PROBABILISTIC STRUCTURAL DYNAMICS
Advanced Theory and Applications

PROBABILISTIC STRUCTURAL DYNAMICS

Advanced Theory and Applications

Y. K. Lin
Florida Atlantic University
G. Q. Cai
Florida Atlantic University

McGraw-Hill, Inc.
New York St. Louis San Francisco Auckland Bogotá
Caracas Lisbon London Madrid Mexico City Milan
Montreal New Delhi San Juan Singapore Sydney Tokyo Toronto

This book was set in Times Roman by Publication Services, Inc.
The editors were B. J. Clark and Margery Luhrs;
the production supervisor was Friederich W. Schulte.
The cover was designed by Carla Bauer.
R. R. Donnelley & Sons Company was printer and binder.

PROBABILISTIC STRUCTURAL DYNAMICS
Advanced Theory and Applications

This book is printed on recycled, acid-free paper
containing 10% postconsumer waste.

1 2 3 4 5 6 7 8 9 0 DOC DOC 9 0 9 8 7 6 5 4

ISBN 0-07-038038-4

Library of Congress Cataloging-in-Publication Data

Lin, Y. K. (Yu-Kweng), (date).
Probabilistic structural dynamics: advanced theory and applications / Y. K. Lin and G. Q. Cai.
p. cm.
Rev. ed. of: Probabilistic theory of structural dynamics, 1967.
Includes bibliographical references (p.) and index.
ISBN 0-07-038038-4
1. Structural dynamics. 2. Stochastic processes. I. Cai, G. Q. (Guo-Qiang) II. Lin, Y. K. (Yu-Kweng), (date). Probabilistic theory of structural dynamics. III. Title.
TA654.L5 1995
624.1'7—dc20 94-13102

About the Authors

Y. K. Lin holds the Charles E. Schmidt Eminent Scholar Chair in Engineering at Florida Atlantic University. He founded the Center for Applied Stochastics Research at Florida Atlantic University and has served as its Director since 1984. He received his Ph.D. in Civil Engineering (Structural) from Stanford University and Doctor of Engineering *honoris causa* from the University of Waterloo. Prior to moving to Florida, he taught for two years in China, one year in Ethiopia, and twenty-four years (1960–1983) at the University of Illinois at Urbana-Champaign. He has served as a consultant for defense, aerospace, and automotive companies and government laboratories. His publications include *Probabilistic Theory of Structural Dynamics* (predecessor of the present volume, McGraw-Hill, 1967, Krieger Publishing Co., 1976) and over 140 technical papers. He is a Fellow of the American Society of Civil Engineers, the American Academy of Mechanics, and the Acoustical Society of America. He is the recipient of the 1984 Alfred M. Freudenthal Medal from the American Society of Civil Engineering for his contribution to stochastic structural dynamics. He is a Registered Professional Engineer in the states of Florida and Illinois, and is listed in *Who's Who in the World, Who's Who in America,* and the *International Who's Who in Education.*

G. Q. Cai holds a joint appointment of Assistant Professor in the Department of Mechanical Engineering and the Center for Applied Stochastics Research at Florida Atlantic University. He received a Ph.D. in Mechanical Engineering from Florida Atlantic University. He has over 40 technical publications that have appeared in the *Journal of Applied Mechanics,* the *AIAA Journal,* the *International Journal of Non-Linear Mechanics,* the *Journal of Engineering Mechanics,* and various conference proceedings.

To Ying-yuh

and to Cong-ying

Contents

Preface

The field of probabilistic structural dynamics has evolved from infancy in the late 1950s to a mature scientific discipline today. Its applications are found in many branches of engineering—aeronautical, astronautical, civil, mechanical, and others. A comprehensive text on the subject requires a balanced treatment of both the mathematical theory of stochastic processes and structural mechanics. This book is a sequel to *Probabilistic Theory of Structural Dynamics,*[*] written with these goals in mind.

The present volume contains some advanced material not generally available at the time of publication of its predecessor volume; thus it is a supplement in several different areas, the most important being the inclusion of multiplicative random excitations on a dynamical system. More thorough treatments of Markov processes are given in Chapters 4, 5, and 7, including the justification of the Markov model idealization from a physical point of view, and the techniques of exact and approximate solutions, applicable to cases of additive random excitations, multiplicative random excitations, or both. Motion stability of dynamical systems due to multiplicative excitations is considered in Chapter 6. Failures due to excursion of the system response into an unsafe region are treated in Chapter 8, again relying strongly on Markov process modeling. Even though the coverage in the earlier volume is still adequate for linear systems under additive random excitations, more rigorous presentations of spectral analysis are given in Chapters 2 and 3 along with recent applications. Chapter 9 is devoted to random uncertainties of system parameters and initial conditions.

[*]Referring to Y. K. Lin, *Probabilistic Theory of Structural Dynamics,* McGraw-Hill, New York, 1967; reprint R. E. Krieger, Melbourne, Florida, 1976.

The progress made in this and related technical fields during the past nearly three decades has been tremendous. It is unavoidable that some important contributions are omitted, or only briefly touched upon, if the materials are considered not essential for the development of the main themes in the book. When quoting a reference, preference is given to the original literary source; thus many excellent textbooks are not included in our citations. Attempts are made to keep the exposition of mathematical principles rigorous and yet comprehensible to engineers with a sound background of mechanics. The combination of the present and the 1967 volumes has been used successfully for a sequence of two graduate courses in structural engineering or engineering mechanics at the University of Illinois at Urbana-Champaign and at Florida Atlantic University. The inclusion of many examples in earthquake and wind engineering also makes the texts suitable references for researchers on these subjects. Nevertheless, the present volume is by and large self-contained; therefore, immediate access to the earlier volume is not necessary.

Much of the material in the book was generated from sponsored research during the past twenty years. We are indebted to the sponsors of our research projects, including National Science Foundation, National Aeronautics and Space Administration, Army Research Office, Air Force Office of Scientific Research, Office of Naval Research, and National Center for Earthquake Engineering Research.

It is a pleasure to acknowledge the help received during the preparation of the manuscript. We are indebted to S. T. Ariaratnam of University of Waterloo, Canada, who made valuable suggestions on Chapters 4 and 6, and to our colleague I. Elishakoff, who commented on Chapters 1 through 3. Constructive criticisms provided by the following reviewers on behalf of McGraw-Hill are also gratefully acknowledged: Mircea Grigoriu, Cornell University; Pol Spanos, Rice University; and Y. K. Wen, University of Illinois.

Y. K. Lin
G. Q. Cai

PROBABILISTIC STRUCTURAL DYNAMICS
Advanced Theory and Applications

CHAPTER 1

INTRODUCTION

Probabilistic or stochastic structural dynamics is a subject dealing with uncertainty in the motion of engineering structures. The cause of motion uncertainty may be the unpredictability of excitations, the imperfection or lack of accurate information in the modeling of physical problems, or a combination of these. Mathematically speaking, modeling a dynamic system is equivalent to setting up the governing equations and specifying the initial and boundary conditions. Thus a probabilistic dynamics problem is posed in probabilistic terms about such equations and conditions, and the problem is solved by providing answers to the ensuing motion also in probabilistic terms.

The idea can best be illustrated by a simple example. Shown in Fig. 1.0.1 is a massless pendulum subjected to two random forces: a vertical force $P[1+\xi_1(t)]$ and a horizontal force $P\xi_2(t)$, where P is a constant; $\xi_1(t)$ and $\xi_2(t)$ are random functions of time or stochastic processes. Assuming that the two hinges shown in the figure are frictionless and the linkage between them rigid, the equation of motion for this dynamic system can be obtained by equating the clockwise and counterclockwise moments about the upper hinge to yield

$$c\frac{d}{dt}[l\sin\theta(t)]l\cos\theta(t) + P[1+\xi_1(t)]l\sin\theta(t) = P\xi_2(t)l\cos\theta(t) \quad (1.0.1)$$

where c = damping coefficient, l = length of the pendulum, and $\theta(t)$ is the angular displacement that describes the motion of the pendulum. The uncertainty in $\theta(t)$ may be due to uncertainties in $\xi_1(t)$, $\xi_2(t)$, c, l, P, and the initial condition $\theta(0)$, individually or in any combinations. Some simplifying assumptions are implied when writing equation (1.0.1), so that the equation is rather simple but still adequate to convey the basic idea. These assumptions include, for example, frictionless hinges, a rigid linkage between the hinges, and negligible inertial properties of the

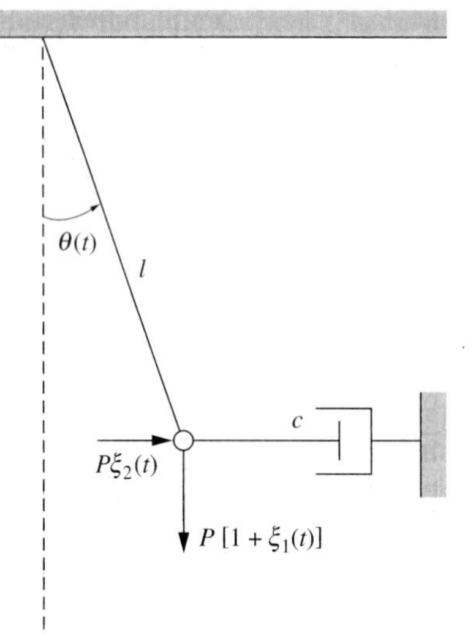

FIGURE 1.0.1
A massless pendulum under random excitations.

system. Otherwise, a partial differential equation in both space and time, with higher derivative terms, will be required to describe the system, and the question of uncertainty in other system parameters in the modeling will also arise.

On the other hand, equation (1.0.1) may be simplified further if the angular displacement $\theta(t)$ is small. We then can use the usual approximations $\sin\theta(t) \approx \theta(t)$ and $\cos\theta(t) \approx 1$, reducing the equation to a linear one:

$$\frac{cl}{P}\frac{d}{dt}\theta(t) + [1 + \xi_1(t)]\theta(t) = \xi_2(t) \tag{1.0.2}$$

In the present form, we have lumped the three system parameters into one factor (cl/P), which can be modeled as one random variable. The initial condition $\theta(0)$ may be another random variable, and the two excitations $\xi_1(t)$ and $\xi_2(t)$ two stochastic processes. The problem is posed by specifying the probabilistic or statistical properties of these random variables and stochastic processes, and the solution is obtained in terms of the corresponding properties for $\theta(t)$.

As seen in equation (1.0.1) or (1.0.2), a random excitation may appear in the coefficient of the unknown, or as an inhomogeneous term on the right-hand side of an equation. These two types of random excitations are described as being multiplicative and additive, respectively, referring directly to their positions in a governing equation. They are also called the parametric and external random excitations, respectively, in the literature, referring more to their physical functions. Even though a parametric random excitation may be generated from an outside source, it causes the basic characteristics of the dynamical system to change randomly with time, whereas an external random excitation does not affect the basic characteristics of the system. In particular, a stable system can become unstable, or vice versa, with the presence of multiplicative excitations.

A more complicated dynamical system may be represented by a set of ordinary differential equations of the type

$$\frac{d}{dt}X_j(t) = f_j[\boldsymbol{X}(t), t] + g_{jk}[\boldsymbol{X}(t), t]\xi_k(t) \qquad j = 1, 2, \ldots, N \quad k = 1, 2, \ldots, M \tag{1.0.3}$$

where $\boldsymbol{X}(t) = \{X_1(t), X_2(t), \ldots, X_N(t)\}$ is a vector of system response variables, $\xi_k(t)$ are random excitations, functions f_j and g_{jk} can be either linear or nonlinear, and a repeated subscript in a product indicates a summation. Such a summation convention will be used elsewhere in this book. The class of dynamical systems represented by equation (1.0.3) is actually quite general. It includes continuous structures of finite size which can be discretized using, for example, a Galerkin or finite-element procedure. The first-order form of the differential equations is not a restriction, because a higher order equation can be replaced by an equivalent set of first-order equations. If uncertainties in the modeling of the system itself are not considered, then the functional forms of f_j and g_{jk} are deterministic. A random excitation $\xi_k(t)$ is multiplicative or additive, depending on whether the associated g_{jk} function does or does not depend explicitly on the components of $\boldsymbol{X}(t)$.

Several solution techniques for systems governed by equations of the type of (1.0.3) are discussed in this book. Only the excitations are assumed to be random, except in Chapter 9.

CHAPTER 2

SPECTRAL ANALYSIS

Spectral analysis in stochastic dynamics is a generalization of Fourier analysis in deterministic dynamics; therefore, it has the same limitations that the dynamic system must be linear and time-invariant. Under these limitations, functions f_j in equation (1.0.3) must be linear in the components of $\boldsymbol{X}(t)$, and functions g_{jk} must be independent of $\boldsymbol{X}(t)$. In particular, spectral analysis cannot be used to treat a system under multiplicative random excitations even when the system is linear.

When a stochastic dynamics problem is solved using spectral analysis, both the random excitations and the system response variables are expressed as Fourier integrals. In the theory of stochastic processes, differentiation and integration may be defined in terms of convergence in several ways, the most commonly used one being convergence in the mean-square or L_2 sense. For convenience, some key results of the L_2 calculus are given below. Readers unfamiliar with these convergence concepts are referred to Lin (1967) for details.

Let $X(t)$ be a continuously valued and continuously parametered stochastic process, with an autocorrelation function

$$\phi_{XX}(t_1, t_2) = E[X(t_1)X^*(t_2)] \tag{2.0.1}$$

where $E[\cdot]$ indicates an ensemble average and an asterisk denotes the complex conjugate. We have the following results.

Continuity. $X(t)$ is continuous at t in the L_2 sense, denoted by

$$\underset{h\to 0}{\text{l.i.m.}}\, X(t+h) = X(t) \tag{2.0.2}$$

if and only if $\phi_{XX}(t_1, t_2)$ is continuous along the diagonal $t_1 = t_2 = t$, where the symbol l.i.m. reads *limit in the mean.*

Differentiability. $X(t)$ is differentiable in the L_2 sense; that is,

$$\dot{X}(t) = \frac{d}{dt}X(t) = \underset{h \to 0}{\text{l.i.m.}} \frac{X(t+h) - X(t)}{h} \tag{2.0.3}$$

exists, if and only if $(\partial^2/\partial t_1 \partial t_2)\phi_{XX}(t_1, t_2)$ exists along the diagonal $t_1 = t_2 = t$.

Integrability. $X(t)$ is Riemann-integrable in the L_2 sense; that is,

$$Y(t) = \int_a^b h(t, \tau)X(\tau)d\tau = \underset{\substack{n \to \infty \\ \Delta_n \to 0}}{\text{l.i.m.}} \sum_{j=1}^{n} h(t, \tau_j')X(\tau_j)(\tau_{j+1} - \tau_j) \tag{2.0.4}$$

exists, where $h(t, \tau)$ is a deterministic weighting function, $a = \tau_0 < \tau_1 < \cdots < \tau_{n+1} = b$, $\tau_j \le \tau_j' \le \tau_{j+1}$, and $\Delta_n = \max(\tau_{j+1} - \tau_j)$, if and only if

$$I(t) = \int_a^b \int_a^b h(t, \tau)h^*(t, u)\phi_{XX}(\tau, u)\, d\tau\, du < \infty \tag{2.0.5}$$

When applied to the spectral analysis of a stochastic process, the L_2 integral in the Riemann form is not entirely satisfactory. Thus we introduce a more general L_2 integral in the Stieltjes form as follows:

$$Y(t) = \int_a^b h(t, \tau)\, dZ(\tau) = \underset{\substack{n \to \infty \\ \Delta_n \to 0}}{\text{l.i.m.}} \sum_{j=1}^{n} h(t, \tau_j')[Z(\tau_{j+1}) - Z(\tau_j)] \tag{2.0.6}$$

Equation (2.0.6) reduces to (2.0.4) if $Z(t)$ is differentiable, so that $dZ(\tau) = X(\tau)\, d\tau$. However, (2.0.6) is meaningful even if $Z(t)$ is not differentiable. The L_2-Stieltjes integral (2.0.6) exists if and only if

$$I(t) = \int_a^b \int_a^b h(t, \tau)h^*(t, u)E[dZ(\tau)\, dZ^*(u)] < \infty \tag{2.0.7}$$

where $$E[dZ(\tau)\, dZ^*(u)] = E\{[Z(\tau + d\tau) - Z(\tau)][Z^*(u + du) - Z^*(u)]\} \tag{2.0.8}$$

Commutability of L_2 limit and ensemble averaging. If an L_2 limit exists, then the order in which the limit and the ensemble average are taken can be interchanged. For example, if $(d/dt)X(t)$ exists as an L_2 derivative, then

$$E\left\{\left[\frac{d}{dt_1}X(t_1)\right]\left[\frac{d}{dt_2}X^*(t_2)\right]\right\} = \frac{\partial^2}{\partial t_1 \partial t_2}E[X(t_1)X^*(t_2)] \tag{2.0.9}$$

or

$$\phi_{\dot{X}\dot{X}}(t_1, t_2) = E[\dot{X}(t_1)\dot{X}^*(t_2)] = \frac{\partial^2}{\partial t_1 \partial t_2}\phi_{XX}(t_1, t_2) \tag{2.0.10}$$

Therefore, the autocorrelation function of the derivative process $\dot{X}(t)$ can be obtained as the mixed second derivative of the autocorrelation function $\phi_{XX}(t_1, t_2)$ of the original process $X(t)$. Moreover,

$$E[\dot{X}^2(t)] = \left[\frac{\partial^2}{\partial t_1 \partial t_2}\phi_{XX}(t_1, t_2)\right]_{t_1 = t_2 = t} \tag{2.0.11}$$

The boundedness of the right-hand side of (2.0.11) is precisely the necessary and sufficient condition for the existence of $\dot{X}(t)$.

It can also be shown that the necessary and sufficient condition (2.0.5) for the L_2-Riemann integral $Y(t)$, defined in (2.0.4), amounts to requiring that the mean-square value $E[Y^2(t)]$ be finite, and that the autocorrelation of $Y(t)$ can be obtained from

$$\phi_{YY}(t_1, t_2) = \int_a^b \int_a^b h(t_1, \tau) h^*(t_2, u) \phi_{XX}(\tau, u) \, d\tau \, du \tag{2.0.12}$$

The case of the L_2-Stieltjes integral is similar.

2.1 STOCHASTIC PROCESSES WITH UNCORRELATED AND ORTHOGONAL INCREMENTS

To explain several subtle points in the Fourier-Stieltjes integral representation of a stochastic process, we require two fundamental concepts.

Let $Z(\omega)$ be a complex-valued stochastic process, defined on $a \leq \omega \leq b$, satisfying

$$E[|Z(\omega_2) - Z(\omega_1)|^2] < \infty \qquad a \leq \omega_1, \omega_2 \leq b \tag{2.1.1}$$

Definition. $Z(\omega)$ is said to be a stochastic process with uncorrelated increments if

$$E\{[Z(\omega_2) - Z(\omega_1)][Z^*(\omega_4) - Z^*(\omega_3)]\} = E[Z(\omega_2) - Z(\omega_1)]E[Z^*(\omega_4) - Z^*(\omega_3)] \tag{2.1.2}$$

for any nonoverlapping intervals $(\omega_1, \omega_2]$ and $(\omega_3, \omega_4]$, where $a \leq \omega_1 < \omega_2 \leq \omega_3 < \omega_4 \leq b$, and $(\omega_j, \omega_k]$ denotes an interval $\omega_j < \omega \leq \omega_k$ which includes point ω_k but not point ω_j. This common practice of representing a closed end of an interval by a bracket and an open end by a parenthesis is followed hereafter.

Definition. $Z(\omega)$ is said to be a stochastic process with orthogonal increments if it has uncorrelated increments, and if the right-hand side of (2.1.2) is equal to zero.

It is clear from the preceding definition that if $Z(\omega)$ is a stochastic process with orthogonal increments, then $Y(\omega) = Z(\omega) - Z_0$ is also a stochastic process with orthogonal increments, where Z_0 is an arbitrary random variable and $E[|Z_0|^2]$ is finite. In particular, Z_0 may be chosen as $Z(\omega_0)$, where ω_0 is an arbitrary reference point on the ω axis.

It can easily be shown that if $Z(\omega)$ is a stochastic process with orthogonal increments, then

$$E\{[Z(\omega_3) - Z(\omega_1)][Z^*(\omega_4) - Z^*(\omega_2)]\} = E\{|Z(\omega_3) - Z(\omega_2)|^2\} \tag{2.1.3}$$

$$a \leq \omega_1 < \omega_2 \leq \omega_3 < \omega_4 \leq b$$

2.2 SPECTRAL REPRESENTATION OF A CORRELATION-STATIONARY STOCHASTIC PROCESS

Let $Z(\omega)$ be a stochastic process with orthogonal increments, and let ω_0 be an arbitrarily chosen reference point on the ω axis. Define a deterministic function

$$\Psi(\omega) = \begin{cases} E\{|Z(\omega) - Z(\omega_0)|^2\} & \omega \geq \omega_0 \\ -E\{|Z(\omega_0) - Z(\omega)|^2\} & \omega < \omega_0 \end{cases} \tag{2.2.1}$$

which implies that $\Psi(\omega_0) = 0$. It can be shown that if $\omega_2 \geq \omega_1$, then regardless of the choice of ω_0,

$$E\{|Z(\omega_2) - Z(\omega_1)|^2\} = \Psi(\omega_2) - \Psi(\omega_1) \tag{2.2.2}$$

We shall now prove that (2.2.2) is valid when $\omega_0 < \omega_1 \leq \omega_2$. Note that the left-hand side of (2.2.2) is equal to

$$\begin{aligned} &E\{[Z(\omega_2) - Z(\omega_1)][Z^*(\omega_2) - Z^*(\omega_1)]\} \\ &\quad = E(\{[Z(\omega_2) - Z(\omega_0)] - [Z(\omega_1) - Z(\omega_0)]\}\{[Z^*(\omega_2) - Z^*(\omega_0)] \\ &\quad - [Z^*(\omega_1) - Z^*(\omega_0)]\}) \end{aligned} \tag{2.2.3}$$

Upon expanding the right-hand side of (2.2.3) and using (2.1.3) and (2.2.1), we obtain

$$\text{rhs} = \Psi(\omega_2) - \Psi(\omega_1) - \Psi(\omega_1) + \Psi(\omega_1) = \Psi(\omega_2) - \Psi(\omega_1) \tag{2.2.4}$$

which agrees with the right-hand side of (2.2.2). The cases of $\omega_1 < \omega_0 < \omega_2$ and $\omega_1 < \omega_2 < \omega_0$ are left as exercises for the reader.

Since the left-hand side of (2.2.2) is nonnegative, $\Psi(\omega)$ must be nondecreasing. In particular, letting $\omega_1 = \omega$, $\omega_2 = \omega + d\omega$ in (2.2.2), we obtain

$$E\{|dZ(\omega)|^2\} = d\Psi(\omega) \tag{2.2.5}$$

Some comments on the result (2.2.5) are in order:

1. If $\Psi(\omega)$ is not differentiable at some ω_j, then $d\Psi(\omega_j)$ can be finite. In this case, $dZ(\omega_j)$ is also finite.
2. If $\Psi(\omega)$ is differentiable, that is, $d\Psi(\omega) = O(d\omega)$, where $O(\cdot)$ denotes the order of magnitude of the quantity within the parentheses, then $dZ(\omega) = O(\sqrt{d\omega})$.

Therefore, a stochastic process with orthogonal increments is always *not* differentiable.

The definition given in Section 2.1 for an orthogonal-increment process is not readily useful in subsequent applications. We next develop an alternative equivalent definition. Write

$$\int_{\omega_1}^{\omega_2} dZ(\omega) = Z(\omega_2) - Z(\omega_1) \tag{2.2.6}$$

$$\int_{\omega_3}^{\omega_4} dZ^*(\omega') = Z^*(\omega_4) - Z^*(\omega_3) \tag{2.2.7}$$

If (2.2.6) and (2.2.7) are interpreted as mean-square integrals, then integration and ensemble-averaging commute. Thus

$$E\{[Z(\omega_2) - Z(\omega_1)][Z^*(\omega_4) - Z^*(\omega_3)]\} = \int_{\omega_1}^{\omega_2}\int_{\omega_3}^{\omega_4} E[dZ(\omega)dZ^*(\omega')] \tag{2.2.8}$$

If

$$E[dZ(\omega)dZ^*(\omega')] = \begin{cases} 0 & \omega \neq \omega' \\ d\Psi(\omega) & \omega = \omega' \end{cases} \tag{2.2.9}$$

then, for any nonoverlapping intervals $(\omega_1, \omega_2]$ and $(\omega_3, \omega_4]$ where $\omega_1 < \omega_2 \leq \omega_3 < \omega_4$, the double integral on the right-hand side of (2.2.8) is zero. Conversely, if, in some range of ω,

$$E[dZ(\omega)\, dZ^*(\omega')] \neq 0 \qquad \omega \neq \omega'$$

then the double integral can be made nonzero by suitably selecting the ω_1, ω_2, ω_3, ω_4 values. Therefore, (2.2.9) is both necessary and sufficient for $Z(\omega)$ to be an orthogonal-increment stochastic process, and it can be adopted as an alternative definition.

The stochastic processes with orthogonal increments can be used to construct certain types of stochastic processes which are important in the study of linear stochastic dynamics. The first type is a correlation-stationary stochastic process, whose correlation function has the form

$$E[X(t_1)X(t_2)] = R_{XX}(\tau) \qquad \tau = t_1 - t_2 \tag{2.2.10}$$

which depends only on the difference $t_1 - t_2$. The relation between such a stochastic process and an orthogonal-increment process is described in the following theorem.

Theorem. Let $X(t)$ be a real-valued stochastic process, $E[X(t)] = 0$, and t is a continuous parameter. A necessary and sufficient condition for $X(t)$ to be a correlation-stationary stochastic process is that it has a Fourier-Stieltjes integral representation

$$X(t) = \int_{-\infty}^{\infty} e^{i\omega t} dZ(\omega) \tag{2.2.11}$$

where $Z(\omega)$ is an orthogonal-increment process.

We next give a heuristic proof of this theorem. Write

$$X(t-\tau) = \int_{-\infty}^{\infty} e^{-i\omega'(t-\tau)} dZ^*(\omega') \tag{2.2.12}$$

and compute the correlation function of $E[X(t)X(t-\tau)]$ by multiplying the right-hand sides of (2.2.11) and (2.2.12) and taking the ensemble averages:

$$E[X(t)X(t-\tau)] = \int_{-\infty}^{\infty}\int_{-\infty}^{\infty} e^{i(\omega-\omega')t+i\omega'\tau} E[dZ(\omega)\, dZ^*(\omega')] \tag{2.2.13}$$

By invoking (2.2.9), the preceding equation can be reduced as follows:

$$R_{XX}(\tau) = \int_{-\infty}^{\infty} e^{i\omega\tau} d\Psi(\omega) \tag{2.2.14}$$

Therefore, the correlation function of a stochastic process $X(t)$, constructed according to (2.2.11), has a stationary form (2.2.10), provided $Z(\omega)$ has orthogonal increments. Conversely, if (2.2.9) does not hold, then the double integral on the right-hand side of (2.2.13) results in a function of both t and τ.

The assumption that $E[X(t)] = 0$ in the preceding theorem can be relaxed. The ensemble mean of a correlation-stationary stochastic process must be a constant, say μ. If this constant is not zero, then it can be added to the right-hand side of (2.2.11). Correspondingly, a constant μ^2 must be added to the right-hand side of (2.2.14).

Equation (2.2.14) shows that the correlation function of a correlation-stationary stochastic process can be expressed as a deterministic Stieltjes-Fourier integral involving a deterministic function $\Psi(\omega)$ defined in (2.2.1). As commented earlier, $Z(\omega)$ is not differentiable, whereas $\Psi(\omega)$ may or may not be differentiable. However, if $\Psi(\omega)$ is indeed differentiable, then we may write $d\Psi(\omega) = \Phi_{XX}(\omega)\, d\omega$, and (2.2.14) becomes

$$R_{XX}(\tau) = \int_{-\infty}^{\infty} e^{i\omega\tau}\Phi_{XX}(\omega)\, d\omega \tag{2.2.15}$$

This expression and the corresponding

$$\Phi_{XX}(\omega) = \frac{1}{2\pi}\int_{-\infty}^{\infty} e^{-i\omega\tau} R_{XX}(\tau)\, d\tau \tag{2.2.16}$$

constitute a Fourier transform pair in the usual Riemann form. If $X(t)$ is a real-valued stochastic process, then an examination of (2.2.15) and (2.2.16) shows that both $R_{XX}(\tau)$ and $\Phi_{XX}(\omega)$ must be even functions of their respective arguments. Therefore, (2.2.15) and (2.2.16) can also be written as cosine Fourier transform pairs. Moreover, since $R_{XX}(0)$ is, by definition, the mean-square value of $X(t)$, substitution of $\tau = 0$ in (2.2.15) yields

$$E[X^2(t)] = \int_{-\infty}^{\infty} \Phi_{XX}(\omega)\, d\omega \tag{2.2.17}$$

This expression shows that integration of $\Phi_{XX}(\omega)$ over the entire ω domain results in the mean-square value of a correlation-stationary stochastic process. For this reason $\Phi_{XX}(\omega)$ is called the *mean-square spectral density* of $X(t)$, or simply the *spectral density* of $X(t)$.

We reiterate the fact that (2.2.14) is more general than (2.2.15) and is applicable even when $\Psi(\omega)$ is not differentiable. Furthermore, according to (2.2.1) the

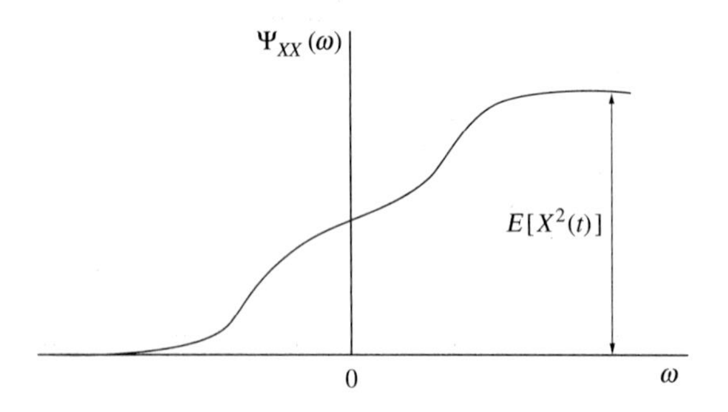

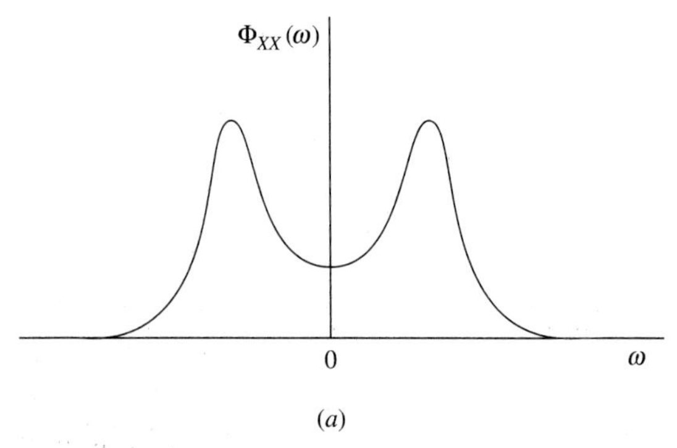

(*a*)

FIGURE 2.2.1
Correlation-stationary stochastic processes. (*a*) Continuous spectral distribution function and the corresponding spectral density function.

value of $\Psi(\omega)$ depends on the choice of the reference point ω_0, and the value can differ by an additive constant if a different reference point is selected. By choosing ω_0 to be $-\infty$, we may write, instead of (2.2.17),

$$E[X^2(t)] = \Psi_{XX}(\infty) - \Psi_{XX}(-\infty) = \Psi_{XX}(\infty) \tag{2.2.18}$$

where $\Psi_{XX}(\omega)$ represents the particular $\Psi(\omega)$ function with $\omega_0 = -\infty$. Function $\Psi_{XX}(\omega)$ is known as the *spectral distribution function* of $X(t)$.

The corresponding spectral distribution function and spectral density function are illustrated in Fig. 2.2.1 for two cases. In Fig. 2.2.1*a*, the spectral distribution function is differentiable everywhere, whereas it is not differentiable at some ω values in Fig. 2.2.1*b*. At those ω values where $\Psi_{XX}(\omega)$ is discontinuous, the mathematically nonexistent $\Phi_{XX}(\omega)$ may be represented formally by the Dirac delta functions.

2.3 LINEAR TIME-INVARIANT SYSTEMS UNDER ADDITIVE EXCITATIONS OF CORRELATION-STATIONARY PROCESSES

When a correlation-stationary stochastic process plays the role of an additive excitation on a time-invariant linear system, the response of the system also tends to a

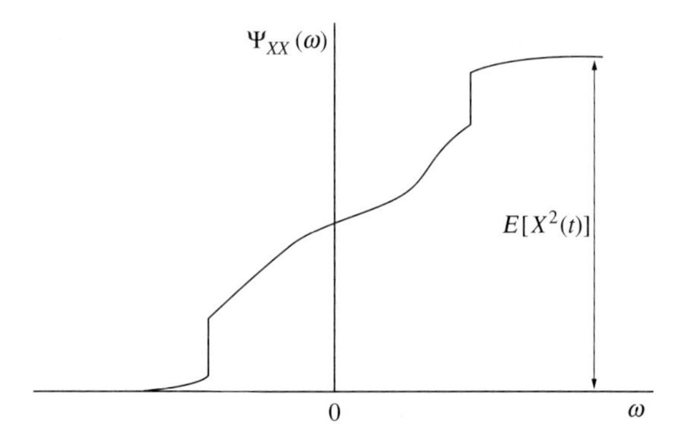

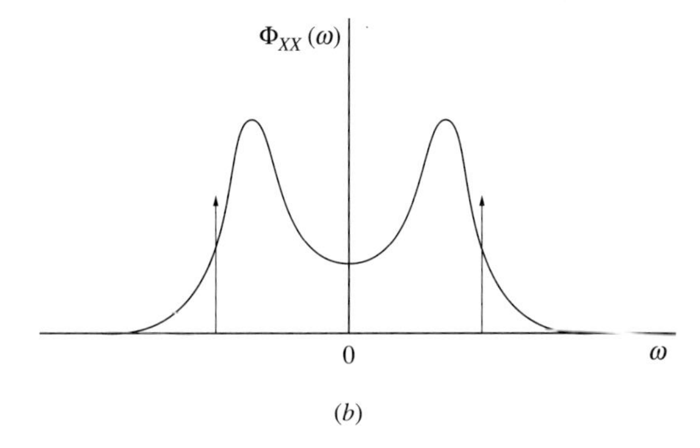

FIGURE 2.2.1 ***(Continued)***
(b) Discontinuous spectral distribution function and the corresponding spectral density function.

correlation-stationary process as the time duration of excitation increases, provided that positive damping is present. The response of interest may be a deflection, a stress, or a strain in a suitable sense. The situation is analogous to the steady-state response in a deterministic vibration analysis.

Since both the excitation and the response are correlation-stationary, they have the Fourier-Stieltjes representations,

$$F(t) = \int_{-\infty}^{\infty} e^{i\omega t} d\tilde{F}(\omega) \tag{2.3.1}$$

$$X(t) = \int_{-\infty}^{\infty} e^{i\omega t} d\tilde{X}(\omega) \tag{2.3.2}$$

where $\tilde{F}(\omega)$ and $\tilde{X}(\omega)$ are two orthogonal-increment processes. Equations (2.3.1) and (2.3.2) show that both $F(t)$ and $X(t)$ are composed of sinusoids with complex coefficients $d\tilde{F}(\omega)$ and $d\tilde{X}(\omega)$, respectively. Since the system is linear and time-invariant, $d\tilde{X}(\omega)$ and $d\tilde{F}(\omega)$ are related as follows:

$$d\tilde{X}(\omega) = H(\omega)\, d\tilde{F}(\omega) \tag{2.3.3}$$

where $H(\omega)$ is the frequency response function, corresponding to a particular pair of excitation (input) and response (output). Multiplying each side of (2.3.3) by its complex conjugate and taking ensemble averages, we obtain

$$d\Psi_{XX}(\omega) = |H(\omega)|^2 \, d\Psi_{FF}(\omega) \tag{2.3.4}$$

where use has been made of the orthogonal-increment property, (2.2.9). When Ψ_{FF} is differentiable, which implies that Ψ_{XX} is also differentiable, we have

$$\Phi_{XX}(\omega) = |H(\omega)|^2 \, \Phi_{FF}(\omega) \tag{2.3.5}$$

Equation (2.3.4) is, of course, more general.

Generalization of the foregoing results to the case of M inputs and N outputs is straightforward. Let the inputs be all additive and correlation-stationary, thus having the representations

$$F_j(t) = \int_{-\infty}^{\infty} e^{i\omega t} d\tilde{F}_j(\omega) \qquad j = 1, 2, \ldots, M \tag{2.3.6}$$

where $\tilde{F}_j(\omega)$ are orthogonal-increment processes satisfying

$$E[d\tilde{F}_j(\omega)\, d\tilde{F}_k^*(\omega')] = \begin{cases} 0 & \omega \neq \omega' \\ d\Psi_{F_jF_k}(\omega) & \omega = \omega' \end{cases} \tag{2.3.7}$$

The function $\Psi_{F_jF_k}(\omega)$ is called the *cross-spectral distribution function* of two jointly correlation-stationary processes $F_j(t)$ and $F_k(t)$, and it becomes a spectral distribution function when $j = k$.

As the exposure time of the system to the inputs increases, the outputs tend also to correlation-stationarity, provided that positive damping exists in the system. Then the outputs also admit the representations

$$X_l(t) = \int_{-\infty}^{\infty} e^{i\omega t} d\tilde{X}_l(\omega) \qquad l = 1, 2, \ldots, N \tag{2.3.8}$$

and

$$E[d\tilde{X}_l(\omega)\, d\tilde{X}_m^*(\omega')] = \begin{cases} 0 & \omega \neq \omega' \\ d\Psi_{X_lX_m}(\omega) & \omega = \omega' \end{cases} \tag{2.3.9}$$

where $\Psi_{X_lX_m}(\omega)$ is the joint spectral distribution function of the two outputs $X_l(t)$ and $X_m(t)$. Since the system is linear and time-invariant, we have

$$d\tilde{X}_l(\omega) = H_{lj}(\omega)\, d\tilde{F}_j(\omega) \tag{2.3.10}$$

where $H_{lj}(\omega)$ is the frequency response function corresponding to a single input $F_j(t)$ and a single output $X_l(t)$, and the repeated subscript in the product signifies a summation over $j = 1, 2, \ldots, M$. Finally, using (2.3.7) and (2.3.9), we obtain

$$d\Psi_{X_lX_m}(\omega) = H_{lj}(\omega) H_{mk}^*(\omega)\, d\Psi_{F_jF_k}(\omega) \tag{2.3.11}$$

This result can be expressed in matrix form:

$$d\boldsymbol{\Psi}_X(\omega) = \boldsymbol{H}(\omega)\, d\boldsymbol{\Psi}_F(\omega) [\boldsymbol{H}^*(\omega)]' \tag{2.3.12}$$

where $\boldsymbol{H}(\omega)$ is an $N \times M$ matrix of frequency response functions, a prime denotes a matrix transposition, and each $d\boldsymbol{\Psi}(\omega)$ is a square matrix of differentials of spectral distributions (appearing on the diagonal) and differentials of cross-spectral distributions (appearing off the diagonal).

If the spectral distributions and cross-spectral distributions of the inputs are differentiable, then (2.3.11) and (2.3.12) may be replaced by their respective relations for spectral densities and cross-spectral densities as follows:

$$\Phi_{X_l X_m}(\omega) = H_{lj}(\omega) H^*_{mk}(\omega) \Phi_{F_j F_k}(\omega) \tag{2.3.13}$$

and

$$\boldsymbol{\Phi}_X(\omega) = \boldsymbol{H}(\omega) \boldsymbol{\Phi}_F(\omega) [\boldsymbol{H}^*(\omega)]' \tag{2.3.14}$$

The case of linear continuous structure under distributed random excitation of the additive type can be treated by replacing summation by integration. The frequency response characteristics are now described by a frequency influence function $H(\boldsymbol{r}, \boldsymbol{\rho}, \omega)$, where $\boldsymbol{\rho}$ denotes an input location and $\boldsymbol{r}$ denotes an output location. Assuming that both distributed excitation $F(\boldsymbol{\rho}, t)$ and distributed response $W(\boldsymbol{r}, t)$ are correlation-stationary in time, they have the representations

$$F(\boldsymbol{\rho}, t) = \int_{-\infty}^{\infty} e^{i\omega t} d\tilde{F}(\boldsymbol{\rho}, \omega) \tag{2.3.15}$$

and

$$W(\boldsymbol{r}, t) = \int_{-\infty}^{\infty} e^{i\omega t} d\tilde{W}(\boldsymbol{r}, \omega) \tag{2.3.16}$$

with

$$E[d\tilde{F}(\boldsymbol{\rho}_1, \omega) d\tilde{F}^*(\boldsymbol{\rho}_2, \omega')] = \begin{cases} 0 & \omega \neq \omega' \\ d\Psi_F(\boldsymbol{\rho}_1, \boldsymbol{\rho}_2, \omega) & \omega = \omega' \end{cases} \tag{2.3.17}$$

and

$$E[d\tilde{W}(\boldsymbol{r}_1, \omega) d\tilde{W}^*(\boldsymbol{r}_2, \omega')] = \begin{cases} 0 & \omega \neq \omega' \\ d\Psi_W(\boldsymbol{r}_1, \boldsymbol{r}_2, \omega) & \omega = \omega' \end{cases} \tag{2.3.18}$$

Then, analogous to (2.3.11), we have

$$d\Psi_W(\boldsymbol{r}_1, \boldsymbol{r}_2, \omega) = \int_R d\boldsymbol{\rho}_1 \int_R d\boldsymbol{\rho}_2 \, H(\boldsymbol{r}_1, \boldsymbol{\rho}_1, \omega) H^*(\boldsymbol{r}_2, \boldsymbol{\rho}_2, \omega) \, d\Psi_F(\boldsymbol{\rho}_1, \boldsymbol{\rho}_2, \omega) \tag{2.3.19}$$

where R represents the region on which the excitation field is acting. The function $\Psi_F(\boldsymbol{\rho}_1, \boldsymbol{\rho}_2, \omega)$ appearing in (2.3.17) and (2.3.19) is the cross-spectral distribution function of the excitation field at two locations $\boldsymbol{\rho}_1$ and $\boldsymbol{\rho}_2$, whereas function $\Psi_W(\boldsymbol{r}_1, \boldsymbol{r}_2, \omega)$ appearing in (2.3.18) and (2.3.19) is the cross-spectral distribution function of the response field at two locations $\boldsymbol{r}_1$ and $\boldsymbol{r}_2$. Assuming differentiability of these cross-spectral distributions, (2.3.19) may be replaced by an expression relating the cross-spectral densities:

$$\Phi_W(\boldsymbol{r}_1, \boldsymbol{r}_2, \omega) = \int_R d\boldsymbol{\rho}_1 \int_R d\boldsymbol{\rho}_2 \, H(\boldsymbol{r}_1, \boldsymbol{\rho}_1, \omega) H^*(\boldsymbol{r}_1, \boldsymbol{\rho}_2, \omega) \Phi_F(\boldsymbol{\rho}_1, \boldsymbol{\rho}_2, \omega) \tag{2.3.20}$$

2.3.1 Along-Wind Motion of a Multistory Building

As an example, consider an N-story building under the excitation of turbulent wind loads. As shown schematically in Fig. 2.3.1a, the structural model of the building

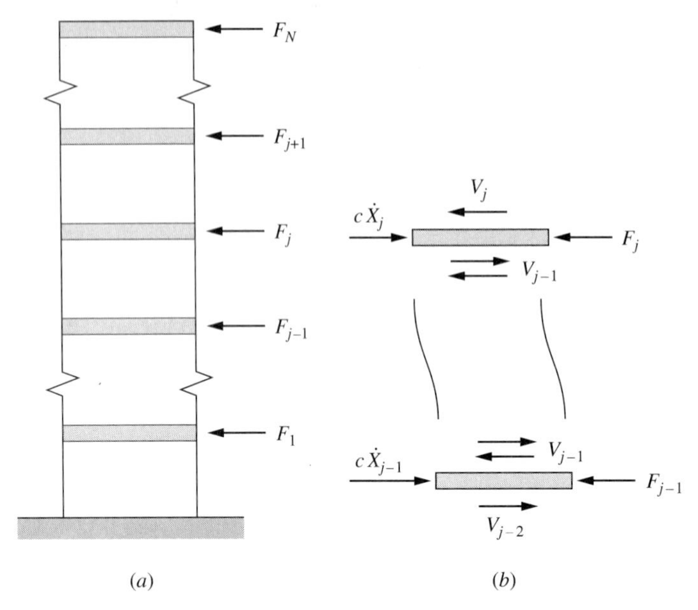

FIGURE 2.3.1
A simplified building-structure model under wind load excitations.

consists of concentrated masses at the floor levels which are capable only of horizontal translation along the wind flow direction. Linear elastic restoring forces are provided by massless shear walls between neighboring floors. The wind loads are also lumped at the floor levels. The entire building structure is clamped to the ground, which is treated as the 0th floor. Free-body diagrams showing forces acting on the jth floor, on the $(j-1)$th floor, and on the shear wall between these two floors are seen in Fig. 2.3.1*b*. These forces include the wind loads, the shear forces in the shear walls, and viscous damping forces. To simplify the subsequent analysis, the entire building structure is assumed to be composed of identically constructed story units, each of which consists of a floor and the shear wall below the floor. A structure composed of identical units in an array is known as a periodic structure, for which a host of analytical techniques is available.

Referring to Fig. 2.3.1*b*, the equations of motion may be written as

$$\begin{aligned} &m\ddot{X}_j + c\dot{X}_j + V_{j-1} - V_j = F_j \\ &V_{j-1} = k(X_j - X_{j-1}) \qquad\qquad j = 1, 2, \ldots, N \end{aligned} \tag{a}$$

subject to the boundary conditions $X_0 = 0$ and $V_N = 0$, where m = concentrated mass at the floor level, X_j = deflection of the jth floor, V_j = shear force above the jth floor, k = elastic constant between neighboring floors, F_j = wind load acting on the jth floor, and each overdot denotes one differentiation with respect to time. Most building structures are sufficiently stiff to justify the assumption that the wind loads F_j are independent of the structural motion. Then F_j are additive random excitations. Moreover, since the mean values of the wind loads are nonzero, so are the mean values of X_j and V_j. In the state of correlation-stationarity, these mean values are constants and they satisfy the algebraic equations

$$\begin{aligned} E[V_{j-1}] - E[V_j] &= E[F_j] \\ E[V_{j-1}] &= k\{E[X_j] - E[X_{j-1}]\} \qquad j = 1, 2, \ldots, N \end{aligned} \tag{b}$$

It is convenient to subtract these mean values from equations (a). The resulting equations after subtraction have the same forms as equations (a), except that X_j, V_j, and F_j now represent, respectively, their random fluctuations from the means. In what follows, these equations are so interpreted.

Restricting our discussion to correlation-stationary excitations and responses, we substitute into equations (a)

$$F_j(t) = \int_{-\infty}^{\infty} e^{i\omega t} d\tilde{F}_j(\omega) \tag{c.1}$$

$$X_j(t) = \int_{-\infty}^{\infty} e^{i\omega t} d\tilde{X}_j(\omega) \tag{c.2}$$

$$V_j(t) = \int_{-\infty}^{\infty} e^{i\omega t} d\tilde{V}_j(\omega) \tag{c.3}$$

where $\tilde{F}_j(\omega)$, $\tilde{X}_j(\omega)$, and $\tilde{V}_j(\omega)$ are stochastic processes with orthogonal increments. The orthogonality of increments in the frequency domain implies that equations (a) must be satisfied at every frequency; thus

$$\begin{aligned} [(i\omega)^2 m + (i\omega)c]\, d\tilde{X}_j(\omega) + d\tilde{V}_{j-1}(\omega) - d\tilde{V}_j(\omega) &= d\tilde{F}_j(\omega) \\ d\tilde{V}_{j-1}(\omega) &= k[d\tilde{X}_j(\omega) - d\tilde{X}_{j-1}(\omega)] \qquad j = 1, 2, \ldots, N \end{aligned} \tag{d}$$

Equation (d) can be cast in the following matrix form:

$$\begin{Bmatrix} d\tilde{X}_j \\ d\tilde{V}_j \end{Bmatrix} = \begin{bmatrix} 1 & k^{-1} \\ (-m\omega^2 + ic\omega) & k^{-1}(-m\omega^2 + ic\omega) + 1 \end{bmatrix} \begin{Bmatrix} d\tilde{X}_{j-1} \\ d\tilde{V}_{j-1} \end{Bmatrix} - \begin{Bmatrix} 0 \\ d\tilde{F}_j \end{Bmatrix} \tag{e}$$

Or, more briefly,

$$d\boldsymbol{Y}_j = \boldsymbol{T} d\boldsymbol{Y}_{j-1} - d\boldsymbol{G}_j \tag{f}$$

where

$$d\boldsymbol{Y}_j = \begin{Bmatrix} d\tilde{X}_j \\ d\tilde{V}_j \end{Bmatrix} \tag{g}$$

and $\boldsymbol{T}$ is known as a transfer matrix. Repeated application of (f) results in a relation between $d\boldsymbol{Y}_0$ and $d\boldsymbol{Y}_m$:

$$d\boldsymbol{Y}_m = \boldsymbol{T}^m d\boldsymbol{Y}_0 - \sum_{j=1}^{m} \boldsymbol{T}^{m-j} d\boldsymbol{G}_j \tag{h}$$

Letting $m = N$ and imposing boundary conditions $d\tilde{X}_0 = 0$ and $d\tilde{V}_N = 0$, we obtain from the second row

$$\tau_{22}(N)\,d\tilde{V}_0 - \sum_{j=1}^{N} \tau_{22}(N-j)\,d\tilde{F}_j = 0 \tag{i}$$

in which $\tau_{pq}(n)$ is the (p, q) element of $\boldsymbol{T}^n$. Solving for $d\tilde{V}_0$:

$$d\tilde{V}_0 = \frac{1}{\tau_{22}(N)} \sum_{j=1}^{N} \tau_{22}(N-j)\,d\tilde{F}_j \tag{j}$$

A great deal of simplification can be achieved by recognizing the fact that eigenvalues of any transfer matrix are reciprocal pairs. Then by denoting the two eigenvalues of $\boldsymbol{T}$ in the present case as $\exp(\pm i\theta)$, it can be shown that

$$\cos\theta = 1 + \frac{-m\omega^2 + ic\omega}{2k} \tag{k}$$

and that (Lin and McDaniel, 1969)

$$\boldsymbol{T}^n = \begin{bmatrix} \cos n\theta & 0 \\ 0 & \cos n\theta \end{bmatrix} + \frac{\sin n\theta}{\sin\theta} \begin{bmatrix} 1-\cos\theta & k^{-1} \\ -2k(1-\cos\theta) & -(1-\cos\theta) \end{bmatrix} \tag{l}$$

In particular,

$$\tau_{22}(n) = \cos n\theta - \frac{\sin n\theta(1-\cos\theta)}{\sin\theta} = \frac{\cos(n+1/2)\theta}{\cos\theta/2} \tag{m}$$

$$\tau_{12}(n) = \frac{\sin n\theta}{k\sin\theta} \tag{n}$$

Note that θ is a complex function of ω as seen from equation (k). It follows from equations (j) and (m) that

$$d\tilde{V}_0 = \frac{1}{\cos(N+1/2)\,\theta} \sum_{j=1}^{N} \cos\left(N - j + \frac{1}{2}\right)\theta\, d\tilde{F}_j \tag{o}$$

Substituting the preceding result into equation (h), we obtain

$$d\boldsymbol{Y}_m = \frac{1}{\cos(N+1/2)\theta} \begin{Bmatrix} \dfrac{\sin m\theta}{k\sin\theta} \\ \dfrac{\cos(m+1/2)\theta}{\cos(\theta/2)} \end{Bmatrix} \sum_{j=1}^{N} \cos\left(N-j+\frac{1}{2}\right)\theta\, d\tilde{F}_j - \sum_{r=1}^{M} \begin{Bmatrix} \dfrac{\sin(m-r)\theta}{k\sin\theta} \\ \dfrac{\cos(m-r+1/2)\theta}{\cos(\theta/2)} \end{Bmatrix} d\tilde{F}_r \tag{p}$$

Finally, using the orthogonal-increment properties of $d\tilde{F}_j$, $d\tilde{V}_j$, and $d\tilde{X}_j$, we can obtain various spectral relations between inputs and outputs. For example, the spectral distribution function of $X_m(t)$, namely, $\Psi_{X_m X_m}(\omega)$, is given by the following expression:

$$
\begin{aligned}
d\Psi_{X_m X_m}(\omega) = {} & \frac{1}{k^2 \left|\cos(N + 1/2)\theta\right|^2} \left|\frac{\sin m\theta}{\sin\theta}\right|^2 \\
& \times \sum_{j=1}^{N} \sum_{l=1}^{N} \cos\left(N - j + \frac{1}{2}\right)\theta \left[\cos\left(N - l + \frac{1}{2}\right)\theta\right]^* d\Psi_{F_j F_l}(\omega) \\
& + \sum_{r=1}^{m} \sum_{p=1}^{m} \frac{1}{k^2 \left|\sin\theta\right|^2} \sin(m - r)\theta[\sin(m - p)\theta]^* d\Psi_{F_r F_p}(\omega) \\
& -2\,\mathrm{Re}\left\{ \frac{\sin m\theta}{k^2 \left|\sin\theta\right|^2 \cos(N + 1/2)\theta} \sum_{j=1}^{N} \sum_{r=1}^{m} \cos\left(N - j + \frac{1}{2}\right)\theta \right. \\
& \left. [\sin(m - r)\theta]^* d\Psi_{F_j F_r}(\omega) \right\} \qquad \text{(q)}
\end{aligned}
$$

where Re{ } denotes the real part of the bracketed expression.

In the foregoing analysis of a wind-excited building, the input-output relations have been obtained in terms of spectral distributions without first obtaining specific expressions for the frequency response functions. This is possible in some cases, such as the simple structural model used in this analysis. However, the presence of frequency response functions is hidden but not eliminated. These functions can be uncovered from equation (p). On the left-hand side of this equation are two types of response which are elements of vector $d\boldsymbol{Y}_m$: the displacement response $d\tilde{X}_m$ and the shear force response $d\tilde{V}_m$. The excitations appearing on the right-hand side of the equation are the wind loads lumped at various floor levels. Since each frequency response function is associated with one input-output pair, namely, a single excitation and a single response quantity due to this excitation, each can be obtained by adding the coefficients of a single $d\tilde{F}_l$ on the right-hand side. For example, the frequency response function for the displacement at the mth floor due to a wind load at the lth floor is given by

$$
H_{X_m F_l}(\omega) = \frac{1}{k \sin\theta} \left[\frac{\cos(N - l + 1/2)\theta}{\cos(N + 1/2)\theta} \sin m\theta - \sin(m - l)\theta \mathrm{II}(m - l) \right] \qquad \text{(r)}
$$

where $\mathrm{II}(\cdot)$ is the *Heaviside unit-step function:*

$$
\mathrm{II}(m - l) = \begin{cases} 1 & m > l \\ 0 & m < l \end{cases} \qquad \text{(s)}
$$

Similarly, the frequency response function for the shear force immediately above the mth floor due to a wind load at the lth floor is given by

$$
H_{V_m F_l}(\omega) = \frac{1}{\cos(\theta/2)} \left[\frac{\cos(N - l + 1/2)\theta}{\cos(N + 1/2)\theta} \cos\left(m + \frac{1}{2}\right)\theta - \cos\left(m - l + \frac{1}{2}\right)\theta \mathrm{II}(m - l) \right] \qquad \text{(t)}
$$

The simplified assumptions for the structural model have been made in order to obtain closed-form solutions, which are more useful for interpreting the analytical results. Since some of these assumptions may be far from being realistic in a practical setting, comments on their relaxations are in order.

The assumption that the building structure is rigidly clamped on the ground may be relaxed by including soil compliancy in the analysis. This can be done quite easily if the soil properties can be modeled by an impedance matrix. The viscous damping depending on the absolute velocities of individual floors may be replaced by other types of damping, resulting in a different construction of the basic transfer matrix. The dimension of the transfer matrices used in the analysis increases from 2×2 to 6×6 if translations of the building in two perpendicular directions and torsion are taken into consideration. The eigenvalues of a 6×6 transfer matrix consist of three reciprocal pairs; however, the analysis remains essentially the same in spite of much greater numerical computations. If the assumption of identically constructed story units is removed so that the structure is no longer spatially periodic, then chain products of transfer matrices will replace powers of one single-transfer matrix, resulting again in much greater numerical computations. Finally, if the shear-wall behavior is not a realistic assumption, then a general-purpose finite-element program may be needed to compute the required frequency response functions.

Numerical calculations have been carried out using equation (r) and the following physical data for the structure: number of stories $N = 40$, individual story height $= 4$ m, lumped mass at individual floor $m = 1{,}290{,}000$ kg, elastic constant for each shear wall $k = 10^9$ N/m, and damping coefficient $c = 21{,}500$ N/(m/s). For the wind environment, the fluctuating part of the wind pressure is assumed to follow a Davenport cross-spectral density of the form

$$\Phi(y_1, y_2, \omega) = \kappa |\omega|^{-1} \left[1 + \left(\frac{600\omega}{\pi u_r}\right)^2\right]^{-4/3} \left(\frac{600\omega}{\pi u_r}\right)^2 \exp\left[-\frac{c_1 |\omega(y_1 - y_2)|}{2\pi u_r}\right] \tag{u}$$

where y_1 and y_2 are two different heights above the ground, $c_1 = 7.7$ is a constant, $\alpha = 0.4$ is a constant, $u_r = 11.46$ m/s is the reference mean wind velocity at 10 m above the ground, and κ is a scaling factor depending on y_1, y_2, and u_r. The cross-spectral density of the concentrated wind loads at two different floors is obtained by multiplying equation (u) by the tributary areas for these floors, which are assumed to be 192 m^2 for each story.

The computed spectral density of the displacement at the top floor is shown as the dashed line in Fig. 2.3.2. Only the positive ω domain is included in the figure; the graph for the negative ω domain can be obtained as its mirror image. The first peak, which occurs at $\omega = 0.04$ rad/s, coincides with the dominant peak in the Davenport spectrum. The second peak, which is taller, is near the fundamental frequency of the structure. Contributions from higher modes are insignificant because the Davenport spectra selected for computation diminish quickly as ω increases. Clearly, if a different model for the wind turbulence, with more energy available in the higher frequency range, is used in the calculation, then more peaks may become visible. Of course, the method of analysis is applicable to any wind models.

The effects of soil compliancy on the structural response have also been investigated, and the result is shown as the solid line in Fig. 2.3.2. The dominant peak is

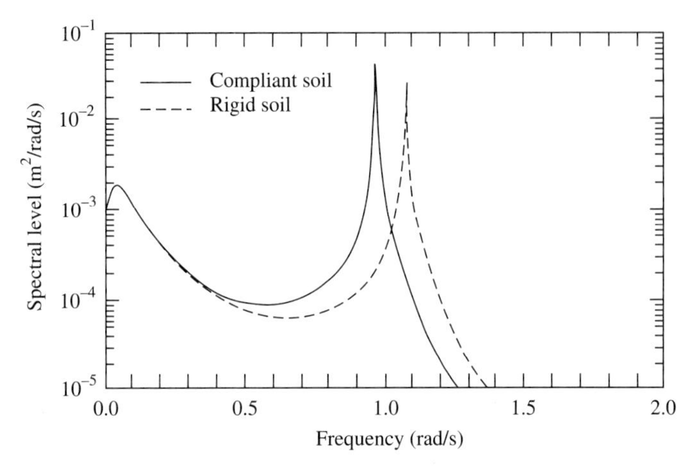

FIGURE 2.3.2
Spectral density of the variance of the top floor displacement of a 40-story building under wind load excitation [after Lin and Wu, 1984b; Copyright ©1984 ASCE, reprinted with permission].

shifted to a slightly lower frequency, as expected. For details, the reader is referred to Lin and Wu (1984b).

In the spectral representation of a correlation-stationary stochastic process, (2.2.11), the symbol ω is used for the parameter of an orthogonal-increment process $Z(\omega)$. If ω has the usual meaning of frequency, then t has the usual meaning of time, and the stochastic process $X(t)$ so constructed is correlation-stationary in time. Analogously, a stochastic process, which is correlation-homogeneous in space, has the representation

$$p(x) = \int_{-\infty}^{\infty} e^{ikx} dZ(k) \tag{2.2.11$'$}$$

where $Z(k)$ is, again, an orthogonal-increment process in k. Equation (2.2.11′) implies that $p(x)$ is composed of sinusoidal components of wavelengths $2\pi/k$, where variable k is known as the wave number. Obviously, the method of spectral analysis in the frequency domain can often be adapted for the wave-number domain, as will be illustrated in two examples.

2.3.2 Response of an Airplane to Atmospheric Turbulence (Lin, 1977)

Shown in Fig. 2.3.3 is an airplane flying into a random turbulence field. We make the following simplified assumptions:

1. The airplane is rigid and the only significant perturbation from its steady forward flight is the up-and-down plunging motion.
2. The forward speed u of the airplane is a constant and is directed toward the $-x_1$ direction.

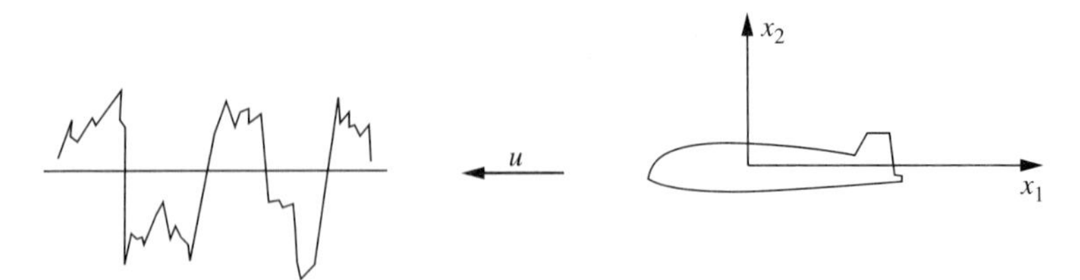

FIGURE 2.3.3
An airplane entering a random turbulence field.

3. Among all three components of the turbulent velocity, only the vertical component V affects the motion of the vehicle.
4. The vertical turbulence velocity field is considered to be *frozen* and stochastically homogeneous in space (Taylor's hypothesis).

With these assumptions, it is sufficient to represent the flow velocity field relative to the unperturbed vehicle as

$$U_i = u\delta_{i1} + V\delta_{i2} \tag{a}$$

where δ_{ij} is the Kronecker delta. The frozen pattern assumption for the turbulence field is reasonable if u is sufficiently high for the air velocities to remain essentially unchanged during a time length $2b/u$, where $2b$ is the chord width of the airplane lifting surface. Clearly, the steady speed u is required to maintain the steady lift. We shall be interested in the vertical acceleration $Z(t)$ of the airplane, the response quantity of interest.

Since the gust field is assumed to be frozen in space, the airplane senses a time-varying excitation which is a function of the difference $(x_1 - ut)$. Such a turbulence field can be represented by a Stieltjes-Fourier integral:

$$V(x_1 - ut) = \int_{-\infty}^{\infty} e^{ik(x_1 - ut)} d\tilde{V}(k) \tag{b}$$

where $\tilde{V}(k)$ is an orthogonal-increment process. This equation implies that V is composed of infinitely many frozen-pattern sinusoids.

Within the framework of a linear analysis, we can also construct the total vehicle response to an arbitrary frozen random velocity field from a basic solution corresponding to a unit frozen sinusoid velocity field. In a certain sense, this basic solution is analogous to the frequency response function. Thus we consider first the problem shown in Fig. 2.3.4, which was considered by Sears (1941) for the case of incompressible fluid. Specifically, Sears considered the change of lift on a two-dimensional airfoil flying into a sinusoidal gust. The result is

$$\eta_1(k) = \sqrt{2}\,\pi\rho b u \phi(k) \tag{c}$$

where ρ is the air density and $\phi(k)$ is a function of the wave number k, known as Sears' function in the aeronautical engineering literature. The exact Sears' function involves Bessel functions, but Liepman (1952) has suggested an approximation:

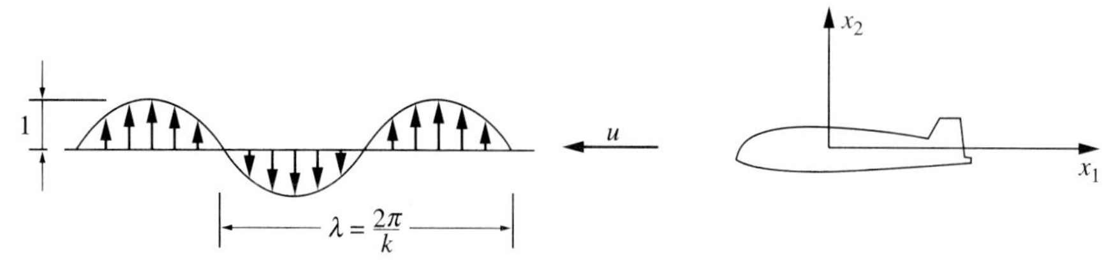

FIGURE 2.3.4
An airplane entering a frozen pattern of a sinusoidal velocity field.

$$\left|\phi(k)\right|^2 = \frac{1}{1 + 2\pi\left|kb\right|} \tag{d}$$

For certain purposes, this approximation leads to unbounded results. Other approximations include one by Howell and Lin (1971):

$$\phi(k) = \left(\frac{0.065}{0.13 + ikb} + \frac{0.5}{1 + ikb}\right)e^{1.10ikb} \tag{e}$$

Additional adjustments are required for compressible flow, but the general form remains similar to (e).

In aeronautical engineering applications, the vehicle acceleration response is of greater interest. For this we require another transformation (Fung, 1953),

$$\eta_2(k) = \sqrt{2}kb\left\{\pi\rho Ab\left[kb(1 + 2\lambda) - 2i\right]\right\}^{-1} \tag{f}$$

where, in addition to the previously defined variables, A is the total lifting surface area, λ is a mass parameter, given by

$$\lambda = \frac{M}{\pi\rho Ab} \tag{g}$$

and M is the total mass of the aircraft.

It is noted that both transfer functions η_1 and η_2 are expressed in terms of the wave number k, which, referring to equation (b), is related to the frequency ω as

$$ku = \omega \tag{h}$$

in the case of a frozen random field. The product kb is, in fact, the well-known Strouhal's number

$$s = kb = \frac{\omega b}{u}$$

In the terminology of aeroelasticity, s is also called the reduced frequency.

The frequency response function from gust velocity to airplane acceleration is then

$$H(\omega) = \eta_2\left(\frac{\omega}{u}\right)\eta_1\left(\frac{\omega}{u}\right) \tag{i}$$

and the acceleration response at the stationary state is

$$Z(t) = \int_{-\infty}^{\infty} H(\omega)e^{-i\omega t} d\tilde{V}\left(\frac{\omega}{u}\right) \tag{j}$$

However, it is not necessary to convert k to ω in actual calculations. Clearly $Z(t)$ can also be computed from

$$Z(t) = \int_{-\infty}^{\infty} \bar{H}(k)e^{-ikut} d\tilde{V}(k) \tag{k}$$

where $\bar{H}(k) = \eta_2(k)\eta_1(k)$, which may be regarded as the wave-number response function analogous to the frequency response function $H(\omega)$. It follows from equation (k) that the correlation function $R_{ZZ}(\tau) = E[Z(t)Z(t+\tau)]$ of the response is given by

$$R_{ZZ}(\tau) = \int_{-\infty}^{\infty} \left|\bar{H}(k)\right|^2 e^{iku\tau} S_{VV}(k)\, dk \tag{l}$$

where $S_{VV}(k)$ is the wave-number spectral density of $V(x)$. Equation (l) has a frequency domain analogue of

$$R_{ZZ}(\tau) = \int_{-\infty}^{\infty} \left|H(\omega)\right|^2 e^{i\omega\tau} \Phi_{VV}(\omega)\, d\omega \tag{m}$$

Comparing (l) and (m), it is seen that

$$\Phi_{VV}(\omega) = \frac{1}{|u|} S_{VV}\left(\frac{\omega}{u}\right) \tag{n}$$

Physically, $\Phi_{VV}(\omega)$ represents the spectral density of the turbulence field, as sensed by an observer on the moving airplane. Although the turbulence is assumed to be frozen in space, it appears to the observer to be fluctuating in time.

In terms of the spectral density of the response, we also have the two forms

$$S_{ZZ}(k) = \left|\bar{H}(k)\right|^2 S_{VV}(k) \tag{o}$$

and

$$\begin{aligned} \Phi_{ZZ}(\omega) &= \left|H(\omega)\right|^2 \Phi_{VV}(\omega) \\ &= \frac{1}{|u|} \left|H(\omega)\right|^2 S_{VV}\left(\frac{\omega}{u}\right) \end{aligned} \tag{p}$$

In practical applications, several different expressions have been used for the spectrum of a homogeneous gust field. The Dryden spectrum (Dryden, 1961) is given by

$$S_{VV}(k) = \frac{\sigma^2 L}{2\pi} \frac{1 + 3(kL)^2}{\left[1 + (kL)^2\right]^2} \tag{q}$$

where σ^2 is the variance of the turbulence (which has a zero mean), and L is called the scale of turbulence, which is a measure of the correlation length in space. Another is the von Karman spectrum (von Karman, 1948), which tends to zero as $k^{-5/3}$ for large k, instead of k^{-2} in the Dryden spectrum.

It is of interest to note that several extensions of the foregoing analysis are possible. If the airplane is modeled as a multi-degree-of-freedom system, but keeping the assumption that only the vertical component of the turbulence field affects the motion of the airplane, then we require the cross-spectrum of the vertical turbulence velocity, for example (Lin, 1967, Eq. 6-102),

$$S_{VV}(x_1, y_1; x_2, y_2; k) = \frac{2\sigma^2}{\pi} e^{-ik\xi} \left\{ k^2 \left[\frac{\eta^2}{4a^2} K_2\left(\frac{a\eta}{L}\right) \right] + \frac{1}{4a^2}\left(\frac{a\eta}{L}\right)\left[3K_1\left(\frac{a\eta}{L}\right) - \frac{a\eta}{L} K_2\left(\frac{a\eta}{L}\right)\right]\right\} \qquad (r)$$

where (x_1, y_1) and (x_2, y_2) are two arbitrary locations, $\xi = x_1 - x_2$, $\eta = y_1 - y_2$, $a = [1 + (Lk)^2]^{1/2}$, and K_m are modified Bessel functions of the second kind. At $\xi = \eta = 0$, (r) reduces to the Dryden spectrum (q) since

$$zK_1(z) \to 1 \qquad z^2 K_2(z) \to 2$$

Another extension is to model the turbulence field as being an inhomogeneous frozen pattern in space, for which we quote the work by Howell and Lin (1971) for the single-degree-of-freedom case and Fujimori and Lin (1973) for the multi-degree-of-freedom case.

2.3.3 Response of an Infinite Beam to Boundary-Layer Turbulence (Lin, 1977)

When a solid object moves slowly within a fluid, the fluid particles in contact with the object acquire the same velocities at various points of the contacting surface, whereas fluid particles at some distances from the contacting surface are essentially undisturbed. The layer of fluid, in which fluid velocity varies from that of the contacting surface to that of undisturbed fluid, is known as the *boundary layer.* If the relative motion between the solid object and the undisturbed fluid is fast, then a stable velocity profile in the boundary layer can no longer be maintained, and turbulence occurs in the layer. The phenomenon is known as *boundary-layer turbulence.*

Vibration and noise induced by boundary-layer turbulence are important considerations to designers of flight and marine vehicles. For the purpose of their computations, experimental measurements have been directed at the pressure fluctuation at the interface generated by the boundary-layer turbulence (e.g., Corcos, 1963). It was found that the cross-spectral density of the pressure field at two locations, separated by a distance ξ in the mean flow direction, can be fitted in the general form of

$$\Phi_{pp}(\xi, \omega) = \Phi_{pp}(0, \omega)\psi(\xi)e^{-i\omega\xi/u_c} \qquad (a)$$

where u_c is a constant, $\psi(\xi)$ is an even function with an absolute maximum equal to unity at $\xi = 0$ and with a nonnegative Fourier transform, and $\Phi_{pp}(0, \omega)$ is the spectral density at an arbitrary point in the pressure field. We have assumed, for simplicity, that the pressure field is one-dimensional and is stochastically homogeneous with respect to a moving frame of reference convected at velocity u_c. In the special case $\psi(\xi) = 1$, equation (a) reduces to the cross-spectral density of a frozen pressure field convected at a velocity u_c. This can be seen by comparing equation (a) with equation (r) in Section 2.3.2, upon substituting $\eta = 0$ and $k = \omega/u$ in that equation.

With the objective of matching the form of the cross-spectral density (a), Lin and Maekawa (1977) proposed the following superposition scheme for the boundary-layer turbulence:

$$p(x,t) = \int_{-\infty}^{\infty} dG(u) \int_{-\infty}^{\infty} e^{i\beta(x-ut)} dF(\beta, u) \tag{b}$$

where $G(u)$ is an orthogonal-increment process in u and $F(\beta, u)$ is an orthogonal-increment process in β. The physical implication of (b) is clear: $p(x,t)$ is composed of infinitely many frozen-pattern components, convected at different velocities. The cross-correlation of $p(x,t)$ is given by

$$E[p(x_1,t_1)p(x_2,t_2)] = \iint^{\infty} \int_{-\infty} \int e^{i(u_1\beta_1 t_1 - u_2\beta_2 t_2)} e^{-i(\beta_1 x_1 - \beta_2 x_2)}$$
$$E[dF(\beta_1, u_1)\, dF^*(\beta_2, u_2)\, dG(u_1)\, dG^*(u_2)] \tag{c}$$

Invoking the properties of orthogonal increments in $F(\beta, u)$ and $G(u)$, the right-hand side of equation (c) simplifies to a function of $\xi = x_2 - x_1$ and $\tau = t_2 - t_1$:

$$R_{pp}(\xi, \tau) = \int^{\infty} \int_{-\infty} e^{i(\beta u\tau - \beta\xi)} S_{pp}(\beta,\ u)\, d\beta\, du \tag{d}$$

Application of the Fourier transform results in

$$\Phi_{pp}(\xi, \omega) = \frac{1}{2\pi} \int_{-\infty}^{\infty} R_{pp}(\xi, \tau) e^{-i\omega\tau} d\tau = \int_{-\infty}^{\infty} \frac{1}{|u|} e^{-i\omega\xi/u} S_{pp}\left(\frac{\omega}{u}, u\right) du \tag{e}$$

Equation (e) is a theoretical pressure cross-spectrum derived under the assumption that the superposition scheme (b) is valid.

We now wish to compare the theoretical cross-spectrum (e) with the experimental one, (a). First we apply the Fourier transform to (a), resulting in

$$\frac{1}{2\pi} \int_{-\infty}^{\infty} \Phi_{pp}(\xi, \omega) e^{i\xi\alpha} d\xi = \Phi_{pp}(0, \omega) \Psi\left(\alpha - \frac{\omega}{u_c}\right) \tag{f}$$

where
$$\Psi(\gamma) = \frac{1}{2\pi} \int_{-\infty}^{\infty} \psi(\xi) e^{i\gamma\xi} d\xi \tag{g}$$

Therefore, the experimental cross-spectrum has a representation of

$$\Phi_{pp}(\xi, \omega) = \Phi_{pp}(0, \omega) \int_{-\infty}^{\infty} \Psi\left(\alpha - \frac{\omega}{u_c}\right) e^{-i\alpha\xi} d\alpha$$

Letting $\alpha = \omega/u$, we arrive at

$$\Phi_{pp}(\xi, \omega) = \Phi_{pp}(0, \omega) \int_{-\infty}^{\infty} \left|\frac{\omega}{u^2}\right| \Psi\left(\frac{\omega}{u} - \frac{\omega}{u_c}\right) e^{-i\xi\omega/u} du \tag{h}$$

Equating (e) and (h):

$$S_{pp}\left(\frac{\omega}{u},\ u\right) = \left|\frac{\omega}{u}\right| \Phi_{pp}(0, \omega)\Psi\left(\frac{\omega}{u} - \frac{\omega}{u_c}\right)$$

$$= \left|\frac{\omega}{u}\right| \Phi_{pp}(0, \omega)\left[\frac{1}{2\pi}\int_{-\infty}^{\infty} \psi(\xi)e^{i\xi(\omega/u-\omega/u_c)}d\xi\right] \tag{i}$$

Equation (i) provides a procedure whereby a measured cross-spectrum, characterized by $\Phi_{pp}(0, \omega)$, the spatial decay function $\psi(\xi)$, and a dominant convection speed u_c, can be incorporated into the theoretical spectrum.

We next show why the theoretical form (e) for the pressure cross-spectrum has an advantage in the structural response computations. By a generalization of equation (p) in Section 2.3.2 to the case of a one-dimensional continuous structure, we obtain

$$\Phi_{WW}(x_1, x_2; \omega) = \int_{-\infty}^{\infty} \frac{1}{|u|}H(x_1, \omega)H^*(x_2, \omega)S_{pp}\left(\frac{\omega}{u}, u\right)du \tag{j}$$

The integration variable u can now be converted to the wave number k, while keeping ω fixed. We note that

$$\frac{\omega}{u} = k \qquad u = \frac{\omega}{k} \qquad du = \left|\frac{\omega}{k^2}\right| dk$$

Consequently,

$$\Phi_{WW}(x_1, x_2; \omega) = \int_{-\infty}^{\infty} \frac{1}{|k|}\bar{H}(x_1, k)\bar{H}^*(x_2, k)S_{pp}\left(k, \frac{\omega}{k}\right)dk \tag{k}$$

Substituting (i) into (k), we obtain an extremely simple result:

$$\Phi_{WW}(x_1, x_2; \omega) = \Phi_{pp}(0, \omega)\int_{-\infty}^{\infty} \bar{H}(x_1, k)\bar{H}^*(x_2, k)\Psi_{pp}\left(k - \frac{\omega}{u_c}\right)dk \tag{l}$$

The simplicity of equation (l) may not be fully appreciated by a casual reader since equation (l) still remains in an integral form. However, remember that for spectral analysis of a one-dimensional structure using the conventional distributed load representation, one would require a double integration over the spatial coordinate. Replacing it by a single integration should result in a reduction in computer time of several orders of magnitude. The fact that (l) is an infinite integral is not a problem. The wave-number response function $\bar{H}(x, k)$ diminishes quickly at large absolute values of k. Furthermore, the range of ω that needs to be considered is restricted within the dominant spectral range of the input spectrum $\Phi_{pp}(0, \omega)$, as is to be expected and clearly demonstrated in equation (l).

For the application to a specific dynamic system, one must know the wave-number response function $\bar{H}(x, k)$ of that system. This wave-number response function is one associated with a unit sinusoidal pressure distribution frozen in space, the computation of which is relatively simple. Again, we choose the simplest model possible, which nevertheless retains important features of the panel vibration problem for airplanes, ships, or other high-speed transportation systems using panel construction on the bodies. The one we choose for the ensuing analysis is an infinite beam

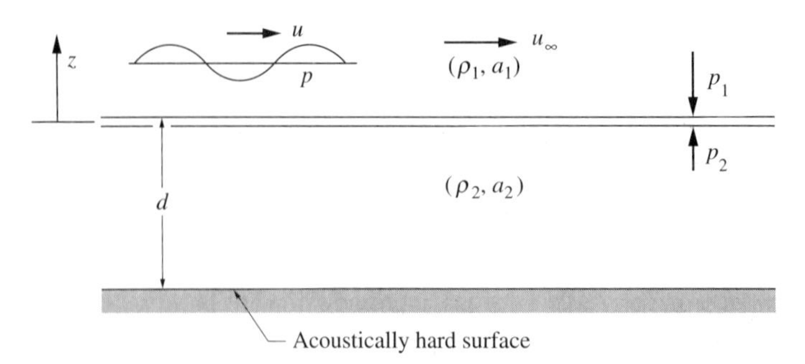

FIGURE 2.3.5
An infinite beam under the excitation of a convected sinusoidal pressure field.

shown in Fig. 2.3.5. On the upper side, the beam is exposed to a fluid with density ρ_1 and sound speed a_1. On the lower side, it forms one boundary of a cavity which contains a fluid with density ρ_2 and sound speed a_2. Fluid 1 has an ambient velocity u_∞ relative to the beam, whereas fluid 2 is initially quiescent with respect to the beam.

The problem of computing $\bar{H}(x, k)$ can be represented symbolically as finding the steady-state solution of the equation

$$\mathscr{L}[\bar{H}(x, k)e^{-i\omega t}] = e^{i(kx-\omega t)} \tag{m}$$

where $\mathscr{L}$ is a linear operator since the problem is assumed to be linear. The equation of motion for the beam is

$$EI\frac{\partial^4 w}{\partial x^4} + m\frac{\partial^2 w}{\partial t} = p + (p_1 - p_2)_{z=0} \tag{n}$$

For the determination of the wave-number response $\bar{H}$, the excitation p is replaced by $e^{i(kx-\omega t)}$ and w is replaced by

$$w = \bar{H}(x, k)e^{-i\omega t} = A(k)e^{i(kx-\omega t)} \tag{o}$$

The second half of equation (o) is derived from the fact that, at the steady state, the exponential factor $e^{i(kx-\omega t)}$ can be canceled from both sides of the equation.

The acoustic pressure p_1, induced by the structural motion on the upper side of the beam, is governed by the convective wave equation

$$\left(\frac{\partial}{\partial t} + u_\infty\frac{\partial}{\partial x}\right)^2 p_1 - a_1^2\left(\frac{\partial^2}{\partial x^2} + \frac{\partial^2}{\partial z^2}\right)p_1 = 0 \tag{p}$$

and satisfies the boundary condition

$$\left(\frac{\partial p_1}{\partial z}\right)_{z=0} = \rho_1\left(\frac{\partial}{\partial t} + u_\infty\frac{\partial}{\partial x}\right)^2 w \tag{q}$$

as well as the radiation condition that it propagates only in the domain $z > 0$. Note that the ambient velocity of the fluid u_∞, appearing in equations (p) and (q), need not be the same as the convection velocity of the excitation, which is equal to $u = \omega/k$

in equation (o). We follow the usual linearization scheme that p_1 and p_2 may be computed without regard to the presence of p.

Because of the ambient velocity u_∞ and the structural motion, the induced pressure p_1 need not propagate exactly in the positive z direction. Let the direction of propagation be at an angle ϕ from the x axis as shown in Fig. 2.3.6. However, the trace of the p_1 wave along the x axis must match that of the structural motion. Therefore,

$$k_t \cos\phi = k \tag{r}$$

Referring to Fig. 2.3.6, p_1 may be expressed as

$$p_1 = P_1 e^{i(kx + kz\tan\phi - \omega t)} \tag{s}$$

Substituting (s) into (p), one obtains, after simplification,

$$\cos\phi = \frac{a_1}{u - u_\infty} \tag{t}$$

Thus

$$\begin{aligned} 0 < \phi \le \frac{\pi}{2} \qquad & u \ge u_\infty \\ \frac{\pi}{2} < \phi < \pi \qquad & u < u_\infty \end{aligned} \tag{u}$$

and

$$k_z = k\tan\phi = \frac{k}{a_1}\sqrt{(u - u_\infty)^2 - a_1^2} \tag{v}$$

Only the positive square root is retained in (v) since p_1 does not propagate in the negative z direction. Furthermore, no propagation is possible when

$$\left| u - u_\infty \right| < a_1$$

since k_z then becomes imaginary.

We now substitute (o), (s), and (v) into the boundary condition (q) and solve for P_1 in terms of the structural motion amplitude A:

$$P_1 = i\rho_1 a_1 \frac{k(u - u_\infty)^2}{\sqrt{(u - u_\infty)^2 - a_1^2}} A \tag{w}$$

Thus the entire p_1 field is determined. In particular, at $z = 0$,

$$(p_1)_{z=0} = i\rho_1 a_1 \frac{k(u - u_\infty)^2}{\sqrt{(u - u_\infty)^2 - a_1^2}} A e^{i(kx - \omega t)} \tag{x}$$

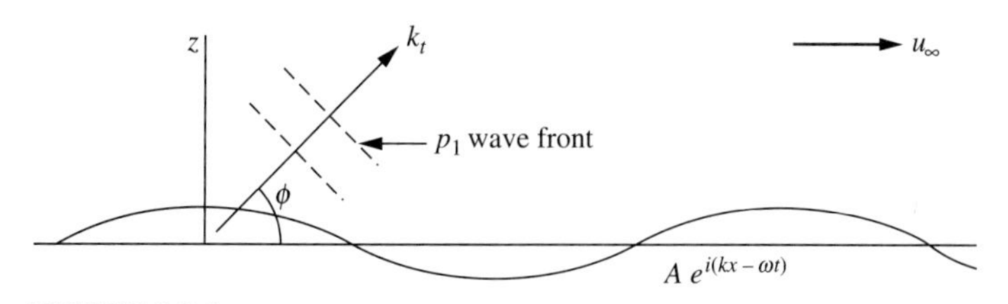

FIGURE 2.3.6
Acoustic pressure p_1 induced by structural motion.

It is of interest to note that equation (x) can also be expressed in the form

$$(p_1)_{z=0} = \frac{i\rho_1 a_1 \omega A e^{i(kx-\omega t)}}{(1 + M\cos\phi)\sin\phi} \tag{y}$$

where $M = u_\infty / a_1$ is the Mach number of the ambient flow.

Next we investigate the other induced pressure p_2, which is governed by

$$\frac{\partial^2 p_2}{\partial t^2} - a_2^2\left(\frac{\partial^2}{\partial x^2} + \frac{\partial^2}{\partial z^2}\right)p_2 = 0 \tag{z}$$

subject to the boundary conditions

$$\left(\frac{\partial p_2}{\partial z}\right)_{z=0} = \rho_2 \ddot{w} \qquad \left(\frac{\partial p_2}{\partial z}\right)_{z=-d} = 0 \tag{aa}$$

In order that the second of the preceding conditions can be satisfied, the solution to equation (z) may be written as

$$p_2 = P_2 \cos[k_z(z + d)]e^{i(kx-\omega t)} \tag{bb}$$

with $$k_z = k\tan\phi \qquad \cos\phi = \frac{a_2}{u} \qquad \tan\phi = \frac{\sqrt{u^2 - a_2^2}}{a_2} \tag{cc}$$

obtained by letting $u_\infty = 0$ from the previous results for p_1. Of course, ϕ must now be measured clockwise from the positive x axis. Substituting (bb) into the first equation in (aa), we obtain

$$P_2 = \frac{\omega^2 \rho_2}{k_z \sin(k_z d)} A \tag{dd}$$

Consequently, $$(p_2)_{z=0} = \frac{\rho_2 \omega^2 a_2 \cot(k_z d)}{k\sqrt{u^2 - a_2^2}} A e^{i(kx-\omega t)} \tag{ee}$$

Finally, substituting (o), (y), and (ee) into (n), we find an equation for A as follows:

$$A = Q^{-1} \tag{ff}$$

where $$Q = EIk^4 - m\omega^2 - i\rho_1 a_1 \frac{k(u - u_\infty)^2}{\sqrt{(u - u_\infty)^2 - a_1^2}} + \rho_2 a_2 \frac{ku^2 \cot(k_z d)}{\sqrt{u^2 - a_2^2}} \tag{gg}$$

The required wave-number response function $\bar{H}(x, k)$ follows from

$$\bar{H}(x, k) = A(k)e^{ikx} = Q^{-1}e^{ikx} \tag{hh}$$

Some comments about the foregoing results are in order:

1. The p_1 term gives rise to damping if $|u - u_\infty| > a_1$.
2. This term gives rise to an apparent mass effect if $|u - u_\infty| < a_1$, in which case, the square root $\sqrt{(u - u_\infty)^2 - a_1^2}$ should be given a positive imaginary value.
3. For a small d (shallow cavity) and $|u| > a_2$, the p_2 term gives rise to an additional spring, and for certain ranges of d value it can change to an additional inertia.

4. When $|u| < a_2$, k_2 becomes imaginary, in which case

$$\frac{\cot(k_z d)}{k_z} = -\frac{\coth|k_z d|}{|k_z|} \tag{ii}$$

Again, the p_2 term adds to the inertia of the system.

In concluding this section, we remark that this analysis is based on an idealized structural model of an infinite beam without supports, which is, of course, not realistic. However, this simple model is capable of bringing into focus various ingredients of the analysis, without being obscured by less essential details. A similar analysis has been carried out for an infinite beam with periodically spaced elastic supports (Lin et al., 1976), and the solution obtained therein is also mathematically exact.

2.3.4 Bridge Response to Wind Excitations

As another example for spectral analysis, the motion of a wind-excited long-span bridge will be considered. The bridge motion at a given location along the span may be decomposed into a vertical component, a horizontal component essentially perpendicular to the span, and a rotational component. Motions of these components are generally coupled both structurally and aerodynamically. However, in presenting the basic ideas in this section, only one component, the rotational component shown in Fig. 2.3.7, is treated.

Wind loads on a bridge may be classified into two types: self-excited loads and buffeting loads (Scanlan, 1978). Buffeting loads are independent of the bridge motion, whereas self-excited loads would disappear if a bridge were held motionless at all times. Assuming that the structure behaves linearly, the equation of motion may be written as

$$\mathcal{L}_x\theta + I\ddot{\theta} + c\dot{\theta} = Q_s(x,t) + Q_b(x,t) \tag{a}$$

where I = mass polar moment of inertia per unit span length, c = structural damping coefficient, Q_s = self-excited torsional moment per unit span length, Q_b = buffeting torsional moment per unit span length, and $\mathcal{L}_x$ = linear operator in the spatial

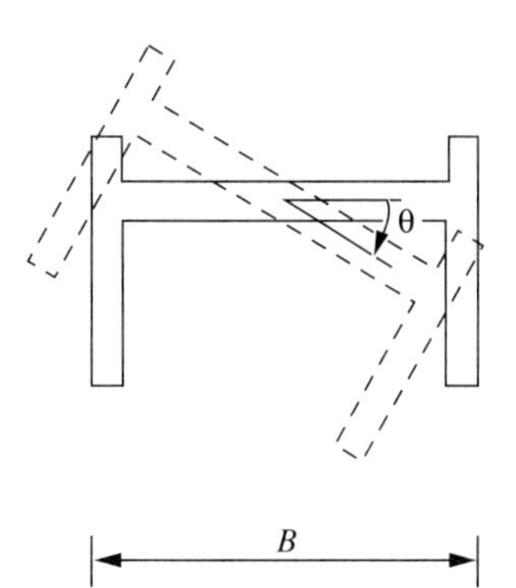

FIGURE 2.3.7
Rotation of a bridge cross section.

variable x, representing the mechanism which generates elastic restoring moment. For instance, if the bridge behaves like a slender bar, then

$$\mathcal{L}_x = \frac{\partial^2}{\partial x^2}\left(E\Gamma\frac{\partial^2}{\partial x^2}\right) - \frac{\partial}{\partial x}\left(GJ\frac{\partial}{\partial x}\right) \tag{b}$$

where E = Young's modulus, G = shear modulus, Γ = warping constant for non-uniform torsion, and J = Saint Venant's constant for torsion. If, in addition, pair-wise stay cables are installed on the two sides of the bridge, at a distance B from each other, then the cable restoring forces must be included in (b):

$$\mathcal{L}_x = \frac{\partial^2}{\partial x^2}\left(E\Gamma\frac{\partial^2}{\partial x^2}\right) - \frac{\partial}{\partial x}\left(GJ\frac{\partial}{\partial x}\right) + \frac{B^2}{2}\sum_{i=1}^{n}\frac{E_i A_i}{l_i}\sin^2\phi_i\delta(x - x_i) \tag{c}$$

in which x_i = location of the ith cable pair; n = number of cable pairs; and E_i, A_i, l_i, ϕ_i are Young's modulus, the cross-sectional area, length, and inclination angle of the ith cable pair, respectively.

The self-excited load depends on the past and present bridge motion, relative to the surrounding wind flow. If the bridge cross section is streamlined, then the generating mechanism for the self-excited load is linear, similar to that of an airfoil, which can be determined theoretically. For bluff cross sections, the mechanism can be strongly nonlinear. However, a linear theory for self-excited load, proposed by Scanlan (1978), is believed to be applicable for bridges with relatively open constructions. This linear theory is adopted in the following discussion.

Near the earth's surface, the average wind velocity is essentially horizontal, with small turbulent perturbations in different directions. The effects of turbulent perturbations on the self-excited loads are important only when the horizontal average wind velocity is high enough so that the bridge motion is close to instability, a subject that is considered in Section 6.6. Here we assume that the bridge motion is far from being unstable, and that the self-excited loads can be obtained without regard to wind turbulence.

Using a linear "strip" theory of aerodynamics (e.g., Bisplinghoff and Ashley, 1962), the self-excited loads may be expressed as

$$Q_s = \frac{1}{2}\rho u^2(2B^2)\left[c_1\theta + \frac{B}{u}c_2\dot{\theta} + \frac{u}{B}\int_{-\infty}^{t} h_s(t-\tau)\theta(\tau)\,d\tau\right] \tag{d}$$

where ρ = air density, B = width of the bridge deck, c_1 and c_2 = constants, and u = the horizontal component of the wind velocity perpendicular to the bridge span, which is now treated as being deterministic, since the effects of turbulence on the self-excited load are neglected here. The physical meaning of the three terms within the brackets in equation (d) are clear. The first two terms account for the effects of the present bridge position $\theta(t)$ and velocity $\dot{\theta}(t)$, and the third term accounts for the effects of the past bridge motion. Function $h_s(t-\tau)$ gives the nondimensionalized aerodynamic pitching moment generated at time t, due to an impulsive change of θ at an earlier time τ. The coefficients B^2, B/u, and u/B in the equation are introduced to nondimensionalize the constants c_1 and c_2 and the "impulse response" function $h_s(t-\tau)$. That function h_s depends only on the time difference $t-\tau$ implies a time-invariant mechanism, which is the case when the effects of wind turbulence on the self-excited loads are negligible.

Unlike the self-excited loads, the buffeting loads are independent of the bridge motion and are caused directly by the wind turbulence, particularly the vertical turbulence. Denote the vertical turbulent velocity by $\eta(x, t)$ and assume that it is stochastically stationary in time. Then, $\eta(x, t)$ has a spectral representation

$$\eta(x, t) = \int_{-\infty}^{\infty} e^{i\omega t} d\tilde{\eta}(x, \omega) \tag{e}$$

where $\tilde{\eta}(x, \omega)$ is an orthogonal increment process in ω:

$$E[d\tilde{\eta}(x_1, \omega_1)\, d\tilde{\eta}^*(x_2, \omega_2)] = \begin{cases} 0 & \omega_1 \neq \omega_2 \\ \Phi_{\eta\eta}(x_1, x_2, \omega)\, d\omega & \omega_1 = \omega_2 = \omega \end{cases} \tag{f}$$

and where $\Phi_{\eta\eta}(x_1, x_2, \omega)$ is the cross-spectral density of $\eta(x_1, t)$ and $\eta(x_2, t)$. Assuming that the buffeting pitching moment on the bridge is also generated by a linear time-invariant mechanism, we may write

$$Q_b(x, t) = \frac{1}{2}\rho u^2 \left(\frac{2B^2}{u}\right) \int_{-\infty}^{t} h_b(t - \tau)\eta(x, \tau)\, d\tau \tag{g}$$

Substituting (e) into (g), we obtain

$$Q_b(x, t) = \frac{1}{2}\rho u^2 \left(\frac{2B^2}{u}\right) \int_{-\infty}^{\infty} e^{i\omega t} H_b(\omega)\, d\tilde{\eta}(x, \omega) \tag{h}$$

where

$$H_b(\omega) = \int_0^{\infty} e^{-i\omega v} h_b(v)\, dv \tag{i}$$

We digress to note that since integral terms are included in equations (d) and (g) for the self-excited and buffeting loads, the governing equation (a) is actually a differential-integral equation. However, it can be converted easily to an equivalent set of simultaneous differential equations without the integral terms, if so desired.

After the onset of the buffeting excitation for some time (usually in the order of $10 \approx 20$ natural period of the fundamental mode), the bridge motion $\theta(x, t)$ practically reaches the state of stochastic stationarity, having the spectral representation

$$\theta(x, t) = \int_{-\infty}^{\infty} e^{i\omega t} d\tilde{\theta}(x, \omega) \tag{j}$$

where

$$E[d\tilde{\theta}(x_1, \omega_1)\, d\tilde{\theta}^*(x_2, \omega_2)] = \begin{cases} 0 & \omega_1 \neq \omega_2 \\ \Phi_{\theta\theta}(x_1, x_2, \omega)\, d\omega & \omega_1 = \omega_2 = \omega \end{cases} \tag{k}$$

Substituting equations (d), (h), and (j) into equation (a), we obtain

$$(\mathscr{L}_x - I\omega^2 + i\omega c)\, d\tilde{\theta}(x, \omega) = \frac{1}{2}\rho u^2 (2B^2) \left[c_1 + i\omega \left(\frac{B}{u}\right) c_2 + \left(\frac{u}{B}\right) H_s(\omega)\right] d\tilde{\theta}(x, \omega) + \frac{1}{2}\rho u^2 \left(\frac{2B^2}{u}\right) H_b(\omega)\, d\tilde{\eta}(x, \omega) \tag{l}$$

where
$$H_s(\omega) = \int_0^\infty e^{-i\omega v} h_s(v)\, dv \tag{m}$$

The aerodynamic constants and functions c_1, c_2, $H_s(\omega)$, and $H_b(\omega)$ can be determined theoretically only if the bridge cross section is streamlined. For bluff cross sections, they must be measured experimentally at the present time. Scanlan and Tomko (1971) measured the self-excited loads in the frequency domain and plotted the results in terms of the so-called flutter derivatives, A_1 and A_3,

$$c_1 + i\omega\left(\frac{B}{u}\right)c_2 + \left(\frac{u}{B}\right)H_s(\omega) = K^2(A_3 + iA_2) \tag{n}$$

where $K = B\omega/u$ is the reduced frequency and A_2 and A_3 are real-valued functions of K. In Scanlan's original notation, A_2 and A_3 were denoted by A_2^* and A_3^*. The asterisk is omitted here to avoid confusion with the notation of a complex conjugate. These measured results are reproduced in Fig. 2.3.8, along with the flutter derivatives for the vertical translation and coupled vertical-rotational motion. Additional measurements were reported by Huston (1986).

Scanlan (1978) proposed that the transfer function $H_b(\omega)$ for the buffeting load may be approximated as

$$H_b(\omega) = \frac{1}{2}\left(\frac{dC_L}{d\theta}\right)_{\theta=\theta_0} \tag{o}$$

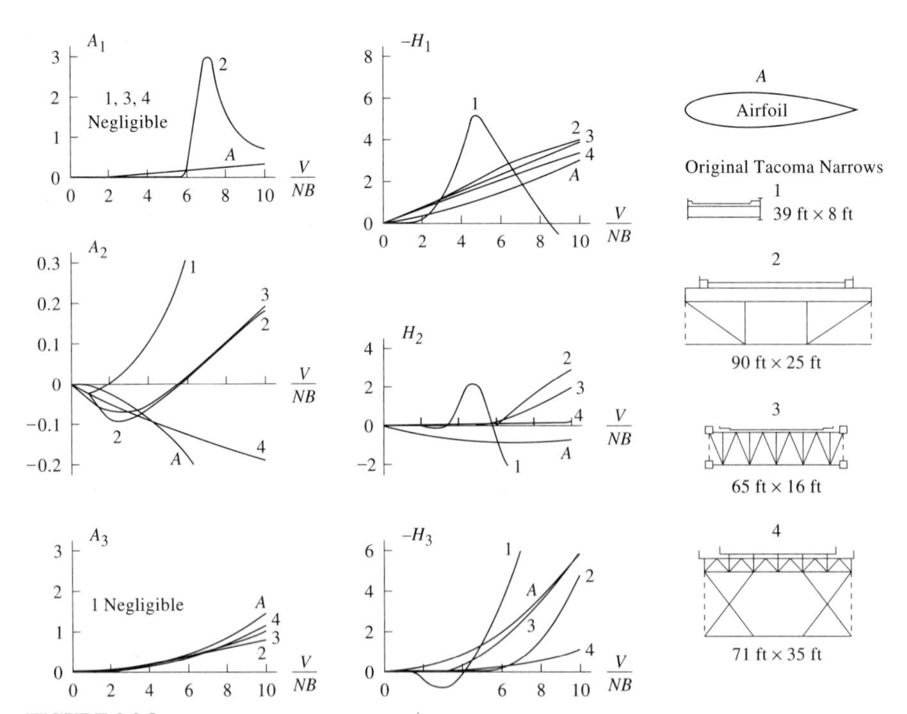

FIGURE 2.3.8
Experimentally determined flutter derivatives for bridge sectional models [after Scanlan and Tomko, 1971; Copyright ©1971 ASCE, reprinted with permission].

where C_L is the lift coefficient and θ_0 is the equilibrium angle of rotation of the bridge under a steady wind velocity u. In this case, $H_b(\omega)$ is treated as a constant, independent of ω. More accurate modelings must again rely on comprehensive experimental measurements, especially for bluff cross sections.

We now expand $d\tilde{\theta}(x, \omega)$ in equation (l):

$$d\tilde{\theta}(x, \omega) = \sum_j \phi_j(x)\, d\tilde{a}_j(\omega) \tag{p}$$

where $\phi_j(x)$ are chosen to be the normal modes of the structure itself. Thus the $\phi_j(x)$ functions satisfy the relation

$$\mathcal{L}_x \phi_j(x) = I\omega_j^2 \phi_j(x) \tag{q}$$

and the orthogonality condition

$$\int_l I\phi_j(x)\phi_k(x)\, dx = \begin{cases} I_j & j = k \\ 0 & j \neq k \end{cases} \tag{r}$$

where ω_j is the natural frequency of the jth mode, and the integral in equation (r) is carried out over the bridge span l. Since structural damping is usually small, we follow the usual practice of neglecting the structure damping coupling, if any. In other words, we assume that

$$\int_l c\phi_j(x)\phi_k(x)\, dx = \begin{cases} 2\zeta_j \omega_j I_j & j = k \\ 0 & j \neq k \end{cases} \tag{s}$$

Using equations (p) through (s) in equation (l), we obtain

$$d\tilde{a}_j(\omega) = \frac{\rho u B}{\Xi_j} H_b(\omega) \int_l \phi_j(x)[d\tilde{\eta}(x, \omega)]\, dx \tag{t}$$

$$\text{where}\, \Xi_j = I_j(\omega_j^2 - \omega^2 + 2i\zeta_j\omega_j\omega) - \rho u^2 B^2 \left[c_1 + i\omega \left(\frac{B}{u}\right) c_2 + \left(\frac{u}{B}\right) H_s(\omega) \right] \tag{u}$$

From equation (p),

$$\Phi_{\theta\theta}(x_1, x_2, \omega)\, d\omega = \sum_j \sum_k \phi_j(x_1)\phi_k(x_2) E[d\tilde{a}_j(\omega)\, d\tilde{a}_k^*(\omega)] \tag{v}$$

and from equation (t),

$$E[d\tilde{a}_j(\omega)\, d\tilde{a}_k^*(\omega)] = d\omega \frac{(\rho u B)^2}{\Xi_j \Xi_k^*} \left|H_b(\omega)\right|^2 \int_l dy_1 \int_l \phi_j(y_1)\phi_k(y_2)\Phi_{\eta\eta}(y_1, y_2, \omega)\, dy_2 \tag{w}$$

It follows upon combining equations (v) and (w) that the cross-spectral densities of the bridge motion and the vertical wind turbulence are related as

$$\Phi_{\theta\theta}(x_1, x_2, \omega)$$

$$= \sum_j \sum_k \frac{(\rho u B)^2}{\Xi_j \Xi_k^*} \left|H_b(\omega)\right|^2 \phi_j(x_1)\phi_k(x_2) \int_l dy_1 \int_l \phi_j(y_1)\phi_k(y_2)\Phi_{\eta\eta}(y_1, y_2, \omega)\, dy_2 \tag{x}$$

TABLE 2.3.1
Roughness length Z_0 of various earth surfaces†

Type of surface	Range of Z_0m
Sea, sand	0.000003–0.004
Snow	0.001–0.006
Mowed grass, to prairie	0.001–0.04
High grass	0.04–0.10
Brush	0.10–0.30
50-ft (12.25-m) pine forest, medium dense	0.90–1.00
Suburbs, outskirts to center	0.20–0.45
Large city centers	0.60–0.80

† After Scanlan (1978).

The source of information for the turbulence spectra is essentially from experiments. Several hypothetical forms have been proposed for the spectral density $\Phi_{\eta\eta}(\omega)$, including one by Lumley and Panofsky (1964):

$$\Phi_{\eta\eta}(\omega) = \frac{168}{2\pi\left[1 + 10(\omega Z/2\pi u)^{5/3}\right]}\left(\frac{u_*}{u}\right)^2\left(\frac{Z}{u}\right) \qquad -\infty < \omega < \infty \tag{y}$$

where Z = the bridge height above the mean terrain level and u_* = the so-called friction velocity. Assuming that the average horizontal wind velocity u varies logarithmically with height, the friction velocity u_* is related to u through (e.g., Scanlan, 1978)

$$u = \frac{u_*}{k}\ln\frac{Z - Z_d}{Z_0} \tag{z}$$

in which $k = 0.4$, $Z_d = \bar{H} - Z_0/k$, $\bar{H}$ is the average height of surrounding buildings (if any), and Z_0 is given in Table 2.3.1.

It has been suggested that the cross-spectral density of the turbulence velocity may be approximated as follows (Davenport, 1962):

$$\Phi_{\eta\eta}(x_1, x_2, \omega) = e^{-(2\pi C\omega/u)|x_1 - x_2|}\Phi_{\eta\eta}(\omega) \tag{aa}$$

where the value of C ranges between 7 and 20 (the lower values being more conservative), and $\Phi_{\eta\eta}(\omega)$ has been given in equation (y). Equation (aa) implies that the turbulence field is not only stochastically stationary in time but also stochastically homogeneous in space. Equations (y) through (aa) furnish the information for the buffeting load spectra required in equation (x).

2.4 CONCLUDING REMARKS

In this chapter we presented methods to obtain the spectral densities and cross-spectral densities of the structural response from those of random excitations. The analysis is applicable only if the excitations are additive and the dynamical system is linear and time-invariant, in the sense that the governing equations of motion are linear differential equations with constant coefficients or convertible to such differential equations. The dynamical system involved may include the structure itself and the surrounding fluids, as illustrated in several examples. By introducing the concept of random process with orthogonal increments, the analysis is made simple, as well as rigorous, and it permits a direct analogy between a random process stochastically stationary in time and one stochastically homogeneous in space. In one example, these two types of random processes are shown to represent the same physical phenomenon, viewed from two different frames of reference.

2.5 EXERCISES

2.1. Complete the proof of equation (2.2.2) for the cases $\omega_1 \leq \omega_0 \leq \omega_2$ and $\omega_1 < \omega_2 < \omega_0$.

2.2. As shown in Fig. P2.2, N particles with equal masses are supported at equal distances by a massless taut string. Let m be the mass of each particle, c the damping coefficient, d the distance between neighboring particles, and T the tension in the string, which is assumed to be unchanged for small particle deflections. The equations of motion are given by

$$m\ddot{W}_j + c\dot{W}_j - \frac{T}{d}(W_{j-1} - 2W_j + W_{j+1}) = F_j(t) \qquad j = 1, 2, \ldots, N$$

with boundary conditions $W_0 = W_{N+1} = 0$.

(a) Find the frequency response function for the displacement $W_m(t)$ of the mth particle due to a force $F_k(t)$ at the kth particle.

(b) Obtain an expression for the spectral density of the displacement $W_m(t)$ in terms of the spectral densities and cross-spectral densities of the excitations $F_j(t)$.

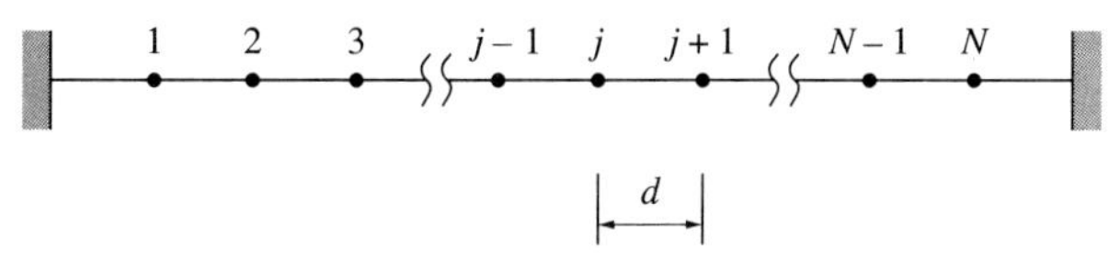

FIGURE P2.2
N particles supported by a massless string.

2.3. Modify the structural model for the wind-excited N-story building in Section 2.3.1 such that the viscous damping forces are dependent on the relative velocities of neighboring floors. Such a damping mechanism is represented schematically in Fig. P2.3.

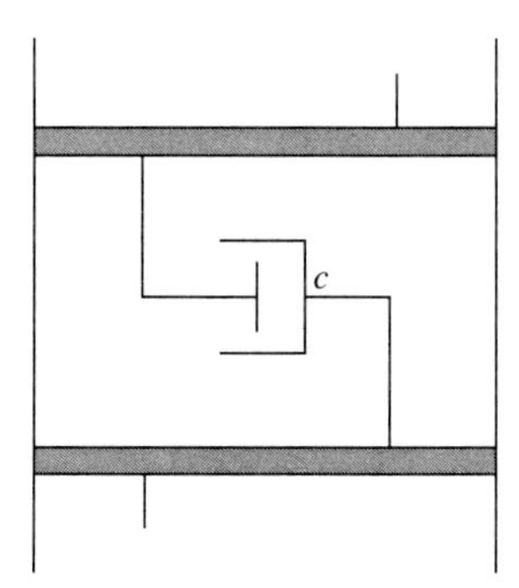

FIGURE P2.3
Schematic of a damping mechanism.

(a) Construct the transfer matrix $\boldsymbol{T}$ for a typical story unit.

(b) Obtain an expression for the spectral density of the base shear $V_0(t)$ in the stationary state.

2.4. As shown in Fig. P2.4, a simply supported beam is exposed to a frozen random pressure field, which is convected at a velocity u_c relative to the beam. Let the spectral density of the pressure field in the wave-number domain be given by

$$S_{pp}(k) = \frac{K\alpha}{\pi(\alpha^2 + k^2)} \qquad K\alpha > 0$$

Assume that the beam motion is dominated by the fundamental mode:

$$W(x, t) \approx X(t) \sin \frac{\pi x}{l}$$

Obtain the spectral density of the beam deflection at the midspan in the frequency domain.

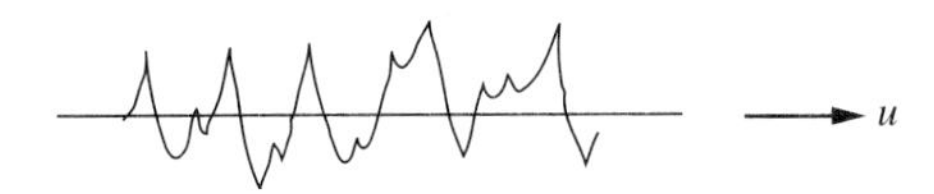

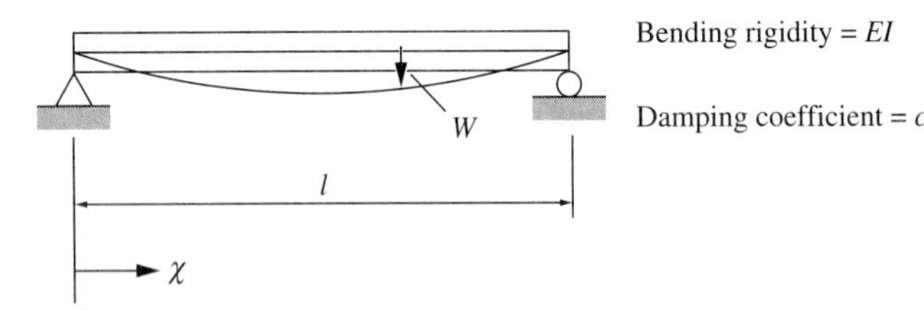

FIGURE P2.4
A simply supported beam in a random pressure field.

2.5. Determine the spectral density of the torsional motion of a bridge under the excitation of a turbulent wind, assuming that the motion is dominated by the fundamental mode

$$\theta(x, t) \approx X(t) \sin \frac{\pi x}{l}$$

The following structural and aerodynamic data are given:

$$\begin{aligned} B &= 16.4 \text{ m} \\ \zeta_1 &= 0.01 \\ \omega_1 &= 0.74\pi \text{ rad/s} \\ I &= 8.98 \times 10^5 \text{ kg} \cdot \text{m}^2/\text{m} \\ \rho &= 1.226 \text{ kg/m}^3 \\ u &= 30 \text{ m/s} \\ c_1 &= 0.0427 \\ c_2 &= -0.3329 \end{aligned}$$

$$H_b(\omega) = 3.14 \text{ (a constant)}$$

$$h_s(t) = -\left\{0.039 \exp\left[-0.1809\left(\frac{u}{B}\right)t\right] + 1.837 \exp\left[-2.1894\left(\frac{u}{B}\right)t\right]\right\}$$

The vertical turbulence is idealized as a uniform white noise along the span length, that is, $\Phi_{\eta\eta}(x_1, x_2, \omega) = K$ (a constant).

CHAPTER 3

EVOLUTIONARY SPECTRAL ANALYSIS

The spectral analysis discussed in Chapter 2 is perhaps the most widely used technique in stochastic dynamics. Although limited to linear systems under covariance-stationary additive excitations, its application is a straightforward extension of the well-known deterministic Fourier analysis. The input-output relationship is simple and transparent in the frequency or wave-number domain, and it is easily generalizable to the case of multiple inputs and multiple outputs. A logical question arises as to whether a similar and almost as simple analysis can be devised for nonstationary excitations. The answer is affirmative for a special type of nonstationary excitations, which is the subject of this chapter.

3.1 EVOLUTIONARY STOCHASTIC PROCESSES

The spectral representation of a correlation-stationary stochastic process (2.2.11) can be extended to a certain class of nonstationary stochastic processes as follows:

$$X(t) = \int_{-\infty}^{\infty} a(t, \omega) e^{i\omega t} dZ(\omega) \tag{3.1.1}$$

where $Z(\omega)$ remains an orthogonal-increment process, and $a(t, \omega)$ is a deterministic function of both t and ω and is generally complex. This extension is due to Priestley (1965), and a stochastic process admissible to such a representation is called an *evolutionary stochastic process.*

Clearly, an evolutionary stochastic process has a zero mean:

$$E[X(t)] = \int_{-\infty}^{\infty} a(t, \omega) e^{i\omega t} E[dZ(\omega)] = 0 \tag{3.1.2}$$

The correlation function of an evolutionary process $X(t)$ is obtained as

$$E[X(t_1)X(t_2)] = \int_{-\infty}^{\infty} \int_{-\infty}^{\infty} a(t_1, \omega_1) a^*(t_2, \omega_2) e^{i(\omega_1 t_1 - \omega_2 t_2)} E[dZ(\omega_1)\, dZ^*(\omega_2)] \tag{3.1.3}$$

Invoking again (2.2.9)

$$E[X(t_1)X(t_2)] = \int_{-\infty}^{\infty} a(t_1, \omega) a^*(t_2, \omega) e^{i\omega(t_1 - t_2)}\, d\Psi(\omega) \tag{3.1.4}$$

which depends on t_1 and t_2 separately. The mean-square value of $X(t)$ is obtained by letting $t_1 = t_2 = t$:

$$E[X^2(t)] = \int_{-\infty}^{\infty} \left| a(t, \omega) \right|^2 d\Psi(\omega) \tag{3.1.5}$$

If $\Psi(\omega)$ is differentiable, then by letting $d\Psi(\omega) = \Phi(\omega)\, d\omega$, the expression for the mean-square value becomes

$$E[X^2(t)] = \int_{-\infty}^{\infty} \left| a(t, \omega) \right|^2 \Phi(\omega)\, d\omega \tag{3.1.6}$$

In equations (3.1.5) and (3.1.6), $\Psi(\omega)$ and $\Phi(\omega)$ are the spectral distribution function and spectral density function, respectively, of *some* suitable stationary process.

The product $|a(t, \omega)|^2 \Phi(\omega)$ is known as the *evolutionary spectral density.* The name is physically meaningful if $a(t, \omega)$ is a slowly varying function of t, such that it can be treated as being nearly independent of t within a time interval considerably longer than $2\pi/\omega$ for any given ω. Otherwise, the name is merely mathematical. Furthermore, since the deterministic function $a(t, \omega)$ in the spectral representation (3.1.1) depends also on ω, the separation between $a(t, \omega)$ and $Z(\omega)$ in the representation need not be unique. However, if an evolutionary process is well-defined, then the evolutionary spectral density $|a(t, \omega)|^2 \Phi(\omega)$ or, more generally, the differential of evolutionary spectral distribution

$$dG(t, \omega) = \left| a(t, \omega) \right|^2 d\Psi(\omega) \tag{3.1.7}$$

must be unique.

A subclass within the class of evolutionary stochastic processes is worthy of special attention. This subclass is obtained by restricting the $a(t, \omega)$ function in the spectral representation (3.1.1) to be dependent only on t, which can then be brought outside the integral:

$$X(t) = a(t) \int_{-\infty}^{\infty} e^{i\omega t} dZ(\omega) \tag{3.1.8}$$

Clearly, such a stochastic process is just a correlation-stationary process of the form of (2.2.11) multiplied by a deterministic function $a(t)$. For practical applications, $a(t)$ may be restricted to being nonnegative without loss of generality, in which case it is called the *envelope function* or the *time-modulation function.* A stochastic process having a spectral representation (3.1.8) is called a *uniformly modulated process.*

3.2 THE RANDOM PULSE TRAIN AND ITS EVOLUTIONARY SPECTRAL REPRESENTATION

The random pulse train is one type of stochastic process which has found applications in a variety of engineering problems. It has the general form of (Lin, 1967, p. 87)

$$X(t) = \sum_{j=1}^{N(T)} Y_j w(t, \tau_j) \qquad 0 < t \leq T \tag{3.2.1}$$

where τ_j is the random time at which the jth pulse is initiated (to be referred to hereafter as the *pulse-arrival time*), $w(t, \tau)$ represents a deterministic pulse shape which satisfies the causality condition $w(t, \tau) = 0$ for $\tau > t$, Y_j is the random magnitude of the jth pulse, and $N(T)$ gives the total number of pulses that arrive in the time interval (0, T].

In most applications, assumptions are made that Y_j are independent and identically distributed random variables which are also independent of the arrival times, and that $w(t, \tau)$ is a function of $t - \tau$. Under the second assumption, equation (3.2.1) is simplified to

$$X(t) = \sum_{j=1}^{N(T)} Y_j w(t - \tau_j) \qquad 0 < t \leq T \tag{3.2.2}$$

The arrival times of the random pulses are generally correlated.

STATISTICAL PROPERTIES OF A RANDOM PULSE TRAIN. Let $\Lambda(t)$ be the random pulse-arrival rate. $\Lambda(t)$ may be characterized by its statistical moments, $E[\Lambda(t)] = f_1(t), E[\Lambda(t_1)\Lambda(t_2)] = f_2(t_1, t_2), E[\Lambda(t_1)\Lambda(t_2)\Lambda(t_3)] = f_3(t_1, t_2, t_3), \ldots$. Stratonovich (1963) has shown that if these moments are continuous functions, then they are related to a generating functional through the expansion

$$\begin{aligned} L_T[v(t)] &= E\left\{\prod_{k=1}^{N(T)}[1 + v(t_k)]\right\} \\ &= 1 + \sum_{m=1}^{\infty} \frac{1}{m!} \int_0^T \cdots \int_0^T f_m(t_1, t_2, \ldots, t_m) v(t_1) \cdots v(t_m)\, dt_1 \cdots dt_m \end{aligned} \tag{3.2.3}$$

where function $v(t)$ belongs to a general class for which the right-hand side of (3.2.3) converges. Alternatively, $\Lambda(t)$ can also be characterized by its cumulants, $g_1(t_1),\ g_2(t_1, t_2),\ g_3(t_1, t_2, t_3),\ \ldots$, which have a generating functional

$$\ln L_T[v(t)] = \sum_{m=1}^{\infty} \frac{1}{m!} \int_0^T \cdots \int_0^T g_m(t_1, t_2, \ldots, t_m) v(t_1) \cdots v(t_m)\, dt_1 \cdots dt_m \tag{3.2.4}$$

where ln denotes the natural logarithm. Equations (3.2.3) and (3.2.4) are changed to the expansions of characteristic functional and log-characteristic functional, respectively, by letting $v(t) = i\theta(t)$. It can be shown that the first cumulant is equal to the first moment, and the second and third cumulants are the same as the second and third central moments, respectively. Thus

$$\begin{aligned} g_1(t) &= f_1(t) \\ g_2(t_1, t_2) &= f_2(t_1, t_2) - f_1(t_1)f_1(t_2) \\ g_3(t_1, t_2, t_3) &= f_3(t_1, t_2, t_3) - f_2(t_1, t_2)f_1(t_3) \\ &\quad - f_2(t_1, t_3)f_1(t_2) - f_2(t_2, t_3)f_1(t_1) \\ &\quad + 2f_1(t_1)f_1(t_2)f_1(t_3) \end{aligned} \tag{3.2.5}$$

A higher order cumulant can also be expressed in terms of the moment of the same order and moments of lower orders, but the relation is more complicated. By invoking the fact that a cumulant function must be symmetric with respect to its arguments, Stratonovich (1963) has given a concise general relationship

$$\begin{aligned} g_n(t_1, t_2, \ldots, t_n) &= f_n(t_1, t_2, \ldots, t_n) - \sum^{ds} g_1(t_1)g_{n-1}(t_2, t_3, \ldots, t_n) \\ &\quad - \sum^{ds} g_2(t_1, t_2)g_{n-2}(t_3, t_4, \ldots, t_n) - \cdots \\ &\quad - g_1(t_1)g_1(t_2)\cdots g_1(t_n) \end{aligned} \tag{3.2.6}$$

where $\sum^{ds}$ indicates the summation of all those terms which are distinctive, but each is of the same form as the one shown behind the symbol. For example, for $n = 4$

$$\begin{aligned} \sum^{ds} g_1(t_1)g_3(t_2, t_3, t_4) &= g_1(t_1)g_3(t_2, t_3, t_4) + g_1(t_2)g_3(t_1, t_3, t_4) \\ &\quad + g_1(t_3)g_3(t_1, t_2, t_4) + g_1(t_4)g_3(t_1, t_2, t_3) \end{aligned} \tag{3.2.7}$$

It should be noted that (3.2.6) is equivalent to the equation given originally by Stratonovich (1963), although it is not in the exact same form. Specifically, the operation $\sum^{ds}$ in (3.2.7) replaces two operations in the original form. The first operation is to take an average of the distinctive terms in a sum, and the second operation is to multiply the averaged result by the number of such distinctive terms.

The cumulant functions of the random arrival rate $\Lambda(t)$ describe its correlation properties at different times. In particular, the equality $g_n(t_1, t_2, \ldots, t_n) = 0$ implies that at least one of the random variables among $\Lambda(t_1), \Lambda(t_2), \ldots, \Lambda(t_n)$ is uncorrelated with all the others. In the special case of independent (Poisson) arrivals, which implies uncorrelated arrivals, all the cumulants are zero except the first cumulant g_1. For this special case, the random pulse train (3.2.1) is also known as a filtered Poisson process (Parzen, 1962).

Return now to $X(t)$ given by (3.2.2). The probabilistic properties of $X(t)$ may be investigated conveniently using its characteristic functional, defined as

$$M_{\{X\}}[\theta(t)] = E\left\{ \exp\left[i\int_0^T X(t)\theta(t)\,dt \right] \right\} \tag{3.2.8}$$

The natural logarithm of (3.2.8) has an expansion

$$\ln M_{\{X\}}[\theta(t)] = \sum_{l=1}^{\infty} \frac{i^l}{l!} \int_0^T \cdots \int_0^T \kappa_l[X(t_1), \ldots, X(t_l)]\theta(t_1)\cdots\theta(t_l)\,dt_1\cdots dt_l \tag{3.2.9}$$

where $\kappa_1[X(t_1), \ldots, X(t_1)]$ is the lth cumulant of $X(t)$. Equation (3.2.9) is analogous to equation (3.2.4). Substituting (3.2.2) into (3.2.8),

$$M_{\{X\}}[\theta(t)] = E\left\{\prod_{j=1}^{N(T)} \exp\left[i\int_0^T Y_j\theta(t)w(t-\tau_j)\,dt\right]\right\}$$

$$= E\left\{\prod_{j=1}^{N(T)}\left[1+\sum_{k=1}^{\infty}\frac{i^k}{k!}Y_j^k\int_0^T\cdots\int_0^T\theta(t_1)\cdots\theta(t_k)w(t_1-\tau_j)\cdots w(t_k-\tau_j)\,dt_1\cdots dt_k\right]\right\} \tag{3.2.10}$$

By equating (3.2.3) and (3.2.10), that is, by letting

$$M_{\{X\}}[\theta(t)] = L_T[v(\tau)] \tag{3.2.11}$$

we obtain

$$v(\tau) = \sum_{k=1}^{\infty}\frac{i^k}{k!}Y_j^k\int_0^T\cdots\int_0^T\theta(t_1)\cdots\theta(t_k)w(t_1-\tau)\cdots w(t_k-\tau)\,dt_1\cdots dt_k \tag{3.2.12}$$

If logarithms are taken of both sides of (3.2.11) and use is made of the expansions (3.2.4) and (3.2.9), we obtain expressions for the cumulants of $X(t)$ by comparing the same number of integrations on the t's on the two sides of the equation. In particular, the first and sccond cumulants are the same as the mean and the covariance, respectively,

$$\kappa_1[X(t)] = E[X(t)] = E[Y]\int_0^T g_1(\tau)w(t-\tau)\,d\tau \tag{3.2.13}$$

$$\kappa_2[X(t_1), X(t_2)] = \mathrm{cov}[X(t_1), X(t_2)] = E[Y^2]\int_0^T g_1(\tau)w(t_1-\tau)w(t_2-\tau)\,d\tau$$
$$+ E^2[Y]\int_0^T\int_0^T g_2(\tau_1,\tau_2)w(t_1-\tau_1)w(t_2-\tau_2)\,d\tau_1\,d\tau_2 \tag{3.2.14}$$

Since $X(t)$ is defined on $0 < t \le T$, we let $g_1(\tau) = 0$ for $\tau < 0$, and $g_2(\tau_1, \tau_2) = 0$ for $\tau_1 < 0$, $\tau_2 < 0$, or both, without loss of generality. The lower integration limits in (3.2.13) and (3.2.14) can therefore be extended to $-\infty$. The upper integration limits can also be extended to $+\infty$, on account of the causality requirement of the pulse shape function $w(t-\tau)$. Taking these into consideration, we may rewrite (3.2.13) and (3.2.14) as

$$E[X(t)] = E[Y]\int_{-\infty}^{\infty} g_1(\tau)w(t-\tau)\,d\tau \tag{3.2.15}$$

$$\mathrm{cov}[X(t_1), X(t_2)] = E[Y^2]\int_{-\infty}^{\infty} g_1(\tau)w(t_1-\tau)w(t_2-\tau)\,d\tau$$
$$+ E^2[Y]\int_{-\infty}^{\infty}\int_{-\infty}^{\infty} g_2(\tau_1,\tau_2)w(t_1-\tau_1)w(t_2-\tau_2)\,d\tau_1\,d\tau_2 \tag{3.2.16}$$

Now comparing (3.2.15) and (3.1.2), we see that $X(t)$ cannot be an evolutionary process if it does not have a zero mean, as would be the case when $E[Y] \neq 0$. However, for most applications, the assumption of $E[X(t)] = 0$ does not result in the loss of generality if a nonzero mean can be treated in a separate deterministic problem. Bearing this in mind, we assume that $E[Y] = 0$. Then (3.2.15) is trivial and (3.2.16) reduces to

$$\operatorname{cov}[X(t_1), X(t_2)] = E[Y^2] \int_{-\infty}^{\infty} g_1(\tau) w(t_1 - \tau) w(t_2 - \tau)\, d\tau \tag{3.2.17}$$

It is interesting to note that with $E[Y] = 0$, the expression for the covariance function of a random pulse train remains the same, whether pulse arrivals are independent or correlated. The effect of correlated arrival is evident only if $E[Y] \neq 0$.

EVOLUTIONARY SPECTRAL REPRESENTATION. We now show that (3.2.17) can be converted to the form of (3.1.4). Since g_1 is nonnegative, we may write

$$b(t, \omega) = \int_{-\infty}^{\infty} w(u) \sqrt{g_1(t-u)}\, e^{-i\omega u} du \tag{3.2.18}$$

Thus

$$w(u) \sqrt{g_1(t-u)} = \frac{1}{2\pi} \int_{-\infty}^{\infty} b(t, \omega) e^{i\omega u} d\omega \tag{3.2.19}$$

Letting $\tau = t - u$, we obtain

$$w(t-\tau) \sqrt{g_1(\tau)} = \frac{1}{2\pi} \int_{-\infty}^{\infty} b(t, \omega) e^{i\omega(t-\tau)} d\omega \tag{3.2.20}$$

Since the left-hand side of (3.2.20) is purely real, we also have

$$w(t-\tau) \sqrt{g_1(\tau)} = \frac{1}{2\pi} \int_{-\infty}^{\infty} b^*(t, \omega) e^{-i\omega(t-\tau)} d\omega \tag{3.2.21}$$

Substituting (3.2.20) and (3.2.21) into (3.2.17) and noting that

$$\frac{1}{2\pi} \int_{-\infty}^{\infty} e^{i(\omega_2 - \omega_1)\tau} d\tau = \delta(\omega_2 - \omega_1) \tag{3.2.22}$$

we arrive at

$$\operatorname{cov}[X(t_1), X(t_2)] = \frac{1}{2\pi} E[Y^2] \int_{-\infty}^{\infty} b(t_1, \omega) b^*(t_2, \omega) e^{i\omega(t_1 - t_2)} d\omega \tag{3.2.23}$$

Comparing this equation with (3.1.4), we see that $(2\pi)^{-1} E[Y^2]\, d\omega$ takes the place of $d\Psi(\omega)$, and $b(t, \omega)$ takes the place of $a(t, \omega)$. The evolutionary spectral density of $X(t)$ is clearly

$$\hat{\Phi}_{XX}(t, \omega) = \frac{1}{2\pi} E[Y^2] \left| b(t, \omega) \right|^2 \tag{3.2.24}$$

where $b(t, \omega)$ is given by (3.2.18).

We caution not to claim that the random pulse-train process given in (3.2.2) is exactly the process

$$\xi(t) = \sqrt{\frac{1}{2\pi} E[Y^2]} \int_{-\infty}^{\infty} b(t, \omega) e^{i\omega t} d\omega \tag{3.2.25}$$

because this is not generally true. What the two processes have in common is the evolutionary spectral density, (3.2.24).

STOCHASTIC EARTHQUAKE MODELING. The random pulse-train model is extremely versatile, capable of simulating a large class of physical phenomena. For example, Lin (1963) and Cornell (1964) proposed independently such a model for earthquake ground motions. Other notable early works included, for example, Tung (1967) and Liu (1970). The model is attractive to engineers and scientists for at least three reasons: (1) the cause for a physical phenomenon being modeled can be incorporated, (2) the response of a linear system to an additive excitation of a random pulse train is another random pulse train, and (3) its evolutionary spectral representation provides clear descriptions of the amplitude variation and the frequency-content variation with time. Interestingly, some other earthquake models proposed in the engineering literature can also be cast in the framework of random pulse train. Since strong earthquakes are among the most severe dynamic loads to civil engineering structures, a brief discussion on these earlier earthquake models is in order; this will also provide a suitable introduction to the more refined modeling to be considered later in this chapter. The references cited in this section, however, are by no means comprehensive.

The occurrence of a future earthquake and the ground motion at a particular site during such an event are not predictable. It is logical that an earthquake ground motion be modeled as a stochastic process, as first suggested by Housner (1947). A typical strong motion earthquake record has a duration from several seconds to one minute, within which time the intensity of the ground motion increases at the beginning and dies down toward the end. Therefore, each complete episode may be treated as a sample function of a nonstationary stochastic process (e.g., Amin and Ang 1968). However, if one is concerned only with the most intense portion of ground shaking, then the use of a stationary process model for such a portion may be justified in some cases. An analysis based on the assumption of a stationary process input is clearly simpler, and in some sense this very assumption may be the key to render an analysis possible. The reasonableness of such an assumption depends primarily on the length of the "stationary" portion, if it exists, in comparison with the so-called relaxation time of the structural system involved. The relaxation time of a dynamical system may be defined as the length of time for the amplitude A of a free motion to reduce to A/e, or to increase to eA, where e is the base of the natural logarithm.

Earlier stochastic earthquake models in the engineering literature were of the phenomenological type; that is, they were attempts to match the appearances of past earthquake records in a given region. One of the earliest nonstationary models, proposed by Bogdanoff, Goldberg, and Bernard (1961), had the form

$$X_0(t) = \begin{cases} 0 & t < 0 \\ \displaystyle\sum_{j=1}^{n} t a_{j0} \exp(-\alpha_{j0} t) \cos(\omega_{j0} t + \Phi_{j0}) & t \geq 0 \end{cases} \tag{3.2.26}$$

where $X_0(t)$ represents the ground acceleration in a given direction; a_{j0}, α_{j0}, and ω_{j0} are given sets of positive numbers with $\omega_{10} < \omega_{20} < \cdots < \omega_{n0}$; and Φ_{j0} are independent random variables uniformly distributed in $[0, 2\pi)$. The implication of (3.2.26) is clear; each term is a sinusoid with a random phase, modulated by a deterministic envelope which rises from $t = 0$, reaches a maximum at $t = \alpha_{j0}^{-1}$, and then dies down at large t. This model can be generalized to

$$X(t) = \sum_{k=0}^{m} X_k(t) \tag{3.2.27}$$

where each $X_k(t)$ is obtained from (3.2.26) by substituting $t - t_k$ for t, and by selecting new sets of a_{jk}, α_{jk}, ω_{jk}, and Φ_{jk}. Apparently, (3.2.26) was motivated by two factors: the general trend of rise and fall of earthquake intensity, describable by the deterministic modulation functions, and the relative ease with which the response of a linear system to each term can be calculated. The generalized version, (3.2.27), represents an attempt to simulate the successive shocks in a typical earthquake, but it results in some loss in the computational expediency.

The Shinozuka-Sato model (1967) is obtained by passing a modulated white noise through a time-invariant linear filter:

$$X(t) = \int_{-\infty}^{\infty} h(t - \tau)\phi(\tau)W(\tau)\,d\tau \tag{3.2.28}$$

where $W(\tau)$ is a white noise, $\phi(\tau)$ is a deterministic modulating function, and $h(t - \tau)$ is the impulse response of the linear filter. It is well-known that a white noise is equivalent to a chain of independently arriving impulses, with independent and identically distributed magnitudes of zero mean (Lin, 1967, Eq. 4-7):

$$W(t) = \sum_{j=1}^{N(T)} Y_j \delta(t - \tau_j) \tag{3.2.29}$$

One can show that the covariance function of $X(t)$ given by (3.2.28) can be computed from (3.2.17) with

$$w(t - \tau_j) = h(t - \tau_j) \tag{3.2.30}$$

and with the average pulse-arrival rate $g_1(\tau)$ modified to $\tilde{g}_1(\tau) = g_1(\tau)\phi^2(\tau)$.

A generalization of the Shinozuka-Sato model is due to Levy and Kozin (1968), with the representation

$$X(t) = \int_{-\infty}^{\infty} h(t - \tau)\phi(\tau)N(\tau)\,d\tau \tag{3.2.31}$$

where

$$N(\tau) = \int_{0}^{c} \psi(\xi)W(\tau - \xi)\,d\xi \tag{3.2.32}$$

and c is a selected constant. In this case, (3.2.31) is a special case of (3.2.1), with

$$w(t, \tau_j) = \int_{-\infty}^{\infty} h(t - u)\phi(u)h_1(u - \tau_j)\,du \tag{3.2.33}$$

and

$$h_1(\xi) = \begin{cases} \psi(\xi) & 0 < \xi < c \\ 0 & \text{otherwise} \end{cases} \tag{3.2.34}$$

The covariance function of $X(t)$ may be obtained as follows:

$$\operatorname{cov}[X(t_1), X(t_2)] = E[Y^2] \int_{-\infty}^{\infty} g_1(\tau) w(t_1, \tau) w(t_2, \tau)\, d\tau \tag{3.2.35}$$

which is a generalization of (3.2.17) for the case of a time-variant pulse shape function.

Another model, proposed by Kozin (1977), involves the following linear equation with variable coefficients:

$$\ddot{X} + a(t)\dot{X} + b(t)X = \phi(t)W(t) \tag{3.2.36}$$

Here a modulated white noise is the input, and $X(t)$ is the output of a time-varying filter. Thus the model represents a generalization of (3.2.28), by replacing the time-invariant impulse response $h(t-\tau)$ with a time-variant one, $h(t, \tau)$. It is also a special case of (3.2.1), having a pulse shape function

$$w(t, \tau_j) = h(t, \tau_j) \tag{3.2.37}$$

On the other hand, earthquake models in the form of uniformly modulated processes have also been used frequently in structural response analyses. A modulating envelope, proposed independently by Bolotin (1965) and Shinozuka and Brant (1969), consists of the difference of two exponential terms, each with a negative exponent. Another envelope, attributable to Amin and Ang (1968), consists of three segments, a quadratically increasing segment followed by a constant segment and then by an exponentially decaying segment. The uniformly modulated process models have been criticized for not simulating the change of frequency contents with time. Improvements in this regard have been suggested by Kameda (1975) and by Scherer, Riera, and Schueller (1982). The evolutionary spectrum of a random pulse-train model discussed earlier provides a useful link between the two entirely different modeling concepts.

3.3 LINEAR TIME-INVARIANT SYSTEMS UNDER ADDITIVE EXCITATIONS OF EVOLUTIONARY STOCHASTIC PROCESSES

When an evolutionary stochastic process plays the role of an additive excitation on a linear time-invariant system, the response of the system is also an evolutionary stochastic process. To show that this is the case, let the excitation process be

$$F(t) = \int_{-\infty}^{\infty} a(t, \omega) e^{i\omega t} d\tilde{G}(\omega) \tag{3.3.1}$$

where $\tilde{G}(\omega)$ is an orthogonal-increment process, namely,

$$E[d\tilde{G}(\omega)\, d\tilde{G}^*(\omega')] = \begin{cases} d\Psi(\omega) & \omega = \omega' \\ 0 & \omega \neq \omega' \end{cases} \tag{3.3.2}$$

The response of a linear time-invariant system to the excitation of $F(t)$ may be obtained as

$$X(t) = \int_0^t h(u) F(t-u)\, du \tag{3.3.3}$$

where $h(u)$ is the impulse response function. Substituting (3.3.1) into (3.3.3) and interchanging the order of integration, we obtain

$$X(t) = \int_{-\infty}^{\infty} \mathcal{M}(t, \omega) e^{i\omega t} d\tilde{G}(\omega) \tag{3.3.4}$$

where
$$\mathcal{M}(t, \omega) = \int_{0}^{t} a(t-u, \omega) h(u) e^{-i\omega u} du \tag{3.3.5}$$

Equation (3.3.4) shows that the response is indeed an evolutionary process.

The correlation function of the response follows as

$$E[X(t_1)X(t_2)] = \int_{-\infty}^{\infty} \mathcal{M}(t_1, \omega)\mathcal{M}^*(t_2, \omega) e^{i\omega(t_1-t_2)} d\Psi(\omega) \tag{3.3.6}$$

If $\Psi(\omega)$ is differentiable, that is, if

$$d\Psi(\omega) = \Phi(\omega)\, d\omega \tag{3.3.7}$$

then the evolutionary spectral density of X(t) is given by

$$\hat{\Phi}_{XX}(t, \omega) = \left|\mathcal{M}(t, \omega)\right|^2 \Phi(\omega) \tag{3.3.8}$$

It is of interest to examine the special case of a constant $a(t, \omega)$, in which the excitation process $F(t)$ becomes a correlation-stationary process. In such a case, function $\mathcal{M}(t, \omega)$ in equation (3.3.5) reduces to

$$\mathcal{H}(t, \omega) = \int_{0}^{t} h(u) e^{-i\omega u} du \tag{3.3.9}$$

where we have let $a(t, \omega) = 1$ without loss of generality. Equation (3.3.9) is the same as Eq. 5-30 of Lin (1967), obtained for the transient response of an initially quiescent linear system, exposed to a correlation-stationary excitation at $t = 0$. As t increases, $\mathcal{H}(t, \omega)$ in (3.3.9) approaches the frequency response function $H(\omega)$, and the response $X(t)$ in (3.3.4) approaches the state of correlation-stationarity, as expected.

Generalization of the foregoing results to the case of m inputs and n outputs is straightforward. Let all the inputs be additive and representable in the form of

$$F_j(t) = \int_{-\infty}^{\infty} a_j(t, \omega) e^{i\omega t} d\tilde{G}_j(\omega) \qquad \text{(no summation on } j) \tag{3.3.10}$$

where
$$E[d\tilde{G}_j(\omega) d\tilde{G}_k^*(\omega')] = \begin{cases} d\Psi_{jk}(\omega) & \omega = \omega' \\ 0 & \omega \neq \omega' \end{cases} \tag{3.3.11}$$

Then
$$X_l(t) = \int_{-\infty}^{\infty} \mathcal{M}_{lj}(t, \omega) e^{i\omega t} d\tilde{G}_j(\omega) \tag{3.3.12}$$

where
$$\mathcal{M}_{lj}(t, \omega) = \int_{0}^{t} a_j(t-u, \omega) h_{lj}(u) e^{-i\omega u} du \tag{3.3.13}$$

Again, the matrix of $\mathcal{M}_{lj}(t, \omega)$ reduces to the matrix of $\mathcal{H}_{lj}(t, \omega)$ in Eq. 6-11 of Lin (1967), when all the $a_j(t, \omega)$ functions are replaced by the same constant unity. The cross-correlation function of $X_l(t)$ and $X_r(t)$ follows as

$$E[X_1(t_1)X_r(t_2)] = \int_{-\infty}^{\infty} \mathcal{M}_{lj}(t_1, \omega)\mathcal{M}_{rk}^*(t_2, \omega)e^{i\omega(t_1-t_2)}d\Psi_{jk}(\omega) \quad (3.3.14)$$

The evolutionary cross-spectrum of $F_j(t)$ and $F_k(t)$, if it exists, is given by

$$\hat{\Phi}_{jk}(t, \omega) = a_j(t, \omega)a_k^*(t, \omega)\Phi_{jk}(\omega) \qquad \text{(no summation on } j \text{ or } k) \quad (3.3.15)$$

where

$$\Phi_{jk}(\omega) = \frac{d}{d\omega}\Psi_{jk}(\omega) \quad (3.3.16)$$

The existence of evolutionary cross-spectra of the excitations implies the same for the responses, given by

$$\hat{\Phi}_{X_lX_r}(t, \omega) = \mathcal{M}_{lj}(t, \omega)\mathcal{M}_{rk}^*(t, \omega)\Phi_{jk}(\omega) \qquad \text{(no summation on } j \text{ or } k) \quad (3.3.17)$$

The case of a continuous structure under distributed random excitation can be treated in an analogous manner to that shown in Section 2.5. The summation over j in (3.3.12) is replaced by an integration over the domain of the structure. However, since the random excitation is now dependent on the spatial location as well as time, it may be an evolutionary stochastic process in space also.

In structural engineering applications, efforts spent to determine the impulse response functions can be greater than what is required subsequently to calculate the evolutionary cross-spectra of the response. In some cases, it is easier to first compute the frequency response functions and then obtain the impulse response functions indirectly from the Fourier transformation, for example,

$$h_{lj}(u) = \int_{-\infty}^{\infty} H_{lj}(\omega)e^{-i\omega u}d\omega \quad (3.3.18)$$

Two illustrative examples are given in the following subsections.

3.3.1 A Multistory Building under Horizontal Earthquake Excitation

As the first example, consider a simplified model for a building structure shown in Fig. 3.3.1. This model has been analyzed previously in Section 2.5.1 for wind excitations; now the same model is subjected to horizontal ground shaking. The problem involves only one input but multiple outputs. The outputs are chosen to be the deflections at different floor levels and the shear forces between neighboring floors. The first step toward solving the problem is to determine the frequency response functions associated with these response variables, which can then be converted to the corresponding impulse response functions according to equation (3.3.18).

Since only one input is present at the ground level, which may be denoted as the zeroth floor, the second subscript j in the impulse response functions $h_{kj}(t)$, or the frequency response functions $H_{kj}(\omega)$, takes the value zero in all cases. For simplicity, the second subscript is omitted in the following discussion.

Referring to Fig. 3.3.1b, the equations of motion may be written as

$$\begin{aligned} &m\ddot{X}_j + c\dot{X}_j + V_{j-1} - V_j = 0 \\ &V_{j-1} = k(X_j - X_{j-1}) \qquad j = 1, 2, \ldots, N \end{aligned} \quad \text{(a)}$$

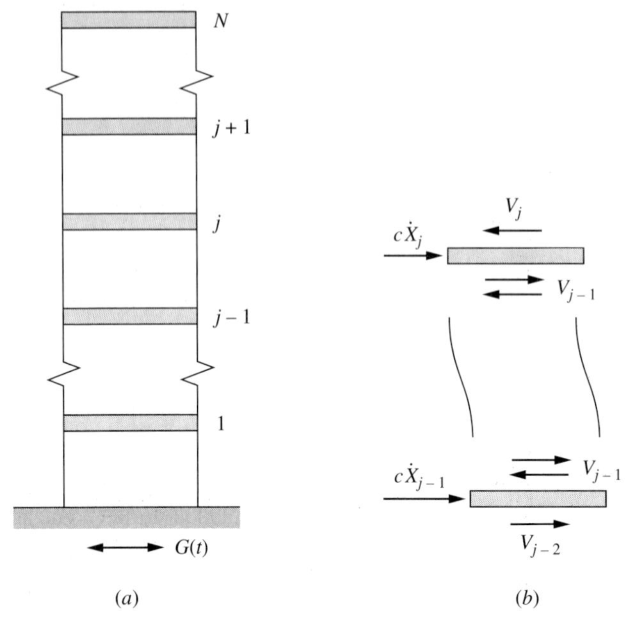

FIGURE 3.3.1
A simplified building structure model subject to horizontal ground shaking.

subject to the boundary conditions $X_0 = G$ and $V_N = 0$. In writing equations (a), the simplifying assumption that all story units are identically constructed was made. In earthquake engineering, we are concerned with motions relative to the ground. Substituting $X_j = Z_j + G$ into equations (a), we obtain

$$\begin{aligned} m\ddot{Z}_j + c\dot{Z}_j + V_{j-1} - V_j &= -m\ddot{G} - c\dot{G} \\ V_{j-1} &= k(Z_j - Z_{j-1}) \qquad j = 1, 2, \ldots, N \end{aligned} \tag{b}$$

subject to the boundary conditions $Z_0 = 0$ and $V_N = 0$. For the purpose of determining the frequency response functions, let

$$G = e^{i\omega t} \tag{c.1}$$

$$Z_j = \bar{Z}_j e^{i\omega t} \tag{c.2}$$

$$V_j = \bar{V}_j e^{i\omega t} \tag{c.3}$$

Equations (b) may be cast into the following matrix form:

$$\boldsymbol{Y}_j = \boldsymbol{T}\boldsymbol{Y}_{j-1} + (-m\omega^2 + ic\omega)\begin{Bmatrix} 0 \\ 1 \end{Bmatrix} \tag{d}$$

where $\boldsymbol{T}$ is the same transfer matrix introduced in Section 2.5.1 for the wind-loading problem, namely,

$$\boldsymbol{T} = \begin{bmatrix} 1 & k^{-1} \\ -m\omega^2 + ic\omega & k^{-1}(-m\omega^2 + ic\omega) + 1 \end{bmatrix} \tag{e}$$

and
$$\boldsymbol{Y}_j = \begin{Bmatrix} \bar{Z}_j \\ \bar{V}_j \end{Bmatrix} \tag{f}$$

Repeated application of (d) results in a relation between $\boldsymbol{Y}_m$ and $\boldsymbol{Y}_0$:

$$\boldsymbol{Y}_m = \boldsymbol{T}^m \boldsymbol{Y}_0 + (-m\omega^2 + ic\omega) \sum_{j=1}^{m} \boldsymbol{T}^{m-j} \begin{Bmatrix} 0 \\ 1 \end{Bmatrix} \tag{g}$$

Letting $m = N$ and imposing the boundary conditions $\bar{Z}_0 = 0$ and $\bar{V}_N = 0$, we find from the second row

$$\bar{V}_0 = \frac{m\omega^2 - ic\omega}{\tau_{22}(N)} \sum_{j=1}^{N} \tau_{22}(N - j) \tag{h}$$

where $\tau_{jk}(n)$ denotes the (j, k) element of matrix $[\boldsymbol{T}]^n$. It is known that the eigenvalues of $\boldsymbol{T}$ are reciprocal pairs. Denoting each pair as $\exp(\pm i\theta)$, where θ is a complex function of ω, the summation shown in equation (h) can be obtained in a closed form:

$$\sum_{j=1}^{N} \tau_{22}(N - j) = \frac{\sin(N\theta)}{\sin\theta} \tag{i}$$

After some additional algebra, we arrive at

$$\bar{V}_0 = \frac{2k \sin(N\theta)\sin(\theta/2)}{\cos[(N + 1/2)\theta]} \tag{j}$$

Then from the second row of (g),

$$\bar{V}_m = \frac{2k \sin(\theta/2)\sin[(N - m)\theta]}{\cos[(N + 1/2)\theta]} \tag{k}$$

To obtain $\bar{Z}_m$ from (g), the following summation is also required:

$$\sum_{j=1}^{m} \tau_{12}(m - j) = \frac{\sin(m\theta/2)\sin[(m - 1)\theta/2]}{k \sin\theta \sin(\theta/2)} \tag{l}$$

Substituting (l) into (g) and simplifying, we obtain

$$\bar{Z}_m = \frac{\cos[(N - m + 1/2)\theta]}{\cos[(N + 1/2)\theta]} - 1 \tag{m}$$

It is interesting to note that the expression for $\bar{V}_m$, equation (k), can also be obtained from

$$\bar{V}_m = k(\bar{Z}_{m+1} - \bar{Z}_m) \tag{n}$$

as expected.

In earthquake engineering practice, the ground acceleration is usually treated as the input. Then the frequency response functions are obtained by letting $\ddot{G} = \exp(i\omega t)$ or $G = -\omega^{-2}\exp(i\omega t)$. It follows that the frequency response functions for the displacement at the mth floor relative to the ground and the shear force above the mth floor due to ground horizontal acceleration are, respectively,

$$H_{Z_m}(\omega) = \frac{1}{\omega^2}\left\{1 - \frac{\cos[(N - m + 1/2)\theta]}{\cos[(N + 1/2)\theta]}\right\} \tag{o}$$

and

$$H_{V_m}(\omega) = \frac{2k \sin(\theta/2) \sin[(N - m)\theta]}{\omega^2 \cos[(N + 1/2)\theta]} \tag{p}$$

Numerical calculations have been carried out for the structural model of an eight-story building ($N = 8$), with the following physical properties: story height $h = 3.6\,\text{m}$, story mass $m = 3.456 \times 10^6$ kg, shear-wall stiffness $k = 3.404 \times 10^9$ N/m, and damping coefficient $c = 1.00 \times 10^6$ N/(m/s). Since we are focusing on the structural aspects of the problem, the model for the horizontal ground acceleration is chosen to be a rather simple uniformly modulated random process:

$$\ddot{G}(t) = a(t) \int_{-\infty}^{\infty} e^{i\omega t} d\tilde{\ddot{G}}(\omega) \tag{q}$$

with an envelope function (Amin and Ang, 1968)

$$a(t) = \begin{cases} 0 & t < 0 \\ \left(\dfrac{t}{t_1}\right)^2 & 0 \leq t \leq t_1 \\ 1 & t_1 \leq t \leq t_2 \\ \exp[-\gamma(t - t_2)] & t > t_2 \end{cases} \tag{r}$$

and a Kanai-Tajimi spectral density function (Kanai, 1957; Tajimi, 1960)

$$\Phi(\omega) = \frac{1 + 4\zeta_g^2(\omega/\omega_g)^2}{[1 - (\omega/\omega_g)^2]^2 + 4\zeta_g^2(\omega/\omega_g)^2} S \tag{s}$$

The parameters in (r) and (s) are chosen to be $t_1 = 3\,\text{s}$, $t_2 = 13\,\text{s}$, $\gamma = 0.26$, $\omega_g = 18.85$ rad/s, $\xi_g = 0.65$, and $S = 4.65 \times 10^{-4}\,\text{m}^2/\text{s}^3$.

The dashed line in Fig. 3.3.2 represents the computed magnitude of the frequency response function for the displacement at the roof, namely, Z_8. The peaks are located near the natural frequencies. However, the eighth peak is not pronounced; it will be more pronounced if a lower damping is used in the computation. Represented by the solid line in the same figure is the computed frequency response amplitude when soil compliancy is taken into account (Lin and Wu, 1984a). The much reduced

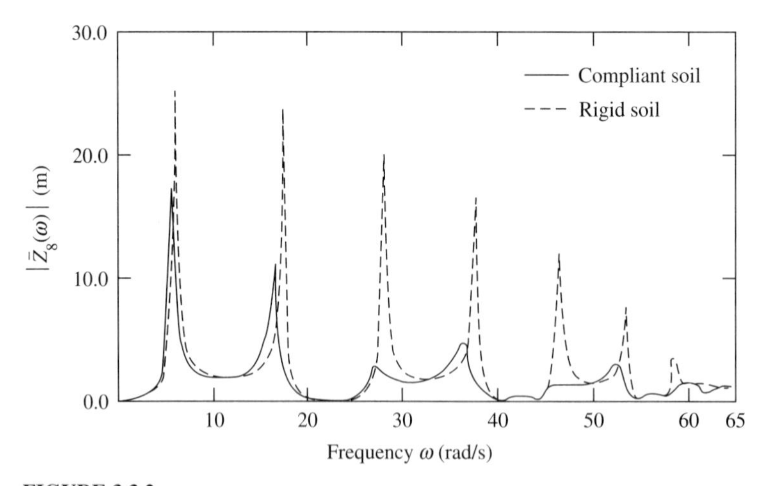

FIGURE 3.3.2
Modulus of frequency response function of the eighth floor displacement, relative to the footing, due to free-field ground displacement input [after Lin and Wu, 1984a; Copyright ©1984 Marcel Dekker Inc., reprinted with permission].

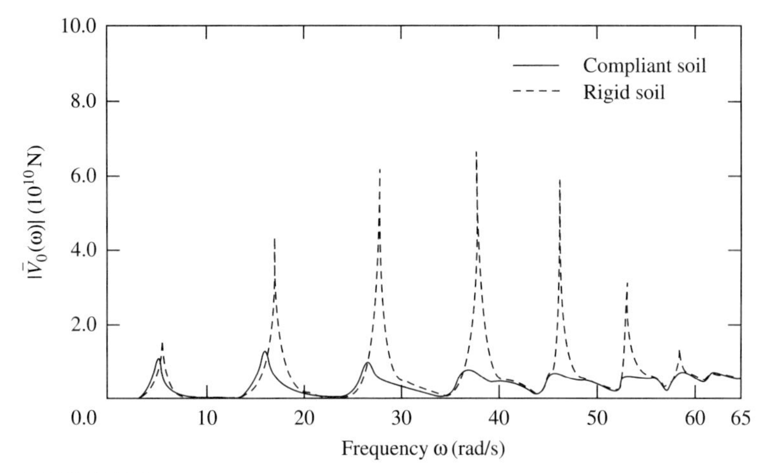

FIGURE 3.3.3
Modulus of frequency response function of the shear force above the footing, due to free-field ground displacement input [after Lin and Wu, 1984a; Copyright ©1984 Marcel Dekker Inc., reprinted with permission].

peak heights and slight shifts of the peak positions indicate a damping effect of the soil, attributable to energy radiation from the foundation back to the soil mass. For a compliant soil, the motion of the foundation is not the same as the so-called free-field ground motion, which would prevail without the building. The solid line in Fig. 3.3.2 refers to the floor displacement relative to the foundation, not the ground.

Figure 3.3.3 shows the computed magnitudes of the frequency response functions for V_0, the shear force immediately above the foundation. Again, the dashed and solid lines represent the cases of rigid and compliant soil, respectively. Interestingly, the first vibration mode is not dominant for this shear force response, contrary to what has been found for the displacement response.

The computed root-mean-square values of the displacement $Z_8(t)$ and the shear force $V_0(t)$ are shown in Figs. 3.3.4 and 3.3.5, respectively.

3.3.2 A Multiply Supported Pipeline under Seismic Surface Wave Excitations (Lin, Zhang, and Yong, 1990)

One important type of seismically generated wave motion in the ground is Rayleigh waves, which are confined closely to the ground surface (Rayleigh, 1887). If the epicenter of an earthquake is sufficiently far from an observer, then a simplified model for the observed Rayleigh waves may be approximated as superposition of decaying plane waves as follows:

$$G(x, t) = \int_{-\infty}^{\infty} e^{-\gamma x} e^{-ik(x-ct)} dZ(k) \tag{a}$$

where $\gamma > 0$, $c > 0$, and $Z(k)$ is an orthogonal-increment process in wave number k. Equation (a) implies that the modeled plane waves originate at $x = 0$. Each k

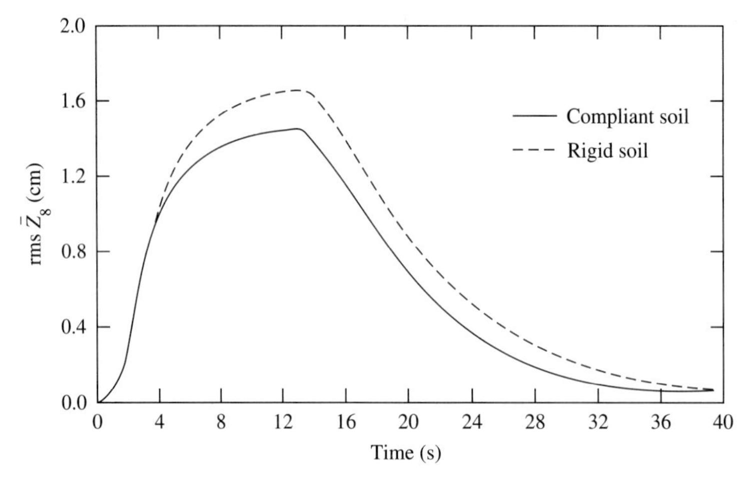

FIGURE 3.3.4
Root-mean-square displacement of the eighth floor, relative to the footing, due to model earthquake excitation [after Lin and Wu, 1984a; Copyright ©1984 Marcel Dekker Inc., reprinted with permission].

identifies a component wave, which has an amplitude $|dZ(k)|$ at $x = 0$ and decays spatially at an exponential rate of γ as the wave propagates in the positive x direction. The speed of wave propagation is c, and the wave length is $2\pi/k$. In general, γ and c may be functions of k. Since $Z(k)$ has orthogonal increments, the ensemble average

$$E[dZ(k_1)\,dZ^*(k_2)] = \begin{cases} \Phi(k)\,dk & k_1 = k_2 = k \\ 0 & k_1 \neq k_2 \end{cases} \tag{b}$$

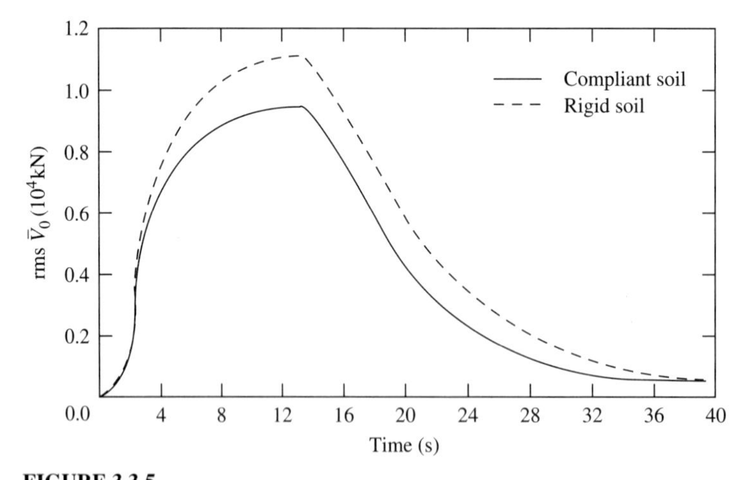

FIGURE 3.3.5
Root-mean-square shear above the footing due to model earthquake excitation [after Lin and Wu, 1984a; Copyright ©1984 Marcel Dekker Inc., reprinted with permission].

in which an asterisk denotes the complex conjugate and $\Phi(k)$ is an even and non-negative function of k, representing the spectral density of the wave group in the wave-number domain at the location $x = 0$.

It is interesting to compare the right-hand sides of equations (a) and (3.1.8), where the integrations are carried out in the wave-number domain and frequency domain, respectively. It is clear from this comparison that, if γ is indeed independent of k, then $G(x, t)$ is a uniformly modulated random process in the spatial variable x, the spatial modulating function being $\exp(-\gamma x)$. However, by relating the time variable t with the space variable x through the propagation speed c, $G(x, t)$ is random in both space and time. Moreover, it is stochastically stationary in time but stochastically inhomogeneous in space.

We now focus our attention on a particular location on a very long pipeline, idealized as being infinitely long. Structurally, we model the pipeline as an Euler-Bernoulli beam, attached to simple supports which are spaced at equal distances l, as shown in Fig. 3.3.6. The beam receives the seismic excitation through the forced displacements of the supports. The ground motion perpendicular to the pipeline is given by equation (a). The effect of the ground motion parallel to the pipeline is disregarded in the simplified analysis in this section.

Within the framework of a linear analysis, the relation between the input given by equation (a) and a suitable output $Y(t)$ may be written symbolically as

$$\mathscr{L}\{Y(t)\} = \int_{-\infty}^{\infty} e^{-\gamma x} e^{-ik(x-ct)} dZ(k) \tag{c}$$

where $\mathscr{L}\{\cdot\}$ represents a linear operator. To facilitate the solution of (c), we consider first the steady-state solution of a fundamental deterministic problem,

$$\mathscr{L}\{\bar{H}_Y(k, x_0, \phi)e^{ikct}\} = e^{-\gamma x} e^{-ik(x-ct)} \tag{d}$$

in which $\bar{H}_Y$ is the wave-number response function. For the problem at hand, $\bar{H}_Y$ is a function of the wave number k, the distance x_0 between the origin of the excitation wave and the location on the structure where the response is of interest, and the incident angle ϕ between the propagation direction of the traveling wave and the orientation of the pipeline. With the knowledge of the wave number response function $\bar{H}_Y$, the solution for Y at the state of stochastic stationarity can be constructed as follows:

$$Y(t) = \int_{-\infty}^{\infty} \bar{H}_Y e^{ikct} dZ(k) \tag{e}$$

The autocorrelation function for $Y(t)$ is given by

$$E[Y(t_1)Y(t_2)] = \int_{-\infty}^{\infty} \left|\bar{H}_Y\right|^2 e^{ikc(t_1-t_2)} \Phi(k)\, dk = R_{YY}(t_1 - t_2) \tag{f}$$

where use has been made of equation (b), the property of orthogonal increments in $Z(k)$. Thus

$$\Phi_{YY}(k) = \left|\bar{H}_Y\right|^2 \Phi(k) \tag{g}$$

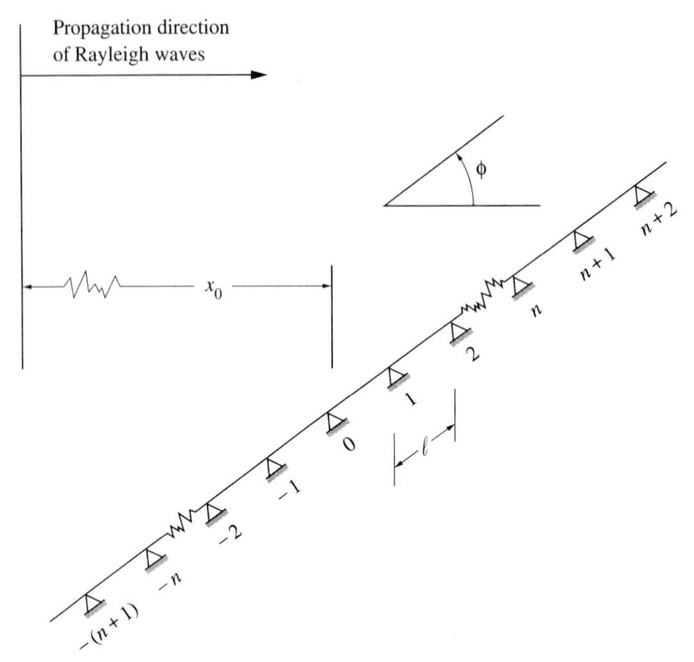

FIGURE 3.3.6
Schematic of mathematical model for multiply supported pipeline.

is the spectral density of $Y(t)$ in the wave-number domain. The mean-square response is obtained by letting $t_1 = t_2$ in equation (f), resulting in

$$E[Y^2(t)] = \int_{-\infty}^{\infty} \left|\bar{H}_Y\right|^2 \Phi(k)\, dk \tag{h}$$

The determination of the wave-number response functions for various types of response is clearly a major task in the present analysis. The structural model shown in Fig. 3.3.6 belongs again to the general class of periodic structures, for which the transfer-matrix formulation is often convenient. Since the structure is infinitely long, the concept of wave propagation in periodic structures can also be used to advantage (Brillouin, 1946; Mead, 1971). A combination of the two techniques is used in the following analysis.

The state vector describing the motion of an Euler-Bernoulli beam is four-dimensional. The four components of such a state vector are commonly chosen to be the complex amplitudes of deflection, slope, bending moment, and transverse shear, denoted by $\bar{w}$, $\bar{\theta}$, $\bar{M}$, and $\bar{V}$, respectively. Now relate the two state vectors at supports n and $n+1$, $\bar{\boldsymbol{Z}}_n = \{\bar{w}_n \bar{\theta}_n \bar{M}_n \bar{V}_n\}'$ and $\bar{\boldsymbol{Z}}_{n+1} = \left\{\bar{w}_{n+1} \bar{\theta}_{n+1} \bar{M}_{n+1} \bar{V}_{n+1}\right\}'$ where a prime denotes a matrix transposition, as follows:

$$\bar{\boldsymbol{Z}}_{n+1} = \boldsymbol{F}\bar{\boldsymbol{Z}}_n \tag{i}$$

where $\boldsymbol{F}$ is a transfer matrix, given by (Pestel and Leckie, 1963)

$$
\boldsymbol{F} = \begin{bmatrix} C_0 & S_1 & \dfrac{C_2}{EI} & \dfrac{S_3}{EI} \\ \sigma^4 S_3 & C_0 & \dfrac{S_1}{EI} & \dfrac{C_2}{EI} \\ \sigma^4 EIC_2 & \sigma^4 EIS_3 & C_0 & S_1 \\ \sigma^4 EIS_1 & \sigma^4 EIC_2 & \sigma^4 S_3 & C_0 \end{bmatrix} \tag{j}
$$

in which EI = bending rigidity of the beam, $\sigma^4 = (kc)^2 m/EI$, m = mass of beam per unit length, and

$$C_0 = \frac{\cosh \sigma l + \cos \sigma l}{2} \tag{k.1}$$

$$C_2 = \frac{\cosh \sigma l - \cos \sigma l}{2\sigma^2} \tag{k.2}$$

$$S_1 = \frac{\sinh \sigma l + \sin \sigma l}{2\sigma} \tag{k.3}$$

$$S_3 = \frac{\sinh \sigma l - \sin \sigma l}{2\sigma^3} \tag{k.4}$$

To account for the material damping of the beam, the bending rigidity EI may be replaced by a complex $EI(1 + i\delta \operatorname{sgn} k)$, where $\delta > 0$ and $\operatorname{sgn} k$ denotes the sign of the wave number k. In this case, σ^4 is complex, but for our purposes any one of its four roots may be used for the value of σ.

We digress to note some common properties of a transfer matrix of an arbitrary order (e.g., see Lin and McDaniel, 1969), which must be an even order for structural applications:

1. The determinant of a transfer matrix is unity.
2. Inversion can be accomplished by rearranging the elements and changing sign for some of the elements.
3. Eigenvalues are in reciprocal pairs.
4. If the structural element represented by a transfer matrix is symmetrical with respect to its midsection, then by a suitable choice of the order and sign convention for the components of state vector $\boldsymbol{Z}$, the transfer matrix can be made symmetrical about its cross-diagonal.

The last property is clearly seen in equation (j).

We now write the upper three rows of equation (i) explicitly as

$$\bar{w}_{n+1} = f_{11}\bar{w}_n + \{f_{12} \quad f_{13}\}\begin{Bmatrix} \bar{\theta} \\ \bar{M} \end{Bmatrix}_n + f_{14}\bar{V}_n \tag{l}$$

$$\begin{Bmatrix} \bar{\theta} \\ \bar{M} \end{Bmatrix}_{n+1} = \begin{Bmatrix} f_{21} \\ f_{31} \end{Bmatrix} \bar{w}_n + \begin{bmatrix} f_{22} & f_{23} \\ f_{32} & f_{33} \end{bmatrix}\begin{Bmatrix} \bar{\theta} \\ \bar{M} \end{Bmatrix} + \begin{Bmatrix} f_{24} \\ f_{34} \end{Bmatrix} \bar{V}_n \tag{m}$$

Eliminating $\bar{V}_n$ from equations (l) and (m), we obtain

$$\boldsymbol{Y}_{n+1} = \boldsymbol{T}\boldsymbol{Y}_n + \boldsymbol{L}\bar{w}_n + \boldsymbol{R}\bar{w}_{n+1} \tag{n}$$

where $\boldsymbol{Y}_j = \{\bar{\theta}_j \, \bar{M}_j\}'$, and the elements of matrices $\boldsymbol{T}$, $\boldsymbol{L}$, and $\boldsymbol{R}$ are given by

$$t_{11} = t_{22} = \frac{\cos\sigma l \sinh\sigma l - \cosh\sigma l \sin\sigma l}{\sinh\sigma l - \sin\sigma l} \tag{o.1}$$

$$t_{12} = \frac{\cosh\sigma l \cos\sigma l - 1}{EI\sigma(\sinh\sigma l - \sin\sigma l)} \tag{o.2}$$

$$t_{21} = -\frac{2EI\sigma \sinh\sigma l \sin\sigma l}{\sinh\sigma l - \sin\sigma l} \tag{o.3}$$

$$l_1 = -\frac{\sigma \sinh\sigma l \sin\sigma l}{\sinh\sigma l - \sin\sigma l} \tag{p.1}$$

$$l_2 = -\frac{EI\sigma^2(\sinh\sigma l \cos\sigma l + \cosh\sigma l \sin\sigma l)}{\sinh\sigma l - \sin\sigma l} \tag{p.2}$$

$$r_1 = \frac{\sigma(\cosh\sigma l - \cos\sigma l)}{\sinh\sigma l - \sin\sigma l} \tag{q.1}$$

$$r_2 = \frac{EI\sigma^2(\sinh\sigma l + \sin\sigma l)}{\sinh\sigma l - \sin\sigma l} \tag{q.2}$$

In the context of the present pipeline problem, $\bar{w}_n$ and $\bar{w}_{n+1}$ are prescribed by the ground motions at supports n and $n+1$, respectively; thus they are inputs to the system. From such a point of view, equation (n) is an inhomogeneous matrix difference equation, with the inputs appearing as inhomogeneous terms. In the absence of the inputs, the equation would become homogeneous, similar to equation (i). Clearly, $\boldsymbol{T}$ in equation (n) is also a transfer matrix. The properties of $|\boldsymbol{T}| = 1$ and reciprocity of the eigenvalues can easily be verified. Equation (n) implies that if deflections at the supports are specified, then the beam motion can be described fully by use of two-dimensional state vectors $\boldsymbol{Y}_j$ at the supports.

Let us focus our attention on a particular support, labeled as support 0. As shown in Fig. 3.3.6, this support is located at a distance x_0 from the origin of the decaying plane wave represented by the right-hand-side of equation (d). From equation (n), we obtain an expression for the response vector $\boldsymbol{Y}_0 = \{\theta_0 \, M_0\}'$ at support 0:

$$\boldsymbol{Y}_0 = \boldsymbol{T}\boldsymbol{Y}_{-1} + \boldsymbol{L}\bar{w}_{-1} + \boldsymbol{R}\bar{w}_0 \tag{r}$$

Similar expressions can also be written for $\boldsymbol{Y}_{-1}, \boldsymbol{Y}_{-2}$, and so on. These expressions can then be combined to yield

$$\boldsymbol{Y}_0 = \boldsymbol{T}^{\infty}\boldsymbol{Y}_{-\infty} + \sum_{n=1}^{N} \boldsymbol{T}^{n-1}(\boldsymbol{L}\bar{w}_{-n} + \boldsymbol{R}\bar{w}_{-n+1}) \qquad N = \text{Integer}\left(\frac{x_0}{l\cos\phi}\right) \tag{s}$$

where Integer ($\cdot$) denotes the integer part of the parenthesized quantity. Note that $\bar{w}_{-N}$ is the last input on the left of support 0; thus the upper limit of the summation is N.

However, an alternative expression for $\boldsymbol{Y}_0$ can also be obtained from equation (n):

$$Y_0 = T^{-1}Y_1 - T^{-1}(L\bar{w}_0 + R\bar{w}_1) \tag{t}$$

and similarly for Y_1, Y_2, and so on. These expressions lead to

$$Y_0 = T^{-\infty}Y_\infty - \sum_{n=1}^{\infty} T^{-n}(L\bar{w}_{n-1} + R\bar{w}_n) \tag{u}$$

Although both equations (s) and (u) are valid, neither can be used for the intended computation alone, since both T^n and T^{-n} diverge as $n \to \infty$.

To circumvent this difficulty, we transform each state vector Y to a corresponding wave vector μ (von Flotow, 1986; Yong and Lin, 1989). Let

$$Y = D\mu \tag{v}$$

where columns in the transformation matrix D are eigenvectors of T. Denoting the two reciprocal eigenvalues of T by $\exp[\mp(\alpha - i\beta)]$, where $\alpha > 0$, we have

$$D^{-1}TD = \begin{bmatrix} e^{-\alpha+i\beta} & 0 \\ 0 & e^{\alpha-i\beta} \end{bmatrix} \tag{w}$$

Thus equations (s) and (u) may be transformed to

$$\begin{aligned} \begin{Bmatrix} \mu_0^r \\ \mu_0^l \end{Bmatrix} &= \begin{bmatrix} e^{-\alpha+i\beta} & 0 \\ 0 & e^{\alpha-i\beta} \end{bmatrix}^{\infty} \mu_{-\infty} \\ &+ \sum_{n=1}^{N} \begin{bmatrix} e^{-\alpha+i\beta} & 0 \\ 0 & e^{\alpha-i\beta} \end{bmatrix}^{n-1} D^{-1}(L\bar{w}_{-n} + R\bar{w}_{-n+1}) \\ &+ \begin{bmatrix} e^{-\alpha+i\beta} & 0 \\ 0 & e^{\alpha-i\beta} \end{bmatrix}^{N} D^{-1}R\bar{w}_{-N} \qquad N = \text{Integer}\left(\frac{x_0}{l\cos\phi}\right) \end{aligned} \tag{x}$$

and

$$\begin{aligned} \begin{Bmatrix} \mu_0^r \\ \mu_0^l \end{Bmatrix} &= \begin{bmatrix} e^{\alpha-i\beta} & 0 \\ 0 & e^{-\alpha+i\beta} \end{bmatrix}^{\infty} \mu_{\infty} \\ &- \sum_{n=1}^{\infty} \begin{bmatrix} e^{\alpha-i\beta} & 0 \\ 0 & e^{-\alpha+i\beta} \end{bmatrix}^{n} D^{-1}(L\bar{w}_{n-1} + R\bar{w}_n) \end{aligned} \tag{y}$$

where the two components of the wave vector μ_0 at support 0, identified as the rightgoing and the leftgoing waves, are denoted by μ_0^r and μ_0^l, respectively. It is clear that equation (x) is suitable for the computation of the rightgoing wave μ_0^r, but not the leftgoing wave μ_0^l, whereas the opposite is true for equation (y). By retaining the useful portion of each equation, μ_0^r and μ_0^l can be calculated, separately. The response vector Y_0 can then be obtained from equation (v).

The required formulas for carrying out the foregoing calculations include the elements of matrices L and R, given in equations (p.1) through (q.2), and the following:

$$\cos(\beta + i\alpha) = t_{11} \tag{z}$$

$$D = \begin{bmatrix} t_{12} & t_{12} \\ i\sin(\beta + i\alpha) & -i\sin(\beta + i\alpha) \end{bmatrix} \tag{aa}$$

$$\boldsymbol{D}^{-1} = \frac{1}{2}\begin{bmatrix} \dfrac{1}{t_{12}} & \dfrac{-i}{\sin(\beta + i\alpha)} \\ \dfrac{1}{t_{12}} & \dfrac{i}{\sin(\beta + i\alpha)} \end{bmatrix} \tag{bb}$$

where t_{jk} are the elements of the transfer matrix $\boldsymbol{T}$, given by equations (o.1) through (o.3). Using these expressions in equations (x) and (y), we obtain

$$\begin{aligned} \mu_0^r = {} & \frac{1}{2}\sum_{n=1}^{N} \exp[(n-1)(-\alpha + i\beta)]\left\{\left[\frac{l_1}{t_{12}} - \frac{il_2}{\sin(\beta + i\alpha)}\right]\bar{w}_{-n} \right. \\ & \left. + \left[\frac{r_1}{t_{12}} - \frac{ir_2}{\sin(\beta + i\alpha)}\right]\bar{w}_{-n+1}\right\} + \frac{1}{2}\exp[N(-\alpha + i\beta)] \\ & \times \left[\frac{r_1}{t_{12}} - \frac{ir_2}{\sin(\beta + i\alpha)}\right]\bar{w}_{-N} \qquad N = \text{Integer}\left(\frac{x_0}{l\cos\phi}\right) \end{aligned} \tag{cc}$$

and

$$\begin{aligned} \mu_0^l = {} & -\frac{1}{2}\sum_{n=1}^{\infty} \exp[n(-\alpha + i\beta)]\left\{\left[\frac{l_1}{t_{12}} - \frac{il_2}{\sin(\beta + i\alpha)}\right]\bar{w}_{n-1} \right. \\ & \left. + \left[\frac{r_1}{t_{12}} + \frac{ir_2}{\sin(\beta + i\alpha)}\right]\bar{w}_n\right\} \end{aligned} \tag{dd}$$

Now, for the computation of wave-number response functions, the input ground motion is given by the right-hand side of equation (d). Thus

$$\bar{w} = \exp[-(\gamma + ik)x] \tag{ee}$$

Referring to Fig. 3.3.6, we obtain

$$\bar{w}_n = \exp[-(\gamma + ik)(x_0 + nl\cos\phi)] \qquad 0 \le \phi \le \frac{\pi}{2} \tag{ff}$$

where ϕ is the incident angle of the seismic waves. The restriction $0 \le \phi \le \pi/2$ is consistent with the physical relationship shown in Fig. 3.3.6. Substituting equation (ff) into equations (cc) and (dd), and summing over n,

$$\begin{aligned} \mu_0^r = {} & \frac{1}{2}\left\{\left[\frac{l_1}{t_{12}} - i\frac{l_2}{\sin(\beta + i\alpha)}\right]\exp[(\gamma + ik)l\cos\phi] + \left[\frac{r_1}{t_{12}} - \frac{ir_2}{\sin(\beta + i\alpha)}\right]\right\} \\ & \times \frac{\exp[-(\gamma + ik)x_0](1 - \exp\{N[-\alpha + i\beta + (\gamma + ik)l\cos\phi]\})}{1 - \exp[-\alpha + i\beta + (\gamma + ik)l\cos\phi]} \\ & + \frac{1}{2}\left[\frac{r_1}{t_{12}} - \frac{ir_2}{\sin(\beta + i\alpha)}\right]\exp[N(-\alpha + i\beta) - (\gamma + ik)(x_0 - Nl\cos\phi)] \\ N = {} & \text{Integer}\left(\frac{x_0}{l\cos\phi}\right) \end{aligned} \tag{gg}$$

$$\begin{aligned} \mu_0^l = {} & -\frac{1}{2}\left\{\left[\frac{l_1}{t_{12}} + \frac{il_2}{\sin(\beta + i\alpha)}\right]\exp[(\gamma + ik)l\cos\phi] + \left[\frac{r_1}{t_{12}} - \frac{ir_2}{\sin(\beta + i\alpha)}\right]\right\} \\ & \times \frac{\exp[-\alpha + i\beta - (\gamma + ik)(x_0 + l\cos\phi)]}{1 - \exp[-\gamma + i\beta - (\gamma + ik)l\cos\phi]} \end{aligned} \tag{hh}$$

Equations (gg) and (hh) give the wave-number response functions for the leftgoing and rightgoing structural waves, respectively, at the support located at $x = x_0$, due to the excitation of a unit decaying sinusoidal surface wave initiated at $x = 0$.

For engineering purposes, the wave vector must be converted to the physical state vector according to equation (v). With the knowledge of the transformation matrix $\boldsymbol{D}$, equation (aa), we obtain the wave-number response functions for the rotation response $\theta(t)$ and the bending-moment response $M(t)$ at support 0, respectively:

$$H_\theta = t_{12}\left[\mu_0^r + \mu_0^l\right] \tag{ii}$$

$$H_M = i\sin(\beta + i\alpha)\left[\mu_0^r - \mu_0^l\right] \tag{jj}$$

The wave-number response function for the shear response, if desired, may be obtained from equation (l). Responses at an arbitrary location on the beam, say, between supports 0 and 1, can be obtained from

$$\bar{\boldsymbol{Z}} = \boldsymbol{F}(\xi)\bar{\boldsymbol{Z}}_0 \tag{kk}$$

where $\boldsymbol{F}(\xi)$ has the same form as equation (j) except that l is replaced by a local coordinate ξ, identifying the location, measured from support 0.

For a periodic structure, large response magnitudes are likely at wave-passage frequencies (Brillouin, 1946). Moreover, since the excitations are generated from a traveling wave, the response can be amplified greatly by a coincidence effect. These concepts are explained next.

Wave-passage frequencies are determined on the basis of an undamped periodic structure. In the present case of an Euler-Bernoulli beam on evenly spaced supports, the wave-passage frequencies ω $(= kc)$ satisfy the inequality (Miles, 1956)

$$|t_{11}| = \left|\frac{\cos\sigma l \sinh\sigma l - \cosh\sigma l \sin\sigma l}{\sinh\sigma l - \sin\sigma l}\right| \leq 1 \tag{ll}$$

which implies that $\alpha = 0$ in equation (z), and the eigenvalues of $\boldsymbol{T}$ can be expressed as $\exp(\pm i\beta)$. At a wave-passage frequency, disturbances can propagate through an undamped infinite periodic structure without attenuation. Such wave-passage frequencies are grouped in distinctive frequency bands. Conversely, a frequency is said to be in a wave-stoppage band if inequality (ll) is not satisfied, indicating that $\alpha > 0$ and $\beta = 0$ if the structure is undamped (β can be nonzero if the transfer matrix $\boldsymbol{T}$ is of an order higher than 2×2). Of course, if damping is present, then α is always positive, but its magnitude is relatively small within a wave-passage band.

In the context of wave motion in a periodic structure, the term "coincidence" is used to indicate a phase matching between the excitation wave and the natural wave in the structure. At a given frequency (or a given wave number) the natural wave in the structure is propagated at a phase difference β from one periodic unit to the next, whereas the excitation wave is propagated at a phase difference $kl\cos\phi$, as seen from Fig. 3.3.6. If the excitation wave and the structure wave are propagating in the same direction, as in the case of μ_0^r, then coincidence occurs when

$$\beta + kl\cos\phi = \pm 2j\pi \qquad j = 0, 1, 2, \ldots \tag{mm}$$

On the other hand, if the excitation and the natural waves are propagated in the opposite directions, as in the case of μ_0^l, then coincidence occurs when

$$\beta - kl\cos\phi = \pm 2j\pi \qquad j = 0, 1, 2, \ldots \tag{nn}$$

Numerical calculations have been carried out using the following physical data:

1. Properties of pipeline:

Bending rigidity $EI = 4.459 \times 10^8 \text{ N} \cdot \text{m}^2$

Loss factor $\delta = 0.005$

Mass per unit length (including fluid inside pipe) $= 757.9$ kg/m

Distance between neighboring supports $l = 22$ m

2. Properties of seismic surface waves:

Origin of seismic wave (distance from pipeline station 0) $x_0 = 4000$ m

Incident angle $\phi = 15°$ or $80°$

Wave-number spectrum at the origin

$$\Phi(k) = \frac{\sigma_g^2}{4\sqrt{\pi}} B^3 k^2 \exp\left(-\frac{B^2}{4} k^2\right) \tag{oo}$$

$\sigma_g^2 =$ mean-square ground displacement, m^2

$B = 65$ m

$$\text{Phase velocity } c = \begin{cases} 650 \text{ m/s} & k < 6.28 \times 10^{-3} \text{ rad/m} \\ 150 \text{ m/s} & k > 6.28 \times 10^{-1} \text{ rad/m} \\ \tilde{c} \text{ m/s} & 6.28 \times 10^{-3} \le k \le 6.28 \times 10^{-1} \text{ rad/m} \end{cases}$$

$\tilde{c} = 650 - 1125u + 750u^2 - 125u^3$, $u = \log(2\pi/k)$

The preceding Rayleigh wave model was proposed by Shinozuka and Deodatis (1988b), assuming that the random process was stochastically stationary in time as well as stochastically homogeneous in space [corresponding to $\gamma = 0$ in equation (a)]. To change the model to a spatially inhomogeneous process, we selected $\gamma = 0.008k$ in our calculation.

Sketched in Fig. 3.3.7 is the wave-number spectrum of the seismic waves according to equation (oo), normalized to $\sigma_g^2 = 1 \text{ m}^2$. Only the positive domain $k > 0$ is shown in the figure, the negative domain being its exact mirror image. For simplicity, the following discussion pertains to $k > 0$. The case of $k < 0$ may be deduced from symmetry. Figure 3.3.7 shows that most seismic energy is concentrated in the range of $0 < k < 0.090$.

The wave-passage bands of the pipeline structure can be determined using inequality (*ll*). The first wave-passage band is found to be $0.0305 \le k \le 0.1168$. Thus seismic energy is available to excite the structural waves approximately within the lower half of this band, as shown in Fig. 3.3.7. These structural waves can

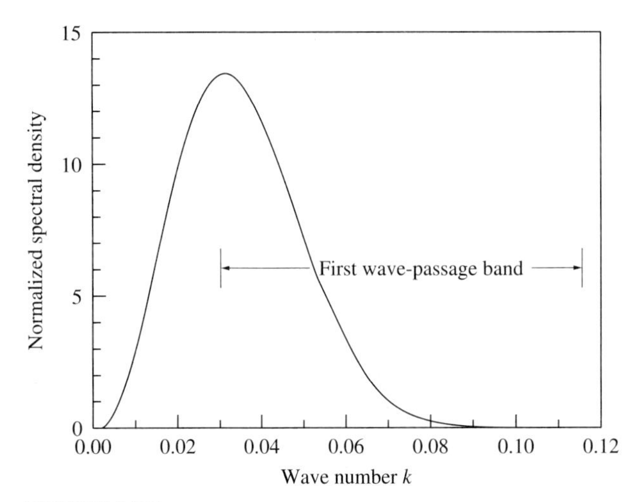

FIGURE 3.3.7
Wave-number spectrum of Rayleigh wave group at the origin $x = 0$, normalized to unit mean-square value [after Lin, Zhang, and Yong, 1990; Copyright ©1990 ASCE, reprinted with permission].

travel to far distances along the structure. Structural waves belonging to the higher wave-passage bands are essentially unexcited.

The magnitude of the wave-number response function for the bending moment at pipeline station 0 is plotted in Fig. 3.3.8 for the case $\phi = 15°$. The location of the first wave-passage band is indicated in the figure. At this incident angle, coincidence

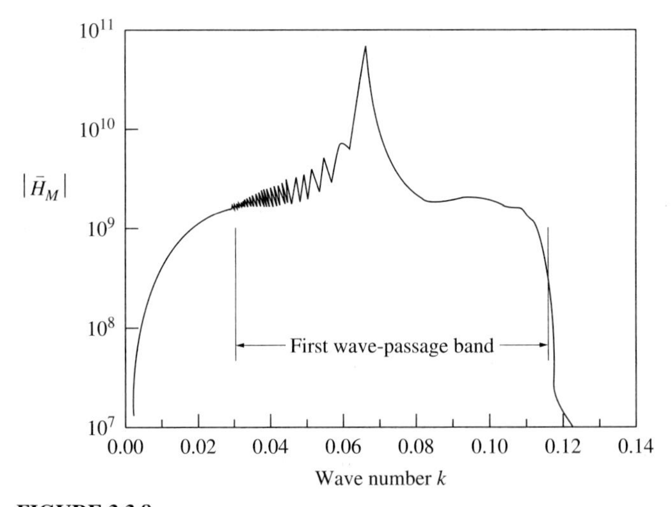

FIGURE 3.3.8
Magnitude of wave-number response for bending moment at support 0, seismic wave incident angle $\phi = 15°$ [after Lin, Zhang, and Yong, 1990; Copyright ©1990 ASCE, reprinted with permission].

between structural and seismic waves occurs at $k = 0.067$, resulting in the highest peak shown in Fig. 3.3.8. It should be noted that the magnitude of the wave-number response is plotted on a logarithmic scale in this figure. Thus the highest peak at coincidence is approximately one order of magnitude greater than that of the second highest peak.

Contributions toward the wave-number response from the rightgoing and left-going structural waves are plotted separately in Fig. 3.3.9, in solid lines and dashed lines, respectively. It is seen that coincidence occurs due to the matching of the seismic wave with the leftgoing structural waves in the neighborhood of $k = 0.067$, traveling in the direction opposite to that of the seismic wave. Furthermore, the cumulative contribution from all the leftgoing waves is much smoother than that from the rightgoing waves because, in our theoretical model, seismic inputs are fed into the structure through an infinite number of supports on the right of station 0, but only through a finite number of supports on the left of station 0. Outside the wave-passage band, the solid and dashed lines nearly overlap, indicating that their contributions are about equal. However, these contributions are local effects, resulting directly from inputs at or near support 0, since disturbances of a frequency outside the wave-passage bands cannot propagate to large distances.

The computed wave-number spectrum for the bending moment at station 0 is plotted in Fig. 3.3.10, corresponding to a normalized $\sigma_g^2 = 1 \text{ m}^2$ for the excitation. The mean-square bending moment obtained by integrating the spectrum over the entire positive and negative wave-number domain is found to be $E[M^2(t)] = 2.672 \times 10^{19} \text{ N}^2 \cdot \text{m}^2$.

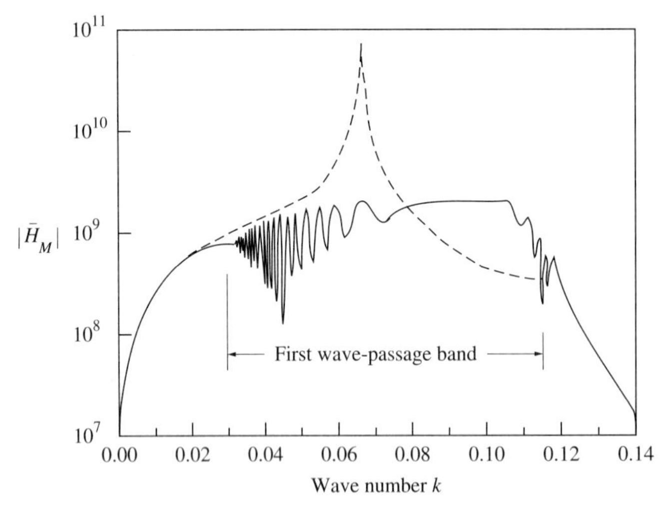

FIGURE 3.3.9
Contributions to wave-number response from rightgoing structural wave (—) and leftgoing structural wave (- - - -), seismic wave incident angle $\phi = 15°$ [after Lin, Zhang, and Yong, 1990; Copyright ©1990 ASCE, reprinted with permission].

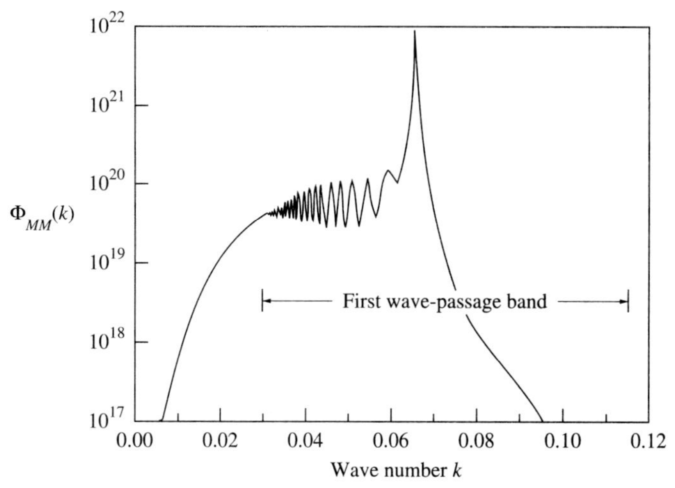

FIGURE 3.3.10
Wave-number spectrum of bending moment response at support 0, seismic wave incident angle $\phi = 15°$ [after Lin, Zhang, and Yong, 1990; Copyright ©1990 ASCE, reprinted with permission].

Figure 3.3.11 shows the magnitude of the wave-number response function for the case of $\phi = 80°$. The highest peak is found at $k = 0.110$. Again, this highest peak results from the coincidence of the rightgoing seismic wave and the leftgoing structural waves, as seen from Fig. 3.3.12, which shows the separate contributions of the leftgoing and rightgoing structural waves. A minor peak is also evident in

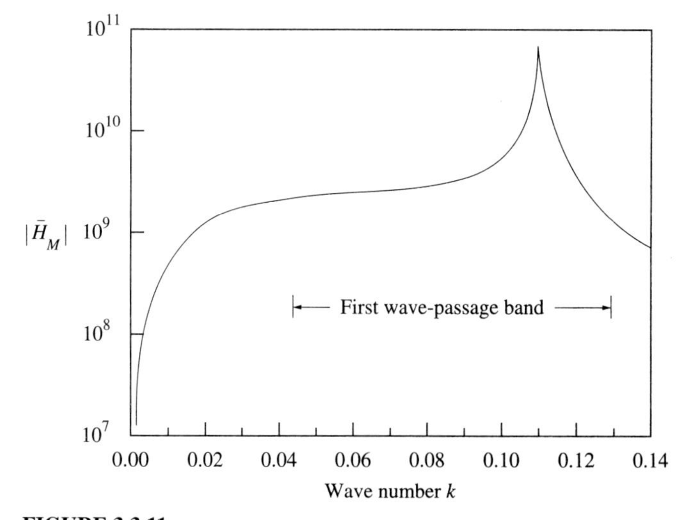

FIGURE 3.3.11
Magnitude of wave-number response for bending moment at support 0, seismic wave incident angle $\phi = 80°$ [after Lin, Zhang, and Yong, 1990; Copyright ©1990 ASCE, reprinted with permission].

Fig. 3.3.12, which is attributable to the rightgoing structural waves, represented by solid lines. However, the latter is not the result of an exact coincidence, for equation (mm) is not exactly satisfied at the peak, although the value on the left-hand side is close to a multiple of 2π. Both the solid and dashed curves in Fig. 3.3.12 are smooth since the seismic wave is propagating toward the pipeline at a nearly perpendicular direction; thus more supports on the left of station 0 are excited as compared with the case of $\phi = 15°$.

The resulting wave-number spectrum for the bending moment response at station 0 is shown in Fig. 3.3.13. The response attributable to coincidence is no longer important due to the low excitation energy in the neighborhood of $k = 0.110$. The mean-square bending moment obtained from integrating the spectrum is $3.174 \times 10^{18}\ \text{N}^2 \cdot \text{m}^2$, which is less than one-eighth the value obtained for $\phi = 15°$.

Numerical results have also been obtained for two additional cases: (1) $l = 25$ m and $\phi = 15°$; and (2) $l = 16$ m and $\phi = 15°$. In case (1), the first wave-passage band is located in $0.0215 < k < 0.0723$, and coincidence occurs near the peak of the seismic excitation spectrum. As expected, the computed mean-square bending moment is a much higher $3.665 \times 10^{20}\ \text{N}^2 \cdot \text{m}^2$. In case (2), the first wave-passage band is found to be $0.0835 < k < 0.4108$. Thus seismic energy is available only to excite the lower wave-stoppage band, and the response is essentially localized. The mean-square bending moment is found to be $1.24 \times 10^{17}\ \text{N}^2 \cdot \text{m}^2$, which is much lower.

The preceding numerical results are based on a normalized input strength of $\sigma_g^2 = 1\ \text{m}^2$. Results for a stronger or weaker input can easily be obtained propor-

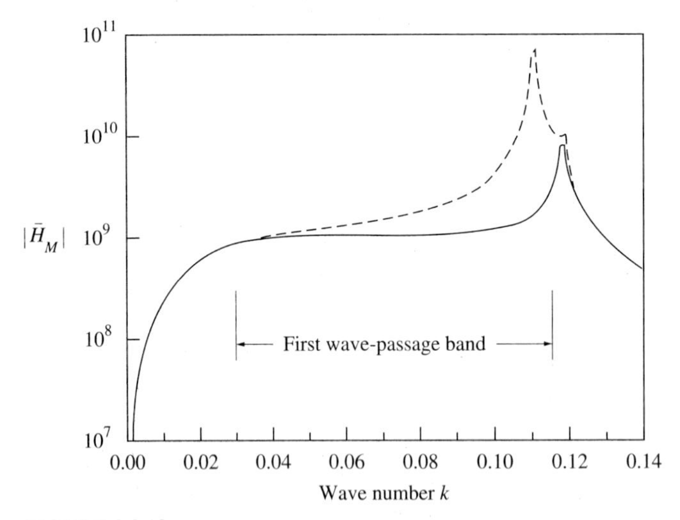

FIGURE 3.3.12
Contributions to wave-number response from rightgoing structural wave (—) and leftgoing structural wave (- - - -), seismic wave incident angle $\phi = 80°$ [after Lin, Zhang, and Yong, 1990; Copyright ©1990 ASCE, reprinted with permission].

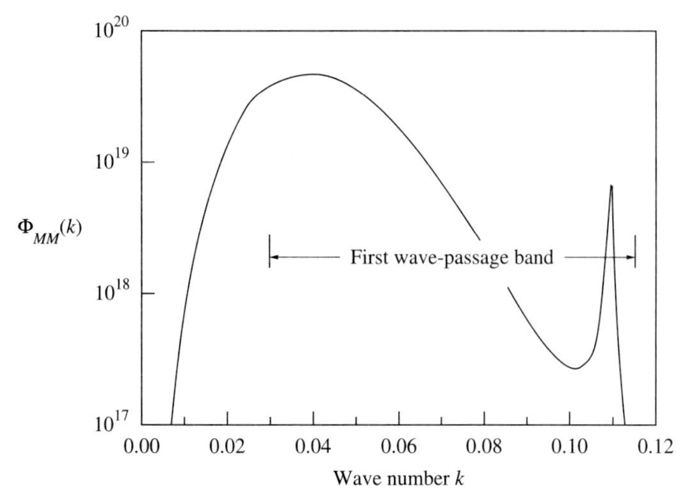

FIGURE 3.3.13
Wave-number spectrum of bending moment response at support 0, seismic wave incident angle $\phi = 80°$ [after Lin, Zhang, and Yong, 1990; Copyright ©1990 ASCE, reprinted with permission].

tionally. Moreover, relatively simple earthquake models are used in the analysis, so that certain features of the structural systems involved can be clearly demonstrated. These earthquake models are essentially phenomenological models. In some cases, more refined models may be required.

The greatest problem with the phenomenological approach of earthquake modeling is the lack of adequate ground motion records for meaningful statistical inference. To compensate for the scarcity of statistical data, the geophysical aspects of earthquake initiation as well as seismic wave propagation, reflection, and refraction may be taken into account. Again, the mathematical framework of random pulse train provides a convenient means to incorporate such geophysical considerations, which are discussed in the order of increasing complexity in Sections 3.4 through 3.6.

3.4 EVOLUTIONARY KANAI-TAJIMI–TYPE EARTHQUAKE MODELS

The Kanai-Tajimi model for ground acceleration (Kanai, 1957; Tajimi, 1960) has been used very widely in the analysis of engineering structures under earthquake excitation. In its original form, the ground acceleration is idealized as a stationary random process, having a spectral density

$$\Phi(\omega) = \frac{\omega_g^4 + (2\zeta_g\omega_g\omega)^2}{(\omega_g^2 - \omega^2)^2 + (2\zeta_g\omega_g\omega)^2}K \tag{3.4.1}$$

This model corresponds to the acceleration of a mass, supported by a linear spring and a dashpot in parallel as shown in Fig. 3.4.1, whose base is undergoing a white

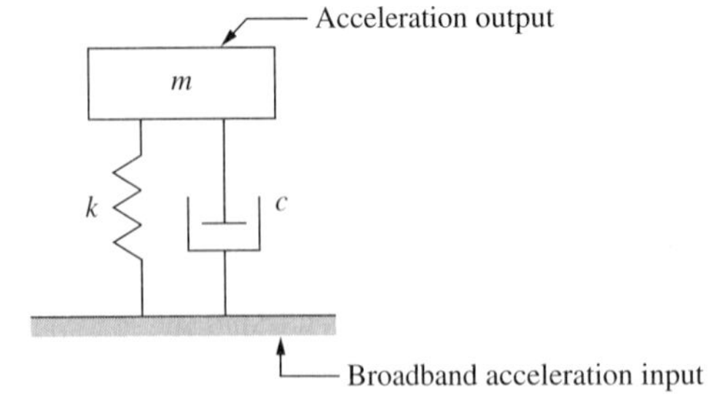

FIGURE 3.4.1
Schematic of Kanai-Tajimi model.

noise acceleration. The three parameters in (3.4.1)—K, ω_g, and ζ_g—represent the spectrum level (normalized to unit mass) of the white noise excitation at the base, the natural frequency of the model, and the ratio of damping to the critical damping, respectively. These parameters can be adjusted according to the earthquake magnitude, ground resonance frequency, and attenuation of seismic waves in the ground.

The most attractive feature of the Kanai-Tajimi model is the ability to simulate ground resonance in a very simple way. Ground resonance arises when a seismic wave propagates through a stratified medium, and it is one distinctive character that should not be disregarded in the modeling (e.g., see Schuëller and Scherer, 1985). The dominant nature of ground resonance has been demonstrated amply in the records of a number of past earthquakes, such as the Mexico City earthquake on September 19, 1985. The records of this event, obtained at sites on a thick clay deposit layer over an old lake bed, show a large percentage of ground shaking energy concentrated around the resonance frequency of the clay layer of approximately 0.5 cps (Beck and Hall, 1986). Another attractive feature of the Kanai-Tajimi model is the relative ease with which the random vibration analysis can be carried out for a structural system under such an excitation.

One unrealistic aspect of the original Kanai-Tajimi model is the implication that an earthquake is a sample function of a stationary random process. This deficiency can be partially remedied by introducing a time-dependent envelope to the original model, but such an envelope describes only the changing intensity of the ground acceleration, not the changing frequency content, which is clearly visible in many available records.

Most earthquakes are caused by slips in a fault zone in the ground (e.g., see Aki and Richards, 1980). When a slip is taking place, the behavior of the material in the vicinity is necessarily inelastic. However, away from the immediate vicinity, the disturbance is transmitted to the ground surface essentially as linear stress waves. Since the exact nature of a slip is largely unknown, an accepted practice among the geophysicists is to treat the earthquake source as being equivalent to a self-equilibrium force system. The early works of Maruyama (1963) and Burridge and Knopoff (1964) established the equivalence of a slip to a pair of equal and opposite moments, known in the literature as a *double couple*.

Furthermore, the existence of high-frequency components in a typical earthquake record suggests that slips in a fault zone occur intermittently rather than smoothly. On the basis of this observation, Aki and his associates (Aki, Bouchon, Chouet, and Das, 1977; Papageorgiou and Aki, 1983) proposed a barrier model in which a fault zone is replaced by a collection of barrier circles, and the breaking of successive circles occurs randomly in time. However, since the actual slipping process in a fault zone is much more complicated and largely unknown at the present time, other forms of modeling are also worthy of consideration.

Alternatively, we assume that slipping occurs in independent short spurts, almost impulsively. A simple mathematical representation of this idea is given by the expression

$$S(t) = \sum_{j=1}^{N(T)} Y_j \delta(t - \tau_j)\delta(\rho - \rho_j) \qquad 0 < t \leq T \tag{3.4.2}$$

where $\delta(\cdot)$ denotes a Dirac delta function, each Y_j represents the random magnitude of a double couple which occurs at (ρ_j, τ_j), that is, at location ρ_j and time τ_j, and $N(T)$ gives the random number of counts in the time interval $(0, T]$. For different j, the magnitudes Y_j are assumed to be independent but have an identical probability distribution. $S(t)$ is a shot noise (Lin, 1967, Eq. 4-104) if the times at which the impulses occur are uncorrelated events, the amplitudes Y_j have a zero mean (Lin, 1986), or both. Here, however, we make the stronger assumption that the impulses occur independently, that is, $N(T)$ is a Poisson random process, except when indicated otherwise.

Let $g_k(\boldsymbol{r}, t; \rho, \tau)$ be a Green's function which describes the ground acceleration in the kth direction at a site location $\boldsymbol{r}$ and time t, due to an impulsive application of a double couple at ρ and time τ. Then the ground acceleration due to a sequence of double couples given by equation (3.4.2) may be obtained from superposition:

$$G_k(\boldsymbol{r}, t) = \sum_{j=1}^{N(T)} Y_j g_k(\boldsymbol{r}, t; \rho, \tau) \tag{3.4.3}$$

For most earthquake engineering purposes, the earth may be considered as a stratified half-space, with generally lighter material in an upper layer than the one below. If the source of an earthquake is reasonably deep, then as seismic waves propagate to the ground surface, their direction of propagation is almost vertically upward, as can be explained by use of Snell's law of refraction. As a first approximation, we may take into account only the uppermost layer between the ground surface and the nearest bedrock and treat the wave propagation in this layer as being one-dimensional and vertical. In the Kanai-Tajimi model, this layer is further approximated by a single-degree-of-freedom linear system shown in Fig. 3.4.1.

The equation of motion for the Kanai-Tajimi model is given by

$$\ddot{G} + 2\zeta_g \omega_g \dot{G} + \omega_g^2 G = 2\zeta_g \omega_g \dot{R} + \omega_g^2 R \tag{3.4.4}$$

where G and R are the accelerations at the mass and the base, respectively. The frequency response function obtained from (3.4.4) is

$$H_1(\omega) = \frac{\omega_g^2 + 2i\zeta_g \omega_g \omega}{\omega_g^2 - \omega^2 + 2i\zeta_g \omega_g \omega} \tag{3.4.5}$$

which is also implied in equation (3.4.1). The impulse response function is essentially the Fourier transform of the frequency response function:

$$h_1(t-\tau) = \frac{1}{2\pi}\int_{-\infty}^{\infty} H_1(\omega)e^{i\omega(t-\tau)}d\omega$$

$$= \begin{cases} 0 & t < \tau \\ \omega_g \exp[-\zeta_g\omega_g(t-\tau)]\left\{ \dfrac{1-2\zeta_g^2}{(1-\zeta_g^2)^{1/2}} \sin[\omega_{gd}(t-\tau)] \right. & \\ \left. \quad + 2\zeta_g \cos[\omega_{gd}(t-\tau) \right\} & t > \tau \end{cases} \tag{3.4.6}$$

where $\omega_{gd} = \sqrt{1-\zeta_g^2}\omega_g$. In the Kanai-Tajimi model, this impulse response function $h_1(t-\tau)$, which is now a function of $t-\tau$, takes the place of Green's function in equation (3.4.3). The designation of the pulse origin ρ_j and the subscript k are dropped, with the understanding that the expression refers to one particular direction of the ground motion, which is propagated vertically upward.

Substituting h_1 in (3.4.6) for Green's function into (3.4.3), we obtain

$$G(t) = \sum_{j=1}^{N(T)} Y_j h_1(t-\tau_j) \tag{3.4.7}$$

Equation (3.4.7) has the same form as (3.2.2); thus its mean and covariance functions may be computed, respectively, from (3.2.13) and (3.2.14), with h_1 substituting for w.

ONE-DIMENSIONAL MODEL OF A LINEAR ELASTIC MEDIUM. The Kanai-Tajimi model is perhaps the simplest representation of the behavior of the uppermost layer between the ground surface and the nearest bedrock, lumping the stiffness, inertia, and dissipative properties of the medium. The question now arises as to whether improvements can be achieved without lumping these properties. We investigate this question within the framework of one-dimensional wave propagation.

We consider first an elastic medium, shown in Fig. 3.4.2. The source of wave motion is located at the bedrock at $y = 0$, and we are concerned with the motion at the free ground surface at $y = l$. The equation for one-dimensional wave propagation is

$$\frac{\partial^2 w}{\partial t^2} - \beta^2 \frac{\partial^2 w}{\partial y^2} = 0 \tag{3.4.8}$$

where β is the wave propagation velocity. The symbol β, which is the standard symbol for shear wave velocity, is used here since the horizontal seismic shear wave is often the most damaging to building structures on the ground. Equation (3.4.8) is, of course, also valid for dilatation waves.

The frequency response function corresponding to the input at $y = 0$ and output at $y = l$ is very simple to derive. Letting $w(0,t) = \exp(i\omega t)$ and $w(y,t) = f(y)\exp(i\omega t)$, we obtain from separation of variables

$$f(y) = C_1\cos(ky) + C_2\sin(ky) \tag{3.4.9}$$

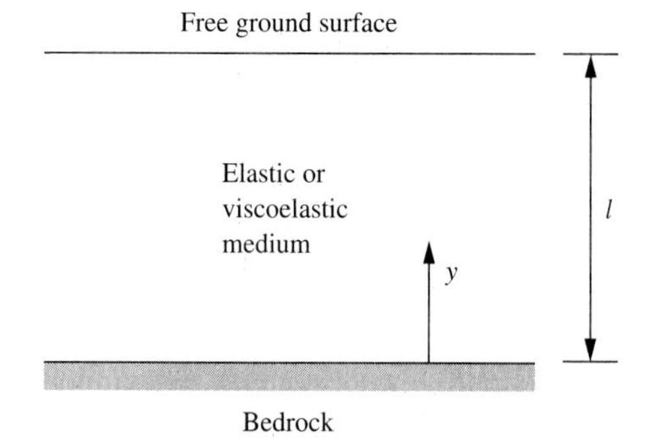

FIGURE 3.4.2
Idealized one-dimensional wave propagation medium.

where $k = \omega/\beta$ is the wave number. Imposition of boundary conditions results in

$$f(y) = \cos(ky) + \tan(kl)\sin(ky) \tag{3.4.10}$$

The frequency response function is obtained as

$$H_2(\omega) = f(l) = \frac{1}{\cos(kl)} = \frac{1}{\cos(\omega l/\beta)} \tag{3.4.11}$$

To account for dissipation in the medium, β is replaced by $\beta(1 + i\gamma \operatorname{sgn}\omega)$:

$$H_2(\omega) = \left[\cos\frac{\omega l}{\beta(1 + i\gamma \operatorname{sgn}\omega)}\right]^{-1} \tag{3.4.12}$$

The impulse response function can be obtained from

$$h_2(t) = \frac{1}{2\pi}\int_{-\infty}^{\infty}\left[\cos\frac{\omega l}{\beta(1 + i\gamma \operatorname{sgn}\omega)}\right]^{-1} e^{i\omega t}\,d\omega \tag{3.4.13}$$

The integration can be evaluated by use of the method of residues, resulting in

$$h_2(t) = \begin{cases} \dfrac{2\beta}{l}\displaystyle\sum_{n=0}^{\infty}(-1)^n \exp\left[-\left(n+\frac{1}{2}\right)\frac{\pi\beta\gamma}{l}t\right] & \\ \quad\times\left\{\gamma\cos\left[\left(n+\frac{1}{2}\right)\frac{\pi\beta}{l}t\right] + \sin\left[\left(n+\frac{1}{2}\right)\frac{\pi\beta}{l}t\right]\right\} & t > 0 \\ 0 & t < 0 \end{cases} \tag{3.4.14}$$

Equations (3.4.12) and (3.4.14) now replace equations (3.4.5) and (3.4.6) for the Kanai-Tajimi model, respectively.

It is clear that each n in (3.4.14) corresponds to one normal mode. For ease of comparison with the Kanai-Tajimi model, the first term in (3.4.14) is cast in the form

$$h_{2,0}(t) = \frac{4}{\pi}\omega_g' \exp(-\gamma\omega_g' t)[\gamma\cos(\omega_g' t) + \sin(\omega_g' t)] \qquad t > 0 \tag{3.4.15}$$

where $\omega_g' = \pi\beta/(2l)$. Equation (3.4.15) bears considerable resemblance to (3.4.6). The other terms in (3.4.14) are higher harmonics, which are damped out faster exponentially.

The one-dimensional model of equation (3.4.8) does not account for geometrical spreading of seismic energy. This can be remedied partially by modifying (3.4.8):

$$\frac{\partial^2 w}{\partial t^2} - \beta^2 \frac{\partial^2 w}{\partial y^2} - \beta^2 \frac{d[\ln A(y)]}{dy} \frac{\partial w}{\partial y} = 0 \tag{3.4.16}$$

Equation (3.4.16) is known as the "horn" equation in acoustics, where $A(y)$ represents the cross section of a horn within $0 \leq y \leq l$. The implied assumption is that the seismic energy is channeled by a horn-shaped device to the ground surface, which is a crude simplification. Nevertheless, the simple model may shed some light on what might be expected of geometric spreading.

An exact solution of equation (3.4.16) is generally difficult, but it can be obtained if the horn is exponential, that is, if

$$A(y) = A_0 \exp(2my) \tag{3.4.17}$$

where A_0 is the cross section at $y = 0$ and m is a positive constant. Then solution (3.4.9) is modified to

$$f(y) = e^{-my}\left[C_1 \cos(\sqrt{k^2 - m^2}\, y) + C_2 \sin(\sqrt{k^2 - m^2}\, y)\right] \tag{3.4.18}$$

Two major effects due to horn-shaped spreading are evident in (3.4.18). First, there is a spatial attenuation characterized by the factor e^{-my}. Second, there exists a low cutoff frequency $\omega_c = m\beta$ below which no seismic wave can be propagated to the ground surface. From the viewpoint of earthquake modeling, spatial attenuation has the same effect as the reduction of source strength. Thus neglecting the spatial attenuation effect can be compensated by lowering the earthquake intensity. The existence of a low cutoff frequency, however, might be a desirable feature since it would eliminate the need for a high-pass filter such as the one proposed by Clough and Penzien (1975).

We return now to (3.4.18) and impose the boundary conditions $f(0) = 1$ and $(df/dy)_{y=1} = 0$. We obtain instead of (3.4.12), the frequency response function,

$$H_3(\omega) = f(l) = e^{-my}\left[\cos(\delta l) - \frac{m}{\delta}\sin(\delta l)\right]^{-1} \tag{3.4.19}$$

where

$$\delta = [\omega^2\beta^{-2}(1 + i\gamma \operatorname{sgn}\omega)^{-2} - m^2]^{1/2} \tag{3.4.20}$$

The impulse response function is related to the frequency response functions as follows:

$$h_3(t) = \frac{1}{2\pi}\int_{-\infty}^{\infty} H_3(\omega)e^{i\omega t}\,d\omega \tag{3.4.21}$$

In principle, the right-hand side of (3.4.21) can be evaluated by applying the method of contour integration, which requires that the residues be evaluated at ω values satisfying

$$\tan(\delta l) = \delta/m \tag{3.4.22}$$

This exact analytical procedure, however, appears complicated. Numerically, $h_3(t)$ can be computed using the standard FFT program.

ONE-DIMENSIONAL MODEL OF A LINEAR MAXWELL MEDIUM. If the soil properties are such that energy is dissipated primarily through viscous damping, then equation (3.4.8) may by modified to read

$$\frac{\partial^2 w}{\partial t^2} + \frac{1}{\tau_r}\frac{\partial w}{\partial t} - \beta^2\frac{\partial^2 w}{\partial y^2} = 0 \tag{3.4.23}$$

where τ_r is known as the relaxation time. Equation (3.4.23) describes the one-dimensional wave propagation in a simple viscoelastic material, called the linear Maxwell material (e.g., Lee, 1962). Another simple viscoelastic model, the linear Kelvin material, will not be considered since it implies an instantaneous elastic response throughout the entire one-dimensional space, similar to the original Kanai-Tajimi model considered earlier. The frequency response function obtained from equation (3.4.23) is given by

$$H_4(\omega) = \left[\cos\left(\frac{l}{\beta}\sqrt{\omega^2 - i\frac{\omega}{\tau_r}}\right)\right]^{-1} \tag{3.4.24}$$

Application of the Fourier transform to (3.4.24) to obtain the impulse response function is again complicated. However, the impulse response function for the *stress* response, which can be converted to that of the *acceleration* response, is available in the literature (Lee and Kanter, 1953). Expressed in the present notation, it reads

$$h_4(t) = \exp\left(-\frac{t}{2\tau_r}\right)\sum_{n=0}^{\infty}(-1)^n\tau_r^{-1}(2n+1)\frac{l}{\beta}\left[t^2 - (2n+1)^2\left(\frac{l}{\beta}\right)^2\right]^{-1/2}$$
$$I_1\left\{(2\tau_r)^{-1}\left[t^2 - (2n+1)^2\left(\frac{l}{\beta}\right)^2\right]^{1/2}\right\} \mathrm{II}\left[t - (2n+1)\frac{l}{\beta}\right] \tag{3.4.25}$$

where $\mathrm{II}[\cdot]$ is the Heaviside unit-step function and I_1 is the first-order Bessel function of the first kind of imaginary argument. It is of interest to note that the right-hand side of equation (3.4.25) can also be obtained by the method of images (Lee and Kanter, 1953), which accounts for multiple reflections of the waves at the two boundaries $y = 0$ and $y = l$.

NUMERICAL SIMULATIONS. Artificial seismograms can be generated easily based on the concept of random pulse train of the form of (3.2.1). In such a model, different pulses are assumed to arrive independently with an average arrival rate $g_1(t)$, and the pulse amplitudes Y_j are independent random variables with an identical probability distribution. Without loss of generality, we may assume further that the Y's have a zero mean. Then the most important statistical property of the model process is the covariance function, obtainable from equation (3.2.17). The form of (3.2.17) suggests that the same ensemble average can be obtained by assuming either g_1 or $E[Y^2]$ to be time-dependent. This means that, when averaging over the entire ensemble, fewer pulse arrivals in a time interval with statistically larger pulse amplitudes have the same effect as that of more pulse arrivals with smaller pulse amplitudes.

Although we have chosen a time-dependent $g_1(t)$ in the foregoing theoretical discussions, it is more convenient, when generating the artificial seismograms, to keep the average pulse-arrival rate constant but change the pulse magnitudes according to the arrival time of each pulse. Artificial seismograms have been simulated (Lin and Yong, 1987) according to

$$G(t) = \sum_{j=1}^{N(T)} Y_j h(t - \tau)$$

using the h functions given by equations (3.4.6), (3.4.14), and (3.4.25), respectively, for the evolutionary Kanai-Tajimi model, one-dimensional elastic model, and one-dimensional Maxwell model. In all three cases, each sample length was chosen to be 60 s and the average pulse-arrival rate $g_1 = 3\text{s}^{-1}$. A total of 180 random numbers were generated between 0 and 60, and these were treated as the pulse-arrival times. In order that the results can be compared, this same set of pulse-arrival times was used in the simulation for all the three cases. For every pulse, a separate Bernoulli trial was conducted to select either a plus or a minus sign for Y_j, and its magnitude was calculated from

$$|Y_j| = \left[\tau_j\left(1 - \cos\frac{\pi\tau_j}{30}\right)\right]^{1/2}.$$

As explained earlier, this leads to the same covariance function that may be obtained by requiring

$$g_1(\tau) = \tau\left(1 - \cos\frac{\pi\tau}{30}\right)$$

and allowing Y_j to be independent identically distributed random variables. Additional parameters selected for different models were

Kanai-Tajimi model: $\omega_g = \pi$ rad/s, $\zeta_g = 0.3$
One-dimensional elastic model: $(\beta/l) = 2\ \text{s}^{-1}$, $\gamma = 0.6$
One-dimensional Maxwell model: $(\beta/l) = 2\ \text{s}^{-1}$, $\tau_r = 0.1$ s

These parametric values were chosen to simulate the dominant ground resonance condition of the 1985 Mexico City earthquake. Typical artificial seismograms are shown in Figs. 3.4.3*a*–*c*. The earthquake record being simulated is shown in Fig. 3.4.3*d*.

From the appearance, the artificial seismograms obtained from both the evolutionary Kanai-Tajimi model and the one-dimensional elastic model, Figs. 3.4.3*a* and *b*, resemble the actual record, Fig. 3.4.3*d*. The seismogram obtained from the one-dimensional Maxwell model, Fig. 3.4.3*c*, is somewhat more jagged, suggesting that the clay deposit over the old lake bed in Mexico City is closer to being elastic than maxwellian. Additional insights are provided in the frequency spectra of the three artificial seismograms shown in Figs. 3.4.4, 3.4.5, and 3.4.6. The ordinates in these figures are the magnitudes of acceleration per radians per second. The units are not indicated since, for the present investigation, only the relative magnitudes are of interest. It appears that, compared with the Kanai-Tajimi model, the one-dimensional elastic model is slightly richer in the intermediate frequency range between 5 and 10

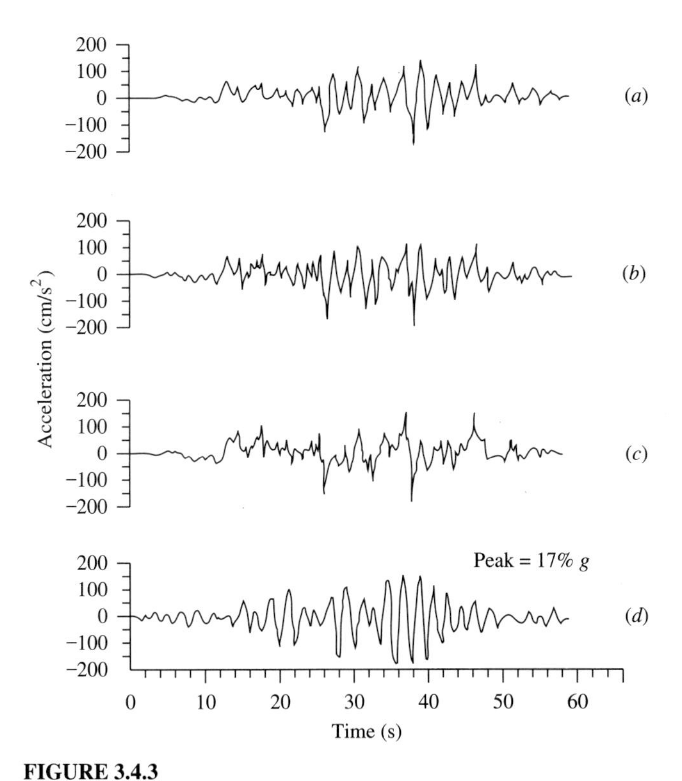

FIGURE 3.4.3
Comparison of artificial seismograms generated from three different models and an actual earthquake record. (*a*) Evolutionary Kanai-Tajimi model; (*b*) one-dimensional elastic model; (*c*) one-dimensional Maxwell model; (*d*) ground acceleration recorded over old lake bed in Mexico City, September 19, 1985, east-west direction [after Lin and Yong, 1987; Copyright ©1987 ASCE, reprinted with permission.]

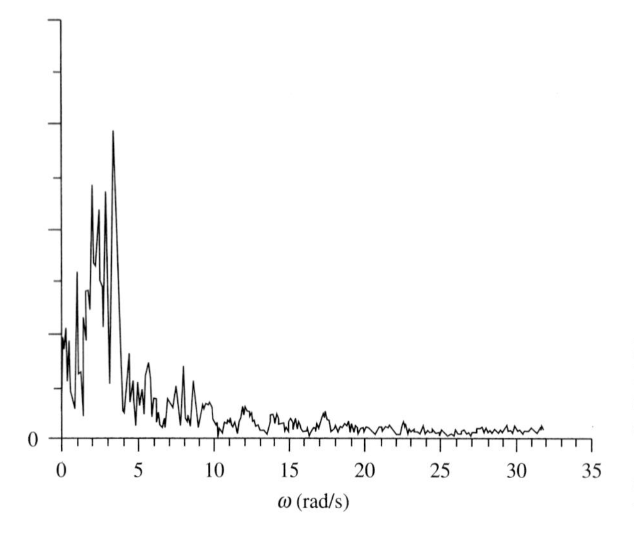

FIGURE 3.4.4
Frequency spectrum of an artificial seismogram generated from evolutionary Kanai-Tajimi model [after Lin and Yong, 1987; Copyright ©1987 ASCE, reprinted with permission].

rad/s, and the one-dimensional Maxwell model has a greater concentration of energy below 2 rad/s. The lower frequency components in the one-dimensional elastic model can be eliminated by incorporating a "spreading horn" in the model.

To illustrate the time variation of the frequency contents of the artificial seismograms, the computed evolutionary spectra of the Kanai-Tajimi model are shown in Fig. 3.4.7. The ordinate in this figure is the mean-square acceleration per radian per second, and the area under each curve represents the mean-square acceleration at a given time, without indicating the unit specifically. The peaks of the spectra are seen to shift only slightly as time changes; they remain near the ground resonance frequency at approximately 3 rad/s. The energy distribution, however,

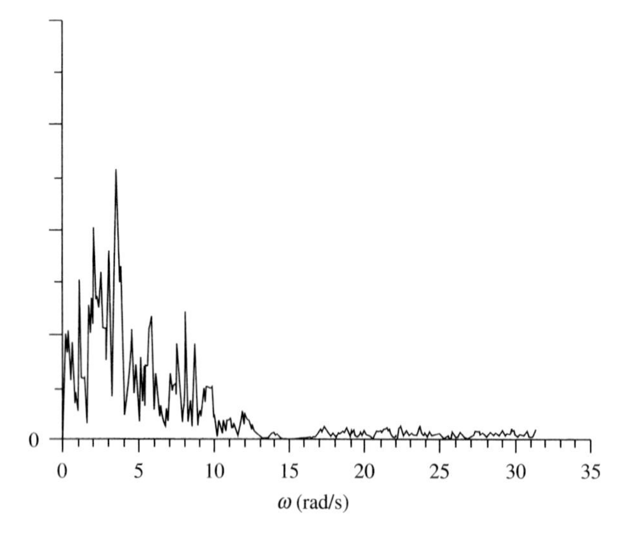

FIGURE 3.4.5
Frequency spectrum of an artificial seismogram generated from one-dimensional elastic model [after Lin and Yong, 1987; Copyright ©1987 ASCE, reprinted with permission].

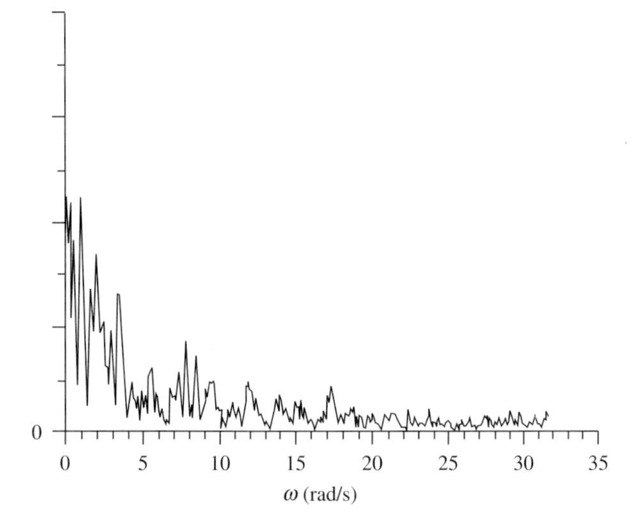

FIGURE 3.4.6
Frequency spectrum of an artificial seismogram generated from one-dimensional Maxwell model [after Lin and Yong, 1987; Copyright ©1987 ASCE, reprinted with permission].

becomes more concentrated around the peak at $t = 40$ s when the maximum mean-square acceleration occurs. Each of these spectra exhibits only one peak. This is to be expected since the original Kanai-Tajimi model has only one degree of freedom, and the one-dimensional elastic or Maxwell model involves a single-layer medium. Either a multi-degree-of-freedom discrete model or a multilayer continuous model is required to produce a multipeak spectrum.

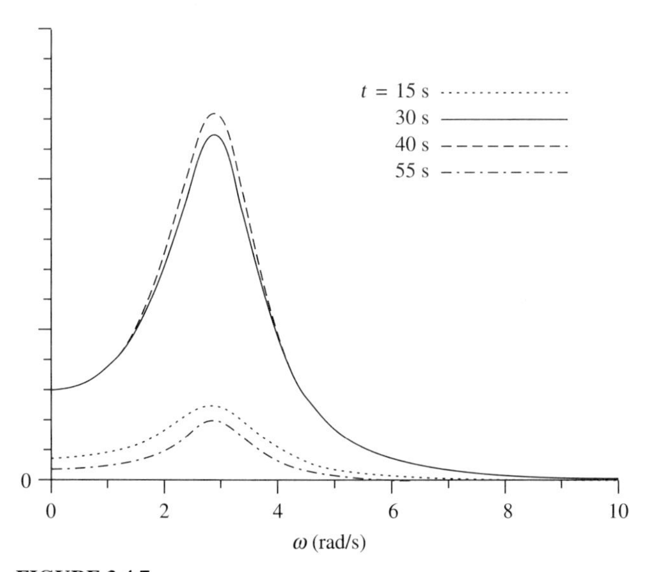

FIGURE 3.4.7
Evolutionary spectra of an evolutionary Kanai-Tajimi earthquake model [after Lin and Yong, 1987; Copyright ©1987 ASCE, reprinted with permission].

One of the criticisms against the original stationary Kanai-Tajimi model for seismic acceleration is the implication that the corresponding velocity and displacement processes are unbounded. However, this is not the case with the evolutionary Kanai-Tajimi model, which is constructed with three basic ingredients: the pulse shape, random pulse amplitude, and random pulse-arrival time. The pulse shape is described by an impulse response function. Using the schematic representation of a mass supported by linear spring and dashpot in parallel, the impulse response function for the acceleration is given by (3.4.6) and those for the velocity and displacement are given, respectively, by

$$h_{1v}(t-\tau) = \begin{cases} 0 & t<0 \\ \exp(-\zeta_g\omega_g t)\left[\zeta_g(1-\zeta_g^2)^{-1/2}\sin(\omega_d t) - \cos(\omega_d t)\right] & t>0 \end{cases} \tag{3.4.26}$$

$$h_{1d}(t-\tau) = \begin{cases} 0 & t<0 \\ -\dfrac{1}{\omega_d}\exp(-\zeta_g\omega_g t)\sin(\omega_d t) & t>0 \end{cases} \tag{3.4.27}$$

As long as the average pulse-arrival rate begins from zero and ends at zero within a finite length of time, the evolutionary spectral densities do not become unbounded at the origin. The acceleration, velocity, and displacement spectra obtained from (3.4.6), (3.4.26), and (3.4.27) are compared in Fig. 3.4.8.

3.5 THE PULSE SHAPE FUNCTION FOR LAYERED EARTH MEDIA

In the preceding section, the earthquake motion was approximated as a one-dimensional wave propagation in the vertical direction within the uppermost earth layer. This is justifiable under two conditions: (1) the geophysical features in the general region in question are characterized essentially by layers, with progressively heavier materials in the layers farther below; and (2) the seismic sources are sufficiently deep so that the propagation directions of the seismic waves become nearly vertical when reaching the ground surface. If these conditions are not met, then a more accurate pulse shape function must be obtained for substitution into equation (3.4.3) for the random pulse chain. The present section is directed at finding such a pulse shape function, following a paper by Zhang, Yong, and Lin (1991a).

WAVE MOTION WITHIN A SINGLE LAYER. As depicted in Fig. 3.5.1, the earth is idealized as a horizontally stratified medium, and each layer in the medium is assumed to be homogeneous, isotropic, and linearly elastic. The layer sequence is ordered from the top downward. A typical layer, say layer i, is characterized by a mass density ρ_i, and two Lame's constants λ_i and μ_i. Let us focus our attention first within one such layer and write the governing equations of motion in the cylindrical coordinates (r, ϕ, z):

$$\rho\frac{\partial^2 u_r}{\partial t^2} = \frac{\partial \tau_{rz}}{\partial z} + \frac{\partial \tau_{rr}}{\partial r} + \frac{1}{r}\frac{\partial \tau_{r\phi}}{\partial \phi} + \frac{1}{r}(\tau_{rr} - \tau_{\phi\phi}) + \rho f_r \tag{3.5.1a}$$

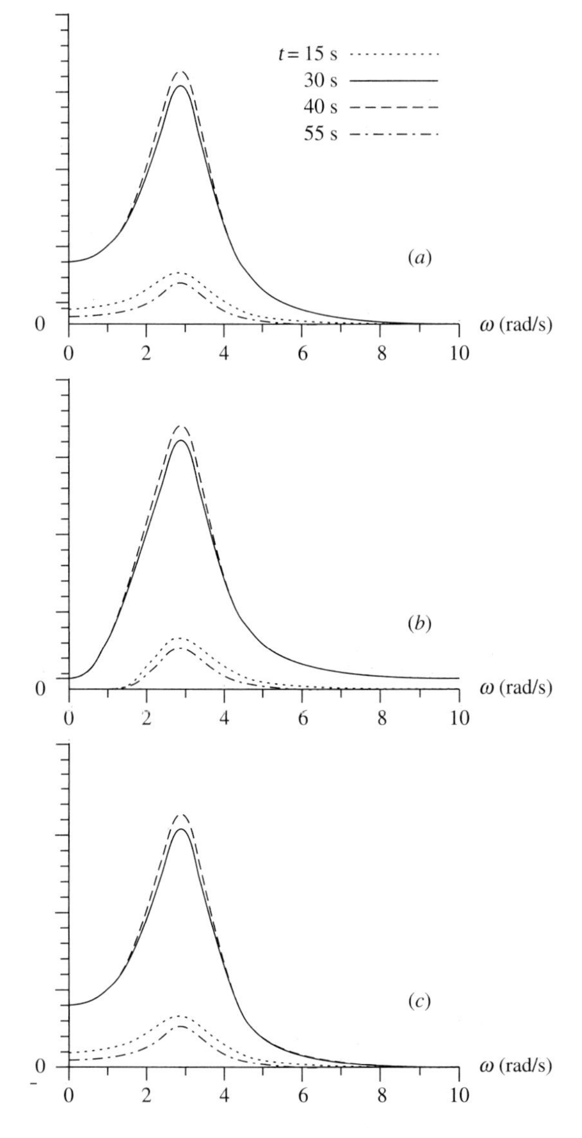

FIGURE 3.4.8
Comparison of (*a*) acceleration, (*b*) velocity and (*c*) displacement spectra of an evolutionary Kanai-Tajimi model [after Lin and Yong, 1987; Copyright ©1989 ASCE, reprinted with permission].

$$\rho\frac{\partial^2 u_\phi}{\partial t^2} = \frac{\partial \tau_{\phi z}}{\partial z} + \frac{\partial \tau_{r\phi}}{\partial r} + \frac{1}{r}\frac{\partial \tau_{\phi\phi}}{\partial \phi} + \frac{2}{r}\tau_{r\phi} + \rho f_\phi \tag{3.5.1b}$$

$$\rho\frac{\partial^2 u_z}{\partial t^2} = \frac{\partial \tau_{zz}}{\partial z} + \frac{\partial \tau_{rz}}{\partial r} + \frac{1}{r}\frac{\partial \tau_{\phi z}}{\partial \phi} + \frac{1}{r}\tau_{rz} + \rho f_z \tag{3.5.1c}$$

where the u's are displacements, the τ's are stresses, and the f's are body forces per unit mass. For simplicity, the identification subscript i for the layer is not indicated. The constitutive relations between displacements and stresses are

$$\tau_{rr} = (\lambda + 2\mu)\frac{\partial u_r}{\partial r} + \lambda\left(\frac{\partial u_z}{\partial z} + \frac{1}{r}\frac{\partial u_\phi}{\partial \phi} + \frac{u_r}{r}\right) \tag{3.5.2a}$$

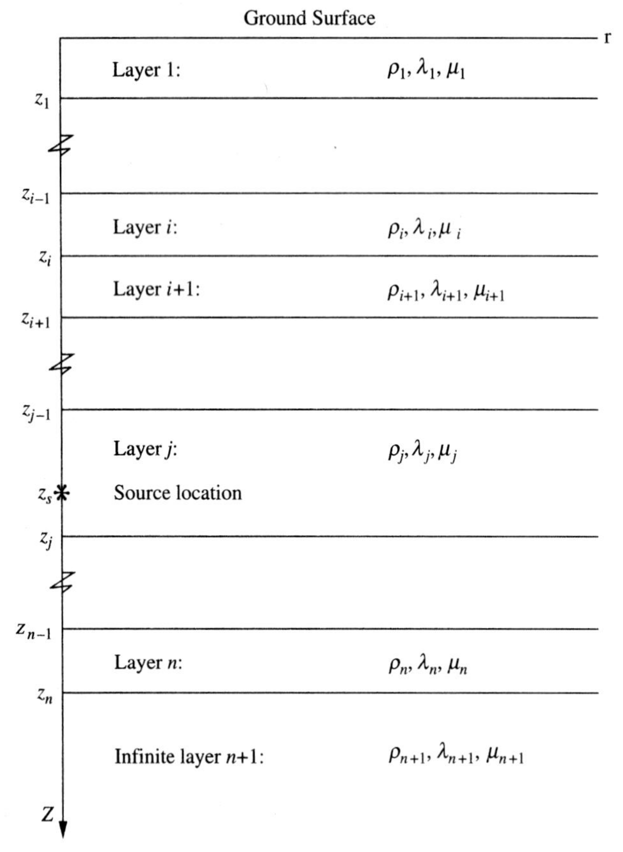

FIGURE 3.5.1
A horizontally stratified medium.

$$\tau_{\phi\phi} = (\lambda + 2\mu)\frac{1}{r}\left(\frac{\partial u_\phi}{\partial \phi} + u_r\right) + \lambda\left(\frac{\partial u_z}{\partial z} + \frac{\partial u_r}{\partial r}\right) \tag{3.5.2b}$$

$$\tau_{zz} = (\lambda + 2\mu)\frac{\partial u_z}{\partial z} + \lambda\left(\frac{\partial u_r}{\partial r} + \frac{1}{r}\frac{\partial u_\phi}{\partial \phi} + \frac{u_r}{r}\right) \tag{3.5.2c}$$

$$\tau_{rz} = \mu\left(\frac{\partial u_r}{\partial z} + \frac{\partial u_z}{\partial r}\right) \tag{3.5.2d}$$

$$\tau_{\phi z} = \mu\left(\frac{1}{r}\frac{\partial u_z}{\partial \phi} + \frac{\partial u_\phi}{\partial z}\right) \tag{3.5.2e}$$

$$\tau_{r\phi} = \mu\left(\frac{\partial u_\phi}{\partial r} - \frac{u_\phi}{r} + \frac{1}{r}\frac{\partial u_r}{\partial \phi}\right) \tag{3.5.2f}$$

It is convenient to introduce the following new variables (Hudson, 1969):

$$u_V = \frac{1}{r}\left[\frac{\partial (r u_r)}{\partial r} + \frac{\partial u_\phi}{\partial \phi}\right] \tag{3.5.3a}$$

$$\tau_{Vz} = \frac{1}{r}\left[\frac{\partial(r\tau_{rz})}{\partial r} + \frac{\partial \tau_{\phi z}}{\partial \phi}\right] \tag{3.5.3b}$$

$$u_H = \frac{1}{r}\left[\frac{\partial(ru_{\phi})}{\partial r} - \frac{\partial u_r}{\partial \phi}\right] \tag{3.5.3c}$$

$$\tau_{Hz} = \frac{1}{r}\left[\frac{\partial(r\tau_{\phi z})}{\partial r} - \frac{\partial \tau_{rz}}{\partial \phi}\right] \tag{3.5.3d}$$

$$f_V = \frac{1}{r}\left[\frac{\partial(rf_r)}{\partial r} + \frac{\partial f_{\phi}}{\partial \phi}\right] \tag{3.5.3e}$$

$$f_H = \frac{1}{r}\left[\frac{\partial(rf_{\phi})}{\partial r} - \frac{\partial f_r}{\partial \phi}\right] \tag{3.5.3f}$$

with which two sets of equations can be obtained. One set corresponds to the coupled dilatational wave (commonly known as the P wave) and vertical shear wave (commonly known as the SV wave), given by

$$\frac{\partial u_z}{\partial z} = -\frac{\lambda}{\lambda + 2\mu}u_V + \frac{1}{\lambda + 2\mu}\tau_{zz} \tag{3.5.4a}$$

$$\frac{\partial u_V}{\partial z} = -\Delta u_z + \frac{1}{\mu}\tau_{V_z} \tag{3.5.4b}$$

$$\frac{\partial \tau_{zz}}{\partial z} = \rho\frac{\partial^2 u_z}{\partial t^2} - \tau_{Vz} - \rho f_z \tag{3.5.4c}$$

$$\frac{\partial \tau_{Vz}}{\partial z} = \rho\frac{\partial^2 u_V}{\partial t^2} - \frac{4\mu(\lambda + \mu)}{\lambda + 2\mu}\Delta u_V - \frac{\lambda}{\lambda + 2\mu}\Delta\tau_{zz} - \rho f_V \tag{3.5.4d}$$

where
$$\Delta = \frac{1}{r}\frac{\partial}{\partial r}\left(r\frac{\partial}{\partial r}\right) + \frac{1}{r^2}\frac{\partial^2}{\partial \phi^2} \tag{3.5.5}$$

is the Laplace operator in the cylindrical coordinates. The particle motion in a P wave coincides with the direction of propagation, whereas the particle motion in an SV wave is perpendicular to the direction of propagation. The other equation set, given by

$$\frac{\partial u_H}{\partial z} = \frac{1}{mu}\tau_{Hz} \tag{3.5.6a}$$

$$\frac{\partial \tau_{Hz}}{\partial z} = \rho\frac{\partial^2 u_H}{\partial t^2} - \mu\Delta u_H - \rho f_H \tag{3.5.6b}$$

corresponds to a horizontal shear wave (commonly known as the SH wave) for which the particle motion is also perpendicular to the direction of propagation. The equation set (3.5.4a–d) is decoupled from the equation set (3.5.6a,b).

A convenient procedure for solving a set of linear partial differential equations is to convert it to a set of ordinary differential equations, using the method of integral transform (e.g., Sneddon, 1951). This method, first applied to the case of a multilayered medium by Haskell (1964), is used in the present analysis. In particular, we apply a combined Fourier transform and truncated Hankel transform (Bouchon, 1979, 1981) to equations (3.5.4a–d) and (3.5.6a,b) to remove the partial differentiations with respect to r, ϕ, and t:

$$\hat{\psi}(k_n, m, z, \omega) = \int_{-\infty}^{\infty} dt\, e^{-i\omega t} \int_0^{r_T} r\, dr\, J_m(k_n r) \frac{1}{2\pi} \int_0^{2\pi} d\phi\, e^{-im\phi} \psi(r, \phi, z, t) \quad (3.5.7)$$

where $\psi(r, \phi, z, t)$ is any one of the original state variables, $\hat{\psi}$ denotes its transformed counterpart, m is the azimuthal order, r_T is the truncation distance in the r direction, J_m is the mth-order Bessel function of the first kind, and k_n is a positive root of the transcendental equation

$$J_m(k_n r_T) = 0 \quad (3.5.8)$$

for a given m. It is of interest to note that as $r_T \to \infty$, equation (3.5.8) is satisfied by any arbitrary k_n. In other words, k_n becomes continuous, which is of course the case with the conventional (untruncated) Hankel transform. The reason for using a truncated Hankel transform will be explained later. The inverse transform corresponding to (3.5.7) is obtained as

$$\psi(r, \phi, z, t) = \frac{1}{2\pi} \int_{-\infty}^{\infty} d\omega\, e^{i\omega t} \sum_{m=-\infty}^{\infty} e^{im\phi} \sum_{n=1}^{\infty} \frac{2J_m(k_n r)}{r_T^2 [J_m'(k_n r_T)]^2} \hat{\psi}(k_n, m, z, \omega) \qquad r \leq r_T \quad (3.5.9)$$

where J_m' represents the derivative of the Bessel function with respect to its argument.

One objective of the integral transform is to remove the laplacian operators on the right-hand sides of equations (3.5.4b), (3.5.4d), and (3.5.6b). For this purpose, use may be made of the identity

$$\widehat{\Delta\psi} = -\frac{k_n r_T}{2\pi} J_m'(k_n r_T) \int_{-\infty}^{\infty} dt\, e^{-i\omega t} \int_0^{2\pi} d\phi\, e^{-im\phi} \psi(r_T, \phi, z, t) - k_n^2 \hat{\psi} \quad (3.5.10)$$

It is seen from equations (3.5.8) and (3.5.9) that

$$\psi(r_T, \phi, z, t) = 0 \quad (3.5.11)$$

Thus (3.5.10) is simplified to

$$\widehat{\Delta\psi} = -k_n^2 \hat{\psi} \quad (3.5.12)$$

However, bearing in mind that ψ represents a state variable at (r, ϕ, z, t), the validity of equation (3.5.11) should be justified. Clearly, (3.5.11) is exact if $t \leq T_w$ where T_w is the time lapse for the fastest seismic wave to propagate from the source to a distance r_T. In practice, equation (3.5.11) always can be rendered approximately valid, if r_T is chosen large enough such that any seismic signal beyond r_T becomes negligible.

When carrying out the inverse Hankel transform numerically according to (3.5.9), the summation over n must be truncated. The number of terms to be kept in the truncation is equal to the number of roots of equation (3.5.8) up to $k_N r_T$ where

k_N is a suitably chosen cutoff wave number, beyond which the ground motion spectrum is considered negligible. Thus more terms must be retained in the truncated series if a larger r_T value is required.

Although the inverse of a conventional (theoretically nontruncated) Hankel transform must be computed numerically as an approximate sum, and the terms associated with large wave numbers must also be truncated, the error involved therein is not the same as that of the truncated Hankel transform approach. The only error associated with the truncated Hankel transform arises from the neglected higher wave-number terms, whereas the error associated with the conventional Hankel transform also includes that of changing from an integral to summation. In view of the highly oscillatory nature of the integrand, the truncated Hankel transform procedure is expected to be numerically more stable, and thus more accurate.

Upon applying the transformation procedure (3.5.7) to equations (3.5.4a–d) for the P-SV wave, the transformed equations may be cast in the form

$$\frac{d}{dz}\begin{Bmatrix} \boldsymbol{w} \\ \boldsymbol{f} \end{Bmatrix} = A\begin{Bmatrix} \boldsymbol{w} \\ \boldsymbol{f} \end{Bmatrix} - \begin{Bmatrix} \boldsymbol{0} \\ \boldsymbol{F} \end{Bmatrix} \tag{3.5.13}$$

where

$$\boldsymbol{A} = \begin{bmatrix} 0 & k_n\left(1 - \dfrac{2\beta^2}{\alpha^2}\right) & \dfrac{\omega}{\rho\alpha^2} & 0 \\ -k_n & 0 & 0 & \dfrac{\omega}{\rho\beta^2} \\ -\rho\omega & 0 & 0 & k_n \\ 0 & \rho\omega\left(\dfrac{\gamma k_n^2}{\omega^2} - 1\right) & -k_n\left(1 - \dfrac{2\beta^2}{\alpha^2}\right) & 0 \end{bmatrix} \tag{3.5.14}$$

and where $\alpha = (\lambda + 2\mu/\rho)^{1/2}$ is the speed of the P wave, $\beta = (\mu/\rho)^{1/2}$ is the speed of the S wave, $\gamma = 4\beta^2(1 - \beta^2/\alpha^2)$, and

$$\boldsymbol{w} = \begin{Bmatrix} \hat{u}_z \\ -k_n^{-1}\hat{u}_V \end{Bmatrix} \quad \boldsymbol{f} = \begin{Bmatrix} \hat{\tau}_{zz} \\ -k_n^{-1}\hat{\tau}_{Vz} \end{Bmatrix} \quad \boldsymbol{F} = \frac{\rho}{\omega}\begin{Bmatrix} \hat{f}_z \\ -k_n^{-1}\hat{f}_V \end{Bmatrix} \tag{3.5.15}$$

For the SH wave, the transformed equations may also be cast in the form of (3.5.13), but with

$$\boldsymbol{A} = \begin{bmatrix} 0 & \dfrac{\omega}{\rho\beta^2} \\ \rho\omega\left(\dfrac{\beta^2 k_n^2}{\omega^2} - 1\right) & 0 \end{bmatrix} \tag{3.5.16}$$

and

$$w = -\frac{1}{k_n}\hat{u}_H \quad f = -\frac{1}{\omega k_n}\hat{\tau}_{Hz} \quad F = -\frac{\rho}{\omega k_n}\hat{f}_H \tag{3.5.17}$$

For earthquake engineering applications, the ground motion at $z = 0$ is of interest. It can be obtained from equation (3.5.13) in the $\omega - k_n - m$ domain, which can then be inverted to yield

$$u_l(r, \phi, z = 0, t) = \frac{1}{2\pi}\int_{-\infty}^{\infty} d\omega e^{i\omega t} \sum_{m=-\infty}^{\infty} e^{im\phi} \sum_{n=1}^{\infty} \frac{2}{r_T^2 J_{m+1}^2(k_n r_T)} b_l \tag{3.5.18}$$

where subscript l refers to the direction of the ground motion and b_l depends on l.

It is often desirable that the ground motion be expressed in cartesian coordinates, namely, $l = x, y$, or z. Expressions for b_x, b_y, and b_z are given, respectively, as

$$b_x = [a_1 \cos\phi - a_2 \sin\phi]\frac{1}{k_n}\hat{u}_V + [a_2 \cos\phi + a_1 \sin\phi]\frac{1}{k_n}\hat{u}_H \quad (3.5.19)$$

$$b_y = [a_1 \sin\phi + a_2 \cos\phi]\frac{1}{k_n}\hat{u}_V + [a_2 \sin\phi - a_1 \cos\phi]\frac{1}{k_n}\hat{u}_H \quad (3.5.20)$$

$$b_z = J_m(k_n r)\hat{u}_z \quad (3.5.21)$$

in which $$a_1 = -\frac{m}{k_n r}J_m(k_n r) + J_{m+1}(k_n r) \qquad a_2 = -\frac{im}{k_n r}J_m(k_n r) \quad (3.5.22)$$

The cartesian and cylindrical coordinates are related as

$$r = (x^2 + y^2)^{1/2} \quad \text{and} \quad \phi = \tan^{-1}\left(\frac{y}{x}\right) \qquad 0 \le r \le r_T \quad (3.5.23)$$

WAVE MOTION IN A MULTILAYER MEDIUM. For a multilayer medium, equation (3.5.13) may be solved for each layer. Consider the ith layer, which is governed by the differential equation

$$\frac{d}{dz}\begin{Bmatrix} \boldsymbol{w}_{i*} \\ \boldsymbol{f}_{i*} \end{Bmatrix} = \boldsymbol{A}_i \begin{Bmatrix} \boldsymbol{w}_{i*} \\ \boldsymbol{f}_{i*} \end{Bmatrix} \qquad i - 1^+ \le i^* \le i^- \quad (3.5.24)$$

where $\boldsymbol{A}_i$ is constructed according to either (3.5.14) or (3.5.16), as the case may be, but with the layer number i so indicated. The subscript i^*, associated with a state variable, denotes the location z_i^*, at which the state variable is evaluated, and $i - 1^+$ and i^- mark the boundaries of the ith layer. Introduce the transformation

$$\begin{Bmatrix} \boldsymbol{w}_{i*} \\ \boldsymbol{f}_{i*} \end{Bmatrix} = \boldsymbol{D}_i \begin{Bmatrix} \boldsymbol{\mu}_u(i*) \\ \boldsymbol{\mu}_d(i*) \end{Bmatrix} \qquad i - 1^+ \le i^* \le i^- \quad (3.5.25)$$

where the columns in $\boldsymbol{D}_i$ are the eigenvectors of $\boldsymbol{A}_i$, with which the state vector $\{\boldsymbol{w}, \boldsymbol{f}\}'$ is transformed to a wave vector $\{\boldsymbol{\mu}_u, \boldsymbol{\mu}_d\}'$, where a prime denotes a matrix transposition, and the subscripts u and d represent the upward and downward propagation directions, respectively, as shown schematically in Figs. 3.5.2 and 3.5.3. It should be noted that the precise propagation directions of these waves are generally not vertical. The adjective upward or downward here refers only to the dependence of a wave motion on z. Its dependences on r and ϕ have been represented in the wave-number and azimuthal domains, respectively.

Substituting (3.5.25) into (3.2.24) and solving for the wave vector at $i - 1^+$ in terms of the wave vector at i^-

$$\begin{Bmatrix} \boldsymbol{\mu}_u(i - 1^+) \\ \boldsymbol{\mu}_d(i - 1^+) \end{Bmatrix} = \boldsymbol{Q}(i^-, i - 1^+) \begin{Bmatrix} \boldsymbol{\mu}_u(i^-) \\ \boldsymbol{\mu}_d(i^-) \end{Bmatrix} \quad (3.5.26)$$

where for the P-SV waves

$$\boldsymbol{Q}(i^-, i - 1^+) = \text{diag}\,[e^{i\omega q_\alpha \Delta z_i}\; e^{i\omega q_\beta \Delta z_i}\; e^{-i\omega q_\alpha \Delta z_i}\; e^{-i\omega q_\beta \Delta z_i}] \quad (3.5.27)$$

and for the SH waves

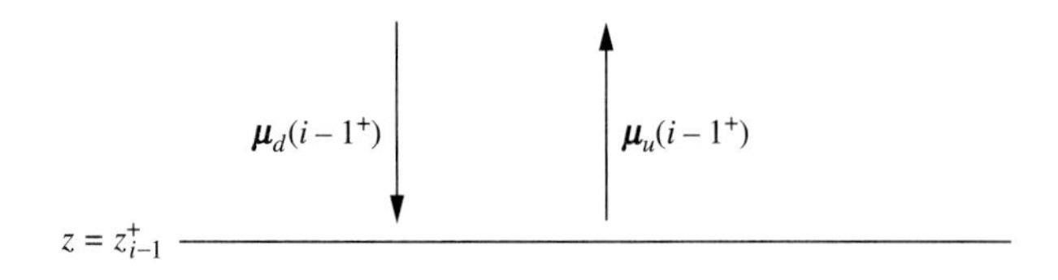

Layer i

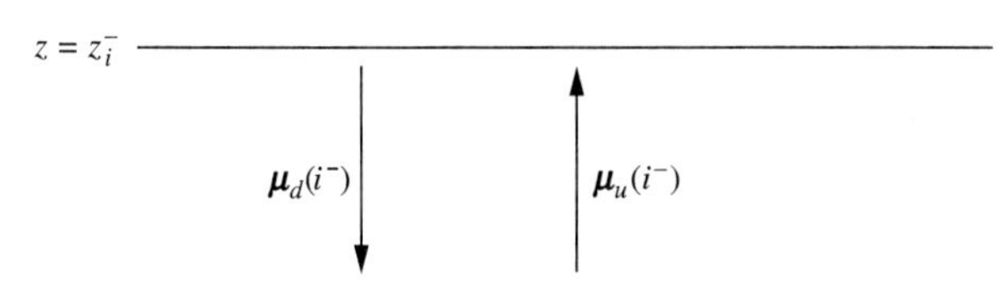

FIGURE 3.5.2
Incoming and outgoing waves to and from a layer.

$$\boldsymbol{Q}(i^-, i-1^+) = \text{diag}\,[e^{i\omega q_\beta \Delta z_i}\; e^{-i\omega q_\beta \Delta z_i}] \tag{3.5.28}$$

In (3.5.27) or (3.5.28), $\boldsymbol{Q}$ is called the wave propagator (or wave transfer matrix), diag indicates a diagonal matrix, and

$$\Delta z_i = |z_i - z_{i-1}| \tag{3.5.29}$$

$$q_\alpha = \left(\alpha^{-2} - \frac{k_n^2}{\omega^2}\right)^{1/2} \qquad q_\beta = \left(\beta^{-2} - \frac{k_n^2}{\omega^2}\right)^{1/2} \tag{3.5.30}$$

To account for damping, the real-valued wave speeds α and β may be replaced by a pair of complex valued wave speeds, $\alpha(1 + i\gamma_\alpha \operatorname{sgn}\omega)$, and $\beta(1 + i\gamma_\beta \operatorname{sgn}\omega)$, respectively. In this case, the branch cuts for the radicals in the expressions for q_α and q_β are taken to be $\text{Im}(\omega q_\alpha) \geq 0$ and $\text{Im}(\omega q_\beta) \geq 0$, respectively.

As shown in Fig. 3.5.2, $\boldsymbol{\mu}_d(i-1^+)$ and $\boldsymbol{\mu}_u(i^-)$ may be treated as incoming wave vectors with respect to layer i, while $\boldsymbol{\mu}_u(i-1^+)$ and $\boldsymbol{\mu}_d(i^-)$ may be treated as outgoing wave vectors. Equation (3.5.26) may then be rearranged according to such an input–output relationship as follows:

$$\begin{Bmatrix} \boldsymbol{\mu}_u(i-1^+) \\ \boldsymbol{\mu}_d(i^-) \end{Bmatrix} = \begin{bmatrix} \boldsymbol{R}(i-1^+, i^-) & \boldsymbol{T}(i^-, i-1^+) \\ \boldsymbol{T}(i-1^+, i^-) & \boldsymbol{R}(i^-, i-1^+) \end{bmatrix} \begin{Bmatrix} \boldsymbol{\mu}_d(i-1^+) \\ \boldsymbol{\mu}_u(i^-) \end{Bmatrix} \tag{3.5.31}$$

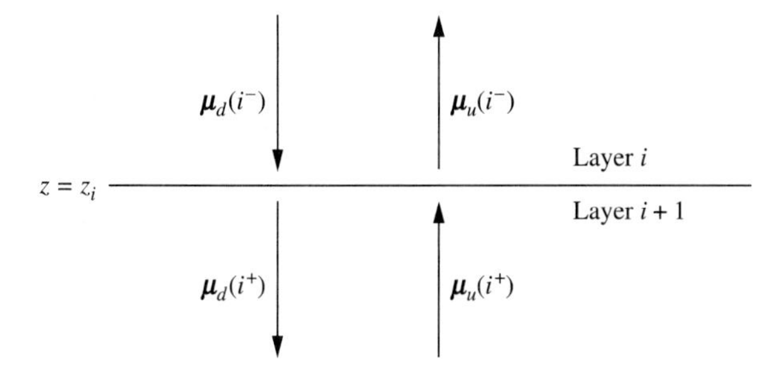

FIGURE 3.5.3
Incoming and outgoing waves to and from an interface between two layers.

where, for the present case,

$$\boldsymbol{R}(i-1^+, i^-) = \boldsymbol{R}(i^-, i-1^+) = \boldsymbol{0} \tag{3.5.32}$$

and

$$\boldsymbol{T}(i-1^+, i^-) = \boldsymbol{T}(i^-, i-1^+) = \begin{cases} \text{diag}\,[e^{i\omega q_\alpha \Delta z_i}\ e^{i\omega q_\beta \Delta z_i}] & \text{for the P-SV waves} \\ e^{i\omega q_\alpha \Delta z_i} & \text{for the SH waves} \end{cases} \tag{3.5.33}$$

The square matrix on the right side of equation (3.5.31) is known as a *wave scattering matrix*, and its submatrices $\boldsymbol{R}$ and $\boldsymbol{T}$ are known as *reflection* and *transmission matrices*, respectively. They characterize the properties of the earth medium between levels z_{i-1^+} and z_{i^-} in response to the incoming waves $\boldsymbol{\mu}_d(i-1^+)$ and $\boldsymbol{\mu}_u(i^-)$. For example, $\boldsymbol{R}(i-1^+, i^-)\boldsymbol{\mu}_d(i-1^+)$ and $\boldsymbol{T}(i-1^+, i^-)\boldsymbol{\mu}_d(i-1^+)$ represent, respectively, the reflected and transmitted parts of the incoming wave $\boldsymbol{\mu}_d(i-1^+)$. The physical meanings of $\boldsymbol{R}(i^-, i-1^+)$ and $\boldsymbol{T}(i^-, i-1^+)$ are similar. By focusing our attention on the layer between levels z_{i-1^+} and z_{i^-}, we have excluded the layer-to-layer interfaces (z_{i-1^-}, z_{i-1^+}) and (z_{i^-}, z_{i^+}) from immediate consideration. Therefore, the two reflection matrices in equation (3.5.31) are null.

Next we focus our attention on the interface (z_{i^-}, z_{i^+}). On the two sides of the interface, the values of the state vectors must be the same, namely,

$$\begin{Bmatrix} \boldsymbol{w}_{i^-} \\ \boldsymbol{f}_{i^-} \end{Bmatrix} = \begin{Bmatrix} \boldsymbol{w}_{i^+} \\ \boldsymbol{f}_{i^+} \end{Bmatrix} \tag{3.5.34}$$

However, the two corresponding wave vectors obtained from applying equation (3.5.25) are different, because the transformation matrices $\boldsymbol{D}_i$ and $\boldsymbol{D}_{i+1}$ are different. These two wave vectors are related as

$$\begin{Bmatrix} \boldsymbol{\mu}_u(i^-) \\ \boldsymbol{\mu}_d(i^-) \end{Bmatrix} = \boldsymbol{Q}(i^+, i^-) \begin{Bmatrix} \boldsymbol{\mu}_u(i^+) \\ \boldsymbol{\mu}_d(i^+) \end{Bmatrix} \tag{3.5.35}$$

where

$$\boldsymbol{Q}(i^+, i^-) = \boldsymbol{D}_i^{-1}\boldsymbol{D}_{i+1} \tag{3.5.36}$$

Recast equation (3.5.35) in an input–output relationship similar to equation (3.5.31):

$$\begin{Bmatrix} \boldsymbol{\mu}_u(i^-) \\ \boldsymbol{\mu}_d(i^+) \end{Bmatrix} = \begin{bmatrix} \boldsymbol{R}(i^-, i^+) & \boldsymbol{T}(i^+, i^-) \\ \boldsymbol{T}(i^-, i^+) & \boldsymbol{R}(i^+, i^-) \end{bmatrix} \begin{Bmatrix} \boldsymbol{\mu}_d(i^-) \\ \boldsymbol{\mu}_u(i^+) \end{Bmatrix} \tag{3.5.37}$$

It can be shown that

$$\boldsymbol{R}(i^-, i^+) = \boldsymbol{Q}_{12}(i^+, i^-)[\boldsymbol{Q}_{22}(i^+, i^-)]^{-1} \tag{3.5.38a}$$

$$\boldsymbol{R}(i^+, i^-) = -[\boldsymbol{Q}_{22}(i^+, i^-)]^{-1}\boldsymbol{Q}_{21}(i^+, i^-) \tag{3.5.38b}$$

$$\boldsymbol{T}(i^-, i^+) = [\boldsymbol{Q}_{22}(i^+, i^-)]^{-1} \tag{3.5.38c}$$

$$\boldsymbol{T}(i^+, i^-) = \boldsymbol{Q}_{11}(i^+, i^-) - \boldsymbol{Q}_{12}(i^+, i^-)[\boldsymbol{Q}_{22}(i^+, i^-)]^{-1}\boldsymbol{Q}_{21}(i^+, i^-) \tag{3.5.38d}$$

in which each $\boldsymbol{Q}_{jk}(j, k = 1, 2)$ is a submatrix of $\boldsymbol{Q}(i^+, i^-)$. The $\boldsymbol{Q}$ matrix constructed according to equation (3.5.36) depends on the physical properties of two layers; therefore, unless the physical properties of the two layers are the same, the reflection matrices in (3.5.37) are not null, and the transmission matrices should account for both energy dissipation and the change of phase, consistent with the well-known Snell's law.

The reflection and transmission matrices, characterizing a stratified medium between any two levels, can be constructed from the basic reflection and transmission matrices for each uniform layer, and the interface between each pair of neighboring layers. This is done according to the following composition rule (Yong and Lin, 1989):

$$\boldsymbol{R}(i,k) = \boldsymbol{R}(i,j) + \boldsymbol{T}(j,i)\boldsymbol{R}(j,k)[\boldsymbol{I} - \boldsymbol{R}(j,i)\boldsymbol{R}(j,k)]^{-1}\boldsymbol{T}(i,j) \tag{3.5.39}$$

$$\boldsymbol{T}(i,k) = \boldsymbol{T}(j,k)[\boldsymbol{I} - \boldsymbol{R}(j,i)\boldsymbol{R}(j,k)]^{-1}\boldsymbol{T}(i,j) \tag{3.5.40}$$

where level j is between levels i and k.

The boundary condition at the ground surface can be characterized by a reflection matrix alone. At the ground surface, the relationship between the state vector and the wave vector can be written as

$$\begin{Bmatrix} \boldsymbol{w}_0 \\ \boldsymbol{f}_0 \end{Bmatrix} = \begin{bmatrix} \boldsymbol{M}_u(1) & \boldsymbol{M}_d(1) \\ \boldsymbol{N}_u(1) & \boldsymbol{N}_d(1) \end{bmatrix} \begin{Bmatrix} \boldsymbol{\mu}_u(0^+) \\ \boldsymbol{\mu}_d(0^+) \end{Bmatrix} \tag{3.5.41}$$

where $\boldsymbol{M}_u(1), \boldsymbol{M}_d(1), \boldsymbol{N}_u(1)$, and $\boldsymbol{N}_d(1)$ are submatrices of $\boldsymbol{D}_1$. Since $\boldsymbol{f}_0 = \{\boldsymbol{0}\}$ at the boundary, it follows from (3.4.41) that

$$\boldsymbol{\mu}_d(0^+) = \boldsymbol{R}(0^+, 0)\boldsymbol{\mu}_u(0^+) \tag{3.5.42}$$

where

$$\boldsymbol{R}(0^+, 0) = [\boldsymbol{N}_d(1)]^{-1}\boldsymbol{N}_u(1) \tag{3.5.43}$$

Equation (3.5.42) characterizes the free ground surface.

The lower part of the earth medium, sufficiently far from the ground surface, may reasonably be modeled as one uniform layer of infinite thickness. The boundary condition at the interface between such an infinite layer and the layer immediately above is examined next. Let this interface be denoted by (n^-, n^+). The wave scattering relationship for the medium (n^-, ∞) can be written as

$$\begin{Bmatrix} \boldsymbol{\mu}_u(n^-) \\ \boldsymbol{\mu}_d(\infty) \end{Bmatrix} = \begin{bmatrix} \boldsymbol{R}(n^-, \infty) & \boldsymbol{T}(\infty, n^-) \\ \boldsymbol{T}(n^-, \infty) & \boldsymbol{R}(\infty, n^-) \end{bmatrix} \begin{Bmatrix} \boldsymbol{\mu}_d(n^-) \\ \boldsymbol{\mu}_u(\infty) \end{Bmatrix} \tag{3.5.44}$$

Clearly, $\boldsymbol{\mu}_u(\infty) = \boldsymbol{0}$ since there exists neither source nor wave reflection at infinity. It follows from (3.5.44) that

$$\boldsymbol{\mu}_u(n^-) = \boldsymbol{R}(n^-, \infty)\boldsymbol{\mu}_d(n^-) \tag{3.5.45}$$

Application of equation (3.5.39) yields

$$\boldsymbol{R}(n^-, \infty) = \boldsymbol{R}(n^-, n^+) + \boldsymbol{T}(n^+, n^-)\boldsymbol{R}(n^+, \infty)[\boldsymbol{I} - \boldsymbol{R}(n^+, n^-)\boldsymbol{R}(n^+, \infty)]^{-1}\boldsymbol{T}(n^-, n^+) \tag{3.5.46}$$

However, $\boldsymbol{R}(n^+, \infty) = \boldsymbol{0}$, since (n^+, ∞) is a uniform medium. Thus

$$\boldsymbol{\mu}_u(n^-) = \boldsymbol{R}(n^-, n^+)\boldsymbol{\mu}_d(n^-) \tag{3.5.47}$$

Equation (3.5.47) characterizes the boundary immediately above an infinitely thick layer.

Most earthquakes are caused by the dislocation of a fault. If the typical dimension of a fault is short compared with its distance from the ground site of interest, then the effect of dislocation may be represented by a set of equivalent body forces in self-equilibrium, known as double couples or collectively a seismic moment (Burridge and Knopoff, 1964). These double couples can be further replaced by equivalent jumps (e.g., Kennett, 1985) in the components of the state vector $\{\boldsymbol{w}, \boldsymbol{f}\}'$ in equation (3.5.13). For example, for a shear dislocation in the x direction, occurring along a fault plane parallel to the ground surface, the seismic moment acts on an x-z plane. In this case, the equivalent jumps corresponding to a unit seismic moment are given by

$$[\hat{u}_z]_-^+ = [\hat{\tau}_{zz}]_-^+ = [\hat{\tau}_{Vz}]_-^+ = [\hat{\tau}_{Hz}]_-^+ = 0 \tag{3.5.48a}$$

$$[\hat{u}_V]_-^+ = \mp\tfrac{1}{2}k_n(\rho\beta^2)^{-1} \qquad m = \pm 1 \tag{3.5.48b}$$

$$[\hat{u}_H]_-^+ = \tfrac{1}{2}ik_n(\rho\beta^2)^{-1} \qquad m = \pm 1 \tag{3.5.48c}$$

where $[\cdot]_-^+$ is an abbreviation for $[\cdot]^+ - [\cdot]^-$. The magnitude of a seismic moment is obtained as (shear modulus) $\times$ (fault area) $\times$ (final slip). The reader is referred to Kennett (1985) for other types of dislocations.

Let the equivalent jumps given in (3.5.48) be located at a source plane z_s within layer j, as shown in Fig. 3.5.1. It is convenient to represent such jumps as a source vector

$$\begin{Bmatrix} \boldsymbol{w}_s \\ \boldsymbol{f}_s \end{Bmatrix} = \begin{Bmatrix} \boldsymbol{w}_s^+ \\ \boldsymbol{f}_s^+ \end{Bmatrix} - \begin{Bmatrix} \boldsymbol{w}_{s-} \\ \boldsymbol{f}_{s-} \end{Bmatrix} \tag{3.5.49}$$

Applying the transformation (3.5.25) to equation (3.5.49) and taking into consideration the reflection matrices above and below the source plane, we obtain

$$\begin{bmatrix} \boldsymbol{M}_u(j) & \boldsymbol{M}_d(j) \\ \boldsymbol{N}_u(j) & \boldsymbol{N}_d(j) \end{bmatrix} \begin{Bmatrix} \boldsymbol{R}(s^+, n^+)\boldsymbol{\mu}_d(s^+) \\ \boldsymbol{\mu}_d(s^+) \end{Bmatrix} - \begin{bmatrix} \boldsymbol{M}_u(j) & \boldsymbol{M}_d(j) \\ \boldsymbol{N}_u(j) & \boldsymbol{N}_d(j) \end{bmatrix} \begin{Bmatrix} \boldsymbol{\mu}_u(s^-) \\ \boldsymbol{R}(s^-, 0)\boldsymbol{\mu}_u(s^-) \end{Bmatrix} = \begin{Bmatrix} \boldsymbol{w}_s \\ \boldsymbol{f}_s \end{Bmatrix} \tag{3.5.50}$$

Solving for $\boldsymbol{\mu}_u(s^-)$ from (3.5.50), and using equations (3.5.25), (3.5.39), (3.5.40), and (3.5.47), the ground motion in the ω-k_n-m domain can be obtained as follows:

$$\boldsymbol{w}_0 = \left[\boldsymbol{M}_u(1) + \boldsymbol{M}_d(1)\boldsymbol{R}(0^+, 0)\right]\left[\boldsymbol{I} - \boldsymbol{R}(0^+, s^-)\boldsymbol{R}(0^+, 0)\right]^{-1}\boldsymbol{T}(s^-, 0^+)\boldsymbol{\mu}_u(s^-) \tag{3.5.51}$$

The ground motion at location (x, y, z) and at time t due to a buried point source at $(0, 0, z_s)$ and initiated at time τ is given by (Aki and Richards, 1980)

$$\begin{aligned} g_l(x, y, z, t; z_s, \tau) &= M_0 \int_0^t m(t', \tau)u_l(x, y, z, t - t')\,dt' \\ &= M_0 \int_{-\infty}^{\infty} m(t', \tau)u_l(x, y, z, t - t')\,dt' \\ &= \frac{M_0}{2\pi}\int_{-\infty}^{\infty} \bar{m}(\omega, \tau)\bar{u}_l(x, y, z, \omega)e^{i\omega t}d\omega \end{aligned} \tag{3.5.52}$$

where $M_0m(t, \tau)$ describes the seismic moment, function u_l has been given in (3.5.18), $\bar{u}_l$ is the Fourier transform of u_l, and use is made of the fact that $u_l(x, y, z, t - t') = 0$ for either $t' < 0$ or $t' > t$. It is clear from (3.5.18) that $\bar{u}_l$ is given by

$$\bar{u}_l(x, y, z, \omega) = \sum_{m=-\infty}^{\infty} e^{im\phi} \sum_{n=1}^{\infty} \frac{2}{r_T^2 J_{m+1}^2(k_n, r_T)} b_l \tag{3.5.53}$$

The time variation of the seismic moment, described by $m(t, \tau)$, is often assumed to be a ramp function (Haskell, 1964) as shown in Fig. 3.5.4, of which the Fourier transform is given by

$$\bar{m}(\omega, \tau) = \frac{1}{\omega^2 T_r}(e^{-i\omega T_r} - 1)e^{-i\omega\tau} + \pi\delta(\omega) \tag{3.5.54}$$

where T_r is the rise time.

For structural engineering applications, the ground acceleration $\ddot{u}_l$ is of greater interest. Its Fourier transform is

$$\bar{\ddot{u}}_l = -\omega^2 \bar{u}_l \tag{3.5.55}$$

Replacing $\bar{u}_l$ by $-\omega^2\bar{u}_l$ in (3.5.52), and using (3.5.54), we obtain the ground acceleration at location (x, y, z) and at time t:

$$\ddot{g}_l(x, y, z, t; z_s, \tau) = \frac{M_0}{2\pi}\int_{-\infty}^{\infty} \frac{1}{T_r}(1 - e^{-i\omega T_r})\bar{u}_l(x, y, z, \omega)e^{i\omega(t-\tau)}d\omega \tag{3.5.56}$$

which arises due to a buried point source at $(0, 0, z_s)$ and initiated at time τ. The pulse shape function for the ground surface acceleration may be obtained from (3.5.56) by letting $M_0 = 1, z = 0$, and replacing x by $x - x_s$ and y by $y - y_s$ to give

$$w_l(x - x_s, y - y_s, 0, t; z_s, \tau) = \frac{1}{2\pi}\int_{-\infty}^{\infty} \frac{1}{T_r}(1 - e^{-i\omega T_r})\bar{u}_l(x - x_s, y - y_s, 0, \omega)e^{i\omega(t-\tau)}d\omega \tag{3.5.57}$$

where use has been made of the assumption of material homogeneity in each layer.

It is of interest to note that in the present analysis, the ground motion is formulated in terms of the fundamental P, SV, and SH waves. These fundamental waves may be reconstituted into various surface waves, known as the Rayleigh wave, Love wave, and Stoneley wave. They are expected to exist in a layered medium, but they are not so identified in the present analysis.

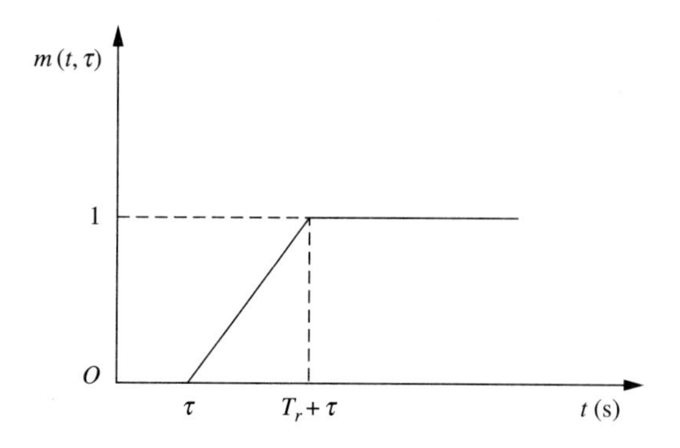

FIGURE 3.5.4
Time variation of an idealized seismic moment.

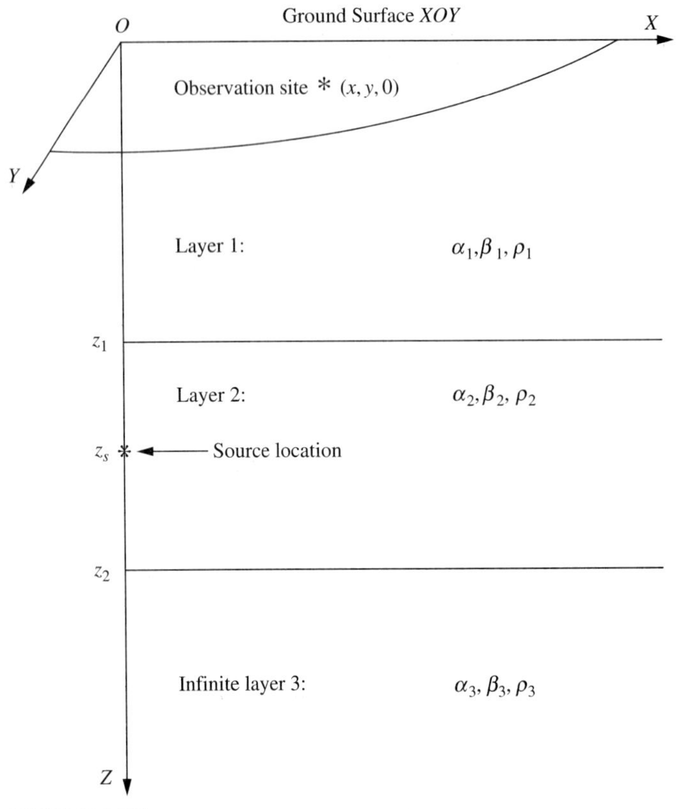

FIGURE 3.5.5
A three-layer half space for numerical computation.

NUMERICAL EXAMPLE. Numeral calculations have been carried out for the pulse shape functions with the earth medium idealized as a three-layer half-space, shown in Fig. 3.5.5. The physical properties of the three layers are given in Table 3.5.1. The other parameters are $T_r = \pi/5$ s, $r_T = 4.2 \times 10^5$ m, $\tau = 0.0$ s. The observation site for the ground motion is selected at $x = 3.0 \times 10^4$ m, $y = 4.0 \times 10^4$ m, $z = 0.0$ m, and the seismic source is assumed to be at $x_s = 0.0$ m, $y_s = 0.0$ m, $z_s = 6000$ m.

Figure 3.5.6 shows the computed pulse shape functions for the ground accelerations in the x, y, and z directions, due to the buried point source described previously. Although multiple wave reflections and refractions at the layer-to-layer

TABLE 3.5.1
Assumed physical properties of earth layers

Layer	ρ, kg/m^3	α, m/s	β, m/s	γ_α	γ_β	z, km
1	1.67×10^3	3000	1730	0.08	0.04	0–0.5
2	2.28×10^3	5000	2890	0.04	0.02	0.5–12
3	2.58×10^3	6000	3460	0.02	0.01	12–∞

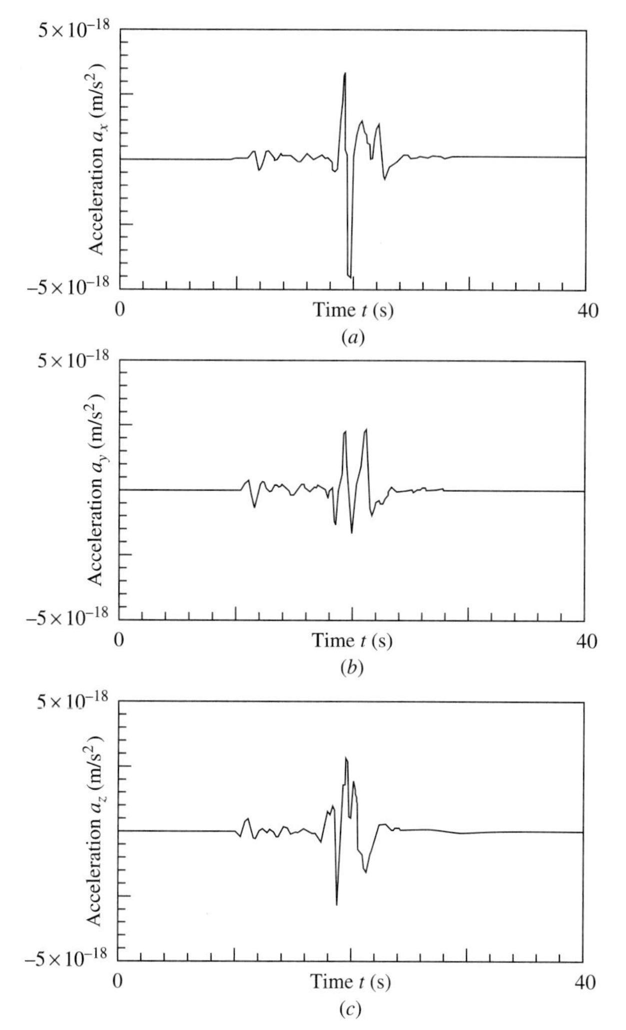

FIGURE 3.5.6
Pulse-shape functions for ground surface accelerations computed for a three-layer medium: (*a*) in the x-direction, (*b*) in the y-direction, (*c*) in the z-direction [after Zhang, Yong and Lin, 1991a; Copyright ©1991 ASCE, reprinted with permission].

interfaces render a detailed interpretation of the entire computed results difficult, some basic characteristics, such as the first arrivals of the P and S wave signals, can still be identified. Referring to Fig. 3.5.5, rough estimates of the times spent for the first P and S waves to travel from the source to the observation site are

$$t_p = \frac{[(x - x_s)^2 + (y - y_s)^2 + (z_1 - z_s)^2]^{1/2}}{\alpha_2} + \frac{z_1}{\alpha_1}$$

and

$$t_s = \frac{[(x - x_s)^2 + (y - y_s)^2 + (z_1 - z_s)^2]^{12}}{\beta_2} + \frac{z_1}{\beta_1}$$

.

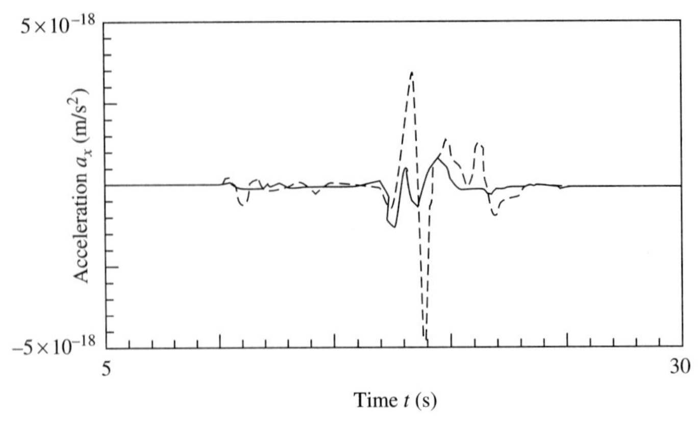

FIGURE 3.5.7
Pulse-shape functions for ground surface accelerations in the x-direction: computed for a uniform medium, (—) computed for a three-layer medium (- - - -) [after Zhang, Yong and Lin, 1991a; Copyright ©1991 ASCE, reprinted with permission].

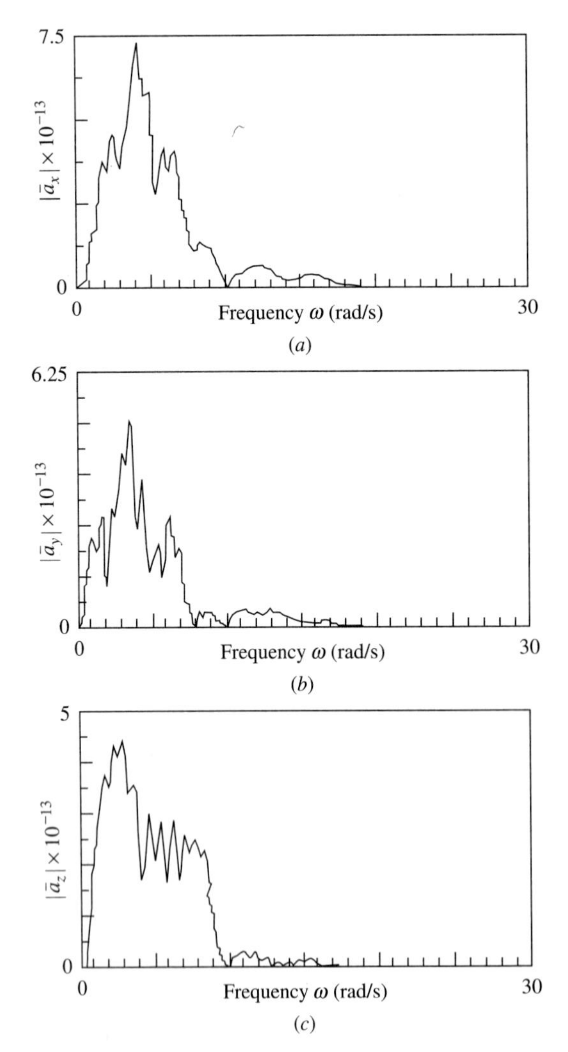

FIGURE 3.5.8
Frequency spectra for the pulse-shape functions: (a) in the x-direction, (b) in the y-direction, (c) in the z-direction [after Zhang, Yong and Lin, 1991a; Copyright ©1991 ASCE, reprinted with permission].

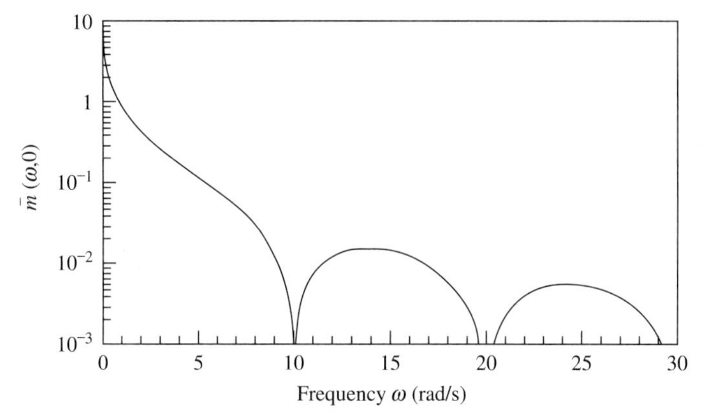

FIGURE 3.5.9
Fourier transform of a hypothetical seismic moment.

Based on the physical properties given in Table 3.5.1, these estimates are obtained as $t_p = 10.23$ s and $t_s = 17.69$ s, respectively. Indeed, as can be seen in Fig. 3.5.6, the time lapse between the initiation of the seismic moment and the first notable ground surface motions in all three directions appears to be about 10.23 s, in agreement with the estimated arrival time of the first P waves. The much larger amplitudes at about 17.69 s are attributable to the arrival of the S waves. Among the three accelerations in the three directions, the peak of a_x is the largest and that of a_y the smallest, due to the fact that the dislocation occurs in the x direction and the equivalent seismic moment acts on the x-z plane.

To gain additional insights, the ground acceleration in the x direction has also been computed for a uniform (unlayered) medium, with the same material properties as those of layer 2 given in Table 3.5.1, and the results are shown as the solid line in Fig. 3.5.7. The dashed line in the figure represents the results obtained earlier for the

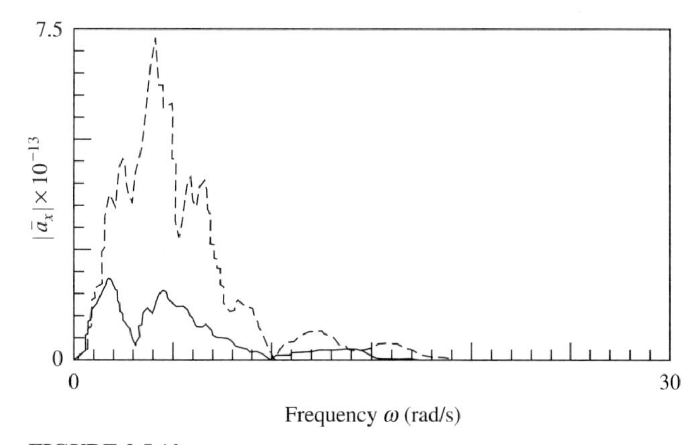

FIGURE 3.5.10
Frequency spectra of pulse-shape functions in the x direction computed for a uniform medium (—), computed for a three-layer medium (- - - -) [after Zhang, Yong and Lin, 1991a; Copyright ©1991 ASCE, reprinted with permission].

three-layer medium. It is seen that the ground surface response of a layered medium can be much higher than that of a uniform medium, due to the amplification effect of stratification. The cases of ground acceleration in the y and z directions are similar.

Figure 3.5.8 depicts the amplitudes of the computed Fourier transforms of the respective pulse shape functions. These frequency spectra are negligible beyond $\omega = 25$ rad/s and are exactly zero at $\omega = 10$ rad/s. The vanishing frequency spectra at $\omega = 10$ rad/s are related to the particular seismic moment $m(t, 0)$ selected for computation. The modulus of the Fourier transform $\bar{m}(\omega, 0)$ of the seismic moment, given by equation (3.5.54), is plotted in Fig. 3.5.9, which shows that $\bar{m}(\omega, t) = 0$ at $\omega T_r = 2n\pi$, or $\omega = 10n$ rad/s for $T_r = \pi/5$. Figure 3.5.8(a) is replotted in Fig. 3.5.10 along with the results obtained for a uniform (unlayered) medium for comparison. The amplification effect of stratification is seen to be quite dramatic.

3.6 EARTHQUAKE GROUND MOTION IN A LAYERED EARTH MEDIUM DUE TO PROPAGATING SEISMIC SOURCES

The pulse shape function obtained in Section 3.5 can now be used to construct an earthquake model. Let $U_l(x, y, 0, t)$ be the ground surface acceleration in the lth direction, and write

$$U_l(x, y, 0, t) = \sum_{k=1}^{N(T)} Y_k w_l(x - x_{sk}, y - y_{sk}, 0, t; z_{sk}, \tau_k) \qquad t \le T \qquad (3.6.1)$$

where w_l is the pulse shape function given in (3.5.57), describing the ground surface acceleration due to a ramp-type seismic moment of unit amplitude. Equation (3.6.1) has the usual pulse-train format (3.2.1). It shows that the kth pulse is caused by a seismic moment of amplitude Y_k that is located at $\boldsymbol{P}_k = (x_{sk}, y_{sk}, z_{sk})$ and is initiated at time τ_k. However, equation (3.6.1) differs from (3.2.1) in one important aspect: the summation index k involves both time and space. Thus equation (3.2.13) does not hold, unless $\boldsymbol{\rho}_k$ is related uniquely to τ_k.

Most seismic faults are narrow, that is, the widths are much shorter than the lengths. For mathematical analysis, it is reasonable to use a fault line approximation, shown in Fig. 3.6.1, and assume that shear dislocation propagates along the fault line, starting from point P and ending at point Q. As shear dislocation propagates, it encounters localized obstacles, or barriers. A hypothesis has been put forward (Das and Aki, 1977) that significant ground motions are caused only by the breakage of such barriers. We follow this hypothesis and assume additionally that (1) such barriers are spaced at equal distances, (2) the strengths of the activated seismic moments Y_k are independent and identically distributed random variables, and independent of τ_k, and (3) the time instants at which the successive barriers are broken are Poisson events; that is, while they are sequential, each occurrence is independent of previous occurrences. These assumptions can now be incorporated in equation (3.6.1). The cumulant functions of $U_l(x, y, 0, t)$ are related to its characteristic functional, defined as

$$M_{\{U_l\}}[\theta(x, y, t)] = E\left\{\exp\left[i\int_0^T dt \int_R \int \theta(x, y, t) U_l(x, y, 0, t)\, dx\, dy\right]\right\} \qquad (3.6.2)$$

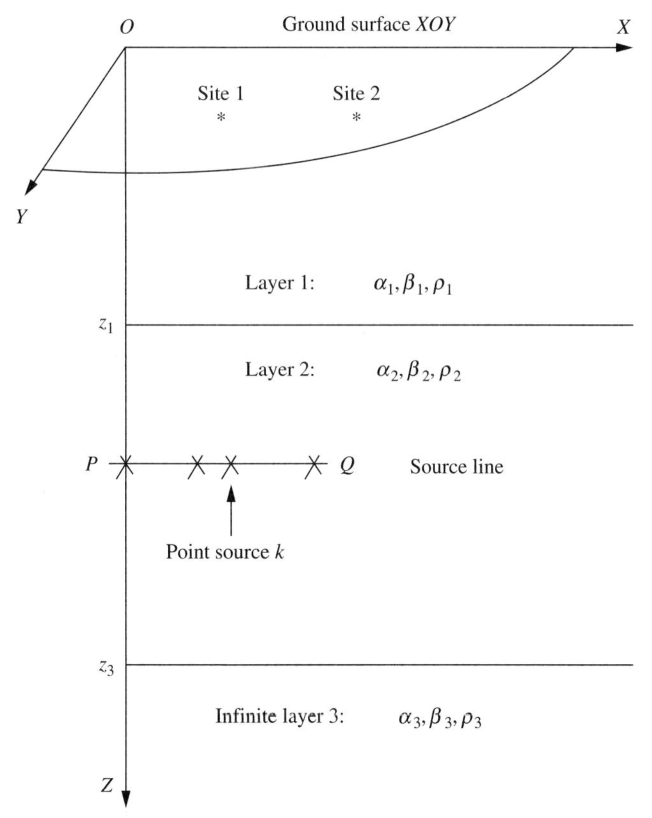

FIGURE 3.6.1
Schematic representation of a numerical example.

where R symbolizes the range of x and y, $0 < t \leq \mathrm{T}$, and function $\theta(x, y, t)$ belongs to a family for which the integral

$$\int_0^T \int_R \int \theta(x, y, t) U_l(x, y, 0, t)\, dx\, dy\, dt \tag{3.6.3}$$

is mean-square convergent. Substitution of (3.6.1) into (3.6.2) yields

$$\begin{aligned} &M_{\{U_l\}}[\theta(x, y, t)] \\ &\quad = E\left\{ \exp\left[i \int_0^T dt \int_R \int \theta(x, y, t) \sum_{k=1}^{N(T)} Y_k w_l(x - x_{sk}, y - y_{sk}, 0, t; z_{sk}, \tau_k)\, dx\, dy \right]\right\} \\ &\quad = E\left(E\left\{ \exp\left[i \int_0^T dt \int_R \int \theta(x, y, t) \sum_{k=1}^{N(T)} Y_k \right.\right.\right. \\ &\qquad\qquad \left.\left.\left. \times\, w_l(x - x_{sk}, y - y_{sk}, 0, t; z_{sk}, \tau_k)\, dx\, dy \right] \Big| N(T) \right\}\right) \end{aligned}$$

$$= \sum_{n=0}^{\infty} P_{\{N\}}(n, T) E \left\{ \exp \left[i \int_0^T dt \int_R \int \theta(x, y, t) \right. \right.$$

$$\left. \left. \times \sum_{k=1}^{N(T)} Y_k w_l(x - x_{sk}, y - y_{sk}, 0, t; z_{sk}, \tau_k) \, dx \, dy \right] \right\} \tag{3.6.4}$$

where $P_{\{N\}}(n, T)$ is the probability for $N(T) = n$.

Let $\lambda(\tau)$ be the average number of barriers broken per unit time. Then the probability for one barrier to be broken in the time interval $(\tau_k, \tau_k + d\tau_k]$ is $\lambda(\tau_k)\,d\tau_k$, and the probability for the generated pulse shape to be $w_l(x - x_{sk}, y - y_{sk}, 0, t; z_{sk}, \tau_k)$ is proportional to $\lambda(\tau_k)\,d\tau_k$. For a Poisson process, the probability $P_{\{N\}}(n, T)$ in equation (3.6.4) is given by (Lin, 1967, Eq. 4-75)

$$P_{\{N\}}(n, T) = \frac{1}{n!} \left[\int_0^T \lambda(\tau)\,d\tau \right]^n \exp \left[- \int_0^T \lambda(\tau)\,d\tau \right] \tag{3.6.5}$$

Moreover, since Y_k and τ_k are independent, we obtain for each n

$$E \left\{ \exp \left[i \int_0^T dt \int_R \int \theta(x, y, t) \sum_{k=1}^{n} Y_k w_l(x - x_{sk}, y - y_{sk}, 0, t; z_{sk}, \tau_k) dx\,dy \right] \middle| N(T) = n \right\}$$

$$= (1 + \alpha)^n \tag{3.6.6}$$

where, as indicated by the statement following a vertical bar, the ensemble average is taken under the condition $N(T) = n$, and where

$$\alpha = \sum_{m=1}^{\infty} \frac{i^m}{m!} E[Y^m] \int_0^T dt_1 \int_R \int \theta(x_1, y_1, t_1)\,dx_1\,dy_1 \cdots$$

$$\int_0^T dt_m \int_R \int \theta(x_m, y_m, t_m)\,dx_m\,dy_m$$

$$\times \frac{\int_0^T w_l(x_1 - x_s, y_1 - y_s, 0, t_1; z_s, \tau) \cdots w_l(x_m - x_s, y_m - y_s, 0, t_m; z_s, \tau) \lambda(\tau)\,d\tau}{\int_0^T \lambda(\tau)\,d\tau} \tag{3.6.7}$$

In equation (3.6.7), Y is a common symbol for any of the independent and identically distributed Y_k. The subscript k associated with τ, x_s, y_s has been removed, since τ is now an integration variable. From combining (3.6.4) and (3.6.7), it follows that

$$M_{\{U_l\}}[\theta(x, y, t)] = \sum_{n=0}^{\infty} \frac{1}{n!} \left[\int_0^T \lambda(\tau)\,d\tau \right]^n \exp \left[- \int_0^T \lambda(\tau)\,d\tau \right] (1 + \alpha)^n$$

$$= \exp \left[\alpha \int_0^T \lambda(\tau)\,d\tau \right] \tag{3.6.8}$$

Taking the natural logarithm of (3.6.8) and using (3.6.7),

$$\ln\left\{M_{\{U_l\}}[\theta(x,y,t)]\right\}$$

$$= \sum_{m=1}^{\infty} \frac{i^m}{m!} E[Y^m] \int_0^T dt_1 \iint_R \theta(x_1, y_1, t_1)\, dx_1\, dy_1 \cdots$$

$$\int_0^T dt_m \iint_R \theta(x_m, y_m, t_m)\, dx_m\, dy_m \int_0^T w_l(x_1 - x_s, y_1 - y_s, 0, t_1; z_s, \tau) \cdots$$

$$w_l(x_m - x_s, y_m - y_s, 0, t_m; z_s, \tau)\lambda(\tau)\, d\tau \tag{3.6.9}$$

It is known (Stratonovich, 1963) that the log-characteristic function for $U_l(x, y, 0, t)$ has the series representation

$$\ln\left\{M_{\{U_l\}}[\theta(x,y,t)]\right\}$$

$$= i\int_0^T dt \iint_R \kappa_1[U_l(x, y, 0, t)]\theta(x, y, t)\, dx\, dy$$

$$+ \frac{i^2}{2!}\int_0^T dt_1 \iint_R dx_1 dy_1 \int_0^T dt_2 \iint_R dx_2 dy_2 \theta(x_1, y_1, t_1)$$

$$\times\ \theta(x_2, y_2, t_2)\kappa_2[U_l(x_1, y_1, 0, t), U_l(x_2, y_2, 0, t_2)] + \cdots \tag{3.6.10}$$

where κ_j denotes the jth cumulant function of $U_l(x, y, 0, t)$. Comparing (3.6.9) and (3.6.10), we obtain

$$\kappa_m[U_l(x_1, y_1, 0, t_1), \ldots, U_l(x_m, y_m 0, t_m)]$$

$$= E[Y^m] \int_0^T w_l(x_1 - x_s, y_1 - y_s, 0, t; z_s, \tau) \cdots$$

$$w_l(x_m - x_s, y_m - y_s, 0, t; z_s, \tau)\lambda(\tau)\, d\tau \tag{3.6.11}$$

The first and the second cumulant functions coincide, respectively, with the mean and the covariance functions:

$$\mu_{U_l(x,y,0,t)} = \mu_Y \int_0^T w_l(x - x_s, y - y_s, 0, t; z_s, \tau)\lambda(\tau)\, d\tau \tag{3.6.12}$$

$$\mathrm{cov}[U_l(x_1, y_1, 0, t_1), U_{l'}(x_2, y_2, 0, t_2)]$$

$$= E[Y^2] \int_0^T w_l(x_1 - x_s, y_1 - y_s, 0, t_1; z_s, \tau) w_{l'}(x_2 - x_s, y_2 - y_s, 0, t_2; z_s, \tau)\lambda(\tau)\, d\tau \tag{3.6.13}$$

where the directions l and l' may be different. When carrying out actual calculations, the causality condition $w_l(x - x_s, y - y_s, 0, t; z_s, \tau) = 0$ for $t < \tau$ must be imposed on equations (3.6.11) through (3.6.13), and (x_s, y_s, z_s) must be expressed in terms of τ.

Although the theme of the foregoing derivation follows that in Section 2.4 for the case where only time dependence is involved, the inclusion of spatial dependence in the pulse shape function raises some subtle issues that must be examined carefully. In this regard, the assumption that $\boldsymbol{\rho}_k$ depends on τ_k is crucial.

When modeling a random process of short duration, such as an earthquake, we let $\lambda(\tau) = 0$ for $\tau < 0$ and $\tau > T$. Equation (3.6.13) can then be rewritten as

$$\operatorname{cov}[U_l(x_1, y_1, 0, t_1), U_{l'}(x_2, y_2, 0, t_2)]$$
$$= E[Y^2]\int_{-\infty}^{\infty} w_l(x_1 - x_s, y_1 - y_s, 0, t_1; z_s, \tau) w_{l'}(x_2 - x_s, y_2 - y_s, 0, t_2; z_s, \tau)\lambda(\tau)\, d\tau \tag{3.6.14}$$

Let

$$b_l(x_1, y_1, 0, t_1; \omega_1)e^{i\omega_1 t_1} = \int_{-\infty}^{\infty} w_l(x_1 - x_s, y_1 - y_s, 0, t_1; z_s, \tau)\sqrt{\lambda(\tau)}e^{i\omega_1\tau}d\tau \tag{3.6.15}$$

Thus

$$w_l(x_1 - x_s, y_1 - y_s, 0, t_1; z_s, \tau)\sqrt{\lambda(\tau)} = \frac{1}{2\pi}\int_{-\infty}^{\infty} b_l(x_1, y_1, 0, t_1; \omega_1)e^{i\omega_1 t_1}e^{-i\omega_1\tau}d\omega_1 \tag{3.6.16}$$

Similarly,

$$w_{l'}(x_2 - x_s, y_2 - y_s, 0, t_2; z_s, \tau)\sqrt{\lambda(\tau)} = \frac{1}{2\pi}\int_{-\infty}^{\infty} b_{l'}^*(x_2, y_2, 0, t_2; \omega_2)e^{-i\omega_2 t_2}e^{i\omega_2\tau}d\omega_2 \tag{3.6.17}$$

The right-hand sides of (3.6.16) and (3.6.17) are written purposely in conjugate forms, but this is permissible since the left-hand sides are real. Substituting (3.6.16) and (3.6.17) into (3.6.14) and noting that

$$\frac{1}{2\pi}\int_{-\infty}^{\infty} e^{i(\omega_2-\omega_1)\tau}d\tau = \delta(\omega_2 - \omega_1) \tag{3.6.18}$$

we arrive at

$$\operatorname{cov}[U_l(x_1, y_1, 0, t_1), U_{l'}(x_2, y_2, 0, t_2)]$$
$$= \frac{1}{2\pi}E[Y^2]\int_{-\infty}^{\infty} b_l(x_1, y_1, 0, t_1; \omega)b_{l'}^*(x_2, y_2, 0, t_2; \omega)e^{i\omega(t_1-t_2)}d\omega \tag{3.6.19}$$

The evolutionary cross-spectral density of $U_l(x_1, y_1, 0, t_1)$ and $U_{l'}(x_2, y_2, 0, t_2)$ is clearly

$$\hat{\Phi}_{UU'}(x_1, y_1, t_1; x_2, y_2, t_2; \omega) = \frac{1}{2\pi}E[Y^2]b_l(x_1, y_1, 0, t_1; \omega)b_{l'}^*(x_2, y_2, 0, t_2; \omega) \tag{3.6.20}$$

The evolutionary spectral density may be obtained by letting $l = l'$, $x_1 = x_2 = x$, $y_1 = y_2 = y$, and $t_1 = t_2 = t$, to yield

$$\hat{\Phi}_{UU}(x, y, t; \omega) = \frac{1}{2\pi}E[Y^2]\left|b_l(x, y, 0, t; \omega)\right|^2 \tag{3.6.21}$$

The derivation from (3.6.15) to (3.6.21) is analogous to that from (3.2.18) to (3.2.24). The evolutionary spectral density and evolutionary cross-spectral density provide the pertinent information for the second-order statistics of the excitation field, required for the computation of the corresponding second-order statistics of linear structural response.

In the context of earthquake modeling, Y_k represents the amplitude of the kth seismic moment, which is equal to the product of the shear modulus, the tributary fault area, and the average slip associated with the kth barrier. We note that the choice of barrier-to-barrier spacing has not been made thus far. Ideally, such a choice should be made according to the geophysical information in the area. In the absence of accurate geophysical information, the choice of a longer spacing will be accompanied by a smaller number of barriers and a larger tributary area for each barrier. If we keep the ratio between the mean-square seismic moment and the barrier spacing constant, then only this ratio affects the calculated results. However, this does not imply that the spacing can be arbitrarily small. Indeed, it should be sufficiently large that the assumption of independent source strength Y_k is deemed reasonable.

We note also that if the barriers are located along a straight line parallel to the ground surface, then the pulses become additive when transformed to the domains of frequency, wave number, and azimuth. In this case, the computation becomes nearly as simple as that of a single point source. The general theory, however, is applicable to the general case, in which the barriers are located on an arbitrary curve, with greater computational efforts.

Numerical calculations have been carried out for an example (Zhang, Yong, and Lin, 1991b), in which the earth medium is modeled as being composed of three layers, with the dislocation occurring in the second layer along a straight line in the x direction, shown in Fig. 3.6.1. Physical parameters of the three layers are the same as those given previously in Table 3.5.1. The other parameters are $T_r = 0.628$ s, $r_T = 4.0\times10^5$ m, source depth $z_s = 6.0\times10^3$ m, length of barrier array $= 1.0\times10^5$ m, and the root-mean-square seismic moment per meter $= 6.0 \times 10^{12}$ N $\cdot$ m/m. The assumed average dislocation propagation speed $\lambda(t)$ is shown in Fig. 3.6.2, which is proportional to the average number of barrier breakages per unit time. The selected observation sites for the ground motion are (1) $x_1 = 3.0 \times 10^4$ m, $y_1 = 4.0 \times 10^4$ m, $z_1 = 0.0$ m, and (2) $x_2 = 4.0 \times 10^4$ m, $y_2 = 5.0 \times 10^4$ m, $z_2 = 0.0$ m.

Although the statistical properties of the stochastic earthquake model can be computed directly, some simulation results are presented first to provide additional insight for the model. For the purpose of simulation, the line source is discretized into 30 point sources, and the associated seismic moments are assumed to be independent random variables uniformly distributed in the interval $\pm\sqrt{3}\,\sigma$, where σ is the root-mean-square value of the seismic moments. Figure 3.6.3a shows the simulated ground acceleration in the x direction at site 1. It can be seen from this figure that the ground acceleration is zero until the first P wave signal arrives, as explained in Section 3.5. The S wave signals, which arrive later, give rise to greater ground acceleration compared with those due to the P waves. The irregular oscillation of the ground acceleration is due to the random variation of the seismic moments and the random activation times. The ground acceleration gradually dies down to zero after the last wave signals arrive from the last point source. Figure 3.6.3b shows the modulus of the corresponding frequency spectrum of the ground acceleration in the x direction. These figures resemble well what have been observed in some real

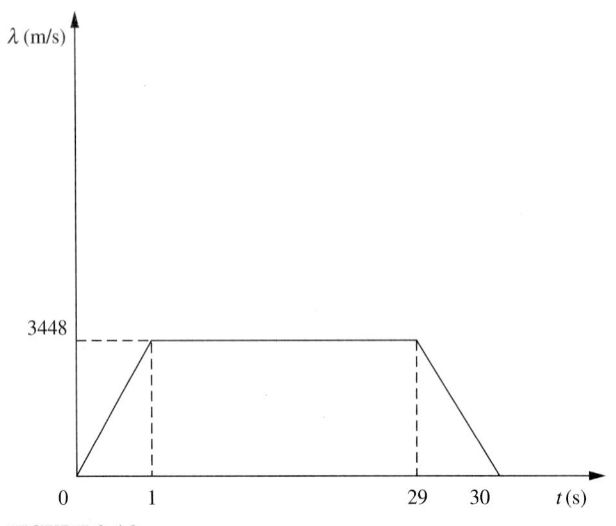

FIGURE 3.6.2
Assumed propagation speed of fault dislocation.

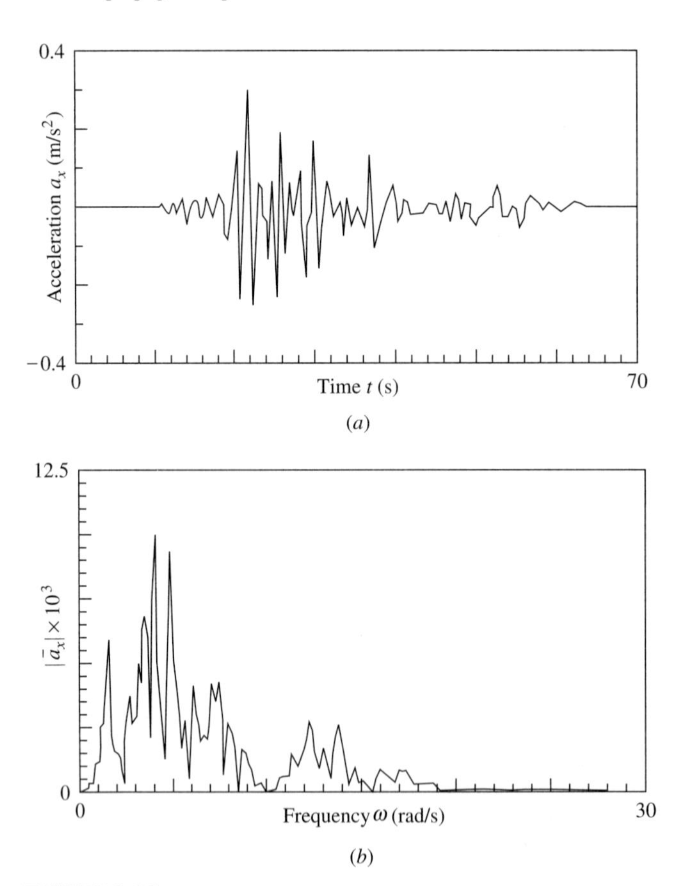

FIGURE 3.6.3
Simulated ground acceleration in the x-direction. (a) Time history; (b) modulus of frequency spectrum. [After Zhang, Yong and Lin, 1991b; Copyright ©1991 ASCE, reprinted with permission].

earthquakes. The simulated ground accelerations in the y and z directions are similar, and they are omitted to save space.

The remaining figures illustrate some key statistical properties of the modeled earthquake, computed directly from the derived formulas without resorting to Monte Carlo simulation. The autocorrelation function of the ground acceleration in the x direction at site 1 is shown in Figs. 3.6.4*a* and *b*. They represent two different views of the same autocorrelation. The mean-square ground acceleration, along the line $t = t_1 = t_2$, becomes noticeable at around $t = 10.23$ s when the first P wave arrives. The first S wave signal arrives at around $t = 17.69$ s. The first arrival times of P and S waves were estimated in Section 3.5. After reaching its maximum, the mean-square acceleration decreases more slowly, due to the contributions from other point sources as dislocation extends along the line. For this particular example, the distance between the dislocation rupture front and the chosen ground site for observation decreases gradually until reaching its minimum and then increases gradually. The energy dissipation of a propagating wave depends, of course, on the distance of propagation. The mean-square acceleration eventually vanishes after all signals have passed the site. Between approximately $t_1 = 10$ s and $t_1 = 18$ s, correlation in the range $t_1 - t_2 > 0$ is almost zero because the signal at t_2 is devoid of the S wave, whereas this is not so in the range $t_1 - t_2 < 0$. The profiles of the correlation function away from $t_1 = t_2$ appear zigzag, but the basic trend of approaching zero at increasing $|t_1 - t_2|$ is quite clear. The zigzag appearance can be smoothed somewhat by taking smaller step size when carrying out the numerical integration in equation (3.6.13), at the expense of greater computer cost.

Figure 3.6.5 depicts the cross-correlation of the ground accelerations at sites 1 and 2, again when viewing from two different directions. The magnitudes near the line $t_1 = t_2$ are still relatively large due to the relative proximity of the two sites. Compared with the autocorrelation at a single site, the cross-correlation is smaller in magnitude along the diagonal $t_1 = t_2$, and has a wider overall spread in its profile, due to the fact that the ground accelerations at the two sites reach their maxima at different times. It also decreases quickly as $|t_1 - t_2|$ becomes large. The autocorrelations and cross-correlations for ground accelerations in the y and z directions are similar.

Figures 3.6.6*a* and *b* are plots of the evolutionary spectral densities of the ground acceleration in the x direction at site 1, seen from two different directions. They show the frequency distributions of the earthquake acceleration at different times. Both the lower and higher frequency components die down more rapidly. As time advances, those components in the intermediate frequency range decay at different rates. The exact reasons for the complicated behavior are unclear at the present time. The stratification of the wave propagation medium and the assumed manner in which fault dislocation takes place and propagates may all be contributing factors. Finally, the evolutionary cross-spectral densities of the ground accelerations in the x direction at sites 1 and 2 are shown in Figs. 3.6.7 through 3.6.9. Again, the results for the ground acceleration in the y and z directions are similar. Figures 3.6.7 and 3.6.8 show the real and imaginary parts of the evolutionary cross-spectrum, respectively, at the two sites and at $t_1 = t_2 = t$. Figure 3.6.9 depicts the absolute value of the preceding evolutionary cross-spectrum.

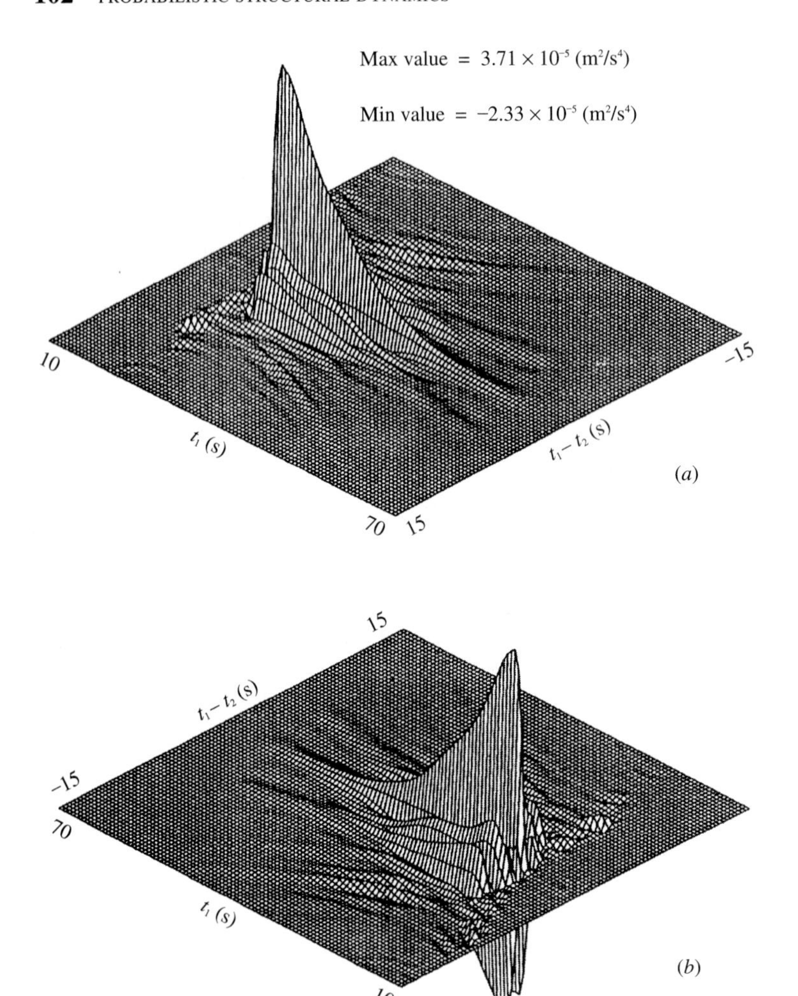

FIGURE 3.6.4
Correlation function of ground acceleration in the *x*-direction at site 1 [after Zhang, Yong and Lin, 1991b; Copyright ©1991 ASCE, reprinted with permission].

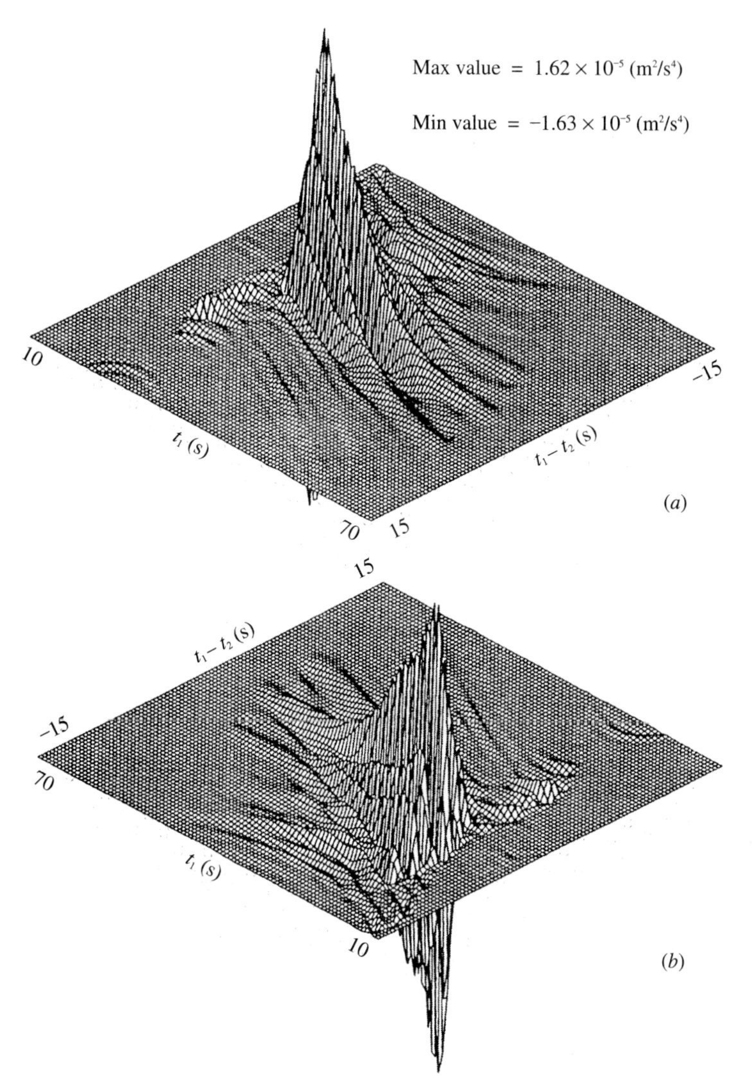

FIGURE 3.6.5
Cross-correlation function of ground accelerations in the x-direction at sites 1 and 2 [after Zhang, Yong and Lin, 1991b].

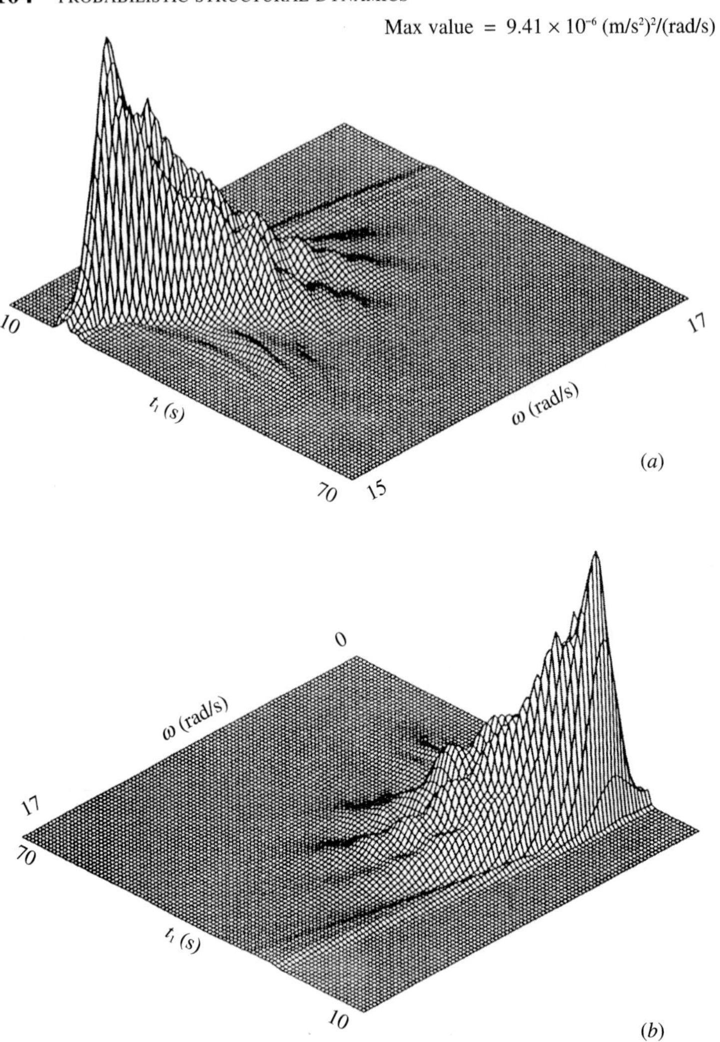

FIGURE 3.6.6
Evolutionary spectrum of ground acceleration in the x-direction at site 1 [after Zhang, Yong and Lin, 1991b; Copyright ©1991 ASCE, reprinted with permission].

Max value = 2.34×10^{-6} $(\text{m/s}^2)^2/(\text{rad/s})$

Min value = -2.03×10^{-6} $(\text{m/s}^2)^2/(\text{rad/s})$

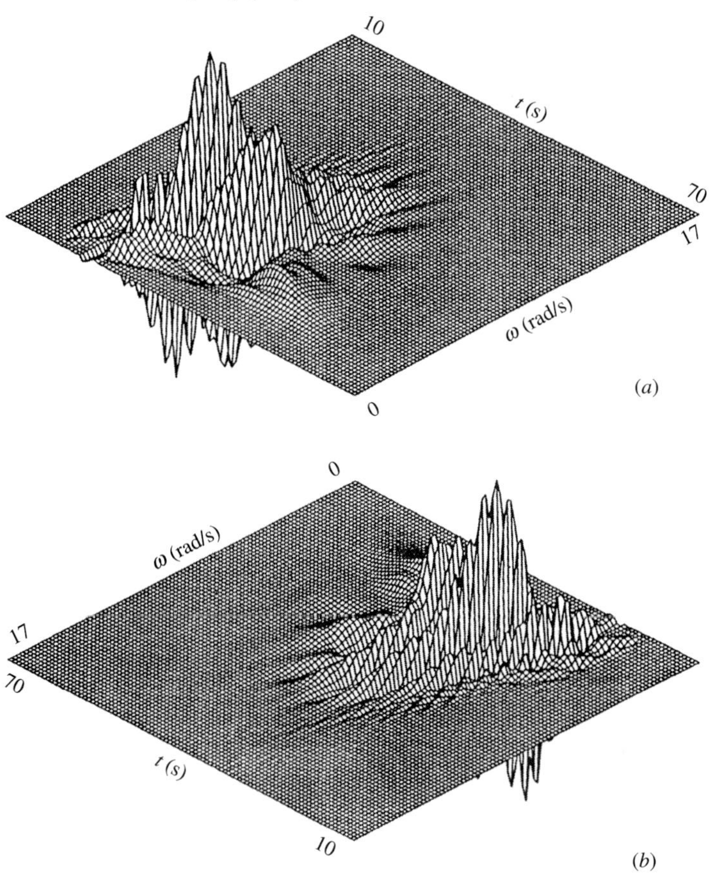

FIGURE 3.6.7
Real part of evolutionary cross-spectrum of ground accelerations in the x-direction at sites 1 and 2 [after Zhang, Yong and Lin, 1991b; Copyright ©1991 ASCE, reprinted with permission].

Max value = 2.31×10^{-6} $(\text{m/s}^2)^2/(\text{rad/s})$

Min value = -1.99×10^{-6} $(\text{m/s}^2)^2/(\text{rad/s})$

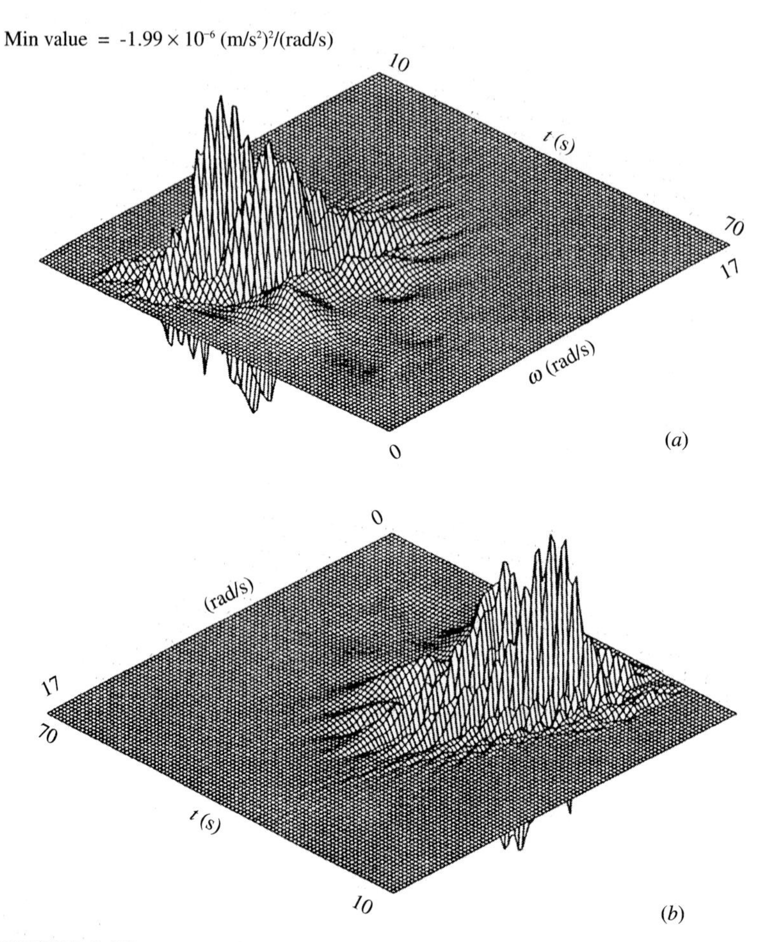

FIGURE 3.6.8
Imaginary part of evolutionary cross-spectrum of ground accelerations in the x-direction at sites 1 and 2 [after Zhang, Yong and Lin, 1991b; Copyright ©1991 ASCE, reprinted with permission].

Max value = 3.17×10^{-6} $(m^2/s^4)^2$(rad/s)

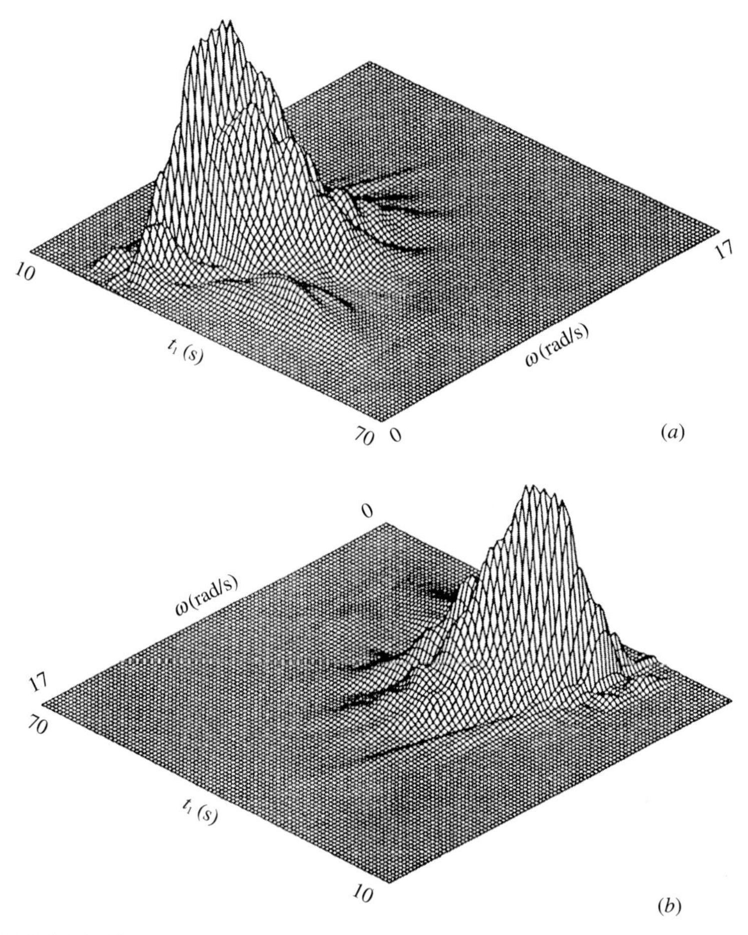

FIGURE 3.6.9
Modulus of evolutionary cross-spectrum of ground accelerations in the x-direction at sites 1 and 2 [after Zhang, Yong and Lin, 1991b; Copyright ©1991 ASCE, reprinted with permission].

3.7 CONCLUDING REMARKS

Provided that a dynamical system is linear and the random excitations are additive, the general input–output relationships are mathematically simple in terms of their respective cumulant functions (as shown in Lin, 1967). However, such simple relationships are seldom used in engineering practice when excitations are stochastically nonstationary, because available statistical data are generally inadequate for a meaningful statistical inference analysis to be conducted of a general nonstationary model. Therefore, practical nonstationary excitation models must incorporate certain physical insights to compensate for the lack of adequate statistical data. The evolutionary process and random pulse-train models are such examples. These two types of models are especially attractive, since their use renders the input–output relationship relatively simple for a linear time-invariant system. In particular, an evolutionary process input gives rise to an evolutionary process output, and a random pulse-train input to a random pulse-train output. An evolutionary process model requires the estimation of a modulation function $a(t, \omega)$ of time and frequency and the spectral contents of the underlying orthogonal-increment process, whereas a random pulse-train model requires the estimation of a pulse shape function and the average pulse-arrival rate. Comparing the two, the insights into the mechanism involved in a physical phenomenon can be incorporated more directly in a random pulse-train model. The fact that a random pulse-train model also has an evolutionary spectral representation further increases its attractiveness in practical applications.

For illustration purposes, we selected examples from earthquake ground motion modeling. Strong earthquakes are among the most severe excitations experienced by engineering structures. The manner in which an earthquake is generated and transmitted to the ground surface can be incorporated quite naturally in a random pulse-train model. These physical characteristics are modeled at various levels of complexity, including the simple mechanism suggested by Kanai and Tajimi, the one-dimensional elastic or viscoelastic medium model, and the three-dimensional stratified medium model. Additional refinements are still possible.

3.8 EXERCISES

3.1. Given an equation of motion

$$\ddot{X} + 2\zeta\omega_0\dot{X} + \omega_0^2 X = a(t)W(t)$$

where $W(t)$ is a gaussian white noise with a spectral density K. Obtain the autocorrelation function of response $X(t)$ for the following cases:

(*a*) $a(t) = e^{-\alpha t}$

(*b*) $a(t) = e^{-\alpha t} - e^{-\beta t}, \qquad \beta > \alpha > 0$

3.2. Given an equation of motion

$$\ddot{X} + 2\zeta\omega_0\dot{X} + \omega_0^2 X = a(t)S(t)$$

where $a(t)$ is a deterministic function and $S(t)$ is a weakly stationary process with zero mean and a spectral density $\Phi_{SS}(\omega)$. Calculate $E[X^2(t)]$ for the following cases:

(*a*) $a(t) = e^{-\alpha t}, \qquad \Phi_{SS}(\omega) = 1/(c^2 + \omega^2)$

(*b*) $a(t) = e^{-\alpha t} - e^{-\beta t}(\beta > \alpha > 0), \qquad \Phi_{SS}(\omega) = 1/(c^2 + \omega^2)$

3.3. Given an equation of motion

$$\ddot{X} + 2\zeta\omega_0\dot{X} + \omega_0^2 X = F(t)$$

where $F(t)$ is an evolutionary process, with a representation

$$F(t) = \int_{-\infty}^{\infty} a(t, \omega)e^{i\omega t} d\tilde{F}(\omega)$$

and where

$$a(t, \omega) = \exp(-\left|\omega^2 - \omega_1^2\right| t)$$

$$E[d\tilde{F}(\omega)d\tilde{F}^*(\omega')] = \begin{cases} K d\omega & \omega = \omega' \\ 0 & \omega \neq \omega' \end{cases}$$

Obtain the evolutionary spectral density of $X(t)$.

3.4. An infinitely long beam shown in Fig. P3.4 is excited by a pressure field

$$p(x, t) = \begin{cases} \int_{-\infty}^{\infty} e^{-\alpha k x} e^{-ik(x-\omega t)} d\tilde{F}(k) & \alpha > 0, x \geq 0 \\ 0 & x < 0 \end{cases}$$

which is a stationary process in time but an evolutionary process in space. The beam is assumed to be uniform in cross-section, undamped, and governed by

$$EI\frac{\partial^4 w}{\partial x^4} + m\frac{\partial^2 w}{\partial t} = p(x, t)$$

Find the cross-spectral density of $w(x_1, t)$ and $w(x_2, t)$.

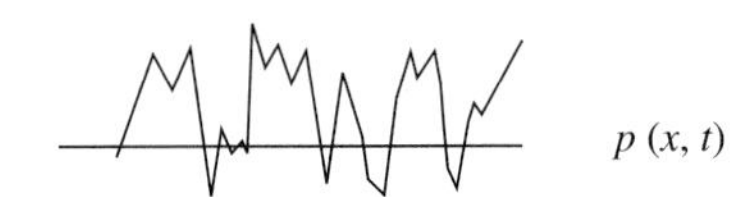

FIGURE P3.4
An infinitely long beam excited by a random pressure field.

3.5. Determine the evolutionary spectral density of the random pulse train

$$X(t) = \sum_{k=1}^{N(T)} Y_k w(t - \tau_k)$$

where the random amplitudes Y_k are independent random variables, with the same probability density

$$p_Y(y) = \begin{cases} (2a)^{-1} & -a \leq y \leq a \\ 0 & \text{otherwise} \end{cases}$$

and τ_k are random pulse arrival times, with an average arrival rate

$$\lambda(\tau) = e^{-\alpha\tau} \qquad \alpha > 0$$

The pulse shape function is given by

$$w(t - \tau) = \begin{cases} \exp[-\beta(t - \tau)] & \beta > 0,\ t > \tau \\ 0 & t < \tau \end{cases}$$

3.6. A modified version of an evolutionary Kanai-Tajimi earthquake model is shown schematically in Fig. P3.6. The model has a random pulse-train representation

$$G(t) = \sum_{k=1}^{N(T)} Y_k w(t - \tau_k)$$

Find the pulse shape function $w(t)$.

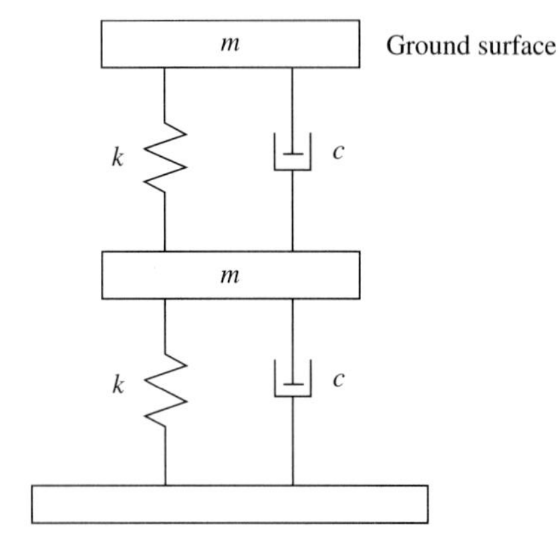

FIGURE P3.6
An improved version of Kanai-Tajimi earthquake model.

3.7. An evolutionary Kanai-Tajimi earthquake model is given by

$$G(t) = \sum_{k=1}^{N(T)} Y_k h_1(t - \tau_k)$$

where Y_k are independent, identically distributed random variables with a zero mean and a finite mean-square value, and the pulse shape function $h_1(t - \tau)$ is given by equation (3.4.6). Obtain the covariance function of the ground surface acceleration, assuming that the average pulse-arrival rate is given by

$$\lambda(\tau) = e^{-\alpha\tau} - e^{-\beta\tau} \qquad \beta > \alpha > 0$$

CHAPTER 4

MARKOV PROCESSES

The spectral analysis and evolutionary spectral analysis described in Chapters 2 and 3 are applicable only if a dynamic system is linear and time-invariant and all random excitations are additive. If a dynamic system is nonlinear or if multiplicative random excitations are present, or both, then a mathematically exact solution is not always obtainable. When such an exact solution is obtained, it is usually based on the assumption that the system response is a Markov stochastic process or related to a Markov process in some sense. Therefore, as a preliminary for later applications, mathematical concepts of Markov processes are discussed in this chapter.

4.1 CHARACTERIZATION OF A MARKOV PROCESS

Definition. A stochastic process $X(t)$ is said to be a scalar Markov process if it has the property

$$\text{Prob}\,[X(t_n) \leq x_n \mid X(t_{n-1}) = x_{n-1}, \ldots, X(t_1) = x_1]$$
$$= \text{Prob}\,[X(t_n) \leq x_n \mid X(t_{n-1}) = x_{n-1}] \qquad t_n > t_{n-1} > \cdots > t_1 \tag{4.1.1}$$

where Prob $[\cdot]$ denotes the probability of an event, and where the statement following a vertical bar specifies certain conditions under which such a probability is defined. In the present case, the conditions are known values of $X(t)$ at earlier time instants $t_1, \ldots, t_{n-1}$. A sufficient condition for $X(t)$ to be a Markov process is that its increments in any two nonoverlapping time intervals are independent; that is, $X(t_2) - X(t_1)$ and $X(t_4) - X(t_3)$ are independent as long as $t_1 < t_2 \leq t_3 < t_4$.

For stochastic dynamics applications, a Markov process $X(t)$ is usually assumed to be continuously valued, and its time parameter t is defined on a continuous space. If values of the parameter t are discrete, it is known as a *Markov series*. If values of both $X(t)$ and the parameter t are discrete, it is called a *Markov chain*. We are concerned here only with continuously parametered and continuously valued Markov processes.

A continuously valued and continuously parametered Markov process is a mathematical idealization. It is usually justified on the basis of how close the sufficient condition of having independent increments is deemed satisfied in an actual physical phenomenon. Since independent increments imply uncorrelated increments, the results obtained in Chapter 2 pertaining to stochastic processes with uncorrelated increments are also valid for scalar Markov processes.

The conditional probability, $\text{Prob}[X(t) \leq x \mid X(t_0) = x_0]$, of a Markov process $X(t)$ is called the *transition probability distribution function*. A Markov process is completely characterized by its transition probability distribution and its probability distribution at an initial time t_0. The latter includes the special case in which the initial state $X(t_0)$ is known deterministically.

If the transition probability distribution function of a Markov process is differentiable, then it is often more convenient to deal with the *transition probability density*, defined as

$$q(x, t \mid x_0, t_0) = \frac{\partial}{\partial x}\text{Prob}\,[X(t) \leq x \mid X(t_0) = x_0] \tag{4.1.2}$$

We assume in all cases considered here that the transition probability density exists.

The concept of a scalar Markov process is readily generalizable to a vector Markov process. Thus $\boldsymbol{X}(t) = \{X_1(t), X_2(t), \ldots, X_m(t)\}'$ is an m-dimensional Markov vector, if it has the property of

$$\text{Prob}\left[\bigcap_{j=1}^{m}\{X_j(t_n) \leq x_j\} \mid \boldsymbol{X}(t_{n-1}) = \boldsymbol{y}_{n-1}, \ldots, \boldsymbol{X}(t_1) = \boldsymbol{y}_1\right]$$

$$= \text{Prob}\left[\bigcap_{j=1}^{m}\{X_j(t_n) \leq x_j\} \mid \boldsymbol{X}(t_{n-1}) = \boldsymbol{y}_{n-1}\right] \quad t_n > t_{n-1} > \cdots > t_1 \tag{4.1.3}$$

where $\cap$ denotes the joint occurrence of multiple events. A sufficient condition for a vectorially valued stochastic process to be a Markov vector is that its vectorial increments be independent in nonoverlapping time intervals. The transition probability density of a vector Markov process is a generalization of (4.1.2):

$$q(\boldsymbol{x}, t \mid \boldsymbol{x}_0, t_0) = \frac{\partial^m}{\partial x_1 \cdots \partial x_m}\text{Prob}\left[\bigcap_{j=1}^{m}\{X_j(t) \leq x_j\} \mid \boldsymbol{X}(t_0) = \boldsymbol{x}_0\right] \tag{4.1.4}$$

It is important to note that the components of a Markov vector may or may not be scalar Markov processes. If a Markov vector has independent increments vectorially, it does not follow necessarily that each of the components has independent increments separately. In particular, some of the components of a vector Markov process may be differentiable, whereas a scalar Markov process is not differentiable.

The higher order probability densities, describing the behavior of a Markov process at several instants of time, can be constructed from the initial probability density and the transition probability density. For example,

$$p(\boldsymbol{x}_1, t_1; \boldsymbol{x}_2, t_2; \ldots; \boldsymbol{x}_n, t_n) = q(\boldsymbol{x}_n, t_n \mid \boldsymbol{x}_{n-1}, t_{n-1}) q(\boldsymbol{x}_{n-1}, t_{n-1} \mid \boldsymbol{x}_{n-2}, t_{n-2}) \cdots$$
$$q(\boldsymbol{x}_2, t_2 \mid \boldsymbol{x}_1, t_1) p(\boldsymbol{x}_1) \qquad t_1 < t_2 < \cdots < t_n \tag{4.1.5}$$

4.2 THE FOKKER-PLANCK EQUATION

In many practical applications, the initial state of a Markov process is known. Then the transition probability density alone characterizes completely the stochastic process. In other cases, only the stationary state behavior of the process is of interest. If such a stationary state does exist, then its stationary probability density can be obtained from

$$p(\boldsymbol{x}) = \lim_{t-t_0 \to \infty} q(\boldsymbol{x}, t \mid \boldsymbol{x}_0, t_0) \tag{4.2.1}$$

The physical implication of (4.2.1) is clear. As the duration of transition time $t-t_0$ increases, the effect of the initial condition $\boldsymbol{X}(t_0) = \boldsymbol{x}_0$ diminishes, and the conditional probability density tends to the unconditional probability density. This unconditional probability density describes the stochastic process at an arbitrary instant of time after it attains the stationary state, and it must be independent of time. Furthermore, at the stationary state, the transition probability density must be a function of the net transition time; that is, $q(\boldsymbol{x}, t \mid \boldsymbol{x}_0, t_0)$ must be a function of $t - t_0$, and the higher order probability densities can be obtained as

$$p(\boldsymbol{x}_1, t_1; \boldsymbol{x}_2, t_2; \ldots; \boldsymbol{x}_n, t_n) = q(\boldsymbol{x}_n, t_n - t_{n-1} \mid \boldsymbol{x}_{n-1}) q(\boldsymbol{x}_{n-1}, t_{n-1} - t_{n-2} \mid \boldsymbol{x}_{n-2}) \cdots$$
$$\times\, q(\boldsymbol{x}_2, t_2 - t_1 \mid \boldsymbol{x}_1) p(\boldsymbol{x}_1) \qquad t_1 < t_2 < \cdots < t_n \tag{4.2.2}$$

where $q(\boldsymbol{x}, \tau \mid \boldsymbol{x}')$ is an abbreviation for $q(\boldsymbol{x}, \tau \mid \boldsymbol{x}', 0)$. Again, the transition probability density alone characterizes a stationary Markov process, since the additional requirement of the unconditional one-time probability density $p(\boldsymbol{x}_1)$ in (4.2.2) is obtainable from the transition probability density, according to (4.2.1).

Now equation (4.1.5) implies that

$$p(\boldsymbol{x}_2, t_2; \boldsymbol{y}, t \mid \boldsymbol{x}_1, t_1) = q(\boldsymbol{x}_2, t_2 \mid \boldsymbol{y}, t) q(\boldsymbol{y}, t \mid \boldsymbol{x}_1, t_1) \qquad t_1 < t < t_2 \tag{4.2.3}$$

Integrating (4.2.3) with respect to $\boldsymbol{y}$, and noting that $p(\boldsymbol{x}_2, t_2 \mid \boldsymbol{x}_1, t_1)$ is in fact a transition probability density, we obtain

$$q(\boldsymbol{x}_2, t_2 \mid \boldsymbol{x}_1, t_1) = \int q(\boldsymbol{x}_2, t_2 \mid \boldsymbol{y}, t) q(\boldsymbol{y}, t \mid \boldsymbol{x}_1, t_1)\, d\boldsymbol{y} \tag{4.2.4}$$

where $d\boldsymbol{y}$ is an abbreviation for $dy_1 dy_2 \ldots dy_m$ and the integration is to cover the entire m-dimensional domain. Equation (4.2.4) is known as the *Chapman-Kolmogorov-Smoluchowski equation.*

Equation (4.2.4) may be considered as an integral equation, which governs the transition probability density of a Markov process. This integral equation may be converted to an equivalent differential equation, called the *Fokker-Planck equation.* We next derive the Fokker-Planck equation using a procedure similar to the one given by Wang and Uhlenbeck (1945) for scalar Markov processes.

Let $\boldsymbol{X}(t)$ be a vector Markov process in an m-dimensional space, and consider the integral

$$I = \int R(\boldsymbol{y}) \frac{\partial}{\partial t} q(\boldsymbol{y}, t \mid \boldsymbol{x}_0, t_0)\, d\boldsymbol{y} \qquad t > t_0 \tag{4.2.5}$$

where $R(\boldsymbol{y})$ is an arbitrary function of y_j, $j = 1, 2, \ldots, m$, which goes to zero sufficiently fast as any y_j approaches either the upper or lower integration limit. Specifically, we assume that for any $s = k + l + \cdots + r$,

$$\frac{\partial^s}{\partial y_1^k \partial y_2^l \cdots \partial y_m^r} R(\boldsymbol{y}) \to 0 \tag{4.2.6}$$

as any y_j approaches an integration limit. We also assume that $R(\boldsymbol{y})$ can be expanded in a Taylor series about a point $\boldsymbol{x}$:

$$\begin{aligned} R(\boldsymbol{y}) &= R(\boldsymbol{x}) + (y_j - x_j)\frac{\partial R(\boldsymbol{x})}{\partial x_j} + \frac{1}{2!}(y_j - x_j)(y_k - x_k)\frac{\partial^2 R(\boldsymbol{x})}{\partial x_j \partial x_k} \\ &\quad + \frac{1}{3!}(y_j - x_j)(y_k - x_k)(y_l - x_l)\frac{\partial^3 R(\boldsymbol{x})}{\partial x_j \partial x_k \partial x_l} + \cdots \end{aligned} \tag{4.2.7}$$

where a repeated subscript in a product indicates a summation over $1, 2, \ldots, m$. Now the integral (4.2.5) may be written as

$$\begin{aligned} I &= \int R(\boldsymbol{y}) \lim_{\Delta t \to 0} \frac{1}{\Delta t} [q(\boldsymbol{y}, t + \Delta t \mid \boldsymbol{x}_0, t_0) - q(\boldsymbol{y}, t \mid \boldsymbol{x}_0, t_0)]\, d\boldsymbol{y} \\ &= \lim_{\Delta t \to 0} \frac{1}{\Delta t} \int R(\boldsymbol{y}) [q(\boldsymbol{y}, t + \Delta t \mid \boldsymbol{x}_0, t_0) - q(\boldsymbol{y}, t \mid \boldsymbol{x}_0, t_0)]\, d\boldsymbol{y} \end{aligned} \tag{4.2.8}$$

obtained by interchanging the order of the integration and the limit operation, which is permissible so long as the integral in the second expression converges uniformly in a neighborhood of t. Using the Chapman-Kolmogorov-Smoluchowski equation,

$$\begin{aligned} I = \lim_{\Delta t \to 0} \frac{1}{\Delta t} &\left[\int d\boldsymbol{y} R(\boldsymbol{y}) \int q(\boldsymbol{y}, t + \Delta t \mid \boldsymbol{x}, t) q(\boldsymbol{x}, t \mid \boldsymbol{x}_0, t_0)\, d\boldsymbol{x} \right. \\ &\left. - \int R(\boldsymbol{y}) q(\boldsymbol{y}, t \mid \boldsymbol{x}_0, t_0)\, d\boldsymbol{y} \right] \end{aligned} \tag{4.2.9}$$

Substituting (4.2.7) into (4.2.9) and integrating first on $\boldsymbol{y}$,

$$\begin{aligned} I = \int &\left[a_j(\boldsymbol{x}, t) \frac{\partial R(\boldsymbol{x})}{\partial x_j} + \frac{1}{2!} b_{jk}(\boldsymbol{x}, t) \frac{\partial^2 R(\boldsymbol{x})}{\partial x_j \partial x_k} \right. \\ &\left. + \frac{1}{3!} c_{jkl}(\boldsymbol{x}, t) \frac{\partial R(\boldsymbol{x})}{\partial x_j \partial x_k \partial x_l} + \cdots \right] q(\boldsymbol{x}, t \mid \boldsymbol{x}_0, t_0)\, d\boldsymbol{x} \end{aligned} \tag{4.2.10}$$

where

$$
\begin{aligned}
a_j(\boldsymbol{x}, t) &= \lim_{\Delta t \to 0} \frac{1}{\Delta t} \int (y_j - x_j) q(\boldsymbol{y}, t + \Delta t \mid \boldsymbol{x}, t)\, d\boldsymbol{y} \\
b_{jk}(\boldsymbol{x}, t) &= \lim_{\Delta t \to 0} \frac{1}{\Delta t} \int (y_j - x_j)(y_k - x_k) q(\boldsymbol{y}, t + \Delta t \mid \boldsymbol{x}, t)\, d\boldsymbol{y} \\
c_{jkl}(\boldsymbol{x}, t) &= \lim_{\Delta t \to 0} \frac{1}{\Delta t} \int (y_j - x_j)(y_k - x_k)(y_l - x_l) q(\boldsymbol{y}, t + \Delta t \mid \boldsymbol{x}, t)\, d\boldsymbol{y} \qquad (4.2.11) \\
&\ \ \vdots
\end{aligned}
$$

In obtaining (4.2.10), use has been made of the fact that

$$
\int q(\boldsymbol{y}, t + \Delta t \mid \boldsymbol{x}, t)\, d\boldsymbol{y} = 1 \tag{4.2.12}
$$

for any $\boldsymbol{x}$, t, and Δt.

The functions defined in (4.2.11) can be expressed more meaningfully as follows:

$$
\begin{aligned}
a_j(\boldsymbol{x}, t) &= \lim_{\Delta t \to 0} \frac{1}{\Delta t} E[X_j(t + \Delta t) - X_j(t) \mid \boldsymbol{X}(t) = \boldsymbol{x}] \\
b_{jk}(\boldsymbol{x}, t) &= \lim_{\Delta t \to 0} \frac{1}{\Delta t} E\{[X_j(t + \Delta t) - X_j(t)][X_k(t + \Delta t) - X_k(t)] \mid \boldsymbol{X}(t) = \boldsymbol{x}\} \\
c_{jkl}(\boldsymbol{x}, t) &= \lim_{\Delta t \to 0} \frac{1}{\Delta t} E\{[X_j(t + \Delta t) - X_j(t)][X_k(t + \Delta t) - X_k(t)] \\
&\qquad \times [X_l(t + \Delta t) - X_l(t)] \mid \boldsymbol{X}(t) = \boldsymbol{x}\} \qquad (4.2.11a) \\
&\ \ \vdots
\end{aligned}
$$

They give the rates of moments of various increments in $\boldsymbol{X}(t)$ at time t, on the condition that $\boldsymbol{X}(t) = \boldsymbol{x}$. They are known as the derivate moments (Moyal, 1949), although other names have been used also in the literature.

The integration in (4.2.10) can now be carried out by parts for each term. By use of the assumption (4.2.6), it leads to

$$
I = \int R(\boldsymbol{x}) \left[-\frac{\partial}{\partial x_j}(a_j q) + \frac{1}{2} \frac{\partial^2}{\partial x_j \partial x_k}(b_{jk} q) - \frac{1}{3!} \frac{\partial^3}{\partial x_j \partial x_k \partial x_l}(c_{jkl} q) + \cdots \right] d\boldsymbol{x} \tag{4.2.13}
$$

where the arguments of $q(\boldsymbol{x}, t \mid \boldsymbol{x}_0, t_0)$, $a_j(\boldsymbol{x}, t)$, $b_{jk}(\boldsymbol{x}, t), \ldots,$ have been omitted for brevity. It follows from a combination of (4.2.5) and (4.2.13) that

$$
\begin{aligned}
\int R(\boldsymbol{x}) \bigg[\frac{\partial}{\partial t} q &+ \frac{\partial}{\partial x_j}(a_j q) - \frac{1}{2} \frac{\partial^2}{\partial x_j \partial x_k}(b_{jk} q) \\
&+ \frac{1}{3!} \frac{\partial^3}{\partial x_j \partial x_k \partial x_l}(c_{jkl} q) - \cdots \bigg] d\boldsymbol{x} = 0 \qquad (4.2.14)
\end{aligned}
$$

Since (4.2.14) must be valid for any choice of $R(\boldsymbol{x})$ which satisfies the rather general restriction (4.2.6) and which can be expanded in a Taylor series, it is necessary that

$$\frac{\partial}{\partial t}q + \frac{\partial}{\partial x_j}(a_j q) - \frac{1}{2}\frac{\partial^2}{\partial x_j \partial x_k}(b_{jk}q) + \frac{1}{3!}\frac{\partial^3}{\partial x_j \partial x_k \partial x_l}(c_{jkl}q) - \cdots = 0 \tag{4.2.15}$$

Equation (4.2.15) is called the Fokker-Planck equation for the transition probability density $q(\boldsymbol{x}, t \mid \boldsymbol{x}_0, t_0)$ of a vector Markov process $\boldsymbol{X}(t)$, the one-dimensional scalar Markov process being a special case.

When the theory is applied to a practical problem in stochastic dynamics, the derivate moments $a_j(\boldsymbol{x}, t)$, $b_{jk}(\boldsymbol{x}, t)$, ... are determined from the equations of motion, and the associated Fokker-Planck equation is solved in conjunction with suitable boundary conditions and the initial condition

$$q(\boldsymbol{x}, t_0 \mid \boldsymbol{x_0}, t_0) = \delta(\boldsymbol{x} - \boldsymbol{x}_0) = \prod_{j=1}^{m} \delta(x_j - x_{j0}) \tag{4.2.16}$$

where x_{j0} are the components of $\boldsymbol{x}_0$.

If a Markov process is also a gaussian process, then by using the gaussian property that higher order moments can be expressed as products of the first- and the second-order moments (Lin, 1967, Eq. 4-50), it can be shown that the derivate moments of an order higher than 2, c_{jkl}, ... are zero. In this special case, the Fokker-Plank equation (4.2.15) reduces to

$$\frac{\partial}{\partial t}q + \frac{\partial}{\partial x_j}(a_j q) - \frac{1}{2}\frac{\partial^2}{\partial x_j \partial x_k}(b_{jk}q) = 0 \tag{4.2.17}$$

Very frequently, the name Fokker-Planck equation refers to the special case (4.2.17) without clarification.

Known exact solutions to Fokker-Planck equations are restricted to the simpler form of (4.2.17), the focus of our attention in the remainder of this chapter and in Chapter 5. An approximate solution technique to account for higher order terms is considered in Chapter 7. We note that the simpler Fokker-Planck equation (4.2.17) is not restricted to gaussian Markov processes.

A *diffusive process* is a Markov process for which the sample functions are continuous with probability 1. A sufficient condition for a Markov process to be diffusive is the Dynkin's condition (Dynkin, 1965),

$$\lim_{\Delta t \to 0} \frac{1}{\Delta t} \text{Prob}\,[\|\boldsymbol{X}(t + \Delta t) - \boldsymbol{X}(t)\| > \varepsilon \mid \boldsymbol{X}(t) = \boldsymbol{x}] = 0 \qquad \varepsilon > 0 \tag{4.2.18}$$

where $\| \, \|$ denotes an euclidean norm of a vector, and the convergence holds uniformly for $\boldsymbol{x}$ in the defining interval of $\boldsymbol{X}$. Furthermore, a sufficient condition for the Dynkin's condition is that (see, e.g., Karlin and Taylor, 1981)

$$\lim_{\Delta t \to 0} \frac{1}{\Delta t} \text{Prob}[\|\boldsymbol{X}(t + \Delta t) - \boldsymbol{X}(t)\|^{2+\delta} \mid \boldsymbol{X}(t) = \boldsymbol{x}] = 0 \qquad \delta > 0 \tag{4.2.19}$$

In other words, its Fokker-Planck equation has the simpler form (4.2.17).

BOUNDARY CONDITIONS FOR THE FOKKER-PLANCK EQUATION. Since (4.2.17) is a second-order differential equation in the state variables x_j, the admissible boundary conditions can be expressed in terms of q, $\partial q/\partial x_j$, or their combinations. To gain some physical insights into suitable boundary conditions, rewrite equation (4.2.17) as

$$\frac{\partial}{\partial t}q + \frac{\partial}{\partial x_j}G_j = 0 \tag{4.2.20}$$

where

$$G_j = a_j q - \frac{1}{2}\frac{\partial}{\partial x_k}(b_{jk}q) \tag{4.2.21}$$

Equation (4.2.20) is analogous to the continuity equation in fluid mechanics, indicating the conservation of mass in a fluid flow. Similarly, equation (4.2.20) may be interpreted as conservation of probability, and G_j as the jth component of the probability flow vector $\boldsymbol{G}(\boldsymbol{x}, t \mid \boldsymbol{x}_0, t_0)$.

Within the domain in which a Markov process is defined, there may exist certain points possessing one of the following two properties:

1. All the second derivate moments b_{jk} vanish.
2. At least one first derivate moment a_j becomes unbounded.

Such points are said to be *singular* of the first and second kinds, respectively. A boundary of the domain is a singular boundary if it is a singular point or consists of a set of singular points. At the present time, singular boundaries are well understood only in the case of the one-dimensional Markov process, which is considered in detail in Section 4.5. For the time being, we restrict our attention to nonsingular boundaries, at which all a_j are finite and at least one b_{jk} is nonzero. A nonsingular boundary is communicative, in the sense that it is accessible by a sample function starting from an interior point, and an interior point is also accessible starting from a nonsingular boundary. In principle, the behavior of a sample function after reaching a nonsingular boundary requires further stipulation. This, in fact, provides some flexibility with which certain physical features can be incorporated, depending on the problem at hand. For engineering applications, the following sample behaviors are often stipulated:

Reflective boundary. A reflective boundary is one at which the passage of probability flow is prohibited:

$$\boldsymbol{n} \cdot \boldsymbol{G}(\boldsymbol{x}, t \mid \boldsymbol{x}_0, t_0) = 0 \qquad \boldsymbol{x} \in S \tag{4.2.22}$$

where S denotes the boundary and $\boldsymbol{n}$ is a unit vector normal to the boundary. A sample function must "leave" a reflective boundary immediately upon "arriving" at the boundary.

Absorbing boundary. A sample function is "removed" from the population of sample functions once it arrives at an absorbing boundary. Therefore, the probability density at an absorbing boundary is always zero:

$$q(\boldsymbol{x}, t \mid \boldsymbol{x}_0, t_0) = 0 \qquad \boldsymbol{x} \in S \tag{4.2.23}$$

An absorbing boundary is also known as an *exit boundary.*

Periodic boundary. The existence of periodic boundaries is clearly indicated in the case of a one-dimensional diffusion process, defined on a circle. In this case, both the probability density and the probability flow must be periodic, with a period $2\pi/n$, where n is an integer. The value n can be determined by examining the periodicity of the first and second derivate moments. More generally, if the first and second derivate moments, $a_j(\boldsymbol{x}, t)$ and $b_{jk}(\boldsymbol{x}, t)$, are periodic functions of x_j with a period L, then we have the periodic boundary conditions

$$q(\boldsymbol{x}, t \mid \boldsymbol{x}_0, t_0)|_{x_j} = q(\boldsymbol{x}, t \mid \boldsymbol{x}_0, t_0)|_{x_j+L} \qquad \boldsymbol{x} \in S \tag{4.2.24a}$$

$$\frac{\partial}{\partial \boldsymbol{x}_k} q(\boldsymbol{x}, t \mid \boldsymbol{x}_0, t_0)\Big|_{\boldsymbol{x}_j} = \frac{\partial}{\partial \boldsymbol{x}_k} q(\boldsymbol{x}, t \mid \boldsymbol{x}_0, t_0)\Big|_{\boldsymbol{x}_j+L} \qquad \boldsymbol{x} \in S \tag{4.2.24b}$$

Boundary at infinity. The probability flow must vanish at a nonsingular boundary at infinity:

$$\lim_{x_j \to \pm\infty} \boldsymbol{G}(\boldsymbol{x}, t \mid \boldsymbol{x}_0, t_0) = 0 \tag{4.2.25}$$

Moreover, since the total probability is finite, we also have

$$\lim_{x_j \to \pm\infty} q(\boldsymbol{x}, t \mid \boldsymbol{x}_0, t_0) = 0 \tag{4.2.26}$$

and it approaches zero at least as fast as $|x_j|^{-\alpha}$, where $\alpha > 1$.

In passing, we note that, in general, conditions at a singular boundary cannot be arbitrarily imposed; instead, they are determined from the values of a_j and b_{jk} at and near the boundary (Section 4.5). When so determined, they can also be any of the preceding types.

KOLMOGOROV BACKWARD EQUATION. The unknown q in (4.2.15) or (4.2.17) is treated as a function of t and x_j, $j = 1, 2, \ldots, m$. Upon imposition of the initial condition, (4.2.16), the resulting solution for q will contain, as parameters, t_0 and x_{j0}, $j = 1, 2, \ldots, m$. Conversely, q may also be considered as a function of t_0 and x_{j0}, with t and x_j playing the roles of parameters. Indeed, an alternative equation for q exists, and it is given by

$$\begin{aligned} \frac{\partial q}{\partial t_0} &+ a_j(\boldsymbol{x}_0, t_0)\frac{\partial q}{\partial x_{j0}} + \frac{1}{2} b_{jk}(\boldsymbol{x}_0, t_0)\frac{\partial^2 q}{\partial x_{j0} \partial x_{k0}} \\ &+ \frac{1}{3!} c_{jkl}(\boldsymbol{x}_0, t_0)\frac{\partial^3 q}{\partial x_{j0} \partial x_{k0} \partial x_{l0}} + \cdots = 0 \end{aligned} \tag{4.2.27}$$

where $a_j, b_{jk}, c_{jkl}, \ldots$ are the same derivate moments, except that they are expressed as functions of $\boldsymbol{x}_0$ and t_0. A proof of equation (4.2.27) is given next.

We write

$$\frac{\partial q}{\partial t_0} = \lim_{\Delta t_0 \to 0} \frac{1}{\Delta t_0} [q(\boldsymbol{x}, t \mid \boldsymbol{x}_0, t_0) - q(\boldsymbol{x}, t \mid \boldsymbol{x}_0, t_0 - \Delta t_0)] \tag{4.2.28}$$

From the Chapman-Kolmogorov-Smoluchowski equation (4.2.4),

$$q(\boldsymbol{x}, t \mid \boldsymbol{x}_0, t_0 - \Delta t_0) = \int q(\boldsymbol{x}, t \mid \boldsymbol{y}, t_0) q(\boldsymbol{y}, t_0 \mid \boldsymbol{x}_0, t_0 - \Delta t_0)\, d\boldsymbol{y} \tag{4.2.29}$$

We can also write

$$q(\boldsymbol{x}, t \mid \boldsymbol{x}_0, t_0) = \int q(\boldsymbol{x}, t \mid \boldsymbol{x}_0, t_0) q(\boldsymbol{y}, t_0 \mid \boldsymbol{x}_0, t_0 - \Delta t_0)\, d\boldsymbol{y} \tag{4.2.30}$$

since

$$\int q(\boldsymbol{y}, t_0 \mid \boldsymbol{x}_0, t_0 - \Delta t_0)\, d\boldsymbol{y} = 1$$

Combination of (4.2.28) through (4.2.30) yields

$$\frac{\partial q}{\partial t_0} = \lim_{\Delta t_0 \to 0} \frac{1}{\Delta t_0} \int [q(\boldsymbol{x}, t \mid \boldsymbol{x}_0, t_0) - q(\boldsymbol{x}, t \mid \boldsymbol{y}, t_0)] q(\boldsymbol{y}, t_0 \mid \boldsymbol{x}_0, t_0 - \Delta t_0)\, d\boldsymbol{y} \tag{4.2.31}$$

Expanding $q(\boldsymbol{x}, t \mid \boldsymbol{y}, t_0)$ in a Taylor series,

$$\begin{aligned} q(\boldsymbol{x}, t \mid \boldsymbol{y}, t_0) = {} & q(\boldsymbol{x}, t \mid \boldsymbol{x}_0, t_0) + (\boldsymbol{y}_j - \boldsymbol{x}_{j0}) \frac{\partial}{\partial x_{j0}} q(\boldsymbol{x}, t \mid \boldsymbol{x}_0, t_0) \\ & + \frac{1}{2!}(y_j - x_{j0})(y_k - x_{k0}) \frac{\partial^2}{\partial x_{j0} \partial x_{k0}} q(\boldsymbol{x}, t \mid \boldsymbol{x}_0, t_0) \\ & + \frac{1}{3!}(y_j - x_{j0})(y_k - x_{k0})(y_l - x_{l0}) \frac{\partial^3}{\partial x_{j0} \partial x_{k0} \partial x_{l0}} q(\boldsymbol{x}, t \mid \boldsymbol{x}_0, t_0) + \cdots \end{aligned} \tag{4.2.32}$$

and substituting (4.2.32) into (4.2.31), we obtain

$$\begin{aligned} \frac{\partial q}{\partial t_0} = {} & \lim_{\Delta t_0 \to 0} \frac{1}{\Delta t_0} \int \Bigg[-(y_j - x_{j0}) \frac{\partial}{\partial x_{j0}} q(\boldsymbol{x}, t \mid \boldsymbol{x}_0, t_0) - \frac{1}{2!}(y_j - x_{j0})(y_k - x_{k0}) \\ & \times \frac{\partial^2}{\partial x_{j0} \partial x_{k0}} q(\boldsymbol{x}, t \mid \boldsymbol{x}_0, t_0) - \frac{1}{3!}(y_j - x_{j0})(y_k - x_{k0})(y_l - x_{l0}) \\ & \times \frac{\partial^3}{\partial x_{j0} \partial x_{k0} \partial x_{l0}} q(\boldsymbol{x}, t \mid \boldsymbol{x}_0, t_0) - \cdots \Bigg] q(\boldsymbol{y}, t_0 \mid \boldsymbol{x}_0, t_0 - \Delta t_0)\, d\boldsymbol{y} \end{aligned} \tag{4.2.33}$$

Now, the definitions (4.2.11) for derivate moments may be rewritten as

$$\begin{aligned} a_j(\boldsymbol{x}_0, t_0) &= \lim_{\Delta t_0 \to 0} \frac{1}{\Delta t_0} \int (y_j - x_{j0}) q(\boldsymbol{y}, t_0 \mid \boldsymbol{x}_0, t_0 - \Delta t_0)\, d\boldsymbol{y} \\ b_{jk}(\boldsymbol{x}_0, t_0) &= \lim_{\Delta t_0 \to 0} \frac{1}{\Delta t_0} \int (y_j - x_{j0})(y_k - x_{k0}) \\ & \qquad \times q(\boldsymbol{y}, t_0 \mid \boldsymbol{x}_0, t_0 - \Delta t_0)\, d\boldsymbol{y} \end{aligned} \tag{4.2.34}$$

$$c_{jkl}(\boldsymbol{x}_0, t_0) = \lim_{\Delta t_0 \to 0} \frac{1}{\Delta t_0} \int (y_j - x_{j0})(y_k - x_{k0})(y_l - x_{l0})$$
$$\times\, q(\boldsymbol{y}, t_0 \mid \boldsymbol{x}_0, t_0 - \Delta t_0)\, d\boldsymbol{y}$$
$$\vdots$$

Equation (4.2.27) is obtained upon substituting (4.2.34) into (4.2.33).

Equation (4.2.27) is known as the *Kolmogorov backward equation* (Kolmogorov, 1931), and the x_{j0} are called the *backward variables,* since they are associated with the earlier time t_0. Correspondingly, equation (4.2.15) is known as the *Kolmogorov forward equation,* and the x_j as the *forward variables,* since they are associated with the later time t. The forward equation is most commonly used for determining the probability density q, since the associated boundary conditions, if nonsingular, can be interpreted analogously to those in fluid mechanics. However, the backward equation is useful for investigating certain system failure problems, as discussed in Chapter 8.

In the case of a diffusive Markov process, (4.2.27) reduces to

$$\frac{\partial}{\partial t_0} q + a_j(\boldsymbol{x}_0, t_0) \frac{\partial}{\partial x_{j0}} q + \frac{1}{2} b_{jk}(\boldsymbol{x}_0, t_0) \frac{\partial^2}{\partial x_{j0} \partial x_{k0}} q = 0 \tag{4.2.35}$$

The initial condition for the backward equation (4.2.35) is still (4.2.16).

BOUNDARY CONDITIONS FOR KOLMOGOROV BACKWARD EQUATION. The four types of nonsingular boundaries associated with (4.2.35) are described as follows:

Reflective boundary. Let τ be a time variable between t_0 and t. Then

$$\frac{\partial}{\partial \tau} q(\boldsymbol{x}, t \mid \boldsymbol{x}_0, t_0) = \frac{\partial}{\partial \tau} \int_R q(\boldsymbol{x}, t \mid \boldsymbol{y}, \tau) q(\boldsymbol{y}, \tau \mid \boldsymbol{x}_0, t_0)\, d\boldsymbol{y} = 0 \tag{4.2.36}$$

where R is an arbitrary region within the domain in which the Markov process $\boldsymbol{X}(t)$ is defined, and where use has been made of the Chapman-Kolmogorov-Smoluchowski equation (4.2.4). Denote

$$q_1 = q(\boldsymbol{y}, \tau \mid \boldsymbol{x}_0, t_0) \qquad q_2 = q(\boldsymbol{x}, t \mid \boldsymbol{y}, \tau)$$

From the Kolmogorov forward and backward equations, we have, respectively,

$$\frac{\partial q_1}{\partial \tau} = \frac{\partial}{\partial \tau} q(\boldsymbol{y}, \tau \mid \boldsymbol{x}_0, t_0) = -\frac{\partial}{\partial y_j}(a_j q_1) + \frac{1}{2} \frac{\partial^2}{\partial y_j \partial y_k}(b_{jk} q_1) \tag{4.2.37}$$

$$\frac{\partial q_2}{\partial \tau} = \frac{\partial}{\partial \tau} q(\boldsymbol{x}, t \mid \boldsymbol{y}, \tau) = -a_j \frac{\partial q_2}{\partial y_j} - \frac{1}{2} b_{jk} \frac{\partial^2 q_2}{\partial y_j \partial y_k} \tag{4.2.38}$$

where a_j and b_{jk} are functions of $\boldsymbol{y}$ and τ. Interchanging the partial differentiation and integration in (4.2.36) and substituting (4.2.37) and (4.2.38), we obtain

$$\int_R \left[-a_j \frac{\partial q_2}{\partial y_j} - \frac{1}{2} b_{jk} \frac{\partial^2 q_2}{\partial y_j \partial y_k} \right] q_1 \, d\boldsymbol{y} + \int_R \left[-\frac{\partial}{\partial y_j}(a_j q_1) + \frac{1}{2} \frac{\partial^2}{\partial y_j \partial y_k}(b_{jk} q_1) \right] q_2 \, d\boldsymbol{y} = 0 \tag{4.2.39}$$

Equation (4.2.39) can be rewritten as

$$\int_R \frac{\partial}{\partial y_j} \left\{ q_2 \left[-a_j q_1 + \frac{1}{2} \frac{\partial}{\partial y_k}(b_{jk} q_1) \right] \right\} d\boldsymbol{y} - \frac{1}{2} \int_R \frac{\partial}{\partial y_j} \left(q_1 b_{jk} \frac{\partial q_2}{\partial y_k} \right) d\boldsymbol{y} = 0 \tag{4.2.40}$$

By applying the Gauss theorem, the volume integrals in (4.2.40) can be transformed into surface integrals as follows:

$$\int_\Gamma q_2 \left[-a_j q_1 + \frac{1}{2} \frac{\partial}{\partial y_k}(b_{jk} q_1) \right] n_j \, d\Gamma - \frac{1}{2} \int_\Gamma q_1 b_{jk} \frac{\partial q_2}{\partial y_k} n_j \, d\Gamma = 0 \tag{4.2.41}$$

where Γ is the surface bounding R and n_j is the jth component of $\boldsymbol{n}$, the unit vector normal to surface Γ. Since Γ can be chosen arbitrarily, it is necessary that

$$q_2 \left[-a_j q_1 + \frac{1}{2} \frac{\partial}{\partial y_k}(b_{jk} q_1) \right] n_j - \frac{1}{2} q_1 b_{jk} \frac{\partial q_2}{\partial y_k} n_j = 0 \tag{4.2.42}$$

everywhere. In particular, on a reflective boundary S, the first part of (4.2.42) vanishes, since

$$\boldsymbol{n} \cdot \boldsymbol{G} = n_j \left[a_j q_1 - \frac{1}{2} \frac{\partial}{\partial y_k}(b_{jk} q_1) \right] = 0 \tag{4.2.43}$$

according to (4.2.22). Therefore, the second part of (4.2.42) also vanishes, leading to

$$n_j b_{jk} \frac{\partial q_2}{\partial y_k} = 0 \tag{4.2.44}$$

Changing the arguments from $\boldsymbol{y}$ and τ to $\boldsymbol{x}_0$ and t_0, we obtain the following reflective boundary condition in terms of backward variables:

$$n_j b_{jk}(\boldsymbol{x}_0, t_0) \frac{\partial}{\partial x_{k0}} q(\boldsymbol{x}, t \mid \boldsymbol{x}_0, t_0) = 0 \qquad \boldsymbol{x}_0 \in S \tag{4.2.45}$$

In the one-dimensional case, (4.2.45) reduces to

$$\frac{\partial}{\partial x_0} q(x, t \mid x_0, t_0) = 0 \qquad x_0 \in S \tag{4.2.46}$$

since we assumed that $b(x_0, t_0)$ is nonzero. It is interesting to note that the first derivate moment is not needed when describing a reflective boundary in terms of backward variables. Yet (4.2.45) is equivalent to (4.2.22).

Absorbing boundary. Since every arriving sample function is removed from an absorbing boundary,

$$q(\boldsymbol{x}, t \mid \boldsymbol{x}_0, t_0) = 0 \qquad \boldsymbol{x}_0 \in S, \quad t > t_0 \tag{4.2.47}$$

Periodic boundary. If the first and second derivate moments $a_j(\boldsymbol{x}_0, t_0)$ and $b_{jk}(\boldsymbol{x}_0, t_0)$ are periodic functions of x_{j0} with a period L, we have

$$q(\boldsymbol{x}, t \mid \boldsymbol{x}_0, t_0)\big|_{x_{j0}} = q(\boldsymbol{x}, t \mid \boldsymbol{x}_0, t_0)\big|_{x_{j0}+L} \qquad \boldsymbol{x}_0 \in S \tag{4.2.48a}$$

$$\frac{\partial}{\partial x_{k0}} q(\boldsymbol{x}, t \mid \boldsymbol{x}_0, t_0)\big|_{x_{j0}} = \frac{\partial}{\partial x_{k0}} q(\boldsymbol{x}, t \mid \boldsymbol{x}_0, t_0)\big|_{x_{j0}+L} \qquad \boldsymbol{x}_0 \in S \tag{4.2.48b}$$

Boundary at infinity. It can be shown by following a similar procedure leading to (4.2.45) that

$$\lim_{x_{j0} \to \pm\infty} \frac{\partial}{\partial x_{k0}} q(\boldsymbol{x}, t \mid \boldsymbol{x}_0, t_0) = 0 \tag{4.2.49}$$

Solutions to the Kolmogorov forward equation (4.2.17) and backward equation (4.2.35) are identical under the same initial condition and equivalent boundary conditions.

If the stationary state of a Markov process exists, then its stationary probability density $p(\boldsymbol{x})$ is the limit of the transition probability density, as shown in equation (4.2.1). This stationary probability density satisfies a reduced Fokker-Planck equation, which is a Fokker-Planck equation without the time derivative term:

$$\frac{\partial}{\partial x_j}(a_j p) - \frac{1}{2}\frac{\partial^2}{\partial x_j \partial x_k}(b_{jk} p) = 0 \tag{4.2.50}$$

Clearly, the first and second derivate moments $a_j$and b_{jk} in (4.2.50) must be independent of t, and there must not be an absorbing part on the boundary; otherwise, the Markov process cannot attain a stationary state. Equation (4.2.50) can also be written as

$$\frac{\partial}{\partial x_j} G_j = 0 \tag{4.2.51}$$

where G_j is the probability flow in the jth direction when the Markov process is in the state of stationarity; that is,

$$G_j = a_j p - \frac{1}{2}\frac{\partial}{\partial x_k}(b_{jk} p) \tag{4.2.52}$$

which depends only on $\boldsymbol{x}$. For nonsingular boundaries, boundary conditions for the reduced Fokker-Planck equation (4.2.50) can be obtained, as the case may be, from (4.2.22) and (4.2.24) to (4.2.26) by replacing $q(\boldsymbol{x}, t \mid \boldsymbol{x}_0, t_0)$ and $\boldsymbol{G}(\boldsymbol{x}, t \mid \boldsymbol{x}_0, t_0)$ by $p(\boldsymbol{x})$ and $\boldsymbol{G}(\boldsymbol{x})$, respectively.

4.3 THE WIENER PROCESS

The simplest example for a Markov process is perhaps the *Wiener process,* also known as the *Brownian motion* process. Denoted by $B(t)$, such a stochastic process is defined as having the following properties: (a) $B(t)$ is a gaussian process, (b)

$B(0) = K$, (c) $E[B(t)] = K$, and (d) it has a covariance function $\text{cov}[B(t_1), B(t_2)] = \sigma^2 \min(t_1, t_2)$; namely,

$$\text{cov}[B(t_1), B(t_2)] = \begin{cases} \sigma^2 t_1 & t_1 < t_2 \\ \sigma^2 t_2 & t_1 > t_2 \end{cases} \tag{4.3.1}$$

where K is a constant and σ^2 is a positive constant. In the following discussion, properties (b), (c), and (d) are replaced, without loss of generality, by (b′) $B(0) = 0$, (c′) $E[B(t)] = 0$, and (d′)

$$E[B(t_1)B(t_2)] = \begin{cases} \sigma^2 t_1 & t_1 < t_2 \\ \sigma^2 t_2 & t_1 > t_2 \end{cases} \tag{4.3.2}$$

Additional important properties are derivable from the preceding basic ones. First, $B(t)$ is continuous in L_2, since its correlation function is continuous along the diagonal $t_1 = t_2$. Furthermore,

$$E\left[\dot{B}(t_1), \dot{B}(t_2)\right] = \frac{\partial^2}{\partial t_1 \partial t_2} E[B(t_1), B(t_2)] = \sigma^2 \delta(t_2 - t_1) \tag{4.3.3}$$

which is unbounded along $t_1 = t_2$. Thus $B(t)$ is not differentiable in L_2. Letting $t_1 < t_2 \leq t_3 < t_4$, it follows from (4.3.2) that

$$E\{[B(t_2) - B(t_1)][B(t_4) - B(t_3)]\} = \sigma^2(t_2 - t_2 - t_1 + t_1) = 0 \tag{4.3.4}$$

Since $B(t)$ has a zero mean,

$$E[B(t_2) - B(t_1)] = E[B(t_4) - B(t_3)] = 0 \tag{4.3.5}$$

Equations (4.3.4) and (4.3.5) show that $B(t)$ is a process with orthogonal increments. Being a gaussian process as well, the increments of $B(t)$ in nonoverlapping time intervals must, in fact, be independent. We then conclude that $B(t)$ satisfies a sufficient condition for a Markov process.

Now write

$$B(t) = \int_0^t dB(\tau) \tag{4.3.6}$$

$$B(t_1)B(t_2) = \int_0^{t_1} dB(\tau_1) \int_0^{t_2} dB(\tau_2) \tag{4.3.7}$$

For $E[B(t)]$ to vanish for any t, we must have $E[dB(\tau)] = dE[B(\tau)] = 0$ for any τ. Taking ensemble averages on the two sides of (4.3.7), we obtain

$$\sigma^2 \min(t_1, t_2) = \int_0^{\min(t_1, t_2)} d\Psi(\tau) \tag{4.3.8}$$

where use has been made of the following property of an orthogonal-increment process discussed in Chapter 2:

$$E[dB(\tau_1)dB(\tau_2)] = \begin{cases} 0 & \tau_1 \neq \tau_2 \\ d\Psi(\tau) & \tau_1 = \tau_2 = \tau \end{cases} \tag{4.3.9}$$

Equation (4.3.8) can be satisfied only if

$$d\Psi(\tau) = \sigma^2 d\tau \tag{4.3.10}$$

The preceding results can now be used to compute the first and second derivate moments of the Markov process $B(t)$. From

$$B(t + \Delta t) - B(t) = \int_t^{t+\Delta t} dB(\tau) \tag{4.3.11}$$

we obtain

$$E\left[B(t + \Delta t) - B(t)\right] = 0 \tag{4.3.12}$$

and

$$E\{[B(t + \Delta t) - B(t)]^2\} = \int_t^{t+\Delta t} \sigma^2 d\tau = \sigma^2 \Delta t \tag{4.3.13}$$

where use has been made of (4.3.9) and (4.3.10). The value of $B(\tau)$ at $\tau = t$ has no effect on (4.3.12) and (4.3.13), on account of the independent increments and $E[dB(\tau)] = 0$. As seen from (4.3.12) and (4.3.13), the first derivate moment a of the Wiener process vanishes, and the second derivate moment $b = \sigma^2$. Since $B(t)$ is gaussian, the higher derivate moments are all zero. Therefore, it satisfies a sufficient condition for a diffusive Markov process, and its sample functions are continuous with probability 1. The Fokker-Planck equation for the transition probability q of $B(t)$ is then

$$\frac{\partial q}{\partial t} - \frac{1}{2}\sigma^2 \frac{\partial^2 q}{\partial z^2} = 0 \tag{4.3.14}$$

Under the initial condition

$$\lim_{t \to t_0} q(z, t \mid z_0, t_0) = \delta(z - z_0) \tag{4.3.15}$$

and the boundary condition $\partial q/\partial z = 0$ as $z \to \pm\infty$, the solution for (4.3.14) is given by

$$q(z, t \mid z_0, t_0) = \frac{1}{\sqrt{2\pi(t - t_0)}\,\sigma} \exp\left[-\frac{(z - z_0)^2}{2\sigma^2(t - t_0)}\right] \tag{4.3.16}$$

which is the probability density of a gaussian random variable, as expected. It has a mean z_0 and a standard deviation $\sigma(t - t_0)^{1/2}$.

The Wiener process possesses the well-known *Lévy's oscillation property* (Lévy, 1948), which is described next. Let $B(t)$ be a unit Wiener process ($\sigma = 1$) defined on the interval $[a, b]$, and let the interval be divided into n subintervals $a = t_0 < t_1 < \cdots < t_{n-1} < t_n = b$. Denote $\Delta t_j = t_j - t_{j-1}$ and $\Delta_n = \max_{1 \le j \le n} \Delta t_j$. Then

$$\underset{\substack{n \to \infty \\ \Delta_n \to 0}}{\text{l.i.m.}} \sum_{j=1}^{n} \left[B(t_j) - B(t_{j-1})\right]^2 = b - a \tag{4.3.17}$$

To prove (4.3.17), let

$$X_j = B(t_j) - B(t_{j-1}) \tag{4.3.18}$$

$$Y_j = X_j^2 - (t_j - t_{j-1}) = X_j^2 - \Delta t_j \tag{4.3.19}$$

$$S_n = \sum_{j=1}^{n} Y_j \tag{4.3.20}$$

From (4.3.13), (4.3.19), and (4.3.20), we have

$$E[S_n] = \sum_{j=1}^{n} \Delta t_j - (b - a) = 0 \tag{4.3.21}$$

Since the X_j are independent gaussian random variables,

$$E[Y_j Y_k] = E[(X_j^2 - \Delta t_j)(X_k^2 - \Delta t_k)] = 0 \qquad j \neq k \tag{4.3.22}$$

$$\begin{aligned} E[Y_j^2] &= E[X_j^4] - 2\Delta t_j E[X_j^2] + \Delta t_j^2 \\ &= 3\Delta t_j^2 - 2\Delta t_j^2 + \Delta t_j^2 = 2\Delta t_j^2 \end{aligned} \tag{4.3.23}$$

Consequently,

$$E[S_n^2] = E\left[\left(\sum_{j=1}^{n} Y_j\right)^2\right] = 2\sum_{j=1}^{n} \Delta t_j^2 \leq 2\Delta_n \sum_{j=1}^{n} \Delta t_j = 2\Delta_n(b - a) \tag{4.3.24}$$

As $n \to \infty$ and $\Delta_n \to 0$, $E[S_n^2]$ also tends to zero, thus proving the validity of the L_2 limit (4.3.17).

It is interesting to note that Lévy's oscillation property is also valid almost surely; that is,

$$\lim_{\substack{n \to \infty \\ \Delta_n \to 0}} \sum_{j=1}^{n} [B(t_j) - B(t_{j-1})]^2 = b - a \tag{4.3.25}$$

with probability 1 (see, e.g., Karlin and Taylor, 1975). This suggests that

$$dB(t_1)dB(t_2) = \begin{cases} 0 & t_1 \neq t_2 \\ dt & t_1 = t_2 = t \end{cases} \tag{4.3.26}$$

which is a much stronger statement than (4.3.9).

Now for every n,

$$\sum_{j=1}^{n} [B(t_j) - B(t_{j-1})]^2 \leq \left[\max_{l \leq k \leq n} \left|B(t_k) - B(t_{k-1})\right|\right] \sum_{j=1}^{n} \left|B(t_j) - B(t_{j-1})\right| \tag{4.3.27}$$

The left-hand side of (4.3.27) tends to $b - a$, whereas $\max_{1 \leq k \leq n} \left|B(t_k) - B(t_{k-1})\right|$ tends to zero, since $B(t)$ is continuous with probability 1. We conclude that

$$\sum_{j=1}^{n} \left|B(t_j) - B(t_{j-1})\right| \to \infty \tag{4.3.28}$$

as $n \to \infty$ and $\Delta_n \to 0$. Therefore, $B(t)$ is of unbounded variation within any finite time interval $[a, b]$. This indicates that the Wiener process is a mathematical idealization. A physical process can be close to a Wiener process, but it can never be exactly a Wiener process.

In certain applications, the asymptotic behaviors of a Wiener process $B(t)$ are of interest, in particular, its behaviors when t is large and when t is small. These can be investigated using the *law of iterative logarithms* (see, e.g., Feller, 1957) as stated below.

Let $X_1, X_2, \ldots, X_n$ be independent and identically distributed random variables, with $E[X_j] = \mu$ and $\text{var}[X_j] = \sigma^2$. Let

$$Z_n = X_1 + X_2 + \cdots + X_n \tag{4.3.29}$$

Then, with probability 1,

$$\limsup_{n\to\infty} \frac{Z_n - n\mu}{\sqrt{2n\ln(\ln n)}} = \sigma \tag{4.3.30}$$

Since (4.3.30) is also valid if we replace X_j by $-X_j$, we also have

$$\liminf_{n\to\infty} \frac{Z_n - n\mu}{\sqrt{2n\ln(\ln n)}} = -\sigma \tag{4.3.31}$$

We now apply (4.3.30) and (4.3.31) to the case of

$$X_j = B(t_j) - B(t_{j-1}) \tag{4.3.18}$$

where $B(t)$ is a unit Wiener process, and $t_j - t_{j-1} = 1$. We have

$$Z_n = B(t_n) - B(0) = B(t_n) \tag{4.3.32}$$

Therefore,

$$\limsup_{n\to\infty} \frac{B(t_n)}{\sqrt{2n\ln(\ln n)}} = 1 \tag{4.3.33}$$

$$\liminf_{n\to\infty} \frac{B(t_n)}{\sqrt{2n\ln(\ln n)}} = -1 \tag{4.3.34}$$

or

$$\limsup_{t\to\infty} \frac{B(t)}{\sqrt{2t\ln(\ln t)}} = 1 \tag{4.3.33a}$$

$$\liminf_{t\to\infty} \frac{B(t)}{\sqrt{2t\ln(\ln t)}} = -1 \tag{4.3.34a}$$

Equations (4.3.33a) and (4.3.34a) show that $B(t)$ grows as $\sqrt{t\ln(\ln t)}$, which is much slower than t.

On the other hand, by letting

$$X_j = jB\left(\frac{1}{j}\right) - (j-1)B\left(\frac{1}{j-1}\right) \tag{4.3.35}$$

it can be shown that $E[X_j] = 0$, $E[X_j^2] = 1$, and $E[X_jX_k] = 0$ if $j \neq k$. In this case,

$$Z_n = nB\left(\frac{1}{n}\right) \tag{4.3.36}$$

and
$$\limsup_{n\to\infty} \frac{nB(1/n)}{\sqrt{2n\ln(\ln n)}} = 1 \tag{4.3.37}$$

Equation (4.3.37) may be rewritten as

$$\limsup_{n\to\infty} \frac{B(1/n)}{\sqrt{2(1/n)\ln\left|\ln(1/n)\right|}} = 1 \tag{4.3.38}$$

or
$$\limsup_{t\to 0^+} \frac{B(t)}{\sqrt{2(t)\ln\left|\ln t\right|}} = 1 \tag{4.3.39}$$

The validity of (4.3.39) also ensures that

$$\liminf_{t\to 0^+} \frac{B(t)}{\sqrt{2(t)\ln\left|\ln t\right|}} = -1 \tag{4.3.40}$$

4.4 THE ITÔ STOCHASTIC DIFFERENTIAL EQUATIONS

In addition to providing a simple example for a Markov process, the Wiener process can be used as a building block to construct other Markov processes. This is accomplished through the Itô stochastic differential equation. According to Itô (1951a), an arbitrary scalar Markov process $X(t)$ may be generated from the stochastic differential equation

$$dX(t) = m(X,t)\,dt + \sigma(X,t)\,dB(t) \tag{4.4.1}$$

where m and σ are called the *drift* and *diffusion coefficients*, respectively, and where $B(t)$ is a unit Wiener process, namely,

$$E\left[B(t_1)B(t_2)\right] = \min(t_1, t_2) \tag{4.4.2}$$

or
$$E\left[dB(t_1)\,dB(t_2)\right] = \begin{cases} 0 & t_1 \neq t_2 \\ dt & t_1 = t_2 = t \end{cases} \tag{4.4.2a}$$

In general, m and σ in (4.4.1) depend on $X(t)$, but they may also depend explicitly on t. Now, equation (4.4.1) is equivalent to

$$X(t) = X(0) + \int_0^t m\left[X(u), u\right] du + \int_0^t \sigma\left[X(u), u\right] dB(u) \tag{4.4.3}$$

The second integral in (4.4.3) is a Stieltjes integral. However, since $B(t)$ is a very unusual stochastic process, being continuous but of unbounded variation within any finite interval, this Stieltjes integral must be interpreted properly. Itô proposed that it be interpreted as a forward L_2 integral:

$$\int_0^t \sigma[X(u), u]\,dB(u) = \underset{\substack{n\to\infty \\ \max\Delta u\to 0}}{\text{l.i.m.}} \sum_{j=1}^{n} \sigma[X(u_j), u_j]\,[B(u_{j+1}) - B(u_j)] \tag{4.4.4}$$

Equation (4.4.4) is known as the Itô integral. Note that in (4.4.4) every difference $B(u_{j+1}) - B(u_j)$ is taken in a forward time interval, following the time instant at which σ is evaluated. Itô's stochastic differential equation implies that (4.4.1) is so interpreted, thus ensuring that $dB(t)$ is independent of $X(t)$ in (4.4.1).

To obtain the first and the second derivate moments, write

$$X(t+\Delta t) - X(t) = \int_t^{t+\Delta t} m\,[X(u), u]\,du + \int_t^{t+\Delta t} \sigma\,[X(u), u]\,dB(u) \quad (4.4.5)$$

If Δt is sufficiently small, and if m and σ are sufficiently smooth, (4.4.5) may be approximated by

$$X(t+\Delta t) - X(t) \approx m\,[X(t), t]\,\Delta t + \sigma\,[X(t), t]\,[B(t+\Delta t) - B(t)] \quad (4.4.6)$$

By using (4.3.12) and (4.3.13), it can be shown that

$$a(x,t) = \lim_{\Delta t \to 0} \frac{1}{\Delta t} E\,[X(t+\Delta t) - X(t) \mid X(t) = x] = m(x,t) \quad (4.4.7)$$

$$b(x,t) = \lim_{\Delta t \to 0} \frac{1}{\Delta t} E\{[X(t+\Delta t) - X(t)]^2 \mid X(t) = x\} = \sigma^2(x,t) \quad (4.4.8)$$

Hence there exist direct correspondences between terms in the Itô stochastic differential equation and the Fokker-Planck equation. Specifically,

$$a(x,t) = [m(X,t)]_{X=x} \quad (4.4.9)$$

$$b(x,t) = [\sigma^2(X,t)]_{X=x} \quad (4.4.10)$$

Conceptually, however, the drift and diffusion coefficients in the Itô equation are functions of the random process $X(t)$, whereas the first and second derivate moments in the corresponding Fokker-Planck equation are functions of the state variable x, which is a possible deterministic value of $X(t)$. By replacing $X(t)$ by its possible value x, we can obtain the first and second derivate moments in the corresponding Fokker-Planck equation, in accordance with (4.4.9) and (4.4.10). At times, we may use the symbols $m(x,t)$ and $\sigma^2(x,t)$ for $a(x,t)$ and $b(x,t)$, respectively, if confusion does not arise.

It is of interest to note that the first term on the right-hand side of equation (4.4.1) is of order dt, while the second term is of order $(dt)^{1/2}$, since $E[|dB(t)|^2] = dt$. Both terms must be retained to describe the differential $dX(t)$ of a Markov process.

Analogously, an arbitrary n-dimensional Markov vector process may be generated from the Itô stochastic differential equations

$$dX_j(t) = m_j(\mathbf{X},t)dt + \sigma_{jk}(\mathbf{X},t)dB_k(t) \qquad j = 1, 2, \ldots, n \quad (4.4.11)$$

where m_j are the drift coefficients, σ_{jk} are the diffusion coefficients, and $B_k(t)$ are mutually independent unit Wiener processes. The first and second derivate moments of the corresponding Fokker-Planck equation can be obtained from

$$a_j(\mathbf{x},t) = [m_j(\mathbf{X},t)]_{\mathbf{X}=\mathbf{x}} \quad (4.4.12)$$

$$b_{jk}(\mathbf{x},t) = [\sigma_{jl}(\mathbf{X},t)\sigma_{kl}(\mathbf{X},t)]_{\mathbf{X}=\mathbf{x}} \quad (4.4.13)$$

Equation (4.4.13) can also be expressed in a matrix form:

$$\mathbf{b}(\mathbf{x},t) = [\boldsymbol{\sigma}(\mathbf{X},t)\boldsymbol{\sigma}'(\mathbf{X},t)]_{\mathbf{X}=\mathbf{x}} \quad (4.4.14)$$

where $\boldsymbol{\sigma}'$ denotes the transpose of matrix $\boldsymbol{\sigma}$. At times, we may use the symbols $m_j(\boldsymbol{x}, t)$ and $\boldsymbol{\sigma}(\boldsymbol{x}, t)\boldsymbol{\sigma}'(\boldsymbol{x}, t)$ for $a_j(\boldsymbol{x}, t)$ and $\boldsymbol{b}(\boldsymbol{x}, t)$, respectively, if confusion does not arise.

One distinctive advantage of using the Itô stochastic differential equations to describe a Markov process lies in the fact that the Itô equation for an arbitrary function of this Markov process can be derived quite simply and without ambiguity. Since the two terms on the right-hand side of (4.4.1), or more generally (4.4.11), are of different orders of magnitude, the usual chain rule of differentiation is not applicable. Itô (1951b) has shown that the differential of an arbitrary scalar function $F(\boldsymbol{X}, t)$ of a Markov vector $\boldsymbol{X}(t)$ and t is given by

$$dF(\boldsymbol{X}, t) = \left(\frac{\partial F}{\partial t} + \mathscr{L}_X F\right)dt + \sigma_{jk}\frac{\partial F}{\partial X_j}\, dB_k(t) \tag{4.4.15}$$

where $\mathscr{L}_X$ is known as the *generating differential operator* of the Markov process $\boldsymbol{X}(t)$, given by

$$\mathscr{L}_X = m_j \frac{\partial}{\partial X_j} + \frac{1}{2}\sigma_{jl}\sigma_{kl}\frac{\partial^2}{\partial X_j \partial X_k} \tag{4.4.16}$$

Equation (4.4.15) can be proved by expanding

$$dF(\boldsymbol{X}, t) = \frac{\partial F}{\partial t}dt + \frac{\partial F}{\partial X_j}dX_j + \frac{1}{2}\frac{\partial^2 F}{\partial X_j \partial X_k}\, dX_j\, dX_k + \cdots \tag{4.4.17}$$

using (4.4.11) and (4.3.26) and keeping terms of the orders dt and $dB_k(t)$. Obviously, (4.4.15) is valid only if $F(\boldsymbol{X}, t)$ is differentiable with respect to t, and twice differentiable with respect to the components of $\boldsymbol{X}(t)$. Equation (4.4.15) is known as *Itô's differential rule*, or *Itô's lemma*.

4.5 THE ONE-DIMENSIONAL DIFFUSION PROCESS

The simple case of the one-dimensional diffusion process is worthy of special attention for at least two reasons: (1) many physical phenomena in science and engineering are essentially one-dimensional, or reducible to one dimension; and (2) a careful examination of the one-dimensional case will provide useful insights which are needed to treat the more complicated cases of higher dimension.

Consider a one-dimensional diffusion process $X(t)$ homogeneous in time; that is, its drift and diffusion coefficients may depend on $X(t)$, but they do not depend explicitly on t. Let $X(t)$ be defined on an interval with a left boundary x_l and a right boundary x_r, and let it be governed by the Itô equation

$$dX(t) = m(X)\, dt + \sigma(X)\, dB(t) \tag{4.5.1}$$

Its transition probability density $q(x, t\,|\,x_0, t_0)$ is governed by the Fokker-Planck equation

$$\frac{\partial q}{\partial t} + \frac{\partial}{\partial x}[a(x)q] - \frac{1}{2}\frac{\partial^2}{\partial x^2}[b(x)q] = 0 \tag{4.5.2}$$

where $a(x)$ and $b(x)$ are related to the drift and diffusion coefficients as in (4.4.9) and (4.4.10). Exact time-dependent solutions for (4.5.2) are known for linear systems and some simple nonlinear systems under additive gaussian white noise excitations (e.g., Caughey, 1971; Caughey and Dienes, 1961). However, the time-independent stationary probability density $p(x)$ of $X(t)$, if it exists, is easy to obtain. This stationary probability density satisfies the reduced Fokker-Planck equation

$$\frac{dG}{dx} = \frac{d}{dx}\left\{a(x)p - \frac{1}{2}\frac{d}{dx}[b(x)p]\right\} = 0 \tag{4.5.3}$$

For the time being, we are concerned only with solving equation (4.5.3).

Integrating (4.5.3) once, we obtain

$$G(x) = a(x)p - \frac{1}{2}\frac{d}{dx}[b(x)p] = \text{constant} = G_c \tag{4.5.4}$$

which indicates that the probability flow must be the same everywhere. The general solution for equation (4.5.4) may be expressed as

$$p(x) = \frac{\psi(x)}{b(x)}\left[C - 2G_c\int_{x_l}^{x}\psi^{-1}(u)\,du\right] \tag{4.5.5}$$

where C is the normalization constant, and

$$\psi(x) = \exp\left[\int \frac{2a(x)}{b(x)}\,dx\right] \tag{4.5.6}$$

We now examine the following two possibilities:

Nonzero probability flow. The case of nonzero probability flow is found most commonly when $a(x)$ and $b(x)$ are periodic, and the state space $[x_l, x_r)$ is, in fact, a closed loop. Then

$$p(x_l) = p(x_r) \tag{4.5.7}$$

$$G(x_l) = G(x_r) = G_c \tag{4.5.8}$$

The constant probability flow G_c can be obtained by substituting (4.5.5) into (4.5.7) to yield

$$G_c = \frac{C}{2}\left[1 - \frac{\psi(x_l)}{\psi(x_r)}\right]\left[\int_{x_l}^{x_r}\psi^{-1}(u)\,du\right]^{-1} \tag{4.5.9}$$

Substituting (4.5.9) into (4.5.5),

$$p(x) = C\frac{\psi(x)}{b(x)}\left\{1 - \left[1 - \frac{\psi(x_l)}{\psi(x_r)}\right]\frac{\displaystyle\int_{x_l}^{x}\psi^{-1}(u)\,du}{\displaystyle\int_{x_l}^{x_r}\psi^{-1}(u)\,du}\right\} \tag{4.5.10}$$

Zero probability flow. If the boundary conditions are such that the probability flow vanishes at both x_l and x_r, then the probability flow must vanish everywhere in the one-dimensional space. If we let $G_c = 0$, solution (4.5.5) is reduced to

$$p(x) = C\frac{\psi(x)}{b(x)} \tag{4.5.11}$$

The case of zero probability flow is quite common if the state space is one-dimensional but uncommon if the state space is multidimensional. Nevertheless, Stratonovich (1963) has obtained stationary solutions for a class of problems which he called "the case of *stationary potential,*" based on the assumption of zero probability flow in all directions. This case and its generalized version are discussed in detail in Chapter 5.

CLASSIFICATION OF BOUNDARIES. According to Feller (1952, 1954), the behavior of a diffusion process at and near a boundary may be classified as follows:

Regular boundary. The process can either reach the boundary from an interior point or reach an interior point from the boundary.

Exit boundary. The process can reach the boundary from an interior point but cannot reach an interior point from the boundary.

Entrance boundary. The process can reach an interior point from the boundary but cannot reach the boundary from an interior point.

Natural boundary. The process cannot reach the boundary from an interior point, nor can it reach an interior point from the boundary.

For the purpose of classifying a boundary, we introduce (Itô and McKean, 1965) a *scale function* $l(x)$ defined as

$$l(x) = \int_{x_0}^{x} \psi^{-1}(u)\,du \tag{4.5.12}$$

and a *speed function* $v(x)$ defined as

$$v(x) = \int_{x_0}^{x} \frac{\psi(u)}{b(u)}\,du \tag{4.5.13}$$

where x_0 is an interior point, that is, $x_0 \in (x_l, x_r)$, and where $\psi(u)$ has been defined previously in (4.5.6). These functions may be used to construct the following two canonical measures:

$$\Sigma(x) = \int_{x_0}^{x} v(u)\,dl(u) \tag{4.5.14}$$

$$N(x) = \int_{x_0}^{x} l(u)\,dv(u) \tag{4.5.15}$$

Physically, $\Sigma(x)$ is a measure of time to reach a point x, starting from an interior point x_0, and $N(x)$ is a measure of time to reach an interior point x_0, starting from a point x. The following relations hold among $l(x)$, $v(x)$, $\Sigma(x)$, and $N(x)$ (Karlin and Taylor, 1981):

1. $l(x) = \infty$ implies $\Sigma(x) = \infty$.
2. $\Sigma(x) < \infty$ implies $l(x) < \infty$.
3. $\nu(x) = \infty$ implies $N(x) = \infty$.
4. $N(x) < \infty$ implies $\nu(x) < \infty$.
5. $\Sigma(x) + N(x) = l(x)\nu(x)$.

The classification of a boundary x_b can now be identified according to these measures, as shown in Table 4.5.1. This table differs from the one given by Karlin and Taylor (1981) in that the class of unattractively natural boundaries is split in two: the class of repulsively natural boundaries and the class of strictly natural boundaries. The additional refinement is useful, as shown in later applications.

Some conclusions can be made readily from the boundary classifications. For example, a nontrivial stationary solution exists within the interval on which the Markov process is defined if each of the two boundaries is either an entrance or repulsively natural. A trivial stationary solution exists and its probability density is a unit delta function at an exit boundary if the other boundary is either an entrance or repulsively natural. A stationary solution does not exist if each of the two boundaries is an exit, or attractively natural, or strictly natural. The presence of a regular boundary renders the solution nonunique without further stipulation on the behaviors of the sample functions after reaching the boundary (Feller, 1952); however, this additional freedom is, in fact, a blessing, since it permits certain assumptions to be incorporated in the context of a physical problem.

SINGULAR BOUNDARIES. Now consider the case in which a boundary x_s is either singular of the first kind—$b(x_s) = 0$, or singular of the second kind—$a(x_s)$ is unbounded. As to be shown in Chapter 6, singular boundaries are quite common, especially when multiplicative random excitations are present. For a singular

TABLE 4.5.1
Classification of a boundary†

Criteria‡				Classifications	
$l(x_b)$	$\nu(x_b)$	$\Sigma(x_b)$	$N(x_b)$		
$< \infty^*$	$< \infty^*$	$< \infty$	$< \infty$	Regular	Accessible
$< \infty$	$= \infty^*$	$< \infty^*$	$= \infty$	Exit	Accessible
$< \infty^*$	$= \infty^*$	$= \infty^*$	$= \infty$	Attractively natural	Inaccessible
$= \infty^*$	$< \infty^*$	$= \infty$	$= \infty^*$	Repulsively natural	Inaccessible
$= \infty^*$	$= \infty^*$	$= \infty$	$= \infty$	Strictly natural	Inaccessible
$= \infty^*$	$< \infty$	$= \infty$	$< \infty^*$	Entrance	Inaccessible

† Modified from the original table of Karlin and Taylor (1981).

‡ The asterisks indicate the minimal sufficient conditions for each type of boundary. For example, the minimal sufficient conditions for a regular boundary are $l(x_b) < \infty$ and $\nu(x_b) < \infty$.

boundary, the integrations shown in (4.5.12) through (4.5.15) for the scale function l, speed function v, and time measures Σ and N are generally difficult (Kozin and Prodromou, 1971). However, only the integrabilities of these integrals are needed for boundary classifications. In the case of a singular boundary of the first kind, Zhang and Kozin (1990) have obtained a set of criteria based on the limiting behaviors of the drift and diffusion coefficients near the boundary. These and additional criteria pertaining to a singular boundary of the second kind are given below.

Singular boundary of the first kind. Consider a singular boundary x_s of the first kind, $\sigma(x_s) = 0$. The boundary is said to be a *shunt* if $m(x_s) \neq 0$, and a *trap* if $m(x_s) = 0$. We introduce the following definitions:

1. α_s is the *diffusion exponent* of x_s if

$$\sigma^2(x) = O\left|x - x_s\right|^{\alpha_s} \qquad \alpha_s \geq 0, \quad \text{as } x \to x_s \tag{4.5.16}$$

2. β_s is the *drift exponent* of x_s if

$$m(x) = O\left|x - x_s\right|^{\beta_s} \qquad \beta_s \geq 0, \quad \text{as } x \to x_s \tag{4.5.17}$$

3. c_s is the *character value* of x_s, given by

$$c_l = \lim_{x \to x_l^+} \frac{2m(x)(x - x_l)^{\alpha_l - \beta_l}}{\sigma^2(x)} \tag{4.5.18}$$

$$c_r = -\lim_{x \to x_r^-} \frac{2m(x)(x_r - x)^{\alpha_r - \beta_r}}{\sigma^2(x)} \tag{4.5.19}$$

In (4.5.16) and (4.5.17), $O|\cdot|$ denotes the order of magnitude of $|\cdot|$, and in (4.5.18) and (4.5.19) the subscripts l and r indicate a left boundary and a right boundary, respectively. The classification of a singular boundary x_s of the first kind can be identified from the values of α_s, β_s, and c_s, as summarized in Table 4.5.2.

Singular boundary of the second kind. If $m(x_s)$ is unbounded and $|x_s| < \infty$, we define the following:

1. α_s is the *diffusion exponent* of x_s if

$$\sigma^2(x) = O\left|x - x_s\right|^{-\alpha_s} \qquad \alpha_s \geq 0, \quad \text{as } x \to x_s \tag{4.5.20}$$

2. β_s is the *drift exponent* of x_s if

$$m(x) = O\left|x - x_s\right|^{-\beta_s} \qquad \beta_s \geq 0, \quad \text{as } x \to x_s \tag{4.5.21}$$

3. c_s is the *character value* of x_s, given by

$$c_l = \lim_{x \to x_l^+} \frac{2m(x)(x - x_l)^{\beta_l - \alpha_l}}{\sigma^2(x)} \tag{4.5.22}$$

$$c_r = -\lim_{x \to x_r^-} \frac{2m(x)(x_r - x)^{\beta_r - \alpha_r}}{\sigma^2(x)} \tag{4.5.23}$$

TABLE 4.5.2

Classification of singular boundary of the first kind[†]

<table>
<tr><th>State</th><th colspan="4">Conditions</th><th>Class</th></tr>
<tr><td rowspan="6">$\sigma(x_s) = 0$
$(\alpha_s > 0)$

$m(x_s) \neq 0$
$(\beta_s = 0)$

(shunt)</td><td>$\alpha_s < 1$</td><td></td><td colspan="2"></td><td>Regular</td></tr>
<tr><td rowspan="3">$\alpha_s = 1$</td><td>$m(x_l) < 0$
or $m(x_r) > 0$</td><td colspan="2"></td><td>Exit</td></tr>
<tr><td rowspan="2">$m(x_l) > 0$
or $m(x_r) < 0$</td><td colspan="2">$0 < c_s < 1$</td><td>Regular</td></tr>
<tr><td colspan="2">$c_s \geq 1$</td><td>Entrance</td></tr>
<tr><td rowspan="2">$\alpha_s > 1$</td><td>$m(x_l) < 0$
or $m(x_r) > 0$</td><td colspan="2"></td><td>Exit</td></tr>
<tr><td>$m(x_l) > 0$
or $m(x_r) < 0$</td><td colspan="2"></td><td>Entrance</td></tr>
<tr><td rowspan="13">$\sigma(x_s) = 0$
$(\alpha_s > 0)$

$m(x_s) = 0$
$(\beta_s > 0)$

(trap)</td><td rowspan="3">$\alpha_s < 1 + \beta_s$</td><td>$\alpha_s < 1$</td><td colspan="2"></td><td>Regular</td></tr>
<tr><td>$1 \leq \alpha_s < 2$</td><td colspan="2"></td><td>Exit</td></tr>
<tr><td>$\alpha_s \geq 2$</td><td colspan="2"></td><td>Attractively natural</td></tr>
<tr><td rowspan="4">$\alpha_s > 1 + \beta_s$</td><td rowspan="2">$\beta_s < 1$</td><td colspan="2">$m(x_l^+) < 0$
or $m(x_r^-) > 0$</td><td>Exit</td></tr>
<tr><td colspan="2">$m(x_l^+) > 0$
or $m(x_r^-) < 0$</td><td>Entrance</td></tr>
<tr><td rowspan="2">$\beta_s \geq 1$</td><td colspan="2">$m(x_l^+) < 0$
or $m(x_r^-) > 0$</td><td>Attractively natural</td></tr>
<tr><td colspan="2">$m(x_l^+) > 0$
or $m(x_r^-) < 0$</td><td>Repulsively natural</td></tr>
<tr><td rowspan="6">$\alpha_s = 1 + \beta_s$</td><td rowspan="3">$\beta_s < 1$</td><td rowspan="2">$c_s > \beta_s$</td><td>$c_s \geq 1$</td><td>Entrance</td></tr>
<tr><td>$c_s < 1$</td><td>Regular</td></tr>
<tr><td>$c_s \leq \beta_s$</td><td></td><td>Exit</td></tr>
<tr><td rowspan="3">$\beta_s \geq 1$</td><td>$c_s > \beta_s$</td><td></td><td>Repulsively natural</td></tr>
<tr><td rowspan="2">$c_s \leq \beta_s$</td><td>$c_s \geq 1$</td><td>Strictly natural</td></tr>
<tr><td>$c_s < 1$</td><td>Attractively natural</td></tr>
</table>

[†] Modified from the original table of Zhang and Kozin (1990).

TABLE 4.5.3
Classification of singular boundary of the second kind ($\lvert x_s \rvert < \infty$)

State	Conditions			Class
$\lvert m(x_s)\rvert = \infty$ ($\beta_s > 0$) $\sigma(x_s) < \infty$ ($\alpha_s = 0$)	$\beta_s < 1$			Regular
	$\beta_s = 1$	$c_s \leq -1$		Exit
		$-1 < c_s < 1$		Regular
		$c_s \geq 1$		Entrance
	$\beta_s > 1$	$m(x_l^+) < 0$ or $m(x_r^-) > 0$		Exit
		$m(x_l) > 0$ or $m(x_r) < 0$		Entrance
$\lvert m(x_s)\rvert = \infty$ ($\beta_s > 0$) $\sigma(x_s) = \infty$ ($\alpha_s > 0$)	$\beta_s < 1 + \alpha_s$			Regular
	$\beta_s > 1 + \alpha_s$	$m(x_l^+) < 0$ or $m(x_r^-) > 0$		Exit
		$m(x_l^+) > 0$ or $m(x_r^-) < 0$		Entrance
	$\beta_s = 1 + \alpha_s$	$c_s \geq -\beta_s$	$c_s \geq 1$	Entrance
			$c_s < 1$	Regular
		$c_s < -\beta_s$		Exit

The classification of a singular boundary x_s of the second kind, $\lvert x_s \rvert < \infty$, is summarized in Table 4.5.3.

Singular boundary of the second kind at infinity. If $m(x_s)$ is unbounded and $\lvert x_s \rvert = \infty$, we define the following:

1. α_s is the *diffusion exponent* of x_s if

$$\sigma^2(x) = O\lvert x\rvert^{\alpha_s} \qquad \alpha_s \geq 0, \quad \text{as } \lvert x\rvert \to \infty \tag{4.5.24}$$

2. β_s is the *drift exponent* of x_s if

$$m(x) = O\lvert x\rvert^{\beta_s} \qquad \beta_s \geq 0, \quad \text{as } \lvert x\rvert \to \infty \tag{4.5.25}$$

3. c_s is the *character value* of x_s, given by

$$c_l = \lim_{x\to-\infty} \frac{2m(x)\lvert x\rvert^{\alpha_l-\beta_l}}{\sigma^2(x)} \tag{4.5.26}$$

$$c_r = -\lim_{x\to\infty} \frac{2m(x)\lvert x\rvert^{\alpha_r-\beta_r}}{\sigma^2(x)} \tag{4.5.27}$$

The classification of a singular boundary x_s of the second kind, $|x_s| = \infty$, is summarized in Table 4.5.4.

In the case of a Markov vector, a singular boundary is generally a line in a two-dimensional state space, and a surface or hypersurface in a higher dimensional state space. Such singular boundaries have not been fully investigated, although their analogy to the one-dimensional case is obvious.

Identification of singular boundaries is now illustrated in an example.

4.5.1 An Example

Consider a one-dimensional diffusion process $X(t)$, defined on $[0, \infty)$, and governed by the Itô equation

$$dX = m(X)\,dt + \sigma(X)\,dB(t) \tag{a}$$

where $B(t)$ is a unit Wiener process and

$$m(X) = (-2\zeta\omega_0 + \pi\omega_0^2 K_1)X + \pi K_2 \tag{b}$$

$$\sigma(X) = (\pi\omega_0^2 K_1 X^2 + 2\pi K_2 X)^{1/2} \tag{c}$$

with $K_1 > 0$ and $K_2 \geq 0$. It can be shown (Section 4.7) that process $X(t)$ is, in fact, the sum of the potential and kinetic energies of a linear oscillator under both additive and multiplicative gaussian white noise excitations. The right boundary at infinity, where the drift coefficient is unbounded, is clearly singular of the second kind. At large X values,

$$m(X) \rightarrow (-2\zeta\omega_0 + \pi\omega_0^2 K_1)X \tag{d}$$

$$\sigma(X) \rightarrow (\pi\omega_0^2 K_1)^{1/2} X \tag{e}$$

The diffusion exponent, drift exponent, and character value for the right boundary at infinity are obtained from (4.5.24), (4.5.25), and (4.5.27) as follows:

$$\alpha_r = 2 \qquad \beta_r = 1 \qquad c_r = \frac{4\zeta}{\pi\omega_0 K_1} - 2 \tag{f}$$

From Table 4.5.4, we conclude that

$$\text{the right boundary at } \infty \text{ is } \begin{cases} \text{repulsively natural} & \zeta > \frac{1}{4}\pi\omega_0 K_1 \\ \text{strictly natural} & \zeta = \frac{1}{4}\pi\omega_0 K_1 \\ \text{attractively natural} & \zeta < \frac{1}{4}\pi\omega_0 K_1 \end{cases}$$

The left boundary $x = 0$ is singular of the first kind since $\sigma(0) = 0$. Two cases, $K_2 > 0$ and $K_2 = 0$, must be considered separately. If $K_2 > 0$, then the left boundary $x = 0$ is a shunt, for which the diffusion exponent and the character value are obtained from (4.5.16) and (4.5.18) as

$$\alpha_l = 1 \qquad c_l = 1 \tag{g}$$

Since $m(0) = \pi K_2 > 0$, the left boundary $x = 0$ is an entrance according to Table 4.5.2.

TABLE 4.5.4
Classification of singular boundary of the second kind at infinity

State	Conditions				Class
$\lvert m(\infty)\rvert = \infty$ $(\beta_s > 0)$ $\sigma(\infty) < \infty$ $(\alpha_s = 0)$	$m(-\infty) < 0$ or $m(+\infty) > 0$	$\beta_s > 1$			Exit
		$\beta_s \le 1$			Attractively natural
	$m(-\infty) > 0$ or $m(+\infty) < 0$	$\beta_s > 1$			Entrance
		$\beta_s \le 1$			Repulsively natural
$\lvert m(\infty)\rvert = \infty$ $(\beta_s > 0)$ $\sigma(\infty) = \infty$ $(\alpha_s > 0)$	$\beta_s > \alpha_s - 1$	$m(-\infty) < 0$ or $m(+\infty) > 0$	$\beta_s > 1$		Exit
			$\beta_s \le 1$		Attractively natural
		$m(-\infty) > 0$ or $m(+\infty) < 0$	$\beta_s > 1$		Entrance
			$\beta_s \le 1$		Repulsively natural
	$\beta_s < \alpha_s - 1$				Regular
	$\beta_s = \alpha_s - 1$	$\beta \le 1$	$c_s > -\beta_s$		Repulsively natural
			$c_s \le -\beta_s$	$c_s \ge -1$	Strictly natural
				$c_s < -1$	Attractively natural
		$\beta_s > 1$	$c_s > -\beta_s$	$c_s \ge -1$	Entrance
				$c_s < -1$	Regular
			$c_s \le -\beta_s$		Exit

On the other hand, if $K_2 = 0$, then $x = 0$ is a trap. Its diffusion exponent, drift exponent, and character value are, respectively,

$$\alpha_l = 2 \qquad \beta_l = 1 \qquad c_l = -\frac{4\zeta}{\pi\omega_0 K_1} + 2 \tag{h}$$

According to Table 4.5.2,

$$\text{the left boundary at } x = 0 \text{ is} \begin{cases} \text{attractively natural} & \zeta > \frac{1}{4}\pi\omega_0 K_1 \\ \text{strictly natural} & \zeta = \frac{1}{4}\pi\omega_0 K_1 \\ \text{repulsively natural} & \zeta < \frac{1}{4}\pi\omega_0 K_1 \end{cases}$$

The foregoing results are summarized as follows:

1. $K_2 > 0, \zeta > \pi\omega_0 K_1/4$. The left boundary at $x = 0$ is an entrance and the right boundary at infinity is repulsively natural. In this case, a stationary probability density exists, and it follows from equation (4.5.11) that

$$p(x) = C\left(1 + \frac{\omega_0^2 K_1}{2K_2} x\right)^{-4\zeta/\pi\omega_0 K_1} \tag{i}$$

   Equation (i) is normalizable only if $\zeta > \pi\omega_0 K_1/4$, which is the same condition for the right boundary to be repulsively natural.
2. $K_2 > 0, \zeta = \pi\omega_0 K_1/4$. The left boundary at $x = 0$ is an entrance, and the right boundary at infinity is strictly natural. In this case, a stationary probability density does not exist, and the behavior of a sample function at very large value is indefinite.
3. $K_2 > 0, \zeta < \pi\omega_0 K_1/4$. The left boundary at $x = 0$ is an entrance, and the right boundary at infinity is attractively natural. In this case, a stationary probability density does not exist, and all sample functions will approach the right boundary at infinity eventually.
4. $K_2 = 0, \zeta > \pi\omega_0 K_1/4$. The left boundary at $x = 0$ is attractively natural, and the right boundary is repulsively natural. In this case, all sample functions will eventually reach and remain at $x = 0$, corresponding to a stationary probability density $p(x) = \delta(x)$.
5. $K_2 = 0, \zeta = \pi\omega_0 K_1/4$. Both boundaries are strictly natural; a stationary probability density does not exist.
6. $K_2 = 0, \zeta < \pi\omega_0 K_1/4$. The left boundary at $x = 0$ is repulsively natural, and the right boundary at infinity is attractively natural. All sample functions will approach the right boundary at infinity eventually, and a stationary probability density does not exist.

It is shown from the preceding analysis that the boundary condition at a singular boundary cannot be imposed arbitrarily, since it must be consistent with the sample behavior dictated by the limiting properties of the first and second derivate moments. On the other hand, the condition at a nonsingular or singular regular boundary can be prescribed according to the context of a physical problem, in much the same way as other boundary value problems. While these conclusions are based on the analysis of a scalar diffusion process, the case of a vector diffusion process should be similar.

4.6 APPROXIMATION OF A PHYSICAL PROCESS BY A MARKOV PROCESS

As indicated earlier, a Markov process is a mathematical idealization; its exact properties cannot be found in a physical process. Therefore, application of the Markov process theory to an engineering problem is an approximation, usually justified on

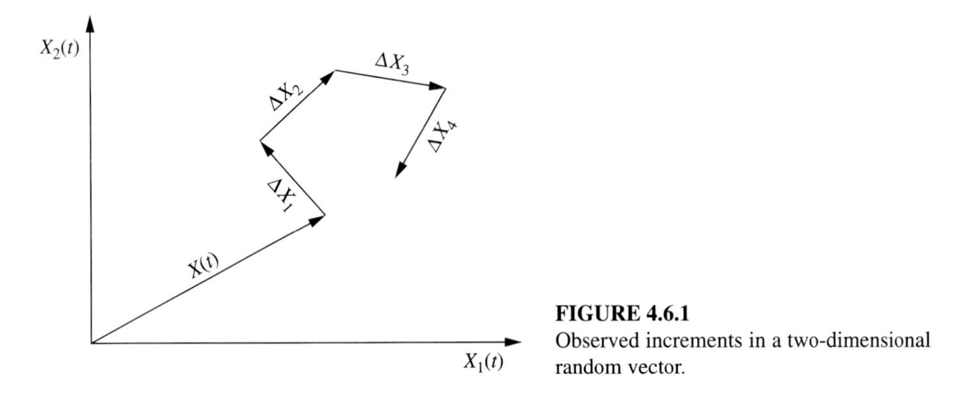

FIGURE 4.6.1
Observed increments in a two-dimensional random vector.

the basis of how close the increments in nonoverlapping time intervals are nearly independent stochastically.

The idea of independent increments is illustrated in Fig. 4.6.1, which shows a two-dimensional vector $\boldsymbol{X}(t)$ whose value changes randomly with time. Let us suppose that $\boldsymbol{X}(t)$ is not being observed continuously, but only at discrete time instants, beginning from t and then at the end of each consecutive time gap Δt_1, Δt_2, and so on. The observed increments are, respectively, $\Delta \boldsymbol{X}_1$, $\Delta \boldsymbol{X}_2$, and so on. The previously mentioned sufficient condition for Markov process requires that $\Delta \boldsymbol{X}_1, \Delta \boldsymbol{X}_2, \ldots$ be independent, regardless of how short $\Delta t_1, \Delta t_2, \ldots$ may be, even when they approach zero (i.e., when the observation becomes continuous). This ideal property can never be found in a real physical process.

On the other hand, if $\Delta t_1, \Delta t_2, \ldots$ are viewed as the time gaps when $\boldsymbol{X}(t)$ is not being observed, then the lengths of these time gaps can be chosen to suit a particular objective, subject, of course, to the loss of some details of the observed results. As long as there is randomness in a real physical process, it is always possible to select long enough time gaps so that the observed increments will appear independent. We shall take this point of view in the following discussion.

If a physical process is to be approximated by a Markov process, then the coefficients in the corresponding Fokker-Planck equation must be obtained from the original equations governing the physical process. This is equivalent to converting the physical equations to the Itô equations, in view of the relations (4.4.12) and (4.4.13). Let the physical equations be of the form

$$\frac{d}{dt}X_j(t) = f_j(\boldsymbol{X}, t) + g_{jr}(\boldsymbol{X}, t)\,\xi_r(t) \tag{4.6.1}$$

where f_j and g_{jr} are deterministic functional forms which can be nonlinear, and $\xi_r(t)$ are random excitations with zero means. To determine a_j and b_{jk} according to (4.2.11a), but regarding Δt as the time gap between two consecutive observations, we require an expression for the increment $X_j(t+\Delta t) - X_j(t)$. This is obtained from (4.6.1) as follows:

$$X_j(t+\Delta t) - X_j(t) = \int_t^{t+\Delta t} f_j(\boldsymbol{X}_u, u)\,du + \int_t^{t+\Delta t} g_{jr}(\boldsymbol{X}_u, u)\,\xi_r(u)\,du \tag{4.6.2}$$

where $\boldsymbol{X}_u$ is an abbreviation for $\boldsymbol{X}(u)$. However, since f_j and g_{jr} on the right-hand side depend on $\boldsymbol{X}_u$ where $t \leq u \leq t + \Delta t$, we have

$$f_j(\boldsymbol{X}_u, u) = f_j(\boldsymbol{X}_t, t) + (u - t)\frac{\partial}{\partial t} f_j(\boldsymbol{X}_t, t) + [X_l(u) - X_l(t)]\frac{\partial}{\partial X_l} f_j(\boldsymbol{X}_t, t) + \cdots \tag{4.6.3}$$

$$g_{jr}(\boldsymbol{X}_u, u) = g_{jr}(\boldsymbol{X}_t, t) + (u - t)\frac{\partial}{\partial t} g_{jr}(\boldsymbol{X}_t, t) + [X_l(u) - X_l(t)]\frac{\partial}{\partial X_l} g_{jr}(\boldsymbol{X}_t, t) + \cdots \tag{4.6.4}$$

In (4.6.3) and (4.6.4) we again substitute $X_l(u) - X_l(t)$ by

$$\int_t^u f_l(\boldsymbol{X}_v, v)\, dv + \int_t^u g_{ls}(\boldsymbol{X}_v, v)\xi_s(v)\, dv \tag{4.6.5}$$

Combining (4.6.2) through (4.6.5) and keeping only the leading terms,

$$\begin{aligned} X_j(t + \Delta t) - X_j(t) &= \int_t^{t+\Delta t} f_j(\boldsymbol{X}_t, t)\, du + \int_t^{t+\Delta t} g_{jr}(\boldsymbol{X}_t, t)\xi_r(u)\, du \\ &+ \int_t^{t+\Delta t} (u - t)\left[\frac{\partial}{\partial t} g_{jr}(\boldsymbol{X}_t, t)\right] \xi_r(u)\, du \\ &+ \int_t^{t+\Delta t} \xi_r(u)\, du \int_t^u \frac{\partial}{\partial X_l} g_{jr}(\boldsymbol{X}_t, t) g_{ls}(\boldsymbol{X}_v, v)\xi_s(v)\, dv + \cdots \end{aligned} \tag{4.6.6}$$

Substituting (4.6.6) into the first equation in (4.2.11a) and recalling that $E[\xi_r(t)] = 0$, we obtain

$$a_j(\boldsymbol{x}_t, t) = f_j(\boldsymbol{x}_t, t) + \frac{1}{\Delta t}\int_t^{t+\Delta t} du \int_t^u g_{ls}(\boldsymbol{x}_v, v)\frac{\partial}{\partial x_l} g_{jr}(\boldsymbol{x}_t, t) E\left[\xi_r(u)\xi_s(v)\right] dv + 0(\Delta t) \tag{4.6.7}$$

where $0(\Delta t)$ indicates the remaining terms of order Δt. In obtaining (4.6.7), an assumption has been made that, within the time span Δt, the variations of the f_j and g_{jr} functions are practically deterministic and are not affected significantly by the randomness in $\boldsymbol{X}(t)$; thus they need not be included in the ensemble-averaging of random excitations during the short time interval. To obtain the second derivate moments from the second equation in (4.2.11a), it is adequate to use (4.6.2), resulting in

$$b_{jk}(\boldsymbol{x}_t, t) = \frac{1}{\Delta t}\int_t^{t+\Delta t} du \int_t^{t+\Delta t} g_{jr}(\boldsymbol{x}_u, u) g_{ks}(\boldsymbol{x}_v, v) E\left[\xi_r(u)\, \xi_s(v)\right]\, dv + 0(\Delta t) \tag{4.6.8}$$

It can be seen from (4.6.7) and (4.6.8) that the effects of random excitations on the first and second derivate moments depend on the respective double integrations, involving correlation functions $E[\xi_r(u)\xi_s(v)]$ in the two equations. To examine their significance more closely, let us consider a special case in which the random excitations are jointly stationary in correlation so that the correlation functions depend

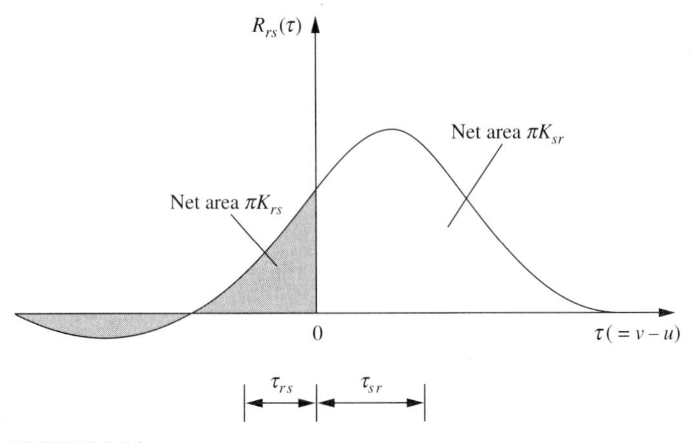

FIGURE 4.6.2
Correlation of random excitations $\xi_r(u)$ and $\xi_s(v)$, jointly stationary in correlation.

only on the difference $v - u$:

$$E\left[\xi_r(u)\,\xi_s(v)\right] = R_{rs}(v-u) \tag{4.6.9}$$

Such a correlation function is sketched in Fig. 4.6.2.

It is convenient to define a *correlation time* as

$$\tau_{rs} = \frac{1}{\sqrt{R_{rr}(0)R_{ss}(0)}}\int_{-\infty}^{0}\left|R_{rs}(\tau)\right|d\tau \tag{4.6.10}$$

which is a measure of "memory" of the present $\xi_r(t)$ with respect to the past $\xi_s(t)$. The double integration indicated in (4.6.7) is restricted to $v \leq u$, the left-hand side of the R_{rs} curve. On this side, the area under the correlation curve is equal to πK_{rs}. Now the domain of integration is the shaded triangular area in Fig. 4.6.3. If Δt is chosen to be much larger than τ_{rs}, then integration on v from t to u will sweep the entire area πK_{rs}, and the value of the double integration is of order Δt. In this case, the double integral makes a significant contribution to a_j.

Now equation (4.6.7) can be recast in the form

$$a_j(\boldsymbol{x},t) = f_j(\boldsymbol{x},t) + \frac{1}{\Delta t}\int_t^{t+\Delta t}du\int_{t-u}^{0} g_{ls}(\boldsymbol{x}_{u+\tau}, u+\tau)\frac{\partial}{\partial x_l}g_{jr}(\boldsymbol{x}_t,t)R_{rs}(\tau)\,d\tau + 0(\Delta t) \tag{4.6.11}$$

If, in addition, functions g_{ls} and $\delta g_{jr}/\partial X_l$ vary slowly within the time interval Δt, then upon integrating first on u, (4.6.11) can be approximated as

$$a_j(\boldsymbol{x},t) = f_j(\boldsymbol{x},t) + \int_{-\Delta t}^{0} g_{ls}(\boldsymbol{x}_{t+\tau}, t+\tau)\frac{\partial}{\partial \boldsymbol{x}_l}g_{jr}(\boldsymbol{x}_t,t)R_{rs}(\tau)\,d\tau + 0(\Delta t) \tag{4.6.12}$$

On the other hand, if Δt is much smaller than τ_{rs}, so that $R_{rs}(v-u)$ remains nearly constant when integrating on v, then the double integral in (4.6.7) yields a quantity

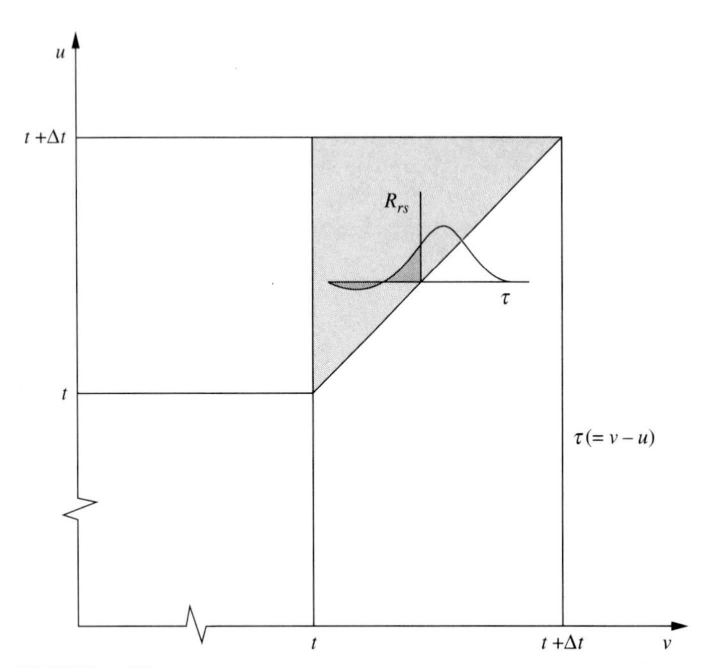

FIGURE 4.6.3
Domain of double integral.

of the order Δt^2, and

$$a_j(\boldsymbol{x}, t) \approx f_j(\boldsymbol{x}, t) + 0(\Delta t) \tag{4.6.13}$$

Thus the random excitations contribute very little compared with what is prescribed by the deterministic term $f_j(\boldsymbol{x}, t)$.

The double integral in (4.6.8) can be interpreted similarly, except that the integration domain is a square, involving both $v \leq u$ and $v \geq u$, and the right-hand side area under the $R_{rs}(v - u)$ curve is also involved in the calculation. Again, if Δt is chosen to be larger than both τ_{rs} and τ_{sr}, then the entire area under the R_{rs} curve will be covered in the integration, and the random excitations contribute significantly, in which case

$$b_{jk}(\boldsymbol{x}, t) = \int_{-\Delta t}^{\Delta t} g_{jr}(\boldsymbol{x}_t, t) g_{ks}(\boldsymbol{x}_{t+\tau}, t + \tau) R_{rs}(\tau)\, dt + 0(\Delta t) \tag{4.6.14}$$

Otherwise, the random excitations contribute little to the second derivate moments:

$$b_{jk}(\boldsymbol{x}, t) \approx 0(\Delta t) \tag{4.6.15}$$

When the random excitations are nonstationary, the correlation functions are no longer functions of the time difference $v - u$ but depend on v and u separately. Such a correlation function is more complicated to sketch, but if its absolute value also tends to be small as $|v - u|$ becomes large, then the arguments given for the stationary excitations still apply in the nonstationary excitation case.

It is important to recall that the correlation $E[\xi_r(u)\,\xi_s(v)]$ is a measure of statistical relation between two random excitation forces at different times u and v. When correlation is zero, the two forces are more likely to be independent (exactly independent if they are gaussian). Since excitations are the cause of change (increments) in the system response, the observation time gap Δt must be much longer than the time difference at which any two random forces become uncorrelated, so that the observed system response can become Markov-like. This single criterion sets a lower limit for the size of Δt in our interpretation. The use of (4.6.7) and (4.6.8) and the retention of the double integral terms imply the choice of a long enough observation time gap to justify a Markov approximation.

As indicated before, Δt cannot be very large or it will lose too much detail in the approximation. Specifically, Δt should not be greater than the so-called *relaxation time* of the system, which is a measure of the "memory" of the system of its earlier state, without taking into account the influence of excitations. The relaxation time, denoted by τ_{rel}, may be suitably defined for either an oscillatory or a nonoscillatory system. In the oscillatory case, τ_{rel} is the time required for the amplitude of oscillation to decrease by a factor e^{-1} or increase by a factor e, where e is the base of the natural logarithm. In the nonoscillatory case, the amplitude is replaced by the motion itself. Too much detail of the motion is lost if the observation intervals are separated longer than τ_{rel}; that is, the upper limit for Δt is τ_{rel}. Therefore, a Markov approximation for the system response is acceptable only if the relaxation time of the system is much greater than the correlation times of all the random excitations.

Very frequently, an analysis involves functions of a Markov process, in which case it is convenient to work with the Itô-type stochastic equations. Equations (4.6.7) and (4.6.8) can also be used to convert the physical equations (4.6.1) to the associated Itô-type equations (4.4.11). According to (4.4.12) and (4.4.13), the drift and diffusion coefficients in the Itô equations are obtained as

$$m_j(\boldsymbol{X}_t, t) = f_j(\boldsymbol{X}_t, t) + \frac{1}{\Delta t}\int_t^{t+\Delta t} du \int_t^u g_{ls}(\boldsymbol{X}_v, v)\frac{\partial}{\partial X_l} g_{jr}(\boldsymbol{X}_t, t) E\left[\xi_r(u)\xi_s(v)\right] dv \tag{4.6.16}$$

$$\sigma_{jl}(\boldsymbol{X}_t, t)\sigma_{kl}(\boldsymbol{X}_t, t) = \frac{1}{\Delta t}\int_t^{t+\Delta t} du \int_t^{t+\Delta t} g_{jr}(\boldsymbol{X}_u, u) g_{ks}(\boldsymbol{X}_v, v) E\left[\xi_r(u)\xi_s(v)\right] dv \tag{4.6.17}$$

The size of Δt must be large enough so that $E[\xi_r(u)\,\xi_s(v)] \approx 0$ when $|u - v| > \Delta t$, and it must be small enough so that it is shorter than the relaxation time of the system.

It is of interest to note that (4.6.17) is an equation for the product of matrices $\boldsymbol{\sigma}$ and $\boldsymbol{\sigma}'$, instead of $\boldsymbol{\sigma}$ itself. However, in actual calculations, only the product $\boldsymbol{\sigma}\boldsymbol{\sigma}'$ is needed, which corresponds to the fact that only the second derivate moments appear in the Fokker-Planck equation.

4.7 STOCHASTIC AVERAGING AND QUASI-CONSERVATIVE AVERAGING

We now consider the special case in which all the random excitations are jointly stationary in correlation, so that the correlation $E[\xi_r(u)\,\xi_s(v)]$ appearing in equations (4.6.16) and (4.6.17) has the form

$$E\,[\xi_r(u)\xi_s(v)] = R_{rs}(v-u) \tag{4.6.9}$$

In this case, equations (4.6.16) and (4.6.17) are reduced, respectively, to

$$m_j(\boldsymbol{X}_t, t) = f_j(\boldsymbol{X}_t, t) + \int_{-\Delta t}^{0} g_{ls}(\boldsymbol{X}_{t+\tau}, t+\tau)\frac{\partial}{\partial X_l} g_{jr}(\boldsymbol{X}_t, t)R_{rs}(\tau)\,d\tau \tag{4.7.1}$$

and
$$\sigma_{jl}(\boldsymbol{X}_t, t)\sigma_{kl}(\boldsymbol{X}_t, t) = \int_{-\Delta t}^{\Delta t} g_{jr}(\boldsymbol{X}_t, t)g_{ks}(\boldsymbol{X}_{t+\tau}, t+\tau)R_{rs}(\tau)\,d\tau \tag{4.7.2}$$

analogous to (4.6.12) and (4.6.14). Now, assuming that $R_{rs}(\tau)$ is nonzero only within a small neighborhood around $\tau = 0$, the lower limits of the integrations on τ in both (4.7.1) and (4.7.2) may be extended to $-\infty$, and the upper limit of the integration on τ in (4.7.2) may be extended to ∞; thus

$$m_j(\boldsymbol{X}_t, t) = f_j(\boldsymbol{X}_t, t) + \int_{-\infty}^{0} g_{ls}(\boldsymbol{X}_{t+\tau}, t+\tau)\frac{\partial}{\partial X_l} g_{jr}(\boldsymbol{X}_t, t)R_{rs}(\tau)\,d\tau \tag{4.7.3}$$

and

$$\sigma_{jl}(\boldsymbol{X}_t, t)\sigma_{kl}(\boldsymbol{X}_t, t) = \int_{-\infty}^{\infty} g_{jr}(\boldsymbol{X}_t, t)g_{ks}(\boldsymbol{X}_{t+\tau}, t+\tau)R_{rs}(\tau)\,d\tau \tag{4.7.4}$$

The physical implication of (4.7.3) and (4.7.4) can be made clear by comparing the physical equation (4.6.1) with the corresponding Itô equation (4.4.11). Each random excitation $dB_k(t)$ in (4.4.11) is independent of the present state $\boldsymbol{X}(t)$, by virtue of Itô's interpretation (4.4.4); however, the random excitations $\xi_r(t)$ in (4.6.1), having some memory with respect to their past histories, are correlated with the present state $\boldsymbol{X}(t)$. This memory affects the tendency for drift. Therefore, when evaluating the drift coefficients for use in an Itô equation, the correlation between the present state of the dynamic system and the past excitation must be taken into consideration, as what is accomplished by the integral in (4.7.3). Without excitations, the drift would be represented by f_j alone.

Similarly, the integral in (4.7.4) sums up the contribution from the past as well as in the future of the random excitations toward diffusion, and lumps the total effect at the present, as implied in the Itô equations (4.4.11).

In the special case in which the random excitations are delta correlated, that is,

$$E\,[\xi_r(u)\xi_s(v)] = R_{rs}(v-u) = 2\pi K_{rs}\delta(v-u) \tag{4.7.5}$$

where K_{rs} are constants, equations (4.7.3) and (4.7.4) reduce to

$$m_j(\boldsymbol{X}_t, t) = f_j(\boldsymbol{X}_t, t) + \pi K_{rs} g_{ls}(\boldsymbol{X}_t, t)\frac{\partial}{\partial X_l} g_{jr}(\boldsymbol{X}_t, t) \tag{4.7.6}$$

and

$$\sigma_{jl}(\boldsymbol{X}, t)\sigma_{kl}(\boldsymbol{X}, t) = 2\pi K_{rs} g_{jr}(\boldsymbol{X}_t, t) g_{ks}(\boldsymbol{X}_t, t) \tag{4.7.7}$$

Equations (4.7.6) and (4.7.7) were first derived by Wong and Zakai (1965). The second term on the right-hand side of (4.7.6) arises when some g_{jr} functions are indeed dependent on $X(t)$; that is, some random excitations are multiplicative. Such terms are known as the *Wong-Zakai correction terms*.

Return again to the general case of correlation-stationary excitations, equations (4.7.3) and (4.7.4), where the excitations may or may not be delta correlated, and consider the possibility of further simplification. Indeed, simplification is possible if the right-hand sides of the physical equations (4.6.1) are small in some sense. Specifically, if (4.6.1) can be represented as

$$\frac{d}{dt} X_j(t) = \varepsilon f_j(\boldsymbol{X}, t) + \varepsilon^{1/2} g_{jr}(\boldsymbol{X}, t)\, \xi_r(t) \tag{4.7.8}$$

where $\varepsilon << 1$ is a small parameter, then the converted Itô equations may be further smoothed by time-averaging the drift and diffusion coefficients. It should be noted that the first and second terms on the right-hand side of (4.7.8) are assumed to be of order ε and $\varepsilon^{1/2}$, respectively. As it will soon become clear this is equivalent to assuming that their contributions to the system response are commensurable. Denote the time-averaged drift and diffusion coefficients by

$$m_j(\boldsymbol{X}) = \varepsilon \left\langle f_j(\boldsymbol{X}_t, t) + \int_{-\infty}^{0} g_{ls}(\boldsymbol{X}_{t+\tau}, t+\tau) \frac{\partial}{\partial X_l} g_{jr}(\boldsymbol{X}_t, t) R_{rs}(\tau) d\tau \right\rangle_t \tag{4.7.9}$$

$$\sigma_{jl}(\boldsymbol{X})\sigma_{kl}(\boldsymbol{X}) = \varepsilon \left\langle \int_{-\infty}^{\infty} g_{jr}(\boldsymbol{X}_t, t) g_{ks}(\boldsymbol{X}_{t+\tau}, t+\tau) R_{rs}(\tau) d\tau \right\rangle_t \tag{4.7.10}$$

where the symbol $\langle[\cdot]\rangle_t$ represents the following time-averaging operation:

$$\langle[\cdot]\rangle_t = \lim_{T\to\infty} \frac{1}{2T} \int_{-T}^{T} [\cdot]\, dt \tag{4.7.11}$$

The smoothed Itô equations, converted from the physical equations (4.7.8), are given by

$$dX_j(t) = m_j(\boldsymbol{X})dt + \sigma_{jk}(\boldsymbol{X})\, dB_k(t) \tag{4.7.12}$$

The procedure indicated in (4.7.9) and (4.7.10), known as *stochastic averaging*, is due originally to Stratonovich (1963), and it may be considered an extension of the well-known averaging method of Bogoliubov and Mitropolski (1961) to stochastic differential equations. This procedure has been provided with a rigorous mathematical interpretation in a limit theorem due to Khasminskii (1966), which states that as ε decreases, $\boldsymbol{X}(t)$ in (4.7.8) approaches a Markov vector governed by (4.7.12), in probability. The solutions obtained from the time-averaged equations are uniformly close to those of the unaveraged equations within a time span of ε^{-1} (Bogoliubov and Mitropolski, 1961).

The term stochastic averaging refers generally to equations (4.7.9) and (4.7.10) in the literature. In a broader sense, it may also refer to the unsmoothed version (4.7.3) and (4.7.4) for correlation-stationary excitations, or even to (4.6.16) and

(4.6.17) for nonstationary excitations. If confusion arises, clarification will be made as whether it is a smoothed or an unsmoothed stochastic averaging.

It is of interest to note that if functions f_j and g_{jr} are indeed explicitly dependent on t, then this t dependence is lost through time-averaging. Thus time-averaging should not be used in cases where certain time-dependent properties of a dynamic system are of primary importance. One example is the rotor blade system of a helicopter in forward flight (Fujimori, Lin, and Ariaratnam, 1979), for which the use of (4.7.3) and (4.7.4) is more appropriate.

NONLINEARLY DAMPED OSCILLATOR. The time-averaging procedure is applicable only if the right-hand sides of the physical equations (4.7.8) are small, which implies that $X_j(t)$ are slowly varying. This may not be the case with the original set of equations but is transformable to such a case. As an example, consider a weakly nonlinear oscillator, governed by the second-order equation

$$\ddot{Y} + \omega_0^2 Y = \varepsilon f(Y, \dot{Y}) + \varepsilon^{1/2} g_k(Y, \dot{Y})\xi_k(t) \tag{4.7.13}$$

where ε is a small parameter, indicating that both the nonlinear and the random excitation terms on the right-hand side are small. Letting $Y = Z_1$ and $\dot{Y} = Z_2$, equation (4.7.13) is replaced by two first-order equations:

$$\dot{Z}_1 = Z_2 \tag{4.7.14}$$

$$\dot{Z}_2 = -\omega_0^2 Z_1 + \varepsilon f(Z_1, Z_2) + \varepsilon^{1/2} g_k(Z_1, Z_2)\xi_k(t) \tag{4.7.15}$$

The right-hand sides of (4.7.14) and (4.7.15) are clearly not small. Now, let

$$Z_1 = A(t)\cos\theta \qquad \theta = \omega_0 t + \phi(t) \tag{4.7.16}$$

$$Z_2 = -A(t)\omega_0 \sin\theta \tag{4.7.17}$$

and substitute (4.7.16) and (4.7.17) into (4.7.14) and (4.7.15) to yield

$$\dot{A}\cos\theta - A\dot{\phi}\sin\theta = 0 \tag{4.7.18}$$

$$\begin{aligned} \dot{A}\sin\theta + A\dot{\phi}\cos\theta = & -\frac{1}{\omega_0}\varepsilon f(A\cos\theta, -A\omega_0\sin\theta) \\ & -\frac{1}{\omega_0}\varepsilon^{1/2} g_k(A\cos\theta, -A\omega_0\sin\theta)\xi_k(t) \end{aligned} \tag{4.7.19}$$

From (4.7.18) and (4.7.19), we obtain

$$\dot{A} = -\frac{\sin\theta}{\omega_0}\left[\varepsilon f(A\cos\theta, -A\omega_0\sin\theta) + \varepsilon^{1/2} g_k(A\cos\theta, -A\omega_0\sin\theta)\,\xi_k(t)\right] \tag{4.7.20}$$

$$\dot{\phi} = -\frac{\cos\theta}{\omega_0 A}\left[\varepsilon f(A\cos\theta, -A\omega_0\sin\theta) + \varepsilon^{1/2} g_k(A\cos\theta, -A\omega_0\sin\theta)\,\xi_k(t)\right] \tag{4.7.21}$$

The right-hand sides of the transformed equations, (4.7.20) and (4.7.21), are indeed small. Physically, $A(t)$ and $\phi(t)$ represent the slowly varying amplitude and phase of a nearly linear oscillator, perturbed by small nonlinearity and small random excitations.

By treating $A(t)$ as $X_1(t)$ and $\phi(t)$ as $X_2(t)$ in equations (4.7.9) and (4.7.10), the smoothed stochastic averaging procedure may be carried out as indicated. It is of interest to note that the time-averaging part of the procedure may sometimes be performed within a "quasi-period" 2π with the same results; that is, replacing (4.7.11) by

$$\langle[\cdot]\rangle_t = \frac{1}{2\pi}\int_0^{2\pi}[\cdot]\,d\theta \tag{4.7.22}$$

NONLINEAR RESTORING FORCE. The procedure indicated in (4.7.22) may be generalized to the case of a strongly nonlinear oscillator, governed by

$$\ddot{Y} + h(Y) = \varepsilon f(Y, \dot{Y}) + \varepsilon^{1/2} g_k(Y, \dot{Y})\,\xi_k(t) \tag{4.7.23}$$

where $h(Y)$ represents a strongly nonlinear restoring force and $\varepsilon f(Y, \dot{Y})$ represents a small damping force. We shall assume that $h(Y)$ is an odd function of Y. Equation (4.7.23) suggests that the system is quasi-periodic, in the sense that it would be periodic if the small damping and the small random excitations on the right-hand side of the equation were absent.

Let us consider, for the time being, the undamped free motion of the system:

$$\ddot{Y} + h(Y) = 0 \tag{4.7.24}$$

or, equivalently,

$$\dot{Y}\frac{d\dot{Y}}{dY} + h(Y) = 0 \tag{4.7.25}$$

since $\ddot{Y} = d\dot{Y}/dt = (d\dot{Y}/dY)(dY/dt)$. Equation (4.7.25) may be integrated to yield

$$\tfrac{1}{2}\dot{Y}^2 + U(Y) = \Lambda \tag{4.7.26}$$

where

$$U(Y) = \int_0^Y h(u)\,du \tag{4.7.27}$$

It is clear that $U(Y)$ is the potential energy, and the constant of integration Λ in (4.7.26) is the total energy of free motion of the undamped system. The amplitude A of free oscillation is reached when the total energy is converted entirely to the potential energy; that is, $\Lambda = U(A)$, or

$$A = U^{-1}(\Lambda) \tag{4.7.28}$$

The period of free oscillation is obtained from

$$T = 4T_{1/4} = 4\int_0^A \frac{dY}{\sqrt{2\,[U(A) - U(Y)]}} \tag{4.7.29}$$

where $T_{1/4}$ is one-quarter of the free period. Equation (4.7.29) shows the well-known fact that the period of free oscillation of a nonlinear oscillator depends on the amplitude, unlike a linear oscillator. The period so obtained may be regarded as the quasi-period of the original system (4.7.23).

Return to equation (4.7.23). Under the influence of damping and random excitations, the total energy Λ becomes varying with time. Differentiating (4.7.26) with respect to time, we obtain

$$\dot{\Lambda} = \dot{Y}\left[\ddot{Y} + h(Y)\right] \tag{4.7.30}$$

Equations (4.7.30) and (4.7.23) can be combined to yield

$$\dot{\Lambda} = \dot{Y}[\varepsilon f(Y, \dot{Y}) + \varepsilon^{1/2} g_k(Y, \dot{Y})\xi_k(t)] \tag{4.7.31}$$

It shows that, indeed, the total energy $\Lambda(t)$ varies slowly with time if damping and random excitations are small.

We assume that the random excitations $\xi_k(t)$ are correlation-stationary and that their correlation times are short compared with ε^{-1}. Then $\Lambda(t)$ may be approximated by a Markov process governed by an Itô equation:

$$d\Lambda(t) = m(\Lambda)dt + \sigma(\Lambda)dB(t) \tag{4.7.32}$$

where the drift and diffusion coefficients, $m(\Lambda)$ and $\sigma(\Lambda)$, can be obtained by another form of averaging procedure, called *quasi-conservative averaging* (Khasminskii, 1964; Landa and Stratonovich, 1962), which will now be described.

First, the velocity $\dot{Y}$ on the right-hand side of (4.7.31) is expressed in terms of Λ and Y according to

$$\dot{Y} = \pm\left\{2\left[\Lambda - U(Y)\right]\right\}^{1/2} \tag{4.7.33}$$

where the positive and negative signs correspond to increasing and decreasing Y, respectively. Next, the "unaveraged" drift and diffusion coefficients analogous to (4.7.3) and (4.7.4) are obtained as follows:

$$\begin{aligned} m(\Lambda) = & \pm \varepsilon \sqrt{2\left[\Lambda - U(Y)\right]}\, f\left(Y, \pm\sqrt{2\left[\Lambda - U(Y)\right]}\right) \\ & + \varepsilon \int_{-\infty}^{0} \left\{ \sqrt{2\left[\Lambda - U(Y)\right]}\, g_r\left(Y, \pm\sqrt{2\left[\Lambda - U(Y)\right]}\right)\right\}_{t+\tau} \\ & \frac{\partial}{\partial\Lambda}\left\{ \sqrt{2\left[\Lambda - U(Y)\right]}\, g_k\left(Y, \pm\sqrt{2\left[\Lambda - U(Y)\right]}\right)\right\}_t R_{kr}(\tau)\,d\tau \end{aligned} \tag{4.7.34}$$

$$\begin{aligned} \sigma^2(\Lambda) = & \varepsilon \int_{-\infty}^{\infty} \left\{ \sqrt{2\left[\Lambda - U(Y)\right]}\, g_r\left(Y, \pm\sqrt{2\left[\Lambda - U(Y)\right]}\right)\right\}_{t+\tau} \\ & \left\{ \sqrt{2\left[\Lambda - U(Y)\right]}\, g_k\left(Y, \pm\sqrt{2\left[\Lambda - U(Y)\right]}\right)\right\}_t R_{kr}(\tau)\,d\tau \end{aligned} \tag{4.7.35}$$

If the correlation times of the excitations are much smaller than a quasi-period, then (4.7.34) and (4.7.35) may be further approximated by time-averaging, or by averaging on Y over a quasi-period T as follows:

$$\langle[\cdot]\rangle_t = \frac{1}{T}\int_{-A}^{A} \frac{\left([\cdot]_{\dot{Y}=\sqrt{2[\Lambda-U(Y)]}} + [\cdot]_{\dot{Y}=-\sqrt{2[\Lambda-U(Y)]}}\right)}{\sqrt{2\left[\Lambda - U(Y)\right]}}\,dY \tag{4.7.36}$$

When evaluating the integral, the indicated integration limits must be converted from A to Λ according to equation (4.7.28). The quasi-period T, obtained from (4.7.29), must also be converted to a function of Λ.

The accuracy of stochastic averaging, or quasi-conservative averaging, can sometimes be improved by first performing the unsmoothed averaging procedures in terms of the original physical variables $Z_1 = Y$ and $Z_2 = \dot{Y}$. These unsmoothed procedures do not require that the right-hand sides be small, since they are carried out prior to the time-averaging. The improved accuracy stems from those Wong-Zakai correction terms, which contribute toward the nondissipative restoring forces, and which can now be combined with the original stiffness terms prior to the variable transformation and time-averaging (Zhu, Yu, and Lin, 1994). Such "pseudo-stiffness" terms, if they exist, modify the natural frequency ω_0 in the weakly nonlinear system (4.7.13), or the $h(Y)$ function in the strongly nonlinear system (4.7.23). It is of interest to note that in the case of a second-order system, equation (4.7.13) or (4.7.23), the Wong-Zakai correction terms arise only if multiplicative random excitations occur in the damping terms in the original equation of motion.

In the case of a multi-degree-of-freedom system under purely multiplicative random excitations, the effect of random loads on the motion stability of the system is of the greatest interest. Near the stability margin, the motion of the system is often dominated by a single mode, called the *critical mode*. The relaxation time of this critical mode increases as the stability margin is approached. Thus the system motion may be approximated by a two-dimensional Markov vector if the correlation times of the excitations are short compared with the relaxation time of the critical mode. In this approximation, the relaxation times of the noncritical modes need not be long compared with the correlation times of the excitations since they are not part of the Markov vector. Nevertheless, the presence of the noncritical modes can affect the stability of the critical mode, on account of energy interchange with the critical modes. A modified stochastic averaging method that has been devised to treat such a problem (Sri Namachchivaya and Lin, 1988) is discussed in Section 6.5.4.

Applications of the Markov process theory to dynamical systems under purely additive excitations of gaussian white noises were discussed in Lin (1967). The case of combined additive and multiplicative excitations is illustrated in the following three examples and in Chapter 5. Stability of single-degree-of-freedom and multi-degree-of-freedom systems is treated in Chapter 6.

4.7.1 A Column under Randomly Varying Axial and Transverse Loads

As shown in Fig. 4.7.1, a simply supported uniform column is subjected to the excitations of an axial compressive force and a transverse concentrated force at the middle of the column. The axial force consists of a static part P and a random fluctuating part $F_1(t)$ with a zero mean. The transverse force $F_2(t)$ is also random with a zero mean and is assumed to be independent of $F_1(t)$. The transverse motion of the column is governed by a partial differential equation,

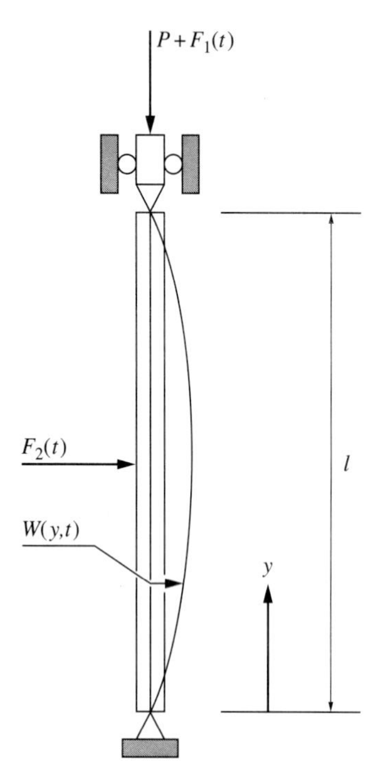

FIGURE 4.7.1
A uniform column under axial and transverse random excitations.

$$EI\frac{\partial^4 W}{\partial y^4} + [P + F_1(t)]\frac{\partial^2 W}{\partial y^2} + m\ddot{W} + c\dot{W} = F_2(t)\delta\left(y - \frac{l}{2}\right) \tag{a}$$

where m = mass of column per unit length and c = damping coefficient. Assume that W is dominated by the first vibration mode:

$$W(y, t) \approx X(t)\sin\frac{\pi y}{l} \tag{b}$$

Equation (a) can be reduced to an ordinary differential equation for the generalized displacement $X(t)$:

$$\ddot{X} + 2\zeta\omega_0\dot{X} + \omega_0^2[1 + \xi_1(t)]X = \xi_2(t) \tag{c}$$

where $\omega_0 = (\pi/l)[(P_{cr} - P)/m]^{1/2}$ is the natural frequency of the dominant mode, $P_{cr} = \pi^2 EI/l^2$ is the static buckling load, $\zeta = c/(2m\omega_0)$, $\xi_1(t) = -F_1(t)/(P_{cr} - P)$, $\xi_x(t) = 2F_2(t)/(ml)$, and where $P < P_{cr}$ is implied. The relaxation time of the dynamic system is of the order $(\zeta\omega_0)^{-1}$. Thus the Markov approximation is acceptable if the correlation times of both $\xi_1(t)$ and $\xi_2(t)$ are much smaller than $(\zeta\omega_0)^{-1}$. We shall assume that this is the case.

To apply the stochastic averaging procedure, the second-order equation (c) must be replaced by an equivalent set of two first-order equations. Now, equation (c) is of the general form of equation (4.7.13) without a multiplicative excitation in the damping term. Therefore, the following transformation is suitable:

$$X(t) = A(t)\cos\theta \qquad \theta = \omega_0 t + \phi(t) \tag{d}$$

$$\dot{X}(t) = -A(t)\omega_0\sin\theta \tag{e}$$

The stochastic processes $A(t)$ and $\phi(t)$ have the physical meaning of randomly varying amplitude and phase, respectively. It is implied in equations (d) and (e) that

$$\dot{A}\cos\theta - A\dot{\phi}\sin\theta = 0 \tag{f}$$

Substituting (d) and (e) into (c), we obtain

$$\dot{A}\sin\theta + A\dot{\phi}\cos\theta = -2\zeta\omega_0 A\sin\theta + \omega_0 A\cos\theta\xi_1(t) - \frac{1}{\omega_0}\xi_2(t) \tag{g}$$

Equations (f) and (g) can be simplified to

$$\dot{A} = -2\zeta\omega_0 A\sin^2\theta + \omega_0 A\sin\theta\cos\theta\xi_1(t) - \frac{1}{\omega_0}\sin\theta\xi_2(t) \tag{h}$$

$$\dot{\phi} = -2\zeta\omega_0\sin\theta\cos\theta + \omega_0\cos^2\theta\xi_1(t) - \frac{1}{A\omega_0}\cos\theta\xi_2(t) \tag{i}$$

Equations (h) and (i) have the same form as (4.7.8) with $A(t)$ and $\phi(t)$ playing the roles of $X_1(t)$ and $X_2(t)$, respectively, in the two physical equations.

Equations (h) and (i) can be converted to the Itô-type stochastic differential equations by stochastic averaging. If $\zeta << 1$, which may be taken as the small parameter ε, and if the excitations $\xi_1(t)$ and $\xi_2(t)$ are correlation-stationary and of order $\varepsilon^{1/2}$, then the smoothed stochastic averaging procedure is applicable. Identifying $A(t)$ as $X_1(t)$ and $\phi(t)$ as $X_2(t)$ when applying the stochastic averaging formulas, (4.7.9) and (4.7.10), we obtain the following results for the drift and diffusion coefficients in the Itô equation for $A(t)$:

$$m_1 = -\left[\zeta\omega_0 - \frac{3\pi}{8}\omega_0^2\Phi_{11}(2\omega_0)\right]A + \frac{\pi}{2\omega_0^2 A}\Phi_{22}(\omega_0) \tag{j}$$

$$(\sigma\sigma')_{11} = \frac{\pi}{4}\omega_0^2\Phi_{11}(2\omega_0)A^2 + \frac{\pi}{\omega_0^2}\Phi_{22}(\omega_0) \tag{k}$$

$$(\sigma\sigma')_{12} = (\sigma\sigma')_{21} = 0 \tag{l}$$

where $\Phi_{11}(\omega)$ and $\Phi_{22}(\omega)$ are, respectively, the spectral density functions of the random excitations $\xi_1(t)$ and $\xi_2(t)$.

It is interesting to note that (j), (k), and (l) are devoid of $\phi(t)$, which means that the probabilistic structure of the smoothed $A(t)$ process is independent of $\phi(t)$. We hasten to add that the drift and diffusion coefficients in the Itô equation for the smoothed $\phi(t)$ process do involve the smoothed $A(t)$ process. Confirmation of this statement will be left as an exercise for the reader. For most engineering applications, however, only the $A(t)$ process is of practical interest.

From (j), (k), and (l) we obtain the following Itô equation for the random amplitude:

$$dA = \left\{-\left[\zeta\omega_0 - \frac{3\pi}{8}\omega_0^2\Phi_{11}(2\omega_0)\right]A + \frac{\pi}{2\omega_0^2 A}\Phi_{22}(\omega_0)\right\}dt + \left[\frac{\pi}{4}\omega_0^2\Phi_{11}(2\omega_0)A^2 + \frac{\pi}{\omega_0^2}\Phi_{22}(\omega_0)\right]^{1/2} dB(t) \tag{m}$$

where $B(t)$ is a unit Wiener process. Equation (m) implies that while $A(t)$ and $\phi(t)$ constitute approximately the components of a two-dimensional Markov vector, one of the components, the $A(t)$ process, is approximately a scalar Markov process itself.

The Fokker-Planck equation for the transient probability density $q(\alpha, t \mid \alpha_0, t_0)$ of $A(t)$, obtained directly from the Itô equation (m), reads

$$\frac{\partial q}{\partial t} - \frac{\partial}{\partial \alpha}\left(\left\{\left[\zeta\omega_0 - \frac{3\pi}{8}\omega_0^2\Phi_{11}(2\omega_0)\right]\alpha - \frac{\pi}{2\omega_0^2\alpha}\Phi_{22}(\omega_0)\right\} q\right)$$
$$- \frac{1}{2}\frac{\partial^2}{\partial \alpha^2}\left\{\left[\frac{\pi}{4}\omega_0^2\Phi_{11}(2\omega_0)\alpha^2 + \frac{\pi}{\omega_0^2}\Phi_{22}(\omega_0)\right] q\right\} = 0 \tag{n}$$

where α is the state variable for $A(t)$. Equation (n) is solved subject to an initial condition and appropriate boundary conditions. However, solution for equation (n) appears difficult to obtain. For the time being, we limit our interest to the stationary probability density $p(\alpha)$ of $A(t)$, governed by the reduced Fokker-Planck equation

$$\frac{d}{d\alpha}G = 0 \tag{o}$$

where
$$G = -\left\{\left[\zeta\omega_0 - \frac{3\pi}{8}\omega_0^2\Phi_{11}(2\omega_0)\right]\alpha - \frac{\pi}{2\omega_0^2\alpha}\Phi_{22}(\omega_0)\right\} p$$
$$- \frac{1}{2}\frac{d}{d\alpha}\left\{\left[\frac{\pi}{4}\omega_0^2\Phi_{11}(2\omega_0)\alpha^2 + \frac{\pi}{\omega_0^2}\Phi_{22}(\omega_0)\right] p\right\} \tag{p}$$

The boundary conditions associated with equation (o) depend on whether the additive random excitation $\xi_2(t)$ is present. If $\xi_2(t)$ is present, then the left boundary at $\alpha = 0$ is singular of the second kind, where the first derivate moment is unbounded. The diffusion exponent, drift exponent, and character value at $\alpha = 0$ are obtained from equations (4.5.20) through (4.5.22) as follows:

$$\alpha_l = 0 \qquad \beta_l = 1 \qquad c_l = 1 \tag{q}$$

From Table 4.5.3, $\alpha = 0$ is identified as an entrance boundary. The right boundary at infinity is also singular of the second kind, and its diffusion exponent, drift exponent, and character value may be obtained from equations (4.5.24), (4.5.25), and (4.5.27) to yield

$$\alpha_r = 2 \qquad \beta_r = 1 \qquad c_r = \frac{8\zeta - 3\pi\omega_0\Phi_{11}(2\omega_0)}{\pi\omega_0\Phi_{11}(2\omega_0)} \tag{r}$$

As shown in Table 4.5.4, the right boundary at infinity is repulsively natural if $c_r > -1$, that is, if

$$\zeta > \frac{\pi}{4}\omega_0\Phi_{11}(2\omega_0) \tag{s}$$

Otherwise, it is strictly natural or attractively natural. A stationary probability density exists under condition (s), in which case it can be obtained from (4.5.11) to yield

$$p(\alpha) = C\alpha(D_1\alpha^2 + D_2)^{-\delta} \tag{t}$$

where C is a normalization constant,

$$D_1 = \frac{\pi}{4}\omega_0^2\Phi_{11}(2\omega_0) \qquad D_2 = \frac{\pi}{\omega_0^2}\Phi_{22}(\omega_0) \tag{u}$$

and

$$\delta = \frac{4\zeta}{\pi\omega_0\Phi_{11}(2\omega_0)} \tag{v}$$

The normalization constant C is given by

$$C = \left[\int_0^\infty \alpha(D_1\alpha^2 + D_2)^{-\delta} d\alpha\right]^{-1} = 2(\delta - 1)D_1 D_2^{\delta-1} \tag{w}$$

It can be shown that condition (s) is also required in order that $p(\alpha)$ in equation (t) is normalizable (i.e., integrable).

It might appear strange that condition (s) would depend only on the spectral density of the multiplicative random excitation at exactly twice the natural frequency. This result can be misleading without considering the fact that the Markov approximation is valid only if the correlation time of $\xi_1(t)$ is short, which is equivalent to requiring a slowly varying spectral density. Therefore, there exists energy input in the general neighborhood of $2\omega_0$, not merely at $2\omega_0$. The excitation energy near $2\omega_0$ is the most crucial, however, in view of the fact that if $\xi_2(t)$ were to be replaced by a sinusoidal excitation, equation (c) would become a damped Mathieu equation and $2\omega_0$ would be the primary resonant frequency.

We digress to remark that if the multiplicative random excitation were to occur at the damping term rather than the stiffness term, then the condition would also involve the excitation spectral density evaluated at $\omega = 0$. This conclusion can be reached by completing exercise 4.5.

If the multiplicative excitation is absent, then condition (s) reduces to $\zeta > 0$. We let $\Phi_{11}(\omega) = 0$ in equation (t) to obtain

$$\begin{aligned} p(\alpha) &= \lim_{\Phi_{11}\to 0} C\alpha(D_1\alpha^2 + D_2)^{-\delta} \\ &= \lim_{\Phi_{11}\to 0} CD_2^{-\delta}\alpha\left\{\left[1 + \frac{\omega_0^4\Phi_{11}}{4\Phi_{22}(\omega_0)}\alpha^2\right]^{\frac{4\Phi_{22}(\omega_0)}{\omega_0^4\Phi_{11}\alpha^2}}\right\}^{-\frac{\zeta\omega_0^3\alpha^2}{\pi\Phi_{22}(\omega_0)}} \end{aligned} \tag{x}$$

It follows upon substituting (w) into (x) and taking the limit

$$p(\alpha) = \frac{2\zeta\omega_0^3}{\pi\Phi_{22}(\omega_0)}\alpha \exp\left[-\frac{\zeta\omega_0^3}{\pi\Phi_{22}(\omega_0)}\alpha^2\right] \tag{y}$$

Thus the $A(t)$ process becomes Rayleigh-distributed, which is a well-known result (see, e.g., Lin, 1967, Eq. 9-71). Of course, this result can also be obtained directly from equation (o) without the $\Phi_{11}(2\omega_0)$ terms.

Next we consider the case in which the additive excitation $\xi_2(t)$ is absent. It can be seen from equation (m) that if $\Phi_{22}(\omega_0) = 0$, then both the drift and diffusion coefficients are zero at the left boundary $\alpha = 0$, indicating a trap. The diffusion exponent α_l, drift exponent β_l, and the character value c_l for the left boundary $\alpha = 0$ can be evaluated from equations (4.5.16)–(4.5.18), resulting in

$$\alpha_l = 2 \qquad \beta_l = 1 \qquad c_l = \frac{-8\zeta + 3\pi\omega_0\Phi_{11}(2\omega_0)}{\pi\omega_0\Phi_{11}(2\omega_0)} \tag{z}$$

As shown in Table 4.5.2,

$$\text{the left boundary } \alpha = 0 \text{ is } \begin{cases} \text{attractively natural} & \zeta > \frac{\pi}{4}\omega_0\Phi_{11}(2\omega_0) \\ \text{strictly natural} & \zeta = \frac{\pi}{4}\omega_0\Phi_{11}(2\omega_0) \\ \text{repulsively natural} & \zeta < \frac{\pi}{4}\omega_0\Phi_{11}(2\omega_0) \end{cases}$$

The behavior of the right boundary at infinity is the same as when both additive and multiplicative excitations are present:

$$\text{the right boundary } \alpha = \infty \text{ is } \begin{cases} \text{repulsively natural} & \zeta > \frac{\pi}{4}\omega_0\Phi_{11}(2\omega_0) \\ \text{strictly natural} & \zeta = \frac{\pi}{4}\omega_0\Phi_{11}(2\omega_0) \\ \text{attractively natural} & \zeta < \frac{\pi}{4}\omega_0\Phi_{11}(2\omega_0) \end{cases}$$

Upon an examination of various combinations of the two boundaries, it is found that a *nontrivial* stationary probability density does not exist without the additive excitation. This conclusion can also be reached from the fact that $p(\alpha)$ in equation (t) is not normalizable if $D_2 = 0$.

The foregoing analysis shows that the existence of a nontrivial stationary probability density can be determined either by identifying the boundaries involved or by examining directly the integrability of the stationary probability density obtained, without the knowledge of boundary conditions. The second approach is sometimes easier. However, if the objective of an analysis is the stability of a system, or if the stationary probability density must be obtained by numerically solving the reduced Fokker-Planck equation, then the knowledge of boundary behaviors is important and necessary, as to be shown in Chapter 6.

4.7.2 Sliding Motion of an Anchored Rigid Block (Lin et al., 1994)

Shown in Fig. 4.7.2 is a rigid block undergoing a sliding displacement $X(t)$ relative to a supporting floor system. The floor system consists of an upper floor, on which sliding of the block takes place, and a subfloor, to which the block is further anchored with elastic ties. The anchorage provides additional resistance against sliding. It is assumed that the anchorage is sufficiently strong to prevent the block from toppling; if it isn't, a toppling analysis can be carried out separately, as shown in Section 8.3.3. The distance between the upper and lower floors is d, and the configuration of the floor system itself is assumed to remain unchanged at all times. When $X(t) = 0$, the elastic ties have a total initial tension S_0. The sliding motion of the block is caused by random shaking of the floor system.

The elongation of each elastic tie due to the sliding motion $X(t)$ of the block is given by

$$\Delta l = d\left[1 + \left(\frac{X}{d}\right)^2\right]^{1/2} - d \tag{a}$$

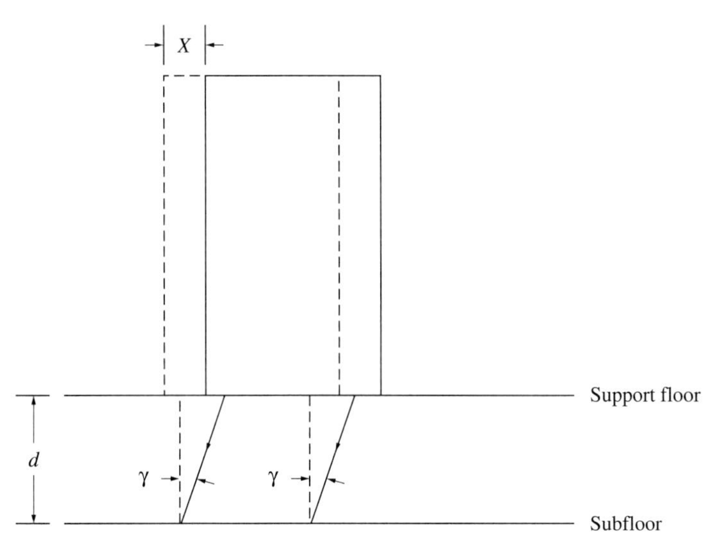

FIGURE 4.7.2
A rigid block undergoing sliding displacement.

Assuming that $X << d$, equation (a) may be approximated as

$$\Delta l \approx \frac{d}{2}\left(\frac{X}{d}\right)^2 \tag{b}$$

The total tension in the elastic ties is then

$$S_T = S_0 + k\Delta l \approx S_0 + \frac{kd}{2}\left(\frac{X}{d}\right)^2 \tag{c}$$

where k is the spring constant of the elastic ties. The horizontal and vertical components of S_T are, respectively, $S_T \sin\gamma$ and $S_T \cos\gamma$, where γ is the angle between a stretched elastic tie and the vertical. Since γ is small, $\cos\gamma \approx 1$ and $\sin\gamma \approx \tan\gamma = X/d$. The horizontal component gives rise to an elastic restoring force against the sliding motion, and the vertical component increases the pressure between the block and the supporting floor.

Making use of these approximations, the equation of motion may be written as

$$M\ddot{X} + \mathrm{sgn}(\dot{X})\mu\left[Mg + S_0 + \frac{1}{2}kd\left(\frac{X}{d}\right)^2 + Mg\ddot{y}_G\right] + S_0\left(\frac{X}{d}\right) + \frac{1}{2}kd\left(\frac{X}{d}\right)^3 = -Mg\ddot{x}_G \tag{d}$$

where M is the mass of the block, g is the gravitational acceleration, μ is the coefficient of friction between the block and the upper floor, $\mathrm{sgn}(\dot{X})$ denotes the sign of $\dot{X}$, and $\ddot{x}_G$ and $\ddot{y}_G$ are the horizontal and vertical base accelerations of the supporting floor system in the units of g. In what follows, $\ddot{x}_G$ and $\ddot{y}_G$ will be idealized as gaussian white noises, with respective spectral densities K_h and K_v and cross-spectral density $\sqrt{K_h K_v}\gamma_{hv}$.

Dividing equation (d) by Md, and letting $Y(t) = X(t)/d$, we obtain the following simplified version:

$$\ddot{Y} + \text{sgn}(\dot{Y})\mu\omega_0^2\left\{1 + \eta^2 Y^2 + \alpha^2\left[1 + \sqrt{2\pi K_v}W_v(t)\right]\right\} + \omega_0^2(Y + \eta^2 Y^3)$$
$$= -\omega_0^2\alpha^2\sqrt{2\pi K_h}W_h(t) \tag{e}$$

where $\omega_0^2 = S_0/(Md)$, $\alpha^2 = Mg/S_0$, $\eta^2 = kd/(2S_0)$, and $W_h(t)$ and $W_v(t)$ are gaussian white noises of unit spectral density. Clearly ω_0 is the natural frequency of the linear system obtained by neglecting all the nonlinear terms. The constants α^2 and η^2 may be redefined as $\alpha^2 = \omega_p^2/\omega_0^2$ and $\eta^2 = \omega_v^2/(2\omega_0^2)$, where $\omega_p = g/d$ is the natural frequency of a pendulum of length d, and $\omega_v = (k/M)^{1/2}$ would be the natural frequency if the mass were suspended vertically by spring k.

Equation (e) represents a system with a strongly nonlinear restoring force. If the damping and excitation terms are small, then the quasi-conservative averaging procedure is applicable. It is of interest to note that although the multiplicative excitation $W_v(t)$ appears in damping, it does not contribute to a pseudo-stiffness term in the averaged equation, since it is associated only with $\text{sgn}(\dot{Y})$.

The potential energy and total energy are, respectively,

$$U(Y) = \omega_0^2(\tfrac{1}{2}Y^2 + \tfrac{1}{4}\eta^2 Y^4) \tag{f}$$

and

$$\Lambda(t) = \tfrac{1}{2}\dot{Y}^2 + \omega_0^2(\tfrac{1}{2}Y^2 + \tfrac{1}{4}\eta^2 Y^4) \tag{g}$$

The quarter-period of free oscillation may be obtained from

$$T_{1/4}(A) = \int_0^A \frac{dY}{\sqrt{2[U(A) - U(Y)]}} = \frac{1}{\omega_0}(1 + \eta^2 A^2)^{-1/2} F\left(\bar{k}, \frac{\pi}{2}\right) \tag{h}$$

where $F(\bar{k}, \pi/2)$ is the complete elliptic integral of the first kind, defined as

$$F\left(\bar{k}, \frac{\pi}{2}\right) = \int_0^{\pi/2} \frac{d\phi}{\sqrt{1 - \bar{k}^2 \sin^2\phi}} \tag{i}$$

and $\bar{k}$ is its modulus. In the present case,

$$\bar{k} = \eta A[2(1 + \eta^2 A^2)]^{-1/2} \tag{j}$$

Since the system is subjected to damping and excitations, $\Lambda(t)$ is now varying with time. Upon differentiating equation (g) with respect to t, and combining the result with equation (e), we obtain

$$\dot{\Lambda} = -\dot{Y}\left\{\text{sgn}(\dot{Y})\mu\omega_0^2[1 + \alpha^2 + \eta^2 Y^2 + \alpha^2\sqrt{2\pi K_v}W_v(t)]\right.$$
$$\left. + \omega_0^2\alpha^2\sqrt{2\pi K_h}W_h(t)\right\} \tag{k}$$

If μ and K_h are small, then $\Lambda(t)$ is a slowly varying random process, which may be approximated by a Markov process governed by the Itô stochastic differential equation

$$d\Lambda = m(\Lambda)dt + \sigma(\Lambda)\,dB(t) \tag{l}$$

where $m(\Lambda)$ and $\sigma(\Lambda)$ are the drift and diffusion coefficients, respectively, and $B(t)$ is a unit Wiener process. The drift and diffusion coefficients may be obtained using the quasi-conservative averaging procedure as follows:

$$\begin{aligned} m(\Lambda) &= \Big\langle -\dot{Y}\mu\omega_0^2(1+\alpha^2+\eta^2Y^2)\,\mathrm{sgn}(\dot{Y}) \\ &\quad + \pi\alpha^4\omega_0^4\frac{d}{d\Lambda}\left\{\dot{Y}^2\left[K_h+\mu^2K_v+2\mu\sqrt{K_hK_v}\,\gamma_{hv}\,\mathrm{sgn}(\dot{Y})\right]\right\}\Big\rangle_t \\ &= -\frac{\mu\omega_0^2Q(\Lambda)}{T_{1/4}} + \pi\alpha^4\omega_0^4K \end{aligned} \tag{m}$$

$$\begin{aligned} \sigma^2(\Lambda) &= 2\pi\alpha^4\omega_0^4\left\langle \dot{Y}^2\left[K_h+\mu^2K_v+2\mu\sqrt{K_hK_v}\,\gamma_{hv}\,\mathrm{sgn}(\dot{Y})\right]\right\rangle_t \\ &= \frac{2\pi\alpha^4\omega_0^4KS(\Lambda)}{T_{1/4}} \end{aligned} \tag{n}$$

where $K = K_h + \mu^2K_v$, which may be regarded as an effective excitation level, and

$$Q(\Lambda) = (1+\alpha^2)A + \tfrac{1}{3}\eta^2A^3 \tag{o}$$

$$S(\Lambda) = \int_0^A \sqrt{2\,[\Lambda - U(Y)]}\,dY = \omega_0A^2\sqrt{1+\eta^2A^2}\int_0^{\pi/2}\sin^2\phi\sqrt{1-\bar{k}^2\sin^2\phi}\,d\phi \tag{p}$$

In obtaining equation (m), use has been made of the fact that

$$\frac{d}{d\Lambda}S(\Lambda) = \int_0^A \frac{dY}{\sqrt{2\,[\Lambda - U(Y)]}} = T_{1/4} \tag{q}$$

We note, in passing, that the white noise idealization for the base excitations is reasonable so long as the correlation times of the excitations are much shorter than the quasi-period T, which is bounded below by $2\pi/\omega_0$. It is also of interest to note that the cross-spectral density of the horizontal and vertical base excitations does not contribute to the effective excitation level K; thus it does not affect the *averaged* motion.

At the state of stochastic stationarity, the probability density $p(\lambda)$ of $\Lambda(t)$ is governed by the reduced Fokker-Planck equation

$$\frac{d}{d\lambda}\left\{a(\lambda)p(\lambda) - \frac{1}{2}\frac{d}{d\lambda}\,[b(\lambda)p(\lambda)]\right\} = 0 \tag{r}$$

where λ is the state variable of $\Lambda(t)$, and the first and second derivate moments $a(\lambda)$ and $b(\lambda)$ may be obtained from equations (m) and (n), by noting that $a(\lambda) = m(\lambda)$ and $b(\lambda) = \sigma^2(\lambda)$. The stationary probability flow, if it exists, must vanish in the one-dimensional state space; therefore, equation (r) may be replaced by

$$a(\lambda)p(\lambda) - \frac{1}{2}\frac{d}{d\lambda}[b(\lambda)p(\lambda)] = 0 \tag{s}$$

Solution to equation (s) is found to be

$$p(\lambda) = CT_{1/4}(\lambda)\exp\left[-\frac{\mu}{\pi\alpha^4\omega_0^2 K}\int\frac{Q(\lambda)}{S(\lambda)}d\lambda\right] \tag{t}$$

where C is the normalization constant. It can be shown that this probability density is always normalizable, indicating that neither boundary can be absorbing or attractively natural. The probability density of the amplitude $A(t)$ is obtained by simple coordinate transformation, to yield

$$\begin{aligned} p_A(\beta) &= p[\lambda(\beta)]\left|\frac{d\lambda}{d\beta}\right| \\ &= CT_{1/4}(\beta)\omega_0^2(\beta + \eta^2\beta^3)\exp\left[-\frac{\mu}{\pi\alpha^4 K}\int\frac{Q(\beta)}{S(\beta)}(\beta+\eta^2\beta^3)\,d\beta\right] \end{aligned} \tag{u}$$

where β is the state variable of $A(t)$. In equation (t), $T_{1/4}$, Q, and S are treated as functions of the total energy λ, whereas in equation (u), they are treated as functions of amplitude β. The relation between λ and β is given by

$$\lambda = \omega_0^2\left(\tfrac{1}{2}\beta^2 + \tfrac{1}{4}\eta^2\beta^4\right) \tag{v}$$

The computed probability densities p_A for the amplitude A are shown in Figs. 4.7.3 and 4.7.4. The results in Fig. 4.7.3 are obtained for $\mu = 0.2$, $\eta^2 = 100$, and $\omega_0 K = 0.002$, and for $\alpha^2 = 0.5$, 1.0, and 2.0. The probability density associated with a larger α^2 value has a flatter shape, with a lower peak located at a higher β value. This is to be expected since a larger α^2 corresponds to a lower initial tension S_0 in the anchorage. In Fig. 4.7.4, the value for α^2 is kept at $\alpha^2 = 1$, while η^2 is allowed to assume three different values, $\eta^2 = 100$, 500, and 1000. A higher η^2 corresponds to a stiffer anchorage spring.

4.7.3 A van der Pol-Type Oscillator

As the final example, consider the following van der Pol-type oscillator under both additive and multiplicative random excitations:

$$\ddot{Y} + \omega_0^2 Y + \delta\dot{Y} + \beta Y^2\dot{Y} + \gamma\dot{Y}^3 = YW_1(t) + \dot{Y}W_2(t) + W_3(t) \tag{a}$$

where δ is a constant, β and γ are positive constants, and $W_1(t)$, $W_2(t)$, and $W_3(t)$ are correlated white noises. Equation (a) is equivalent to a pair of first-order equations:

$$\dot{Z}_1 = Z_2 \tag{b}$$

$$\dot{Z}_2 = -\omega_0^2 Z_1 - \delta Z_2 - \beta Z_1^2 Z_2 - \gamma Z_2^3 + Z_1 W_1(t) + Z_2 W_2(t) + W_3(t) \tag{c}$$

which can be converted to the Itô-type stochastic differential equations as follows, using equations (4.7.6) and (4.7.7):

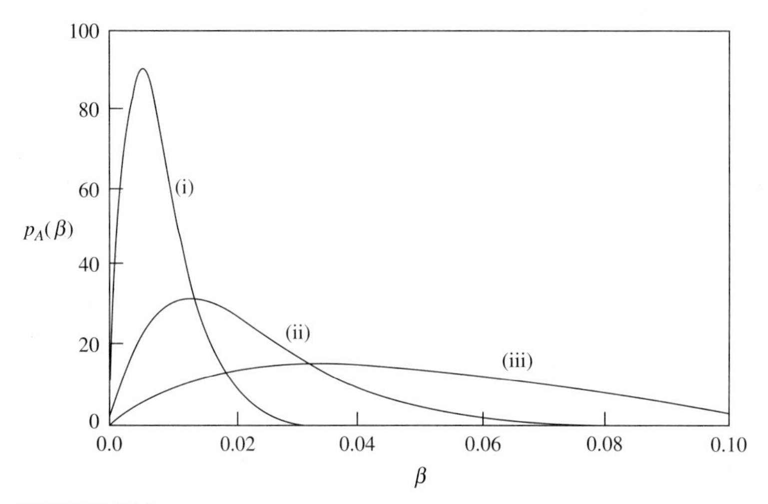

FIGURE 4.7.3
Probability density of amplitude A. System parameters: $\mu = 0.2, \eta^2 = 100$, and (i) $\alpha^2 = 0.5$, (ii) $\alpha^2 = 1.0$, (iii) $\alpha^2 = 2.0$. Excitation parameter $\omega_0 K = 0.002$ [after Lin et al., 1994; Copyright ©1994 Elsevier Science Ltd., reprinted with permission].

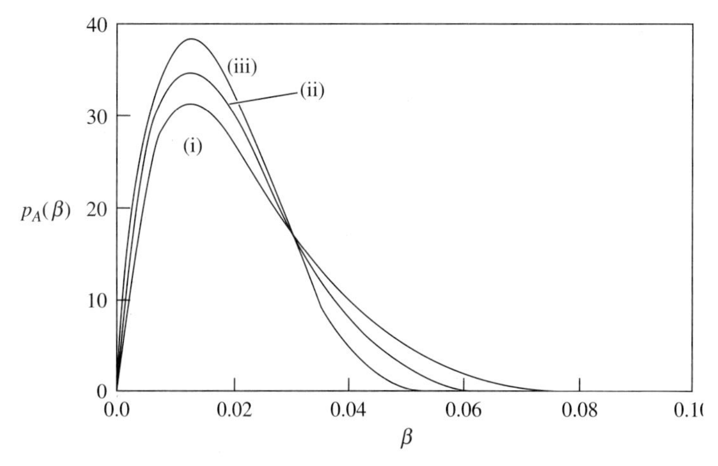

FIGURE 4.7.4
Probability density of amplitude A. System parameters: $\mu = 0.2, \alpha^2 = 1$, and (i) $\eta^2 = 100$, (ii) $\eta^2 = 500$, (iii) $\eta^2 = 1000$. Excitation parameter $\omega_0 K = 0.002$ [after Lin *et al.,* 1994; Copyright ©1994 Elsevier Science Ltd., reprinted with permission].

$$dZ_1 = Z_2 dt \tag{d}$$

$$\begin{aligned} dZ_2 &= \left[-\omega_0^2 Z_1 - \delta Z_2 - \beta Z_1^2 Z_2 - \gamma Z_2^3 + \pi(K_{12}Z_1 + K_{22}Z_2 + K_{23})\right] \\ &\quad + \left[2\pi(K_{11}Z_1^2 + K_{22}Z_2^2 + K_{33} + 2K_{12}Z_1Z_2 + 2K_{13}Z_1 + 2K_{23}Z_2)\right]^{1/2} dB(t) \end{aligned} \tag{e}$$

where $B(t)$ is a unit Wiener process, and the $K_{ij}(i, j = 1, 2, 3)$ are the spectral and cross-spectral densities of the white noises. Appearing in equation (e) are three Wong-Zakai correction terms:

$$\pi(K_{12}Z_1 + K_{22}Z_2 + K_{23}) \tag{f}$$

The first term, $\pi K_{12}Z_1$, modifies the conservative restoring force; the second term, $\pi K_{22}Z_2$, contributes to a negative damping; and the third term, πK_{23}, provides an "equivalent" static load.

Rewrite equation (e) in the abbreviated form of

$$dZ_2 = [-\bar{\omega}_0^2 Z_1 + \pi K_{23} - f(Z_1, Z_2)]\, dt + \sqrt{g(Z_1, Z_2)}\, dB(t) \tag{e$'$}$$

where

$$f(Z_1, Z_2) = \bar{\delta} Z_2 + \beta Z_1^2 Z_2 + \gamma Z_2^3 \tag{g}$$

$$g(Z_1, Z_2) = 2\pi(K_{11}Z_1^2 + K_{22}Z_2^2 + K_{33} + 2K_{12}Z_1Z_2 + 2K_{13}Z_1 + 2K_{23}Z_2) \tag{h}$$

and where $\bar{\omega}_0^2 = \omega_0^2 - \pi K_{12}$ and $\bar{\delta} = \delta - \pi K_{22}$. If $\bar{\delta}$, β, γ, and K_{ij} are small, then (d) and (e$'$) describe a weakly nonlinear system.

To render the time-averaging procedure applicable, let

$$\begin{aligned} Z_1 &= A(t)\cos\theta + z_{10} \qquad \theta = \bar{\omega}_0 t + \phi(t) \\ Z_2 &= -A(t)\bar{\omega}_0 \sin\theta \end{aligned} \tag{i}$$

where $z_{10} = \pi K_{23}\bar{\omega}_0^{-2}$, which accounts for the equivalent static deflection. The transformation defined in equations (i) may be rewritten as

$$\begin{aligned} A(t) &= \left[(Z_1 - z_{10})^2 + \left(\frac{Z_2}{\bar{\omega}_0}\right)^2\right]^{1/2} \\ \phi(t) &= -\tan^{-1}\left[\frac{Z_2}{\bar{\omega}_0(Z_1 - z_{10})}\right] - \bar{\omega}_0 t \end{aligned} \tag{j}$$

Using Itô's differential rule (4.4.15), we obtain

$$\begin{aligned} dA &= \left[\frac{\sin\theta}{\bar{\omega}_0} f(A\cos\theta + z_{10}, -A\bar{\omega}_0\sin\theta) + \frac{\cos^2\theta}{2\bar{\omega}_0^2 A} g\left[(A\cos\theta + z_{10}, -A\bar{\omega}_0\sin\theta)\right]\right] dt \\ &\quad - \frac{\sin\theta}{\bar{\omega}_0}\sqrt{g(A\cos\theta + z_{10}, -A\bar{\omega}_0\sin\theta)}\, dB(t) \end{aligned} \tag{k}$$

$$d\phi = \left[\frac{\cos\theta}{A\bar{\omega}_0} f(A\cos\theta + z_{10}, -A\bar{\omega}_0 \sin\theta) - \frac{\sin 2\theta}{2A^2\bar{\omega}_0^2} g(A\cos\theta + z_{10}, -A\bar{\omega}_0 \sin\theta)\right] dt$$

$$- \frac{\cos\theta}{A\bar{\omega}_0} \sqrt{g(A\cos\theta + z_{10}, -A\bar{\omega}_0 \sin\theta)}\, dB(t) \tag{l}$$

If the values of the spectral densities and cross-spectral densities of $W_1(t)$, $W_2(t)$, and $W_3(t)$ are small, then from equations (g) and (h), the functions f and g are small. The right-hand sides of (k) and (l) can now be time-averaged to yield

$$dA = m_A dt + \sigma_{AA} dB_1(t) + \sigma_{A\phi} dB(t) \tag{m}$$

$$d\phi = m_\phi dt + \sigma_{\phi A} dB_1(t) + \sigma_{\phi\phi} dB_2(t) \tag{n}$$

where $B_1(t)$ and $B_2(t)$ are independent unit Wiener processes, and

$$m_A = \left\langle \frac{\sin\theta}{\bar{\omega}_0} f(A\cos\theta + z_{10}, -A\bar{\omega}_0 \sin\theta) + \frac{\cos^2\theta}{2A\bar{\omega}_0^2} g(A\cos\theta + z_{10}, -A\bar{\omega}_0 \sin\theta)\right\rangle_t$$

$$= (-\frac{\delta}{2} - \frac{\beta}{2} z_{10}^2 + \frac{3\pi}{8\bar{\omega}_0^2} K_{11} + \frac{5\pi}{8} K_{22})A + \frac{\pi}{2A\bar{\omega}_0^2}(K_{33} + 2K_{13}z_{10}) - \frac{1}{8}(\beta + 3\gamma\bar{\omega}_0^2)A^3 \tag{o}$$

$$\sigma_{AA}^2 = \left\langle \frac{\sin^2\theta}{\bar{\omega}_0^2} g(A\cos\theta + z_{10}, -A\bar{\omega}_0 \sin\theta)\right\rangle_t$$

$$= \frac{\pi}{4\bar{\omega}_0^2}(K_{11} + 3\bar{\omega}_0^2 K_{22})A^2 + \frac{\pi}{\bar{\omega}_0^2}(K_{33} + 2K_{13}z_{10}) \tag{p}$$

$$m_\phi = \left\langle \frac{\cos\theta}{A\bar{\omega}_0} f(A\cos\theta + z_{10}, -A\bar{\omega}_0 \sin\theta) - \frac{\sin 2\theta}{2A^2\bar{\omega}_0^2} g(A\cos\theta + z_{10}, -A\bar{\omega}_0 \sin\theta)\right\rangle_t$$

$$= -\frac{\pi}{2\bar{\omega}_0} K_{12} \tag{q}$$

$$\sigma_{\phi\phi}^2 = \left\langle \frac{\cos^2\theta}{A^2\bar{\omega}_0^2} g(A\cos\theta + z_{10}, -A\bar{\omega}_0 \sin\theta)\right\rangle_t$$

$$= \frac{\pi}{4\bar{\omega}_0^2}(3K_{11} + \bar{\omega}_0^2 K_{22}) + \frac{\pi}{A^2\bar{\omega}_0^2}(K_{33} + 2K_{13}z_{10}) \tag{r}$$

$$\sigma_{A\phi} = \sigma_{\phi A} = 0 \tag{s}$$

Thus the time-averaged amplitude $A(t)$ is a one-dimensional Markov process, on which our attention is henceforth focused.

The stationary probability density $p(\alpha)$ of $A(t)$ is governed by the reduced Fokker-Planck equation

$$\frac{\partial}{\partial\alpha}G = 0 \tag{t}$$

where
$$G = -\frac{1}{2}\frac{\partial}{\partial\alpha}[(\tilde{\gamma}\alpha^2 + \tilde{\beta})p] - \left(\tilde{\lambda}\alpha^3 + \tilde{\delta}\alpha - \frac{\tilde{\beta}}{2\alpha}\right)p \tag{u}$$

and
$$\begin{aligned}
\tilde{\beta} &= \frac{\pi}{\bar{\omega}_0^2}(K_{33} + 2K_{13}z_{10} + K_{11}z_{10}^2) \\
\tilde{\delta} &= \frac{\delta}{2} + \frac{\beta}{2}z_{10}^2 - \frac{3\pi K_{11}}{8\bar{\omega}_0^2} - \frac{5\pi}{8}K_{22} \\
\tilde{\gamma} &= \frac{\pi}{4\bar{\omega}_0^2}(K_{11} + 3\bar{\omega}_0^2 K_{22}) \\
\tilde{\lambda} &= \tfrac{1}{8}(\beta + 3\gamma\bar{\omega}_0^2)
\end{aligned} \tag{v}$$

If a stationary probability density exists, the total probability flow must vanish in the one-dimensional space,

$$G = 0 \tag{w}$$

which is readily solved to yield

$$p(\alpha) = C\alpha(\tilde{\gamma}\alpha^2 + \tilde{\beta})^{\kappa}\exp\left(-\frac{\tilde{\lambda}}{\tilde{\gamma}}\alpha^2\right) \tag{x}$$

where C is the normalization constant, and

$$\kappa = \frac{\tilde{\beta}\tilde{\lambda}}{\tilde{\gamma}^2} - \frac{\tilde{\delta}}{\tilde{\gamma}} - \frac{3}{2} \tag{y}$$

It is of interest to note that $p(\alpha)$ in equation (x) is always normalizable as long as $\tilde{\beta} \neq 0$, which is the case when the additive excitation $W_3(t)$ is present. However, if the additive excitation is absent, then $\tilde{\beta} = 0$, and $p(\alpha)$ is normalizable only if $\kappa > -1$, that is, only if

$$\delta < \frac{\pi}{2}\left(\frac{K_{11}}{\omega_0^2} + K_{22}\right) \tag{z}$$

4.8 THE STRATONOVICH STOCHASTIC DIFFERENTIAL EQUATIONS

The use of equations (4.7.3) and (4.7.4) or the time-averaged version (4.7.9) and (4.7.10) has the primary objective of obtaining the drift and diffusion coefficients for the Itô-type stochastic differential equation (4.4.11). Mathematically speaking, the Itô-type equation defines clearly a diffusive Markov process, and it has the advantage that the drift and diffusion coefficients are related directly to the first and second derivate moments of the Fokker-Planck equation.

However, a Markov process may also be defined by other types of stochastic differential equations, with other suitable interpretations. The Stratonovich-type stochastic differential equation has the form of

$$dX(t) = \tilde{m}(X,t)dt + \tilde{\sigma}(X,t) \circ dB(t) \tag{4.8.1}$$

where $B(t)$ is also a unit Wiener process. For the ease of discussion, (4.8.1) is written in the one-dimensional form; a multidimensional generation is straightforward. We use the same symbol $X(t)$ in (4.8.1) and (4.4.1) to indicate that both the Stratonovich and Itô equations define the same Markov process. A small circle is inserted between $\tilde{\sigma}(X,t)$ and $dB(t)$ to distinguish the Stratonovich type from the Itô type.

Equation (4.8.1) is equivalent to

$$X(t) = X(0) + \int_0^t \tilde{m}\,[X(u),u]\,du + \int_0^t \tilde{\sigma}\,[X(u),u] \circ dB(u) \tag{4.8.2}$$

The second integral on the right-hand side, called a *Stratonovich integral*, is interpreted as

$$\int_0^t \tilde{\sigma}\,[X(u),u] \circ dB(u)$$
$$= \underset{\substack{n\to\infty \\ \max\Delta u\to 0}}{\text{l.i.m.}} \sum_{j=1}^{n} \tilde{\sigma}\left[\frac{X(u_j)+X(u_{j+1})}{2}, \frac{u_j+u_{j+1}}{2}\right]\left[B(u_{j+1})-B(u_j)\right] \tag{4.8.3}$$

To obtain the first and second derivate moments from (4.8.1), write

$$X(t+\Delta t) - X(t) = \int_t^{t+\Delta t} \tilde{m}\,[X(u),u]\,du + \int_t^{t+\Delta t} \tilde{\sigma}\,[X(u),u] \circ dB(u) \tag{4.8.4}$$

If Δt is sufficiently small, and if $\tilde{m}$ and $\tilde{\sigma}$ are sufficiently smooth, (4.8.4) may be approximated by

$$\begin{aligned}
&X(t+\Delta t) - X(t) \\
&= \tilde{m}_t\Delta t + \left\{\tilde{\sigma}_t + \left(\frac{\partial\tilde{\sigma}}{\partial X}\right)_t \left[\frac{X(t)+X(t+\Delta t)}{2} - X(t)\right]\right. \\
&\quad \left. + \left(\frac{\partial\tilde{\sigma}}{\partial t}\right)_t \left(\frac{t+t+\Delta t}{2} - t\right) + \cdots\right\}[B(t+\Delta t) - B(t)] \\
&= \tilde{m}_t\Delta t + \left\{\tilde{\sigma}_t + \left(\frac{\partial\tilde{\sigma}}{\partial X}\right)_t \left[\frac{X(t+\Delta t)-X(t)}{2}\right]\right. \\
&\quad \left. + \left(\frac{\partial\tilde{\sigma}}{\partial t}\right)_t \frac{\Delta t}{2} + \cdots\right\}[B(t+\Delta t) - B(t)] \\
&= \tilde{m}_t\Delta t + \tilde{\sigma}_t\,[B(t+\Delta t) - B(t)] \\
&\quad + \frac{1}{2}\left(\frac{\partial\tilde{\sigma}}{\partial X}\right)_t \tilde{\sigma}_t\,[B(t+\Delta t) - B(t)]^2 + 0(\Delta t^{3/2})
\end{aligned} \tag{4.8.5}$$

where

$$\tilde{m}_t = \tilde{m}\,[X(t), t] \tag{4.8.6a}$$

$$\tilde{\sigma}_t = \tilde{\sigma}\,[X(t), t] \tag{4.8.6b}$$

$$\left(\frac{\partial \tilde{\sigma}}{\partial X}\right)_t = \frac{\partial \tilde{\sigma}\,[X(t), t]}{\partial X(t)} \tag{4.8.6c}$$

$$\left(\frac{\partial \tilde{\sigma}}{\partial t}\right)_t = \frac{\partial \tilde{\sigma}\,[X(t), t]}{\partial t} \tag{4.8.6d}$$

It follows that

$$a(x, t) = \lim_{\Delta t \to 0} \frac{1}{\Delta t} E\,[X(t + \Delta t) - X(t) \mid X(t) = x] = \tilde{m}(x, t) + \frac{1}{2}\tilde{\sigma}(x, t)\frac{\partial \tilde{\sigma}(x, t)}{\partial x} \tag{4.8.7}$$

$$b(x, t) = \lim_{\Delta t \to 0} \frac{1}{\Delta t} E\left\{[X(t + \Delta t) - X(t)]^2 \mid X(t) = x\right\} = \tilde{\sigma}^2(x, t) \tag{4.8.8}$$

Comparing (4.8.7) with (4.4.9), and (4.8.8) with (4.4.10), we obtain

$$m(X, t) = \tilde{m}(X, t) + \frac{1}{2}\tilde{\sigma}(X, t)\frac{\partial \tilde{\sigma}(X, t)}{\partial X} \tag{4.8.9}$$

$$\sigma^2(X, t) = \tilde{\sigma}^2(X, t) \tag{4.8.10}$$

In the case of a Markov vector, (4.8.9) and (4.8.10) are generalized to

$$m_j(\mathbf{X}, t) = \tilde{m}_j(\mathbf{X}, t) + \frac{1}{2}\tilde{\sigma}_{kl}(\mathbf{X}, t)\frac{\partial \tilde{\sigma}_{jl}(\mathbf{X}, t)}{\partial X_k} \tag{4.8.11}$$

$$\sigma_{jl}(\mathbf{X}, t)\sigma_{kl}(\mathbf{X}, t) = \tilde{\sigma}_{jl}(\mathbf{X}, t)\tilde{\sigma}_{kl}(\mathbf{X}, t) \tag{4.8.12}$$

Thus the stochastic differential equation of Itô and that of Stratonovich have the same set of diffusion coefficients but different sets of drift coefficients, which can be converted from one to the other according to equation (4.8.11).

The Stratonovich integral and stochastic differential equation have the advantage that they can be treated in the same way as nonstochastic equations. To illustrate, consider first the following Stratonovich integral:

$$\int_a^b B(t) \circ dB(u) = \underset{\substack{n \to \infty \\ \Delta_n \to 0}}{\text{l.i.m.}} \sum_{j=1}^{n} B\left(\frac{t_{j+1} + t_j}{2}\right)[B(t_{j+1}) - B(t_j)] \tag{4.8.13}$$

where $\Delta_n = \max(t_{j+1} - t_j)$. Since $B(t)$ is continuous with probability 1,

$$B\left(\frac{t_{j+1} + t_j}{2}\right) \approx \frac{1}{2}[B(t_{j+1}) - B(t_j)] \tag{4.8.14}$$

as j_{j+1} and t_j become close. Substituting (4.8.14) into (4.8.13), and taking the limit,

$$\int_a^b B(t) \circ dB(u) \approx \tfrac{1}{2}[B^2(b) - B^2(a)] \tag{4.8.15}$$

The same result can be obtained if the integral is a regular integral.

As another example, consider the Itô differential equation

$$dX(t) = KX(t)\,dB(t) \tag{4.8.16}$$

where K is a constant. If (4.8.16) were to be treated as a regular equation, it would be replaceable by

$$d \ln X(t) = K dB(t) \tag{4.8.17}$$

However, (4.8.17) is incorrect, since according to the Itô differential rule

$$d \ln X(t) = -\tfrac{1}{2} K^2 \, dt + K \, dB(t) \tag{4.8.18}$$

which can then be integrated to obtain

$$X(t) = X(0) \exp\left[-\frac{1}{2} K^2 t + K B(t) \right] \tag{4.8.19}$$

On the other hand, (4.8.16) can be converted first to a Stratonovich stochastic differential equation in the form of (4.8.1). Using (4.8.10) and (4.8.9),

$$\tilde{\sigma}^2(X, t) = \sigma^2(X, t) = K^2 X^2(t) \tag{4.8.20}$$

$$\tilde{m}(X, t) = m(X, t) - \frac{1}{2} \tilde{\sigma}(X, t) \frac{\partial \tilde{\sigma}(X, t)}{\partial X} = -\frac{1}{2} K^2 X(t) \tag{4.8.21}$$

Thus

$$dX(t) = -\tfrac{1}{2} K^2 X(t) \, dt + K X(t) \circ dB(t) \tag{4.8.22}$$

This equation can be integrated in the usual way to yield the same result (4.8.19).

4.9 CONCLUDING REMARKS

If a randomly excited dynamical system is nonlinear, or some of the random excitations are multiplicative, or both, then an exact probabilistic solution is possible only if the system response can be modeled as a diffusive Markov process. While such exact solutions are obtainable only under the ideal situation of gaussian white noise excitations, they provide a basis for obtaining approximate solutions for realistic excitations. Therefore, a thorough understanding of the fundamentals of the diffusion processes, including various mathematical implications, is important.

The key to the Markov process modeling is the derivation of the Fokker-Planck equation, which governs the probability density of a Markov process. Given a set of physical equations, the way in which the basic ingredients in the Fokker-Planck equation are determined is contained in the concise procedure of stochastic averaging. The conditions under which this procedure is valid are explained in physical terms. Distinction is made between the unsmoothed and the smoothed versions of the stochastic averaging procedure. The later version, in which additional time-averaging is performed, requires that the response variables vary slowly with time. In most cases, the original response variables must be transformed to an equivalent set of variables to satisfy such a requirement. Additional accuracy may sometimes be gained by performing the unsmoothed stochastic averaging in terms of the original response variables, prior to the transformation of variables and time-averaging.

Since there exists simple one-to-one correspondence between a Fokker-Planck equation governing the probability density of a Markov process and the Itô-type differential equations describing the process, the stochastic averaging procedure may be viewed as converting a given set of physical equations to a corresponding set of Itô equations. The Itô equation representation of a Markov process has a distinct advantage in that the investigation of an arbitrary function of the Markov process becomes very simple. By means of Itô's differential rule, applicable under rather general differentiability conditions, the Itô equation for such an arbitrary function can be obtained simply, which leads immediately to its associated Fokker-Planck equation. Conversion from one Fokker-Planck equation to another would be much more complicated.

The stochastic averaging procedure is illustrated in the three examples in Section 4.7. In each example, a transformation is introduced to render the time-averaging applicable. The question of motion stability of a randomly excited system is touched upon in the first example. This question will be explored further in Chapter 6.

Finally, a comment on the use of the Itô equation versus the Stratonovich equation is in order. While conversion between the two types is straightforward, the Itô equation has the advantage of relating more directly to the Fokker-Planck equation; thus it is preferred by many analysts. On the other hand, since the Stratonovich equation can be treated in the same way as a regular differential equation, it is better suited when performing Monte Carlo simulations.

4.10 EXERCISES

4.1. Let $B(t)$ be a Wiener process with a correlation function

$$E\left[B(t_1)B(t_2)\right] = \sigma^2 \min(t_1, t_2)$$

Show that the correlation function of the mean-square derivative $\dot{B}(t) = dB(t)/dt$ is given by

$$E[\dot{B}(t_1)\dot{B}(t_2)] = \sigma^2\delta(t_1 - t_2)$$

4.2. A Markov vector $\boldsymbol{X} = \{X_1, X_2\}'$ is governed by the following two Itô differential equations:

$$dX_j = m_j(\boldsymbol{X}, t)\, dt + \sigma_{jk}(\boldsymbol{X}, t)\, dB_k(t) \qquad j = 1, 2$$

where $B_k(t)$ are independent unit Wiener processes. Find the Itô stochastic differential equations for A and θ which are related to X_1 and X_2 as $X_1 = A\cos\theta$ and $X_2 = A\sin\theta$.

4.3. The equation of motion for the system shown in Fig. 1.0.1 is given by

$$\dot{\Theta} + \frac{P}{cl}[1 + W_1(t)]\Theta = \frac{P}{cl}W_2(t)$$

where $W_1(t)$ and $W_2(t)$ are correlated white noises, and

$$E[W_i(t)W_j(t+\tau)] = 2\pi K_{ij}\delta(\tau)$$

Find the Itô stochastic differential equation for $\Theta(t)$.

4.4. A one-dimensional diffusion process $X(t)$, defined on $[0, \infty)$, is governed by the Itô equation

$$dX = \left[(-\zeta\omega_0 + \frac{3\pi}{8}\omega_0^2 K_1)X + \frac{\pi K_2}{2\omega_0^2 X} - \delta X^3\right]dt + \left(\frac{\pi}{4}\omega_0^2 K_1 X^2 + \frac{\pi K_2}{\omega_0^2}\right)^{1/2} dB(t)$$

where $\zeta > 0$, $\delta > 0$, and $B(t)$ is a unit Wiener process. Identify the left boundary $x = 0$ and the right boundary at infinity according to Feller's classifications for (a) $K_1 = 0$ and $K_2 > 0$, (b) $K_1 > 0$ and $K_2 = 0$, and (c) $K_1 > 0$ and $K_2 > 0$.

4.5. Consider the system

$$\ddot{X} + 2\zeta\omega_0 [1 + \xi(t)] \dot{X} + \omega_0^2 X = 0$$

where $\xi(t)$ is a correlation-stationary random excitation with a broadband spectral density $\Phi(\omega)$.

(a) Replace the original equation with two first-order equations for $A(t)$ and $\phi(t)$, using the transformation

$$X(t) = A(t) \cos\theta \qquad \theta = \omega_0 t + \phi(t)$$
$$\dot{X}(t) = -A(t)\omega_0 \sin\theta$$

(b) Obtain an Itô stochastic differential equation for $A(t)$ by employing the smoothed stochastic averaging procedure.

(c) Obtain an Itô stochastic differential equation for $A^n(t)$.

4.6. Obtain the Itô equation for $\phi(t)$ for the column system discussed in Section 4.7.1, using the following approximations:

$$\int_{-\infty}^{0} R_{11}(\tau) \sin\omega\tau \, d\tau \approx 0 \qquad \int_{-\infty}^{0} R_{22}(\tau) \sin\omega\tau \, d\tau \approx 0$$

4.7. Consider the nonlinear system

$$\ddot{X} + \alpha\dot{X} + \beta X^2 \dot{X} + \gamma \dot{X}^3 + X = W(t)$$

where $W(t)$ is a white noise with spectral density K, and α, β, and γ are small positive parameters.

(a) Replace the original equation with two first-order equations for $A(t)$ and $\phi(t)$, using the transformation

$$X(t) = A(t) \cos\theta \qquad \theta = \omega_0 t + \phi(t)$$
$$\dot{X}(t) = -A(t)\omega_0 \sin\theta$$

(b) Derive the Fokker-Planck equation for the Markov vector $\{A(t), \phi(t)\}'$, using the smoothed stochastic averaging procedure.

(c) Find the stationary probability density $p(\alpha)$ of $A(t)$.

(d) Prove that the stationary phase process $\phi(t)$ is uniformly distributed on a unit circle.

(e) Obtain the stationary probability density $p(x, \dot{x})$ of the original variables $X(t)$ and $\dot{X}(t)$.

4.8. Consider an oscillator with a linear viscous damping and a nonlinear power law spring governed by the equation of motion

$$\ddot{X} + \alpha\dot{X} + k\left|X\right|^{\rho} \operatorname{sgn}(X) = W(t)$$

where α and ρ are positive constants and $W(t)$ is a white noise with spectral density K.

(a) Derive the Fokker-Planck equation for the total energy

$$\Lambda = \tfrac{1}{2}\dot{X}^2 + k \int \left|X\right|^{\rho} \operatorname{sgn}(X) \, dX$$

using the quasi-conservative averaging procedure.

(b) Determine the stationary probability density for $\Lambda(t)$.

4.9. A linear oscillator is subjected to combined harmonic and random excitations and is governed by the equation of motion

$$\ddot{X} + 2\varepsilon\zeta\omega_0\dot{X} + \omega_0^2 X = \varepsilon\lambda \sin \nu t + \varepsilon^{1/2} W(t)$$

where ε is a small parameter and $W(t)$ is a gaussian white noise with spectral density K. Using the transformation

$$X(t) = A(t)\cos\theta \qquad \theta = \gamma t + \phi(t)$$

$$\dot{X}(t) = -A(t)\gamma \sin\theta$$

and the smoothed stochastic averaging procedure, derive the Itô type stochastic differential equations for the amplitude process $A(t)$ and phase process $\phi(t)$. Two cases must be considered separately: the resonant case when $|\nu - \omega_0|$ is of order ε, and the nonresonant case when $|\nu - \omega_0|$ is not small, for which the proper choices of the γ values are $\gamma = \nu$ and $\gamma = \omega_0$, respectively.

4.10. Let $B(t)$ be a unit Wiener process defined on $[a, b]$. Show that as $n \to \infty$ and $\max\limits_{1 \le j \le n} (t_{j+1} - t_j) \to 0$,

$$I_1 = \sum_{j=1}^{n} B(t_j)\left[B(t_{j+1}) - B(t_j)\right] \to \tfrac{1}{2}\left[B^2(b) - B^2(a)\right] - \tfrac{1}{2}(b - a)$$

$$I_2 = \sum_{j=1}^{n} B(t_{j+1})\left[B(t_{j+1}) - B(t_j)\right] \to \tfrac{1}{2}\left[B^2(b) - B^2(a)\right] + \tfrac{1}{2}(b - a)$$

$$I_3 = \sum_{j=1}^{n} B\left[rt_{j+1} + (1 - r)t_j\right]\left[B(t_{j+1}) - B(t_j)\right] \to \tfrac{1}{2}\left[B^2(b) - B^2(a)\right] + \left(r - \tfrac{1}{2}\right)(b - a)$$

where $0 \le r \le 1$.

CHAPTER 5

EXACT SOLUTIONS FOR MULTIDIMENSIONAL NONLINEAR SYSTEMS

In Chapter 4, much emphasis was placed on those randomly excited systems for which a crucial response variable may be modeled as a one-dimensional Markov process, since the stationary probability density of such a process, if it exists, can easily be obtained. This crucial response variable may be the amplitude or the total energy of a single-degree-of-freedom system or of the critical mode of a multi-degree-of-freedom system. If the response must be modeled as a multidimensional Markov process, then the possibility of an exact probabilistic solution becomes much more limited. This chapter is devoted to obtaining such exact solutions in certain cases.

Consider a nonlinear system, governed by the equations of motion

$$\frac{d}{dt}X_i = f_i(\boldsymbol{X}, t) + g_{ij}(\boldsymbol{X}, t)\,\xi_j(t) \qquad i = 1, 2, \ldots, n, \quad j = 1, 2, \ldots, m \tag{5.0.1}$$

where X_i are components of the system response vector $\boldsymbol{X}$, and $\xi_j(t)$ are random excitations. Moreover, each X_i is assumed to be distributed on the entire range of real values $(-\infty, \infty)$, unless stated otherwise. Functions f_i and g_{ij} are generally nonlinear; however, their functional forms are assumed to be deterministic. An exact probabilistic solution for such a system is difficult to obtain. The possibility for exact solution does exist, however, when random excitations are gaussian white noises, with the cross-correlations

$$E\left[\xi_j(t)\xi_k(t+\tau)\right] = 2\pi K_{jk}\delta(\tau) \tag{5.0.2}$$

where each K_{jk} is a constant, representing the cross-spectral density of $\xi_j(t)$ and $\xi_k(t)$. For this special case, the system response $\boldsymbol{X}$ is a Markov vector. As shown in Chapter 4, the probability density of a Markov vector is governed by the Fokker-Planck equation

$$\frac{\partial}{\partial t}q + \frac{\partial}{\partial x_i}G_i = 0 \tag{5.0.3}$$

where $q = q(\boldsymbol{x},t \mid \boldsymbol{x}_0, t_0)$, and G_i is the probability flow in the ith direction given by

$$G_i = a_i(\boldsymbol{x}, t)q - \frac{1}{2}\frac{\partial}{\partial x_j}[b_{ij}(\boldsymbol{x}, t)q] \tag{5.0.4}$$

In equation (5.0.4), the first and second derivate moments a_i and b_{ij} can be derived from the equation of motion (5.0.1) as follows:

$$a_i(\boldsymbol{x}, t) = f_i(\boldsymbol{x}, t) + \pi K_{ls} g_{rs}(\boldsymbol{x}, t)\frac{\partial}{\partial x_r} g_{il}(\boldsymbol{x}, t) \tag{5.0.5}$$

$$b_{ij}(\boldsymbol{x}, t) = 2\pi K_{ls} g_{il}(\boldsymbol{x}, t) g_{js}(\boldsymbol{x}, t) \tag{5.0.6}$$

Equation (5.0.3) is solved subject to the initial condition

$$q(\boldsymbol{x}, t_0 \mid \boldsymbol{x}_0, t_0) = \prod_{j=1}^{n} \delta(x_j - x_{j,0}) \tag{5.0.7}$$

and some appropriate boundary conditions. In this chapter, we are concerned only with either reflective or natural boundaries at which the probability flows vanish:

$$G_i = 0 \qquad \boldsymbol{x} \in S \tag{5.0.8}$$

As indicated in Chapter 4, a natural boundary is inaccessible from any interior point. A solution of the Fokker-Planck equation (5.0.3), satisfying the initial condition (5.0.7) and boundary conditions (5.0.8), will show how the probabilistic structure of the system response evolves with time. Unfortunately, for nonlinear systems, complete analytical solutions are known only for some special first-order systems (e.g., Caughey, 1971; Caughey and Dienes, 1961; Gardiner, 1983). Although numerical procedures have been developed for higher order cases (e.g., Naess and Johnsen, 1993; Spencer and Bergman, 1993; Sun and Hsu, 1990) the demand for computer time is generally high.

For higher order nonlinear systems, known analytical solutions have been restricted to the reduced Fokker-Planck equation

$$\frac{\partial}{\partial x_i}G_i = 0 \tag{5.0.9}$$

where

$$G_i = a_i(\boldsymbol{x})p(\boldsymbol{x}) - \frac{1}{2}\frac{\partial}{\partial x_j}\left[b_{ij}(\boldsymbol{x})p(\boldsymbol{x})\right] \tag{5.0.10}$$

Equation (5.0.9) is obtained by dropping the time-derivative term in (5.0.3) and is valid only for the special case in which the first and second derivate moments, a_i and b_{ij}, are independent of t. It is solved subject to the boundary conditions (5.0.8). The unknown $p(\boldsymbol{x})$ in equation (5.0.9) is the unconditional probability density of the system response when it reaches the state of statistical stationarity. Of course, a stationary response of system (5.0.1) exists only if every random excitation is a stationary process (in the present case a gaussian white noise), if f_i and g_{ij} do not depend explicitly on t, and if some energy dissipation mechanism exists in the system such that the energy input from the random excitations is balanced statistically by energy output from dissipation. A stationary solution, if obtainable, is still very useful. It is needed, for example, to compute the average rate at which a system response variable crosses a specified threshold (Lin, 1967, Chapter 9). The existence of a stationary solution is also related to the stability conditions of the system response in some sense (Chapter 6).

The usefulness of exact stationary solutions has motivated many attempts to identify the class of problems for which they are obtainable. Andronov et al. (1933) were the first to show that a second-order system with a nonlinear spring and under an additive white noise excitation was solvable. The same conclusion was reached independently by Kramers (1940). This class of solvable problems was extended by Caughey (1964), and further extended by Caughey and Ma (1982a) to include certain types of nonlinear damping. However, for some systems in the solvable class, a special relationship was required between the spectral density level of the additive white noise excitation and system parameters.

The cases involving multiplicative excitations are more difficult. The first success was reported in a paper by Dimentberg (1982), in which a specific second-order oscillator was considered. Yong and Lin (1987) made use of the concept of detailed balance (e.g., Graham and Haken, 1971; van Kampen, 1957) and developed a general procedure to obtain the exact stationary solutions for nonlinear systems under both additive and multiplicative white noise excitations. It was shown that the class of detailed balance included those problems of Caughey and Ma (1982a), as well as the specific multiplicative excitation problem of Dimentberg (1982). This solvable class was further extended by Lin and Cai (1988a) to include certain systems not in detailed balance. The solution procedure of Lin and Cai is also a generalization of Stratonovich's method of stationary potential (Stratonovich, 1963), which, in its original form, is not applicable to oscillatory systems.

In the following sections, the concepts of stationary potential and detailed balance are introduced to provide the needed physical insights for the development of a systematic procedure for obtaining exact stationary solutions. This procedure is then applied to single- and multi-degree-of-freedom oscillatory systems.

5.1 STATIONARY POTENTIAL

To explain the concept of stationary potential (Stratonovich, 1963), let us consider first the case of a one-dimensional scalar Markov process, say $X(t)$, for which the reduced Fokker-Planck equation reads

$$\frac{dG}{dx} = \frac{d}{dx}\left\{a(x)p(x) - \frac{1}{2}\frac{d}{dx}[b(x)p(x)]\right\} = 0 \tag{5.1.1}$$

Integrate (5.1.1) once to yield

$$G = a(x)p(x) - \frac{1}{2}\frac{d}{dx}[b(x)p(x)] = G_c = \text{constant} \tag{5.1.2}$$

According to (5.0.8), the probability flow vanishes at the boundaries; thus it must vanish everywhere: $G_c = 0$. The solution of equation (5.1.2) may be expressed as

$$p(x) = Ce^{-\phi(x)} \tag{5.1.3}$$

where C is a normalization constant, and

$$\phi(x) = \ln b - 2\int \frac{a}{b}\,dx \tag{5.1.4}$$

Function ϕ is called the *probability potential*. The exponential form of the solution, equation (5.1.3), guarantees that $p(x)$ is nonnegative for a positive normalization constant C.

In the case of a Markov vector, a steady pattern of probability flow may exist in the multidimensional state space without crossing the boundaries. Therefore, vanishing probability flow at every boundary does not guarantee that it vanishes everywhere. The term *stationary potential* refers to a special case in which the probability flow does vanish everywhere, not only at the boundaries (Stratonovich, 1963):

$$G_i = a(\boldsymbol{x})p(\boldsymbol{x}) - \frac{1}{2}\frac{\partial}{\partial x_j}[b_{ij}(\boldsymbol{x})p(\boldsymbol{x})] = 0 \tag{5.1.5}$$

for every i. Write the probability density $p(\boldsymbol{x})$ again in the form

$$p(\boldsymbol{x}) = C\exp[-\phi(\boldsymbol{x})] \tag{5.1.6}$$

Substituting (5.1.6) into (5.1.5), one obtains

$$b_{ij}\frac{\partial\phi}{\partial x_j} = \frac{\partial}{\partial x_j}b_{ij} - 2a_i \tag{5.1.7}$$

These are n equations for ϕ. The stochastic system is said to belong to the class of stationary potential if a consistent ϕ can be found to satisfy all these equations.

If matrix $\boldsymbol{B} = [b_{ij}]$ is nonsingular, so that its inverse $\boldsymbol{B}^{-1} = \boldsymbol{D} = [d_{ij}]$ exists, then (5.1.7) may be simplified to

$$\frac{\partial\phi}{\partial x_l} = d_{li}\left(\frac{\partial b_{ij}}{\partial x_j} - 2a_i\right) \tag{5.1.8}$$

A consistent ϕ function is obtainable if the following compatibility conditions are satisfied:

$$\frac{\partial}{\partial x_m}d_{li}\left(\frac{\partial b_{ij}}{\partial x_j} - 2a_i\right) = \frac{\partial}{\partial x_l}d_{mi}\left(\frac{\partial b_{ik}}{\partial x_k} - 2a_i\right) \tag{5.1.9}$$

Specializing further to the case of isotropic diffusion (Stratonovich, 1963), defined as

$$b_{ij}(\boldsymbol{x}) = \begin{cases} K(\boldsymbol{x}) & i = j \\ 0 & i \neq j \end{cases} \tag{5.1.10}$$

equation (5.1.8) becomes

$$\frac{\partial\phi}{\partial x_i} = \frac{1}{K}\left(\frac{\partial K}{\partial x_i} - 2a_i\right) \tag{5.1.11}$$

and the compatibility conditions (5.1.9) reduce to

$$\frac{\partial}{\partial x_m}\left(\frac{a_l}{K}\right) = \frac{\partial}{\partial x_l}\left(\frac{a_m}{K}\right) \tag{5.1.12}$$

When these conditions are satisfied, the following stationary probability potential is obtained from (5.1.11):

$$\phi = \ln K - \int \frac{2a_i}{K} dx_i \tag{5.1.13}$$

which has the same form as (5.1.4). The stationary probability density is then given by

$$p(\boldsymbol{x}) = \frac{C}{K(\boldsymbol{x})} \exp\left[\int \frac{2a_i(\boldsymbol{x})}{K(\boldsymbol{x})} dx_i\right] \tag{5.1.14}$$

provided that $p(\boldsymbol{x})$ is integrable over the state space $\boldsymbol{x}$.

In passing, we note that under the condition of vanishing probability flow at the boundaries, every first-order system must belong to the class of stationary potential. However, many higher order systems do not belong to this class, particularly the oscillatory systems, for which the $\boldsymbol{B}$ matrix is singular. Therefore, a more general solution scheme is required.

5.2 STATE OF DETAILED BALANCE

Consider a dynamical system under gaussian white noise excitations, governed by a set of differential equations of the form of (5.0.1). The response vector $\boldsymbol{X}(t)$ is a Markov vector process. Following Graham and Haken (1971), the components of $\boldsymbol{X}(t)$ may be classified as either odd or even variables, according to the transformation from X_i to $\tilde{X}_i$ upon a time reversal $t \to -t$. The even variables do not change their signs when time is reversed, whereas the odd variables do. Thus we denote

$$\tilde{X}_i = \varepsilon_i X_i \qquad \text{(no summation on } i) \tag{5.2.1}$$

where $\varepsilon_i = \pm 1$, corresponding to the even and odd variables, respectively. That the summation rule does not apply to the subscript of ε will be understood in the remainder of this section. For a structural system, the displacements are even variables, while the velocities are odd variables.

Definition. A Markov process is said to be in the state of detailed balance if

$$p_2(\boldsymbol{x}, t; \boldsymbol{x}', t') = p_2(\tilde{\boldsymbol{x}}', t; \tilde{\boldsymbol{x}}, t') \qquad t > t' \tag{5.2.2}$$

where p_2 is the joint probability density of $\boldsymbol{X}(t)$ at two different times. Equation (5.2.2) implies that each transition from one state to another is balanced statistically by a reversed transition.

Equation (5.2.2) may be rewritten as

$$q(\boldsymbol{x}, t \mid \boldsymbol{x}', t')p_1(\boldsymbol{x}', t') = q(\tilde{\boldsymbol{x}}', t \mid \tilde{\boldsymbol{x}}, t')p_1(\tilde{\boldsymbol{x}}, t') \tag{5.2.3}$$

where q is the transition probability density and p_1 is the first-order probability density. When a Markov vector reaches the state of statistical stationarity, p_1 becomes time-invariant and q depends only on the transition time $\tau = t - t'$. Then equation (5.2.3) simplifies to

$$q(\boldsymbol{x}, \tau \mid \boldsymbol{x}')p(\boldsymbol{x}') = q(\tilde{\boldsymbol{x}}', \tau \mid \tilde{\boldsymbol{x}})p(\tilde{\boldsymbol{x}}) \tag{5.2.4}$$

where $p(\boldsymbol{x})$ is the stationary probability density.

Definition. A stationary Markov process is said to be in the state of detailed balance if equation (5.2.4) is satisfied.

Theorem. The necessary and sufficient conditions for a stationary Markov process to be in the state of detailed balance are

$$a_i(\boldsymbol{x})p(\boldsymbol{x}) + \varepsilon_i a_i(\tilde{\boldsymbol{x}})p(\boldsymbol{x}) - \frac{\partial}{\partial x_j}[b_{ij}(\boldsymbol{x})p(\boldsymbol{x})] = 0 \tag{5.2.5}$$

and

$$b_{ij}(\boldsymbol{x}) - \varepsilon_i\varepsilon_j b_{ij}(\tilde{\boldsymbol{x}}) = 0 \tag{5.2.6}$$

Proof of necessity. Integrating both sides of equation (5.2.4) at $\tau = 0$, and noting that $q(\boldsymbol{x}, 0 \mid \boldsymbol{x}') = \delta(\boldsymbol{x} - \boldsymbol{x}')$ and $q(\tilde{\boldsymbol{x}}', 0 \mid \tilde{\boldsymbol{x}}) = \delta(\tilde{\boldsymbol{x}}' - \tilde{\boldsymbol{x}})$, it is seen readily that $p(\tilde{\boldsymbol{x}}) = p(\boldsymbol{x})$. Thus equation (5.2.4) may be rearranged to read

$$q(\boldsymbol{x}, \tau \mid \boldsymbol{x}') = \frac{p(\boldsymbol{x})}{p(\boldsymbol{x}')} q(\tilde{\boldsymbol{x}}', \tau \mid \tilde{\boldsymbol{x}}) \tag{5.2.7}$$

Now, the transition probability density q satisfies the Fokker-Planck equation

$$\frac{\partial}{\partial \tau} q(\boldsymbol{x}, \tau \mid \boldsymbol{x}') = -\frac{\partial}{\partial x_i}[a_i(\boldsymbol{x})q] + \frac{1}{2}\frac{\partial^2}{\partial x_i \partial x_j}[b_{ij}(\boldsymbol{x})q] \tag{5.2.8}$$

and the stationary probability density p satisfies the reduced Fokker-Planck equation

$$\frac{\partial}{\partial x_i}[a_i(\boldsymbol{x})p(\boldsymbol{x})] - \frac{1}{2}\frac{\partial^2}{\partial x_i \partial x_j}[b_{ij}(\boldsymbol{x})p(\boldsymbol{x})] = 0 \tag{5.2.9}$$

Substituting (5.2.7) into (5.2.8) and imposing (5.2.9), we obtain

$$\frac{\partial q(\tilde{\boldsymbol{x}}', \tau \mid \tilde{\boldsymbol{x}})}{\partial \tau} + a_i(\boldsymbol{x})\frac{\partial q(\tilde{\boldsymbol{x}}', \tau \mid \tilde{\boldsymbol{x}})}{\partial x_i} - \frac{1}{p(\boldsymbol{x})}\frac{\partial q(\tilde{\boldsymbol{x}}', \tau \mid \tilde{\boldsymbol{x}})}{\partial x_i}\frac{\partial [b_{ij}(\boldsymbol{x})p(\boldsymbol{x})]}{\partial x_j}$$
$$-\frac{1}{2}b_{ij}(\boldsymbol{x})\frac{\partial^2 q(\tilde{\boldsymbol{x}}', \tau \mid \tilde{\boldsymbol{x}})}{\partial x_i \partial x_j} = 0 \tag{5.2.10}$$

The transition probability density q is also a solution of the Kolmogorov backward equation, which may be cast in the form

$$\frac{\partial q(\tilde{\boldsymbol{x}}', \tau \mid \tilde{\boldsymbol{x}})}{\partial \tau} = a_i(\tilde{\boldsymbol{x}})\frac{\partial q(\tilde{\boldsymbol{x}}', \tau \mid \tilde{\boldsymbol{x}})}{\partial \tilde{x}_i} + \frac{1}{2}b_{ij}(\tilde{\boldsymbol{x}})\frac{\partial^2 q(\tilde{\boldsymbol{x}}', \tau \mid \tilde{\boldsymbol{x}})}{\partial \tilde{x}_i \partial \tilde{x}_j} \tag{5.2.11}$$

Equation (5.2.11) differs from the usual backward equation in that the backward time variable has been replaced by the forward time variable τ. Noticing that

$$\frac{\partial}{\partial \tilde{x}_i} = \varepsilon_i \frac{\partial}{\partial x_i} \quad \text{and} \quad \frac{\partial^2}{\partial \tilde{x}_i \partial \tilde{x}_j} = \varepsilon_i \varepsilon_j \frac{\partial^2}{\partial x_i \partial x_j}$$

combination of (5.2.10) and (5.2.11) yields

$$\left\{\frac{1}{p(\boldsymbol{x})}\frac{\partial[b_{ij}(\boldsymbol{x})p(\boldsymbol{x})]}{\partial x_j} - a_i(\boldsymbol{x}) - \varepsilon_i a_i(\tilde{\boldsymbol{x}})\right\}\frac{\partial q(\tilde{\boldsymbol{x}}', \tau \mid \tilde{\boldsymbol{x}})}{\partial x_i}$$

$$+\frac{1}{2}[b_{ij}(\boldsymbol{x}) - \varepsilon_i\varepsilon_j b_{ij}(\tilde{\boldsymbol{x}})]\frac{\partial^2 q(\tilde{\boldsymbol{x}}', \tau \mid \tilde{\boldsymbol{x}})}{\partial x_i \partial x_j} = 0 \tag{5.2.12}$$

Multiplying equation (5.2.12) by an arbitrary, twice-differentiable function $F(\tilde{\boldsymbol{x}}')$ and integrating over $\tilde{\boldsymbol{x}}'$ for the special case $\tau = 0$ when $q(\tilde{\boldsymbol{x}}', 0 \mid \tilde{\boldsymbol{x}} = \delta(\tilde{\boldsymbol{x}}' - \tilde{\boldsymbol{x}})$, we obtain

$$\{\frac{1}{p(\boldsymbol{x})}\frac{\partial[b_{ij}(\boldsymbol{x})p(\boldsymbol{x})]}{\partial x_i} - a_i(\boldsymbol{x}) - \varepsilon_i a_i(\tilde{\boldsymbol{x}})\}\frac{\partial F(\tilde{\boldsymbol{x}})}{\partial x_i} + \frac{1}{2}[b_{ij}(\boldsymbol{x}) - \varepsilon_i\varepsilon_j b_{ij}(\tilde{\boldsymbol{x}})]\frac{\partial^2 F(\tilde{\boldsymbol{x}})}{\partial x_i\, \partial x_j} = 0 \tag{5.2.13}$$

Since $F(\tilde{\mathbf{x}})$ is arbitrary, the coefficients for each derivative of $F(\tilde{\boldsymbol{x}})$ must vanish; that is, equations (5.2.5) and (5.2.6) must be satisfied.

Proof of sufficiency. Let

$$S(\boldsymbol{x}, \boldsymbol{x}', \tau) = \frac{p(\boldsymbol{x})}{p(\boldsymbol{x}')}q(\tilde{\boldsymbol{x}}', \tau \mid \tilde{\boldsymbol{x}}) \tag{5.2.14}$$

It can be shown that

$$-\frac{\partial}{\partial x_i}[a_i(\boldsymbol{x})S] + \frac{1}{2}\frac{\partial^2}{\partial x_i\, \partial x_j}[b_{ij}(\boldsymbol{x})S]$$

$$= \frac{1}{p(\boldsymbol{x}')}\left\{-a_i(\boldsymbol{x})p(\boldsymbol{x}) + \frac{\partial[b_{ij}(\boldsymbol{x})p(\boldsymbol{x})]}{\partial x_j}\right\}\frac{\partial q(\tilde{\boldsymbol{x}}', \tau \mid \tilde{\boldsymbol{x}})}{\partial x_i}$$

$$+\frac{1}{2}b_{ij}(\boldsymbol{x})\frac{p(\boldsymbol{x})}{p(\boldsymbol{x}')}\frac{\partial^2 q(\tilde{\boldsymbol{x}}', \tau \mid \tilde{\boldsymbol{x}})}{\partial x_i \partial x_j} \tag{5.2.15}$$

Upon imposing (5.2.5) and (5.2.6), the right-hand side of equation (5.2.15) becomes

$$\varepsilon_i\frac{a_i(\tilde{\boldsymbol{x}})p(\boldsymbol{x})}{p(\boldsymbol{x}')}\frac{\partial q(\tilde{\boldsymbol{x}}', \tau \mid \tilde{\boldsymbol{x}})}{\partial x_i} + \frac{1}{2}\varepsilon_i\varepsilon_j\frac{b_{ij}(\tilde{\boldsymbol{x}})p(\boldsymbol{x})}{p(\boldsymbol{x}')}\frac{\partial^2 q(\tilde{\boldsymbol{x}}', \tau \mid \tilde{\boldsymbol{x}})}{\partial x_i \partial x_j}$$

$$= \left[a_i(\tilde{\boldsymbol{x}})\frac{\partial q(\tilde{\boldsymbol{x}}', \tau \mid \tilde{\boldsymbol{x}})}{\partial \tilde{x}_i} + \frac{1}{2}b_{ij}(\tilde{\boldsymbol{x}})\frac{\partial^2 q(\tilde{\boldsymbol{x}}', \tau \mid \tilde{\boldsymbol{x}})}{\partial \tilde{x}_i \partial \tilde{x}_j}\right]\frac{p(\boldsymbol{x})}{p(\boldsymbol{x}')}$$

$$= \left[\frac{\partial q(\tilde{\boldsymbol{x}}', \tau \mid \tilde{\boldsymbol{x}})}{\partial \tau}\right]\frac{p(\boldsymbol{x})}{p(\boldsymbol{x}')} = \frac{\partial S(\boldsymbol{x}, \boldsymbol{x}', \tau)}{\partial \tau} \tag{5.2.16}$$

By comparing the left-hand side of (5.2.15) and the right-hand side of (5.2.16), it is clear that $S(\boldsymbol{x}, \boldsymbol{x}', \tau)$, satisfies the Fokker-Planck equation. If $S(\boldsymbol{x}, \boldsymbol{x}', \tau)$ is normalized for a total probability measure 1, then it must be the unique solution for the Fokker-Planck equation:

$$S(\boldsymbol{x}, \boldsymbol{x}', \tau) = q(\boldsymbol{x}, \tau \mid \boldsymbol{x}') \tag{5.2.17}$$

Combination of equations (5.2.14) and (5.2.17) leads to equation (5.2.4), the definition of the state of detailed balance of a Markov process in the stationary state.

We have proved that (5.2.5) and (5.2.6) constitute a set of sufficient and necessary conditions for detailed balance. Since each of the equations in (5.2.5) contains the stationary probability density $p(\boldsymbol{x})$, each may be used to obtain $p(\boldsymbol{x})$, instead of the original Fokker-Planck equation. Of course, a solution is acceptable only if it satisfies all the equations in (5.2.5). To facilitate such a solution scheme, Graham and Haken (1971) developed an alternative set of equations, equivalent to those in (5.2.5) as follows.

Separate each first derivate moment into the reversible and irreversible parts,

$$a_i(\boldsymbol{x}) = a_i^R(\boldsymbol{x}) + a_i^I(\boldsymbol{x}) \tag{5.2.18}$$

where

$$a_i^R(\boldsymbol{x}) = \tfrac{1}{2}\left[a_i(\boldsymbol{x}) - \varepsilon_i a_i(\tilde{\boldsymbol{x}})\right] \tag{5.2.19}$$

$$a_i^I(\boldsymbol{x}) = \tfrac{1}{2}\left[a_i(\boldsymbol{x}) + \varepsilon_i a_i(\tilde{\boldsymbol{x}})\right] \tag{5.2.20}$$

The adjectives "reversible" and "irreversible" refer to the fact that

$$a_i^R(\boldsymbol{x}) = -\varepsilon_i a_i^R(\tilde{\boldsymbol{x}}) \tag{5.2.21}$$

$$a_i^I(\boldsymbol{x}) = \varepsilon_i a_i^I(\tilde{\boldsymbol{x}}) \tag{5.2.22}$$

where $\varepsilon_i = 1$ or -1, depending on whether $\boldsymbol{X}_i$, with which a_i is associated, is an even or odd variable. Expressing $p(\boldsymbol{x})$ in the exponential form (5.1.6) and substituting (5.2.18) and (5.1.6) into (5.2.5), we obtain

$$a_i^I(\boldsymbol{x}) = \frac{1}{2}\left[\frac{\partial}{\partial x_j} b_{ij}(\boldsymbol{x}) - b_{ij}(\boldsymbol{x})\frac{\partial \phi}{\partial x_j}\right] \tag{5.2.23}$$

Equations (5.2.23) are equivalent to (5.2.5), but with the probability potential $\phi(\boldsymbol{x})$ as the unknown. Now $\phi(\boldsymbol{x})$ must also satisfy

$$\frac{\partial}{\partial x_i} a_i^R(\boldsymbol{x}) = a_i^R(\boldsymbol{x})\frac{\partial \phi}{\partial x_i} \tag{5.2.24}$$

which is obtained by substituting (5.1.6), (5.2.18), and (5.2.23) into the reduced Fokker-Planck equation (5.2.9). The combination of equations (5.2.23) and (5.2.24) is a set of $n + 1$ first-order differential equations. If a stationary Markov process is in detailed balance, then (5.2.23) and (5.2.24) must be consistent, and they can be used to determine the probability potential $\phi(\boldsymbol{x})$.

Equation (5.2.23) is of the same form as (5.1.7) except that a_i in (5.1.7) is replaced by a_i^I in (5.2.23). Thus the case of detailed balance is more general than that of stationary potential, since only the irreversible parts of the first derivate moments are associated with the vanishing probability flow (Yong and Lin, 1987).

The application of equations (5.2.23) and (5.2.24) is illustrated in the following two examples. We note in passing that the concept of detailed balance has been applied by van Kampen (1957) to a more general class of equations, called the master equations, of which the Fokker-Planck equation is a special case.

5.2.1 A Nonlinear System under Additive White Noise Excitation

Consider the system

$$\ddot{X} + h(\Lambda)\dot{X} + u(X) = W(t) \tag{a}$$

where $W(t)$ is a gaussian white noise with a spectral density K, $u(X)$ is a nonlinear spring force, $h(\Lambda)$ is an arbitrary function which must be positive when its argument becomes large, and Λ is the would-be total energy of the system if the damping and the excitation were absent, that is,

$$\Lambda = \tfrac{1}{2}\dot{X}^2 + \int_0^X u(z)\,dz \tag{b}$$

Let $X_1 = X$ and $X_2 = \dot{X}$. Equation (a) is equivalent to a set of two Itô-type stochastic differential equations

$$dX_1 = X_2\,dt \tag{c}$$

$$dX_2 = [-h(\Lambda)X_2 - u(X_1)]\,dt + \sqrt{2\pi K}\,dB(t) \tag{d}$$

The corresponding reduced Fokker-Planck equation is then

$$x_2\frac{\partial p}{\partial x_1} - \frac{\partial}{\partial x_2}\{[h(\lambda)x_2 + u(x_1)]\,p\} - \pi K\frac{\partial^2 p}{\partial x_2^2} = 0 \tag{e}$$

The first and second derivate moments are

$$a_1 = x_2 \qquad a_2 = -h(\lambda)x_2 - u(x_1) \tag{f}$$

and

$$b_{11} = b_{12} = b_{21} = 0 \qquad b_{22} = 2\pi K \tag{g}$$

Clearly, one of the conditions for detailed balance, equation (5.2.6), is satisfied. As noted before, the displacement X_1 is an even variable and the velocity X_2 is an odd variable: $\varepsilon_1 = 1$ and $\varepsilon_2 = -1$. The reversible and irreversible parts of the first derivate moments can then be obtained from (5.2.19) and (5.2.20) as follows:

$$a_1^R = x_2 \qquad a_1^I = 0 \qquad a_2^R = -u(x_1) \qquad a_2^I = -h(\lambda)x_2 \tag{h}$$

Substituting (g) and (h) into (5.2.23) and (5.2.24), we obtain

$$\pi K\frac{\partial \phi}{\partial x_2} = h(\lambda)x_2 \tag{i}$$

and

$$x_2\frac{\partial \phi}{\partial x_1} - u(x_1)\frac{\partial \phi}{\partial x_2} = 0 \tag{j}$$

The general solution for equation (i) is

$$\phi = \frac{1}{\pi K}\int_0^\lambda h(u)\,du + g(x_1) \tag{k}$$

In order that (j) is satisfied also, function g must be a constant, which may be taken to be zero without loss of generality. Thus

$$p(x_1, x_2) = C\exp\left[-\frac{1}{\pi K}\int_0^\lambda h(u)\,du\right] \tag{l}$$

This result was obtained earlier by Caughey (1971), using a different procedure.

5.2.2 A Nonlinearly Damped System under Both Multiplicative and Additive White Noise Excitations

Next we consider a nonlinear system governed by

$$\ddot{X} + (\alpha + \beta X^2)\dot{X} + \omega_0^2[1 + W_1(t)]X = W_2(t) \tag{a}$$

where $W_1(t)$ and $W_2(t)$ are independent gaussian white noises with spectral densities K_{11} and K_{22}, respectively. Equation (a) is replaced by the following Itô equations:

$$dX_1 = X_2 dt \tag{b}$$

$$dX_2 = -[(\alpha + \beta X_1^2)X_2 + \omega_0^2 X_1]\,dt + [2\pi(\omega_0^4 K_{11} X_1^2 + K_{22})]^{1/2} dB(t) \tag{c}$$

The multiplicative excitation $W_1(t)$ does not give rise to the Wong-Zakai correction, since it is associated with the spring term. The reduced Fokker-Planck equation, corresponding to (b) and (c), is given by

$$x_2\frac{\partial p}{\partial x_1} - \frac{\partial}{\partial x_2}\left\{\left[(\alpha + \beta x_1^2)x_2 + \omega_0^2 x_1\right]p\right\} - \pi(\omega_0^4 K_{11} x_1^2 + K_{22})\frac{\partial^2 p}{\partial x_2^2} = 0 \tag{d}$$

The first and second derivate moments are

$$a_1 = x_2 \qquad a_2 = -(\alpha + \beta x_1^2)x_2 - \omega_0^2 x_1 \tag{e}$$

and

$$b_{11} = b_{12} = b_{21} = 0 \qquad b_{22} = 2\pi(\omega_0^4 K_{11} x_1^2 + K_{22}) \tag{f}$$

The second derivate moments satisfy equation (5.2.6). Separate the first derivate moments into

$$a_1^R = x_2 \qquad a_1^I = 0 \qquad a_2^R = -\omega_0^2 x_1 \qquad a_2^I = -(\alpha + \beta x_1^2)x_2 \tag{g}$$

Substituting (f) and (g) into (5.2.23) and (5.2.24), we obtain

$$\pi(K_{11}\omega_0^4 x_1^2 + K_{22})\frac{\partial \phi}{\partial x_2} = (\alpha + \beta x_1^2)x_2 \tag{h}$$

$$x_2\frac{\partial \phi}{\partial x_1} - \omega_0^2 x_1 \frac{\partial \phi}{\partial x_2} = 0 \tag{i}$$

The general solution for equation (i) is

$$\phi(x_1, x_2) = \phi(\lambda) \tag{j}$$

where

$$\lambda = \tfrac{1}{2}x_2^2 + \tfrac{1}{2}\omega_0^2 x_1^2 \tag{k}$$

Combining (h), (j), and (k),

$$\frac{d\phi}{d\lambda} = \frac{\alpha + \beta x_1^2}{\pi(K_{22} + K_{11}\omega_0^4 x_1^2)} \tag{l}$$

Since the left-hand side of equation (l) is a function of λ, the right-hand side must also be a function of λ regardless of the value of x_1, which can be true only if

$$\frac{\alpha}{\beta} = \frac{K_{22}}{K_{11}\omega_0^4} \tag{m}$$

Equation (m) imposes a condition for system (a) to be in detailed balance. Under this condition, the probability potential function can be obtained as

$$\phi(\lambda) = \frac{\alpha}{\pi K_{22}}\lambda \tag{n}$$

namely, the stationary probability density is given by

$$p(x_1, x_2) = C\exp\left[-\frac{\alpha}{2\pi K_{22}}(\omega_0^2 x_1^2 + x_2^2)\right] \tag{o}$$

Equation (o) shows that the displacement $X(t) = X_1(t)$ and the velocity $\dot{X}(t) = X_2(t)$ at the same t are jointly gaussian distributed. The same gaussian distribution is obtained if we set $\beta = 0$ and $W_1(t) = 0$, that is, if the system is linear and under only the additive white noise excitation. Therefore, under a suitable combination of additive and multiplicative gaussian random excitations, as indicated in equation (m), the stationary response of a nonlinear system can also be gaussian. Indeed, different stochastic systems can sometimes share a common probability distribution, to be discussed in more detail in Section 5.4.

5.3 GENERALIZED STATIONARY POTENTIAL

The solvable class of detailed balance will now be expanded further to include those systems that are not in detailed balance. For this purpose, we split *both* the first and second derivate moments as follows:

$$a_i = a_i^{(1)} + a_i^{(2)} \tag{5.3.1}$$

$$b_{ij} = b_{ij}^{(i)} + b_{ji}^{(j)} \tag{5.3.2}$$

where the separation between $a_i^{(1)}$ and $a_i^{(2)}$ in (5.3.1) is not restricted to irreversible and reversible, and the superscript (i) or (j) in (5.3.2) does not imply a summation. However, equation (5.3.2) does retain the symmetric property of the second derivate moments; that is, $b_{ij} = b_{ji}$. Substituting (5.3.1) and (5.3.2) into (5.0.9), we obtain

$$\frac{\partial}{\partial x_i}\left[a_i^{(1)}p - \frac{\partial}{\partial x_j}b_{ij}^{(i)}p\right] + \frac{\partial}{\partial x_i}\left[a_i^{(2)}p\right] = 0 \tag{5.3.3}$$

Equation (5.3.3) is satisfied under the following more restrictive sufficient conditions:

$$a_i^{(1)}p - \frac{\partial}{\partial x_j}b_{ij}^{(i)}p = 0 \tag{5.3.4}$$

$$\frac{\partial}{\partial x_i}\left[a_i^{(2)}p\right] = 0 \tag{5.3.5}$$

It is interesting to note that equation (5.3.4) resembles equation (5.1.5) obtained earlier under the assumption of vanishing probability flow. Thus $a_i^{(1)}$ in (5.3.4) represents a portion of the drift in the ith direction which is required to balance the differential diffusion and maintain a zero probability flow in the ith direction. The remaining portion $a_i^{(2)}$ must then cause a steady circulatory probability flow, since the distribution of the total probability remains unchanged in the stationary state.

Now let p be expressed in terms of the potential function ϕ. Equations (5.3.4) and (5.3.5) are replaced by

$$a_i^{(1)}(\boldsymbol{x}) = \frac{\partial}{\partial x_j} b_{ij}^{(i)}(\boldsymbol{x}) - b_{ij}^{(i)}(\boldsymbol{x}) \frac{\partial \phi}{\partial x_j} \tag{5.3.6}$$

and

$$\frac{\partial}{\partial x_i} a_i^{(2)}(\boldsymbol{x}) = a_i^{(2)}(\boldsymbol{x}) \frac{\partial \phi}{\partial x_i} \tag{5.3.7}$$

respectively. There are a total of $n + 1$ equations for ϕ in the set of (5.3.6) and (5.3.7). The problem is solvable if a consistent ϕ function can be obtained from these equations. Such a solvable system is said to belong to the class of generalized stationary potential.

In the case of detailed balance, the required splitting of a_i and b_{ij} can be identified as

$$a_i^{(1)} = a_i^I \tag{5.3.8}$$

$$a_i^{(2)} = a_i^R \tag{5.3.9}$$

$$b_{ij}^{(i)} = b_{ji}^{(j)} = \tfrac{1}{2} b_{ij} \tag{5.3.10}$$

Therefore, the class of generalized stationary potential is more general than that of detailed balance. Moreover, the additional requirement for detailed balance, equation (5.2.6), is superfluous, since it is not needed when obtaining an exact solution.

In the following discussion, equations (5.3.6) and (5.3.7) are applied to single- and multi-degree-of-freedom systems.

5.3.1 Single-Degree-of-Freedom Systems

The equation of motion for a single-degree-of-freedom second-order stochastic system has the general form

$$\ddot{X} + h(X, \dot{X}) = g_i(X, \dot{X}) W_i(t) \tag{a}$$

where $h(\cdot)$ and $g_i(\cdot)$ are generally nonlinear, $W_i(t)$ are gaussian white noises, and

$$E[W_i(t) W_j(t + \tau)] = 2\pi K_{ij} \delta(\tau) \tag{b}$$

The first and second derivate moments can be obtained from (5.0.5) and (5.0.6), yielding

$$\begin{aligned} a_1 &= x_2 \\ a_2 &= -h(x_1, x_2) + \pi K_{ij} g_i(x_1, x_2) \frac{\partial}{\partial x_2} g_j(x_1, x_2) \\ b_{11} &= b_{12} = b_{21} = 0 \qquad b_{22} = 2\pi K_{ij} g_i(x_1, x_2) g_j(x_1, x_2) \end{aligned} \tag{c}$$

where x_1 and x_2 are the state variables of $X(t)$ and $\dot{X}(t)$, respectively. The reduced Fokker-Planck equation is then given by

$$x_2 \frac{\partial}{\partial x_1} p + \frac{\partial}{\partial x_2} \left\{ \left[-h(x_1, x_2) + \pi K_{ij} g_i(x_1, x_2) \frac{\partial g_j(x_1, x_2)}{\partial x_2} \right] p \right\}$$
$$- \pi K_{ij} \frac{\partial^2}{\partial x_2^2} [g_i(x_1, x_2) g_j(x_1, x_2) p] = 0 \tag{d}$$

subject to the condition that the probability flow vanishes either as $|x_1| \to \infty$, or as $|x_2| \to \infty$, or both; thus,

$$p \to 0 \qquad \frac{\partial p}{\partial x_2} \to 0 \qquad \text{as } x_1^2 + x_2^2 \to +\infty \tag{e}$$

Now split the first and second derivate moments into

$$\begin{aligned}
&a_1^{(1)} = 0 \qquad a_1^{(2)} = x_2 \\
&a_2^{(1)} = -h(x_1, x_2) + \pi K_{ij} g_i(x_1, x_2) \frac{\partial}{\partial x_2} g_j(x_1, x_2) + \frac{\lambda_x}{\lambda_y} \qquad a_2^{(2)} = -\frac{\lambda_x}{\lambda_y} \\
&b_{11}^{(1)} = 0 \qquad b_{12}^{(1)} = -b_{21}^{(2)} \qquad b_{22}^{(2)} = \tfrac{1}{2} b_{22}
\end{aligned} \tag{f}$$

where $x = x_1$, $y = x_2^2/2$, $\lambda(x, y)$ is an arbitrary function differentiable with respect to x and twice differentiable with respect to y, and a subscript x or y represents a partial differentiation. The physical meaning of the term λ_x/λ_y is explained later. Application of equations (5.3.6) and (5.3.7) leads to

$$b_{12}^{(1)} \frac{\partial \phi}{\partial x_2} = \frac{\partial}{\partial x_2} b_{12}^{(1)} \tag{g}$$

$$b_{21}^{(2)} \frac{\partial \phi}{\partial x_1} + \pi K_{ij} g_i g_j \frac{\partial \phi}{\partial x_2} = \frac{\partial b_{21}^{(2)}}{\partial x_1} + h(x_1, x_2) + \pi K_{ij} g_i \frac{\partial g_j}{\partial x_2} - \frac{\lambda_x}{\lambda_y} \tag{h}$$

$$-x_2 \frac{\partial \phi}{\partial x_1} - \frac{\partial}{\partial x_2} \left(\frac{\lambda_x}{\lambda_y} \right) + \frac{\lambda_x}{\lambda_y} \frac{\partial \phi}{\partial x_2} = 0 \tag{i}$$

We next seek the condition under which a consistent ϕ function can be obtained from equations (g), (h), and (i).

The general solution of equation (i) is

$$\phi(x_1, x_2) = -\ln \lambda_y + \phi_0(\lambda) \tag{j}$$

where $\phi_0(\lambda)$ is an arbitrary function of λ. Equation (g) may be considered a first-order differential equation for $b_{12}^{(1)}$, which may be solved to obtain

$$b_{12}^{(1)} = A(x_1) \exp(\phi) \tag{k}$$

where $A(x_1)$ is an arbitrary function of x_1. Substitution of (j) and (k) into (h) leads to

$$h(x_1, x_2) = \pi x_2 K_{ij} g_i(x_1, x_2) g_j(x_1, x_2) \left[\lambda_y \frac{d\phi_0(\lambda)}{d\lambda} - \frac{\lambda_{yy}}{\lambda_y} \right]$$
$$- \pi K_{ij} g_i(x_1, x_2) \frac{\partial g_j(x_1, x_2)}{\partial x_2} + \frac{\lambda_x}{\lambda_y} + \frac{B(x_1)}{\lambda_y} e^{\phi_0(\lambda)} \tag{l}$$

where $B(x_1)$ is also an arbitrary function of x_1. System (a) is said to belong to the class of generalized stationary potential if function h can be cast in the form of (l). In this case, the stationary probability density of the response of system (a) is given by

$$p(x_1, x_2) = C\exp(-\phi) = C\lambda_y \exp(-\phi_0) \tag{m}$$

where C is a normalization constant.

It is of interest to note that the $B(x_1)$ term in equation (l) is a direct consequence of letting $b_{12}^{(1)} = -b_{21}^{(2)} \neq 0$. However, within the scope of a single degree of freedom, we have yet to find a practical use of this additional generality. Thus we shall let $B(x_1) = 0$ in the remainder of this section.

The physical meaning of the term λ_x/λ_y in equation (l) can be revealed by considering the case of undamped free motion. Letting $g_i = g_j = 0$ in equation (l), we obtain

$$h(x_1, x_2) = \frac{\lambda_x}{\lambda_y} \tag{n}$$

This h function corresponds to a deterministic system

$$\ddot{x} + \frac{\lambda_x}{\lambda_y} = 0 \tag{o}$$

Noting that $\ddot{x} = \dot{x}\,d\dot{x}/dx = dy/dx$, equation (o) may be replaced by

$$d\lambda = \lambda_x\,dx + \lambda_y\,dy = 0 \tag{p}$$

Therefore, $\lambda(x, y)$ is not truly arbitrary; it represents the total energy in a conservative system. Moreover, equation (o) shows that the conservative force in the system, which gives rise to the acceleration $\ddot{x}$, is equal to $-\lambda_x/\lambda_y$ per unit mass.

Return now to the randomly excited system, and note the particular way in which the second derivate moment a_2 is split, as indicated in (f). The splitting is to render $a_2^{(1)}$ devoid of the conservative force. This is accomplished by subtracting $-\lambda_x/\lambda_y$ from (or adding λ_x/λ_y to) the original a_2 to obtain $a_2^{(1)}$. It is important to note that if multiplicative random excitations appear in damping, then the Wong-Zakai correlation terms

$$\pi K_{ij} g_i(x_1, x_2) \frac{\partial}{\partial x_2} g_j(x_1, x_2)$$

may produce a contribution toward the conservative force, in which case the added λ_x/λ_y term must be such as to remove the total "effective" conservative force from a_2.

For most structural or mechanical engineering applications, we may let $\lambda = y + \int u(x_1)dx_1$. Then $\lambda_y = 1$, $\lambda_x/\lambda_y = u(x_1)$, $\phi(\lambda) = \phi_0(\lambda)$, and equation ($l$) is simplified to

$$\begin{aligned} h(x_1, x_2) &= \pi x_2 K_{ij}\, g_i(x_1, x_2) g_j(x_1, x_2) \phi'(\lambda) \\ &\quad - \pi K_{ij}\, g_i(x_1, x_2) \frac{\partial g_j(x_1, x_2)}{\partial x_2} + u(x_1) \end{aligned} \tag{q}$$

where a prime denotes one differentiation with respect to the argument.

Equation (q) or more generally (l) provides a restriction between the constitutive law of the dynamic system, as represented by $h(x_1, x_2)$ and $g_i(x_1, x_2)$ on one hand, and the spectral densities of the random excitations K_{ij} on the other, so that an exact solution for the stationary probability density is obtainable using the present procedure. Application of the procedure is illustrated in the following examples.

Consider first the following nonlinear system under an additive random excitation:

$$\ddot{X} + \alpha\left[\Lambda_Y f(\Lambda) - \frac{\Lambda_{YY}}{\Lambda_Y}\right]\dot{X} + \frac{\Lambda_X}{\Lambda_Y} = W(t) \tag{r}$$

where $W(t)$ is a gaussian white noise with a spectral density K, Λ is an arbitrary function of X and $Y = \dot{X}^2/2$, and $f(\Lambda)$ is an arbitrary function of Λ. This system has been considered previously by Caughey and Ma (1982a). Comparing equations (a) and (r), we find that

$$h(x_1, x_2) = \alpha x_2\left[\lambda_y f(\lambda) - \frac{\lambda_{yy}}{\lambda_y}\right] + \frac{\lambda_x}{\lambda_y} \tag{s}$$

However, in order for system (r) to be in the class of generalized stationary potential, it is necessary that

$$h(x_1, x_2) = \pi K x_2\left[\lambda_y \phi_0'(\lambda) - \frac{\lambda_{yy}}{\lambda_y}\right] + \frac{\lambda_x}{\lambda_y} \tag{t}$$

which is obtained from equation (l) without the $B(x_1)$ term. This is satisfied if $\alpha = \pi K$. The stationary probability density for the response is therefore

$$p(x_1, x_2) = C\lambda_y \exp\left[-\int_0^{\lambda} f(v)\,dv\right] \tag{u}$$

according to equation (m). Equation (u) agrees with the result obtained earlier by Caughey and Ma (1982a). When λ depends linearly on y, λ_y becomes a constant, in which case equations (s) and (u) reduce, respectively, to equations (a) and (l) in Section 5.2.1.

Next let us examine a case involving both additive and multiplicative random excitations, described by the equation of motion

$$\ddot{X} + [f(\Lambda) + W_2(t)]\dot{X} + \omega_0^2[1 + W_1(t)]X = W_3(t) \tag{v}$$

where $\Lambda = (\dot{X}^2 + \omega_0^2 X^2)/2$ and $W_1(t)$, $W_2(t)$, and $W_3(t)$ are independent gaussian white noises. Comparing (v) with (a), we see that $g_1 = -\omega_0^2 x_1$, $g_2 = -x_2$, $g_3 = 1$, and

$$h(x_1, x_2) = x_2 f(\lambda) + \omega_0^2 x_1 \tag{w}$$

According to (q), however,

$$h(x_1, x_2) = \pi x_2(K_{11}\omega_0^4 x_1^2 + K_{22}x_2^2 + K_{33})\phi'(\lambda) - \pi K_{22}x_2 + \omega_0^2 x_1 \tag{x}$$

Equating (w) and (x), we find that

$$\phi'(\lambda) = \frac{f(\lambda) + \pi K_{22}}{\pi(K_{11}\omega_0^4 x_1^2 + K_{22}x_2^2 + K_{33})} \tag{y}$$

The right-hand side of equation (y) can be a function of λ only if

$$\omega_0^2 K_{11} = K_{22} \tag{z}$$

which is a condition for the system to belong to the class of generalized stationary potential. Under condition (z), equation (y) becomes

$$\phi'(\lambda) = \frac{f(\lambda) + \pi K_{22}}{\pi(2K_{22}\lambda + K_{33})} \tag{aa}$$

Upon integrating (aa), we obtain the probability potential ϕ, and hence the stationary probability density

$$p(x_1, x_2) = C(2K_{22}\lambda + K_{33})^{-1/2} \exp\left[-\int_0^{\lambda} \frac{f(v)}{\pi(2K_{22}v + K_{33})} dv\right] \tag{bb}$$

In the special case in which f is a linear function of λ, namely, $f(\lambda) = \alpha + \beta\lambda$, we obtain

$$p(x_1, x_2) = C(2K_{22}\lambda + K_{33})^{-\gamma} \exp\left(-\frac{\beta}{2\pi K_{22}}\lambda\right) \tag{cc}$$

where

$$\gamma = \frac{1}{2}\left(\frac{\alpha}{\pi K_{22}} + 1 - \frac{\beta}{2\pi K_{22}^2} K_{33}\right) \tag{dd}$$

Equation (cc) was obtained earlier by Dimentberg (1982) in different notations. In the absence of external excitation, $K_{33} = 0$, the right-hand side of (cc) is integrable only if $\gamma < 1$, or if

$$K_{22} > \frac{\alpha}{\pi} \tag{ee}$$

In this case, a nontrivial response exists, and the trivial solution $X = \dot{X} = 0$ is unstable in some sense (see Chapter 6).

An interesting nonlinear oscillator which is capable of exhibiting limit cycle and bifurcation behaviors is governed by the equation

$$\ddot{X} + X^2\left(\beta + \frac{4\alpha}{X^2 + \dot{X}^2}\right)\dot{X} + [1 + W(t)]X = 0 \qquad \beta > 0 \tag{ff}$$

We note that without the random multiplicative excitation, the equation has a stable trivial solution $X = \dot{X} = 0$, if $\alpha > 0$; it has a nontrivial solution corresponding to a stable limit cycle if $\alpha < 0$. With the multiplicative white noise $W(t)$ in place, the system possesses a probability potential

$$\phi = \frac{\beta}{2\pi K}(x_1^2 + x_2^2) + \frac{2\alpha}{\pi K} \ln(x_1^2 + x_2^2) \tag{gg}$$

where x_1 and x_2 are the state variables of $X(t)$ and $\dot{X}(t)$, respectively; this can be shown by using equation (q). Thus the stationary probability density of the response is

$$p(x_1, x_2) = C(x_1^2 + x_2^2)^{-2\alpha/\pi K} \exp\left[-\frac{\beta}{2\pi K}(x_1^2 + x_2^2)\right] \tag{hh}$$

In the case of $\alpha < 0$, the right-hand side of (hh) is always normalizable, indicating that a nontrivial stationary probability density always exists. However, in the case of $\alpha > 0$, a nontrivial stationary probability density exists only if $K > 2\alpha/\pi$. Thus $2\alpha/\pi$ represents a critical spectral level for the multiplicative excitation, at which bifurcation occurs in some sense, and below which the stationary solution is trivial $X = \dot{X} = 0$.

When a nontrivial stationary probability density of the system response exists ($K > 2\alpha/\pi$), its shape depends on whether $\alpha \geq 0$ or $\alpha < 0$. It can be shown that if $\alpha \geq 0$, the potential function ϕ has a minimum at $x_1 = x_2 = 0$. A minimum ϕ corresponds to a peak of the probability density, and its location is sometimes referred to as the most probable value. On the other hand, if $\alpha < 0$, the probability potential ϕ has a minimum at $x_1^2 + x_2^2 = 4|\alpha|/\beta$, and it tends to infinity as $x_1^2 + x_2^2$ approaches to either zero or infinity. In this case, the probability density has the shape shown in Fig. 5.3.1, and the most probable values for the response lie on the circle $x_1^2 + x_2^2 = 4|\alpha|/\beta$, which is the stochastic analogue of a limit cycle.

As the last example, consider a system governed by

$$\ddot{X} + \delta\dot{X} + \alpha(X + \beta\dot{X})^2\dot{X} + f(X) = (aX + b\dot{X})W_1(t) + W_2(t) \tag{ii}$$

where $\alpha > 0, \delta > 0, ab > 0$, and $W_1(t)$ and $W_2(t)$ are independent gaussian white noises. It can be shown that condition (5.2.6) is not satisfied. Therefore, the system is not in detailed balance. By using equation (q), we obtain

$$\begin{aligned} h(x_1, x_2) &= \delta x_2 + \alpha(x_1 + \beta x_2)^2 x_2 + f(x_1) \\ &= \pi x_2[K_{11}(ax_1 + bx_2)^2 + K_{22}]\phi'(\lambda) - \pi K_{11}b(ax_1 + bx_2) + u(x_1) \end{aligned} \tag{jj}$$

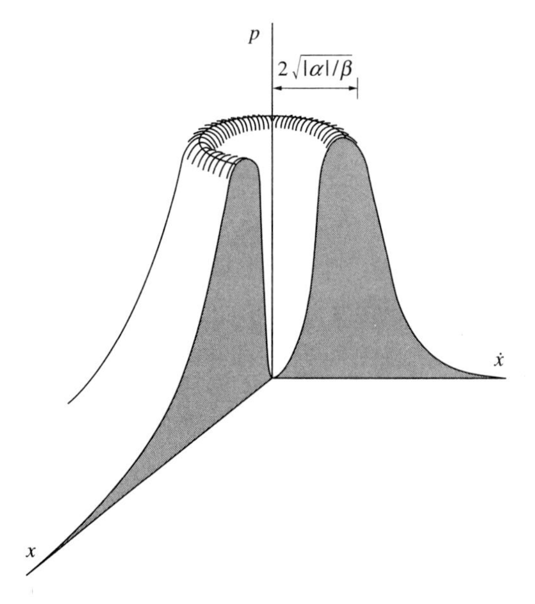

FIGURE 5.3.1
Stationary probability density (with the right-front quarter removed) of nonlinear system $\ddot{X} + X^2[\beta + 4\alpha/(X^2 + \dot{X}^2)]\dot{X} + [1 + W(t)]X = 0, \beta > 0, \alpha < 0$.

The system belongs to the class of generalized stationary potential if we let

$$u(x_1) = f(x_1) + \pi K_{11} abx_1 \tag{kk}$$

and if the following conditions are satisfied:

$$\beta = \frac{b}{a} \qquad \alpha = \frac{K_{11}}{K_{22}} a^2(\delta + \pi K_{11} b^2) \tag{ll}$$

In this case,

$$\lambda = \tfrac{1}{2} x_2^2 + \tfrac{1}{2} \pi K_{11} abx_1^2 + \int_0^{x_1} f(v)\, dv \tag{mm}$$

and $$p(x_1, x_2) = C \exp\left\{ -\frac{\delta + \pi K_{11} b^2}{2\pi K_{22}} \left[x_2^2 + \pi K_{11} abx_1^2 + 2\int_0^{x_1} u_1(v)\, dv \right] \right\} \tag{nn}$$

It is of interest to note that the term $\pi K_{11} abx_1$ in equation (kk) arises from the Wong and Zakai correction and contributes to the effective spring force.

5.3.2 Multi-Degree-of-Freedom Systems

The results obtained in Section 5.3.1 can be generalized for the case of multi-degree-of-freedom oscillatory systems. Let the system equations be

$$\ddot{Z}_i + h_i(\mathbf{Z}, \dot{\mathbf{Z}}) = g_{ij}(\mathbf{Z}, \dot{\mathbf{Z}}) W_j(t) \qquad i = 1, 2, \ldots, n, \quad j = 1, 2, \ldots, m \tag{a}$$

where $\mathbf{Z} = \{Z_1, \ldots, Z_n\}'$ and $\dot{\mathbf{Z}} = \{\dot{Z}_1, \ldots, \dot{Z}_n\}'$ are vectors of displacements and velocities, respectively. Equations (a) are equivalent to

$$\begin{aligned} \dot{X}_{2i-1} &= X_{2i} \\ \dot{X}_{2i} &= -h_i(\mathbf{X}) + g_{ij}(\mathbf{X}) W_j(t) \end{aligned} \tag{b}$$

where $X_{2i-1} = Z_i$, $X_{2i} = \dot{Z}_i$, and $\boldsymbol{X} = \{X_1, \ldots, X_{2n}\}'$. The first and second derivate moments are identified as follows:

$$\begin{gathered} a_{2i-1} = x_{2i} \qquad a_{2i} = -h_i + \pi K_{ls} g_{js} \frac{\partial g_{il}}{\partial x_{2j}} \\ b_{2i-1,2i-1} = b_{2i-1,j} = b_{j,2i-1} = 0 \qquad b_{2i,2j} = 2\pi K_{ls} g_{js} g_{il} \end{gathered} \tag{c}$$

Split them into

$$\begin{gathered} a_{2i-1}^{(1)} = 0 \qquad a_{2i-1}^{(2)} = x_{2i} \\ a_{2i}^{(1)} = -h_j + \pi K_{ls} g_{js} \frac{\partial g_{il}}{\partial x_{2j}} + \frac{\lambda_{x_{2i-1}}}{\lambda_{y_i}} \qquad a_{2i}^{(2)} = -\frac{\lambda_{x_{2i-1}}}{\lambda_{y_i}} \\ b_{2i-1,2i-1}^{(2i-1)} = 0 \qquad b_{2i-1,j}^{(2i-1)} = -b_{j,2i-1}^{(j)} \qquad b_{2i,2j}^{(2i)} + b_{2j,2i}^{(2j)} = 2\pi K_{ls} g_{js} g_{il} \end{gathered} \tag{d}$$

where $y_i = x_{2i}^2/2$ and $\lambda(x_1, x_3, \ldots, x_{2n-1}, y_1, \ldots, y_n)$ is an arbitrary function, differentiable with respect to $x_1, x_3, \ldots, x_{2n-1}$ and twice differentiable with respect to $y_1, y_2, \ldots, y_n$. Equations (5.3.6) and (5.3.7) lead to

$$b_{2i-1,j}^{(2i-1)} \frac{\partial \phi}{\partial x_j} = \frac{\partial}{\partial x_j} b_{2i-1,j}^{(2i-1)} \tag{e}$$

$$b_{2i,j}^{(2i)} \frac{\partial \phi}{\partial x_j} = \frac{\partial}{\partial x_j} b_{2i,j}^{(2i)} + h_i - \pi K_{ls} f_{js} \frac{\partial f_{il}}{\partial x_{2j}} - \frac{\lambda_{x_{2i-1}}}{\lambda_{y_i}} \tag{f}$$

$$-x_{2i} \frac{\partial \phi}{\partial x_{2i-1}} - \frac{\partial}{\partial x_{2i}} \left(\frac{\lambda_{x_{2i-1}}}{\lambda_{y_i}} \right) + \frac{\lambda_{x_{2i-1}}}{\lambda_{y_i}} \frac{\partial \phi}{\partial x_{2i}} = 0 \tag{g}$$

The next step is to find the conditions under which a consistent ϕ function can be found to satisfy the preceding three sets of equations. Unfortunately, this task is much more difficult for a multi-degree-of-freedom system. Two additional requirements will now be added to render the task simpler; however, the solvable class will become somewhat smaller than what would be the case without these additional requirements.

The first additional condition to be imposed is

$$\lambda_{y_i} = G(x_1, x_3, \ldots, x_{2n-1}) \qquad \text{for all } i \tag{h}$$

which amounts to requiring that λ depend linearly on each y_i in exactly the same manner. In this case, equation set (g) can be solved to yield

$$\phi(\boldsymbol{x}) = -\ln G + \phi_0(\lambda) \tag{i}$$

Next, we restrict

$$b_{2i-1,j}^{(2i-1)} = b_{j,2i-1}^{(j)} = 0 \tag{j}$$

thus removing condition (e) from further consideration. Finally, substitution of (i) into (f) results in

$$h_i = b_{2i,2j}^{(2i)} x_{2j} G \phi_0'(\lambda) - \frac{\partial}{\partial x_{2j}} b_{2i,2j}^{(2i)} + \pi K_{ls} g_{js} \frac{\partial g_{il}}{\partial x_{2j}} + \frac{\lambda_{x_{2i-1}}}{G} \tag{k}$$

If the h_i functions in (a) can be expressed in the form of (k), then the response of system (a) has the stationary probability density

$$p(\boldsymbol{x}) = CG(x_1, x_3, \ldots, x_{2n-1}) \exp[-\phi_0(\lambda)] \tag{l}$$

Equation (l) shows that the kinetic energies are identically distributed among different modes, known as equipartitioning of energy in statistical mechanics, which is a consequence of condition (h).

As the first example, consider the following two-degree-of-freedom system:

$$\begin{aligned} \ddot{Z}_1 + h_1(Z_1, \dot{Z}_1, Z_2, \dot{Z}_2) &= g_{11} W_1(t) + g_{12} W_2(t) \\ \ddot{Z}_2 + h_2(Z_1, \dot{Z}_1, Z_2, \dot{Z}_2) &= g_{21} W_1(t) + g_{22} W_2(t) \end{aligned} \tag{m}$$

where g_{ij} are generally functions of Z_1, $\dot{Z}_1$, Z_2, and $\dot{Z}_2$ without being so indicated, and $W_1(t)$ and $W_2(t)$ are gaussian white noises. These two second-order equations are equivalent to four first-order equations:

$$
\begin{aligned}
\dot{X}_1 &= X_2 \\
\dot{X}_2 &= -h_1(X_1, X_2, X_3, X_4) + g_{11}W_1(t) + g_{12}W_2(t) \\
\dot{X}_3 &= X_4 \\
\dot{X}_4 &= -h_2(X_1, X_2, X_3, X_4) + g_{21}W_1(t) + g_{22}W_2(t)
\end{aligned}
\tag{n}
$$

In this case, equation (k) leads to

$$
\begin{aligned}
h_1 &= \left[b_{22}^{(2)}x_2 + b_{24}^{(2)}x_4\right]G\phi_0'(\lambda) - \frac{\partial}{\partial x_2}b_{22}^{(2)} - \frac{\partial}{\partial x_4}b_{24}^{(2)} \\
&\quad + \pi\left[\frac{\partial g_{11}}{\partial x_2}(K_{11}g_{11} + K_{12}g_{12}) + \frac{\partial g_{12}}{\partial x_2}(K_{21}g_{11} + K_{22}g_{12})\right. \\
&\quad \left. + \frac{\partial g_{11}}{\partial x_4}(K_{11}g_{21} + K_{12}g_{22}) + \frac{\partial g_{12}}{\partial x_4}(K_{21}g_{21} + K_{22}g_{22})\right] + \frac{\lambda_{x_1}}{G}
\end{aligned}
\tag{o}
$$

$$
\begin{aligned}
h_2 &= \left[b_{42}^{(4)}x_2 + b_{44}^{(4)}x_4\right]G\phi_0'(\lambda) - \frac{\partial}{\partial x_2}b_{42}^{(4)} - \frac{\partial}{\partial x_4}b_{44}^{(4)} \\
&\quad + \pi\left[\frac{\partial g_{21}}{\partial x_2}(K_{11}g_{11} + K_{12}g_{12}) + \frac{\partial g_{22}}{\partial x_2}(K_{21}g_{11} + K_{22}g_{12})\right. \\
&\quad \left. + \frac{\partial g_{21}}{\partial x_4}(K_{11}g_{21} + K_{12}g_{22}) + \frac{\partial g_{22}}{\partial x_4}(K_{21}g_{21} + K_{22}g_{22})\right] + \frac{\lambda_{x_3}}{G}
\end{aligned}
\tag{p}
$$

System (m) belongs to the class of generalized stationary potential if a consistent $\phi_0'(\lambda)$ function can be obtained from (o) and (p).

Scheurkogel and Elishakoff (1988) have treated a special case in which

$$
g_{11} = g_{22} = 1 \qquad g_{12} = g_{21} = 0 \tag{q}
$$

and

$$
\phi(\lambda) = \phi_0(\lambda) = \mu\lambda \qquad \lambda = y_1 + y_2 + H(x_1, x_3) \tag{r}
$$

where μ is a positive constant and H is nonnegative. Then $G = 1$, and the system belongs to the class of generalized stationary potential if

$$
h_1 = \pi\mu(K_{11}x_2 + 2\beta K_{12}x_4) + \frac{\partial}{\partial x_1}H(x_1, x_3) \tag{s}
$$

$$
h_2 = \pi\mu[2(1-\beta)K_{12}x_2 + K_{22}x_4] + \frac{\partial}{\partial x_3}H(x_1, x_3) \tag{t}
$$

where β is an arbitrary constant. In this case, the system admits a stationary solution

$$
p(x_1, x_2, x_3, x_4) = C\exp(-\mu\lambda) = C\exp\left\{-\mu\left[\tfrac{1}{2}(x_2^2 + x_4^2) + H(x_1, x_3)\right]\right\} \tag{u}
$$

independent of the choice of β. The same result was obtained by Scheurkogel and Elishakoff (1988) using a different procedure. It is of interest to note that the system is in detailed balance only if $\beta = \frac{1}{2}$.

5.3.3 Stochastically Perturbed Hamiltonian Systems (Zhu, Cai, and Lin, 1991)

The equations of motion for a structural or mechanical system can often be formulated simply in the hamiltonian form. Therefore, randomly excited hamiltonian systems are of special importance. Consider such a system governed by

$$\dot{Q}_j = \frac{\partial H}{\partial P_j} \qquad j = 1, 2, \ldots, n \tag{a}$$

$$\dot{P}_j = -\frac{\partial H}{\partial Q_j} - h(H)c_{jk}(\boldsymbol{Q}, \boldsymbol{P})\frac{\partial H}{\partial P_k} + f(H)g_{jl}(\boldsymbol{Q}, \boldsymbol{P})W_l(t) \tag{b}$$

$$j, k = 1, 2, \ldots, n, \qquad l = 1, 2, \ldots, m$$

where Q_j and P_j are generalized displacements and generalized momenta, respectively, $H(\boldsymbol{Q}, \boldsymbol{P})$ is a hamiltonian with continuous first-order derivatives, and $W_l(t)$ are gaussian white noises. In addition, it is assumed that functions $c_{ij}(\boldsymbol{Q}, \boldsymbol{P})$ are differentiable, and $h(H)$, $f(H)$, and $g_{jl}(\boldsymbol{Q}, \boldsymbol{P})$ are twice-differentiable. The system is generally nonlinear and includes both additive and multiplicative random excitations. It is a generalization of the one considered previously by Soize (1988) by introducing the variable coefficients $c_{ij}(\boldsymbol{Q}, \boldsymbol{P})$ and $g_{jl}(\boldsymbol{Q}, \boldsymbol{P})$, and by allowing the random excitations to be correlated.

Equations (a) and (b) are in the same first-order form as equations (5.0.1). Identifying Q_j with X_{2j-1} and P_j with X_{2j}, the first and second derivate moments for the hamiltonian system are obtained from equations (5.0.4) and (5.0.5) as follows:

$$a_{2j-1} = \frac{\partial H}{\partial p_j} \qquad a_{2j} = -\frac{\partial H}{\partial q_j} - hc_{jk}\frac{\partial H}{\partial p_k} + \pi f K_{ls} g_{ks}\frac{\partial}{\partial p_k}(fg_{jl})$$
$$b_{2j-1,k} = b_{k,2j-1} = 0 \qquad b_{2j,2k} = f^2 B_{jk} \tag{c}$$

where $B_{jk} = 2\pi K_{ls} g_{jl} g_{ks}$, and where the arguments for H, h, f, c_{jk}, and g_{jl} have been replaced by their counterpart deterministic variables p_j and q_j. By splitting these derivate moments into

$$a_{2j-1}^{(1)} = 0 \qquad a_{2j-1}^{(2)} = \frac{\partial H}{\partial p_j}$$
$$a_{2j}^{(1)} = -hc_{jk}\frac{\partial H}{\partial p_k} + \pi f K_{ls} g_{ks}\frac{\partial}{\partial p_k}(fg_{jl}) \qquad a_{2j}^{(2)} = -\frac{\partial H}{\partial q_j} \tag{d}$$
$$b_{2j-1,k}^{(2j-1)} = b_{k,2j-1}^{(k)} = 0 \qquad b_{2j,2k}^{(2j)} = f^2 B_{jk}^{(j)}$$

where $B_{jk}^{(j)} + B_{kj}^{(k)} = B_{jk}$, and substituting them into equations (5.3.6) and (5.3.7), we obtain

$$\frac{\partial H}{\partial q_j}\frac{\partial \phi}{\partial p_j} - \frac{\partial H}{\partial p_j}\frac{\partial \phi}{\partial q_j} = 0 \tag{e}$$

$$f^2 B_{jk}^{(j)} \frac{\partial \phi}{\partial p_k} = \frac{\partial}{\partial p_k}[f^2 B_{jk}^{(j)}] + hc_{jk}\frac{\partial H}{\partial p_k} - \pi f K_{ls} g_{ks} \frac{\partial}{\partial p_k}(f g_{jl}) \tag{f}$$

The general solution for equation (e) may be written as

$$\phi(\boldsymbol{q}, \boldsymbol{p}) = \phi[H(\boldsymbol{q}, \boldsymbol{p})] \tag{g}$$

where ϕ is an arbitrary function of H. Substitution of (g) into (f) leads to

$$f^2 B_{jk}^{(j)} \frac{d\phi}{dH}\frac{\partial H}{\partial p_k} = hc_{jk}\frac{\partial H}{\partial p_k} - \pi f K_{ls} g_{ks} \frac{\partial}{\partial p_k}(f g_{jl}) + \frac{\partial}{\partial p_k}[f^2 B_{jk}^{(j)}] \tag{h}$$

These are n equations for ϕ. If a consistent function ϕ can be found satisfying all these equations, then the exact stationary probability density is given by

$$p(\boldsymbol{q}, \boldsymbol{p}) = C \exp[-\phi(H)] \tag{i}$$

Two special cases will now be investigated.

Case 1. In the system equations (a) and (b), f is a constant and c_{jk} and g_{jl} are functions only of $\boldsymbol{q}$. In this case, equation (h) is reduced to

$$f^2 B_{jk}^{(j)} \frac{d\phi}{dH}\frac{\partial H}{\partial p_k} = hc_{jk}\frac{\partial H}{\partial p_k} \tag{j}$$

with a common solution

$$\frac{d\phi}{dH} = \frac{\eta h(H)}{f^2} \tag{k}$$

under the condition $c_{jk} = \eta B_{jk}^{(j)}$ or, equivalently, $c_{jk} + c_{kj} = \eta B_{jk}$, where η is a constant. The stationary probability density is then

$$p(\boldsymbol{q}, \boldsymbol{p}) = C \exp\left[-\frac{\eta}{f^2}\int_0^H h(u)du\right] \tag{l}$$

Case 2. In the system equations (a) and (b), f is a function of H and c_{jk} and g_{jl} remain functions only of $\boldsymbol{q}$. In this case, equation (h) may be rearranged to read

$$\frac{d\phi}{dH} = \frac{c_{jk}\dfrac{\partial H}{\partial p_k}}{B_{jk}^{(j)}\dfrac{\partial H}{\partial p_k}}\frac{h}{f^2} - \frac{B_{jk}\dfrac{\partial H}{\partial p_k}}{B_{jk}^{(j)}\dfrac{\partial H}{\partial p_k}}\frac{1}{2f}\frac{df}{dH} + \frac{2}{f}\frac{df}{dH} \tag{m}$$

In order for the right-hand side of (m) to be just a function of H, it is required that $c_{jk} = c_{kj} = \eta B_{jk}/2$. Under this restriction, we obtain

$$p(\boldsymbol{q}, \boldsymbol{p}) = C f^{-1}(H) \exp\left[-\eta \int_0^H \frac{h(u)}{f^2(u)}du\right] \tag{n}$$

Equation (n) is reduced to equation (l) if f is a constant.

On the other hand, equations (a) and (b) may be generalized to

$$\dot{Q}_j = D(\boldsymbol{Q})\frac{\partial H}{\partial P_j} \qquad j = 1, 2, \ldots, n \tag{o}$$

$$\dot{P}_j = -D(\boldsymbol{Q})\frac{\partial H}{\partial Q_j} - h(H)c_{jk}(\boldsymbol{Q}, \boldsymbol{P})\frac{\partial H}{\partial P_k} + f(H)g_{jl}(\boldsymbol{Q}, \boldsymbol{P})W_l(t)$$
$$j, k = 1, 2, \ldots, n, \quad l = 1, 2, \ldots, m \tag{p}$$

where $D(\boldsymbol{Q})$ is an arbitrary function of $\boldsymbol{Q}$. The hamiltonian system described by equations (a) and (b) is a special case with $D(\boldsymbol{Q}) = 1$. Following a similar procedure and denoting the probability potential by $\phi^*(\boldsymbol{q}, \boldsymbol{p})$, we obtain, instead of equations (e) and (f),

$$-\frac{\partial D}{\partial q_j}\frac{\partial H}{\partial p_j} + D\frac{\partial \phi^*}{\partial q_j}\frac{\partial H}{\partial p_j} + \frac{\partial D}{\partial p_j}\frac{\partial H}{\partial q_j} - D\frac{\partial \phi^*}{\partial p_j}\frac{\partial H}{\partial q_j} = 0 \tag{q}$$

$$f^2 B_{jk}^{(j)}\frac{\partial \phi^*}{\partial p_k} = \frac{\partial}{\partial p_k}[f^2 B_{jk}^{(j)}] + hc_{jk}\frac{\partial H}{\partial p_k} - \pi f K_{ls}g_{ks}\frac{\partial}{\partial p_k}(f g_{jl}) \tag{r}$$

Equation (q) may be rewritten as

$$\frac{\partial H}{\partial p_j}\left[\frac{\partial}{\partial q_j}(\phi^* - \ln D)\right] = \frac{\partial H}{\partial q_j}\left[\frac{\partial}{\partial p_j}(\phi^* - \ln D)\right] \tag{s}$$

which possesses a general solution

$$\phi^*(\boldsymbol{q}, \boldsymbol{p}) = \ln D(\boldsymbol{q}) + \phi[H(\boldsymbol{q}, \boldsymbol{p})] \tag{t}$$

where ϕ is an arbitrary function of H. Substituting (t) into (r), we obtain the same equation (h). Therefore, if ϕ is the stationary probability potential for the hamiltonian system governed by (a) and (b), then ϕ^* is the stationary probability potential for the more general system governed by (o) and (p). The stationary probability density corresponding to (t) is

$$p^*(\boldsymbol{q}, \boldsymbol{p}) = \frac{C\exp[-\phi(H)]}{D(\boldsymbol{q})} \tag{u}$$

or, in view of (i),

$$p^*(\boldsymbol{q}, \boldsymbol{p}) = \frac{p(\boldsymbol{q}, \boldsymbol{p})}{D(\boldsymbol{q})} \tag{v}$$

If, in the system equations (o) and (p), f is a constant, c_{jk} and g_{jl} depend only on $\boldsymbol{q}$, and $c_{jk} + c_{kj} = \eta B_{jk}$, then

$$p^*(\boldsymbol{q}, \boldsymbol{p}) = \frac{C}{D(\boldsymbol{q})}\exp\left[-\frac{\eta}{f^2}\int_0^H h(u)\,du\right] \tag{w}$$

Equation (w) reduces to equation (l) when $D = 1$.

If, in equations (o) and (p), f is a function of H, c_{jk} and g_{jl} depend only on $\boldsymbol{q}$, and $c_{jk} = c_{kj} = \eta B_{jk}/2$, then

$$p^*(\boldsymbol{q}, \boldsymbol{p}) = \frac{C}{D(\boldsymbol{q})f(H)}\exp\left[-\eta\int_0^H \frac{h(u)}{f^2(u)}du\right] \tag{x}$$

Equation (x) reduces to equation (n) when $D = 1$.

Application of this procedure is illustrated in the following examples. First consider a hamiltonian

$$H = \sum_{j=1}^{n} \int_{0}^{P_j} r_j(u)\,du + U(Q_1, Q_2, \ldots, Q_n) \tag{y}$$

The equations of motion are obtained by substituting H into equations (a) and (b) to yield

$$\dot{X}_{2j-1} = r_j(X_{2j}) \tag{z}$$

$$\dot{X}_{2j} = -h(H)c_{jk}r_k(X_{2k}) - \frac{\partial U}{\partial X_{2j-1}} + f(H)g_{jl}W_l(t) \tag{aa}$$

where $X_{2j-1} = Q_j$, $X_{2j} = P_j$, and where c_{jk}, g_{jl}, and U are functions of $X_1, X_3, \ldots, X_{2n-1}$. The stationary probability density is given by equation (n) under the condition $c_{jk} = c_{kj} = \eta\pi K_{ls}g_{jl}g_{ks}$. If f is a constant, then the stationary probability density reduces to (l) under a less restrictive condition $c_{jk} + c_{kj} = 2\eta\pi K_{ls}g_{jl}g_{ks}$. In the special case

$$f = 1 \qquad c_{jk} = \begin{cases} 1 & j \neq k \\ 0 & j = k \end{cases} \qquad g_{jl} = \begin{cases} 1 & j \neq l \\ 0 & j = l \end{cases} \qquad K_{ls} = \begin{cases} K & l \neq s \\ 0 & l = s \end{cases} \tag{bb}$$

the stationary probability density obtained from (l) is given by

$$p(\boldsymbol{x}) = C \exp\left[-\frac{1}{\pi K}\int_{0}^{H} h(u)\,du\right] \tag{cc}$$

under the condition $\eta = 1/\pi K$. This case was investigated earlier by Caughey and Ma (1982b).

Consider another hamiltonian

$$H = \tfrac{1}{2}[\boldsymbol{m}^{-1}(\boldsymbol{Q})]_{jk}P_jP_k + U(\boldsymbol{Q}) \tag{dd}$$

where Q_j are the generalized displacements, $P_j = m_{jk}\dot{Q}_k$ are the generalized momenta, and $\boldsymbol{m}(\boldsymbol{Q})$ is a symmetric matrix. Substituting this hamiltonian into equation (a), we obtain

$$\dot{Q}_j = \frac{\partial H}{\partial P_j} = [\boldsymbol{m}^{-1}(\boldsymbol{Q})]_{jk}P_k \tag{ee}$$

Equations (ee) and (b) govern a hamiltonian system perturbed by gaussian white noises, and the stationary probability density $p(\boldsymbol{q}, \boldsymbol{p})$ can be obtained if the system belongs to the class of generalized stationary potential, using the foregoing procedure.

It is interesting to note that this particular hamiltonian system is equivalent to

$$\frac{d}{dt}[m_{jk}(\boldsymbol{X})\dot{X}_k] + h(H)c_{jk}(\boldsymbol{X}, \dot{\boldsymbol{X}})\dot{X}_k + \frac{\partial H}{\partial X_j} = f(H)g_{jl}(\boldsymbol{X}, \dot{\boldsymbol{X}})W_l(t) \tag{ff}$$

which represents a system with varying inertia coupling. The stationary probability density of $\boldsymbol{X}$ and $\dot{\boldsymbol{X}}$ can be obtained from that of $\boldsymbol{Q}$ and $\boldsymbol{P}$ as follows:

$$p(\boldsymbol{x}, \dot{\boldsymbol{x}}) = |J|\, p(\boldsymbol{q}, \boldsymbol{p}) \tag{gg}$$

where J is a jacobian given by

$$J = \begin{vmatrix} \dfrac{\partial q_1}{\partial x_1} & \dfrac{\partial q_1}{\partial \dot{x}_1} & \dfrac{\partial q_1}{\partial x_2} & \dfrac{\partial q_1}{\partial \dot{x}_2} & \cdots \\ \dfrac{\partial p_1}{\partial x_1} & \dfrac{\partial p_1}{\partial \dot{x}_1} & \dfrac{\partial p_1}{\partial x_2} & \dfrac{\partial p_1}{\partial \dot{x}_2} & \cdots \\ \dfrac{\partial q_2}{\partial x_1} & \dfrac{\partial q_2}{\partial \dot{x}_1} & \dfrac{\partial q_2}{\partial x_2} & \dfrac{\partial q_2}{\partial \dot{x}_2} & \cdots \\ \dfrac{\partial p_2}{\partial x_1} & \dfrac{\partial p_2}{\partial \dot{x}_1} & \dfrac{\partial p_2}{\partial x_2} & \dfrac{\partial p_2}{\partial \dot{x}_2} & \cdots \\ \cdots & \cdots & \cdots & \cdots & \cdots \end{vmatrix}$$

$$= \begin{vmatrix} 1 & 0 & 0 & 0 & 0 & 0 & \cdots \\ 0 & m_{11} & 0 & m_{12} & 0 & m_{13} & \cdots \\ 0 & 0 & 1 & 0 & 0 & 0 & \cdots \\ 0 & m_{21} & 0 & m_{22} & 0 & m_{23} & \cdots \\ 0 & 0 & 0 & 0 & 1 & 0 & \cdots \\ 0 & m_{31} & 0 & m_{32} & 0 & m_{33} & \cdots \\ \cdots & \cdots & \cdots & \cdots & \cdots & \cdots & \cdots \end{vmatrix} = \det(\boldsymbol{m}) \tag{hh}$$

and where $\det(\boldsymbol{m})$ denotes the determinant of matrix $\boldsymbol{m}$. A special case of system (ff) with constant c_{jk} and g_{jl} was investigated by Soize (1988).

The last example is the hamiltonian

$$H = \frac{P_j P_j}{2D(\boldsymbol{Q})} + U(\boldsymbol{Q}) \tag{ii}$$

In this case, the system described by equations (o) and (p) is equivalent to

$$\ddot{X}_j + \frac{1}{D(\boldsymbol{X})} h(H) c_{jk}(\boldsymbol{X}, \dot{\boldsymbol{X}}) \dot{X}_k + D(\boldsymbol{X}) \frac{\partial H}{\partial X_j} = f(H) g_{jl}(\boldsymbol{X}, \dot{\boldsymbol{X}}) W_l(t) \tag{jj}$$

where $Q_j = X_j, P_j = \dot{X}_j$. Thus the stationary probability density $p(\boldsymbol{q}, \boldsymbol{p})$ may be obtained from either (l) or (n) under the respective conditions. The results obtained previously by Caughey and Ma (1982a) and Zhu (1990) were special cases.

5.4 EQUIVALENT STOCHASTIC SYSTEMS

As shown in Section 5.2.2, the same gaussian probability density was obtained for two different systems: a nonlinearly damped system under simultaneous multiplicative and additive random excitations and a linear system under an additive random excitation alone. The possibility for different stochastic systems sharing an identical probability distribution is next investigated in more general terms (Lin and Cai, 1988b).

Let two different dynamical systems be governed, respectively, by the following equations of motion:

$$\dot{X}_i = f_i(\boldsymbol{X}, t) + g_{ij}(\boldsymbol{X}, t)\xi_j(t) \qquad i = 1, \ldots, n, \quad j = 1, \ldots, m \tag{5.4.1}$$

$$\dot{Y}_l = F_l(\boldsymbol{Y}, t) + G_{lr}(\boldsymbol{Y}, t)\eta_r(t) \qquad l = 1, \ldots, n, \quad r = 1, \ldots, s \tag{5.4.2}$$

where X_i are components of $\boldsymbol{X}$, Y_l are components of $\boldsymbol{Y}$, and $\xi_j(t)$ and $\eta_r(t)$ are random excitations. The two dynamical systems are said to be stochastically equivalent if the two system responses $\boldsymbol{X}(t)$ and $\boldsymbol{Y}(t)$ share the same probability distribution.

If $\boldsymbol{X}(t)$ and $\boldsymbol{Y}(t)$ tend to statistical stationarity as t increases, then it is possible that $\boldsymbol{X}(t)$ and $\boldsymbol{Y}(t)$ share only the same stationary probability distribution, but possess different transient nonstationary distributions under identical initial conditions. In such a case, the two systems (5.4.1) and (5.4.2) are said to be stochastically equivalent in the wide (or weak) sense. On the other hand, if $\boldsymbol{X}(t)$ and $\boldsymbol{Y}(t)$ also share the same transient nonstationary distribution, then systems (5.4.1) and (5.4.2) are said to be stochastically equivalent in the strict (or strong) sense. Since only the stationary probability densities are presently obtainable analytically for the second- and higher order nonlinear systems under gaussian white noise excitations, the following discussion focuses on weakly equivalent stochastic systems. The qualification phrase "weakly" or "in the wide sense" is omitted for convenience, unless there is a need for clarification. However, the condition under which strong equivalency is possible is indicated.

The assertion that different dynamical systems can share the same probabilistic solution, either in the wide sense or in the strict sense, may seem strange to readers accustomed to deterministic analysis. However, this is possible in the case of stochastic systems, since the assertion merely means that the probabilistic distribution of all sample solutions is the same for either system, while individual sample solutions can be all different. In fact, they are generally different even for one stochastic system.

We recall that in Section 5.3 our objective was to obtain a consistent function ϕ from equations (5.3.6) and (5.3.7) in order to construct an exact solution for a given stochastic system. Instead, our present objective is to identify a group of dynamical systems which are stochastically equivalent to a given system with known solution. From the present point of view, the ϕ function is known. We now attempt to find the possible first and second derivate moments, a_i and b_{ij}, such that equations (5.3.6) and (5.3.7) can be satisfied.

In order to satisfy (5.3.7), we let

$$a_i^{(2)} = \frac{\partial}{\partial x_j}\delta_{ij}^{(1)} - \delta_{ij}^{(1)}\frac{\partial \phi}{\partial x_j} \tag{5.4.3}$$

where $\delta_{ij}^{(1)}$ is antisymmetric, that is, $\delta_{ji}^{(1)} = -\delta_{ij}^{(1)}$, as can be verified by substituting (5.4.3) into (5.3.7). Adding (5.3.6) and (5.4.3),

$$a_i = \frac{\partial}{\partial x_j}[b_{ij}^{(i)} + \delta_{ij}^{(1)}] - [b_{ij}^{(1)} + \delta_{ij}^{(1)}]\frac{\partial \phi}{\partial x_j} \tag{5.4.4}$$

Now, since b_{ij} is symmetric, it is always possible to write

$$b_{ij}^{(i)} = \tfrac{1}{2}b_{ij} + \delta_{ij}^{(2)} \tag{5.4.5}$$

where $\delta_{ij}^{(2)}$ are also antisymmetric. Then equation (5.4.4) becomes

$$a_i = \frac{\partial}{\partial x_j}\left[\frac{1}{2}b_{ij} + \delta_{ij}\right] - \left[\frac{1}{2}b_{ij} + \delta_{ij}\right]\frac{\partial \phi}{\partial x_j} \tag{5.4.6}$$

where
$$\delta_{ij} = \delta_{ij}^{(1)} + \delta_{ij}^{(2)} \tag{5.4.7}$$
is again antisymmetric. The set of n equations in (5.4.6), which replaces the set of $n+1$ equations in (5.3.6) and (5.3.7), is simpler for the purpose of finding stochastically equivalent systems, since the splitting of a_i and b_{ij} is now replaced by a single process of selecting δ_{ij}. With the knowledge of the solution of one stochastic system, namely, the knowledge of ϕ, we can construct a large number of equivalent stochastic systems by selecting physically meaningful combinations of b_{ij} and δ_{ij} matrices.

Given a probability density, there are two ways to construct a set of equivalent stochastic systems. We can select different b_{ij} and different δ_{ij}, and then obtain a_i from (5.4.6) for each set of b_{ij} and δ_{ij}. The systems so obtained have different Fokker-Planck equations, although they share a common stationary solution. On the other hand, if we keep the same b_{ij} but vary δ_{ij}, and determine the a_i for different sets of δ_{ij}, then the equivalent stochastic systems obtained possess the same Fokker-Planck equation, and they must be strictly equivalent.

The constitutive relation (5.4.6) for equivalent stochastic systems is expressed in terms of the first and second derivate moments. For practical purposes, it is more convenient to operate directly on the equations of motion. Consider a stochastic system governed by
$$\dot{X}_i = f_i(\boldsymbol{X}) + g_{ij}(\boldsymbol{X})W_j(t) \tag{5.4.8}$$
which is the same as (5.4.1), except that functions f_i and g_{ij} are restricted to being time-invariant, and the random excitations are now gaussian white noises. Suppose that the system belongs to the class of generalized stationary potential, and its stationary probability potential ϕ is known. Then another system governed by
$$\dot{X}_i = \bar{f}_i(\boldsymbol{X}, t) + \bar{g}_{ij}(\boldsymbol{X}, t)W_j(t) \tag{5.4.9}$$
is stochastically equivalent to (5.4.8) if
$$\begin{aligned}\bar{f}_i(\boldsymbol{X}) &= \pi K_{ls}\bar{g}_{js}(\boldsymbol{X})\frac{\partial}{\partial X_j}\bar{g}_{il}(\boldsymbol{X}) - \pi K_{ls}\bar{g}_{il}(\boldsymbol{X})\bar{g}_{js}(\boldsymbol{X})\frac{\partial \phi}{\partial X_j} \\ &\quad + \frac{\partial}{\partial X_j}\delta_{ij}(\boldsymbol{X}) - \delta_{ij}(\boldsymbol{X})\frac{\partial \phi}{\partial X_j}\end{aligned} \tag{5.4.10}$$
The validity of the preceding statement can be verified by using (5.0.5), (5.0.6), and (5.4.6). Equation (5.4.10) may be used to obtain an equivalent stochastic system by varying δ_{ij} and letting $\bar{g}_{ij} = g_{ij}$, or by varying both $\bar{g}_{ij}$ and δ_{ij}.

Structural and mechanical systems are often oscillatory, and they are often described by a set of second-order equations of the form
$$\ddot{Z}_i + h_i(\boldsymbol{Z}, \dot{\boldsymbol{Z}}) = g_{ij}(\boldsymbol{Z}, \dot{\boldsymbol{Z}})W_j(t) \tag{5.4.11}$$
Letting $X_{2i-1} = Z_i$ and $X_{2i} = \dot{Z}_i$, and applying (5.4.10) to the present case, we obtain
$$X_{2i} = \frac{\partial}{\partial X_r}\delta_{2i-1,r} - \delta_{2i-1,r}\frac{\partial}{\partial X_r}\phi$$
$$i = 1, 2, \ldots, n, \;\; r = 1, 2, \ldots, 2n, \;\; r \neq 2i-1 \tag{5.4.12}$$

$$\bar{h}_i = -\pi K_{ls}\bar{g}_{js}\frac{\partial}{\partial X_{2j}}\bar{g}_{il} + \pi K_{ls}\bar{g}_{il}\bar{g}_{js}\frac{\partial \phi}{\partial X_{2j}} - \frac{\partial}{\partial X_r}\delta_{2i,r} + \delta_{2i,r}\frac{\partial \phi}{\partial X_r}$$
$$i = 1, 2, \ldots, n,\ j = 1, 2, \ldots, n,\ l = 1, 2, \ldots, m,$$
$$s = 1, 2, \ldots, m,\ r = 1, 2, \ldots, 2n;\ r \neq 2i \tag{5.4.13}$$

It is of interest to note that the choice of the antisymmetric δ_{ij} functions in (5.4.12) and (5.4.13) cannot be entirely arbitrary. For example, in view of equation (5.4.13), the choice of every $\delta_{\text{pr}} = 0$ would imply a totally vanishing probability flow. However, this would also require that all $X_{2i} = 0$, that is, the system would be motionless, as can be concluded from (5.4.12). Therefore, an oscillatory system does not belong to the class of stationary potential, because a portion of its probability flow must be circulatory.

Additional insight can be gained by examining in detail the case of single degree of freedom, for which equations (5.4.12) and (5.4.13) become

$$X_2 = \frac{\partial}{\partial X_2}\delta - \delta\frac{\partial \phi}{\partial X_2} \tag{5.4.14}$$

$$\bar{h} = -\pi K_{ls}\bar{g}_s\frac{\partial \bar{g}_l}{\partial X_2} + \pi K_{ls}\bar{g}_l\bar{g}_s\frac{\partial \phi}{\partial X_2} + \frac{\partial \delta}{\partial X_1} - \delta\frac{\partial \phi}{\partial X_1} \tag{5.4.15}$$

where $\delta = \delta_{12} = -\delta_{21}$. The most general δ function that satisfies (5.4.14) is

$$\delta = \left[D(X_1) + \int X_2 \exp(-\phi)\, dX_2\right]\exp(\phi) \tag{5.4.16}$$

where $D(X_1)$ is an arbitrary function of X_1. Substituting (5.4.16) into (5.4.15),

$$\begin{aligned}\bar{h} = &-\pi K_{ls}\bar{g}_s\frac{\partial \bar{g}_l}{\partial X_2} + \pi K_{ls}\bar{g}_l\bar{g}_s\frac{\partial \phi}{\partial X_2}\\ &-\exp(\phi)\int X_2\frac{\partial \phi}{\partial X_1}\exp(-\phi)\, dX_2 + D'(X_1)\exp(\phi)\end{aligned} \tag{5.4.17}$$

Equation (5.4.17) can be used to construct an unlimited number of second-order equivalent stochastic systems sharing the same stationary probability density

$$p(\mathbf{x}) = C\exp(-\phi) \tag{5.4.18}$$

Of special practical interest is the case $D'(X_1) = 0$, $\phi = \phi(\Lambda)$, where $\Lambda = \frac{1}{2}X_2^2 + \int_0^{X_1} u(v)dv$. In this case,

$$\bar{h} = -\pi K_{ls}\bar{g}_s\frac{\partial \bar{g}_l}{\partial X_2} + \pi X_2 K_{ls}\bar{g}_l\bar{g}_s\frac{d}{d\Lambda}\phi(\Lambda) + u(X_1) \tag{5.4.19}$$

As examples, we recall first the nonlinear system considered previously in Section 5.2.2 governed by

$$\ddot{X} + (\alpha + \beta X^2)\dot{X} + \omega_0^2[1 + W_1(t)]X = W_2(t) \qquad \alpha > 0, \quad \beta > 0 \tag{5.4.20}$$

where the multiplicative excitation $W_1(t)$ and the additive excitation $W_2(t)$ are independent gaussian white noises, with spectral densities K_{11} and K_{22}, respectively. It has been shown that the system is in detailed balance under the condition $\beta/\alpha =$

$\omega_0^4 K_{11}/K_{22}$, in which case the joint stationary probability density of $X_1 = X$ and $X_2 = \dot{X}$ is given by

$$p(x_1, x_2) = C \exp\left[-\frac{\beta}{2\pi K_{11}}(x_2^2 + \omega_0^2 x_1^2)\right] \tag{5.4.21}$$

Equation (5.4.21) shows that $X(t)$ and $\dot{X}(t)$ are independent and jointly gaussian distributed, the same as the stationary response of the linear system

$$\ddot{X} + \beta\dot{X} + \omega_0^2 X = W_1(t) \qquad \beta > 0 \tag{5.4.22}$$

under an additive random excitation $W_1(t)$ alone. Thus when in detailed balance, system (5.4.20) is stochastically equivalent to system (5.4.22). It appears that the restriction $\beta/\alpha = \omega_0^4 K_{11}/K_{22}$ imposed on (5.4.20) provides the necessary balance, so that the same probabilistic solution can be obtained as that for the linear system (5.4.22).

Another pair of weakly equivalent stochastic systems can be constructed by adding a nonlinear spring term to both equations (5.4.20) and (5.4.22):

$$\ddot{X} + (\alpha + \beta X^2)\dot{X} + \omega_0^2[1 + W_1(t)]X + u_1(X) = W_2(t) \qquad \alpha > 0, \beta > 0 \tag{5.4.23}$$

$$\ddot{X} + \beta\dot{X} + \omega_0^2 X + u_1(X) = W_1(t) \qquad \beta > 0 \tag{5.4.24}$$

It can be shown by using (5.4.19) that under the same condition $\beta/\alpha = \omega_0^4 K_{11}/K_{22}$, the two stochastic systems (5.4.23) and (5.4.24) share a stationary probability density

$$p(x_1, x_2) = C \exp\left\{-\frac{\beta}{\pi K_{11}}\left[\frac{1}{2}x_2^2 + \int_0^{x_1} u(v)\, dv\right]\right\} \tag{5.4.25}$$

where $u(v) = \omega_0^2 v + u_1(v)$. Among numerous stochastic systems sharing the same stationary probability density (5.4.25), we cite the following:

$$\ddot{X} + \beta X^2\dot{X} + u(X) = XW_1(t) \tag{5.4.26}$$

$$\ddot{X} - \pi K_{11}\dot{X} + \beta\dot{X}^3 + u(X) = \dot{X}W_1(t) \tag{5.4.27}$$

$$\ddot{X} + (\beta - \pi K_{33})\dot{X} + \beta\dot{X}\left(\frac{K_{22}}{K_{11}}X^2 + \frac{K_{33}}{K_{11}}\dot{X}^2\right) + u(X) = W_1(t) + XW_2(t) + \dot{X}W_3(t) \tag{5.4.28}$$

where $W_1(t)$, $W_2(t)$, and $W_3(t)$ are independent gaussian white noises.

Finally, we consider briefly the existence and the forms of strictly equivalent stochastic systems, that is, systems sharing an identical probability distribution in both the stationary and nonstationary states. We begin with the full Fokker-Planck equation (5.0.2), which must be shared by stochastic systems strictly equivalent to each other. This equation remains unchanged if antisymmetric elements are added to the matrix of the second derivate moments. In other words, the equation

$$\frac{\partial q}{\partial t} + \frac{\partial}{\partial x_i}\left\{a_i q - \frac{1}{2}\frac{\partial}{\partial x_j}[(b_{ij} + \delta_{ij})q]\right\} = 0 \tag{5.4.29}$$

is the same as equation (5.0.2), since the antisymmetric terms δ_{ij} cancel upon differentiations with respect to x_j and x_i. However, by its very definition, the b_{ij} matrix must be symmetric. This means that strictly equivalent stochastic systems must share the same b_{ij} matrix, and that equation (5.4.29) must be interpreted in another sense. Indeed, it should be interpreted as changing the first derivate moments from a_i to $\bar{a}_i$ such that

$$\bar{a}_i q = a_i q - \frac{1}{2}\frac{\partial}{\partial x_j}(\delta_{ij} q) \tag{5.4.30}$$

Now, the nonstationary probability density q in (5.4.30) is time-dependent, which can be cast in the form of

$$q(\boldsymbol{x}, t) = C(t)\exp[-\phi(\boldsymbol{x}, t)] \tag{5.4.31}$$

Substitution of (5.4.31) into (5.4.30) yields

$$\bar{a}_i = a_i - \frac{1}{2}\left(\frac{\partial}{\partial x_j}\delta_{ij} - \delta_{ij}\frac{\partial \phi}{\partial x_j}\right) \tag{5.4.32}$$

Equation (5.4.32) provides a set of conditions for stochastically equivalent systems in the strict sense. For example, given a stochastic system

$$\ddot{X} + h(X, \dot{X}) = g_i(X, \dot{X})W_i(t) \tag{5.4.33}$$

let us be required to find another stochastic system

$$\ddot{X} + \bar{h}(X, \dot{X}) = \bar{g}_i(X, \dot{X})\bar{W}_i(t) \tag{5.4.34}$$

such that both systems share the same Fokker-Planck equation. Since the matrices of the second derivate moments must be identical,

$$\bar{K}_{ls}\bar{g}_l\bar{g}_s = K_{ls}g_l g_s \tag{5.4.35}$$

where $\bar{K}_{ls}$ is the cross-spectral density of $\bar{W}_l(t)$ and $\bar{W}_s(t)$. For most practical purposes, (5.4.35) may be replaced by the more restrictive $\bar{K}_{ls} = K_{ls} = \bar{g}_l = g_l$. Imposition of (5.4.32) leads to

$$\frac{\partial}{\partial X_2}\delta - \delta\frac{\partial \phi}{\partial X_2} = 0 \tag{5.4.36}$$

and

$$\bar{h} = h - \frac{\partial}{\partial X_1}\delta + \delta\frac{\partial \phi}{\partial X_1} \tag{5.4.37}$$

where $X_1 = X$ and $X_2 = \dot{X}$. Equation (5.4.36) is satisfied if

$$\delta = F(X_1, t)\exp(\phi) \tag{5.4.38}$$

where F is an arbitrary function of X_1 and t. Then from (5.4.37),

$$\bar{h} = h + \bar{F}(X_1, t)\exp(\phi) \tag{5.4.39}$$

where $\bar{F}$ is also an arbitrary function of X_1 and t. A slightly more complicated $\bar{h}$ function is obtained if a condition less restrictive than (5.4.35) is imposed.

It should be noted that exact nonstationary probabilistic solutions are largely unknown for the second- and higher order systems, except for linear systems under purely additive gaussian excitations. Thus results for strictly equivalent stochastic systems given here are perhaps only of mathematical interest at the present time.

5.5 CONCLUDING REMARKS

When a system is linear and the random excitations are additive, the elementary spectral analysis and the evolutionary spectral analysis are applicable. Therefore, the Markov process modeling is truly needed only for treating nonlinear systems. At the present time, however, exact probabilistic solutions for nonlinear systems are obtainable only in limited cases, and only if the random excitations are gaussian white noises.

The simplest nonlinear systems, in which only the conservative forces are nonlinear and only additive random excitations are present, are discussed in Lin (1967). The method of generalized stationary potential given in this chapter has the primary objective of treating nonlinear damping, multiplicative excitations, or both. It is shown that solvability of such more complicated problems generally requires a restrictive relationship between the system parameters and the spectral levels of the excitations, a condition rarely satisfied in practice. However, the exact solution technique developed herein is not merely academic; it provides the very foundation for the development of approximate solution techniques which are the subjects of Chapter 7. In general, practical problems must be solved with approximation techniques.

5.6 EXERCISES

5.1. Show that an oscillator governed by an equation of the form

$$\ddot{X} + f(X, \dot{X}) = g_j(X, \dot{X})W_j(t)$$

does not belong to the class of *stationary* potential, where $W_j(t)$ are gaussian white noises.

5.2. Consider the Itô equations

$$dX_1 = [-\zeta\omega_0(1-r)X_1 - \delta X_2]\,dt + \sqrt{2\pi K}dB_1(t)$$

$$dX_2 = [\delta X_1 - \zeta\omega_0(1+r)X_2]\,dt + \sqrt{2\pi K}dB_2(t)$$

where $B_1(t)$ and $B_2(t)$ are independent unit Wiener processes. The reduced Fokker-Planck equation for the stationary probability density $p(x_1, x_2)$ is given by

$$\frac{\partial G_1}{\partial x_1} + \frac{\partial G_2}{\partial x_2} = 0$$

where

$$G_1 = [-\zeta\omega_0(1-r)x_1 - \delta x_2]p - \pi K\frac{\partial p}{\partial x_1}$$

$$G_2 = [\delta x_1 - \zeta\omega_0(1+r)x_2]p - \pi K\frac{\partial p}{\partial x_2}$$

(*a*) Show that the probability flows G_1 and G_2 do not satisfy the conditions for stationary potential except when $\delta = 0$.

(*b*) The reduced Fokker-Planck equation can also be written in the form

$$\frac{\partial G_1^*}{\partial x_1} + \frac{\partial G_2^*}{\partial x_2} = 0$$

where

$$G_1^* = G_1 - D\frac{\partial p}{\partial x_2} = [-\zeta\omega_0(1-r)x_1 - \delta x_2]p - \pi K\frac{\partial p}{\partial x_1} - D\frac{\partial p}{\partial x_2}$$

$$G_2^* = G_2 + D\frac{\partial p}{\partial x_1} = [\delta x_1 - \zeta\omega_0(1+r)x_2]p + D\frac{\partial p}{\partial x_1} - \pi K\frac{\partial p}{\partial x_2}$$

and where D is a constant. Show that the probability flows G_1^* and G_2^* satisfy the conditions for stationary potential for a certain constant D.

(c) Find the stationary probability density $p(x_1, x_2)$.

5.3. Consider the nonlinear system

$$\ddot{X} + \alpha\dot{X} + \beta\dot{X}^3 + \gamma X^2\dot{X} + X = W(t)$$

where $W(t)$ is a gaussian white noise with spectral density K.

(a) Find the first and second derivate moments for the system, and separate the first derivate moments into reversible and irreversible parts.

(b) Determine the condition for the system to be in detailed balance, and the stationary probability density $p(x, \dot{x})$ if the condition is satisfied.

5.4. Consider the system

$$\ddot{X} + 2\zeta\omega_0\dot{X} + \omega_0^2[1 + W_1(t)]X = W_2(t)$$

where $W_1(t)$ and $W_2(t)$ are independent gaussian white noises with spectral densities K_{11} and K_{22}, respectively. Show that the system belongs to the class of generalized stationary potential only when $K_{11} = 0$.

5.5. Consider the system

$$\ddot{X} + \alpha\dot{X} + \beta\dot{X}^3 + X = W_1(t) + \dot{X}W_2(t)$$

where $W_1(t)$ and $W_2(t)$ are gaussian white noises with the correlation functions

$$E[W_i(t)W_j(t+\tau)] = 2\pi K_{ij}\delta(\tau) \qquad i, j = 1, 2$$

(a) Letting $X = X_1, \dot{X} = X_2$, obtain a Fokker-Planck equation for the Markov vector $\{X_1, X_2\}$.

(b) Determine the condition under which the system belongs to the class of generalized stationary potential.

(c) Obtain the exact stationary probability density when (b) is satisfied.

5.6. Determine the conditions under which the following systems belong to the class of generalized stationary potential:

(a) $\ddot{X} + 2\zeta\omega_0\dot{X}(1 + \alpha X^2) + \omega_0^2[1 + W_1(t)]X + f(X) = W_2(t)$

(b) $\ddot{X} + 2\zeta\omega_0\dot{X}(1 + \alpha X^2 + \beta\dot{X}^2 + \gamma\left|X^3\right|) + \omega_0^2X(1 + \delta\left|X\right|) = W(t)$

where $W_1(t)$, $W_2(t)$, and $W(t)$ are independent gaussian white noises, and find the exact probability density functions $p(x, \dot{x})$ for these systems.

5.7. Find at least three stochastic systems weakly equivalent to the system

$$\ddot{X} + \alpha\dot{X} + \beta(X^2 + \dot{X}^2)\dot{X} + X = W(t)$$

where $W(t)$ is a gaussian white noise with a spectral density K.

CHAPTER 6

STABILITY OF STOCHASTIC SYSTEMS

We learn in elementary mechanics that a perfectly straight uniform column can buckle under a static axial load if the load is greater than the critical Euler load. When the load is not purely static but contains a time-varying component, buckling can still occur under certain conditions, even if the total load never exceeds the static Euler load at any given time. The two cases are referred to as static and dynamic instabilities, respectively. This chapter is concerned with dynamic stability when time-varying loads are stochastic. However, in order to explain the fundamental concepts, a brief discussion of dynamic stability of deterministic systems is appropriate.

Perhaps the most extensively studied dynamic instability problem is described by the well-known Mathieu-Hill equation

$$\ddot{x} + 2\zeta\omega_0\dot{x} + (\omega_0^2 + \varepsilon\cos\omega t)x = 0 \tag{6.0.1}$$

which may be regarded as a simplified equation for the column problem, by assuming that the lateral deflection of the column is dominated by one mode and the time-varying component of the axial load is sinusoidal with a frequency ω. In this case, $x(t)$ represents the amplitude of the dominant mode. The term $\varepsilon\cos\omega t$ in equation (6.0.1) plays the role of a multiplicative (parametric) excitation which renders the bending stiffness of the column dependent on time.

Obviously, the trivial solution $x = \dot{x} = 0$ satisfies equation (6.0.1). The purpose of a stability analysis is to determine whether a nontrivial solution under initial

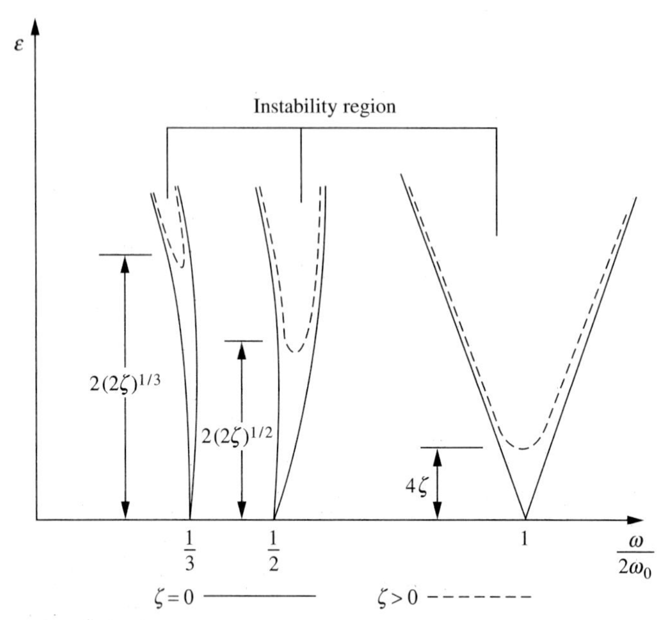

FIGURE 6.0.1
Strutt diagram.

condition $x(0) \neq 0$ and/or $\dot{x}(0) \neq 0$ will be bounded and, if so, whether it will tend to the trivial solution as time increases. These are qualitative properties of the solution, not the detailed quantitative values of the solution. Interestingly enough, such a qualitative question can often be answered without knowing the quantitative solution.

Figure 6.0.1 shows the stability boundaries for the Mathieu-Hill equation (6.0.1) plotted on the parameter plane (ω, ε) of the excitation for the cases of $\zeta = 0$ and $\zeta > 0$. Such a plot is known as the Strutt diagram (e.g., Metler, 1962). By scaling the frequency parameter ω by the unit of $2\omega_0$, successive instability regions are found near $\omega/2\omega_0 = 1/n$ where n is a positive integer. It is seen that the instability region is widest near $n = 1$. Furthermore, the instability boundaries are raised from the $\omega/2\omega_0$ axis by $2(2\zeta)^{1/n}$ when damping is present. Thus for $\zeta \ll 1$, the instability region near $n = 1$ is of the greatest practical concern.

If a lateral load is also present on the column, then an inhomogeneous term will appear on the right-hand side of equation (6.0.1), playing the role of an additive excitation on the system, as seen in Section 4.7.1 for the case of random excitations. If the response of a linear system is bounded under both multiplicative and additive excitations, then it must be stable without the additive excitation, and vice versa. However, this may or may not be true with nonlinear systems.

In general, the state of a dynamic system may be represented by an n-dimensional vector $\boldsymbol{x}(t)$, measured from a reference solution, called the trivial solution in the following discussion. The boundedness and convergence of $\boldsymbol{x}(t)$ can be defined rigorously in terms of a suitable norm of $\boldsymbol{x}(t)$, denoted by $\|\boldsymbol{x}(t)\|$.

Examples of such norms are

$$\|\boldsymbol{x}(t)\| = \sum_{i=1}^{n} |x_i(t)| \qquad \|\boldsymbol{x}(t)\| = \left[\sum_{i=1}^{n} x_i^2(t)\right]^{1/2}$$

$$\|\boldsymbol{x}(t)\| = \left[\sum_{i=1}^{n}\sum_{j=1}^{n} a_{ij} x_i(t) x_j(t)\right]^{1/2}$$

In the last example, a_{ij} are the elements of a positive definite square matrix. The following concepts, attributable to Lyapunov (1892), are fundamental in the investigation of stability of deterministic systems.

Lyapunov stability. The trivial solution is said to be stable if for every $\varepsilon > 0$, there exists a $\delta(\varepsilon, t_0) > 0$ such that

$$\sup_{t \geq t_0} \|\boldsymbol{x}(t; \boldsymbol{x}_0, t_0)\| < \varepsilon \tag{6.0.2}$$

provided $\|\boldsymbol{x}_0\| \leq \delta$ where $\boldsymbol{x}_0 = \boldsymbol{x}(t_0)$. The stability is said to be uniform if (6.0.2) holds for any t_0.

Lyapunov asymptotic stability. The trivial solution is said to be asymptotically stable, if it is stable, and if there exists a $\delta'(t_0) > 0$ such that

$$\lim_{t \to \infty} \|\boldsymbol{x}(t; \boldsymbol{x}_0, t_0)\| < 0 \tag{6.0.3}$$

provided $\|\boldsymbol{x}_0\| \leq \delta'$. The trivial solution is said to be asymptotically stable in the large if (6.0.3) holds for any $\boldsymbol{x}_0$.

Stability investigations are sometimes formulated in terms of a Lyapunov function $V(\boldsymbol{x}, t) \geq \alpha \|\boldsymbol{x}\|$ where $\alpha > 0$, which includes the norm itself as a special case.

6.1 CONCEPTS OF STOCHASTIC STABILITY

Under the excitation of stochastic processes, the response of a dynamic system is generally an n-dimensional stochastic vector $\boldsymbol{X}(t)$ measured from a referenced solution. Stability of such a system must be defined in terms of boundedness and convergence of $\|\boldsymbol{X}(t)\|$. Since convergence of a sequence of random variables can be interpreted in more than one way [see, e.g., Lin (1967)], different definitions for stochastic stability are possible. In a survey article, Kozin (1969) presented a set of definitions, but other definitions for stochastic stability have also appeared in the literature. The following definitions are used in this book.

Lyapunov stability with probability 1. The trivial solution is said to be stable in the Lyapunov sense with probability 1 if, for every pair of $\varepsilon_1, \varepsilon_2 > 0$, there exists a $\delta(\varepsilon_1, \varepsilon_2, t_0) > 0$ such that

$$\text{Prob}\left\{\bigcup_{\|\boldsymbol{x}_0\| \leq \delta}\left[\sup_{t \geq t_0} \|\boldsymbol{X}(t; \boldsymbol{x}_0, t_0)\| \geq \varepsilon_1\right]\right\} \leq \varepsilon_2 \tag{6.1.1}$$

where $x_0 = X(t_0)$ is deterministic. The stability is said to be uniform if (6.1.1) holds for any t_0.

Inequality (6.1.1) states that except for those sample functions which are associated with an arbitrarily small probability of occurrence ε_2, all the remaining sample functions are stable in the Lyapunov sense. Since ε_1 and ε_2 are arbitrarily small, this type of stability is also known as the almost sure sample stability, or simply sample stability. It is concerned with all the sample functions in the time interval $[t_0, t]$.

Lyapunov asymptotic stability with probability 1. The trivial solution is said to be asymptotically stable in the Lyapunov sense with probability 1, if (6.1.1) holds and if, for every $\varepsilon > 0$, there exists a $\delta'(\varepsilon, t_0) > 0$ such that

$$\lim_{t_1 \to \infty} \text{Prob}\left[\sup_{t \geq t_1} \left\| X(t; x_0, t_0) \right\| \geq \varepsilon \right] = 0 \tag{6.1.2}$$

provided $\|x_0\| \leq \delta'$. The stability is said to be in the large if (6.1.2) holds for any x_0. Since ε is arbitrarily small, inequality (6.1.2) is also known as the almost sure asymptotic sample stability, or simply asymptotic sample stability.

In principle, the almost sure sample stability and asymptotic sample stability are ideally suited to describe the qualitative behavior of stochastic systems, since they describe the boundedness and convergence of the greatest excursions of all the sample functions on the entire semi-infinite time domain $t_0 \leq t < \infty$. However, inequality (6.1.1) is often difficult to prove when applied to practical dynamic systems.

Stability in probability. The trivial solution is said to be stable in probability if, for every pair of $\varepsilon_1, \varepsilon_2 > 0$, there exists a $\delta(\varepsilon_1, \varepsilon_2, t_0) > 0$ such that

$$\text{Prob}\left[\left\| X(t; x_0, t_0) \right\| \geq \varepsilon_1 \right] \leq \varepsilon_2 \qquad t \geq t_0 \tag{6.1.3}$$

provided $\|x_0\| \leq \delta$, where $x_0 = X(t_0)$ is deterministic. The stability is said to be uniform if (6.1.3) holds for any t_0.

Asymptotic stability in probability. The trivial solution is said to be asymptotically stable in probability if (6.1.3) holds, and if for every $\varepsilon > 0$ there exists a $\delta'(\varepsilon, t_0) > 0$ such that

$$\lim_{t \to \infty} \text{Prob}\left[\left\| X(t; x_0, t_0) \right\| \geq \varepsilon \right] = 0 \tag{6.1.4}$$

provided $\|x_0\| \leq \delta'$. The stability is said to be in the large if (6.1.4) holds for any x_0.

It is important to note that in inequality (6.1.3) the probability associated with the boundedness of $\|X(t; x_0, t_0)\|$ is specified for each value of t, not for the maximum value in the time interval $[t_0, t]$. Thus the stability indicated in (6.1.3) is not of a Lyapunov type. It is clear that (6.1.1) implies (6.1.3), and (6.1.2) implies (6.1.4). However, if the system is linear, then the reverse is also true, and the two types of stability become equivalent.

Stability in the *m*th moment. The trivial solution is said to be stable in the *m*th moment if, for every $\varepsilon > 0$, there exists a $\delta(\varepsilon, t_0) > 0$ such that

$$E\left[\left\|X(t; x_0, t_0)\right\|^m\right] \leq \varepsilon \qquad m > 0, t \geq t_0 \tag{6.1.5}$$

provided $\|x_0\| \leq \delta$, where $x_0 = X(t_0)$ is deterministic. The stability is said to be uniform if (6.1.5) holds for any t_0.

In (6.1.5), the *m*th moment of $\|X(t; x_0, t_0)\|$ is again specified for each value of t; thus (6.1.5) is not a Lyapunov-type definition. However, it is not immediately clear whether (6.1.5) is weaker than (6.1.1) or vice versa.

Asymptotic stability in the *m*th moment. The trivial solution is said to be asymptotically stable in the *m*th moment if (6.1.5) holds and if there exists a $\delta'(\varepsilon, t_0) > 0$ such that

$$\lim_{t\to\infty} E\left[\left\|X(t; x_0, t_0)\right\|^m\right] = 0 \qquad m > 0 \tag{6.1.6}$$

provided $\|x_0\| \leq \delta'$. The stability is said to be in the large if (6.1.6) holds for any x_0.

It is of interest to note that although the asymptotic moment stability (6.1.6) is concerned only with the convergence of the *m*th moment of $\|X(t; x_0, t_0)\|$ for each t as t increases, it is more stringent than the asymptotic sample stability condition (6.1.2) for linear systems under multiplicative gaussian white noise excitations (Arnold, 1984; Kozin and Sugimoto, 1977). Moreover, in the special case $m = 2$, condition (6.1.6) is always more stringent than (6.1.2) for linear systems, since convergence in mean square implies convergence in probability, and since stability in probability is equivalent to stability with probability 1 for linear systems.

In Section 6.2, our attention will be focused on the sample stability of linear systems under multiplicative excitations of well-behaved nonwhite ergodic processes. Sample and moment stabilities for linear systems under multiplicative excitations, which are either white noises or replaceable by white noises, are considered in Sections 6.3 through 6.6. It will be shown that inequality (t) of Section 4.7.1 is, in fact, the condition for asymptotic stability in probability of a linear system, obtained by considering the existence of the stationary probability. This condition also guarantees that the system is asymptotically stable with probability 1, since the system is linear. Stability conditions in probability and in moments for some nonlinear systems for which such conditions can be determined are discussed in Section 6.7.

6.2 ASYMPTOTIC SAMPLE STABILITY OF LINEAR SYSTEMS UNDER ERGODIC RANDOM EXCITATIONS WITH ZERO MEANS AND BOUNDED VARIANCES

For nonwhite stochastic excitations, even when they are ergodic, only sufficient conditions have been obtained for the asymptotic sample stability and only for linear systems. We are concerned, of course, with structural systems for which the state space is at least two-dimensional. Since stochastic stability of a multidimensional system is defined in terms of a norm, and since there are different choices for such a norm, the sufficient stability conditions so obtained are generally different. Therefore, early efforts were directed at increasing the sharpness of sufficient conditions, notably the works of Kozin (1963), Caughey and Gray (1965), and Ariaratnam (1967). In principle, the "sharpest" condition, if obtainable, must also be the necessary condition.

An important advance in the study of sample stability of linear systems under arbitrary ergodic random excitations was due to Infante (1968), who made use of the "pencil" of quadratic form. To explain Infante's fundamental work, we require the following concepts and definitions.

Let $\boldsymbol{D}$ and $\boldsymbol{B}$ be $n \times n$ real symmetric matrices, and let $\boldsymbol{B}$ be positive definite. Then for an n-dimensional vector, the expression $\boldsymbol{x}'\boldsymbol{D}\boldsymbol{x} - \boldsymbol{x}'\boldsymbol{B}\boldsymbol{x}$ is called a *regular pencil of quadratic form*, where a prime denotes the transpose of a matrix. The equation $|\boldsymbol{D} - \lambda \boldsymbol{B}| = 0$ or $|\boldsymbol{D}\boldsymbol{B}^{-1} - \lambda \boldsymbol{I}| = 0$ is called the *characteristic equation*, and its roots λ the *eigenvalues of the pencil*. These eigenvalues are real since $\boldsymbol{D}\boldsymbol{B}^{-1}$ is symmetrical. By arranging the eigenvalues in the order of $\lambda_1 \leq \lambda_2 \leq \cdots \leq \lambda_n$, it can be shown that

$$\lambda_1[\boldsymbol{D}\boldsymbol{B}^{-1}] = \min_{x} \frac{\boldsymbol{x}'\boldsymbol{D}\boldsymbol{x}}{\boldsymbol{x}'\boldsymbol{B}\boldsymbol{x}} \tag{6.2.1}$$

$$\lambda_n[\boldsymbol{D}\boldsymbol{B}^{-1}] = \max_{x} \frac{\boldsymbol{x}'\boldsymbol{D}\boldsymbol{x}}{\boldsymbol{x}'\boldsymbol{B}\boldsymbol{x}} \tag{6.2.2}$$

where $\lambda_j[\boldsymbol{Q}]$ indicates the jth eigenvalue of matrix $\boldsymbol{Q}$.

In mechanics, the right-hand sides of (6.2.1) and (6.2.2) may be interpreted as the minimum and maximum values, respectively, for Rayleigh's quotient of an n-degree-of-freedom undamped linear system, with a mass matrix $\boldsymbol{B}$ and a stiffness matrix $\boldsymbol{D}$, obtained by allowing the assumed mode $\boldsymbol{x}$ to vary in an n-dimensional space. Thus the preceding minimum and maximum properties of eigenvalues are well-known.

Now consider an n-dimensional linear stochastic system

$$\dot{\boldsymbol{X}} = [\boldsymbol{A} + \boldsymbol{F}(t)]\boldsymbol{X} \tag{6.2.3}$$

where $\boldsymbol{X}(t)$ is the state vector, $\boldsymbol{A}$ is a constant matrix, and $\boldsymbol{F}(t)$ is a matrix whose nonzero elements $f_{ij}(t)$ are ergodic processes with zero means. The zero-mean assumption of the $\boldsymbol{F}(t)$ matrix is made without loss of generality, since a nonzero mean, which must be a constant for an ergodic process, can be added to matrix $\boldsymbol{A}$. To investigate the sample stability of $\boldsymbol{X}(t)$, choose a Lyapunov function $V(\boldsymbol{X}) = \boldsymbol{X}'\boldsymbol{B}\boldsymbol{X}$ which is positive for any nontrivial $\boldsymbol{X}(t)$ and is zero only if $\boldsymbol{X} = \boldsymbol{0}$. Then along the trajectories of (6.2.3), define

$$\lambda(t) = \frac{\dot{V}(\boldsymbol{X})}{V(\boldsymbol{X})} = \frac{\boldsymbol{X}'[(\boldsymbol{A}' + \boldsymbol{F}')\boldsymbol{B} + \boldsymbol{B}(\boldsymbol{A} + \boldsymbol{F})]\boldsymbol{X}}{\boldsymbol{X}'\boldsymbol{B}\boldsymbol{X}} \tag{6.2.4}$$

which gives the exponential rate of growth of the Lyapunov function at time t. Letting $\boldsymbol{D} = [(\boldsymbol{A}' + \boldsymbol{F}')\boldsymbol{B} + \boldsymbol{B}(\boldsymbol{A} + \boldsymbol{F})]$ in (6.2.2), we obtain

$$\lambda(t) \leq \lambda_{\max}[\boldsymbol{A}' + \boldsymbol{F}' + \boldsymbol{B}(\boldsymbol{A} + \boldsymbol{F})\boldsymbol{B}^{-1}] \tag{6.2.5}$$

Now, from (6.2.4)

$$V[\boldsymbol{X}(t)] = V(\boldsymbol{x}_0)\exp\left[\int_0^t \lambda(\tau)\,d\tau\right] = V(\boldsymbol{x}_0)\exp\left\{t\left[\frac{1}{t}\int_0^t \lambda(\tau)\,d\tau\right]\right\} \tag{6.2.6}$$

where $\boldsymbol{x}_0 = \boldsymbol{X}(0)$. Since $\boldsymbol{F}(t)$ is a matrix of ergodic elements, $\lambda(t)$ also tends to be ergodic as t increases. Thus

$$\lim_{t\to\infty}\frac{1}{t}\int_0^t \lambda(\tau)\,d\tau = E[\lambda(t)] \tag{6.2.7}$$

It follows from (6.2.6) that $V[\boldsymbol{X}(t)] \to 0$ as $t \to \infty$ provided

$$E[\lambda(t)] \le -\varepsilon \qquad \text{for some } \varepsilon > 0 \tag{6.2.8}$$

Since $\lambda(t)$ is bounded above as shown in (6.2.5), we obtain a sufficient condition for asymptotic sample stability as follows:

$$E\left\{\lambda_{\max}[\boldsymbol{A}' + \boldsymbol{F}' + \boldsymbol{B}(\boldsymbol{A} + \boldsymbol{F})\boldsymbol{B}^{-1}]\right\} \le -\varepsilon \tag{6.2.9}$$

The preceding sufficient condition is known as *Infante's theorem,* and it can be optimized by varying the elements of matrix $\boldsymbol{B}$, subject to the constraints that $\boldsymbol{B}$ remains symmetrical and positive definite.

In passing, we note that the use of a Lyapunov function for the stability investigation of a stochastic system was first made by Bertram and Sarachik (1959) in the sense of stability in the mean.

Although (6.2.9) is valid for arbitrary linear systems, it is practical only for lower order systems for two reasons. First, the maximum eigenvalue must be obtained algebraically before taking the ensemble average, and second, the optimization procedure can be extremely complex if the matrices involved are large. In the following discussion we consider only the two-dimensional systems.

The governing equation for a single-degree-of-freedom linear system under multiplicative random excitations may be written in the nondimensional form

$$\ddot{Y} + 2[\zeta + \eta(t)]\dot{Y} + [1 + \xi(t)]Y = 0 \tag{6.2.10}$$

where an overdot denotes one differentiation with respect to the nondimensional time, which is so selected that the normalized natural frequency of the system is unity. Two multiplicative random excitations appear in equation (6.2.10), $\eta(t)$ in the damping term and $\xi(t)$ in the stiffness term. In the original work of Infante (1968), optimized stability conditions were obtained for the special cases $\eta(t) = 0$ and $\xi(t) = 0$ separately, but for the general case of $\eta(t) \neq 0$ and $\xi(t) \neq 0$, similar optimization was unsuccessful. Subsequently, Ariaratnam and Ly (1989) extended Infante's work to include independent nonzero $\eta(t)$ and $\xi(t)$, and they showed that optimization is possible by taking the intensities of $\eta(t)$ and $\xi(t)$ into consideration. Further extension was reported by Ariaratnam and Xie (1989) to the case of correlated $\eta(t)$ and $\xi(t)$. The following discussion follows closely the development in the last two references.

Equation (6.2.10) can be simplified using a transformation

$$Y = Ze^{-\zeta t} \tag{6.2.11}$$

Substitution of (6.2.11) into (6.2.10) yields

$$\ddot{Z} + 2\eta(t)\dot{Z} + [c + \gamma(t)]Z = 0 \tag{6.2.12}$$

where $c = 1 - \zeta^2$ and

$$\gamma(t) = \xi(t) - 2\zeta\eta(t) \tag{6.2.13}$$

Equation (6.2.13) is equivalent to

$$\frac{d}{dt}\begin{Bmatrix} X_1 \\ X_2 \end{Bmatrix} = \begin{bmatrix} 0 & 1 \\ -c & 0 \end{bmatrix}\begin{Bmatrix} X_1 \\ X_2 \end{Bmatrix} + \begin{bmatrix} 0 & 0 \\ -\gamma(t) & -2\eta(t) \end{bmatrix}\begin{Bmatrix} X_1 \\ X_2 \end{Bmatrix} \tag{6.2.14}$$

which has the form of equation (6.2.3); thus the corresponding $\boldsymbol{A}$ and $\boldsymbol{F}(t)$ matrices can be identified.

We now select a Lyapunov function $V = \boldsymbol{X}'\boldsymbol{B}\boldsymbol{X}$ with

$$\boldsymbol{B} = \begin{bmatrix} \alpha_1^2 & \alpha_2 \\ \alpha_2 & 1 \end{bmatrix} \qquad \alpha_1^2 - \alpha_2^2 > 0 \tag{6.2.15}$$

For the purpose of our analysis, this $\boldsymbol{B}$ matrix is of the most general form of a symmetric positive definite matrix of order 2×2. Evaluating V along the trajectory of (6.2.14) yields

$$\dot{V} = \boldsymbol{X}'\boldsymbol{D}\boldsymbol{X} \tag{6.2.16}$$

where

$$\boldsymbol{D} = \begin{bmatrix} -2\alpha_2(c+\gamma) & \alpha_1^2 - 2\alpha_2\eta - (c+\gamma) \\ \alpha_1^2 - 2\alpha_2\eta - (c+\gamma) & 2\alpha_2 - 4\eta \end{bmatrix} \tag{6.2.17}$$

The eigenvalues of $\boldsymbol{D}\boldsymbol{B}^{-1}$ may be computed from

$$|\boldsymbol{D} - \lambda\boldsymbol{B}| = \begin{vmatrix} -2\alpha_2(c+\gamma) - \alpha_1^2\lambda & \alpha_1^2 - 2\alpha_2\eta - (c+\gamma) - \alpha_2\lambda \\ \alpha_1^2 - 2\alpha_2\eta - (c+\gamma) - \alpha_2\lambda & 2\alpha_2 - 4\eta - \lambda \end{vmatrix} = 0 \tag{6.2.18}$$

The maximum eigenvalue obtained from (6.2.18) is

$$\lambda_{\max}(t) = -2\eta + \left(4\eta^2 + \frac{G}{\alpha_1^2 - \alpha_2^2}\right)^{1/2} \tag{6.2.19}$$

where

$$G = (\alpha_1^2 - c)^2 + 4c\alpha_2^2 - 4\alpha_2(\alpha_1^2 + c)\eta + 4\alpha_2^2\eta^2 - 4\alpha_2\eta\gamma + (4\alpha_2^2 + 2c - 2\alpha_1^2)\gamma + \gamma^2 \tag{6.6.20}$$

Thus a sufficient condition for the asymptotic sample stability for equation (6.2.12) is given by

$$E\left[\left(4\eta^2 + \frac{G}{\alpha_1^2 - \alpha_2^2}\right)^{1/2}\right] \leq -\varepsilon \qquad \varepsilon > 0 \tag{6.2.21}$$

where use has been made of the fact that $E[\eta(t)] = 0$. Of course, inequality (6.2.21) cannot be satisfied since the left-hand side is the ensemble average of a nonnegative random variable. This is not surprising, in view of the fact that equation (6.2.12) for the $Z(t)$ process does not have a deterministic damping term. Now, from (6.2.11)

$$|Y|^2 = |Z|^2 e^{-2\zeta t} \tag{6.2.22}$$

It follows that for the original system (6.2.10) a corresponding sufficient condition for asymptotic sample stability is

$$E\left[\left(4\eta^2 + \frac{G}{\alpha_1^2 - \alpha_2^2}\right)^{1/2}\right] \le 2\zeta - \varepsilon \qquad \varepsilon > 0 \tag{6.2.23}$$

6.2.1 Multiplicative Excitations with Known Variances and Covariance

The ensemble averaging in (6.2.23) cannot be carried out if only the variances and covariance of the excitations are given. In this case, we resort to the Schwarz inequality and obtain

$$E\left[4\eta^2 + \frac{G}{\alpha_1^2 - \alpha_2^2}\right] \ge \left\{E\left[\left(4\eta^2 + \frac{G}{\alpha_1^2 - \alpha_2^2}\right)^{1/2}\right]\right\}^2 \tag{a}$$

Then the condition

$$E\left[4\eta^2 + \frac{G}{\alpha_1^2 - \alpha_2^2}\right] \le 4\zeta^2 - \varepsilon' \qquad \varepsilon' > 0 \tag{b}$$

implies condition (6.2.23), but it is generally not as sharp as (6.2.23).

Carrying out the ensemble averaging in (b), we obtain the boundary of a sufficient condition for asymptotic sample stability as follows:

$$F = -4\zeta^2(\alpha_1^2 - \alpha_2^2) + 4(\alpha_1^2 - \alpha_2^2)\sigma_\eta^2 + E[G] = 0 \tag{c}$$

where, from (6.2.20) and (6.2.13),

$$E[G] = (\alpha_1^2 - c)^2 + 4c\alpha_2^2 + 4\alpha_2^2\sigma_\eta^2 - 4\alpha_2 E[\eta\gamma] + E[\gamma^2] \tag{d}$$

$$E[\eta\gamma] = \rho\sigma_\xi\sigma_\eta - 2\zeta\sigma_\eta^2 \tag{e}$$

$$E[\gamma^2] = \sigma_\xi^2 - 4\zeta\rho\sigma_\xi\sigma_\eta + 4\zeta^2\sigma_\eta^2 \tag{f}$$

and where ρ is the correlation coefficient of $\xi(t)$ and $\eta(t)$. The stability boundary (c) can now be optimized by varying α_1^2 and α_2 such that σ_ξ and σ_η attain their maximum values. This is accomplished by requiring

$$\frac{\partial\sigma_\xi}{\partial\alpha_1^2} = \frac{\partial\sigma_\xi}{\partial\alpha_2} = \frac{\partial\sigma_\eta}{\partial\alpha_1^2} = \frac{\partial\sigma_\eta}{\partial\alpha_2} = 0 \tag{g}$$

These conditions result in four equations for the two unknowns α_1^2 and α^2; therefore, only two equations can be used at each time. In practice, either σ_η or σ_ξ can be held fixed while the other is maximized. By so doing, two of the equations in (g) are effectively discarded. However, since

$$\begin{aligned} dF = {} & \frac{\partial F}{\partial\alpha_1^2}d\alpha_1^2 + \frac{\partial F}{\partial\alpha_2}d\alpha_2 + \frac{\partial F}{\partial\sigma_\xi}\left(\frac{\partial\sigma_\xi}{\partial\alpha_1^2}d\alpha_1^2 + \frac{\partial\sigma_\xi}{\partial\alpha_2}d\alpha_2\right) \\ & + \frac{\partial F}{\partial\sigma_\eta}\left(\frac{\partial\sigma_\eta}{\partial\alpha_1^2}d\alpha_1^2 + \frac{\partial\sigma_\eta}{\partial\alpha_2}d\alpha_2\right) = 0 \end{aligned} \tag{h}$$

two of the conditions in (g) can be replaced by

$$\frac{\partial F}{\partial \alpha_1^2} = \frac{\partial F}{\partial \alpha_2} = 0 \tag{i}$$

Using (c) through (f) in (i), we obtain

$$-4\zeta^2 + 4\sigma_\eta^2 + 2(\alpha_1^2 - c) = 0 \tag{j}$$

$$8(\zeta^2 + c)\alpha_2 - 4(\rho\sigma_\xi\sigma_\eta - 2\zeta\sigma_\eta^2) = 0 \tag{k}$$

These lead to a pair of optimal α_1^2 and α_2:

$$\alpha_1^2 = 1 + \zeta^2 - 2\sigma_\eta^2 \tag{l}$$

$$\alpha_2 = \tfrac{1}{2}\rho\sigma_\xi\sigma_\eta - \zeta\sigma_\eta^2 \tag{m}$$

noticing that $c = 1-\zeta^2$. Of course, these α_1^2 and α_2 values constitute a true optimal pair only if $\alpha_1^2 - \alpha_2^2 > 0$ holds, which must be checked in each case. Substitution of (l) and (m) into (c) yields

$$(1 - \rho^2\sigma_\eta^2)\sigma_\xi^2 - 4\zeta\rho(1 - \sigma_\eta^2)\sigma_\eta\sigma_\xi + 4(1 - \sigma_\eta^2)[(1 + \zeta^2)\sigma_\eta^2 - \zeta^2] = 0 \tag{n}$$

In passing, we note that this equation should be interpreted as either describing a sufficient stability boundary for σ_η given a value σ_ξ or, conversely, describing a sufficient stability boundary for σ_ξ given a value σ_η, because (i) can replace only two of the four conditions specified in (g).

Three special cases are examined next.

Case 1. For the case $\eta(t) = 0$, equation (6.2.10) reduces to

$$\ddot{Y} + 2\zeta\dot{Y} + [1 + \xi(t)]Y = 0 \tag{o}$$

Letting $\sigma_\eta = 0$ and $\rho = 0$ in (n) we obtain the boundary of a sufficient condition for asymptotic sample stability

$$\sigma_\xi^2 - 4\zeta^2 = 0 \tag{p}$$

Case 2. For the case $\xi(t) = 0$, equation (6.2.10) reduces to

$$\ddot{Y} + 2[\zeta + \eta(t)]\dot{Y} + Y = 0 \tag{q}$$

A sufficient-condition boundary for asymptotic sample stability is obtained by letting $\sigma_\xi = 0$ and $\rho = 0$ in (n), resulting in

$$\sigma_\eta^2 - \frac{\zeta^2}{1 + \zeta^2} = 0 \tag{r}$$

Both (p) and (r) were first obtained by Infante (1968). They are illustrated in Figs. 6.2.1 and 6.2.2.

Case 3. For the case of uncorrelated $\eta(t)$ and $\xi(t)$, namely, $\rho = 0$, the boundary of sufficient condition (n) reduces to

$$\sigma_\xi^2 + 4(1 - \sigma_\eta^2)[(1 + \zeta^2)\sigma_\eta^2 - \zeta^2] = 0 \tag{s}$$

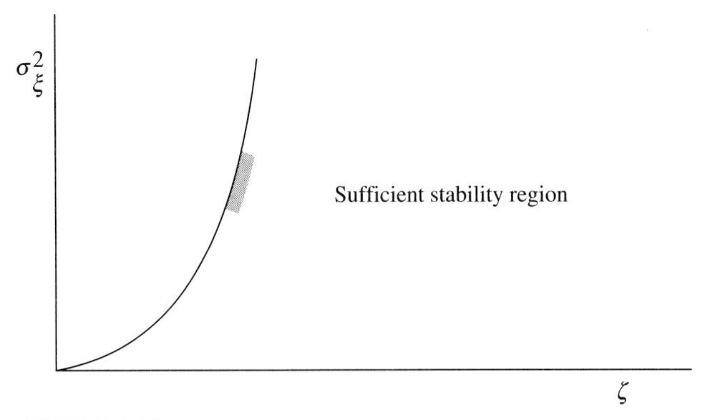

FIGURE 6.2.1
Sufficient condition for asymptotic sample stability for system $\ddot{Y} + 2\zeta\dot{Y} + [1 + \xi(t)]Y = 0$.

This result, obtained by Ariaratnam and Ly (1989), is plotted in Fig. 6.2.3 as contours of σ_η vs. σ_ξ for different values of ζ.

Return now to the general case, represented by (n). Contours similar to those of Fig. 6.2.3 can be plotted for different values of the correlation coefficient ρ. The cases of $\rho = \pm 0.5$ and $\rho = \pm 1$ are shown in Fig. 6.2.4 and Fig. 6.2.5, respectively. Compared with the case of two uncorrelated multiplicative excitations, a positive correlation between the two excitations tends to enlarge the sufficient stability region obtained from the foregoing procedure, whereas the opposite is true for a negative correlation.

6.2.2 Multiplicative Excitations with Known Probability Distributions

If the joint probability density of the multiplicative excitations $\eta(t)$ and $\xi(t)$ is known, then, in principle, the ensemble averages indicated in the sufficient condition

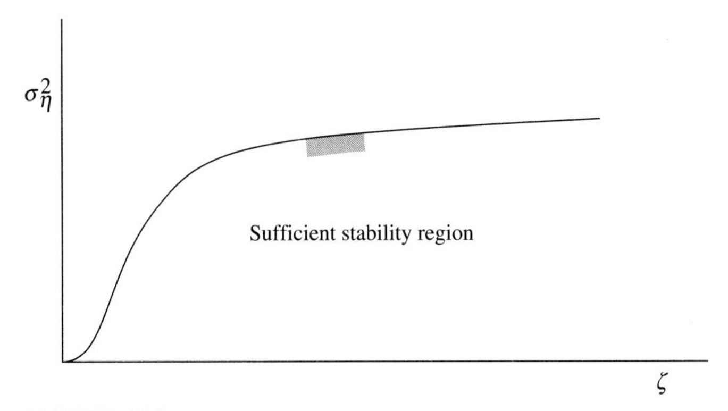

FIGURE 6.2.2
Sufficient condition for asymptotic sample stability for system $\ddot{Y} + 2[\zeta + \eta(t)]\dot{Y} + Y = 0$.

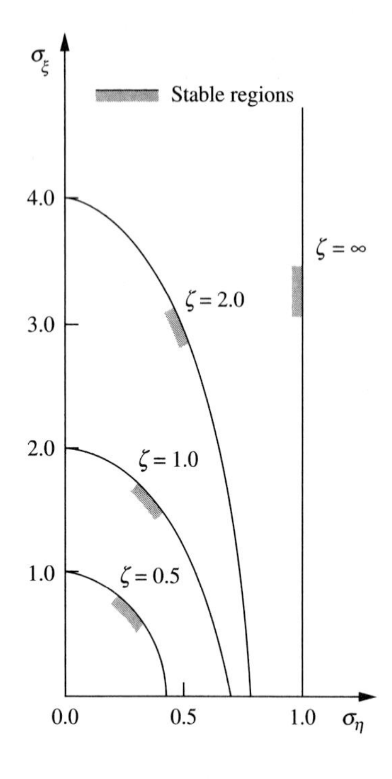

FIGURE 6.2.3
Regions of almost-sure asymptotic stability for system $\ddot{Y} + 2[\zeta + \eta(t)]\dot{Y} + [1 + \xi(t)]Y = 0$ via the Schwarz inequality ($\rho = 0$) [after Ariaratnam and Ly, 1989].

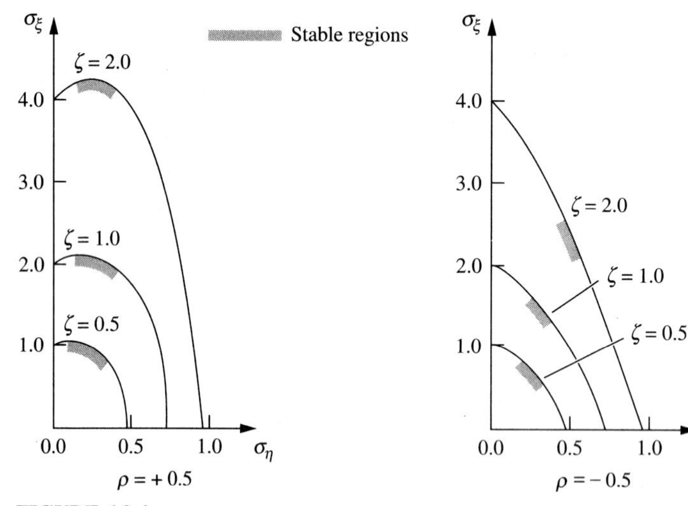

FIGURE 6.2.4
Regions of almost-sure asymptotic stability for system $\ddot{Y} + 2[\zeta + \eta(t)]\dot{Y} + [1 + \xi(t)]Y = 0$ via the Schwarz inequality [after Ariaratnam and Xie, 1989].

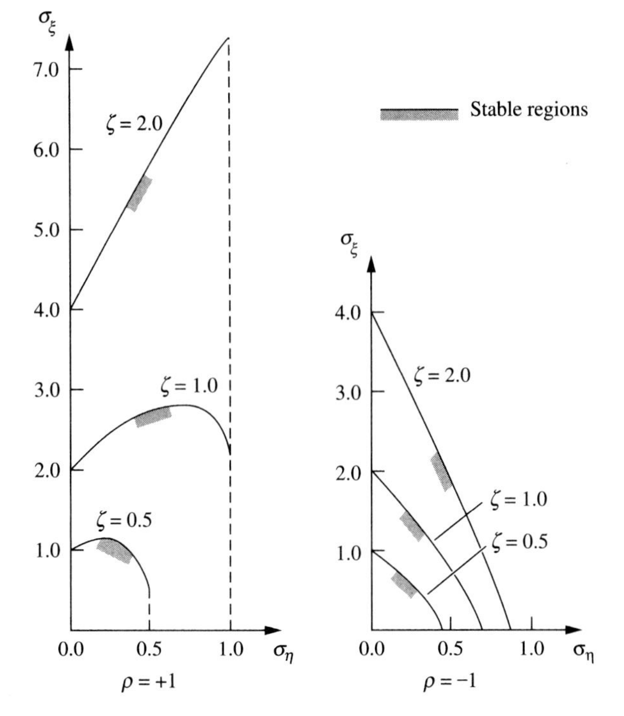

FIGURE 6.2.5
Regions of almost-sure asymptotic stability for system $\ddot{Y} + 2[\zeta + \eta(t)]\dot{Y} + [1 + \xi(t)]Y = 0$ via the Schwarz inequality [after Ariaratnam and Xie, 1989].

for asymptotic sample stability (6.2.23) can be calculated. The boundary for the sufficient stability condition so determined and optimized is expected to be sharper in the sense that the indicated stability region is larger and contains the smaller region obtained with the knowledge of only σ_ξ, σ_η, and ρ.

Return now to inequality (6.2.23), and specify the boundary for the sufficient asymptotic stability condition as

$$H = E\{[4\eta^2(\alpha_1^2 - \alpha_2^2) + G]^{1/2}\} - 2\zeta(\alpha_1^2 - \alpha_2^2)^{1/2} = 0 \tag{a}$$

where G has been given in (6.2.20):

$$G = (\alpha_1^2 - c)^2 + 4c\alpha_2^2 - 4\alpha_2(\alpha_1^2 + c)\eta + 4\alpha_2^2\eta^2 - 4\alpha_2\eta\gamma + (4\alpha_2^2 + 2c - 2\alpha_1^2)\gamma + \gamma^2 \tag{6.2.20}$$

First consider two special cases.

Case 1. For the case $\eta(t) = 0$, we have $\gamma = \xi(t)$. Then (a) and (6.2.20) are reduced to

$$H = E[G^{1/2}] - 2\zeta(\alpha_1^2 - \alpha_2^2)^{1/2} = 0 \tag{b}$$

$$G = (\alpha_1^2 - c)^2 + 4c\alpha_2^2 + (4\alpha_2^2 + 2c - 2\alpha_1^2)\xi + \xi^2 \tag{c}$$

The optimal α_1^2 and α_2 are obtained from

$$\frac{\partial H}{\partial \alpha_1^2} = E[(\alpha_1^2 - c - \xi)G^{-1/2}] - \zeta(\alpha_1^2 - \alpha_2^2)^{-1/2} = 0 \tag{d}$$

$$\frac{\partial H}{\partial \alpha_2} = \alpha_2\{2E[(c+\xi)G^{-1/2}] + \zeta(\alpha_1^2 - \alpha_2^2)^{-1/2}\} = 0 \tag{e}$$

Equation (e) is satisfied by letting $\alpha_2 = 0$. It follows from (c) that

$$G = (\alpha_1^2 - c - \xi)^2 \tag{f}$$

and from (b) and (d) that

$$E[|\alpha_1^2 - c - \xi|] - 2\zeta\alpha_1 = 0 \tag{g}$$

$$E[\operatorname{sgn}(\alpha_1^2 - c - \xi)] - \frac{\zeta}{\alpha_1} = 0 \tag{h}$$

Equations (g) and (h) were obtained by Ariaratnam and Ly (1989), and equation (g) was obtained by Kozin and Wu (1973) earlier in different notations.

Case 2. For the case $\xi(t) = 0$, we have $\gamma = -2\zeta\eta(t)$. Then (6.2.20) has the form of

$$\begin{aligned} G = {} & (\alpha_1^2 - c)^2 + 4c\alpha_2^2 - 4\alpha_2(\alpha_1^2 + c)\eta + 4\alpha_2^2\eta^2 + 8\zeta\alpha_2\eta^2 \\ & - 2\zeta\eta(4\alpha_2^2 + 2c - 2\alpha_1^2) + 4\zeta^2\eta^2 \end{aligned} \tag{i}$$

For optimal α_1^2 and α_2,

$$\begin{aligned} \frac{\partial H}{\partial \alpha_1^2} = {} & E\{[2\eta^2 + \alpha_1^2 - c - 2\alpha_2\eta + 2\zeta\eta][4\eta^2(\alpha_1^2 - \alpha_2^2) + G]^{-1/2}\} \\ & - \zeta(\alpha_1^2 - \alpha_2^2)^{-1/2} = 0 \end{aligned} \tag{j}$$

$$\begin{aligned} \frac{\partial H}{\partial \alpha_2} = {} & E\{[2c\alpha_2 - (\alpha_1^2 + c)\eta + 2\zeta\eta^2 - 4\zeta\alpha_2\eta][4\eta^2(\alpha_1^2 - \alpha_2^2) + G]^{-1/2}\} \\ & + \zeta\alpha_2(\alpha_1^2 - \alpha_2^2)^{-1/2} = 0 \end{aligned} \tag{k}$$

Multiplying (j) by $2(\zeta\alpha_2 - c)$ and (k) by $-2(\alpha_2 + \zeta)$, and adding the results to (a), we obtain

$$\begin{aligned} (\alpha_1^2 + 2\zeta\alpha_2 - c)\big(E\{[4\eta^2 - 2(\alpha_2 - 3\zeta)\eta + \alpha_1^2 - 3c][4\eta^2(\alpha_1^2 - \alpha_2^2) + G]^{-1/2}\} \\ - 2\zeta(\alpha_1^2 - \alpha_2^2)^{-1/2}\big) = 0 \end{aligned} \tag{l}$$

Equation (l) is satisfied if

$$\alpha_1^2 + 2\zeta\alpha_2 - c = 0 \tag{m}$$

which leads, after some algebra, to

$$[4\eta^2(\alpha_1^2 - \alpha_2^2) + G]^{1/2} = 2\,|\eta - \alpha_2| \tag{n}$$

Thus equation (a) is reduced to

$$E[|\eta - \alpha_2|] - \zeta[1 - (\zeta + \alpha_2)^2]^{1/2} = 0 \tag{o}$$

Multiplying equation (j) by $-\zeta$ and adding the result to equation (k), we obtain

$$E[\operatorname{sgn}(\alpha_2 - \eta)] - \zeta(\alpha_2 + \zeta)[1 - (\alpha_2 + \zeta)^{1/2}]^{-1/2} = 0 \tag{p}$$

Again, equations (o) and (p) were obtained by Ariaratnam and Ly (1989), and equation (o) was obtained by Kozin and Wu (1973) earlier in different notations.

For the case of $\eta(t) \neq 0$ and $\xi(t) \neq 0$, the required optimization process is not quite straightforward, and for its implementation, Ariaratnam and Xie (1989) used a general algorithm, called the complex method for constrained optimization, due to Box (1965) and Richardson and Kuester (1973).

The complex method for constrained optimization provides a sequential search for the maximum of a multivariable nonlinear function subject to nonlinear constraints. Specifically, it is used to maximize an objective function

$$F(X_1, X_2, \ldots, X_N) \tag{q}$$

under the following constraints:

$$G_k \leq X_k \leq H_k \qquad k = 1, 2, \ldots, M, \quad M > N \tag{r}$$

The variables $X_{N+1}, \ldots, X_M$ are implicit, and they are functions of the explicit independent variables $X_1, X_2, \ldots, X_N$. The upper and lower constraints H_k and G_k are either constants or functions of independent variables. The algorithm proceeds as follows:

1. An original "complex" of $K \geq N + 1$ points is obtained by selecting a reasonable starting point X_i and $K - 1$ additional points generated from

$$X_{i,j} = G_i + r_{i,j}(H_i - G_i) \qquad i = 1, 2, \ldots, N, j = 1, 2, \ldots, K - 1 \tag{s}$$

where $r_{i,j}$ are random numbers between 0 and 1.

2. The above complex of points is then tested against constraints (r). If an explicit constraint is violated at any step, the point is moved a small distance δ inside the violated limit. If an implicit constraint is violated, the point is moved one-half the distance toward the centroid $\bar{X}_{i,c}$ of the remaining points:

$$X_{i,j}(\text{new}) = \frac{X_{i,j}(\text{old}) + \bar{X}_{i,c}}{2} \qquad i = 1, 2, \ldots, N \tag{t}$$

where
$$\bar{X}_{i,c} = \frac{1}{K - 1}\left[\sum_{j=1}^{K} X_{i,j} - X_{i,j}(\text{old})\right] \tag{u}$$

This process is repeated until all the implicit constraints are satisfied.

3. The objective function is evaluated at each point. The point corresponding to the lowest objective function is replaced by its reflection:

$$X_{i,j}(\text{new}) = \gamma[\bar{X}_{i,c} - X_{i,j}(\text{old})] + \bar{X}_{i,c} \qquad i = 1, 2, \ldots, N \tag{v}$$

where
$$\bar{X}_{i,c} = \frac{1}{K - 1}\left[\sum_{j=1}^{K} X_{i,j} - X_{i,j}(\text{lowest})\right] \qquad i = 1, 2, \ldots, N \tag{w}$$

and a recommended value for γ is 1.3.

4. If a point gives repeatedly the lowest function value on consecutive trials, it is moved one-half the distance toward the centroid of the remaining points.

5. The new point is checked against the constraints and is adjusted as before if the constraints are violated.

6. The search ends when the value of the objective function at each point is below a preset limit ε in m consecutive iterations.

The complex method for constrained optimization just described will now be applied to maximize the sufficient sample stability condition (6.2.23). Assume that the excitation processes $\xi(t)$ and $\eta(t)$ are jointly gaussian ergodic processes, with a joint probability density

$$p(\xi, \eta) = (2\pi\sigma_\xi\sigma_\eta)^{-1}(1-\rho^2)^{-1/2} \times \exp\left\{-\frac{1}{2}(1-\rho^2)^{-1}\left[\left(\frac{\xi}{\sigma_\xi}\right)^2 + \left(\frac{\eta}{\sigma_\eta}\right)^2 - 2\rho\left(\frac{\xi}{\sigma_\xi}\right)\left(\frac{\eta}{\sigma_\eta}\right)\right]\right\} \quad \text{(x)}$$

This probability density is used to compute the ensemble average shown in (6.2.21):

$$E[F(\xi, \eta)] = \int_{-\infty}^{\infty} d\eta \int_{-\infty}^{\infty} F(\xi, \eta) p(\xi, \eta)\, d\xi \quad \text{(y)}$$

where

$$F(\xi, \eta) = \left[4\eta^2 + \frac{G}{\alpha_1^2 - \alpha_2^2}\right]^{1/2} \quad \text{(z)}$$

and G is defined according to (6.2.13) and (6.2.20). Substituting (z) into (y), and changing ξ and η to new variables

$$\bar{\xi} = \xi\sigma_\xi^{-1}[2(1-\rho^2)]^{-1/2} \quad \text{and} \quad \bar{\eta} = \eta\sigma_\eta[2(1-\rho^2)]^{-1/2} \quad \text{(aa)}$$

we obtain

$$E[F(\xi, \eta)] = \pi^{-1}(1-\rho^2)^{1/2}\int_{-\infty}^{\infty} \exp(-\bar{\xi}^2)\, d\bar{\xi} \int_{-\infty}^{\infty} \bar{F}(\bar{\xi}, \bar{\eta}) \exp(2\rho\bar{\xi}\bar{\eta} - \bar{\eta}^2)\, d\bar{\eta} \quad \text{(bb)}$$

where $\bar{F}(\bar{\xi}, \bar{\eta})$ is the same as $F(\xi, \eta)$, except that the original variables ξ and η are converted to the new ones $\bar{\xi}$ and $\bar{\eta}$. The sufficient stability condition (6.2.23) can be restated as

$$-2\pi\zeta(1-\rho^2)^{-1/2} + \int_{-\infty}^{\infty} \exp(-\xi^2)\, d\xi \int_{-\infty}^{\infty} \bar{F}(\bar{\xi}, \bar{\eta}) \exp(2\rho\bar{\xi}\bar{\eta} - \bar{\eta}^2)\, d\bar{\eta} < -\varepsilon \qquad \varepsilon > 0 \quad \text{(cc)}$$

We can now construct an optimization model as follows:

Given:	σ_η and ρ
Maximize:	σ_ξ
Subject to constraints:	$0.01 \le \sigma_\xi = X_1 \le 10.0$
	$0.01 \le \alpha_1 = X_2 \le 5.0$
	$-\alpha_1 + 0.01 \le \alpha_2 = X_3 \le \alpha_1 - 0.01$
	$0.0 \le X_4 \le 2\pi\zeta(1-\rho^2)^{-1/2}$

where

$$X_4 = 2\pi\zeta(1-\rho^2)^{-1/2} - \int_{-\infty}^{\infty} \exp(-\xi^2)\, d\xi \int_{-\infty}^{\infty} \bar{F}(\bar{\xi}, \bar{\eta}; \sigma_\xi, \sigma_\eta, \rho) \exp(2\rho\bar{\xi}\bar{\eta} - \bar{\eta}^2)\, d\bar{\eta} \quad \text{(dd)}$$

The values of the upper and lower constraints in this model are chosen for convenience of computation. When implementing this optimization procedure, an initial estimate for $X_1 = \sigma_\xi$ may be obtained from (n) of Section 6.2.1, which is derived with only known variances and covariance of the excitation processes.

The results obtained from the constrained optimization procedure are plotted in Figs. 6.2.6 through 6.2.8 for the cases of $\rho = 0, +0.5$, and -0.5.

6.2.3 Additional Knowledge of Multiplicative Excitation Processes

In principle, the sufficient stability boundary of a stochastic system can be sharpened further with additional knowledge of the multiplicative excitation processes. Indeed, such a possibility was demonstrated by Ariaratnam and Xie (1988). Return to the equation

$$\ddot{Y} + 2[\zeta + \eta(t)]\dot{Y} + [1 + \xi(t)]Y = 0 \tag{a}$$

By letting

$$Y = X \exp\left[-\zeta t - k\int_0^t \eta(\tau)\,d\tau\right] \tag{b}$$

equation (a) is transformed to

$$\ddot{X} + 2(1 - k)\eta(t)\dot{X} + [c + \phi(t)]X = 0 \tag{c}$$

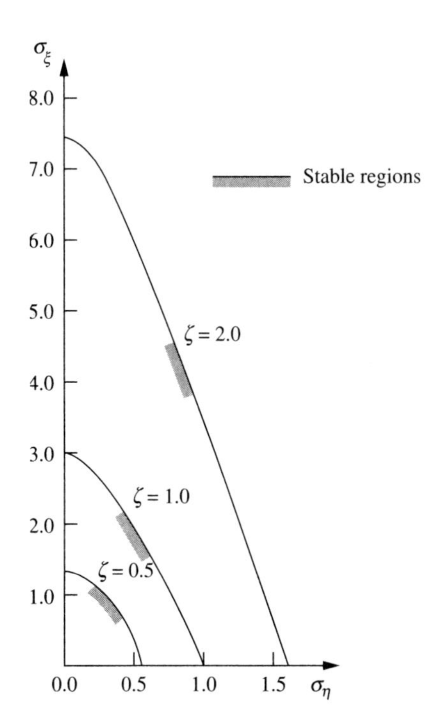

FIGURE 6.2.6
Regions of almost-sure asymptotic stability for system $\ddot{Y}+2[\zeta+\eta(t)]\dot{Y}+[1+\xi(t)]Y = 0$ via the optimization method; $\rho = 0; \eta(t), \zeta(t)$ gaussian [after Ariaratnam and Xie, 1989].

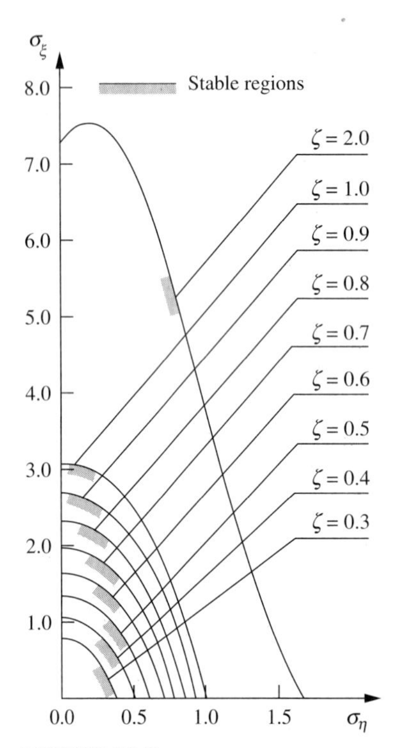

FIGURE 6.2.7
Regions of almost-sure asymptotic stability for system $\ddot{Y} + 2[\zeta + \eta(t)]\dot{Y} + [1 + \xi(t)]Y = 0$ via the optimization method; $\rho = +0.5$; $\eta(t), \zeta(t)$ gaussian [after Ariaratnam and Xie, 1989].

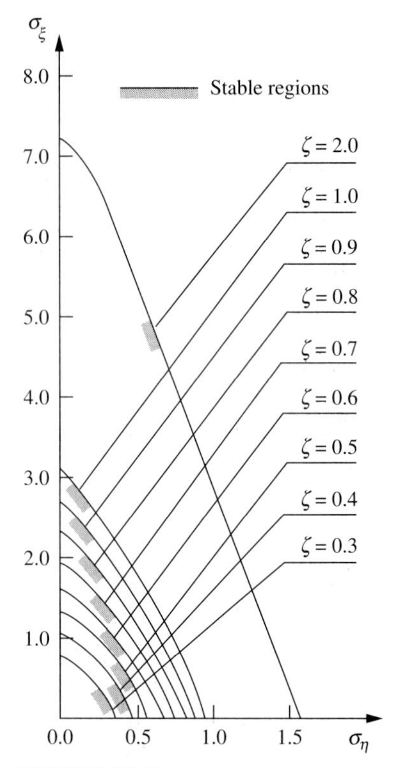

FIGURE 6.2.8
Regions of almost-sure asymptotic stability for system $\ddot{Y} + 2[\zeta + \eta(t)]\dot{Y} + [1 + \xi(t)]Y = 0$ via the optimization method; $\rho = -0.5$; $\eta(t), \zeta(t)$ gaussian [after Ariaratnam and Xie, 1989].

where $c = 1 - \zeta^2$ and

$$\phi(t) = \xi(t) - 2\zeta\eta(t) - k(2-k)\eta^2(t) - k\dot{\eta}(t) \tag{d}$$

Equation (d) suggests that the additional knowledge of the derivative of $\eta(t)$ may be useful in improving the sufficient stability boundary.

Rewrite (c) in the form of

$$\frac{d}{dt}\begin{Bmatrix} X_1 \\ X_2 \end{Bmatrix} = \begin{bmatrix} 0 & 1 \\ -(c+\phi) & -2(1-k)\eta \end{bmatrix}\begin{Bmatrix} X_1 \\ X_2 \end{Bmatrix} = \boldsymbol{Q}\begin{Bmatrix} X_1 \\ X_2 \end{Bmatrix} \tag{e}$$

To investigate the sample stability of $\boldsymbol{X} = \{X_1, X_2\}'$, we select again a Lyapunov function $V(\boldsymbol{X}) = \boldsymbol{X}'\boldsymbol{B}\boldsymbol{X}$, where

$$\boldsymbol{B} = \begin{bmatrix} \alpha_1^2 & \alpha_2 \\ \alpha_2 & 1 \end{bmatrix} \qquad \alpha_1^2 - \alpha_2^2 > 0 \tag{f}$$

The time derivative of $V(\boldsymbol{X})$ along the trajectory of (e) is

$$\dot{V}(\boldsymbol{X}) = \boldsymbol{X}'(\boldsymbol{Q}'\boldsymbol{B} + \boldsymbol{B}\boldsymbol{Q})\boldsymbol{X} \tag{g}$$

A sufficient condition for the asymptotic sample stability of system (e) is given by

$$E[\lambda(t)] < 2\zeta \tag{h}$$

where $\lambda(t)$ is the larger root of the characteristic equation

$$|Q'B + BQ - \lambda B| = \begin{vmatrix} -2\alpha_2(c+\phi) - \alpha_1^2\lambda & \alpha_1^2 - 2\alpha_2(1-k)\eta - (c+\phi) - \alpha_2\lambda \\ \alpha_1^2 - 2\alpha_2(1-k)\eta - (c+\phi) - \alpha_2\lambda & 2\alpha_2 - 4(1-k)\eta - \lambda \end{vmatrix} = 0 \tag{i}$$

or

$$\lambda^2 + 4(1-k)\eta\lambda - \frac{G}{\alpha_1^2 - \alpha_2^2} = 0 \tag{j}$$

where

$$G = (\alpha_1^2 - c)^2 + 4c\alpha_2^2 - 4\alpha_2(\alpha_1^2 + c)(1-k)\eta + 4\alpha_2^2(1-k)^2\eta^2 - 4\alpha_2(1-k)\eta\phi + (4\alpha_2^2 + 2c - 2\alpha_1^2)\phi + \phi^2 \tag{k}$$

The larger root of equation (j) is clearly

$$\lambda = -2(1-k)\eta + \left[4(1-k)^2\eta^2 + \frac{G}{\alpha_1^2 - \alpha_2^2}\right]^{1/2} \tag{l}$$

By substituting (l) into (h) and noticing that $E[\eta(t)] = 0$, the condition for asymptotic sample stability reduces to

$$E\left\{\left[4(1-k)^2\eta^2 + \frac{G}{\alpha_1^2 - \alpha_2^2}\right]^{1/2}\right\} < 2\zeta \tag{m}$$

In principle, the ensemble average in inequality (m) may be computed if the joint probability density of $\xi(t)$, $\eta(t)$, and $\dot{\eta}(t)$ is known, and the best stability boundary is obtained by optimizing α_1^2, α_2, and k. If such a boundary is defined in terms of σ_ξ, σ_η, and $\sigma_{\dot{\eta}}$, then the objective is to determine the maximum value for one of the three quantities, while fixing the values of the other two.

The optimization procedure can, again, be carried out using the complex method for constrained optimization, discussed in Section 6.2.2. Specifically, let

$$F(\xi, \eta, \dot{\eta}) = \left[4(1-k)^2\eta^2 + \frac{G}{\alpha_1^2 - \alpha_2^2}\right]^{1/2} \tag{n}$$

and let $V = \sigma_\eta$ be maximized, subject to the constraints

$$\alpha_1 > 0 \qquad -\alpha_1 < \alpha_2 < \alpha_1 \qquad 0 \le k \le 1 \tag{o}$$

$$E[F(\xi, \eta, \dot{\eta})] < 2\zeta \tag{p}$$

To illustrate, assume that $\xi(t)$ and $\eta(t)$ are stationary ergodic gaussian random processes, and that the joint probability density of $\xi(t)$, $\eta(t)$, and $\dot{\eta}(t)$ is given by

$$p(\xi, \eta, \dot{\eta}) = (2\pi)^{-3/2}(\sigma_\xi\sigma_\eta\sigma_{\dot{\eta}})^{-1} \exp\left\{-\frac{1}{2}\left[\left(\frac{\xi}{\sigma_\xi}\right)^2 + \left(\frac{\eta}{\sigma_\eta}\right)^2 + \left(\frac{\dot{\eta}}{\sigma_{\dot{\eta}}}\right)^2\right]\right\} \tag{q}$$

The ensemble average on the left-hand side of (p) is obtained as

$$E[F(\xi, \eta, \dot{\eta})] = \int_{-\infty}^{\infty} d\eta \int_{-\infty}^{\infty} d\xi \int_{-\infty}^{\infty} F(\xi, \eta, \dot{\eta}) p(\xi, \eta, \dot{\eta})\, d\dot{\eta} \tag{r}$$

By the change of variables $u = \xi/\sqrt{2}\sigma_\xi$, $v = \eta/\sqrt{2}\sigma_\eta$, and $w = \dot{\eta}/\sqrt{2}\sigma_{\dot{\eta}}$, equation (r) becomes

$$E[F(\xi, \eta, \dot{\eta})] = \pi^{-3/2} I \tag{s}$$

$$\text{where} \quad I = \int_{-\infty}^{\infty} \exp(-u^2)\, du \int_{-\infty}^{\infty} \exp(-v^2)\, dv \int_{-\infty}^{\infty} \exp(-w^2) F(u, v, w)\, dw \tag{t}$$

The triple integral may be calculated numerically by use of the following approximation:

$$I \approx \sum_{i=1}^{n} W_i \sum_{j=1}^{n} W_j \sum_{k=1}^{n} W_k F(u_i, u_j, u_k) \tag{u}$$

where u_i, u_j, and $u_k (i, j, k = 1, 2, \ldots, n)$ are the zeros of the Hermite polynomial (Davis and Rabinowitz, 1984)

$$H_n(z) = (-1)^n \exp(z^2) \frac{d^n}{dz^n} \exp(-z^2) = \sum_{k=0}^{[n/2]} \frac{(-1)^k n!}{k!(n-2k)!} (2z)^{n-2k} \tag{v}$$

and where the coefficients W_i, W_j, and W_k $(i, j, k = 1, 2, \ldots, n)$ are obtained from

$$W_k = \frac{2^{n+1} n! \sqrt{\pi}}{[H_{n+1}(z_k)]^2} \tag{w}$$

The approximate expression (u) is known as the Gauss-Hermite formula.

The computed results are illustrated in Figs. 6.2.9 through 6.2.11. In Fig. 6.2.9, the damping ratio is chosen to be 0.2, and the boundary for the sufficient condi-

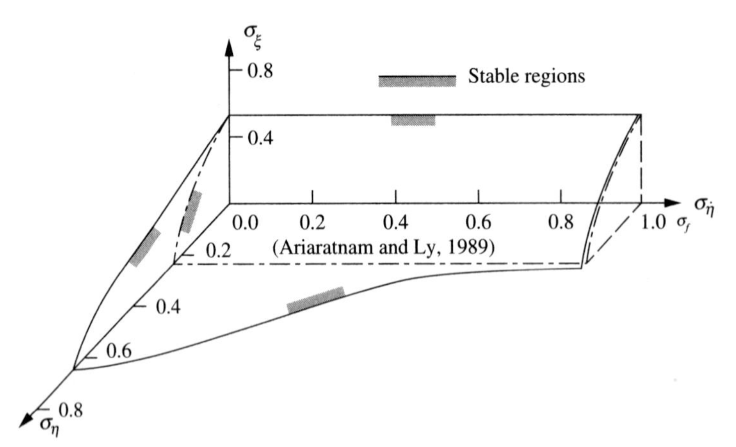

FIGURE 6.2.9
Regions of almost-sure asymptotic stability for system $\ddot{Y} + 2[\zeta + \eta(t)]\dot{Y} + [1 + \xi(t)]Y = 0, \zeta = 0.2$, [after Ariaratnam and Xie, 1988].

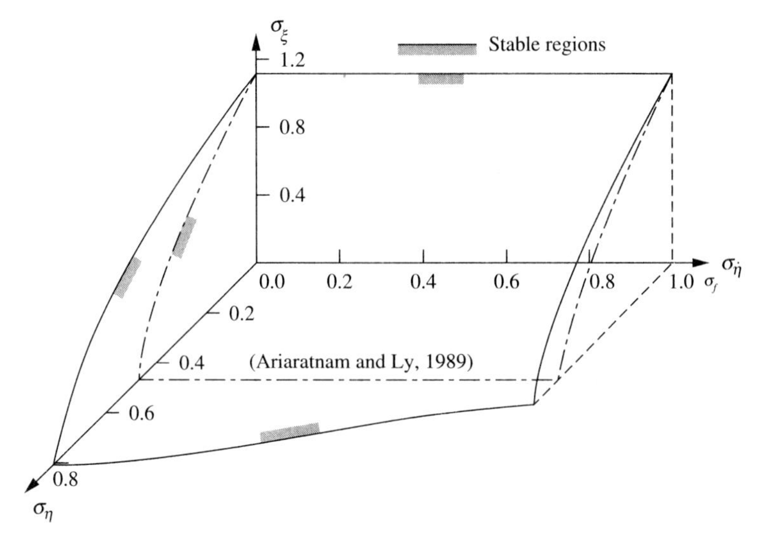

FIGURE 6.2.10
Regions of almost-sure asymptotic stability for system $\ddot{Y} + 2[\zeta + \eta(t)]\dot{Y} + [1 + \xi(t)Y = 0, \zeta = 0.4$ [after Ariaratnam and Xie, 1988].

tion for asymptotic sample stability is plotted as a three-dimensional surface in the σ_ξ–σ_η–$\sigma_{\dot{\eta}}$ spacc. Thc boundary obtained without the additional knowledge of $\dot{\eta}$ (Ariaratnam and Ly, 1989) is also shown for comparison. The improvement due to the additional knowledge of $\dot{\eta}(t)$ is seen to be quite significant. Figure 6.2.10 is similar to Fig. 6.2.9, except that a higher ζ value has been selected for computation. Figure 6.2.11 is plotted for the special case $\xi(t) = 0$, and for five different ζ values. The sufficient conditions for asymptotic sample stability are again sharper than the ones obtained by Kozin and Wu (1973) without the knowledge of $\dot{\eta}(t)$.

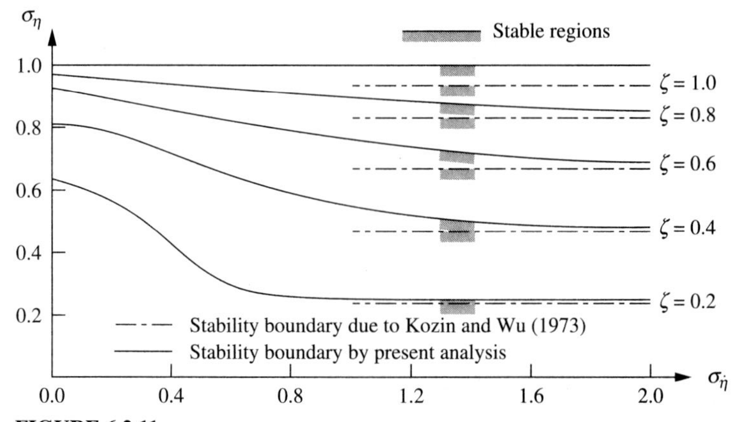

FIGURE 6.2.11
Regions of almost-sure asymptotic stability for system $\ddot{Y} + 2[\zeta + \eta(t)]\dot{Y} + Y = 0$ [after Ariaratnam and Xie, 1988].

6.2.4 Method of Digital Simulation

Early attempts to obtain stability boundaries by digital simulation were unsuccessful, because numerical difficulties invariably arise when a sample function is near a stability boundary (Lin et al., 1986). To circumvent such difficulties, Wedig (1990) devised a new scheme which is applicable for a general n-dimensional system.

Let an n-dimensional linear system be governed by

$$\dot{X}_j = a_{jk}X_k + X_l F_{jl}(t) \tag{a}$$

which is equivalent to equation (6.2.3). Now introduce the normalized variables

$$U_j = \frac{X_j}{\|\boldsymbol{X}\|} \tag{b}$$

where $\|\boldsymbol{X}\|$ is the euclidean norm defined as

$$\|\boldsymbol{X}\| = (X_k X_k)^{1/2} = (X_1^2 + X_2^2 + \cdots + X_n^2)^{1/2} \tag{c}$$

The vector process $\boldsymbol{U}(t)$ is the projection of the original vector process $\boldsymbol{X}(t)$ on a multidimensional unit sphere. However, the n components of $\boldsymbol{U}(t)$ are not independent, since they must satisfy the constraint

$$(U_k U_k) = 1 \tag{d}$$

Another variable is required to complete the transformation, for which we choose

$$\Upsilon(t) = \ln \|\boldsymbol{X}\| \tag{e}$$

It can be shown that

$$\dot{U}_j = (a_{jk} - a_{kr}U_j U_r)U_k + (F_{jl} - U_j U_k F_{kl})U_l \tag{f}$$

$$\dot{\Upsilon} = \frac{1}{\|\boldsymbol{X}\|}\frac{d}{dt}\|\boldsymbol{X}\| = a_{jk}U_j U_k + U_j U_l F_{jl} \tag{g}$$

The preceding transformation scheme is originally due to Khasminskii (1967), in his investigation of the stability of linear systems under random multiplicative excitations of gaussian white noises, to be discussed in Sections 6.3 and 6.4. Clearly, this scheme is equally useful in the case of nonwhite noise excitations.

Unlike the original process $\boldsymbol{X}(t)$, the $\boldsymbol{U}(t)$ process governed by equation (f) is always well-behaved since it is confined to a unit sphere. If the multiplicative forcing is stationary and ergodic, then $\boldsymbol{U}(t)$ also tends to stationary and ergodic. Thus equation (f) provides a suitable framework for digital simulation.

As noted by Wedig (1990), geometrical singularities exist in equation (f), and they must be avoided in the simulation. To explain the nature of such singularities, consider the following second-order system:

$$\ddot{Y} + 2\zeta\dot{Y} + [1 + \xi(t)]Y = 0 \tag{h}$$

Replacing equation (h) by

$$\frac{d}{dt}\begin{Bmatrix} X_1 \\ X_2 \end{Bmatrix} = \begin{bmatrix} 0 & 1 \\ -1 & -2\zeta \end{bmatrix}\begin{Bmatrix} X_1 \\ X_2 \end{Bmatrix} + \begin{bmatrix} 0 & 0 \\ -\xi(t) & 0 \end{bmatrix}\begin{Bmatrix} X_1 \\ X_2 \end{Bmatrix} \tag{i}$$

we identify $a_{11} = 0$, $a_{12} = 1$, $a_{21} = -1$, $a_{22} = -2\zeta$, $F_{21} = -\xi(t)$, and $F_{11} = F_{12} = F_{22} = 0$. The projection of the two-dimensional $\boldsymbol{X}(t)$ vector on a unit circle is governed by

$$\dot{U}_1 = U_2[1 + 2\zeta U_1 U_2 + U_1^2 \xi(t)] \tag{j}$$

$$\dot{U}_2 = -U_1 - 2\zeta U_2(1 - U_2^2) - U_1(1 - U_2^2)\xi(t) \tag{k}$$

Without excitation and for $0 < \zeta < 1$, the $\boldsymbol{U}(t)$ vector would rotate in the clockwise direction. Equation (j) cannot be used for simulation near $U_1 = 1$, where $U_2 = \dot{U}_1 = 0$. However, only one equation is needed for simulation, and in the neighborhood of $U_1 = 1$, equation (k) becomes more effective. Wedig (1990) proposed that equation (j) be used if the projection is in the upper or lower part of the unit circle and equation (k) be used in the right or left part of the unit circle. The extension to the general case of an n-dimensional unit sphere is obvious. Once the absolute value of a particular U_k becomes greater than the others, the equation associated with $\dot{U}_k$ is set aside in the simulation process.

Return now to equation (g). Since the $U_j(t)$ and $F_{jl}(t)$ are stationary and ergodic, the right-hand side of (g) is also stationary and ergodic. Its long time average is then equal to the ensemble average:

$$\lambda = E[\dot{Y}(t)] = \lim_{T\to\infty} \frac{1}{T}\left[a_{jk}\int_0^T U_j(t)U_k(t)\,dt + \int_0^T U_j(t)U_l(t)F_{jl}(t)\,dt\right] \tag{l}$$

A computer program for the simulation of equations (f) and (l) has been provided by Wedig (1990).

6.3 ASYMPTOTIC SAMPLE STABILITY OF LINEAR SYSTEMS UNDER GAUSSIAN WHITE NOISE EXCITATIONS

In Section 6.2, sufficient conditions for asymptotic sample stability were obtained under the assumption that variances of the multiplicative random excitations were finite. Therefore, the procedures given there are not applicable if the multiplicative excitations are modeled as white noises. However, a multiplicative white noise excitation does not always lead to instability for at least two reasons. First, the white noise idealization is expected to be reasonably accurate if the real noise has a correlation time much shorter than the relaxation time of the dynamical system. Second, the conditions obtained in Section 6.2 are sufficient, not necessary conditions.

The present section is concerned with multiplicative excitations which may be idealized as gaussian white noises, so that the system response is markovian. As it turns out, an asymptotic sample stability condition, which is both necessary and sufficient, can be obtained if the dynamical system is linear. A general procedure for obtaining such a condition, due to Khasminskii (1967), is explained next.

Consider an n-dimensional linear system, whose free motion is governed by a set of ordinary differential equations with constant coefficients. Some of the coefficients are changed to stochastic processes when multiplicative excitations are introduced. Let the random parts of the coefficients be white noises or be replaced by white noises upon stochastic averaging. We obtain the Itô-type stochastic differential equations as follows:

$$dX_j = \alpha_{jl} X_l \, dt + \gamma_{jkr} X_k \, dB_r(t) \tag{6.3.1}$$

where α_{jl} and γ_{jkr} are constants, and the $B_r(t)$ are independent unit Wiener processes. Equation (6.3.1) shows that the drift coefficients $\alpha_{jl} X_l$ and the diffusion coefficients $\gamma_{jkr} X_k$ are linear functions of $\boldsymbol{X}(t)$. The generating operator $\mathscr{L}$ of the Markov vector $\boldsymbol{X}(t)$ is given by

$$\mathscr{L} = m_j \frac{\partial}{\partial X_j} + \frac{1}{2} \sigma_{jr} \sigma_{qr} \frac{\partial^2}{\partial X_j \partial X_q} \tag{6.3.2}$$

where

$$m_j = \alpha_{jl} X_l \tag{6.3.3}$$

$$\sigma_{jr} \sigma_{qr} = \gamma_{jkr} \gamma_{qlr} X_k X_l \tag{6.3.4}$$

The sample stability of (6.3.1) may be interpreted in terms of the euclidean norm

$$\|\boldsymbol{X}\| = (X_j X_j)^{1/2} \tag{6.3.5}$$

Introduce the normalized version of X_s as follows:

$$U_s = \frac{X_s}{\|\boldsymbol{X}\|} \tag{6.3.6}$$

This describes the projection of the original vector process $\boldsymbol{X}(t)$ on an n-dimensional unit sphere. The Itô equation for U_s is given by

$$dU_s = \mathscr{L}[U_s(t)] \, dt + \gamma_{jkr} X_k \frac{\partial U_s}{\partial X_j} dB_r(t) \tag{6.3.7}$$

Upon carrying out the operations on the right-hand side,

$$\begin{aligned} dU_s = {} & \left[\alpha_{jl} U_l (\delta_{js} - U_j U_s) + \tfrac{1}{2} \gamma_{jkr} \gamma_{qlr} U_k U_l (3 U_j U_s U_q - U_s \delta_{jq} - U_q \delta_{qs})\right] dt \\ & + \gamma_{jkr} U_k (\delta_{js} - U_j U_s) \, dB_r(t) \end{aligned} \tag{6.3.8}$$

where each δ_{js} is a Kronecker delta. It is of interest to note that the coefficients in (6.3.8) depend only on $\boldsymbol{U}$; therefore, vector $\boldsymbol{U}$ is also a Markov vector. Under certain conditions (Khasminskii, 1967), the $\boldsymbol{U}(t)$ process tends to a stationary ergodic vector process as t increases. However, this process is at most $(n-1)$-dimensional, because the n components are subject to the constraint

$$\|\boldsymbol{U}\| = 1 \tag{6.3.9}$$

The associated Fokker-Planck equation for $\boldsymbol{U}$ can be constructed from the Itô equations (6.3.8). In principle, it may be solved to obtain the probability density of $\boldsymbol{U}$.

The growth (or decay) of the euclidean norm $\|\boldsymbol{X}\|$ is characterized by its logarithm:

$$\Upsilon = \ln \|X\| \tag{6.3.10}$$

The Itô equation for Υ is found to be

$$d\Upsilon(t) = \mathcal{L}[\Upsilon(t)]\,dt + \gamma_{jkr}U_jU_k\,dB_r(t) \tag{6.3.11}$$

where $$\mathcal{L}[\Upsilon(t)] = Q(U) = \alpha_{jl}U_jU_l + \tfrac{1}{2}\gamma_{jkr}\gamma_{qlr}U_kU_l(\delta_{jq} - 2U_jU_q) \tag{6.3.12}$$

Remarkably, the right-hand side of (6.3.11) is again dependent only on $\boldsymbol{U}$. Therefore, while $\Upsilon(t)$ is not a Markov process, the combination of $\Upsilon(t)$ and $\boldsymbol{U}(t)$ is a Markov vector.

Integrating (6.3.11) from 0 to t, we obtain

$$\begin{aligned}\Upsilon(t) - \Upsilon(0) &= \ln\|\boldsymbol{X}(t)\| - \ln\|\boldsymbol{X}(0)\| \\ &= \int_0^t Q[\boldsymbol{U}(\tau)]\,d\tau + \gamma_{jkr}\int_0^t U_j(\tau)U_k(\tau)\,dB_r(\tau)\end{aligned} \tag{6.3.13}$$

Dividing (6.3.13) by t,

$$\frac{1}{t}\ln\|\boldsymbol{X}(t)\| = \frac{1}{t}\ln\|\boldsymbol{X}(0)\| + \frac{1}{t}\int_0^t Q[\boldsymbol{U}(\tau)]\,d\tau + \gamma_{jkr}\frac{1}{t}\int_0^t U_j(\tau)U_k(\tau)\,dB_r(\tau) \tag{6.3.14}$$

As $t \to \infty$, the first term in (6.3.14) vanishes. The last term also vanishes, since $U_j(\tau)$ and $U_k(\tau)$ are bounded, and the $B_r(\tau)$ grow as $[\tau\ln(\ln\tau)]^{1/2}$. If $\boldsymbol{U}(t)$ is stationary and ergodic, then equation (6.3.14) becomes

$$\lambda = \lim_{t\to\infty}\frac{1}{t}\ln\|\boldsymbol{X}(t)\| = E\{Q[\boldsymbol{U}(t)]\} \tag{6.3.15}$$

The computation of the ensemble average in (6.3.15) requires the knowledge of the stationary probability density of the $\boldsymbol{U}(t)$ process.

Equation (6.3.15) implies that as t increases

$$\|\boldsymbol{X}(t)\| \approx \exp\left(tE\{Q[\boldsymbol{U}(\tau)]\}\right) \tag{6.3.16}$$

Therefore, in terms of $\|\boldsymbol{X}(t)\|$, the necessary and sufficient condition for the asymptotic sample stability is

$$\lambda = E\{Q[\boldsymbol{U}(t)]\} < 0 \tag{6.3.17}$$

This result is due to Khasminskii (1967).

The ensemble average λ given in (6.3.17) is known as the Lyapunov exponent, which characterizes the exponential growth (if $\lambda > 0$) or decay (if $\lambda < 0$) of a dynamical system as time increases. For an n-dimensional system, there exists a finite number of finite Lyapunov exponents, called the *Lyapunov spectrum* (see, e.g., Arnold and Wihstutz, 1984). These Lyapunov exponents may be arranged as $\lambda_1 \le \lambda_2 \le \cdots \le \lambda_n = \lambda_T$. The asymptotic behavior of a deterministic system, characterized by one of the Lyapunov exponents, is dependent on the initial condition. However, in the case of a stochastic linear system (6.3.1), equation (6.3.15) converges to the largest Lyapunov exponent λ_T, provided that the normalized process $\boldsymbol{U}(t)$ is ergodic on the *entire* unit sphere.

The convergence of (6.3.15) to the largest Lyapunov exponent has the theoretical basis of the *multiplicative ergodic theorem* of Oseledec (1968). The essence of the theorem may be explained heuristically as follows. Let the transformation from

$\boldsymbol{X}(t_j)$ to $\boldsymbol{X}(t_{j+1})$ be represented as

$$\boldsymbol{X}(t_{j+1}) = \boldsymbol{T}(t_j, \Delta t)\boldsymbol{X}(t_j) \qquad \Delta t = t_{j+1} - t_j \tag{6.3.18}$$

where $\boldsymbol{T}(t_j, \Delta t)$ is a transformation matrix. If $\boldsymbol{T}(t_j, \Delta t)$ for different j are independent and identically distributed, then sequential transformations tend to an ergodic Markov chain. Since sequential transformations are equivalent to successive multiplications of the transformation matrices, the result is dominated by the largest eigenvalues of these matrices. The ergodic assumption on the normalized state space ensures that the subspace associated with the largest Lyapunov exponent λ_T will be visited sooner or later. At times, we may refer to λ_T merely as the Lyapunov exponent if there is no danger of confusion.

Unfortunately, the normalized process $\boldsymbol{U}(t)$ may or may not be ergodic over the entire unit sphere. In the latter case, the unit sphere is divided into regions by singular boundaries. Some regions may not admit ergodic solutions, while some others may be noncommunicative and admit different ergodic solutions. Such complexities are now explained in terms of a two-dimensional system.

Consider the following pair of linear Itô equations:

$$\begin{aligned} dX_1 &= \alpha_{1j}X_j\,dt + \gamma_{1kr}X_k\,dB_r(t) \\ dX_2 &= \alpha_{2j}X_j\,dt + \gamma_{2kr}X_k\,dB_r(t) \end{aligned} \tag{6.3.19}$$

where the indices j and k may assume the values of 1 and 2. The projection of $\boldsymbol{X}(t) = \{X_1(t), X_2(t)\}'$ on a unit circle may be represented by an angle $\Theta(t)$. This representation is equivalent to choosing the pair $U_1 = \cos\Theta$ and $U_2 = \sin\Theta$, which satisfies $(U_1^2 + U_2^2)^{1/2} = 1$. The Itô equation for the $\Theta(t)$ process has the usual form

$$d\Theta(t) = m_\Theta\,dt + \sigma_\Theta\,dB(t) \tag{6.3.20}$$

By applying Itô's differential rule to $\Theta = \tan^{-1}(X_2/X_1)$, it can be shown that both m_Θ and σ_Θ are functions of $\cos 2\Theta$ and $\sin 2\Theta$.

The stationary probability density $p(\theta)$ of $\Theta(t)$, if it exists, is governed by the reduced Fokker-Planck equation

$$\frac{1}{2}\frac{d^2}{d\theta^2}[b_\theta p(\theta)] - \frac{d}{d\theta}[a_\theta p(\theta)] = 0 \tag{6.3.21}$$

where a_θ and b_θ are obtained, respectively, from m_Θ and σ_Θ^2 by replacing $\Theta(t)$ by its deterministic counterpart θ. In the present case,

$$a_\theta = a_\theta(\cos 2\theta, \sin 2\theta) \tag{6.3.22}$$

$$b_\theta = b_\theta(\cos 2\theta, \sin 2\theta) \tag{6.3.23}$$

which are periodic with a period π. Equation (6.3.21) may be integrated once to yield

$$\frac{1}{2}\frac{d}{d\theta}[b_\theta p(\theta)] - a_\theta p(\theta) = -G_c \tag{6.3.24}$$

where G_c has the meaning of the negative probability flow [see equation (4.5.4)]. Note that G_c may be different for different ergodic regions.

The following comments on the one-dimensional diffusion process $\Theta(t)$ are in order:

1. The stationary probability density $p(\theta)$ is a periodic function of θ, with a period π.
2. There may exist singular points of the first kind in the interval $[0, 2\pi)$, where b_θ vanishes. For every such singular point θ_s, there exists another singular point at $\theta_s + \pi$. These singular points divide the unit circle into separate intervals. In general, the reduced Fokker-Planck equation should be solved for each interval bounded by two neighboring singular points.
3. If a singular boundary θ_s dividing two ergodic intervals is a shunt, then the direction of sample paths near the boundary is determined by the sign of $a_\theta(\theta_s)$. A sample path will cross the boundary in the direction from $\theta < \theta_s$ to $\theta > \theta_s$ if $a_\theta(\theta_s)$ is positive, and in the opposite direction if $a_\theta(\theta_s)$ is negative. The behavior of sample trajectories is more complicated when approaching a trap-type singular boundary. It depends on the drift exponent, the diffusion exponent, and the character value, as summarized in Table 4.5.2.

The asymptotic sample stability of system (6.3.19) may now be investigated in terms of

$$\Upsilon(t) = \frac{1}{2}\ln(X_1^2 + X_2^2) \tag{6.3.25}$$

The drift coefficient for $\Upsilon(t)$ is obtained as

$$m_\Upsilon = \mathscr{L}[\Upsilon(t)] \tag{6.3.26}$$

where $\mathscr{L}$ is the generating differential operator defined in (6.3.2). It can be shown that m_Υ is also a function of $\cos 2\theta$ and $\sin 2\theta$:

$$m_\Upsilon = Q(\cos 2\theta, \sin 2\theta) \tag{6.3.27}$$

which is the two-dimensional version of (6.3.12). The asymptotic behavior of $\boldsymbol{X}(t)$ can be examined by taking the ensemble average of Q over θ. Three different cases are now considered; these cases depend on whether singularities exist on the unit circle and the nature of such singularities.

Case 1. No singularity exists on $[0, 2\pi)$ and the entire unit circle is one ergodic region. As shown in Section 4.5, the constant G_c in (6.3.24) can be determined by use of the periodicity condition $p(0) = p(2\pi)$. From equation (4.5.10),

$$p(\theta) = C\frac{\psi(\theta)}{b_\theta(\theta)}\left\{1 - \left[1 - \frac{\psi(2\pi)}{\psi(0)}\right]\frac{\int_0^\theta \psi^{-1}(\phi)\,d\phi}{\int_0^{2\pi}\psi^{-1}(\phi)\,d\phi}\right\} \qquad 0 \leq \theta < 2\pi \tag{6.3.28}$$

where C is the normalization constant and

$$\psi(\theta) = \exp\left[\int \frac{2a_\theta(\theta)}{b_\theta(\theta)}d\theta\right] \tag{6.3.29}$$

The largest Lyapunov exponent is obtained from

$$\lambda_T = \int_0^{2\pi} Q(\theta)p(\theta)\,d\theta \tag{6.3.30}$$

Case 2. All the singular points in $[0, 2\pi)$ are shunts in the same direction. In this case, the entire unit circle is also one ergodic region, but the Fokker-Planck equation (6.3.21) must be solved separately for each interval, bounded by two neighboring singular points. If all the singular points are left shunts, then four types of intervals are possible: (exit, entrance), (regular, entrance), (exit, regular), and (regular, regular). Note that the left boundary can be either an exit or a regular boundary, while the right boundary can be either an entrance or a regular boundary. For each interval on the unit circle, bounded by two neighboring left shunts θ_l and θ_r, we obtain (Itô and McKean, 1965)

$$p(\theta) = C\frac{\psi(\theta)}{b_\theta(\theta)}\int_{\theta_l}^{\theta} \psi^{-1}(\phi)\,d\phi \qquad \theta_l < \theta < \theta_r \tag{6.3.31}$$

The values of $p(\theta)$ at the boundaries can be obtained by taking the limit of (6.3.31) as θ approaches θ_l or θ_r, to yield (Zhang, 1991)

$$p(\theta_l) = \infty \qquad \theta_l \ \text{regular} \tag{6.3.32}$$

$$\lim_{\theta\to\theta_l} p(\theta) = \begin{cases} \lim\limits_{\theta\to\theta_l} \dfrac{2C}{b'_\theta - 2a_\theta} & \theta_l \ \text{exit}, \alpha_l = 1 \\[2ex] \lim\limits_{\theta\to\theta_l} \dfrac{-C}{a_\theta} & \theta_l \ \text{exit}, \alpha_l > 1 \end{cases} \tag{6.3.33}$$

$$p(\theta_r) = \infty \qquad \theta_r \ \text{regular} \tag{6.3.34}$$

$$\lim_{\theta\to\theta_r} p(\theta) = \begin{cases} \lim\limits_{\theta\to\theta_r} \dfrac{2C}{b'_\theta - 2a_\theta} & \theta_r \ \text{entrance}, \alpha_r = 1 \\[2ex] \lim\limits_{\theta\to\theta_r} \dfrac{-C}{a_\theta} & \theta_r \ \text{entrance}, \alpha_r > 1 \end{cases} \tag{6.3.35}$$

where b'_θ denotes the derivative of b_θ with respect to θ, and α_l and α_r are the diffusion exponents of the boundaries θ_l and θ_r respectively, defined in (4.5.16). In the case of a regular boundary, further stipulation is required on the behaviors of the sample functions after reaching the boundary, as indicated in Chapter 4. In what follows, it will be assumed that all sample functions have a zero holding-time at a regular boundary.

On the other hand, if all the singular points are right shunts, then the possible intervals are (entrance, exit), (regular, exit), (entrance, regular), and (regular, regular). The probability density in the interval bounded by θ_l and θ_r is then given by (Itô and McKean, 1965)

$$p(\theta) = -C\frac{\psi(\theta)}{b_\theta(\theta)}\int_{\theta}^{\theta_r} \psi^{-1}(\phi)\,d\phi \qquad \theta_l < \theta < \theta_r \tag{6.3.36}$$

The values of $p(\theta)$ at the boundaries are obtained as (Zhang, 1991)

$$p(\theta_l) = \infty \qquad \theta_l \quad \text{regular} \tag{6.3.37}$$

$$\lim_{\theta\to\theta_l} p(\theta) = \begin{cases} \lim_{\theta\to\theta_l} \dfrac{2C}{b'_\theta - 2a_\theta} & \theta_l \quad \text{entrance}, \alpha_l = 1 \\ \lim_{\theta\to\theta_l} \dfrac{-C}{a_\theta} & \theta_l \quad \text{entrance}, \alpha_l > 1 \end{cases} \tag{6.3.38}$$

$$p(\theta_r) = \infty \qquad \theta_r \quad \text{regular} \tag{6.3.39}$$

$$\lim_{\theta\to\theta_r} p(\theta) = \begin{cases} \lim_{\theta\to\theta_r} \dfrac{2C}{b'_\theta - 2a_\theta} & \theta_r \quad \text{exit}, \alpha_r = 1 \\ \lim_{\theta\to\theta_r} \dfrac{-C}{a_\theta} & \theta_r \quad \text{exit}, \alpha_r > 1 \end{cases} \tag{6.3.40}$$

If the entire unit circle is one ergodic region and is divided into N intervals, then the stationary probability density of $\Theta(t)$ is given by

$$p(\theta) = p_i(\theta) \qquad \text{for} \quad \theta_{li} < \theta < \theta_{ri}, \qquad i = 1, 2, \ldots, N \tag{6.3.41}$$

where θ_{li} and θ_{ri} are the two boundaries of the ith interval, and where $p_i(\theta)$ is given by (6.3.31) for the case of left shunt and by (6.3.36) for the case of right shunt. The normalization constant C is obtained from

$$\int_0^{2\pi} p(\phi)\, d\phi = \sum_{i=1}^{N} \int_{\theta_{li}}^{\theta_{ri}} p_i(\phi)\, d\phi = 1 \tag{6.3.42}$$

The largest Lyapunov exponent is then evaluated as

$$\lambda_T = \int_0^{2\pi} Q(\phi) p(\phi)\, d\phi = \sum_{i=1}^{N} \int_{\theta_{li}}^{\theta_{ri}} Q(\phi) p_i(\phi)\, d\phi \tag{6.3.43}$$

Case 3. The unit circle is divided into several intervals, each of which is bounded by opposite shunts (entrance or regular), entrance traps, or repulsively natural traps. In this case, the process $\Theta(t)$ will remain in an interval, once it enters that interval. Therefore, every interval is a separate ergodic region. The probability density for an interval bounded by θ_l and θ_r can be obtained from (Tanaka, 1957)

$$p(\phi) = C \frac{\psi(\theta)}{b_\theta(\theta)} \qquad \theta_l < \theta < \theta_r \tag{6.3.44}$$

where the normalization constant C is given by

$$C = \left[\int_{\theta_l}^{\theta_r} \frac{\psi(\phi)}{b_\theta(\phi)} d\phi \right]^{-1} \tag{6.3.45}$$

The values of $p(\theta)$ at the boundaries are (Zhang, 1991)

$$p(\theta_s) = \infty \qquad \theta_s \quad \text{regular shunt} \tag{6.3.46}$$

$$p(\theta_s) = \begin{cases} \text{finite} & \theta_s \quad \text{entrance shunt}, \quad \alpha_s = 1, c_s = 1 \\ 0 & \theta_s \quad \text{entrance shunt}, \quad \alpha_s = 1, c_s > 1 \quad \text{or} \quad \alpha_s > 1 \\ 0 & \theta_s \quad \text{entrance trap or repulsively natural trap} \end{cases} \tag{6.3.47}$$

Since the entire unit circle is divided into separate ergodic intervals, the stationary probability density for $\Theta(t)$ depends on the interval in which the initial state $\Theta(0) = \theta_0$ is located. Consequently, the Lyapunov exponent obtained from ensemble averaging also depends on θ_0. If θ_0 is located in the ith ergodic region bounded by θ_{li} and θ_{ri}, then the Lyapunov exponent is given by

$$\lambda_i = \int_{\theta_{li}}^{\theta_{ri}} Q(\phi)p(\phi)\,d\phi \tag{6.3.48}$$

where $p(\theta)$ is obtained from (6.3.44).

Case 4. If one boundary, say θ_l, of an interval $[\theta_l, \theta_r]$ is an exit trap, and the other boundary θ_r is an entrance or a repulsively natural boundary, then a sample path initiated within this interval will eventually reach this exit boundary and remain there. Therefore, the stationary probability density is given by

$$p(\theta) = \delta(\theta_l) \qquad \theta_l \leq \theta \leq \theta_r \tag{6.3.49}$$

and the Lyapunov exponent is given by

$$\lambda = \int_{\theta_l}^{\theta_r} Q(\phi)p(\phi)\,d\phi = Q(\theta_l) \tag{6.3.50}$$

provided that the initial θ_0 is within this interval.

6.4 NUMERICAL METHODS FOR OBTAINING LYAPUNOV EXPONENTS

In spite of its simple appearance, the application of Khasminskii's result (6.3.17) is complicated numerically. For the two-dimensional linear Itô system (6.3.19), the stationary probability density for the phase diffusion is given by (6.3.28), (6.3.31), (6.3.36), or (6.3.44). In any case, the computation of the Lyapunov exponent according to (6.3.30), (6.3.43), or (6.3.48) involves multiple integrations. In some cases it may be simpler to solve the reduced Fokker-Planck equation (6.3.24) directly by numerical means, making use of the boundary conditions described by (6.3.33), (6.3.35), (6.3.38), (6.3.40), or the last two of (6.3.47). These numerical procedures are discussed in this section. Obviously, a numerical approach is not suitable if the probability density is unbounded at a boundary, such as those described by (6.3.32), (6.3.34), (6.3.37), and (6.3.46).

First, consider the second case in Section 6.3 in which all singular points are shunts in the same direction. In this case, the Laplace method of asymptotic expansion (e.g., see Copson, 1965) is sometimes applicable to obtain an approximate value for λ. As seen from equation (6.3.29), the expression for the ψ function is exponential. By changing the integration variable, the integral in equation (6.3.31) or (6.3.36) may be cast in the following form (Xie, 1990):

$$\int_{\theta_0}^{\theta} \psi^{-1}(\phi)d\phi = \int_{\alpha}^{\beta} f(x)e^{\nu h(x)}dx \tag{6.4.1}$$

If ν is a large positive constant, if $h(x)$ attains its largest value at $x = \beta$, and if $h'(\beta) > 0$, then

$$\int_{\alpha}^{\beta} f(x)e^{\nu h(x)}dx \approx \frac{f(\beta)}{\nu h'(\beta)}e^{\nu h(\beta)} \tag{6.4.2}$$

The approximate stationary probability density so obtained can then be substituted into the expression for the Lyapunov exponent and the remaining integration can be carried out. This asymptotic method is nonetheless restricted by those conditions imposed on (6.4.2).

In principle, the integration to obtain $p(\theta)$ and the integration to obtain λ can also be combined in another asymptotic procedure. By changing both integration variables, (6.3.43) may be cast in the form (Xie, 1990)

$$\lambda = \frac{\int_D \int P(u)e^{\nu h(u,v)}du\,dv}{\int_D \int e^{\nu h(u,v)}du\,dv} \tag{6.4.3}$$

If v is a large positive constant, and if $h(u, v)$ attains its extremum at (u_0, v_0) within the domain of integration D, then (Hsu, 1948)

$$\int_D \int f(u, v)e^{\nu h(u,v)}du\,dv \approx \frac{\pi}{\nu}\frac{f(u_0, v_0)e^{\nu h(u_0,v_0)}}{\left|f_{uu}(u_0, v_0)f_{vv}(u_0, v_0) - h_{uv}(u_0, v_0)\right|^{1/2}} \tag{6.4.4}$$

However, Xie (1990) has found that the accuracy of this more general asymptotic procedure is not always satisfactory.

The stationary probability density $p(\theta)$ can also be obtained by directly solving the reduced Fokker-Planck equation (6.3.24) numerically (Wedig, 1989; Wedig, Lin, and Cai, 1990; Xie, 1990). Consider again the case in which all singular points are shunts in the same direction, and none is a regular boundary. The value of the stationary probability density at each singular point θ_s may be obtained from one of the equations in (6.3.33), (6.3.35), (6.3.38), (6.3.40), or the last two of (6.3.47), depending on the nature of the singular point θ_s. This value provides a starting point for the numerical solution of (6.3.24).

The values of $p(\theta)$ at other θ locations can be obtained as follows. Since the coefficients in (6.3.24) are periodic with a period π, the calculation needs to be carried out only between θ_s and $\theta_s + \pi$. Let $p_n = p(\theta_n)$, $\theta_n = \theta_s + n\Delta\theta$, $\Delta\theta = \pi/N$, and $n = 1, 2, \ldots, N$. Replace (6.3.24) by its backward difference version

$$\sigma_\theta^2(\theta_n)\frac{p_n - p_{n-1}}{\Delta\theta} + 2[\sigma_\theta(\theta_n)\sigma_\theta'(\theta_n) - m_\theta(\theta_n)]p_n = -G_c \tag{6.4.5}$$

which may be rearranged to read

$$p_n = \frac{\dfrac{\sigma_\theta^2(\theta_n)}{\Delta\theta}p_{n-1} - G_c}{\dfrac{\sigma_\theta^2(\theta_n)}{\Delta\theta} + 2[\sigma_\theta(\theta_n)\sigma_\theta'(\theta_n) - m_\theta(\theta_n)]} \tag{6.4.6}$$

The constant G_c is determined from the normalization condition

$$\sum_{n=1}^{N} p_n \Delta\theta = \tfrac{1}{2} \tag{6.4.7}$$

The largest Lyapunov exponent is obtained from

$$\lambda_T = 2\sum_{n=1}^{N} Q(\theta_n) p_n \Delta\theta \tag{6.4.8}$$

Another successive approximation method for computing Lyapunov exponents is due to Wedig (1988), based on the concept of linear transformation and the relationship between Lyapunov exponents of the original process and its pth moment. Return to (6.3.19), and define the pth norm of $\boldsymbol{X}(t)$ as

$$A_p(t) = \|\boldsymbol{X}\|^p = (X_1^2 + X_2^2)^{p/2} \qquad p > 0 \tag{6.4.9}$$

To obtain an Itô equation for A_p, we note that

$$\frac{\partial}{\partial X_i} A_p(t) = p\,\|\boldsymbol{X}\|^{p-2} X_i \tag{6.4.10}$$

and

$$\frac{\partial^2}{\partial X_i \partial X_j} A_p(t) = \begin{cases} p(p-2)\,\|\boldsymbol{X}\|^{p-4} X_i X_j & i \neq j \\ p(p-2)\,\|\boldsymbol{X}\|^{p-4} X_i^2 + p\,\|\boldsymbol{X}\|^{p-2} & i = j \end{cases} \tag{6.4.11}$$

It can be shown that

$$\begin{aligned} dA_p = {} & pA_p\left[\alpha_{jl} U_j U_l + \tfrac{1}{2}(p-2)\gamma_{jkr}\gamma_{qlr} U_j U_k U_q U_l + \gamma_{jkr}\gamma_{jir} + U_j U_k\right] dt \\ & + pA_p \gamma_{jkr} U_j U_k \, dB_r(t) \end{aligned} \tag{6.4.12}$$

where $U_1 = \cos\Theta$ and $U_2 = \sin\Theta$. Clearly, the drift and the diffusion coefficients in (6.4.12) are again functions of $\cos 2\Theta$ and $\sin 2\Theta$. Equation (6.4.12) is equivalent to the following more concise form:

$$dA_p = pA_p f(\Theta)\, dt + pA_p g(\Theta)\, dB(t) \tag{6.4.13}$$

where $g(\Theta) = (\gamma_{jkr}\gamma_{lqr} U_j U_k U_l U_q)^{1/2}$. Equation (6.4.13) and the Itô equation for Θ,

$$d\Theta = m_\Theta\, dt + \sigma_\Theta\, dB(t) \tag{6.3.20}$$

describe a two-dimensional Markov process.

Following Wedig (1988), a linear transformation

$$S = T(\Theta) A_p \qquad A_p = T^{-1}(\Theta) S \tag{6.4.14}$$

is applied to the vector process $\{A_p, \Theta\}$. The resulting Itô equation for S reads

$$dS = \left[\tfrac{1}{2}\sigma_\Theta^2 T'' + (pg\sigma_\Theta + m_\Theta)T' + pfT\right] T^{-1} S\, dt + (\sigma_\Theta T'' + pgT) T^{-1} S\, dB(t) \tag{6.4.15}$$

where each prime associated with T denotes one differentiation with respect to Θ. As long as the transformation $T(\Theta)$ is bounded and nonsingular, the $S(t)$ process has the same stability behavior as that of the $A_p(t)$ process. For the purpose of investigating the stability behavior of the $S(t)$ process, the most desirable transformation is one that renders the drift coefficient in (6.4.15) independent of Θ. Let

$$\left[\tfrac{1}{2}\sigma_\Theta^2 T'' + (pg\sigma_\Theta + m_\Theta)T' + pfT\right]T^{-1} = \bar{\lambda}_p \qquad (6.4.16)$$

and rewrite (6.4.15) as

$$dS = \bar{\lambda}_p S\,dt + (\sigma_\theta T'' + pgT)T^{-1}S\,dB(t) \qquad (6.4.17)$$

Taking the ensemble average of (6.4.17) results in

$$\frac{d}{dt}E[S] = \bar{\lambda}_p E[S] \qquad (6.4.18)$$

It is clear that $\bar{\lambda}_p$ is the Lyapunov exponent of $E[S]$, which is the pth moment of the norm $\|\boldsymbol{X}(t)\|$.

Now equation (6.4.16) may be treated as the following eigenvalue problem:

$$\tfrac{1}{2}\sigma_\Theta^2 T'' + (pg\sigma_\Theta + m_\Theta)T' + pfT = \bar{\lambda}_p T \qquad (6.4.19)$$

Since the coefficients in the second-order differential equation (6.4.19) are periodic with a period π, the eigenfunction T may be expressed as a Fourier series

$$T = \sum_{n=0}^{\infty}(a_n \cos 2n\Theta + b_n \sin 2n\Theta) \qquad (6.4.20)$$

Substituting (6.4.20) into (6.4.19), and equating the coefficients of like trigonometric terms, we obtain an infinite set of linear homogeneous algebraic equations for the unknown a_n and b_n. For nontrivial a_n and b_n, the determinant of the coefficients of these algebraic equations must be zero, giving rise to an equation for $\bar{\lambda}_p$. In practice, the Fourier representation (6.4.20) of the eigenfunction T must be truncated, and the determinant so obtained is then of a finite size. Generally, a five-term approximation, in which only a_0, a_1, a_2, b_1, and b_2 are retained, is sufficiently accurate.

According to a theorem due to Kozin and Sugimoto (1977) and Arnold (1984), the largest Lyapunov exponent of a linear system, under gaussian white noise excitations, is related to that of the pth moment of its norm as follows:

$$\lambda_T = \lim_{p\to 0}\frac{1}{p}\ln E[\|\boldsymbol{X}\|^p] \qquad (6.4.21)$$

Equation (6.4.21) suggests that λ_T may be obtained as the slope of $\bar{\lambda}_p$ at $p = 0$:

$$\lambda_T = \lim_{p\to 0}\frac{1}{p}\bar{\lambda}_p = \left(\frac{d}{dp}\bar{\lambda}_p\right)_{p=0} \qquad (6.4.22)$$

If an analytical solution for $\bar{\lambda}_p$ can be obtained from the preceding approximate procedure (Wedig, 1988), then the largest Lyapunov exponent for the original linear system may be determined from (6.4.22).

The Khasminskii method has been applied to two-dimensional systems or those reducible to two-dimensional systems by Kozin and Prodromou (1971), Mitchell and Kozin (1974), Nishioka (1976), Pardoux and Wihstutz (1988), Wedig (1988, 1989, 1990), Ariaratnam and Xie (1990, 1992), Xie (1990), and others. An example is given in Section 6.4.1.

6.4.1 A Column under Randomly Varying Axial Load

As the first example, return to the column problem considered in Section 4.7.1. Since we are interested in the stability of the trivial solution of this linear system, the additive excitation term associated with the lateral load will be ignored. The equation of motion can then be expressed as

$$\ddot{Y} + 2\zeta\dot{Y} + [1 + W(t)]Y = 0 \tag{a}$$

obtained from equation (c) in Section 4.7.1 by dropping the additive excitation $\xi_2(t)$, replacing $\xi_1(t)$ by a white noise $W(t)$, normalizing ω_0 to 1, and denoting the normalized displacement by Y. This equation is also a special case of equation (6.2.10). Letting

$$Y = Ze^{-\zeta t} \tag{b}$$

equation (a) is simplified to

$$\ddot{Z} + [1 - \zeta^2 + W(t)]Z = 0 \tag{c}$$

The Lyapunov exponents λ_Y and λ_Z of systems (a) and (c), respectively, are related as

$$\lambda_Y = -\zeta + \lambda_Z \tag{d}$$

Let $Z = X_1$ and $\dot{Z} = X_2$. Equation (c) may be replaced by two first-order equations, which then can be converted to the Itô-type stochastic differential equations as follows:

$$dX_1 = X_2\,dt \tag{e}$$

$$dX_2 = -(1 - \zeta^2)X_1 + \sqrt{2\pi K}X_1\,dB(t) \tag{f}$$

where $B(t)$ is a unit brownian motion process and K is the spectral density of $W(t)$. For this two-dimensional case, we use the transformation

$$\Upsilon = \ln\|X(t)\| = \ln(X_1^2 + X_2^2)^{1/2} \tag{g}$$

$$\Theta = \tan^{-1}\left(\frac{X_2}{X_1}\right) \tag{h}$$

Equation (h) is equivalent to the pair $U_1 = \cos\Theta$ and $U_2 = \sin\Theta$, which satisfies $U_1^2 + U_2^2 = 1$. The Itô equations for $\Upsilon(t)$ and $\Theta(t)$ can be obtained from (e) through (h), to yield

$$d\Upsilon(t) = m_\Upsilon\,dt + \sigma_\Upsilon\,dB(t) \tag{i}$$

$$d\Theta(t) = m_\Theta\,dt + \sigma_\Theta\,dB(t) \tag{j}$$

where

$$m_\Upsilon = \tfrac{1}{2}\zeta^2\sin 2\Theta + \pi K\cos 2\Theta\cos^2\Theta \tag{k}$$

$$\sigma_\Upsilon = \sqrt{\frac{\pi K}{2}}\sin 2\Theta \tag{l}$$

$$m_\Theta = -1 + \zeta^2\cos^2\Theta - \pi K\sin 2\Theta\cos^2\Theta \tag{m}$$

$$\sigma_\Theta = \sqrt{2\pi K}\cos^2\Theta \tag{n}$$

Thus $\Theta(t)$ is a diffusive Markov process on a unit circle. The stationary probability density $p(\theta)$ of $\Theta(t)$ is governed by the reduced Fokker-Planck equation

$$\frac{1}{2}\frac{d^2}{d\theta^2}(b_\theta p) - \frac{d}{d\theta}(a_\theta p) = 0 \tag{o}$$

where $b_\theta = (\sigma_\Theta^2)_{\Theta=\theta}$ and $a_\theta = (m_\Theta)_{\Theta=\theta}$. Equation (o) is singular at $\theta = \pi/2$ and $\theta = 3\pi/2$ where b_Θ vanishes. However, a_θ is negative at these points, indicating that both are left shunts. The probability flow can then pass from one region in $-\pi/2 < \theta < \pi/2$ to another in $\pi/2 < \theta < 3\pi/2$ at these singular points, permitting the $\Theta(t)$ process to become ergodic within the entire domain of $[0, 2\pi)$. Therefore, the Lyapunov exponent obtained from equation (6.3.15) is, indeed, the largest within the Lyapunov spectrum.

The solution for equation (o) can be obtained from (6.3.31) as (Nishioka, 1976; Xie, 1990)

$$p(\theta) = \begin{cases} Cf(\theta) & -\pi/2 < \theta < \pi/2 \\ Cf(\theta - \pi) & \pi/2 < \theta < 3\pi/2 \end{cases} \tag{p}$$

where C is the normalization constant and

$$f(\theta) = \frac{\psi(\theta)}{b_\theta(\theta)}\int_{-\pi/2}^{\theta} \psi^{-1}(\phi)\,d\phi \tag{q}$$

in which function ψ is given by (6.3.29). This solution has a period π, dictated by the fact that the coefficients in equation (o) have a period π, as can be seen in (m) and (n). Upon carrying out the indicated integrations in equation (q),

$$p(\theta) = \begin{cases} Ce^{-g(\theta)}\sec^2(\theta)\displaystyle\int_{-2/\pi}^{\theta} e^{g(\phi)}\sec^2\phi\,d\phi & -\pi/2 < \theta < \pi/2 \\ Ce^{-g(\theta-\pi)}\sec^2(\theta - \pi)\displaystyle\int_{-2/\pi}^{\theta-\pi} e^{g(\phi)}\sec^2\phi\,d\phi & \pi/2 < \theta < 3\pi/2 \end{cases} \tag{r}$$

where
$$g(\theta) = \frac{1}{3\pi K}(3 - 3\zeta^2 + \tan^2\theta)\tan\theta \tag{s}$$

The largest Lyapunov exponent for the Z process is obtained as

$$\lambda_Z = 2\int_{-\pi/2}^{\pi/2} (\tfrac{1}{2}\zeta^2\sin 2\theta + \pi K\cos 2\theta\cos^2\theta)p(\theta)\,d\theta \tag{t}$$

The right-hand side of equation (t) can be evaluated using one of the numerical procedures discussed in Section 6.4. Letting $x = \tan\theta$, equation (t) is changed to

$$\lambda_Z = \frac{\int_{-\infty}^{\infty} J(x)e^{-\nu h(x)}dx\int_{-\infty}^{x} e^{\nu h(u)}du}{\int_{-\infty}^{\infty} e^{-\nu h(x)}dx\int_{-\infty}^{x} e^{\nu h(u)}du} \tag{u}$$

where $\nu = (\pi k)^{-1}$ and

$$h(x) = (1 - \zeta^2)x + \frac{x^3}{3} \tag{v}$$

$$J(x) = \frac{\zeta^2 x}{1 + x^2} + \frac{\pi K(1 - x^2)}{(1 + x^2)^2} \tag{w}$$

For $0 < \zeta < 1$, the inner integrals in equation (u) may be evaluated approximately by using the Laplace method of asymptotic expansion (Xie, 1990). In this case, $h(u)$ attains its largest value at $u = x$ and $h'(x) > 0$. Thus, from (6.4.2),

$$\int_{-\infty}^{x} e^{\nu h(u)} du \approx \frac{1}{\nu h'(x)} e^{\nu h(x)} \tag{x}$$

Substituting (x) into (u),

$$\lambda_Z \approx \frac{\displaystyle\int_{-\infty}^{\infty} \frac{\pi K}{1 - \zeta^2 + x^2} \left[\frac{\zeta^2 x}{1 + x^2} + \frac{\pi K(1 - x^2)}{(1 + x^2)^2} \right] dx}{\displaystyle\int_{-\infty}^{\infty} \frac{\pi K}{1 - \zeta^2 + x^2} dx} = \frac{\pi K}{\left(1 + \sqrt{1 - \zeta^2}\right)^2} \tag{y}$$

The Lyapunov exponent for the original system (a) is then

$$\lambda_Y \approx -\zeta + \frac{\pi K}{\left(1 + \sqrt{1 - \zeta^2}\right)^2} \tag{z}$$

For $\zeta \ll 1$, equation (z) can be further approximated as

$$\lambda_Y \approx -\zeta + \frac{\pi K}{4} \left(1 + \frac{1}{2}\zeta^2\right) \tag{aa}$$

indicating an approximate asymptotic sample stability condition

$$\zeta > \frac{\pi K}{4} \left(1 + \frac{1}{2}\zeta^2\right) \tag{bb}$$

We recall that the column problem was treated in Section 4.7.1, where both additive and multiplicative excitations were assumed to be present. The two-dimensional problem was reduced to one-dimensional, represented by the average amplitude process $A(t)$, using the smoothed stochastic averaging procedure. It is of interest to obtain the asymptotic sample stability condition for $A(t)$ and compare with inequality (bb).

Let

$$\begin{aligned} Y(t) &= A(t) \cos\theta \qquad \theta = t + \phi(t) \\ \dot{Y}(t) &= -A(t) \sin\theta \end{aligned} \tag{cc}$$

Equation (a) may be replaced by two first-order equations:

$$\dot{A} = -2\zeta A \sin^2\theta + A \sin\theta \cos\theta W(t) \tag{dd}$$

$$\dot{\phi} = -2\zeta \sin\theta \cos\theta + \cos^2\theta W(t) \tag{ee}$$

Applying the smooth procedure of stochastic averaging, we obtain an Itô equation for the averaged amplitude process as follows:

$$dA = -\left(\zeta - \frac{3\pi}{8} K\right) A\, dt + \sqrt{\frac{\pi}{4}} K\, A\, dB(t) \tag{ff}$$

It follows from the Itô differential rule (4.4.15),

$$d(\ln A) = -\left(\zeta - \frac{1}{4}\pi K\right)dt + \sqrt{\frac{\pi}{4}K}\,dB(t) \tag{gg}$$

$$\zeta > \frac{\pi K}{4} \tag{hh}$$

Inequalities (bb) and (hh) differ by an order of ζ^2, which is beyond the accuracy of the stochastic averaging procedure. Of course, the applicability of the stochastic averaging procedure requires only that the excitation in equation (a) be a wideband stationary process, not necessarily a white noise. In this more general case, the constant K in inequality (kk) is replaced by the spectral density of the wideband excitation, evaluated at twice the natural frequency of the system.

6.5 ASYMPTOTIC MOMENT STABILITY OF LINEAR SYSTEMS

As shown in Sections 6.2 through 6.4, the analytical and numerical procedures for obtaining the sample stability condition for a linear stochastic system are complicated; therefore, these procedures are generally practical only for a lower order system. On the other hand, the moment stability conditions are much easier to determine if the excitations can be modeled as gaussian white noises or are reducible to gaussian white noises, so that the equations of motion can be converted to the Itô-type stochastic differential equations.

Let the Itô equations so obtained for an n-dimensional system be given by

$$dX_j(t) = m_j(\boldsymbol{X}, t)\,dt + \sigma_{jk}(\boldsymbol{X}, t)\,dB_k(t) \tag{6.5.1}$$

where m_j are the drift coefficients, σ_{jk} are the diffusion coefficients, and $B_k(t)$ are independent unit Wiener processes. The Itô equation for a scalar function $F(\boldsymbol{X})$ of $\boldsymbol{X}(t)$ can be obtained by applying the Itô differential rule:

$$dF(\boldsymbol{X}) = (\mathscr{L}_X F)\,dt + \sigma_{jk}\frac{\partial F}{\partial X_j}dB_k(t) \tag{6.5.2}$$

where $\mathscr{L}_X$ is the generating differential operator of $\boldsymbol{X}(t)$ given by

$$\mathscr{L}_X = m_j\frac{\partial}{\partial X_j} + \frac{1}{2}\sigma_{jl}\sigma_{kl}\frac{\partial^2}{\partial X_j \partial X_k} \tag{6.5.3}$$

Equation (6.5.2) is valid, provided $F(\boldsymbol{X})$ is twice differentiable with respect to the components of $\boldsymbol{X}(t)$. Taking the expectation of equation (6.5.2), we obtain

$$\frac{d}{dt}E[F(t)] = E[\mathscr{L}_X F(\boldsymbol{X})] \tag{6.5.4}$$

Now, letting

$$F(\boldsymbol{X}) = X_1^{r_1}X_2^{r_2}\cdots X_n^{r_n} \tag{6.5.5}$$

where r_m are nonnegative integers and $N = r_1 + r_2 + \cdots + r_n$ is a positive integer, we obtain a set of deterministic equations for various Nth moments by keeping the same N but changing the composition of r_1 through r_n. It can be shown that if the Itô equations (6.5.1) are linear and all random excitations are multiplicative, then each of these equations is linear and contains only moments of order N. Such moment equations are said to be closed for a given N. By letting $N = 1, 2, \ldots$, the stability conditions for moments of different orders can be investigated separately.

If the excitations are wideband stochastic processes, then under certain conditions (see Sections 4.6 and 4.7), the original equations of motion may be converted to the Itô-type equations using the stochastic averaging or the quasi-conservative averaging procedure. If the original equations are linear and without the additive excitation terms, then the Itô equations obtained are also linear. The equations for moments can again be derived in the same manner as in equations (6.5.4) and (6.5.5), and they are closed for each order N. Some examples are now given.

6.5.1 A Column under Axial White Noise Excitation

The case of column stability was discussed in Section 6.4.1 from the viewpoint of sample stability. It is illuminating to reexamine the problem in terms of statistical moments. Again, we assume that the transverse motion of the column in question is dominated by the fundamental mode, for which the equation of motion is given by

$$\ddot{X} + 2\zeta\omega_0\dot{X} + \omega_0^2[1 + W(t)]X = 0 \tag{a}$$

which is obtained from equation (c) of Section 4.7.1 by dropping the additive excitation $\xi_2(t)$ and changing the multiplicative excitation $\xi_1(t)$ to a gaussian white noise $W(t)$. The additive term is not retained since it does not affect the stability of a linear system. The parameters ζ and ω_0 in equation (a) are positive constants, as defined in Section 4.7.1.

Equation (a) may be replaced by the following Itô equations:

$$dX_1 = X_2\,dt \tag{b}$$

$$dX_2 = (-2\zeta\omega_0X_2 - \omega_0^2X_1)\,dt - \omega_0^2X_1\sqrt{2\pi K}\,dB(t) \tag{c}$$

where K is the spectral density of $W(t)$. Let $\mu_{ij} = E[X_1^iX_2^j]$. The equations for the first-order moments μ_{10} and μ_{01} are obtained as follows by taking ensemble averages of equations (b) and (c):

$$\frac{d}{dt}\begin{Bmatrix}\mu_{10}\\ \mu_{01}\end{Bmatrix} = \begin{bmatrix}0 & 1\\ -\omega_0^2 & -2\zeta\omega_0\end{bmatrix}\begin{Bmatrix}\mu_{10}\\ \mu_{01}\end{Bmatrix} \tag{d}$$

The eigenvalues of the square matrix on the right-hand side of (d) are found to be

$$\lambda = -\zeta\omega_0 \pm i\sqrt{1-\zeta^2}\,\omega_0 \tag{e}$$

The stability condition for the first-order moments is that the real parts of the preceding eigenvalues be negative, or

$$\zeta > 0 \tag{f}$$

since $\omega_0 > 0$.

The Itô equations for X_1^2, X_1X_2, and X_2^2 are derived by applying the Itô differential rule (6.5.2), resulting in

$$dX_1^2 = 2X_1X_2\,dt \tag{g}$$

$$dX_1X_2 = (-\omega_0^2X_1^2 - 2\zeta\omega_0X_1X_2 + X_2^2)dt - \omega_0^2X_1^2\sqrt{2\pi K}\,dB(t) \tag{h}$$

$$dX_2^2 = (2\pi K\omega_0^4X_1^2 - 2\omega_0^2X_1X_2 - 4\zeta\omega_0X_2^2)dt - 2\omega_0^2X_1X_2\sqrt{2\pi K}\,dB(t) \tag{i}$$

The equations for the second-order moments, μ_{20}, μ_{11}, and μ_{02}, are obtained by taking the ensemble averages of (g), (h), and (i), and they may be cast in the following matrix form:

$$\frac{d}{dt}\begin{Bmatrix}\mu_{20}\\ \mu_{11}\\ \mu_{02}\end{Bmatrix} = \begin{bmatrix}0 & 2 & 0\\ -\omega_0^2 & -2\zeta\omega_0 & 1\\ 2\pi K\omega_0^4 & -2\omega_0^2 & -4\zeta\omega_0\end{bmatrix}\begin{Bmatrix}\mu_{20}\\ \mu_{11}\\ \mu_{02}\end{Bmatrix} \tag{j}$$

The characteristic equation of (j) is given by

$$\begin{vmatrix}\lambda & -2 & 0\\ \omega_0^2 & \lambda+2\zeta\omega_0 & -1\\ -2\pi K\omega_0^4 & 2\omega_0^2 & \lambda+4\zeta\omega_0\end{vmatrix} = \lambda^3 + 6\zeta\omega_0\lambda^2 + 4\omega_0^2\lambda + 4\omega_0^3(2\zeta - \pi K\omega_0) = 0 \tag{k}$$

According to the well-known Routh-Hurwitz criteria (see, e.g., Chetayev, 1961), the real parts of all the roots of equation (k) are negative if and only if

$$\Delta_1 = 6\zeta\omega_0 > 0 \tag{l}$$

$$\Delta_2 = \begin{vmatrix}6\zeta\omega_0 & 1\\ 4\omega_0^3(2\zeta-\pi K\omega_0) & 4\omega_0^2\end{vmatrix} > 0 \tag{m}$$

$$\Delta_3 = \begin{vmatrix}6\zeta\omega_0 & 1 & 0\\ 4\omega_0^3(2\zeta-\pi K\omega_0) & 4\omega_0^2 & 6\zeta\omega_0\\ 0 & 0 & 4\omega_0^3(2\zeta-\pi K\omega_0)\end{vmatrix} > 0 \tag{n}$$

These inequalities can all be satisfied only if

$$\zeta > \frac{\pi\omega_0}{2}K \tag{o}$$

Thus inequality (o) is the necessary and sufficient condition for the stability of the second-order moments μ_{20}, μ_{11}, and μ_{02}.

The stability conditions for higher order moments can be obtained in a similar way.

6.5.2 A Column under Axial Wideband Excitation

We use, once again, a one-mode approximation described in Section 4.7.1. The governing equation is now given by

$$\ddot{X} + 2\zeta\omega_0\dot{X} + \omega_0^2[1 + \xi_1(t)]X = 0 \tag{a}$$

which is obtained by letting $\xi_2(t) = 0$ in equation (c) of Section 4.7.1. Let

$$X(t) = A(t)\cos\theta \qquad \theta = \omega_0 t + \phi(t) \tag{b}$$

$$\dot{X}(t) = -\omega_0 A(t)\sin\theta \tag{c}$$

Equation (a) is replaced by two first-order equations:

$$\dot{A} = -2\zeta\omega_0 A\sin^2\theta + A\omega_0\sin\theta\cos\theta\xi_1(t) \tag{d}$$

$$\dot{\phi} = -2\zeta\omega_0\sin\theta\cos\theta + \omega_0\cos^2\theta\xi_1(t) \tag{e}$$

We assume that the right-hand sides of equations (d) and (e) are small, so that the time-averaged version of the stochastic averaging procedure is applicable. The averaged amplitude process $A(t)$ is itself a Markov process, with an Itô equation

$$dA = \left[-\zeta\omega_0 + \frac{3\pi}{8}\omega_0^2\Phi_{11}(2\omega_0)\right]A\,dt + \left[\frac{\pi}{4}\omega_0^2\Phi_{11}(2\omega_0)\right]^{1/2} A\,dB(t) \tag{f}$$

which agrees with equation (n) of Section 4.7.1 upon setting the spectral density $\Phi_{22}(\omega_0)$ of the additive excitation to zero.

Applying the Itô differential rule to $F(A) = A^n$, we obtain

$$dA^n = n\left[-\zeta\omega_0 + \frac{\pi}{8}(n+2)\omega_0^2\Phi_{11}(2\omega_0)\right]A^n\,dt + n\left[\frac{\pi}{4}\omega_0^2\Phi_{11}(2\omega_0)\right]^{1/2} A^n\,dB(t) \tag{g}$$

The equation for the nth moment follows from taking the ensemble average of equation (g):

$$\frac{d}{dt}E[A^n] = n\left[-\zeta\omega_0 + \frac{\pi}{8}(n+2)\omega_0^2\Phi_{11}(2\omega_0)\right]E[A^n] \tag{h}$$

The stability of the nth moment requires that

$$\zeta > \frac{\pi}{8}(n+2)\omega_0\Phi_{11}(2\omega_0) \tag{i}$$

This condition is seen to be increasingly more stringent, as the order n of the moment increases. In the special case $n = 2$, it agrees with inequality (o) in Section 6.5.1.

It is of interest to show that the moment stability condition (i) can be obtained by an alternative approach. For this purpose, we modify the equation of motion (a) by adding, for the time being, an additive excitation term to the right-hand side to read

$$\ddot{X} + 2\zeta\omega_0\dot{X} + \omega_0^2[1 + \xi_1(t)]X = \xi_2(t) \tag{j}$$

where $\xi_2(t)$ is another wideband excitation, with a spectral density $\Phi_{22}(\omega)$, and independent of $\xi_1(t)$. This is the same equation as equation (c) in Section 4.7.1. The drift and diffusion coefficients in the Itô equation for the averaged amplitude are now

$$m(A) = \left[-\zeta\omega_0 + \frac{3\pi}{8}\omega_0^2\Phi_{11}(2\omega_0)\right]A + \frac{\pi\Phi_{22}(\omega_0)}{2\omega_0^2 A} \tag{k}$$

$$\sigma(A) = \left[\frac{\pi}{4}\omega_0^2\Phi_{11}(2\omega_0)A^2 + \frac{\pi\Phi_{22}(\omega_0)}{\omega_0^2}\right]^{1/2} \tag{l}$$

as seen in equation (m) of Section 4.7.1. The stationary probability density, if it exists, can be obtained from equation (4.5.11) to yield

$$p(\alpha) = C\alpha\left[\alpha^2 + \frac{4\Phi_{22}(\omega_0)}{\omega_0^4\Phi_{11}(2\omega_0)}\right]^{-\delta} \tag{m}$$

where α is the state variable for $A(t)$, and

$$\delta = \frac{4\zeta}{\pi\omega_0\Phi_{11}(2\omega_0)} \qquad C = 2(\delta - 1)\left[\frac{4\Phi_{22}(\omega_0)}{\omega_0^4\Phi_{11}(2\omega_0)}\right]^{\delta-1} \tag{n}$$

The nth moment is then obtained as

$$E[A^n] = \int_0^\infty \alpha^n p(\alpha)\,d\alpha = (\delta - 1)\left[\frac{4\Phi_{22}(\omega_0)}{\omega_0^4\Phi_{11}(2\omega_0)}\right]^{n/2} \frac{\Gamma\left(\dfrac{n+2}{2}\right)\Gamma\left(\delta - \dfrac{n+2}{2}\right)}{\Gamma(\delta)} \tag{o}$$

which is bounded under the condition

$$\delta - \frac{n+2}{2} > 0 \tag{p}$$

It can be shown that (p) is exactly (i). Once this condition is satisfied, we find the nth moment for the original system, without the additive excitation $\xi_2(t)$, as follows:

$$\lim_{\Phi_{22}(\omega_0)\to 0} E[A^n] = 0 \tag{q}$$

in agreement with the original definition for the moment stability. The foregoing alternative approach is obviously more cumbersome for linear systems, for which the use of Itô's differential rule is quite straightforward. However, it is an effective means to investigate the moment stability conditions for some nonlinear systems, to be shown in Section 6.7.

The condition for asymptotic stability in probability of the averaged amplitude can also be determined by requiring that the probability density $p(\alpha)$, given in equation (m), be integrable, or by letting $n = 0$ in inequality (p). In terms of the spectral density of the multiplicative excitation, it reads

$$\zeta > \frac{\pi}{4}\omega_0\Phi_{11}(2\omega_0) \tag{r}$$

which is reducible to inequality (kk) in Section 6.4.1, upon letting $\omega_0 = 1$ and $\Phi_{11}(\omega) = K$. This is also the condition for the asymptotic sample stability since the system is linear.

6.5.3 Coupled Bending-Torsional Stability of a Beam (Ariaratnam and Srikantaiah, 1978)

As the third example, consider a uniform elastic beam of narrow rectangular cross section, shown in Fig. 6.5.1. The beam is simply supported and is subjected to randomly varying end couples $M(t)$ at $z = 0$ and $z = l$. The two end couples are assumed to be identical wideband stationary processes.

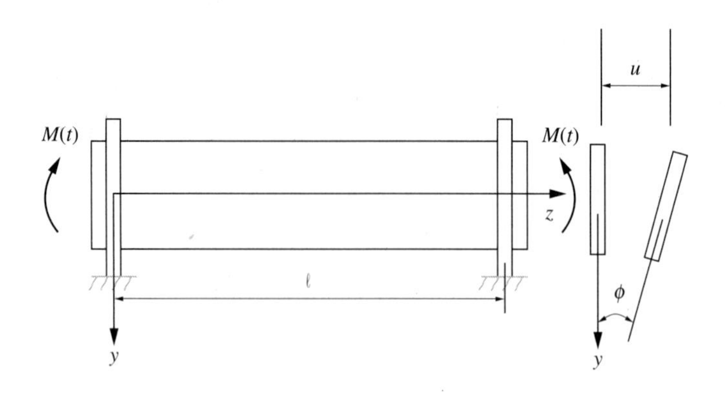

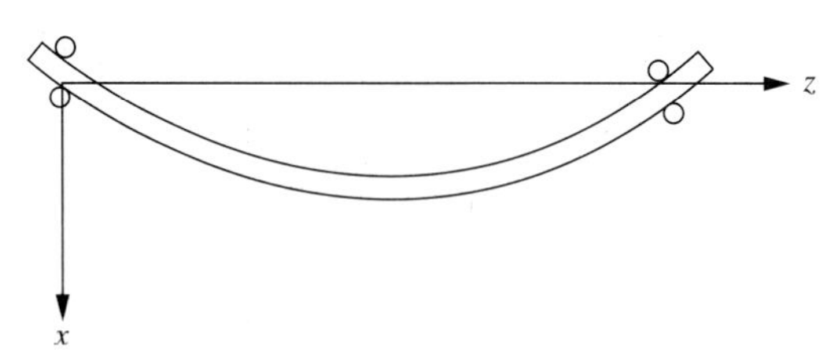

FIGURE 6.5.1
A beam subject to random end couples.

The motion of the beam is governed by

$$EI_x \frac{\partial^4}{\partial z^4} v + m\ddot{v} + \alpha \dot{v} = 0 \tag{a.1}$$

$$EI_y \frac{\partial^4}{\partial z^4} u + M(t) \frac{\partial^2}{\partial z^2} \phi + m\ddot{u} + \beta \dot{u} = 0 \tag{a.2}$$

$$M(t) \frac{\partial^2}{\partial z^2} u - GJ \frac{\partial^2}{\partial z^2} \phi + m\rho^2 \ddot{\phi} + \gamma \dot{\phi} = 0 \tag{a.3}$$

with boundary conditions

$$EI_x \frac{\partial^2}{\partial z^2} v(0, t) = EI_x \frac{\partial^2}{\partial z^2} v(l, t) = -M(t) \tag{b.1}$$

$$u(0, t) = u(l, t) = \frac{\partial^2}{\partial z^2} u(0, t) = \frac{\partial^2}{\partial z^2} u(l, t) = 0 \tag{b.2}$$

$$\phi(0, t) = \phi(l, t) = 0 \tag{b.3}$$

In equations (a.1) through (b.3), v = deflection in the y direction, u = deflection in the x direction, ϕ = angle of twist, EI_x = bending stiffness about the x axis, EI_y = bending stiffness about the y axis, GJ = torsional stiffness, m = beam mass per unit length, ρ = polar radius of gyration of the beam cross section, and α, β, and γ are viscous damping coefficients. It is seen that the u and ϕ motions are coupled, but they are uncoupled with the v motion. The case of deterministic $M(t)$ was considered by Bolotin (1964).

Since the beam is much more rigid against the bending deformation about the x axis, the coupled u and ϕ motions are of the greatest practical concern. Considering only the fundamental modes, described by

$$u(z, t) = \rho Q_1(t) \sin \frac{\pi z}{l} \tag{c.1}$$

$$\phi(z, t) = Q_2(t) \sin \frac{\pi z}{l} \tag{c.2}$$

we obtain, upon substituting (c.1) and (c.2) into (a.2) and (a.3),

$$\ddot{Q}_1 + \omega_1^2 Q_1 + 2\varepsilon\zeta_1\omega_1\dot{Q}_1 - \varepsilon^{1/2}\omega_1\omega_2 Q_2\xi(t) = 0 \tag{d.1}$$

$$\ddot{Q}_2 + \omega_2^2 Q_2 + 2\varepsilon\zeta_2\omega_2\dot{Q}_2 - \varepsilon^{1/2}\omega_1\omega_2 Q_1\xi(t) = 0 \tag{d.2}$$

where $\omega_1^2 = \pi^2 EI_y/(ml^4)$, $\omega_2^2 = \pi^2 GJ/(ml^2\rho^2)$, $2\varepsilon\zeta_1\omega_1 = \beta/m$, $2\varepsilon\zeta_2\omega_2 = \gamma/(m\rho^2)$, $\varepsilon^{1/2}\xi(t) = M(t)/M_{\text{cr}}$, and where $M_{\text{cr}} = \pi(EI_y GJ)^{1/2}/l$ is the static buckling moment, that is, the beam would buckle under a pair of static couples M_{cr} at the ends. It is implied in (d.1) and (d.2) that β and γ are of order ε, and $M(t)$ is of order $\varepsilon^{1/2}$. Now, let

$$Q_j = A_j \cos\theta_j \qquad \theta_j = \omega_j t + \phi_j(t) \tag{e.1}$$

$$\dot{Q}_j = -A_j\omega_j \sin\theta_j \tag{e.2}$$

where $j = 1, 2$, and where the repeated subscript j does not indicate a summation operation. Equations (d.1) and (d.2) are transformed into four first-order equations:

$$\dot{A}_1 = -2\varepsilon\zeta_1\omega_1 A_1 \sin^2\theta_1 - \varepsilon^{1/2}\omega_2 A_2 \sin\theta_1 \cos\theta_2 \xi(t) \tag{f.1}$$

$$\dot{\phi}_1 = -2\varepsilon\zeta_1\omega_1 \sin\theta_1 \cos\theta_1 - \varepsilon^{1/2}\omega_2 \frac{A_2}{A_1} \cos\theta_1 \cos\theta_2 \xi(t) \tag{f.2}$$

$$\dot{A}_2 = -2\varepsilon\zeta_2\omega_2 A_2 \sin^2\theta_2 - \varepsilon^{1/2}\omega_1 A_1 \sin\theta_2 \cos\theta_1 \xi(t) \tag{f.3}$$

$$\dot{\phi}_2 = -2\varepsilon\zeta_2\omega_2 \sin\theta_2 \cos\theta_2 - \varepsilon^{1/2}\omega_1 \frac{A_1}{A_2} \cos\theta_1 \cos\theta_2 \xi(t) \tag{f.4}$$

The four-dimensional vector $(A_1, \phi_1, A_2, \phi_2)$ is approximately a Markov vector, if the correlation time of $M(t)$ is of an order smaller than ε^{-1}.

Since the right-hand sides of (f.1) through (f.4) are assumed to be small, the Itô equations for the system may be obtained by applying the smoothed version of the stochastic averaging procedure, (4.7.9) and (4.7.10). Again, the averaged amplitude processes $A_1(t)$ and $A_2(t)$ are found to be uncoupled with $\phi_1(t)$ and $\phi_2(t)$, and their Itô equations are given by

$$dA_1 = m_1 dt + \sigma_{11} dB_1(t) + \sigma_{12} dB_2(t) \tag{g.1}$$

$$dA_2 = m_2 dt + \sigma_{21} dB_1(t) + \sigma_{22} dB_2(t) \tag{g.2}$$

where
$$m_1 = -\varepsilon\zeta_1\omega_1 A_1 + \frac{\varepsilon\pi\omega_2}{8A_1}[2\omega_1 A_1^2\Phi^- + \omega_2 A_2^2\Phi^+] \tag{h.1}$$

$$m_2 = -\varepsilon\zeta_2\omega_2 A_2 + \frac{\varepsilon\pi\omega_1}{8A_2}[\omega_1 A_1^2\Phi^+ + 2\omega_2 A_2^2\Phi^-] \tag{h.2}$$

$$(\sigma\sigma')_{11} = \frac{\varepsilon\pi}{4}\omega_2^2 A_2^2\Phi^+ \tag{h.3}$$

$$(\sigma\sigma')_{22} = \frac{\varepsilon\pi}{4}\omega_1^2 A_1^2\Phi^+ \tag{h.4}$$

$$(\sigma\sigma')_{12} = (\sigma\sigma')_{21} = \frac{\varepsilon\pi}{4}\omega_1\omega_2 A_1 A_2\Phi^- \tag{h.5}$$

in which
$$\Phi^+ = \Phi(\omega_1 + \omega_2) + \Phi(\omega_1 - \omega_2) \tag{i.1}$$

$$\Phi^- = \Phi(\omega_1 + \omega_2) - \Phi(\omega_1 - \omega_2) \tag{i.2}$$

and $\Phi(\omega)$ is the spectral density of $\xi(t)$. Equations (g.1) and (g.2) show that the averaged $A_1(t)$ and $A_2(t)$ constitute a two-dimensional Markov vector.

We now derive the second moment equations for $A_1(t)$ and $A_2(t)$. Applying, in turn, the Itô differential rule to A_1^2 and A_2^2,

$$dA_1^2 = \varepsilon\left[\left(-2\zeta_1\omega_1 + \frac{\pi}{2}\omega_1\omega_2\Phi^-\right)A_2^2 + \frac{\pi}{2}\omega_2^2\Phi^+ A_2^2\right]dt$$
$$+ 2A_1[\sigma_{11}dB_1(t) + \sigma_{12}dB_2(t)] \tag{j.1}$$

$$dA_2^2 = \varepsilon\left[\left(-2\zeta_2\omega_2 + \frac{\pi}{2}\omega_1\omega_2\Phi^-\right)A_2^2 + \frac{\pi}{2}\omega_1^2\Phi^+ A_1^2\right]dt$$
$$+ 2A_2[\sigma_{21}dB_1(t) + \sigma_{22}dB_2(t)] \tag{j.2}$$

Taking the ensemble averages of (j.1) and (j.2), respectively, we obtain the following matrix equation for $E[A_1^2]$ and $E[A_2^2]$:

$$\frac{d}{dt}\begin{Bmatrix} E[A_1^2] \\ E[A_2^2] \end{Bmatrix} = \varepsilon\begin{bmatrix} -2\zeta_1\omega_1 + \frac{\pi}{2}\omega_1\omega_2\Phi^- & \frac{\pi}{2}\omega_2^2\Phi^+ \\ \frac{\pi}{2}\omega_1^2\Phi^+ & -2\zeta_2\omega_2 + \frac{\pi}{2}\omega_1\omega_2\Phi^- \end{bmatrix}\begin{Bmatrix} E[A_1^2] \\ E[A_2^2] \end{Bmatrix} \tag{k}$$

According to the Routh-Hurwitz criteria, we obtain the stability conditions in the second moments as follows:

$$\zeta_1 > \frac{\pi}{4}\omega_2\Phi^- \qquad \zeta_2 > \frac{\pi}{4}\omega_1\Phi^-$$
$$\left(\zeta_1 - \frac{\pi}{4}\omega_2\Phi^-\right)\left(\zeta_2 - \frac{\pi}{4}\omega_1\Phi^-\right) > \frac{\pi^2}{16}\omega_1\omega_2(\Phi^+)^2 \tag{l}$$

The regions of stability are shown in Fig. 6.5.2 for two cases of $\Phi^- > 0$ and $\Phi^- < 0$. It may be noted that when $\Phi^- < 0$, that is, $\Phi(\omega_1 + \omega_2) < \Phi(\omega_1 - \omega_2)$, the system can be stable even when one of the damping coefficients is negative, provided that the other is sufficiently large and positive. In the case when the spectral density of $\xi(t)$ is constant and of magnitude K_0 over a wide frequency range, conditions (l)

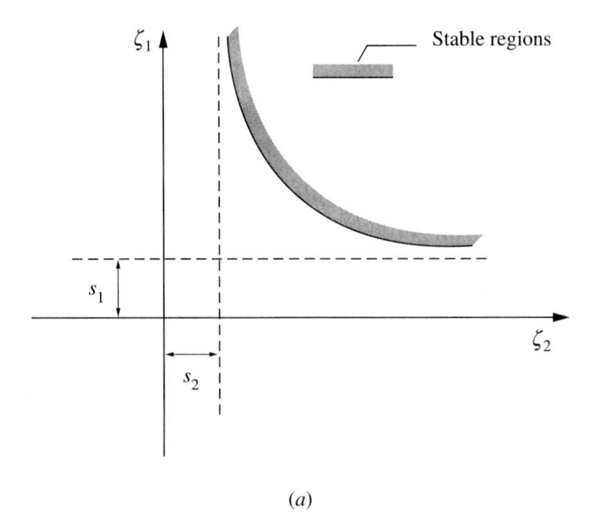

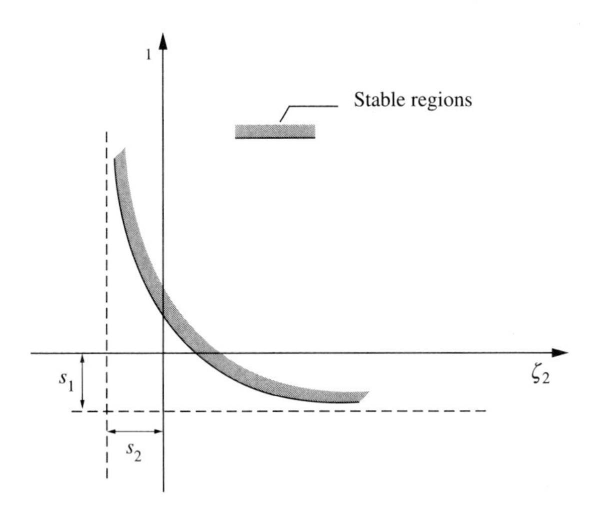

(*b*)

$s_1 = \frac{\pi}{4}\,\omega_2\Phi-$ $\qquad$ $s_2 = \frac{\pi}{4}\,\omega_1\Phi-$

FIGURE 6.5.2
Stability regions (*a*) $\phi > 0$; (*b*) $\phi < 0$ [after Ariaratnam and Srikantaiah, 1978; Copyright ©1978 Marcel Dekker Inc., reprinted with permission].

reduce to

$$\zeta_1 > 0 \qquad \zeta_2 > 0 \qquad \zeta_1\zeta_2 > \frac{\pi^2}{4}K_0^2 \tag{m}$$

The foregoing analysis is based on the assumptions that the lateral deflection u and torsion ϕ are each dominated by one mode, and that both modes are lightly damped. It is clear that the same analysis may be generalized to the case of arbitrary N modes, as long as all modes involved are lightly damped (Ariaratnam and

Srikantaiah, 1978). However, if only some of the modes are lightly damped, then a modified stochastic averaging procedure should be employed, as shown in the next example.

6.5.4 Stability of a Coupled System with Both Low-Damping and High-Damping Modes

One key assumption in the smoothed version of the stochastic averaging procedure is that damping in the system is low, so that the relaxation time of the system is long compared with the correlation time of the random excitations. In such a case, the response of the system is approximately a Markov vector. If both low-damping and high-damping modes are present in the system, then recourse may be sought from a modified stochastic averaging procedure (Papanicolaou and Kohler, 1975; Sri Namachchivaya and Lin, 1988).

Return to the problem of coupled bending-torsional motion of a thin beam considered in Section 6.5.3. Let the governing equations be given by

$$\ddot{Q}_1 + \omega_1^2 Q_1 + 2\varepsilon\zeta_1\omega_1\dot{Q}_1 - \varepsilon^{1/2}\rho\omega_1\omega_2 Q_2\xi(t) = 0 \tag{a.1}$$

$$\ddot{Q}_2 + \omega_2^2 Q_2 + 2\zeta_2\omega_2\dot{Q}_2 - \varepsilon^{1/2}\rho\omega_1\omega_2 Q_1\xi(t) = 0 \tag{a.2}$$

These are taken from equations (d.1) and (d.2) in Section 6.5.3 with only one exception: while the damping coefficient ζ_2 for the Q_2 mode is less than 1, it is not very much smaller than 1. Again, $\xi(t)$ is assumed to be a wideband stationary random process, with a correlation function $E[\xi(t)\xi(t+\tau)] = R_{\xi\xi}(\tau)$. We use the same transformation for the low-damping mode:

$$Q_1 = A_1\cos\theta_1 \qquad \theta_1 = \omega_1 t + \phi_1(t) \tag{b.1}$$

$$\dot{Q}_1 = -A_1\omega_1\sin\theta_1 \tag{b.2}$$

to arrive at

$$\dot{A}_1 = -2\varepsilon\zeta_1\omega_1 A_1\sin^2\theta_1 - \varepsilon^{1/2}\rho\omega_2\sin\theta_1 Q_2\xi(t) \tag{c.1}$$

$$\dot{\phi}_1 = -2\varepsilon\zeta_1\omega_1\sin\theta_1\cos\theta_1 - \varepsilon^{1/2}\omega_2 A_1^{-1}\cos\theta_1 Q_2\xi(t) \tag{c.2}$$

However, for the high-damping mode, we replace equation (a.2) by

$$\frac{d}{dt}\begin{Bmatrix} U_3 \\ U_4 \end{Bmatrix} = \boldsymbol{B}\begin{Bmatrix} U_3 \\ U_4 \end{Bmatrix} + \begin{bmatrix} 0 & 0 \\ \varepsilon^{1/2}\rho\omega_1\omega_2 & 0 \end{bmatrix}\begin{Bmatrix} Q_1\xi(t) \\ 0 \end{Bmatrix} \tag{d}$$

where $U_3 = Q_2$, $U_4 = \dot{Q}_2$, and

$$\boldsymbol{B} = \begin{bmatrix} 0 & 1 \\ -\omega_2^2 & -2\zeta_2\omega_2 \end{bmatrix} \tag{e}$$

The eigenvalues of $\boldsymbol{B}$ are $\lambda_{1,2} = -\zeta_2\omega_2 \pm i\omega_{2d}$, where $\omega_{2d} = \omega_2(1-\zeta_2^2)^{1/2}$. Let

$$\begin{Bmatrix} U_3 \\ U_4 \end{Bmatrix} = \begin{bmatrix} 1 & 1 \\ \lambda_1 & \lambda_2 \end{bmatrix}\begin{Bmatrix} X_3 \\ X_4 \end{Bmatrix} \tag{f}$$

Equation (d) is diagonized to read

$$\frac{d}{dt}\begin{Bmatrix} X_3 \\ X_4 \end{Bmatrix} = \begin{bmatrix} \lambda_1 & 0 \\ 0 & \lambda_2 \end{bmatrix}\begin{Bmatrix} X_3 \\ X_4 \end{Bmatrix} + \frac{i\varepsilon^{1/2}\rho\omega_1}{2\left(1-\zeta_2^2\right)^{1/2}}\begin{Bmatrix} -1 \\ 1 \end{Bmatrix} Q_1\xi(t) \tag{g}$$

By applying another transformation $X_3 = Y_3 \exp(\lambda_1 t)$ and $X_4 = Y_4 \exp(\lambda_2 t)$, equation (g) is simplified to

$$\frac{d}{dt}\begin{Bmatrix} Y_3 \\ Y_4 \end{Bmatrix} = +\frac{i\varepsilon^{1/2}\rho\omega_1}{2\left(1-\zeta_2^2\right)^{1/2}}\begin{Bmatrix} -\exp(-\lambda_1 t) \\ \exp(-\lambda_2 t) \end{Bmatrix} Q_1\xi(t) \tag{h}$$

Since λ_1 and λ_2 are complex conjugates, Y_3 and Y_4 are also complex conjugates. Finally, substituting

$$Q_2 = U_3 = X_3 + X_4 = Y_3 e^{\lambda_1 t} + Y_4 e^{\lambda_2 t} \tag{i}$$

into equation (c.1) and (c.2), and $Q_1 = A_1 \cos\theta_1$ into equation (h), we obtain

$$\dot{A}_1 = -2\varepsilon\zeta_1\omega_1 A_1 \sin^2\theta_1 + \varepsilon^{1/2} g_1 \xi(t) \tag{j.1}$$

$$\dot{\phi}_1 = -2\varepsilon\zeta_1\omega_1 \sin\theta_1 \cos\theta_1 + \varepsilon^{1/2} g_2 \xi(t) \tag{j.2}$$

$$\dot{Y}_3 = \varepsilon^{1/2} g_3 \xi(t) \tag{j.3}$$

$$\dot{Y}_4 = \varepsilon^{1/2} g_4 \xi(t) \tag{j.4}$$

where

$$g_1 = -\rho\omega_2 \sin\theta_1 (Y_3 e^{\lambda_1 t} + Y_4 e^{\lambda_2 t}) \tag{k.1}$$

$$g_2 = -\rho\omega_2 A_1^{-1} \cos\theta_1 (Y_3 e^{\lambda_l t} + Y_4 e^{\lambda_2 t}) \tag{k.2}$$

$$g_3 = -\frac{i\rho\omega_1}{2(1-\zeta_2^2)^{1/2}} e^{-\lambda_1 t} A_1 \cos\theta_1 \tag{k.3}$$

$$g_4 = \frac{i\rho\omega_1}{2(1-\zeta_2^2)^{1/2}} e^{-\lambda_2 t} A_1 \cos\theta_1 \tag{k.4}$$

If the correlation time of $\xi(t)$ is short compared with $(\varepsilon\zeta_1\omega_1)^{-1}$, then the two-dimensional vector (A_1, ϕ_1) may be approximated by a Markov vector, governed by the Itô equations

$$dA_1 = m_1\,dt + \sigma_{11}\,dB_1(t) + \sigma_{12}\,dB_2(t) \tag{l.1}$$

$$d\phi_1 = m_2\,dt + \sigma_{21}\,dB_1(t) + \sigma_{22}\,dB_2(t) \tag{l.2}$$

where

$$m_1 = -\varepsilon\zeta_1\omega_1 A_1 + \varepsilon\left\langle \int_{-\infty}^{0} \left[\frac{\partial g_1(t)}{\partial A_1} g_1(t+\tau) + \frac{\partial g_1(t)}{\partial \phi_1} g_2(t+\tau) + \frac{\partial g_1(t)}{\partial Y_3} g_3(t+\tau) + \frac{\partial g_1(t)}{\partial Y_4} g_4(t+\tau) \right] R_{\xi\xi}(\tau)\,d\tau \right\rangle_t \tag{m.1}$$

$$m_2 = \varepsilon\left\langle \int_{-\infty}^{0} \left[\frac{\partial g_2(t)}{\partial A_1} g_1(t+\tau) + \frac{\partial g_2(t)}{\partial \phi_1} g_2(t+\tau) + \frac{\partial g_2(t)}{\partial Y_3} g_3(t+\tau) + \frac{\partial g_2(t)}{\partial Y_4} g_4(t+\tau) \right] R_{\xi\xi}(\tau)\,d\tau \right\rangle_t \tag{m.2}$$

$$(\boldsymbol{\sigma\sigma}')_{11} = \varepsilon\left\langle\int_{-\infty}^{\infty} g_1(t)g_1(t+\tau)R_{\xi\xi}(\tau)\,d\tau\right\rangle_t \tag{n.1}$$

$$(\boldsymbol{\sigma\sigma}')_{12} = (\boldsymbol{\sigma\sigma}')_{21} = \varepsilon\left\langle\int_{-\infty}^{\infty} g_1(t)g_2(t+\tau)R_{\xi\xi}(\tau)\,d\tau\right\rangle_t \tag{n.2}$$

$$(\boldsymbol{\sigma\sigma}')_{22} = \varepsilon\left\langle\int_{-\infty}^{\infty} g_2(t)g_2(t+\tau)R_{\xi\xi}(\tau)\,d\tau\right\rangle_t \tag{n.3}$$

Carrying out the operations indicated in equations (m.1), (n.1), and (n.2), we obtain

$$m_1 = \varepsilon\left[-\zeta_1\omega_1 + \frac{\pi\rho^2\omega_1\omega_2}{4(1-\zeta_2^2)^{1/2}}(\Phi_d^+ - \Phi_d^-)\right]A_1 = -\varepsilon\gamma A_1 \tag{o.1}$$

$$(\boldsymbol{\sigma\sigma}')_{11} = (\boldsymbol{\sigma\sigma}')_{12} = (\boldsymbol{\sigma\sigma}')_{21} = 0 \tag{o.2}$$

where

$$\Phi_d^+ = \frac{1}{\pi}\int_0^\infty e^{-\zeta_2\omega_2\tau}\cos[(\omega_1+\omega_{2d})\tau]R_{\xi\xi}(\tau)\,dt \tag{p.1}$$

$$\Phi_d^- = \frac{1}{\pi}\int_0^\infty e^{-\zeta_2\omega_2\tau}\cos[(\omega_1-\omega_{2d})\tau]R_{\xi\xi}(\tau)\,dt \tag{p.2}$$

Therefore, the $A_1(t)$ process is itself governed by

$$dA_1 = -\varepsilon\gamma A_1 dt \tag{q}$$

The equation for the nth moment of $A_1(t)$ is obtained readily as

$$\frac{d}{dt}E[A_1^n] = -\varepsilon n\gamma E[A_1^n] \tag{r}$$

The stability condition for the nth moment is clearly $\gamma > 0$, or

$$\zeta_1 > \frac{\pi\rho^2\omega_2}{4(1-\zeta_2^2)^{1/2}}(\Phi_d^+ - \Phi_d^-) \tag{s}$$

which is independent of n.

It is interesting to note that since the diffusion term in equation (q) is zero, the Fokker-Planck equation associated with the averaged $A_1(t)$ process reduces to the so-called *Liouville equation*

$$\frac{\partial p}{\partial t} + \frac{\partial}{\partial\alpha}(\varepsilon\gamma\alpha p) = 0 \tag{t}$$

where the unknown $p(\alpha, t)$ is the nonstationary probability density of $A_1(t)$. Solution to equation (t), subject to the initial condition

$$p(\alpha, 0) = p_0(\alpha) \tag{u}$$

is given by (e.g., see Soong, 1973)

$$p(\alpha, t) = p_0(e^{\gamma t}\alpha)e^{\gamma t} \tag{v}$$

If $\gamma > 0$, then the right-hand side of (v) tends to $\delta(\alpha)$ as t increases. Therefore, inequality (s) is also the almost-sure stability condition. In passing, we note that the Liouville equation is a useful analytical tool for treating random initial condition problems in general. Another example is given in Section 9.1.

The absence of a diffusion term in the averaged equation (q) indicates that the averaged $A_1(t)$ process is "quasi-deterministic" and that its randomness arises from the possible random initial condition. It is therefore not surprising that the same condition is found for the almost-sure stability and the moment stability of any order. As shown in equation (a.1), the random excitation $\xi(t)$ affects the Q_1 motion by way of the Q_2 motion, which is quickly damped out. If a multiplicative random excitation term associated with Q_1 were to be introduced in equation (a.1), it would result in a diffusion term in equation (q).

The scope of engineering problems treatable with the stochastic averaging procedure is greatly expanded by adding the modified version of the procedure. It has been applied by Bucher and Lin (1988) in the investigation of moment stability of long-span bridges in turbulent wind. The case of sample stability is discussed in the following section.

6.6 MOTION STABILITY OF LONG-SPAN BRIDGES IN TURBULENT FLOW

The response of a long-span bridge to turbulent wind excitations was analyzed in Section 2.3.4, without considering the possibility of motion instability. In particular, the random fluctuation in the horizontal wind velocity was ignored, so that the bridge could be modeled as a time-invariant dynamical system, subjected only to the additive buffeting loads, and the spectral analysis became applicable. Such an analysis is valid only if both the mean horizontal wind velocity and the horizontal turbulence intensity are sufficiently low that the bridge motion is far from being unstable. In the present section, our attention is focused instead on those cases in which motion stability is a major concern.

The stochastic stability theory for bridges, to be discussed in this section, is a randomized version of a linear deterministic theory, due essentially to Scanlan and his co-workers (e.g., Scanlan, Beliveau, and Budlong, 1974), which is now reviewed briefly. For simplicity, we assume that the bridge in question is dominated by one torsional mode. Ignoring the presence of turbulence in the wind, the governing equation may be written as

$$\begin{aligned} &I(\ddot{\alpha} + 2\zeta_\alpha \omega_\alpha \dot{\alpha} + \omega_\alpha^2 \alpha) \\ &\quad = \frac{1}{2}\rho(2B^2)\left\{ u^2\left[c_1\alpha + c_2\left(\frac{B}{u}\right)\dot{\alpha}\right] + \int_{-\infty}^{t} u^2[X_{M\alpha}(t-\tau) - 1]\frac{\partial \alpha(\tau)}{\partial \tau}d\tau \right\} \end{aligned} \tag{6.6.1}$$

where α = modal torsional angle, I = modal moment of inertia, ω_α = natural frequency, ζ_α = ratio of structural damping to critical damping, ρ = air density, B = width of bridge deck, u = horizontal wind velocity, which is treated as a constant in the deterministic theory, $X_{M\alpha}(t)$ is known as the aerodynamic indicial function, and c_1 and c_2 are aerodynamic constants. In the case of an unstreamlined bridge cross section, $X_{M\alpha}(t)$, c_1, and c_2 must be determined experimentally at the present time. Equation (6.6.1) may be obtained from the governing equation in Section 2.3.4 by discarding the buffeting load term and representing the torsional displacement by a single mode. Note that the derivative of the indicial function $X_{M\alpha}(t)$ is equal to function $h_s(t)$ in equation (d) of Section 2.3.4. Equation (6.6.1) is cast in a form suggested by Lin and Ariaratnam (1980) for ease of extension to the stochastic theory.

It has been found (Beliveau, Vaicaitis, and Shinozuka, 1977; Scanlan, Beliveau, and Budlong, 1974) that the experimentally obtained data for the indicial function $X_{M\alpha}$ can be fitted in the form

$$X_{M\alpha}(t) = 1 + c_3 \exp\left[-d_3\left(\frac{u}{B}\right)t\right] + c_4 \exp\left[-d_4\left(\frac{u}{B}\right)t\right] \tag{6.6.2}$$

in which c_3, c_4, d_3, and d_4 are constants. Letting $y_1 = \alpha$, $y_2 = \dot{\alpha}$, and

$$y_3 = \frac{\rho u^2 B^2 c_3}{I} \int_{-\infty}^{t} \exp\left[-d_3\left(\frac{u}{B}\right)(t-\tau)\right] \frac{\partial \alpha(\tau)}{\partial \tau} d\tau \tag{6.6.3}$$

$$y_4 = \frac{\rho u^2 B^2 c_4}{I} \int_{-\infty}^{t} \exp\left[-d_4\left(\frac{u}{B}\right)(t-\tau)\right] \frac{\partial \alpha(\tau)}{\partial \tau} d\tau \tag{6.6.4}$$

equations (6.6.1) through (6.6.4) may be replaced by the following matrix equation:

$$\frac{d}{dt}\boldsymbol{y} = \boldsymbol{A}\boldsymbol{y} \tag{6.6.5}$$

where $\boldsymbol{y} = (y_1, y_2, y_3, y_4)'$ and

$$\boldsymbol{A} = \begin{bmatrix} 0 & 1 & 0 & 0 \\ -\left(\omega_\alpha^2 - \dfrac{\rho u^2 B^2 c_1}{I}\right) & -\left(2\zeta_\alpha \omega_\alpha - \dfrac{\rho u^2 B^2 c_2}{I}\right) & 1 & 1 \\ 0 & \dfrac{\rho u^2 B^2 c_3}{I} & -\dfrac{d_3 u}{B} & 0 \\ 0 & \dfrac{\rho u^2 B^2 c_4}{I} & 0 & -\dfrac{d_4 u}{B} \end{bmatrix} \tag{6.6.6}$$

The eigenvalues of the system are obtained from the characteristic equation

$$\left|\boldsymbol{A} - \lambda \boldsymbol{I}\right| = 0 \tag{6.6.7}$$

and they are dependent on the flow velocity u. If u is smaller than a critical velocity u_c, then the eigenvalues are either (1) two complex-conjugate pairs, $\lambda_{1,2} = -\zeta_1 \pm i\omega_1$ and $\lambda_{3,4} = -\zeta_3 \pm i\omega_3$, or (2) one complex-conjugate pair, $\lambda_{1,2} = -\zeta_1 \pm i\omega_1$, and two real values, $\lambda_3 = -\zeta_3$ and $\lambda_4 = -\zeta_4$, where ζ_1, ζ_3, ζ_4, ω_1, and ω_3 are positive constants. In what follows, we assume that $\omega_1 \neq 0$ and $\zeta_1 < \zeta_3 < \zeta_4$, so that the impending instability is of an oscillatory type.

The critical velocity u_c is the smallest u at which a pair of complex-conjugate eigenvalues becomes purely imaginary: $\zeta_1 = 0$. Physically speaking, $u = u_c$ is the flow velocity, enabling the combined structure-fluid system to execute an undamped sinusoidal motion. The "mode shape" of this coupled structure-fluid motion is described by the eigenvector associated with either $i\omega_1$ or $-i\omega_1$ (the two possible choices are associated with a complex-conjugate pair of eigenvectors). Mathematically, other critical u values greater than u_c may exist, corresponding to other unstable modes. For the system described by equations (6.6.5) and (6.6.6), these additional mathematical solutions include either one oscillatory mode or two divergence modes, but they are of no practical concern.

The equation of motion (6.6.5) can now be "randomized" by replacing the constant flow velocity u by

$$U(t) = u[1 + F(t)] \tag{6.6.8}$$

where $F(t)$ is a random process with a zero mean and u is now the average flow velocity. Implicit in this replacement is the assumption that the physical laws governing the conversion of fluid flow to flow forces acting on the structure remain unchanged, which is a reasonable assumption if the length scale of the turbulent fluctuation $uF(t)$ is much larger than the dimensions of the bridge cross section. There has been some experimental evidence (Jancauskas and Melbourne, 1985) to suggest that the flow pattern around a bridge cross section may change due to the presence of small eddies; however, this effect cannot be accounted for theoretically at the present time.

Since the root-mean-square value of the turbulence rarely exceeds 30 percent of the mean velocity, it is reasonable to approximate

$$U^2(t) \approx u^2[1 + 2F(t)] \tag{6.6.9}$$

Thus the randomized version of equation (6.6.5) reads

$$\frac{d\boldsymbol{Y}}{dt} = \boldsymbol{AY} + \boldsymbol{DY}F(t) \tag{6.6.10}$$

where $\boldsymbol{Y} = (Y_1, Y_2, Y_3, Y_4)'$ and, in addition to the previously defined matrix $\boldsymbol{A}$,

$$\boldsymbol{D} = \begin{bmatrix} 0 & 0 & 0 & 0 \\ \dfrac{\rho u^2 B^2 c_1}{I} & \dfrac{\rho u^2 B^2 c_2}{I} & 0 & 0 \\ 0 & \dfrac{\rho u^2 B^2 c_3}{I} & -\dfrac{d_3 u}{B} & 0 \\ 0 & \dfrac{\rho u^2 B^2 c_4}{I} & 0 & -\dfrac{d_4 u}{B} \end{bmatrix} \tag{6.6.11}$$

It is implied in equations (6.6.9) and (6.6.10) that the average wind velocity u is now used in the scaling of parameters in equations (6.6.1) and (6.6.2).

To proceed, we let $\boldsymbol{Y} = \boldsymbol{TZ}$ where the matrix of transformation $\boldsymbol{T}$ is constructed from the eigenvectors of $\boldsymbol{A}$. Specifically, if the eigenvalues consist of a complex-conjugate pair $\lambda_{1,2} = -\zeta_1 \pm i\omega_1$ and two real values $\lambda_3 = -\zeta_3$ and $\lambda_4 = -\zeta_4$, and if the associated eigenvectors are $\boldsymbol{V}_{1R} \pm i\boldsymbol{V}_{1I}$, $\boldsymbol{V}_3$, $\boldsymbol{V}_4$, respectively, we choose

$$\boldsymbol{T} = [\boldsymbol{V}_{1R} \quad \boldsymbol{V}_{1I} \quad \boldsymbol{V}_3 \quad \boldsymbol{V}_4] \tag{6.6.12}$$

The transformation yields

$$\frac{d\boldsymbol{Z}}{dt} = \boldsymbol{QZ} + \boldsymbol{RZ}F(t) \tag{6.6.13}$$

where

$$\boldsymbol{Q} = \boldsymbol{T}^{-1}\boldsymbol{AT} = \begin{bmatrix} -\zeta_1 & \omega_1 & 0 & 0 \\ -\omega_1 & -\zeta_1 & 0 & 0 \\ 0 & 0 & -\zeta_3 & 0 \\ 0 & 0 & 0 & -\zeta_4 \end{bmatrix} \tag{6.6.14}$$

$$\boldsymbol{R} = \boldsymbol{T}^{-1}\boldsymbol{D}\boldsymbol{T} = \begin{bmatrix} R_{11} & R_{12} & R_{13} & R_{14} \\ R_{21} & R_{22} & R_{23} & R_{24} \\ R_{31} & R_{32} & R_{33} & R_{34} \\ R_{41} & R_{42} & R_{43} & R_{44} \end{bmatrix} \tag{6.6.15}$$

If the eigenvalues of $\boldsymbol{A}$ consist of two complex-conjugate pairs, $\lambda_{1,2} = -\zeta_1 \pm i\omega_1$ and $\lambda_{3,4} = -\zeta_3 \pm i\omega_3$, associated with the eigenvectors $\boldsymbol{V}_{1R} \pm i\boldsymbol{V}_{1I}$ and $\boldsymbol{V}_{3R} \pm i\boldsymbol{V}_{3I}$, then the matrix of transformation should be $\boldsymbol{T} = [\boldsymbol{V}_{1R} \quad \boldsymbol{V}_{1I} \quad \boldsymbol{V}_{3R} \quad \boldsymbol{V}_{3I}]$. However, this second case will not be discussed separately, since it is quite similar.

In passing, we note that through the transformation, the state space is divided into two subspaces as shown in (6.6.14): a critical subspace represented by Z_1 and Z_2, in which possible instability may occur, and a stable subspace, represented by Z_3 and Z_4. Moreover, the transformation is based on matrix $\boldsymbol{A}$ from the deterministic formulation.

The random process $F(t)$, introduced in equation (6.6.8), represents the turbulent fluctuation in the wind flow, normalized and nondimensionalized with respect to the average wind velocity u. Typical strong wind records suggest that within a time span of the order of 1 hour, $F(t)$ may be treated as a correlation-stationary stochastic process, with a magnitude smaller than 1 and a smooth spectral density. A versatile mathematical model possessing these properties may be expressed as

$$F(t) = \varepsilon \sin[\nu t + \Theta(t)] \tag{6.6.16a}$$

$$\Theta(t) = \sigma B(t) + \chi \tag{6.6.16b}$$

where ε, ν, σ are positive constants, $\varepsilon < 1$, χ is a random variable uniformly distributed in $[0, 2\pi)$, and $B(t)$ is a unit Wiener process. Such a random process model has been used by Dimentberg (1988) and Wedig (1989) in the investigations of Mathieu-type stochastic systems, without a random phase χ shown in equation (6.6.16b). The inclusion of a uniformly distributed random phase renders the process correlation-stationary at all times.

The correlation function of $F(t)$ is given by

$$\begin{aligned} E[F(t_1)F(t_2)] &= \frac{\varepsilon^2}{2} E\{\cos[\nu(t_2 - t_1) + \sigma B(t_2) - \sigma B(t_1)] \\ &\quad - \cos[\nu(t_2 + t_1) + \sigma B(t_2) + \sigma B(t_1) + 2\chi]\} \end{aligned} \tag{6.6.17}$$

The ensemble average of the second cosine term in equation (6.6.17) is zero, since χ is uniformly distributed. The remaining averaging may be carried out by noticing that $B(t_2) - B(t_1)$ is a gaussian random variable with a zero mean and a standard deviation $\sqrt{|t_2 - t_1|}$. Then a simple calculation yields

$$E[F(t_1)F(t_2)] = R_{FF}(t_1 - t_2) = \frac{\varepsilon^2}{2} \cos \nu(t_1 - t_2) \exp\left(-\frac{\sigma^2}{2} |t_1 - t_2|\right) \tag{6.6.18}$$

which is an even function of $\tau = t_1 - t_2$, as expected. The two-sided spectral density of $F(t)$ is given by

$$\Phi_{FF}(\omega) = \frac{1}{\pi} \int_0^\infty R_{FF}(\tau) \cos \omega\tau \, d\tau = \frac{(\varepsilon\sigma)^2}{2\pi} \left[\frac{1}{4(\omega - \nu)^2 + \sigma^4} + \frac{1}{4(\omega + \nu)^2 + \sigma^4}\right] \tag{6.6.19}$$

The following comments about this turbulence model are in order:

1. The normalized turbulence intensity, represented by its root-mean-square value, is fixed at $\varepsilon/\sqrt{2}$ regardless of the values of ν and σ.
2. The shape of the spectral density is determined by the ν and σ values. In general, it has two symmetrical peaks, symmetrically located in the positive and the negative frequency domains; their bandwidth depends on σ, and their locations depend largely on ν and, to a lesser degree, also on σ. The two peaks merge into one at $\omega = 0$, when $\nu/\sigma^2 \ll 1$. When $\nu/\sigma^2 \gg 1$, $F(t)$ becomes a narrow-band random process.
3. In the limit as σ approaches infinity, the random process becomes a "white noise" of constant spectral density. However, since the mean-square value must still be $\varepsilon^2/2$, this constant spectral level reduces to zero in the limit.

Hypothetical spectral densities of $F(t)$ corresponding to several pairs of ν and σ values are plotted in Fig. 6.6.1, while the mean-square value—the area under every spectral curve—is kept the same. One of the hypothetical spectral curves is compared in Fig. 6.6.2 with a Dryden spectrum (Dryden, 1961) having the same mean-square value and the same height of spectral peaks at the same locations. The Dryden spectrum is obtainable under the assumption of homogeneous and isotropic turbulence. In Fig. 6.6.3, another hypothetical spectral curve is compared with a von Karman spectrum (von Karman, 1948) for longitudinal turbulence, again by matching the mean-square value and the height of spectral peak. The peak of the von Karman spectrum is located at $\omega = 0$, in accord with the energy-cascade hypothesis of turbulence. Other well-known spectra in wind engineering (e.g., see Simiu and Scanlan, 1986) are similar. However, in some recent experiments (Su and Lian, 1992), it was found that not all turbulence vortices broke down as implied in the cascade hypothesis, but some would grow in size by merging with one another when

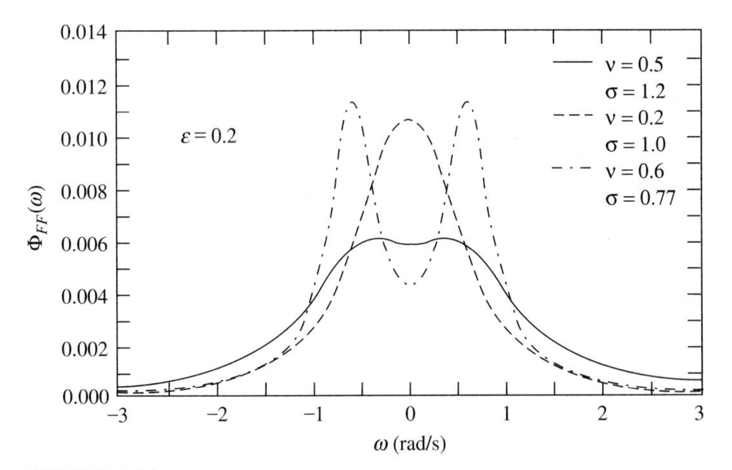

FIGURE 6.6.1
Possible spectral shapes of proposed wind turbulence model [after Li, 1993].

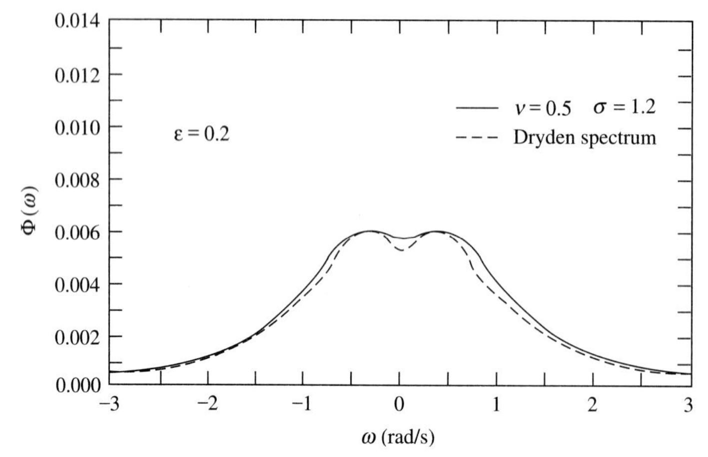

FIGURE 6.6.2
Comparison of spectral density of turbulence model and targeted Dryden spectrum [after Li, 1993].

being transported at slower speeds. It is therefore reasonable to assume that under certain topographical and meteorological conditions, the initial instability in an atmospheric boundary layer may be dominated by eddies of a certain size, and that these eddies can either break down or merge, depending on the speeds at which they are convected by the surrounding flow. They tend to break down at higher convection speeds and merge at lower convection speeds. In a state of statistical equilibrium, the entire eddy population has a smooth distribution of sizes, and the peak frequency of the spectral density corresponds to the dominant eddy size in the population. The present turbulence model, given in equations (6.6.16a) and (6.6.16b), captures these

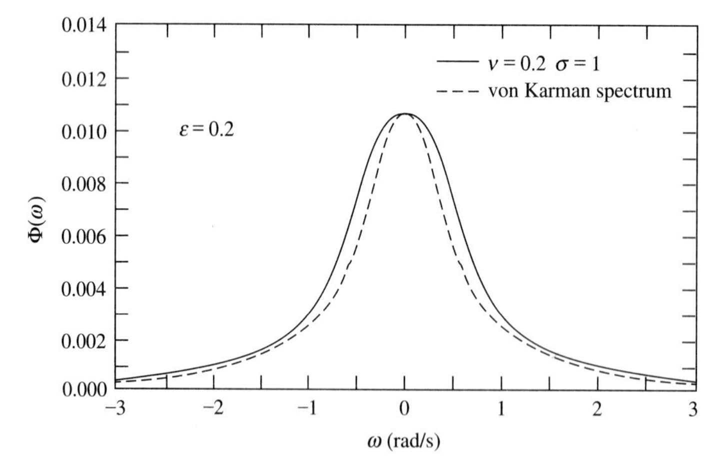

FIGURE 6.6.3
Comparsion of spectral density of turbulence model and targeted von Karman spectrum [after Li, 1993].

physical features. In addition, the level of turbulence, or its rms value, can be preassigned independent of its spectral shape.

Return now to equation (6.6.13), and introduce the transformation

$$Z_1 = A(t)\cos\beta \qquad \beta = \Omega t + \phi(t) \tag{6.6.20a}$$

$$Z_2 = -A(t)\sin\beta \tag{6.6.20b}$$

where Ω is a positive constant to be chosen later. This transformation changes Z_1 and Z_2 to A and β but keeps the same critical subspace. It follows upon substituting (6.6.20a) and (6.6.20b) into equation (6.6.13),

$$\begin{aligned}\dot{A} &= [-\zeta_1 A\cos\beta - \omega_1 A\sin\beta + (R_{11}A\cos\beta - R_{12}A\sin\beta + R_{13}Z_3 \\ &\quad + R_{14}Z_4)F(t)]\cos\beta - [-\omega_1 A\cos\beta + \zeta_1 A\sin\beta + (R_{21}A\cos\beta \\ &\quad - R_{22}A\sin\beta + R_{23}Z_3 + R_{24}Z_4)F(t)]\sin\beta\end{aligned} \tag{6.6.21a}$$

$$\begin{aligned}\dot{\beta} &= -\frac{1}{A}\{[-\zeta_1 A\cos\beta - \omega_1 A\sin\beta + (R_{11}A\cos\beta - R_{12}A\sin\beta \\ &\quad + R_{13}Z_3 + R_{14}Z_4)F(t)]\sin\beta + [-\omega_1 A\cos\Theta + \zeta_1 A\sin\beta \\ &\quad + (R_{21}A\cos\beta - R_{22}A\sin\beta + R_{23}Z_3 + R_{24}Z_4)F(t)]\cos\beta\}\end{aligned} \tag{6.6.21b}$$

$$\dot{Z}_3 = -\zeta_3 Z_3 + [A(R_{31}\cos\beta - R_{32}\sin\beta) + R_{33}Z_3 + R_{34}Z_4]F(t) \tag{6.6.21c}$$

$$\dot{Z}_4 = -\zeta_4 Z_4 + [A(R_{41}\cos\beta - R_{42}\sin\beta) + R_{43}Z_3 + R_{44}Z_4]F(t) \tag{6.6.21d}$$

Since we are concerned with the asymptotic behavior of the system, it is convenient to remove the fast-decaying parts from Z_3 and Z_4. For this purpose, we replace Z_3 and Z_4 with two new variables $X_3 = Z_3\exp(\zeta_3 t)$ and $X_4 = Z_4\exp(\zeta_4 t)$, and we recast equations (6.6.21a) through (6.6.21d) as follows:

$$\begin{aligned}\dot{A} &= -\zeta_1 A + \varepsilon\left\{\frac{A}{2}[R_{11} + R_{22} + (R_{11} - R_{22})\cos(2\Omega t + 2\phi) - (R_{12} + R_{21})\right. \\ &\quad \times\sin(2\Omega t + 2\phi)] + [R_{13}\cos(\Omega t + \phi) - R_{23}\sin(\Omega t + \phi)]e^{-\zeta_3 t}X_3 \\ &\quad \left. + [R_{14}\cos(\Omega t + \phi) - R_{24}\sin(\Omega t + \phi)]e^{-\zeta_4 t}X_4\right\}\sin(\nu t + \Theta)\end{aligned} \tag{6.6.22a}$$

$$\begin{aligned}\dot{\phi} &= \omega_1 - \Omega + \varepsilon\left\{\frac{1}{2}[R_{12} - R_{21} - (R_{12} + R_{21})\cos(2\Omega t + 2\phi) + (R_{22} - R_{11})\right. \\ &\quad \times\sin(2\Omega t + 2\phi)] - \frac{1}{A}[R_{13}\sin(\Omega t + \phi) + R_{23}\cos(\Omega t + \phi)]e^{-\zeta_3 t}X_3 \\ &\quad \left. - \frac{1}{A}[R_{14}\sin(\Omega t + \phi) + R_{24}\cos(\Omega t + \phi)]e^{-\zeta_4 t}X_4\right\}\sin(\nu t + \Theta)\end{aligned} \tag{6.6.22b}$$

$$\begin{aligned}\dot{X}_3 &= \varepsilon\{A[R_{31}\cos(\Omega t + \phi) - R_{32}\sin(\Omega t + \phi)]e^{\zeta_3 t} \\ &\quad + R_{33}X_3 + R_{34}e^{(\zeta_3 - \zeta_4)t}X_4\}\sin(\nu t + \Theta)\end{aligned} \tag{6.6.22c}$$

$$\begin{aligned}\dot{X}_4 &= \varepsilon\{A[R_{41}\cos(\Omega t + \phi) - R_{42}\sin(\Omega t + \phi)]e^{\zeta_4 t} \\ &\quad + R_{43}e^{(\zeta_4 - \zeta_3)t}X_3 + R_{44}X_4\}\sin(\nu t + \Theta)\end{aligned} \tag{6.6.22d}$$

in which $F(t)$ has been replaced by the right-hand side of equation (6.6.16a) and β by $\Omega t + \phi(t)$. If the average wind velocity u is close to the deterministic flutter velocity u_c, then ζ_1 is small, and $A(t)$ is a slowly varying random process. Moreover, $\phi(t)$ fluctuates slowly about a constant rotation of $(\omega_1 - \Omega)t$. Therefore, equations (6.6.22a) and (6.6.22b) for $A(t)$ and $\phi(t)$ may be simplified by averaging. The effects of X_3 and X_4 on A and ϕ can be accounted for in the averaging process (Sri Namachchivaya and Lin, 1988; see also Section 6.5.4). However, in order to carry out the averaging procedure, a suitable Ω value must be selected. Three different cases must now be treated separately (Li, 1993).

NARROW-BAND TURBULENCE NEARLY TUNED TO $2\omega_1$. As seen in equation (6.6.19), the bandwidth of the turbulence model is controlled by the value of the parameter σ. For the present analysis, a turbulence is considered to be a narrow-band random process when $\sigma^2/\nu \ll 1$. In such a case, the peak location of the turbulence spectrum is largely determined by the value of ν. The turbulence is said to be tuned to the resonance frequency of the flutter mode if $\nu = 2\omega_1$. The extent of detuning may be measured by the difference $\Delta = \omega_1 - \nu/2$ between the deterministic flutter frequency and one-half the peak frequency of the turbulence spectral density.

For the case of near-tuning—that is, when $|1 - \nu/(2\omega_1)|$ is small—the proper choice for Ω is $\Omega = \nu/2$. The reason for this choice will be explained later. Then equations (6.6.22a) through (6.6.22d) may be rewritten in the form

$$\frac{d}{dt}\begin{pmatrix} A \\ \phi \\ X_3 \\ X_4 \end{pmatrix} = \begin{pmatrix} f_A \\ f_\phi \\ f_3 \\ f_4 \end{pmatrix} \tag{6.6.23}$$

where

$$\begin{aligned} f_A = {} & -\zeta_1 A + \varepsilon\left\{ \frac{A}{2}[R_{11} + R_{22} + (R_{11} - R_{22})\cos(\nu t + 2\phi) - (R_{12} + R_{21}) \right. \\ & \times \sin(\nu t + 2\phi)] + \left[R_{13}\cos\left(\frac{\nu}{2}t + \phi\right) - R_{23}\sin\left(\frac{\nu}{2}t + \phi\right)\right] e^{-\zeta_3 t} X_3 \\ & \left. + \left[R_{14}\cos\left(\frac{\nu}{2}t + \phi\right) - R_{24}\sin\left(\frac{\nu}{2}t + \phi\right)\right] e^{-\zeta_4 t} X_4 \right\} \sin(\nu t + \Theta) \end{aligned} \tag{6.6.24a}$$

$$\begin{aligned} f_\phi = {} & \omega_1 - \frac{\nu}{2} + \varepsilon\left\{ \frac{1}{2}[R_{12} - R_{21} - (R_{12} + R_{21})\cos(\nu t + 2\phi) + (R_{22} - R_{11}) \right. \\ & \times \sin(\nu t + 2\phi)] - \frac{1}{A}\left[R_{13}\sin\left(\frac{\nu}{2}t + \phi\right) + R_{23}\cos\left(\frac{\nu}{2}t + \phi\right)\right] e^{-\zeta_3 t} X_3 \\ & \left. - \frac{1}{A}\left[R_{14}\sin\left(\frac{\nu}{2}t + \phi\right) + R_{24}\cos\left(\frac{\nu}{2}t + \phi\right)\right] e^{-\zeta_4 t} X_4 \right\} \sin(\nu t + \Theta) \end{aligned} \tag{6.6.24b}$$

$$\begin{aligned} f_3 = {} & \varepsilon\left\{ A\left[R_{31}\cos\left(\frac{\nu}{2}t + \phi\right) - R_{32}\sin\left(\frac{\nu}{2}t + \phi\right)\right] e^{\zeta_3 t} \right. \\ & \left. + R_{33}X_3 + R_{34}e^{(\zeta_3 - \zeta_4)t} X_4 \right\} \sin(\nu t + \Theta) \end{aligned} \tag{6.6.24c}$$

$$f_4 = \varepsilon\left\{A\left[R_{41}\cos\left(\frac{\nu}{2}t+\phi\right) - R_{42}\sin\left(\frac{\nu}{2}t+\phi\right)\right]e^{\zeta_4 t} + R_{43}e^{(\zeta_4-\zeta_3)t}X_3 + R_{44}X_4\right\}\sin(\nu t+\Theta) \qquad (6.6.24d)$$

Applying the first-order approximation procedure of Bogoliubov and Mitropolsky (1961), we obtain a pair of averaged equations for $A(t)$ and $\phi(t)$, and we express in the Itô form as

$$dA = \langle f_A\rangle_t\, dt \qquad (6.6.25a)$$

$$d\phi = \langle f_\phi\rangle_t\, dt \qquad (6.6.25b)$$

where $< \cdot >_t$ indicates the time-averaging procedure; that is, for a general function $g(\boldsymbol{X}, t)$ of t and $\boldsymbol{X}$,

$$\langle g(\boldsymbol{X}, t)\rangle_t = \lim_{T\to\infty}\frac{1}{T}\int_0^T g(\boldsymbol{X}, t)\, dt \qquad (6.6.26)$$

where $\boldsymbol{X}$ is treated as being independent of time. Upon carrying out the averaging procedure in (6.6.25a) and (6.6.25b), this yields

$$dA = A[-\zeta_1 + k_1\sin(2\phi-\Theta) - k_2\cos(2\phi-\Theta)]\, dt \qquad (6.6.27a)$$

$$d\phi = [\Delta + k_2\sin(2\phi-\Theta) + k_1\cos(2\phi-\Theta)]\, dt \qquad (6.6.27b)$$

where $\Delta = \omega_1 - \nu/2$ is the detuning parameter and

$$k_1 = \frac{\varepsilon}{4}(R_{22} - R_{11}) \qquad (6.6.28a)$$

$$k_2 = \frac{\varepsilon}{4}(R_{12} + R_{21}) \qquad (6.6.28b)$$

We digress to note that the sinusoidal terms in equations (6.6.27a) and (6.6.27b) would disappear if a value other than $\nu/2$ were selected for Ω.

Introducing

$$\kappa = \sqrt{k_1^2 + k_2^2} = \frac{\varepsilon}{4}\sqrt{(R_{22}-R_{11})^2 + (R_{12}+R_{21})^2} \qquad (6.6.29a)$$

$$\cos q = -\frac{k_2}{\kappa} \qquad (6.6.29b)$$

$$\sin q = -\frac{k_1}{\kappa} \qquad (6.6.29c)$$

equations (6.6.27a) and (6.6.27b) can be further simplified to

$$dA = A[-\zeta_1 + \kappa\cos(2\phi - \Theta + q)]\, dt \qquad (6.6.30a)$$

$$d\phi = [\Delta - \kappa\sin(2\phi - \Theta + q)]\, dt \qquad (6.6.30b)$$

Finally, letting $\Psi = \phi - \Theta/2 + q/2$, we obtain

$$dA = A[-\zeta_1 + \kappa\cos(2\Psi)]\, dt \qquad (6.6.31a)$$

$$d\Psi = [\Delta - \kappa\sin 2\Psi]\, dt - \frac{\sigma}{2}\, dB(t) \qquad (6.6.31b)$$

in which we have used equation (6.6.16b) to obtain

$$d\Theta = \sigma\, dB(t) \tag{6.6.32}$$

It is interesting to note that equations (6.6.31a) and (6.6.31b) are obtained using the *deterministic* averaging procedure of Bogoliubov and Mitropolsky, without resorting to stochastic averaging. Moreover, Ψ is uncoupled from Λ in equation (6.6.31b); therefore, it is a one-dimensional Markov process. At the state of statistical stationarity, the probability density $p(\psi)$ of Ψ is governed by the following reduced Fokker-Planck equation:

$$\frac{d}{d\psi}[(\Delta - \kappa \sin 2\psi)p(\psi)] - \frac{1}{2}\frac{d^2}{d\psi^2}\left[\left(-\frac{\sigma}{2}\right)^2 p(\psi)\right] = 0 \tag{6.6.33}$$

Equation (6.6.33) is solved to yield

$$p(\psi) = C\int_0^{\pi} \exp[-\delta\mu + 2\gamma \sin\mu \sin(2\psi + \mu)]\, d\mu \tag{6.6.34}$$

where $\delta = 8\Delta/\sigma^2$, $\gamma = 4\kappa/\sigma^2$, and C is a normalization constant given by

$$C = \left\{2\int_0^{\pi}\int_0^{\pi} \exp[-\delta\mu + 2\gamma \sin\mu \sin(2\psi + \mu)]\, d\psi\, d\mu\right\}^{-1} \tag{6.6.35}$$

The probability density (6.6.34) is periodic with a period π; it was obtained earlier by Stratonovich (1967).

From (6.6.31a), the largest Lyapunov exponent λ_T of the system is obtained as

$$\lambda_T = E[-\zeta_1 + \kappa \cos 2\Psi] = -\zeta_1 + \kappa E[\cos 2\Psi] \tag{6.6.36}$$

In the special case of exact tuning, namely, $\Delta = 0$, equation (6.6.36) is simplified to

$$\lambda_T = -\zeta_1 + \kappa\frac{\mathrm{I}_1(\gamma)}{\mathrm{I}_0(\gamma)} \tag{6.6.37}$$

where I_1 and I_0 are the modified Bessel functions with imaginary arguments. However, an exact tuning should be rare for a wind-excited bridge.

NARROW-BAND DETUNED TURBULENCE. If the spectral peak of a narrow-band turbulence is not close to $2\omega_1$, where ω_1 is the deterministic flutter frequency, then the proper choice for the Ω value is $\Omega = \omega_1$ when carrying out the averaging procedure. The equations of motion can again be written in the form of (6.6.23),

$$\frac{d}{dt}\begin{pmatrix} A \\ \phi \\ X_3 \\ X_4 \end{pmatrix} = \begin{pmatrix} f_A \\ f_\phi \\ f_3 \\ f_4 \end{pmatrix} \tag{6.6.38}$$

where the right-hand sides are now given by

$$\begin{aligned} f_A = -\zeta_1 A + \varepsilon\Big\{&\frac{A}{2}[R_{11} + R_{22} + (R_{11} - R_{22})\cos(2\omega_1 t + 2\phi) - (R_{12} + R_{21}) \\ &\times \sin(2\omega_1 t + 2\phi)] + [R_{13}\cos(\omega_1 t + \phi) - R_{23}\sin(\omega_1 t + \phi)]e^{-\zeta_3 t}X_3 \\ &+ [R_{14}\cos(\omega_1 t + \phi) - R_{24}\sin(\omega_1 t + \phi)]e^{-\zeta_4 t}X_4\Big\}\sin(\nu t + \Theta) \end{aligned} \tag{6.6.39a}$$

$$f_\phi = \varepsilon\left\{\frac{1}{2}[R_{12} - R_{21} - (R_{12} + R_{21})\cos(2\omega_1 t + 2\phi)\right.$$
$$+ (R_{22} - R_{11})\sin(2\omega_1 t + 2\phi)] - \frac{1}{A}[R_{13}\sin(\omega_1 t + \phi)$$
$$+ R_{23}\cos(\omega_1 t + \phi)]e^{-\zeta_3 t}X_3 - \frac{1}{A}[R_{14}\sin(\omega_1 t + \phi)$$
$$\left. + R_{24}\cos(\omega_1 t + \phi)]e^{-\zeta_4 t}X_4\right\}\sin(\nu t + \Theta) \tag{6.6.39b}$$

$$f_3 = \varepsilon\{A[R_{31}\cos(\omega_1 t + \phi) - R_{32}\sin(\omega_1 t + \phi)]e^{\zeta_3 t}$$
$$+ R_{33}X_3 + R_{34}e^{(\zeta_3 - \zeta_4)t}X_4\}\sin(\nu t + \Theta) \tag{6.6.39c}$$

$$f_4 = \varepsilon\{A[R_{41}\cos(\omega_1 t + \phi) - R_{42}\sin(\omega_1 t + \phi)]e^{\zeta_4 t}$$
$$+ R_{43}e^{(\zeta_4 - \zeta_3)t}X_3 + R_{44}X_4\}\sin(\nu t + \Theta) \tag{6.6.39d}$$

In this case, however, terms of order ε vanish upon applying the first-order averaging procedure. It is then necessary to apply the second-order averaging procedure of Bogoliubov and Mitropolsky (1961), in order that the effects of turbulence can be revealed. The averaged $A(t)$ and $\phi(t)$ equations so obtained can again be expressed in the Itô form:

$$dA = \left\langle f_A + \tilde{f}_A\frac{\partial f_A}{\partial A} + \tilde{f}_\phi\frac{\partial f_A}{\partial \phi} + \tilde{f}_3\frac{\partial f_A}{\partial x_3} + \tilde{f}_4\frac{\partial f_A}{\partial x_4}\right\rangle_t dt \tag{6.6.40a}$$

$$d\phi = \left\langle f_\phi + \tilde{f}_A\frac{\partial f_\phi}{\partial A} + \tilde{f}_\phi\frac{\partial f_\phi}{\partial \phi} + \tilde{f}_3\frac{\partial f_\phi}{\partial x_3} + \tilde{f}_4\frac{\partial f_\phi}{\partial x_4}\right\rangle_t dt \tag{6.6.40b}$$

where $< \cdot >_t$ indicates the time-averaging procedure, defined in (6.6.26), and $\tilde{}$ denotes the following integration operator:

$$\tilde{g}(\boldsymbol{X}, t) = \int [g(\boldsymbol{X}, t) - \langle g(\boldsymbol{X}, t)\rangle_t]\,dt \tag{6.6.41}$$

In (6.6.41), $\boldsymbol{X}$ is again treated as being independent of time. Upon performing the second approximation procedure indicated in equations (6.6.40a) and (6.6.40b), we obtain

$$dA = -\zeta A\,dt \tag{6.6.42a}$$

$$d\phi = \eta\,dt \tag{6.6.42b}$$

where

$$\zeta = \zeta_1 - \frac{\varepsilon^2}{8}\sum_{j=3}^{4}\left[\frac{\zeta_j(R_{1j}R_{j1} + R_{2j}R_{j2}) + (\omega_1 + \nu)(R_{1j}R_{j2} - R_{2j}R_{j1})}{\zeta_j^2 + (\omega_1 + \nu)^2}\right.$$
$$\left. + \frac{\zeta_j(R_{1j}R_{j1} + R_{2j}R_{j2}) + (\omega_1 - \nu)(R_{1j}R_{j2} - R_{2j}R_{j1})}{\zeta_j^2 + (\omega_1 - \nu)^2}\right] \tag{6.6.43}$$

$$\eta = -\frac{\varepsilon^2 \omega_1}{4(4\omega_1^2 - \nu^2)} \left[(R_{22} - R_{11})^2 + (R_{12} + R_{21})^2 \right]$$
$$- \frac{\varepsilon^2}{8} \sum_{j=3}^{4} \left[\frac{\zeta_j (R_{2j} R_{j1} - R_{1j} R_{j2}) + (\omega_1 + \nu)(R_{1j} R_{j1} + R_{2j} R_{j2})}{\zeta_j^2 + (\omega_1 + \nu)^2} \right.$$
$$\left. + \frac{\zeta_j (R_{2j} R_{j1} - R_{1j} R_{j2}) + (\omega_1 - \nu)(R_{1j} R_{j1} + R_{2j} R_{j2})}{\zeta_j^2 + (\omega_1 - \nu)^2} \right] \tag{6.6.44}$$

It is seen that the averaged equations for $A(t)$ and $\phi(t)$ are decoupled from each other and are devoid of noise terms. Therefore, these equations are, in fact, deterministic. The largest Lyapunov exponent of the system is clearly ζ, as seen from equation (6.6.42a). The effects of a detuned narrow-band turbulence are accounted for in the summation terms on the right-hand side of equation (6.6.43). The bandwidth parameter σ has no effect on the averaged motion of the bridge as long as it is small.

WIDEBAND TURBULENCE. As seen in equation (6.6.18), the correlation time of the wind turbulence is of order σ^{-2}, whereas the relaxation time of the system is of order ζ_1^{-1}. The turbulence is considered to be a wideband random process if $\zeta_1^{-1} \gg \sigma^{-2}$. In this case, the stochastic averaging procedure is applicable, and ω_1 is the proper choice for the value of Ω. Return to equations (6.6.22a) through (6.6.22d). These equations can be recast in the form

$$\dot{X}_j = h_j + g_j F(t) \qquad j = 1, 2, 3, 4 \tag{6.6.45}$$

where $X_1 = A$, $X_2 = \phi$, and the turbulence component in the wind is denoted once again by the original symbol $F(t)$. The h_j and g_j functions are specifically

$$h_1 = -\zeta_1 A \tag{6.6.46a}$$

$$h_2 = h_3 = h_4 = 0 \tag{6.6.46b}$$

$$g_1 = \frac{A}{2}[R_{11} + R_{22} + (R_{11} - R_{22})\cos(2\omega_1 t + 2\phi) - (R_{12} + R_{21})\sin(2\omega_1 t + 2\phi)]$$
$$+ [R_{13}\cos(\omega_1 t + \phi) - R_{23}\sin(\omega_1 t + \phi)]e^{-\zeta_3 t} X_3$$
$$+ [R_{14}\cos(\omega_1 t + \phi) - R_{24}\sin(\omega_1 t + \phi)]e^{-\zeta_4 t} X_4 \tag{6.6.47a}$$

$$g_2 = \frac{1}{2}[R_{12} - R_{21} - (R_{12} + R_{21})\cos(2\omega_1 t + 2\phi) + (R_{22} - R_{11})\sin(2\omega_1 t + 2\phi)]$$
$$- \frac{1}{A}[R_{13}\sin(\omega_1 t + \phi) + R_{23}\cos(\omega_1 t + \phi)]e^{-\zeta_3 t} X_3$$
$$- \frac{1}{A}[R_{14}\sin(\omega_1 t + \phi) + R_{24}\cos(\omega_1 t + \phi)]e^{-\zeta_4 t} X_4 \tag{6.6.47b}$$

$$g_3 = A[R_{31}\cos(\omega_1 t + \phi) - R_{32}\sin(\omega_1 t + \phi)]e^{\zeta_3 t}$$
$$+ R_{33} X_3 + R_{34} e^{(\zeta_3 - \zeta_4)t} X_4 \tag{6.6.47c}$$

$$g_4 = A[R_{41}\cos(\omega_1 t + \phi) - R_{42}\sin(\omega_1 t + \phi)]e^{\zeta_4 t}$$
$$+ R_{43} e^{(\zeta_4 - \zeta_3)t} X_3 + R_{44} X_4 \tag{6.6.47d}$$

The averaged Itô equations for $A(t)$ and $\phi(t)$ are given by

$$\begin{Bmatrix} dA \\ d\phi \end{Bmatrix} = \begin{Bmatrix} m_1 \\ m_2 \end{Bmatrix} dt + \begin{bmatrix} \sigma_{11} & \sigma_{12} \\ \sigma_{21} & \sigma_{22} \end{bmatrix} \begin{Bmatrix} dB_1(t) \\ dB_2(t) \end{Bmatrix} \tag{6.6.48}$$

where $B_1(t)$ and $B_2(t)$ are independent unit Wiener processes, and where the drift and diffusion coefficients can be obtained in the same manner as shown in equations (m.1) through (n.3) in Section 6.5.4. After some algebra, we obtain

$$m_1 = \gamma_1 A \tag{6.6.49a}$$

$$m_2 = \gamma_2 \tag{6.6.49b}$$

$$(\sigma\sigma')_{11} = \gamma_3 A^2 \tag{6.6.50a}$$

$$(\sigma\sigma')_{12} = (\sigma\sigma')_{21} = \gamma_4 A \tag{5.6.50b}$$

$$(\sigma\sigma')_{22} = \gamma_5 \tag{6.6.50c}$$

where

$$\gamma_1 = -\zeta_1 + \frac{\pi}{8}\Bigg\{ 2(R_{11}+R_{22})^2\Phi(0) + 3[(R_{11}-R_{22})^2 + (R_{12}+R_{21})^2]\Phi(2\omega_1) + 4\sum_{j=3}^{4}[(R_{1j}R_{j1}+R_{2j}R_{j2})\Phi_j(\omega_1) + (R_{2j}R_{j1} - R_{1j}R_{j2})\Psi_j(\omega_1)]\Bigg\} \tag{6.6.51a}$$

$$\gamma_2 = \frac{\pi}{2}\Bigg\{ [(R_{11}-R_{22})^2 + (R_{12}+R_{21})^2]\Psi(2\omega_1) + \sum_{j=3}^{4}[(R_{1j}R_{j2} - R_{2j}R_{j1})\Phi_j(\omega_1) + (R_{1j}R_{j1} + R_{2j}R_{j2})\Psi_j(\omega_1)]\Bigg\} \tag{6.6.51b}$$

$$\gamma_3 = \frac{\pi}{4}\{2(R_{11}+R_{22})^2\Phi(0) + [(R_{11}-R_{22})^2 + (R_{12}+R_{21})^2]\Phi(2\omega_1)\} \tag{6.6.51c}$$

$$\gamma_4 = \frac{\pi}{2}(R_{11}+R_{22})(R_{12}-R_{21})\Phi(0) \tag{6.6.51d}$$

$$\gamma_5 = \frac{\pi}{4}\{2(R_{12}-R_{21})^2\Phi(0) + [(R_{11}-R_{22})^2 + (R_{12}+R_{21})^2]\Phi(2\omega_1)\} \tag{6.6.51e}$$

and where

$$\Phi(\omega) = \frac{1}{\pi}\int_{-\infty}^{0} \cos\omega\tau\, R_{FF}(\tau)\, d\tau = \frac{(\varepsilon\sigma)^2}{2\pi}\left[\frac{1}{4(\omega-\nu)^2+\sigma^4} + \frac{1}{4(\omega+\nu)^2+\sigma^4}\right] \tag{6.6.52a}$$

$$\Psi(\omega) = \frac{1}{\pi}\int_{-\infty}^{0} \sin\omega\tau\, R_{FF}(\tau)\, d\tau = \frac{\varepsilon^2}{\pi}\left[\frac{\omega-\nu}{4(\omega-\nu)^2+\sigma^4} + \frac{\omega+\nu}{4(\omega+\nu)^2+\sigma^4}\right] \tag{6.6.52b}$$

$$\begin{aligned}\Phi_j(\omega) &= \frac{1}{\pi}\int_{-\infty}^{0} \exp(-\zeta_j\tau)\cos\omega\tau\, R_{FF}(\tau)\, d\tau \\ &= \frac{\varepsilon^2(\sigma^2+2\zeta_j)}{2\pi}\left[\frac{1}{4(\omega-\nu)^2+(\sigma^2+2\zeta_j)^2} + \frac{1}{4(\omega+\nu)^2+(\sigma^2+2\zeta_j)^2}\right]\end{aligned} \tag{6.6.52c}$$

$$\Psi_j(\omega) = \frac{1}{\pi}\int_{-\infty}^{0} \exp(-\zeta_j\tau)\sin\omega\tau R_{FF}(\tau)\,d\tau$$

$$= \frac{\varepsilon^2}{\pi}\left[\frac{\omega-\nu}{4(\omega-\nu)^2+(\sigma^2+2\zeta_j)^2} + \frac{\omega+\nu}{4(\omega+\nu)^2+(\sigma^2+2\zeta_j)^2}\right] \qquad (6.6.52d)$$

It is interesting to note that the drift coefficients m_j and the diffusion coefficients σ_{jk}, obtained in equations (6.6.49a) through (6.6.50c), are independent of ϕ. Therefore, the joint probability density of the $A(t)$ and $\phi(t)$ processes, if it exists, must also be independent of ϕ.

Now the Itô equation for the $A(t)$ process is given by

$$dA = m_1\,dt + \sigma_{11}\,dB_1(t) + \sigma_{12}\,dB_2(t) \qquad (6.6.53)$$

It can be shown that equation (6.6.53) is equivalent to

$$dA = m_1\,dt + \sqrt{(\boldsymbol{\sigma}\boldsymbol{\sigma}')_{11}}\,dB(t) \qquad (6.6.54)$$

where $B(t)$ is another unit Wiener process, in the sense that both Itô equations correspond to the same Fokker-Planck equation. Substituting (6.6.49a) and (6.6.50a) into (6.6.54),

$$dA = \gamma_1 A\,dt + \sqrt{\gamma_3}\,AB(t) \qquad (6.6.55)$$

To determine the stability condition for this simple Markov process, we apply Itô's differential rule to $Y = \ln A$ to obtain

$$dY = \left(\gamma_1 - \tfrac{1}{2}\gamma_3\right)dt + \sqrt{\gamma_3}\,dB(t) \qquad (6.6.56)$$

It is then clear that the necessary and sufficient condition for the asymptotic sample stability is

$$\frac{2\gamma_1}{\gamma_3} < 1 \qquad (6.6.57)$$

Inequality (6.6.57) is equivalent to

$$-4\zeta_1 + \pi[(R_{11}-R_{22})^2 + (R_{12}+R_{21})^2]\Phi(2\omega_1)$$

$$+ 2\pi\sum_{j=3}^{4}[(R_{1j}R_{j1} + R_{2j}R_{j2})\Phi_j(\omega_1) + (R_{2j}R_{j1} - R_{1j}R_{j2})\Psi_j(\omega_1)] < 0 \qquad (6.6.58)$$

It is of interest to examine the limiting case as the bandwidth parameter σ of the turbulence tends to infinity. In the limit the turbulence model becomes a "white noise," but with a finite mean-square value. Thus functions $\Phi(\omega)$, $\Psi(\omega)$, $\Phi_j(\omega)$, and $\Psi_j(\omega)$, defined in equations (6.6.52a) through (6.6.52d), vanish, and inequality (6.6.58) reduces to

$$\zeta_1 > 0 \qquad (6.6.59)$$

the same as the one without turbulence. This is to be expected, since the spectral density of such a white noise vanishes at all frequencies.

Numerical computation has been carried out for the preceding three cases of turbulence, using the following structural and aerodynamic data:

$$\omega_\alpha = 0.74\pi \text{ rad/s} \qquad \zeta_\alpha = 0.01 \qquad B = 16.4 \text{ m} \qquad I = 8.95 \times 10^5 \text{ kg m}^2$$
$$\rho = 1.226 \text{ kg/m}^3 \qquad c_1 = 0.043 \qquad c_2 = -0.33 \qquad c_3 = 0.22$$
$$c_4 = 0.84 \qquad d_3 = 0.18 \qquad d_4 = 2.19$$

The structural data were taken from a paper by Scanlan and Tomko (1971), identified as bridge deck no. 2 in their Fig. 8, and the aerodynamic data were converted from the measured flutter derivatives available in the same paper, using a least-square difference procedure. The accuracy of the system identification procedure was checked by converting the foregoing aerodynamic data back to the flutter derivatives and comparing with the original experimental data, as shown in Fig. 6.6.4. The deterministic (without turbulence) critical velocity for this bridge model is found to be $u_c = 60.8$ m/s, and the corresponding frequency $\omega_1 = 2.23$ rad/s.

The computed stability boundaries are shown in Fig. 6.6.5 for three combinations of the ν and σ values. The vertical coordinate u/u_c is the ratio of the mean wind velocity and the deterministic critical velocity, and the horizontal coordinate $\varepsilon/\sqrt{2}$ is the root-mean-square value of the nondimensional turbulence velocity, known alternatively as the turbulence intensity in the literature. It is seen that the nearly tuned narrow-band turbulence ($\nu = 4.45, \sigma = 0.2$) has a destabilizing effect, in the sense that u/u_c decreases as ε increases, while the detuned narrow-band turbulence ($\nu = 1.40, \sigma = 0.2$) has a stabilizing effect. The wideband turbulence ($\nu = 0.20, \sigma = 10$) is stabilizing, but the effect is quite small. This last conclusion would be different if the spectral level of the wideband turbulence, instead of the

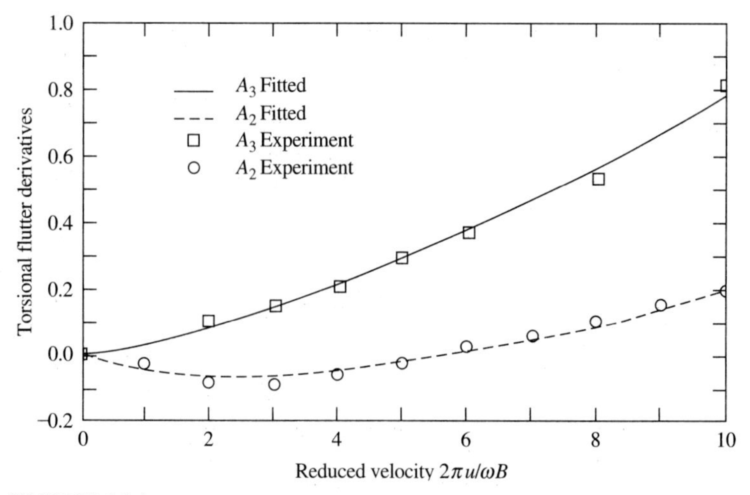

FIGURE 6.6.4
Torsional flutter derivative: curve fitting vs. experimental data of Scanlan and Tomko (1971).

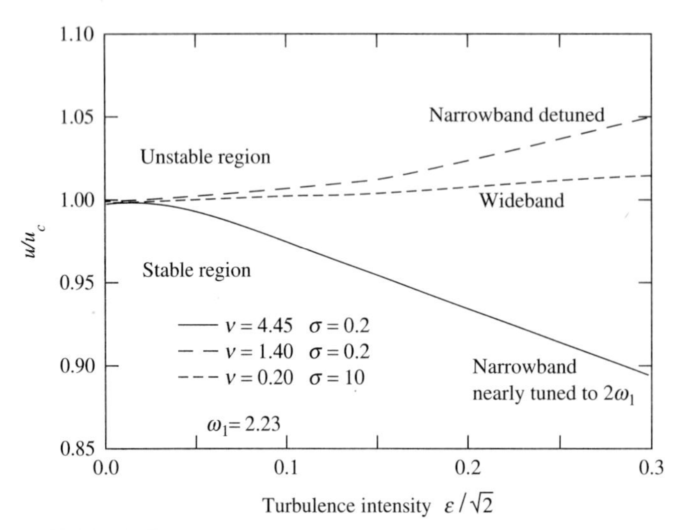

FIGURE 6.6.5
Effect of turbulence intensity on asymptotic sample stability boundary for torsional bridge motion in turbulent wind.

mean-square value, were to be kept constant, as was done in an earlier analysis (Lin and Ariaratnam, 1980).

Although the preceding numerical results were obtained only for a particular bridge model and for a single structural mode, some general trends can be observed. Without wind turbulence, the onset of instability is marked by a critical wind velocity, which is associated with an eigenvector describing a combined structural and fluid mode. This particular composition of structural and fluid components strikes a balance between the energy inflow from fluid to structure and the energy outflow from structure to fluid and structure damping, so that an undamped structural motion becomes sustainable. Two things take place at the introduction of turbulence. First, the same composition of the structural and fluid components is being changed continuously, permitting energy flow from the critical subspace to the stable subspace. Second, the wind velocity fluctuates, permitting frequent excursions to higher values than the mean velocity. The first effect is stabilizing, whereas the second effect is destabilizing. If turbulence is of a narrow band and is tuned to the resonance frequency $2\omega_1$ of the system, then the second effect dominates, since essentially the same composition of the structural and the fluid components as that of the deterministic critical mode is maintained. On the other hand, a detuned narrow-band turbulence has a greater influence on changing the composition; therefore, it can be stabilizing.

6.7 ASYMPTOTIC STABILITY OF NONLINEAR SYSTEMS

As shown in Sections 6.3 and 6.4, the necessary and sufficient condition for asymptotic sample stability of a multidimensional linear system under gaussian white noise excitations can be obtained by employing a general procedure due to Khasminskii.

This procedure is also applicable to a special type of nonlinear systems, for which the drift and diffusion coefficients of the response Markov vector are homogeneous (Zhang and Kozin, 1990), in the sense that $f(k\boldsymbol{x}) = kf(\boldsymbol{x})$, where $f(\cdot)$ is the drift or diffusion coefficient in question. Such a homogeneity requirement is met, for example, in the case of two coupled amplitude processes, described by equations (g.1) through (i.2) in Section 6.5.3, but it is quite restrictive for most other nonlinear systems.

However, if a system is one-dimensional, or reducible to one dimension, then the problem becomes much simpler. The reduction of dimensionality can often be accomplished by using the stochastic averaging procedure or quasi-conservative averaging procedure. For a one-dimensional diffusive process, the necessary and sufficient conditions for asymptotic stability in probability can be obtained by examining the sample behaviors at the boundaries. These boundaries are generally singular in the case of multiplicative excitation, and they can be identified in terms of their diffusion exponents, drift exponents, and character values, as described in Section 4.5. The trivial solution is asymptotically stable if and only if it is an exit or attractively natural, and the other boundary is an entrance or repulsively natural. This procedure has been applied to some nonlinear systems by Kozin and Sunahara (1987), Sri Namachchivaya (1989), and Zhang (1991).

For nonlinear systems, the stability conditions of the trivial solution in statistical moments cannot be determined from the moment equations, since such equations constitute an infinite hierarchy, and the application of a closure scheme is often unreliable for the stability analysis (Ariaratnam, 1980; Brückner and Lin, 1987). If a stationary probability density exists and is of a non-delta-type without an additive excitation, then all statistical moments of positive orders must be asymptotically unstable. However, if a non-delta-type stationary probability density does not exist, then the following alternative approach is applicable. First, the system is modified by adding an additive excitation in order to obtain a non-delta-type stationary probability density. Then, the condition for the boundedness of a nontrivial statistical moment of the modified system is determined. Finally, limit is taken by letting the additive excitation tend to zero. The validity of the alternative approach was demonstrated in Section 6.5.2 for a linear oscillator.

These procedures are illustrated in the following three examples.

6.7.1 A Nonlinearly Damped Oscillator

Consider a nonlinearly damped oscillator under a multiplicative random excitation, governed by

$$\ddot{X} + 2\zeta\omega_0\dot{X} + \eta|\dot{X}|^{\delta}\,\mathrm{sgn}\,\dot{X} + \omega_0^2[1 + W(t)]X = 0 \qquad \zeta > 0, \eta > 0, \delta \geq 0, \delta \neq 1 \tag{a}$$

where $W(t)$ is a gaussian white noise with a spectral density K. Letting

$$X(t) = A(t)\cos\theta \qquad \theta = \omega_0 t + \phi(t) \tag{b}$$

$$\dot{X}(t) = -\omega_0 A(t)\sin\theta \tag{c}$$

we obtain a set of differential equations for the amplitude process $A(t)$ and the phase process $\phi(t)$:

$$\dot{A} = -2\zeta A \sin^2\theta - \eta\omega_0^{\delta-1}|\sin\theta|^{\delta+1}A^{\delta} + \omega_0 A \sin\theta\cos\theta W(t) \tag{d}$$

$$\dot{\phi} = -2\zeta\omega_0 \sin\theta\cos\theta - \eta\omega_0^{\delta-1}|\sin\theta|^{\delta}\cos\theta\,\mathrm{sgn}(\sin\theta)A^{\delta-1} + \omega_0\cos^2\theta W(t) \tag{e}$$

We assume that both the linear and nonlinear damping terms and the white noise spectral density are small, so that the smoothed stochastic averaging procedure can be applied to obtain

$$dA = m(A)\,dt + \sigma(A)\,dB(t) \tag{f}$$

where

$$m(A) = \left(-\zeta\omega_0 + \frac{3\pi}{8}\omega_0^2 K\right)A - DA^{\delta} \tag{g}$$

$$\sigma(A) = \left(\frac{\pi}{4}\omega_0^2 K\right)^{1/2} A \tag{h}$$

and where D is a positive constant given by

$$D = \frac{2}{\pi}\eta\omega_0^{\delta-1}\int_0^{\pi/2}\sin^{\delta+1}\theta\,d\theta = \frac{\eta\omega_0^{\delta-1}\Gamma\left(\dfrac{\delta}{2}+1\right)}{\sqrt{\pi}\,\Gamma\left(\dfrac{\delta+3}{2}\right)} \tag{i}$$

Equation (f) for the amplitude process $A(t)$ is devoid of the phase process $\phi(t)$; therefore, $A(t)$ itself is a Markov process. It can be seen from equations (g) and (h) that the left boundary $a = 0$ is singular of the first kind, and the right boundary $a = \infty$ is singular of the second kind, where a is the state variable of A. The stability analysis, however, must be conducted separately for two cases, $0 \le \delta < 1$ and $\delta > 1$, as described next.

Case 1: $0 \le \delta < 1$. The left boundary at $a = 0$ is a shunt exit if $\delta = 0$, since $m(0) = -2\eta/\pi\omega_0 < 0$. It becomes a trap if $0 < \delta < 1$, and the diffusion and drift exponents are given, respectively, by

$$\alpha_l = 2 \qquad \beta_l = \delta \tag{j}$$

which can be obtained from equations (4.5.16) and (4.5.17). In this case, $\alpha_l > \beta_l + 1$ and $m(0^+) < 0$; therefore, the left boundary is also an exit, as shown in Table 4.5.2.

The diffusion exponent α_r, drift exponent β_r, and character value c_r at the right boundary at infinity can be obtained from equations (4.5.24), (4.5.25), and (4.5.27), yielding

$$\alpha_r = 2 \qquad \beta_r = 1 \qquad c_r = \frac{8\zeta - 3\pi\omega_0 K}{\pi\omega_0 K} \tag{k}$$

According to Table 4.5.4, the right boundary at infinity is attractively natural if $c_r < -1$, strictly natural if $c_r = -1$, and repulsively natural if $c_r > -1$.

The behaviors of sample functions near the two boundaries are represented schematically in Fig. 6.7.1. The left boundary is an exit regardless of the ζ value. The necessary and sufficient condition for the trivial solution to be asymptotically

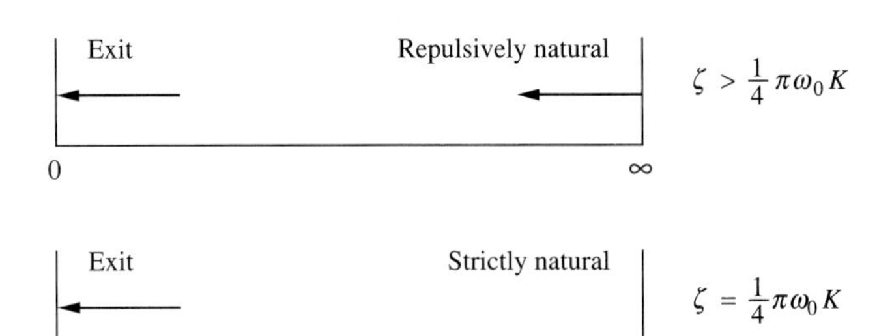

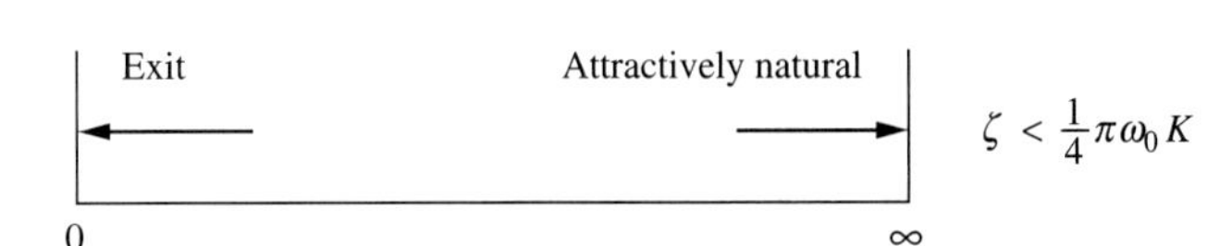

FIGURE 6.7.1
Boundary behaviors of sample functions of a nonlinear system (adding a nonlinear damping term to a linear oscillator, $0 \leq \delta < 1$).

$$\zeta > \tfrac{1}{4}\pi\omega_0 K \tag{l}$$

stable in probability is since the right boundary is then repulsively natural. Under condition (*l*), the stationary probability density is the delta function $\delta(a)$. A stationary probability density does not exist if $\zeta < \frac{1}{4}\pi\omega_0 K$, since the right boundary becomes attractively natural even though the left boundary remains an exit. It is of interest to note that at the bifurcation point $\zeta = \frac{1}{4}\pi\omega_0 K$, the right boundary is strictly natural, in which case a stationary probability density also does not exist, and the trivial solution $a = 0$ is not stable in probability in the sense that instability is not associated with a vanishing probability.

To investigate the asymptotic stability condition of the trivial solution in statistical moments, let equation (a) be modified by adding an additive excitation to the right-hand side to read

$$\ddot{X} + 2\zeta\omega_0\dot{X} + \eta|\dot{X}|^{\delta}\operatorname{sgn}\dot{X} + \omega_0^2[1 + W(t)]X = W_1(t) \tag{m}$$

where $W_1(t)$ is another gaussian white noise, with a spectral density K_1, and independent of $W(t)$. The averaged Itô differential equation for the amplitude process is now

$$dA = \left[\left(-\zeta\omega_0 + \frac{3\pi}{8}\omega_0^2 K\right)A - DA^{\delta} + \frac{\pi K_1}{2\omega_0^2 A}\right]dt + \left(\frac{\pi K_1}{\omega_0^2} + \frac{\pi}{4}\omega_0^2 K A^2\right)^{1/2} dB(t) \tag{n}$$

The reduced Fokker-Planck equation corresponding to equation (n) is

$$\frac{d}{da}\left\{\left[\left(-\zeta\omega_0 + \frac{3\pi}{8}\omega_0^2 K\right)a - Da^{\delta} + \frac{\pi K_1}{2\omega_0^2 a}\right]p\right\} - \frac{d^2}{da^2}\left[\left(\frac{\pi K_1}{2\omega_0^2} + \frac{\pi K}{8}\omega_0^2 a^2\right)p\right] = 0 \tag{o}$$

If a stationary probability density exists, then the probability flow must vanish in the one-dimensional state space:

$$\left[\left(-\zeta\omega_0 + \frac{3\pi}{8}\omega_0^2 K\right)a - Da^\delta + \frac{\pi K_1}{2\omega_0^2 a}\right]p - \frac{d}{da}\left[\left(\frac{\pi K_1}{2\omega_0^2 a} + \frac{\pi K}{8}\omega_0^2 a^2\right)p\right] = 0 \quad \text{(p)}$$

Equation (p) is solved to yield

$$p(a) = C\frac{1}{4\pi K_1 + \pi\omega_0^4 K a^2}\exp\left[-\int \frac{(8\zeta - 3\pi\omega_0 K)\omega_0^3 a + 8\omega_0^2 D a^\delta - 4\pi K_1/a}{4\pi K_1 + \pi\omega_0^4 K a^2}da\right] \quad \text{(q)}$$

The nth moment of $A(t)$ is then obtained as

$$\begin{aligned} E[A^n] &= \int_0^\infty a^n p(a)\,da \\ &= C\int_0^\infty \frac{a^n}{4\pi K_1 + \pi\omega_0^4 K a^2} \\ &\quad \times \exp\left[-\int \frac{(8\zeta - 3\pi\omega_0 K)\omega_0^3 a + 8\omega_0^2 D a^\delta - 4\pi K_1/a}{4\pi K_1 + \pi\omega_0^4 K a^2}da\right]da \end{aligned} \quad \text{(r)}$$

The limiting behavior of the integrand $a^n p(a)$ can be inferred from (r) as follows:

$$a^n p(a) \to \frac{C}{\pi\omega_0^4 K}a^{-(8\zeta/\pi\omega_0 K)+n+1} \quad \text{as} \quad a \to \infty \quad \text{(s)}$$

$$a^n p(a) \to \frac{C}{4\pi K}a^{n+1} \quad \text{as} \quad a \to 0 \quad \text{(t)}$$

The integrability condition for equation (r) is controlled by the limiting behavior of the integrand at the right boundary at infinity, leading to

$$\zeta > \frac{\pi}{8}(n+2)\omega_0 K \quad \text{(u)}$$

This is the same as inequality (i) in Section 6.5.2 when the multiplicative excitation on a linear system is a gaussian white noise. Since inequality (u) is a sufficient condition for inequality (l), the stationary probability density $p(a)$ in equation (r) tends to $\delta(a)$, and $E[A^n]$ vanishes as K_1 tends to zero. Therefore, the asymptotic stability condition of the trivial solution in statistical moments remains unchanged by adding a nonlinear damping term $\eta|\dot{X}|^\delta \operatorname{sgn}\dot{X}$ to a linear system, if $0 \le \delta < 1$.

Case 2: $\boldsymbol{\delta} > 1$. Return now to equations (f), (g), and (h). The left boundary at $a = 0$ is a trap. Its diffusion exponent α_l, drift exponent β_l, and character value c_l can be obtained from equations (4.5.16) through (4.5.18), resulting in

$$\alpha_l = 2 \qquad \beta_l = 1 \qquad c_l = \frac{-8\zeta + 3\pi\omega_0 K}{\pi\omega_0 K} \quad \text{(v)}$$

According to Table 4.5.2, the left boundary is attractively natural if $c_l < 1$, strictly natural if $c_l = 1$, and repulsively natural if $c_l > 1$.

For the right boundary at infinity, it can be shown that

$$\alpha_r = 2 \qquad \beta_r = \delta > 1 \qquad m(+\infty) < 0 \tag{w}$$

indicating an entrance boundary, according to Table 4.5.4.

These results are represented schematically in Fig. 6.7.2. If $\zeta > \frac{1}{4}\pi\omega_0 K$, then $a = 0$ is an attractively natural boundary. Since the right boundary is an entrance, regardless of the ζ value, the trivial solution is asymptotically stable in probability. However, a non-delta-type stationary probability density exists if $\zeta < \frac{1}{4}\pi\omega_0 K$, since both boundaries are now unreachable if a sample path begins from an interior point. This probability density is governed by the reduced Fokker-Planck equation

$$\frac{d}{da}\left\{\left[\left(-\zeta\omega_0 + \frac{3\pi}{8}\omega_0^2 K\right)a - Da^\delta\right]p\right\} - \frac{\pi}{8}\omega_0^2 K \frac{d^2}{da^2}(a^2 p) = 0 \tag{x}$$

which is solved to yield

$$p(a) = Ca^{-(8\zeta/\pi\omega_0 K)+1} \exp\left[-\frac{8D}{\pi(\delta-1)\omega_0^2 K}a^{\delta-1}\right] \qquad \delta > 1 \tag{y}$$

where

$$C = \frac{\delta-1}{\Gamma(\mu)}\left[\frac{8D}{\pi(\delta-1)\omega_0^2 K}\right]^\mu \tag{z}$$

$$\mu = \frac{2}{\delta-1}\left(-\frac{4\zeta}{\pi\omega_0 K} + 1\right) \tag{aa}$$

It can be shown that the condition for the right-hand side of equation (y) to be integrable is exactly $\zeta < \frac{1}{4}\pi\omega_0 K$. The existence of a non-delta-type stationary probability density implies that the trivial solution is unstable in probability.

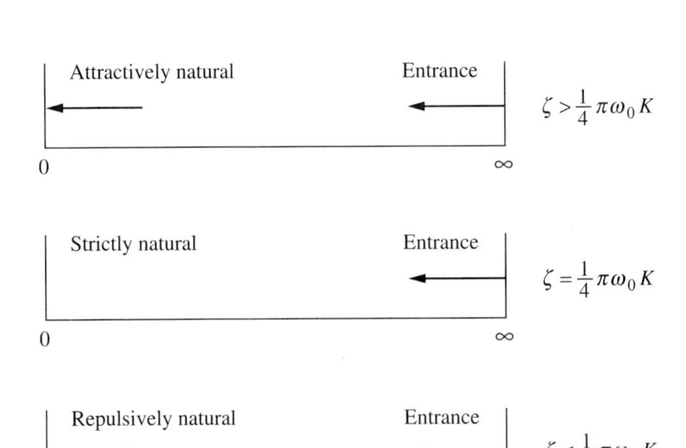

FIGURE 6.7.2
Boundary behaviors of sample functions of a nonlinear system (adding a nonlinear damping or stiffness term to a linear oscillator, $\delta > 1$).

At the bifurcation point $\zeta = \frac{1}{4}\pi\omega_0 K$, the left boundary is strictly natural where the sample behavior is indefinite. Again, the trivial solution is neither stable in probability nor unstable in probability. A stationary probability also does not exist.

Conditions for the asymptotic stability of the trivial solution in statistical moments can be investigated in this case without the aid of an additive excitation. The nth moment obtained from equation (y) is given by

$$E[A^n] = \int_0^\infty a^n p(a)\,da = \left[\frac{\pi(\delta-1)\omega_0^2 K}{8D}\right]^{n/(\delta-1)} \frac{\Gamma\left(\mu + \dfrac{n}{\delta-1}\right)}{\Gamma(\mu)} \qquad n > 0 \qquad \text{(bb)}$$

The condition for a nontrivial $E[A^n]$ is $\mu > 0$, or, equivalently, $\zeta < \frac{1}{4}\pi\omega_0 K$. As ζ approaches $\frac{1}{4}\pi\omega_0 K$,

$$\lim_{\zeta \to \frac{\pi}{4}\omega_0 K} E[A^n] = 0 \qquad n > 0 \qquad \text{(cc)}$$

independent of n. However, the same condition $\zeta > \frac{1}{4}\pi\omega_0 K$ as that of the asymptotic stability in probability is required for the asymptotic stability in statistical moments, since the former is a prerequisite of the latter.

Figure 6.7.3a shows the first-, second-, third-, and fourth-order moments of A, computed from equation (bb) with $\delta = 3$, $\eta = 0.1$, $K = 0.1$, and $\omega_0 = 1$. The computed second- and fourth-order moments are compared with the results obtained from Monte Carlo simulations in Fig. 6.7.3b. Both the original equation (a) and the averaged equation (f) have been used in the simulation. These results demonstrate that $\zeta = \frac{1}{4}\pi\omega_0 K = 0.0786$ is the same bifurcation point for all the moments, independent of order n.

In summary, the condition, $\zeta > \frac{1}{4}\pi\omega_0 K$ for asymptotic stability in probability remains unchanged upon adding a nonlinear damping term to a linear oscillator. The asymptotic stability condition in the nth moment is also unchanged if $0 \leq \delta < 1$, but it becomes independent of n if $\delta > 1$. In the latter case ($\delta > 1$), we obtain a single condition $\zeta > \frac{1}{4}\pi\omega_0 K$ for the trivial solution to be asymptotically stable both in probability and in statistical moments of all orders.

6.7.2 An Oscillator with Nonlinear Restoring Force

To investigate the effect of nonlinearity in the restoring force, consider the equation of motion

$$\ddot{X} + 2\zeta\omega_0\dot{X} + \omega_0^2[1 + W(t)]X + \eta\,|X|^\delta \operatorname{sgn} X = 0$$
$$\zeta > 0, \eta > 0, \delta \geq 0, \delta \neq 1 \qquad \text{(a)}$$

where the multiplicative random excitation is a gaussian white noise with a spectral density K. We assume that both damping and the spectral density of the excitation are sufficiently small and that the quasi-conservative averaging procedure described in Section 4.7 is applicable. The total energy of the oscillator is given by

$$\Lambda = \tfrac{1}{2}\dot{X}^2 + U(X) \qquad \text{(b)}$$

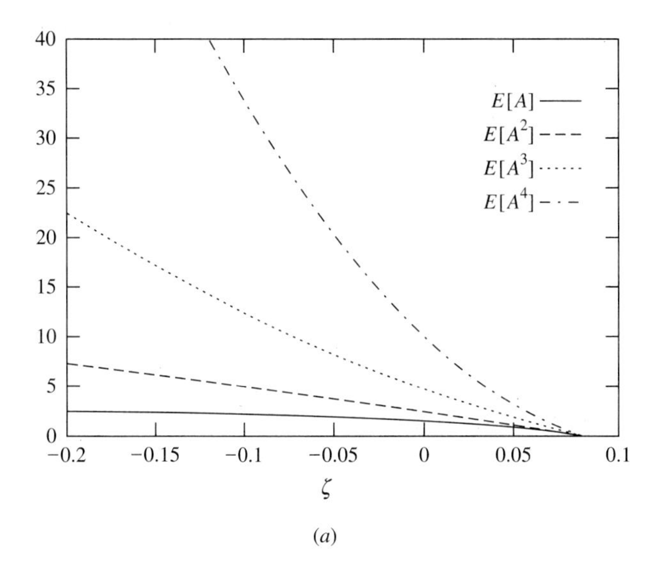

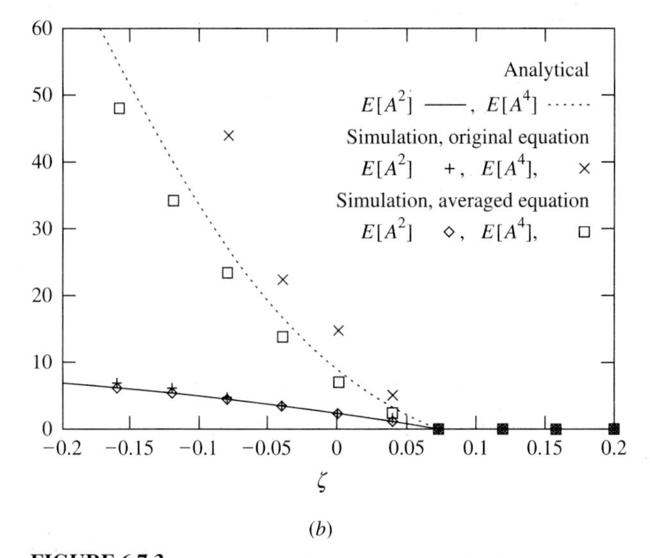

FIGURE 6.7.3
Statistical moments of amplitude A of a nonlinearly damped system, $\delta > 1$. (*a*) The first four moments, analytical results; (*b*) the second and fourth moments, analytical and simulation results.

where
$$U(X) = \frac{1}{2}\omega_0^2 X^2 + \frac{\eta}{\delta + 1}\left|X\right|^{\delta+1} \tag{c}$$

Differentiating equation (b) with respect to t, and combining the result with equation (a), we obtain

$$\dot{\Lambda} = -2\zeta\omega_0 \dot{X}^2 - \omega_0^2 X\dot{X}W(t) \tag{d}$$

Equation (d) may be approximated by the Itô stochastic differential equation

$$d\Lambda = m(\Lambda) + \sigma(\Lambda)\, dB(t) \tag{e}$$

where the drift and diffusion coefficients are evaluated as follows, using the quasi-conservative averaging procedure:

$$\begin{aligned} m(\Lambda) &= \langle -2\zeta\omega_0\dot{X}^2 + \pi\omega_0^4 K X^2\rangle_t \\ &= \frac{1}{T_{1/4}}\left\{-2\zeta\omega_0\int_0^A [2\Lambda - 2U(x)]^{1/2} dX + \pi\omega_0^4 K\int_0^A X^2[2\Lambda - 2U(X)]^{-1/2} dX\right\} \end{aligned} \tag{f}$$

$$\sigma^2(\Lambda) = \langle 2\pi\omega_0^4 K X^2 \dot{X}^2\rangle_t = \frac{2\pi\omega_0^4 K}{T_{1/4}}\int_0^A X^2[2\Lambda - 2U(X)]^{1/2} dX \tag{g}$$

In (f) and (g), each time-averaging $< \cdot >_t$ has been replaced by averaging on X over a quarter-period $T_{1/4}$, the upper limit A of each integration is the positive root of the equation $U(A) = \Lambda$, and the quarter-period is obtained from

$$T_{1/4} = \int_0^A [2\Lambda - 2U(X)]^{-1/2} dX \tag{h}$$

Now, the total energy $\Lambda(t)$ is a Markov process defined on $[0, +\infty)$. It can be shown from equations (f) and (g) that $m(0) = 0$, $\sigma(0) = 0$, $|m(\infty)| = \infty$, and $\sigma(\infty) = \infty$, indicating that $\lambda = 0$ and $\lambda = \infty$ are singular boundaries of the first and second kinds, respectively. Two cases of $0 \le \delta < 1$ and $\delta > 1$ are discussed next.

Case 1: $0 \le \delta < 1$. When Λ tends to infinity, the nonlinear restoring force can be neglected when compared with the linear restoring force; thus

$$m(\Lambda) \to (-2\zeta\omega_0 + \pi\omega_0^2 K)\Lambda \qquad \text{as } \Lambda \to \infty \tag{i}$$

$$\sigma^2(\Lambda) \to \pi\omega_0^2 K\Lambda^2 \qquad \text{as } \Lambda \to \infty \tag{j}$$

It can be shown that the diffusion exponent, drift exponent, and character value at the right boundary at infinity are given by

$$\alpha_r = 2 \qquad \beta_r = 1 \qquad c_r = \frac{4\zeta - 2\pi\omega_0 K}{\pi\omega_0 K} \tag{k}$$

The right boundary is repulsively natural if $\zeta > \frac{1}{4}\pi\omega_0 K$, attractively natural if $\zeta < \frac{1}{4}\pi\omega_0 K$, and strictly natural if $\zeta = \frac{1}{4}\pi\omega_0 K$. On the other hand, when Λ tends to zero, the linear restoring force is negligible; thus

$$m(\Lambda) \to -4\zeta\omega_0\frac{\delta + 1}{\delta + 3}\Lambda \qquad \text{as } \Lambda \to 0 \tag{l}$$

$$\sigma^2(\Lambda) \to 4\pi\omega_0^4 K\left[\frac{2(\delta + 1)}{\eta}\right]^{2/(\delta+1)} \frac{B\left(\frac{3}{\delta+1}, \frac{3}{2}\right)}{B\left(\frac{1}{\delta+1}, \frac{1}{2}\right)}\Lambda^{(\delta+3)/(\delta+1)} \qquad \text{as } \Lambda \to 0 \tag{m}$$

where $B(\cdot, \cdot)$ denotes the beta function. It follows that at the left boundary $\lambda = 0$, the diffusion and drift exponents are, respectively,

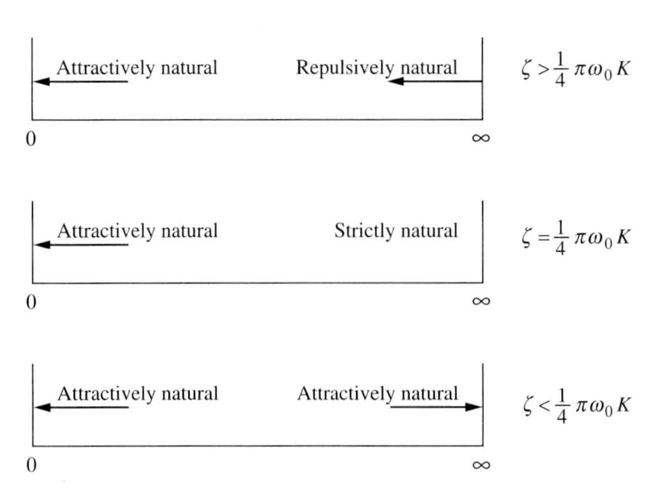

FIGURE 6.7.4
Boundary behaviors for sample functions of a nonlinear system (adding a nonlinear stiffness term to a linear oscillator, $0 \leq \delta < 1$).

$$\alpha_l = \frac{\delta + 3}{\delta + 1} \qquad \beta_l = 1 \tag{n}$$

Since $\alpha_l > 1 + \beta_l = 2$ and $m(0^+) < 0$, $\lambda = 0$ is an attractively natural boundary, independent of the ζ value.

The boundary behaviors are shown schematically in Fig. 6.7.4. The trivial solution is asymptotically stable if $\zeta > \frac{1}{4}\pi\omega_0 K$, although it is only approachable, not accessible from an interior point. A non-delta-type stationary probability density does not exist irrespective of the ζ value.

For obtaining the asymptotic moment stability conditions, we add an additive excitation term to equation (a) to read

$$\ddot{X} + 2\zeta\omega_0\dot{X} + \omega_0^2[1 + W(t)]X + \eta\left|X\right|^{\delta}\operatorname{sgn} X = W_1(t) \tag{o}$$

Upon performing the quasi-conservative averaging, the drift and diffusion coefficients for the energy process $\Lambda(t)$ of system (o) are obtained as follows:

$$m_1(\Lambda) = \pi K_1 + \frac{1}{T_{1/4}}\left\{-2\zeta\omega_0\int_0^A [2\Lambda - 2U(X)]^{1/2} dX \right.$$
$$\left. + \pi\omega_0^4 K\int_0^A X^2[2\Lambda - 2U(X)]^{-1/2} dX\right\} \tag{p}$$

$$\sigma_1^2(\Lambda) = \frac{1}{T_{1/4}}\left\{2\pi\omega_0^4 K\int_0^A X^2[2\Lambda - 2U(X)]^{1/2} dX \right.$$
$$\left. + 2\pi K_1\int_0^A [2\Lambda - 2U(X)]^{1/2} dX\right\} \tag{q}$$

The stationary probability density of $\Lambda(t)$ is given by

$$p(\lambda) = C\frac{1}{\sigma_1^2(\lambda)}\exp\left[\int \frac{2m_1(\lambda)}{\sigma_1^2(\lambda)}\,d\lambda\right] \tag{r}$$

The nth moment of the energy process $\Lambda(t)$ is now obtained from

$$E[\Lambda^n] = C\int_0^\infty \frac{\lambda^n}{\sigma_1^2(\lambda)}\exp\left[\int \frac{2m_1(\lambda)}{\sigma_1^2(\lambda)}\,d\lambda\right]d\lambda \tag{s}$$

It can be shown that

$$\begin{aligned} m_1(\Lambda) &\to \pi K_1 && \text{as } \Lambda \to 0 \\ \sigma_1^2(\Lambda) &\to 4\pi K_1\frac{\delta+1}{\delta+3}\Lambda && \text{as } \Lambda \to 0 \end{aligned} \tag{t}$$

$$\begin{aligned} m_1(\Lambda) &\to (-2\zeta\omega_0 + \pi\omega_0^2 K)\Lambda && \text{as } \Lambda \to \infty \\ \sigma_1^2(\Lambda) &\to \pi\omega_0^2 K\Lambda^2 && \text{as } \Lambda \to \infty \end{aligned} \tag{u}$$

The integrability of equation (s) requires that $\zeta > \frac{1}{4}(n+1)\pi\omega_0 K$. If this condition is satisfied, then $E[A^n] \to 0$ as $K_1 \to 0$, indicating that the trivial solution of the original system (a) is asymptotically stable in the nth moment.

Case 2: $\delta > 1$. Return to equations (e), (f), and (g). In this case, the nonlinear component of the restoring force is negligible when Λ is small but predominant when Λ is large. It can be shown that at the right boundary at infinity, the diffusion exponent, drift exponent, and character value are given by

$$\alpha_r = \frac{\delta+3}{\delta+1} \qquad \beta_r = 1 \qquad m(+\infty) < 0 \tag{v}$$

indicating that the right boundary at infinity is repulsively natural, independent of the value of ζ. On the other hand, at the left boundary $\lambda = 0$, they are

$$\alpha_l = 2 \qquad \beta_l = 1 \qquad c_l = \frac{-4\zeta + 2\pi\omega_0 K}{\pi\omega_0 K} \tag{w}$$

The classifications of the two boundaries are shown schematically in Fig. 6.7.5. The trivial solution is asymptotically stable in probability if $\zeta > \frac{1}{4}\pi\omega_0 K$. A non-delta-type stationary probability exists if $\zeta < \frac{1}{4}\pi\omega_0 K$, and it can be determined from

$$p(\lambda) = C\frac{1}{\sigma^2(\lambda)}\exp\left[\int \frac{2m(\lambda)}{\sigma^2(\lambda)}\,d\lambda\right] \tag{x}$$

where $m(\lambda)$ and $\sigma(\lambda)$ are given by (f) and (g), respectively. The nth moment is then obtained as

$$E[\Lambda^n] = C\int_0^\infty \frac{\lambda^n}{\sigma^2(\lambda)}\exp\left[\int \frac{2m(\lambda)}{\sigma^2(\lambda)}\,d\lambda\right]d\lambda \tag{y}$$

By examining the limiting behaviors of $m(\lambda)$ and $\sigma(\lambda)$ as $\lambda \to 0$ and $\lambda \to \infty$, it can be shown that (y) is integrable if $\zeta < \frac{1}{4}\pi\omega_0 K$. It also can be shown that $E[\Lambda^n] \to 0$ as $\zeta \to \frac{1}{4}\pi\omega_0 K$. Therefore, $\zeta > \frac{1}{4}\pi\omega_0 K$ is the necessary and sufficient condition for the trivial solution to be stable in all the statistical moments of positive orders.

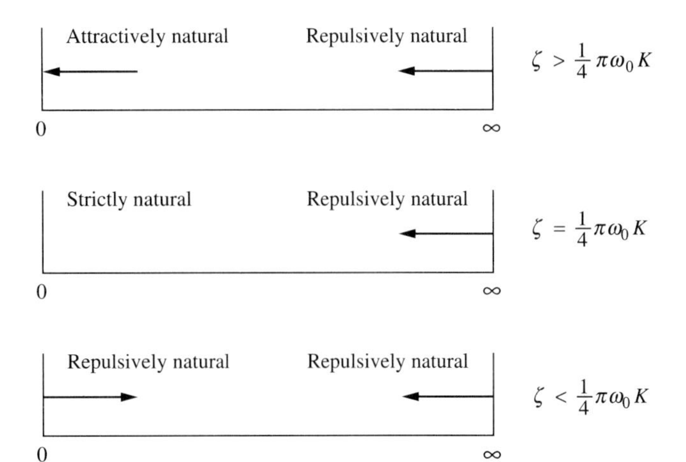

FIGURE 6.7.5
Boundary behaviors for sample functions of a nonlinear system (adding a nonlinear stiffness term to a linear oscillator, $\delta > 1$).

We conclude that the effects of adding a nonlinear spring term $\eta|X|^\sigma \operatorname{sgn} X$ to an originally linear system are the same as those of adding a nonlinear damping term $\eta|\dot{X}|^\delta \operatorname{sgn} \dot{X}$.

6.7.3 A Nonlinear System in the Class of Generalized Stationary Potential

Consider the nonlinear system

$$\ddot{X} + X^2\left[\beta + \frac{4\alpha}{X^2 + \dot{X}^2}\right]\dot{X} + [1 + W(t)]X = 0 \qquad \beta > 0 \tag{a}$$

where $W(t)$ is a gaussian white noise with a spectral density K. As shown in Section 5.3.1, the exact joint probability density of $X(t)$ and $\dot{X}(t)$ in the stationary state is obtainable, and it is given by

$$p(x, \dot{x}) = C(x^2 + \dot{x}^2)^{-2\alpha/\pi K} \exp\left[-\frac{\beta}{2\pi K}(x^2 + \dot{x}^2)\right] \tag{b}$$

where C is a normalization constant given by

$$C = \frac{1}{\pi}\left(\frac{2\pi K}{\beta}\right)^{2\alpha/\pi K - 1}\left[\Gamma\left(-\frac{2\alpha}{\pi K} + 1\right)\right]^{-1} \tag{c}$$

The integrability of the right-hand side of equation (b) requires

$$\alpha < \tfrac{1}{2}\pi K \tag{d}$$

which is the condition for the existence of a non-delta-type stationary probability density. Since equation (a) is devoid of additive excitation, inequality (d) is also a sufficient condition for the trivial solution to become unstable in probability.

To obtain a condition that is both necessary and sufficient for the asymptotic stability in probability is not always simple for a two-dimensional nonlinear system. However, in the present case, equation (b) can be used to obtain the following probability density for the amplitude $A = (X^2 + \dot{X}^2)^{1/2}$:

$$p(a) = C_1 a^{1-4\alpha/\pi K} \exp\left(-\frac{\beta}{2\pi K}a^2\right) \tag{e}$$

Equation (e) corresponds to an Itô equation

$$dA = \left[\left(-\frac{1}{2}\alpha + \frac{3}{8}\pi K\right)A - \frac{1}{8}\beta A^3\right]dt + \frac{\sqrt{\pi K}}{2}A\,dB(t) \tag{f}$$

It can be shown that

$$\begin{aligned} a = 0 \quad & \begin{cases} \text{attractively natural} & \alpha > \pi K/2 \\ \text{strictly natural} & \alpha = \pi K/2 \\ \text{repulsively natural} & \alpha < \pi K/2 \end{cases} \\ a = \infty \quad & \text{entrance} \end{aligned} \tag{g}$$

Thus, the trivial solution $A = 0$ is asymptotically stable in probability only if

$$\alpha > \tfrac{1}{2}\pi K \tag{h}$$

It is clear that $A = 0$ is equivalent to $X = 0$ and $\dot{X} = 0$.

The conditions for the asymptotic stability in the statistical moments can be determined directly from (b). The nth moment of the norm $\|\boldsymbol{X}(t)\| = [X^2(t)+\dot{X}^2(t)]^{1/2}$ is obtained as follows:

$$E[\|\boldsymbol{X}\|^n] = \int_{-\infty}^{\infty}\int_{-\infty}^{\infty}(x^2 + \dot{x}^2)^{n/2}p(x,\dot{x})\,dx\,d\dot{x} = \left(\frac{2\pi K}{\beta}\right)^{n/2}\frac{\Gamma(n/2 + 1 - 2\alpha/\pi K)}{\Gamma(1 - 2\alpha/\pi K)} \tag{i}$$

Equation (i) is valid under the same condition (d). As α approaches $\pi K/2$, $\Gamma(1 - 2\alpha/\pi K)$ tends to infinity, leading to

$$\lim_{\alpha\to\frac{1}{2}\pi K} E[\|\boldsymbol{X}\|^n] = 0 \quad n > 0 \tag{j}$$

Since stability in probability is a prerequisite for stability in the statistical moments of positive orders, we obtain again a single condition (h) under which the trivial solution is asymptotically stable in the statistical moments of all positive orders. Since equation (b) is an exact stationary probability density, this single condition is exact.

6.8 CONCLUDING REMARKS

Since motion stability is defined in terms of boundedness and convergence, and since convergence of a stochastic sequence can be interpreted in several different ways, stability of a randomly excited system also can be interpreted in several different ways. The main objective of the present chapter is to determine the condition or conditions under which the motion stability of a dynamical system can be assured on the basis of each of these interpretations.

The Lyapunov stability with probability 1 (or almost sure stability, or sample stability) is the most relevant physically. It assures that almost all sample functions are bounded in the entire semi-infinite time interval $[0, \infty)$. Its asymptotic version assures additionally that almost all sample functions tend to zero as time increases. For linear systems, sufficient conditions for asymptotic sample stability can be obtained if the multiplicative random excitations are ergodic and if the variances of these excitations are finite. Such sufficient conditions can be improved (or "sharpened") with additional knowledge of the excitations, for example, their probability distributions.

Necessary and sufficient conditions for asymptotic sample stability are obtainable for linear systems under gaussian white noise excitations. In this case, the system response is a Markov vector, and a method due to Khasminskii is applicable. The key step is to determine the stationary probability density of the projection of the system response on a unit sphere, which is needed in the computation of the Lyapunov exponent for the norm of the system response.

Stability in probability and its asymptotic version are concerned with the convergence properties of sample functions at an arbitrary instant of time $t \geq 0$, not with the sample functions in the entire semi-infinite time interval $[0, \infty)$. Therefore, they are not of the Lyapunov type and are generally less stringent. However, when applied to a linear system, convergence of the solution at an arbitrary time instant guarantees the same in the entire semi-infinite time interval beyond this time instant, and the two types of asymptotic stability conditions coincide. In the case of a one-dimensional system under gaussian white noise excitations, the asymptotic stability condition in probability can be inferred from the sample behaviors at the boundaries. The trivial solution is asymptotically stable in probability if it is either an exit or attractively natural, while the other boundary is either an entrance or repulsively natural.

Stability in statistical moments of various positive orders are also concerned with an arbitrary instant of time and are not of the Lyapunov type. It can be shown, by using the Schwarz inequality and induction, that moment stability of a higher positive order implies moment stability of a lower positive order. Furthermore, since the integrability of the probability density is a prerequisite for the existence of any positive order moment, moment stability of a positive order is generally more stringent than stability in probability. The asymptotic stability in probability coincides with the asymptotic sample stability in the special case of linear systems. Indeed, Kozin and Sugimoto (1977) and Arnold (1984) have shown that in this case the condition for the asymptotic sample stability can be obtained in a limiting process by letting the order of the asymptotic statistical moment approach zero.

Equations for the statistical moments of positive integral orders are easy to derive if the physical equations of a system can first be converted to the corresponding Itô equations. If a system is linear, then such moment equations are closed, in the sense that equations derived for moments of order n are linear, and they do not contain moments of higher order; therefore, stability conditions can be determined sequentially for moments of ascending orders. These moment stability conditions depend on the orders of the moments. However, the same method is not applicable to nonlinear systems, for which the equations for statistical moments form an infinite hierarchy. While several truncation or closure schemes have been proposed to solve such moment equations (to be discussed in Section 7.1), they are generally unsuitable for determining the stability conditions. An alternative approach can be applied in which the original system is modified by adding an additive excitation, and the statistical moment in question is obtained for the modified system. Then, the asymptotic stability condition of the trivial solution of the original system in terms of this statistical moment can be determined by taking the limit as the additive excitation approaches zero. Of course, modification is unnecessary if a non-delta-type stationary probability density exists without an additive excitation.

6.9 EXERCISES

6.1. Consider the system

$$\ddot{X} + [2\zeta + W(t)]\dot{X} + X = 0$$

where $W(t)$ is a gaussian white noise with spectral density K.

(*a*) Derive the Itô stochastic differential equations for $\Upsilon(t)$ and $\Theta(t)$ using the transformation

$$\Upsilon = \frac{1}{2}\ln(X^2 + \dot{X}^2) \qquad \Theta = \tan^{-1}\frac{\dot{X}}{X}$$

(*b*) Show that $\Theta(t)$ is a Markov process and obtain the Fokker-Planck equation for process $\Theta(t)$.

(*c*) Identify the singular points for the $\Theta(t)$ process in the interval of $[0, 2\pi)$, and find the stationary probability density $p(\theta)$ for $\Theta(t)$.

(*d*) Obtain an expression for the largest Lyapunov exponent for the $X(t)$ process.

6.2. Given the equation of motion

$$\ddot{X} + [2\zeta\omega_0 + W(t)]\dot{X} + \omega_0^2 X = 0$$

where $W(t)$ is a gaussian white noise, obtain equations for the first- and second-order moments of the system response $X(t)$ and $\dot{X}(t)$. Determine the stability conditions for the first- and second-order moments.

6.3. Consider the system

$$\ddot{X} + [2\zeta\omega_0 + \xi_2(t)]\dot{X} + \omega_0^2 X = \xi_1(t)$$

where $\xi_1(t)$ and $\xi_2(t)$ are correlation-stationary random excitations with broadband spectral densities $\Phi_{ij}(\omega)$; $i, j = 1, 2$.

(*a*) Derive the Itô stochastic differential equation for the averaged amplitude process $A(t)$ using the transformation

$$X = A\cos\theta \qquad \theta = \omega_0 t + \phi$$
$$\dot{X} = -A\omega_0 \sin\theta$$

and the smoothed version of the stochastic averaging procedure.

(*b*) Solve the corresponding reduced Fokker-Planck equation for the stationary probability density $p(a)$ of $A(t)$, and find the integrability condition for the ensemble average of $A^n(t)$.

(*c*) Obtain the Itô stochastic differential equation for $A^n(t)$ and find the stability condition for $E[A^n]$.

(*d*) Compare the results obtained from two alternative procedures, (*b*) and (*c*).

6.4. Consider the linear system

$$\ddot{X} = [2\zeta\omega_0 + \xi_2(t)]\dot{X} + \omega_0^2[1 + \xi_1(t)]X = 0$$

where $\xi_1(t)$ and $\xi_2(t)$ are correlation-stationary random processes with broadband spectral densities $\Phi_{ij}(\omega)$; $i, j = 1, 2$. Using the transformation

$$X = A\cos\theta \qquad \theta = \omega_0 t + \phi$$
$$\dot{X} = -A\omega_0 \sin\theta$$

and the smoothed version of the stochastic averaging procedure, obtain an Itô equation for the averaged amplitude process $A(t)$. Determine the stability conditions for the first- and second-order moments of process $A(t)$.

6.5. Consider the nonlinear system

$$\ddot{X} + \alpha\dot{X} + \beta\dot{X}^3 + X = \dot{X}W(t) \qquad \beta > 0$$

where $W(t)$ is a gaussian white noise with spectral density K.

(*a*) Using the transformation

$$X = A\cos\theta \qquad \theta = \omega_0 t + \phi$$
$$\dot{X} = -A\omega_0 \sin\theta$$

and the smoothed stochastic version of the averaging procedure, obtain an Itô stochastic differential equation for the averaged amplitude process $A(t)$ and the corresponding Fokker-Planck equation.

(*b*) Solve the reduced Fokker-Planck equation for a non-delta-type stationary probability density $p(a)$ of $A(t)$, and determine its integrability condition.

(*c*) Determine the type of singular boundary at $a = 0$, find the asymptotic stability condition in probability for the trivial solution $A(t) = 0$, and compare the result with that of (*b*).

6.6. A nonlinearly damped oscillator, subjected to a multiplicative gaussian white noise excitation, is governed by

$$\ddot{X} + \eta|\dot{X}|^\delta \operatorname{sgn}\dot{X} + \omega_0^2 X = \dot{X}W(t) \qquad \eta > 0, \delta \geq 0$$

(*a*) Derive an Itô stochastic differential equation for the amplitude process

$$A(t) = \left[X^2(t) + \frac{\dot{X}^2(t)}{\omega_0^2}\right]^{1/2}$$

using the smoothed version of the stochastic averaging procedure.

(*b*) Investigate the sample behaviors of the amplitude process $A(t)$ at the two boundaries.

(*c*) Find the condition for the asymptotic stability in probability for the trivial solution $A(t) = 0$.

(*d*) Find the condition for the asymptotic stability in the nth moment of the trivial solution $A(t) = 0$.

6.7. An oscillator with a nonlinear spring and subjected to a multiplicative gaussian white noise excitation is governed by

$$\ddot{X} + 2\eta\dot{X} + k|X|^\delta \operatorname{sgn}X = \dot{X}W(t) \qquad k > 0,\ \delta \geq 0$$

(*a*) Derive an Itô stochastic differential equation for the energy process

$$\Lambda(t) = \frac{1}{2}\dot{X}^2(t) + \frac{k}{\delta + 1}|X(t)|^{\delta+1}$$

using the quasi-conservative averaging procedure.

(*b*) Investigate the sample behaviors of the energy process $\Lambda(t)$ at the two boundaries.

(*c*) Find the asymptotic stability condition in probability for the trivial solution $\Lambda(t) = 0$.

(*d*) Find the asymptotic stability condition for the nth moment of the trivial solution $\Lambda(t) = 0$.

CHAPTER 7

APPROXIMATE SOLUTIONS FOR MULTIDIMENSIONAL NONLINEAR SYSTEMS

It was shown in Chapter 5 that when a multidimensional nonlinear system is subjected to both additive and multiplicative excitations of gaussian white noises, the reduced Fokker-Planck equation can be solved in closed form only with certain highly restrictive relations between the system parameters and the spectral densities of the excitations. Rarely can such restrictive requirements be met in practical cases; therefore, approximate solution techniques are generally needed.

The most frequently used approximation scheme is the *equivalent linearization procedure* (Atalik and Utku, 1976; Booton, 1954; Caughey, 1959a,b; Roberts and Spanos, 1990; Spanos, 1981), in which the original system is replaced by an equivalent linear system. The parameters of the replacement linear system are determined using the statistical criterion of the least-mean-square difference. The theoretical basis of the equivalent linearization procedure and its application were elucidated in Lin (1967). The procedure is versatile, without the limitation of gaussian white noise excitations, and it can easily be applied to multi-degree-of-freedom systems. It has been used mostly to compute the second-order statistical moments of the stationary response to stationary random excitations, for which the moment equations to be solved are algebraic. Extension to the case of nonstationary transient response, however, is possible. In the special case of gaussian white noise excitations, this procedure is equivalent to another scheme, called *gaussian closure* (Crandall, 1978).

Therefore, it is considered unsuitable when the system is highly nonlinear, or when multiplicative random excitations are present, because in either case the probability distribution of the system response is usually far from being gaussian.

To improve the accuracy of an approximate solution, several nongaussian closure schemes have been proposed; these include additional terms describing the nongaussian features in increasingly greater detail. In one approach, the unknown probability density of the system response is expressed as a truncated Gram-Charlier or Edgeworth series in which the first term is the gaussian distribution. The coefficients in the truncated series are then determined using the governing equations of the dynamical system. Once the approximate probability distribution is known, the required statistical moments can be computed easily. This approach has been used by Assaf and Zirkie (1976) and Crandall (1980). However, it becomes rather tedious when the transient nonstationary moments are required or, in the case of a multi-degree-of-freedom system, when a multidimensional Gram-Charlier series is required.

Another generalization of the gaussian closure procedure focuses on the properties of cumulants (also called semi-invariants). Since the third and higher order cumulants (including joint cumulants) of gaussian random variables are zero, gaussian closure is equivalent to neglecting those cumulants above the second order. A more general scheme, known as *cumulant-neglect closure*, is based on the premise that successive improvements can be achieved by retaining additionally the third, fourth, fifth, and higher order cumulants. It was used earlier in the studies of turbulence (e.g., Beran, 1968), and it was then applied to dynamical systems by Wu and Lin (1984) and Ibrahim, Soundararajan, and Heo (1985). The scheme is also quite versatile, applicable to multi-degree-of-freedom systems and to finding the statistical moments of transient nonstationary response, provided that the excitations are gaussian white noises. The results so obtained appear to be accurate when a system is subjected only to additive excitations. When multiplicative excitations are also present, its use may lead to erroneous results, as shown by Sun and Hsu (1987).

The method of equivalent nonlinear system was first proposed by Lutes (1970), in an effort to improve the accuracy of approximate solutions for highly nonlinear hysteretic systems under additive excitations. In Lutes's paper, the original hysteretic system was replaced by an equivalent nonlinear system for which the response probability density was known. The criteria for selecting an equivalent system were devised specifically for hysteretic systems, and the energy dissipation due to the hysteresis was assumed to be small. Subsequently, the method was applied by Caughey (1986) to other nonlinearly damped systems under additive random excitations, using the criterion of least-mean-square difference. However, the form of the replacement system must be preselected, which may be crucial for the accuracy of the solution.

Based on the knowledge of the exact solutions for the class of generalized stationary potential (Chapter 5), Cai and Lin (1988b) developed another approximation procedure, called dissipation energy balancing, in which a given nonlinear system is replaced by another nonlinear system belonging to the class of generalized stationary potential. The replacement system is selected with the criterion that the average dissipated energy remains the same. In some sense, this procedure is a variation of the method of equivalent nonlinear systems, but it is applicable to systems under additive and/or multiplicative excitations of gaussian white noises. Moreover, the form of the replacement system need not be preselected, which is an important advantage.

The method of dissipation energy balancing may be viewed as an application of the general scheme of weighted residuals; thus further generalization is possible (Cai, Lin, and Elishakoff, 1992). Within this general framework, a set of weighting functions is selected, and the replacement system is required to satisfy the constraint that each weighted residual be zero. One of the constraints is equivalent to dissipation energy balancing. In principle, the accuracy of the approximate solution may be improved by judiciously choosing the other constraints.

In this chapter, the cumulant-neglect closure is introduced first. Next, the method of weighted residuals is described in detail, illustrated by examples, and applied to randomly excited hysteretic structures. Since the method is applicable only if the excitations are gaussian white noises, relaxation of the gaussianity assumption in the excitations is then investigated. Finally, the case of combined harmonic and broadband random excitations is considered.

7.1 CUMULANT-NEGLECT CLOSURE

The nth cumulant of random variables $X_1, X_2, \ldots, X_n$ is defined as (e.g., see Lin, 1967)

$$\kappa_n [X_1, X_2, \ldots, X_n] = \frac{1}{i^n} \left[\frac{\partial^n}{\partial\theta_1 \partial\theta_2 \cdots \partial\theta_n} \ln M_{X_1 X_2 \cdots X_n}(\theta_1, \theta_2, \cdots, \theta_n) \right]_{\theta_1 = \theta_2 = \ldots = \theta_n = 0} \tag{7.1.1}$$

where $i = \sqrt{-1}$, and $M_{X_1 X_2 \cdots X_n}(\theta_1, \theta_2, \ldots, \theta_n)$ is the joint characteristic function of random variables $X_1, X_2, \ldots, X_n$, defined as

$$\begin{aligned} M_{X_1 X_2 \cdots X_n}(\theta_1, \theta_2, \ldots, \theta_n) &= E\{\exp[i(\theta_1 X_1 + \theta_2 X_2 + \cdots + \theta_n X_n)]\} \\ &= \int_{-\infty}^{\infty} \cdots \int_{-\infty}^{\infty} p_{X_1 X_2 \cdots X_n}(x_1, x_2, \ldots, x_n) \\ &\quad \times \exp[i(\theta_1 x_1 + \theta_2 x_2 + \cdots + \theta_n x_n)]\, dx_1\, dx_2 \cdots dx_n \end{aligned} \tag{7.1.2}$$

and where the logarithm of a complex variable is taken to be the principal value. Equation (7.1.1) implies that the natural logarithm of a characteristic function has a series expansion

$$\begin{aligned} \ln M_{X_1 X_2 \cdots X_n}(\theta_1, \theta_2, \ldots, \theta_n) &= i\theta_j \kappa_1[X_j] + \frac{1}{2!}(i\theta_j)(i\theta_k)\kappa_2[X_j, X_k] \\ &\quad + \frac{1}{3!}(i\theta_j)(i\theta_k)(i\theta_l)\kappa_3[X_j, X_k, X_l] + \cdots \end{aligned} \tag{7.1.3}$$

that all the cumulants involved exist, and that the series converges. The left-hand side of (7.1.3) is called the log-characteristic function of $X_1, X_2, \ldots, X_n$. The cumulants are related to the statistical moments as follows (Stratonovich, 1963):

$$\begin{aligned} E[X_j] &= \kappa_1[X_j] \\ E[X_jX_k] &= \kappa_2[X_j, X_k] + \kappa_1[X_j]\kappa_1[X_k] \\ E[X_jX_kX_l] &= \kappa_3[X_j, X_k, X_l] + 3\{\kappa_1[X_j]\kappa_2[X_k, X_l]\}_s \\ &\quad + \kappa_1[X_j]\kappa_1[X_k]\kappa_1[X_l] \\ E[X_jX_kX_lX_m] &= \kappa_4[X_j, X_k, X_l, X_m] + 3\{\kappa_2[X_j, X_k]\kappa_2[X_l, X_m]\}_s \\ &\quad + 4\{\kappa_1[X_j]\kappa_3[X_k, X_l, X_m]\}_s + 6\{\kappa_1[X_j]\kappa_1[X_k]\kappa_2[X_l, X_m]\}_s \\ &\quad + \kappa_1[X_j]\kappa_1[X_k]\kappa_1[X_l]\kappa_1[X_m] \end{aligned} \tag{7.1.4}$$

where $\{\cdot\}_s$ denotes a symmetrizing operation with respect to all its arguments; that is, an operation that takes the arithmetic mean of *different* permuted terms similar to the one within the braces. For example,

$$\begin{aligned} \{\kappa_1[X_j]\kappa_2[X_k, X_l]\}_s = \tfrac{1}{3}\{&\kappa_1[X_j]\kappa_2[X_k, X_l] \\ &+\kappa_1[X_k]\kappa_2[X_l, X_j] + \kappa_1[X_l]\kappa_2[X_j, X_k]\} \end{aligned} \tag{7.1.5}$$

It is of interest to note that in (7.1.4) the coefficient associated with each symmetrizing operation is exactly equal to the total number of terms being averaged. A general relationship between the nth-order moment and the nth- and lower order cumulants can be expressed more conveniently in an alternative form, as shown in equation (3.2.6).

If the third expression in (7.1.4) is altered to read

$$\kappa_3[X_j, X_k, X_l] = E[X_jX_kX_l] - E[X_j]E[X_k]E[X_l] - 3\{E[X_j]\kappa_2[X_k, X_l]\}_s \tag{7.1.6}$$

then the third cumulant is seen to be the result of subtracting the following from the third moment: (1) the value of a third moment if all three random variables are mutually uncorrelated, and (2) terms attributable to pairwise correlations. Therefore, if a third cumulant is equal to zero, then at least one random variable is uncorrelated with the other two; hence the alternative name, triple correlation, has been used elsewhere for the third cumulant, for example, in the book by Stratonovich (1963). Similarly, if $\kappa_n[X_1, X_2, \ldots, X_n] = 0$, it implies that at least one of the n random variables is uncorrelated with any of the others.

If all the first moments are zero, then the remaining expressions in (7.1.4) are simplified to

$$\begin{aligned} E[X_jX_k] &= \kappa_2[X_j, X_k] \\ E[X_jX_kX_l] &= \kappa_3[X_j, X_k, X_l] \\ E[X_jX_kX_lX_m] &= \kappa_4[X_j, X_k, X_l, X_m] + 3\{\kappa_2[X_j, X_k]\kappa_2[X_l, X_m]\}_s \\ E[X_jX_kX_lX_mX_p] &= \kappa_5[X_j, X_k, X_l, X_m, X_p] \\ &\quad + 10\{\kappa_2[X_j, X_k]\kappa_3[X_l, X_m, X_p]\}_s \end{aligned}$$

$$
\begin{aligned}
E[X_jX_kX_lX_mX_pX_r] = \; & \kappa_6[X_j, X_k, X_l, X_m, X_p, X_r] \\
& + 15\left\{\kappa_2[X_j, X_k]\kappa_4[X_l, X_m, X_p, X_r]\right\}_s \\
& + 15\left\{\kappa_2[X_j, X_k]\kappa_2[X_l, X_m]\kappa_2[X_p, X_r]\right\}_s \\
& + 10\left\{\kappa_3[X_j, X_k, X_l]\kappa_3[X_m, X_p, X_r]\right\}_s
\end{aligned}
\tag{7.1.7}
$$

Let $X(t)$ be a random process, and let $X_j = X(t_j)$ where $j = 1, 2, \ldots, n$. Then (7.1.3) is a series expansion of the nth-order log-characteristic function of $X(t)$. Therefore, a random process can also be described by a complete set of cumulants which are functions of various times. Since a cumulant function of an order higher than 1 represents a measure of correlation of the random process at different times, its value tends to zero as the t values are separated further from each other. Furthermore, it is expected that the physical significance of a cumulant decreases as the order increases, and the most important properties of a random process are revealed in the lower order cumulants.

Now, consider a nonlinear system

$$
\frac{d}{dt}X_i = f_i(\mathbf{X}) + g_{ij}(\mathbf{X})W_j(t) \qquad i = 1, 2, \ldots, n, \quad j = 1, 2, \ldots, m \tag{7.1.8}
$$

where $W_j(t)$ are gaussian white noises. Equation (7.1.8) is equivalent to the following set of Itô-type stochastic differential equations:

$$
dX_i = \left(f_i + \pi K_{ls}g_{rs}\frac{\partial g_{il}}{\partial X_r}\right)dt + \sqrt{2\pi K_{ls}g_{il}g_{js}}\,dB_j(t) \tag{7.1.9}
$$

where $B_j(t)$ are independent unit Wiener processes, and K_{ls} are the cross-spectral density of $W_l(t)$ and $W_s(t)$. Let $M(X) = X_1^{k_1}X_2^{k_2}\cdots X_n^{k_n}$ where the superscripts $k_1, k_2, \ldots, k_n$ are nonnegative integers. According to the Itô differential rule,

$$
\begin{aligned}
dM = & \left[\left(f_i + \pi K_{ls}g_{rs}\frac{\partial g_{il}}{\partial X_r}\right)\frac{\partial M}{\partial X_i} + \pi K_{ls}g_{il}g_{js}\frac{\partial^2 M}{\partial X_i \partial X_j}\right]dt \\
& + \sqrt{2\pi K_{ls}g_{il}g_{js}}\frac{\partial M}{\partial X_i}dB_j(t)
\end{aligned}
\tag{7.1.10}
$$

It follows upon taking the ensemble average of equation (7.1.10)

$$
\frac{d}{dt}E[M] = E\left[\left(f_i + \pi K_{ls}g_{rs}\frac{\partial g_{il}}{\partial X_r}\right)\frac{\partial M}{\partial X_i}\right] + \pi K_{ls}E\left[g_{il}g_{js}\frac{\partial^2 M}{\partial X_i \partial X_j}\right] \tag{7.1.11}
$$

The left-hand side of (7.1.11) is the time derivative of a statistical moment of order N, where $N = k_1 + k_2 + \cdots + k_n$, whereas the right-hand side depends on the functional forms of f_i and g_{jk}. If functions f_i and g_{jk} are linear, then the right-hand side contains only the Nth- and lower order statistical moments. In this case, equations for statistical moments of the type of (7.1.11) can be solved recursively, beginning from $N = 1$. However, if at least one of the f_i and g_{jk} functions is a nonlinear polynomial, then the right-hand side of (7.1.11) also contains moments of orders higher than N. An exact solution is no longer obtainable since the moment equations for $N = 1, 2, \ldots,$ constitute an infinite hierarchy.

Now, let all the cumulants higher than a given order p be set to zero. Then a statistical moment of an order higher than p can be expressed in terms of those moments of order p and lower than p, using relations in (7.1.4) or (7.1.7). For example, if all the random variables in question, $X_1, X_2, \ldots, X_n$, have a zero mean, and if the truncation level is set at $p = 2$, then we obtain

$$
\begin{aligned}
E[X_j X_k X_l] &= 0 \\
E[X_j X_k X_l X_m] &= 3\{E[X_j X_k] E[X_l X_m]\}_s \\
E[X_j X_k X_l X_m X_p] &= 0 \\
E[X_j X_k X_l X_m X_p X_r] &= 15\{E[X_j X_k]\{3E[X_p X_r] E[X_l X_m]\}_s\}_s \\
&\quad + 15\{E[X_j X_k] E[X_l X_m] E[X_p X_r]\}_s \\
&\cdots\cdots\cdots
\end{aligned}
\tag{7.1.12}
$$

These are the same relations as those of gaussian random variables. On the other hand, by letting $p = 4$, we have for moments of orders higher than the fourth

$$
\begin{aligned}
E[X_j X_k X_l X_m X_p] &= 10\{E[X_j X_k] E[X_l X_m X_p]\}_s \\
E[X_j X_k X_l X_m X_p X_r] &= 15\{E[X_j X_k](E[X_l X_m X_p X_r] \\
&\quad - 3\{E[X_l X_m] E[X_p X_r]\}_s)\}_s \\
&\quad + 15\{E[X_j X_k] E[X_l X_m] E[X_p X_r]\}_s \\
&\quad + 10\{E[X_j X_k X_l] E[X_m X_p X_r]\}_s \\
&\cdots\cdots\cdots
\end{aligned}
\tag{7.1.13}
$$

By selecting a truncation level p, the total number of equations of the form of (7.1.11) required for computing unknown statistical moments up to the pth order becomes finite. Although these equations are nonlinear, they can be solved numerically.

If the system is asymptotically stable, then the motion tends to statistical stationarity. At the stationary state, the statistical moments are independent of time, and equation (7.1.11) reduces to an algebraic equation.

The following example illustrates the application of the cumulant-neglect closure procedure.

7.1.1 A Duffing Oscillator under Additive Random Excitation (Wu and Lin, 1984)

Consider the Duffing oscillator investigated previously by Crandall (1980):

$$\ddot{X} + \eta \dot{X} + (X + \varepsilon X^3) = \sqrt{\eta}\, W(t) \tag{a}$$

where ε is a parameter representing the degree of nonlinearity and $W(t)$ is a gaussian white noise with a correlation function

$$E[W(t)W(t+\tau)] = 2\delta(\tau) \tag{b}$$

Equations (a) and (b) have been normalized so that when $\varepsilon = 0$ the stationary mean-square displacement and velocity are both equal to 1.

The stationary probability density of the displacement $X = X_1$ and velocity $\dot{X} = X_2$ is known to be

$$p(x_1, x_2) = C \exp(-\tfrac{1}{2}x_2^2 - \tfrac{1}{2}x_1^2 - \tfrac{1}{4}\varepsilon x_1^4) \tag{c}$$

where C is the normalization constant given by

$$C = \sqrt{\frac{\varepsilon}{\pi}} \left[\exp\left(\frac{1}{8\varepsilon}\right) K_{1/4}\left(\frac{1}{8\varepsilon}\right) \right]^{-1} \tag{d}$$

in which $K_{1/4}$ is a modified Bessel function (Gradshteyn and Ryzhik, 1980). It is seen that X_1 and X_2 are independent, each with a zero mean. The mean-square velocity $E\left[X_2^2\right] = 1$ can be calculated readily. The mean-square displacement is found to be

$$\begin{aligned} E\left[X_1^2\right] &= \sqrt{2\pi} C \int_{-\infty}^{\infty} x_1^2 \exp\left(-\frac{1}{2}x_1^2 - \frac{1}{4}\varepsilon x_1^4\right) dx_1 \\ &= C\pi 2^{1/4} \varepsilon^{-3/4} \exp\left(\frac{1}{8\varepsilon}\right) D_{-3/2}\left(\frac{1}{\sqrt{2\varepsilon}}\right) \end{aligned} \tag{e}$$

where $D_{-3/2}$ is a parabolic cylinder function (Abramowitz and Stegun, 1972).

Corresponding to system (a), equation (7.1.11) reads

$$\frac{d}{dt} E[M] = E\left[X_2 \frac{\partial M}{\partial X_1}\right] - E\left[(\eta X_2 + X_1 + \varepsilon X_1^3) \frac{\partial M}{\partial X_2}\right] + \eta E\left[\frac{\partial^2 M}{\partial^2 X_2}\right] \tag{f}$$

Letting $M = X_1^i X_2^j (i + j = 1, 2, 3, 4)$ and denoting $\mu_{ij} = E[X_1^i X_2^j]$, we obtain the following differential equations for the statistical moments:

$$\left\{ \begin{aligned} \frac{d\mu_{10}}{dt} &= \mu_{01} \\ \frac{d\mu_{01}}{dt} &= -\eta\mu_{01} - \mu_{10} - \varepsilon\mu_{30} \end{aligned} \right. \tag{g.1}$$

$$\left\{ \begin{aligned} \frac{d\mu_{20}}{dt} &= 2\mu_{11} \\ \frac{d\mu_{11}}{dt} &= \mu_{02} - \eta\mu_{11} - \mu_{20} - \varepsilon\mu_{40} \\ \frac{d\mu_{02}}{dt} &= -2\eta\mu_{02} - 2\mu_{11} - 2\varepsilon\mu_{31} + 2\eta \end{aligned} \right. \tag{g.2}$$

$$\left\{ \begin{aligned} \frac{d\mu_{30}}{dt} &= 3\mu_{21} \\ \frac{d\mu_{21}}{dt} &= 2\mu_{12} - \eta\mu_{21} - \mu_{30} - \varepsilon\mu_{50} \\ \frac{d\mu_{12}}{dt} &= \mu_{03} - 2\eta\mu_{12} - 2\mu_{21} - 2\varepsilon\mu_{41} + 2\eta\mu_{10} \\ \frac{d\mu_{03}}{dt} &= -3\eta\mu_{03} - 3\mu_{12} - 3\varepsilon\mu_{32} + 6\eta\mu_{01} \end{aligned} \right. \tag{g.3}$$

$$
\left\{
\begin{aligned}
\frac{d\mu_{40}}{dt} &= 4\mu_{31} \\
\frac{d\mu_{31}}{dt} &= 3\mu_{22} - \eta\mu_{31} - \mu_{40} - \varepsilon\mu_{60} \\
\frac{d\mu_{22}}{dt} &= 2\mu_{13} - 2\eta\mu_{22} - 2\mu_{31} - 2\varepsilon\mu_{51} + 2\eta\mu_{20} \\
\frac{d\mu_{13}}{dt} &= \mu_{04} - 3\eta\mu_{13} - 3\mu_{22} - 3\varepsilon\mu_{42} + 6\eta\mu_{11} \\
\frac{d\mu_{04}}{dt} &= -4\eta\mu_{04} - 4\mu_{13} - 4\varepsilon\mu_{33} + 12\eta\mu_{02}
\end{aligned}
\right. \tag{g.4}
$$

The groups of equations (g.1) and (g.2) are required for gaussian closure, or the second-order cumulant-neglect closure (neglecting those cumulants above the second order). Group (g.3) must be added for the third-order cumulant-neglect closure, group (g.4) further added for the fourth-order cumulant-neglect closure. Additional groups are required if higher order moments are included.

For stationary solutions, the statistical moments are constants, and their time derivatives are zero. Equations (g.1), (g.2), and so on, then reduce to algebraic equations. Furthermore, it is obvious that $E[X_1] = \mu_{10} = 0$ since, in the stationary state, the probability density of the displacement governed by the equation of motion (a) must be an even function. Thus equations (7.1.7), (7.1.12), and (7.1.13) are applicable. Letting $\kappa_n = 0$ for $n \geq 3$, we obtain

$$
\mu_{30} = 0 \qquad \mu_{40} = 3(\mu_{20})^2 \qquad \mu_{31} = 3\mu_{20}\mu_{11} \tag{h}
$$

Substitution of (h) into (g.1) and (g.2) yields the gaussian closure result

$$
E[X_1^2] = \mu_{20} = \frac{-1 + \sqrt{1 + 12\varepsilon}}{6\varepsilon} \tag{i}
$$

which is an approximation, as well as $E[X_2^2] = \mu_{02} = 1$ and $E[X_1X_2] = \mu_{11} = 0$, which are exact.

Focusing our attention on $\mu_{20} = E[X_1^2]$, we can show that, by neglecting cumulants above the fourth order, we obtain for μ_{20}

$$
30\varepsilon^2\mu_{20}^3 + 15\varepsilon\mu_{20}^2 + (1 - 12\varepsilon)\mu_{20} - 1 = 0 \tag{j}
$$

and, by ignoring cumulants beyond the sixth,

$$
714\varepsilon^3\mu_{20}^4 + 420\varepsilon^2\mu_{20}^3 + (63 - 336\varepsilon)\varepsilon\mu_{20}^2 + (1 - 90\varepsilon)\mu_{20} - (1 - 30\varepsilon) = 0 \tag{k}
$$

It is of interest to compare the estimates for $E[X_1^2]$ computed from equations (i), (j), and (k) with the exact solution (e). For a small ε, we obtain the following asymptotic expansions:

From gaussian closure:

$$
E[X_1^2] \approx 1 - 3\varepsilon + 18\varepsilon^2 - \cdots \tag{l}
$$

From the fourth-order cumulant-neglect closure:

$$
E[X_1^2] \approx 1 - 3\varepsilon + 24\varepsilon^2 - 297\varepsilon^3 + 4536\varepsilon^4 - \cdots \tag{m}
$$

From the sixth-order cumulant-neglect closure:

$$E[X_1^2] \approx 1 - 3\varepsilon + 24\varepsilon^2 - 297\varepsilon^3 + 4986\varepsilon^4 - 100{,}278\varepsilon^5 + \cdots \tag{n}$$

From exact solution:

$$E[X_1^2] \approx 1 - 3\varepsilon + 24\varepsilon^2 - 297\varepsilon^3 + 4986\varepsilon^4 - 100{,}278\varepsilon^5 + \cdots \tag{o}$$

Equation (m) is the same as the fourth-order approximation obtained by Crandall (1980) using a truncated Gram-Charlier expansion. Equation (n) is slightly closer to the exact solution than the sixth-order approximation of Crandall (1980). For a large ε, the present analysis yields the following equations:

From gaussian closure:

$$E[X_1^2] \approx \frac{0.5774}{\sqrt{\varepsilon}} \tag{p}$$

From the fourth-order cumulant-neglect closure:

$$E[X_1^2] \approx \frac{0.6325}{\sqrt{\varepsilon}} \tag{q}$$

From the sixth-order cumulant-neglect closure:

$$E[X_1^2] \approx \frac{0.6480}{\sqrt{\varepsilon}} \tag{r}$$

From cxact solution:

$$E[X_1^2] \approx \frac{0.6760}{\sqrt{\varepsilon}} \tag{s}$$

Equation (r) is not as accurate as Crandall's sixth-order result of $0.6606/\sqrt{\varepsilon}$. Nevertheless, the cumulant-neglect procedure does have a tendency to converge to the exact solution for all values of ε as shown in Fig. 7.1.1.

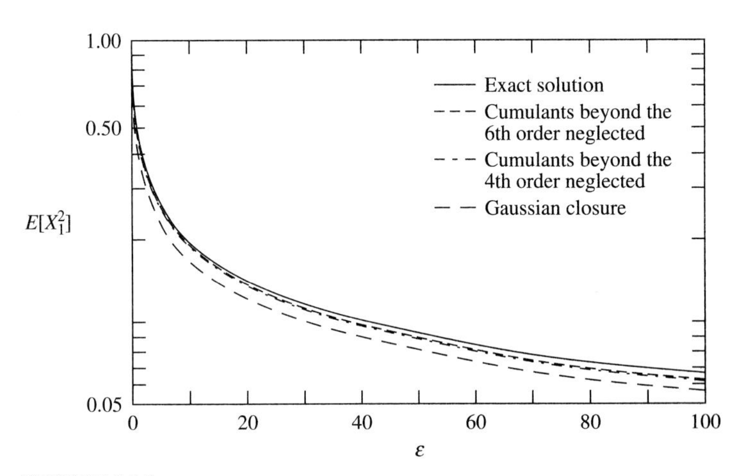

FIGURE 7.1.1
Stationary mean-square displacement of a Duffing oscillator [after Wu and Lin, 1984; Copyright ©1984 Elsevier Science Ltd., reprinted with permission].

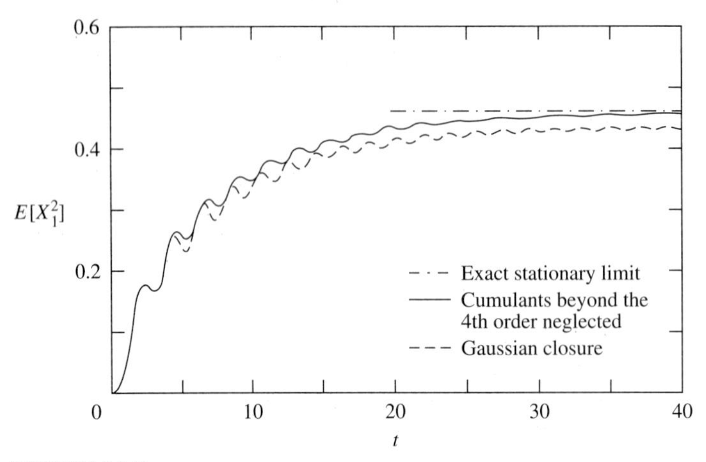

FIGURE 7.1.2
Transient mean-square displacement of a Duffing oscillator [after Wu and Lin, 1984; Copyright ©1984 Elsevier Science Ltd., reprinted with permission].

When transient nonstationary results are required, the time derivatives in equations (g.1), (g.2), and so on, must be retained. Application of a cumulant-neglect closure to the right-hand sides of these differential equations results in nonlinear terms, and, in general, numerical integrations must be performed to obtain the solutions. Several efficient subroutines are available for this purpose (e.g., Gear, 1971). The results for $E[X_1^2]$ obtained from gaussian closure and from the fourth-order cumulant-neglect closure are plotted in Fig. 7.1.2 for the Duffing system governed by equation (a) with $\varepsilon = 1$, $\eta = 0.1$, and a deterministic initial condition $X_1(0) = X_2(0) = 0$.

We note in passing that the use of a closure scheme may lead to large errors when the system response is near a stability boundary. Therefore, it should be avoided if multiplicative excitations are present, particularly if the question of system stability is of primary concern. Illustrative examples in this regard may be found in papers by Sun and Hsu (1987) and by Brückner and Lin (1987).

7.2 METHOD OF WEIGHTED RESIDUALS

Instead of calculating approximate statistical moments, the method of weighted residuals focuses directly on the unknown probability density. Consider a nonlinear system

$$\frac{d}{dt}X_i = F_i(\boldsymbol{X}) + g_{ij}(\boldsymbol{X})W_j(t) \qquad i = 1, 2, \ldots, n, \quad j = 1, 2, \ldots, m \tag{7.2.1}$$

where $W_j(t)$ are gaussian white noises. The reduced Fokker-Planck equation associated with system (7.2.1) is given by

$$\frac{\partial}{\partial x_i}\left[\left(F_i + \pi K_{ls} g_{rs}\frac{\partial g_{il}}{\partial x_r}\right)\tilde{p}\right] - \pi K_{ls}\frac{\partial^2}{\partial x_i \partial x_j}(g_{il}g_{js}\tilde{p}) = 0 \tag{7.2.2}$$

where K_{ls} is the cross-spectral density of $W_l(t)$ and $W_s(t)$, and $\tilde{p}(\boldsymbol{x})$ is the stationary probability density of the response vector $\boldsymbol{X}$. Suppose that an exact solution for equation (7.2.2) is not obtainable analytically, and we wish to find a replacement system within the solvable class of generalized stationary potential, which is closest to the original system in some statistical sense. Then the exact solution of the replacement system may be considered as an approximate solution for the replaced original system.

Let the replacing system be governed by

$$\frac{d}{dt}X_i = f_i(\boldsymbol{X}) + g_{ij}(\boldsymbol{X})W_j(t) \qquad i = 1, 2, \ldots, n, \quad j = 1, 2, \ldots, m \tag{7.2.3}$$

and let $p(\boldsymbol{x})$ be the exact stationary probability density of its response; that is, $p(\boldsymbol{x})$ satisfies the reduced Fokker-Planck equation

$$\frac{\partial}{\partial x_i}\left[\left(f_i + \pi K_{ls} g_{rs} \frac{\partial g_{il}}{\partial x_r}\right)p\right] - \pi K_{ls} \frac{\partial^2}{\partial x_i \partial x_j}(g_{il} g_{js} p) = 0 \tag{7.2.4}$$

Our objective is to select a set of appropriate $f_i(x)$ functions such that system (7.2.3) is the closest to system (7.2.1) in some statistical sense. Replacing $\tilde{p}(\boldsymbol{x})$ by $p(\boldsymbol{x})$ in equation (7.2.2) results in a residual error

$$\delta = \frac{\partial}{\partial x_i}\left[\left(F_i + \pi K_{ls} g_{rs} \frac{\partial g_{il}}{\partial x_r}\right)p\right] - \pi K_{ls} \frac{\partial^2}{\partial x_i \partial x_j}(g_{il} g_{js} p) \tag{7.2.5}$$

Subtracting equation (7.2.5) from (7.2.4), we obtain

$$\delta = \frac{\partial}{\partial x_i}\left\{[F_i(\boldsymbol{x}) - f_i(\boldsymbol{x})]\, p(\boldsymbol{x})\right\} \tag{7.2.6}$$

The residual δ, as a function of $\boldsymbol{x}$, is a measure of error of the approximate solution $p(\boldsymbol{x})$ from the unknown true solution $\tilde{p}(\boldsymbol{x})$.

The method of weighted residuals (e.g., Finlayson, 1972) is a global minimization scheme, in which functions f_i, hence $p(\boldsymbol{x})$, are chosen such that the integral

$$\Delta_M = \int M(\boldsymbol{x})\delta\, d\boldsymbol{x} = \int M(\boldsymbol{x})\frac{\partial}{\partial x_i}\left\{[F_i(\boldsymbol{x}) - f_i(\boldsymbol{x})]\, p(\boldsymbol{x})\right\} d\boldsymbol{x} = 0 \tag{7.2.7}$$

for some selected weighting function $M(\boldsymbol{x})$. In equation (7.2.7), the subscript M of Δ indicates that the value of Δ depends on the choice of the weighting function M. For the present purpose, we shall impose rather general restrictions that the M function is differentiable with respect to x_i and that, at the boundaries of the $\boldsymbol{x}$ domain,

$$M(F_i - f_i)p = 0 \tag{7.2.8}$$

Upon integration by parts on every x_i, equation (7.2.7) is changed to read

$$\Delta_M = E\left\{[F_i(\boldsymbol{X}) - f_i(\boldsymbol{X})]\frac{\partial M(\boldsymbol{x})}{\partial X_i}\right\} = 0 \tag{7.2.9}$$

where $E[\cdot]$ denotes the ensemble average with respect to the probability density $p(\boldsymbol{x})$. By judiciously selecting a set of M functions, we obtain a corresponding set of constraints of the type of (7.2.9). The $f_i(\boldsymbol{x})$ functions for the replacement system can then be determined from these constraints. Clearly, the required number of constraints, hence the required number of weighting functions, depends on the number of parameters to be determined in functions $f_i(\boldsymbol{x})$.

We digress to show that the foregoing approximate solution procedure is self-consistent in terms of certain statistical moments. Let the weighting functions M be chosen in the form of $M(\mathbf{X}) = X_1^{k_1} X_2^{k_2} \cdots X_n^{k_n}$, where $k_1, k_2, \ldots, k_n$ are positive integers. The equation for $E[M]$ for each set of $k_1, k_2, \ldots, k_n$ may be obtained as follows:

$$\frac{d}{dt}E[M] = E\left[\left(f_i + \pi K_{ls} g_{rs} \frac{\partial g_{il}}{\partial X_r}\right)\frac{\partial M}{\partial X_i}\right] + \pi K_{ls} E\left[g_{il} g_{rs} \frac{\partial^2 M}{\partial X_i \partial X_j}\right] \tag{7.2.10}$$

which is the same as equation (7.1.11). After exposure to the white noise excitations for a sufficiently long time, the system response attains the state of statistical stationarity, at which time (7.2.10) reduces to the algebraic equation

$$E\left[\left(f_i + \pi K_{ls} g_{rs} \frac{\partial g_{il}}{\partial X_r}\right)\frac{\partial M}{\partial X_i}\right] + \pi K_{ls} E\left[g_{il} g_{js} \frac{\partial^2 M}{\partial X_i \partial X_j}\right] = 0 \tag{7.2.11}$$

Substitution of (7.2.9) into (7.2.11) results in

$$E\left[\left(F_i + \pi K_{ls} g_{rs} \frac{\partial g_{il}}{\partial X_r}\right)\frac{\partial M}{\partial X_i}\right] + \pi K_{ls} E\left[g_{il} g_{js} \frac{\partial^2 M}{\partial X_i \partial X_j}\right] = \Delta_M \tag{7.2.12}$$

If constraint (7.2.9) is satisfied, then

$$E\left[\left(F_i + \pi K_{ls} g_{rs} \frac{\partial g_{il}}{\partial X_r}\right)\frac{\partial M}{\partial X_i}\right] + \pi K_{ls} E\left[g_{il} g_{js} \frac{\partial^2 M}{\partial X_i \partial X_j}\right] = 0 \tag{7.2.13}$$

Thus the statistical moments computed from the approximate probability density $p(\mathbf{x})$ satisfy equation (7.2.13). However, it can be shown that (7.2.13) is an exact moment equation which could have been derived directly from the original system (7.2.1); therefore, the method of weighted residuals is self-consistent insofar as those statistical moments appearing in (7.2.13) are concerned.

In particular, let the method be applied to a single-degree-of-freedom second-order system governed by

$$\ddot{X} + H(X, \dot{X}) = g_i(X, \dot{X}) W_i(t) \tag{7.2.14}$$

We assume that this system does not belong to the exactly solvable class of generalized stationary potential. Let equation (7.2.14) be replaced by a solvable

$$\ddot{X} + h(X, \dot{X}) = g_i(X, \dot{X}) W_i(t) \tag{7.2.15}$$

The function $h(X, \dot{X})$ is required to satisfy

$$\Delta_M = E\left\{[H(X_1, X_2) - h(X_1, X_2)]\,\frac{\partial M(X_1, X_2)}{\partial X_2}\right\} = 0 \tag{7.2.16}$$

where $X_1 = X$ and $X_2 = \dot{X}$. A set of constraints can be obtained by selecting a set of suitable weighting functions $M(X_1, X_2)$. In principle, the accuracy of approximation can be improved by adding more constraints, as long as they are compatible.

The method of weighted residuals is next applied under three different settings, by selecting the replacement system from increasingly larger classes of systems which are solvable exactly.

7.2.1 Equivalent Linearization

The method of equivalent linearization was discussed, e.g., in Lin (1967). It is an adequate approximation scheme for a weakly nonlinear system under purely additive random excitations. If the additive excitations are gaussian white noises, then it yields the same results as those of gaussian closure. In this section we show that this method may be cast in the framework of weighted residuals by restricting the replacement system to be linear and the approximate probability density to be gaussian.

Consider the multi-degree-of-freedom system

$$\begin{aligned} \ddot{Y}_j + H_j(Y_1, \ldots, Y_n; \dot{Y}_1, \ldots, \dot{Y}_n) &= c_{jk} W_k(t) \\ j = 1, 2, \ldots, n, \qquad k &= 1, 2, \ldots, m \end{aligned} \tag{a}$$

where c_{jk} are constants and $W_k(t)$ are gaussian white noises. Equation (a) may be rewritten in the first-order form

$$\begin{aligned} \dot{X}_j &= X_{j+n} \\ \dot{X}_{j+n} &= -H_j(X_1, X_2, \ldots, X_{2n}) + c_{jk} W_k(t) \end{aligned} \tag{b}$$

where $X_j = Y_j$ and $X_{j+n} = \dot{Y}_j$. Replace (b) by the linear systems

$$\begin{aligned} \dot{X}_j &= X_{j+n} \\ \dot{X}_{j+n} &= -k_{jl} X_l - \beta_{jl} X_{l+n} + c_{jk} W_k(t) \end{aligned} \tag{c}$$

The coefficients k_{jl} and β_{jl} are chosen to satisfy the constraints

$$E\left[(H_j - k_{jl} X_l - \beta_{jl} X_{l+n}) \frac{\partial M(\boldsymbol{X})}{\partial X_{j+n}}\right] = 0 \tag{d}$$

Letting $M = X_r X_s$, where $r, s = 1, 2, \ldots, 2n$, we obtain

$$E\,[X_s\,(H_s - k_{sl} X_l - \beta_{sl} X_{l+n})] = 0 \qquad \text{(no summation on } s\text{)} \tag{e}$$

$$E\,[X_{s+n}\,(H_s - k_{sl} X_l - \beta_{sl} X_{l+n})] = 0 \qquad \text{(no summation on } s\text{)} \tag{f}$$

Equations (e) and (f) may be combined and cast in the matrix form

$$E[\boldsymbol{X}\boldsymbol{X}'] \begin{bmatrix} \boldsymbol{k}' \\ \boldsymbol{\beta}' \end{bmatrix} = \boldsymbol{E}[\boldsymbol{X}\boldsymbol{H}'] \tag{g}$$

where a prime denotes a matrix transposition, $\boldsymbol{X} = \{X_1, \ldots, X_{2n}\}'$, $\boldsymbol{k} = [k_{ij}]$, $\boldsymbol{\beta} = [\beta_{ij}]$, and $\boldsymbol{H} = \{H_1, \ldots, H_n\}'$. The set of equations (g) can be solved for the unknowns k_{ij} and β_{ij}.

In the case of a single-degree-of-freedom nonlinear oscillator

$$\ddot{X} + H(X, \dot{X}) = W(t) \tag{h}$$

the linear approximation has the form

$$\ddot{X} + \beta_e \dot{X} + k_e X = W(t) \tag{i}$$

Equation (g) can then be written specifically as follows:

$$\begin{bmatrix} E[X_1^2] & E\,[X_1 X_2] \\ E\,[X_1 X_2] & E[X_2^2] \end{bmatrix} \begin{Bmatrix} k_e \\ \beta_e \end{Bmatrix} = \begin{Bmatrix} E\,[X_1 H(X_1, X_2)] \\ E\,[X_2 H(X_1, X_2)] \end{Bmatrix} \tag{j}$$

Equation (j) is solved to yield

$$k_e = \frac{E\left[X_1 H(X_1, X_2)\right]}{E[X_1^2]} \tag{k}$$

$$\beta_e = \frac{E\left[X_2 H(X_1, X_2)\right]}{E[X_2^2]} \tag{l}$$

The stationary solution for the linear system (i) has the well-known gaussian distribution

$$p(x_1, x_2) = C \exp\left[-\frac{\beta_e}{2\pi K}(x_2^2 + k_e x_1^2)\right] \tag{m}$$

where K is the spectral density of $W(t)$. This probability density is now treated as an approximate solution for the original system (h). From (m), we obtain $E[X_1^2] = \pi K/\beta_e k_e$ and $E[X_2^2] = \pi K/\beta_e$, with which equations (k) and (l) are reduced to

$$E\left[X_2 H\left(X_1, X_2\right)\right] = \pi K \tag{n}$$

$$E\left[X_1 H\left(X_1, X_2\right)\right] = \frac{\pi K}{\beta_e} \tag{o}$$

Equations (n) and (o) are a set of nonlinear algebraic equations for the unknowns β_e and k_e. In particular, if $H\left(X_1, X_2\right)$ is a polynomial of X_1 and X_2, then the ensemble averages indicated on the left-hand sides of these equations can be carried out analytically using the approximate probability density given in (m), and β_e and k_e can be solved in closed form.

Equation (j) may be derived also on the condition of the least-mean-square difference:

$$E\{[H\left(X_1, X_2\right) - \beta_e X_2 - k_e X_1]^2\} = \text{minimum} \tag{p}$$

Condition (p) is the very basis for the original equivalent linearization procedure (Lin, 1967), without requiring that the excitation be a gaussian white noise.

7.2.2 Partial Linearization

In Section 7.2.1, an approximate solution was obtained for a nonlinear oscillator by seeking its replacement within the class of solvable linear oscillators. It is reasonable to expect that the accuracy of approximation may be improved if the replacement system is drawn from a larger class, rather than being restricted to linear systems. Let the original system be described by

$$\ddot{X} + H_1(X, \dot{X}) + g(X) = W(t) \tag{a}$$

where $H_1(X, \dot{X})$ represents the damping force, $g\left(X\right)$ represents the conservative force dependent only on the displacement, and $W(t)$ is a gaussian white noise. Then a closer replacement system is obviously

$$\ddot{X} + \beta_e \dot{X} + g\left(X\right) = W(t) \tag{b}$$

in which the conservative force is kept unchanged. Equation (b) possesses an exact stationary solution

$$p(x_1, x_2) = C \exp\left\{ -\frac{\beta_e}{\pi K} \left[\frac{1}{2} x_2^2 + \int_0^{x_1} g(u)\, du \right] \right\} \tag{c}$$

In this case, only the dissipative force needs to be linearized, and the equivalent linear damping coefficient β_e can be determined from

$$\beta_e = \frac{E\left[X_2 H_1 (X_1, X_2)\right]}{E[X_2^2]} \tag{d}$$

Equation (d) has the same form as equation (l) in Section 7.2.1; however, $H_1(X_1, X_2)$ here does not include the nonlinear restoring force term. This method, called partial linearization (Elishakoff and Cai, 1992), is more accurate than the full linearization procedure since it retains the nonlinear restoring force of the original nonlinear system.

It is interesting to note that by letting the weighting function be $M = X_2^2$ in equation (7.2.16), we obtain

$$\int_{-\infty}^{\infty} \int_{-\infty}^{\infty} x_2 \left[H_1(x_1, x_2) - \beta_e x_2\right] \exp\left\{ -\frac{\beta_e}{\pi K} \left[\frac{x_2^2}{2} + \int_0^{x_1} g(u)\, du \right] \right\} dx_1\, dx_2 = 0 \tag{e}$$

Equations (d) and (e) are, in fact, identical; therefore, partial linearization can also be classified under the method of weighted residuals.

Equation (e) is a nonlinear algebraic equation for β_e, which can be simplified to read

$$\int_{-\infty}^{\infty} \int_{-\infty}^{\infty} \left[x_2 H_1(x_1, x_2) - \pi K\right] \exp\left\{ -\frac{\beta_e}{\pi K} \left[\frac{x_2^2}{2} + \int_0^{x_1} g(u)\, du \right] \right\} dx_1\, dx_2 = 0 \tag{f}$$

It can be solved either analytically or numerically, depending on the form of the original system. Once β_e is known, the approximate joint probability density, equation (c), is determined. If the nonlinear damping force depends only on the velocity—that is, if $H_1(x_1, x_2) = H_1(x_2)$—then equation (f) is reduced to

$$\int_{-\infty}^{\infty} \left[x_2 H_1(x_2) - \pi K\right] \exp\left(-\frac{\beta_e x_2^2}{2\pi K} \right) dx_2 = 0 \tag{g}$$

Equation (g) can often be solved analytically for β_e.

It should be noted that if the damping force depends on both the displacement and velocity, then the equivalent damping coefficient calculated from (f) is different from that which would have been obtained from equivalent linearization [from equation (n) of Section 7.2.1] because the displacement here is not approximated as a gaussian random variable.

Although partial linearization is generally more accurate than full linearization, both methods are considered unsuitable if multiplicative random excitations are present.

7.2.3 Dissipation Energy Balancing

In the equivalent linearization or the partial linearization procedure, the replacement system is selected from a subclass of the exactly solvable class of generalized stationary potential. By enlarging the selection pool to include the entire solvable class of generalized stationary potential, the accuracy of the approximate solution obtained can be further improved. Moreover, it is desirable that an approximate solution technique is applicable when both additive and multiplicative random excitations are present, not just additive excitations.

Let the original system be governed by the second-order equation (7.2.14), and let it be replaced by equation (7.2.15). It was shown in Section 5.3.1 that system (7.2.15) belongs to the class of generalized stationary potential if function h can be expressed in the form

$$h(X_1, X_2) = \pi X_2 K_{ij} g_i g_j \phi'(\Lambda) - \pi K_{ij} g_i \frac{\partial g_j}{\partial X_2} + u(X_1) \tag{a}$$

where $u(X_1)$ is the *effective* conservative force, Λ is the *effective* total energy given by

$$\Lambda = \tfrac{1}{2} X_2^2 + \int u(X_1)\, dX_1 \tag{b}$$

and where function ϕ is the probability potential. The probability density of the system response is given by

$$p(x_1, x_2) = C \exp(-\phi) \tag{c}$$

The effective conservative force may include contributions from the Wang and Zakai correction terms. As implied in the notation, the expression for $u(X_1)$ does not contain X_2 , but it may contain a constant. Our objective is to find a replacement system for the original system (7.2.14) from such a general class, subject to the constraints (7.2.16) of vanishing weighted residuals.

Substitution of (a) into (7.2.16) leads to

$$\Delta_M = E\left\{\left[H - \pi X_2 K_{ij} g_i g_j \phi'(\Lambda) + \pi K_{ij} g_i \frac{\partial g_i}{\partial X_2} - u(X_1)\right] \frac{\partial M}{\partial X_2}\right\} = 0 \tag{d}$$

By choosing a set of appropriate weighting functions M, we obtained a corresponding set of statistical conditions for finding the total effective conservative force $u(X_1)$ and the probability potential ϕ for the replacement system.

First, we let $M = X_2 X_1^k$ where k is a nonnegative integer, and we obtain from (d)

$$E\left\{X_1^k \left[H - \pi X_2 K_{ij} g_i g_j \phi'(\Lambda) + \pi K_{ij} g_i \frac{\partial g_j}{\partial X_2} - u(X_1)\right]\right\} = 0 \tag{e}$$

We assume that

$$\pi K_{ij} g_i g_j p = 0 \qquad |x_2| \to +\infty \tag{f}$$

which is a reasonable assumption in most practical cases. Thus

$$\begin{aligned} &E[X_1^k \pi K_{ij} X_2 g_i g_j \phi'(\Lambda)] \\ &= \int_{-\infty}^{\infty} x_1^k \left[\int_{-\infty}^{\infty} \pi K_{ij} x_2 g_i g_j \phi'(\lambda)\, p\, dx_2\right] dx_1 \\ &= \int_{-\infty}^{\infty} x_1^k \left\{\left[-\pi K_{ij} g_i g_j p\right]_{x_2=-\infty}^{x_2=\infty} + \int_{-\infty}^{\infty} 2\pi K_{ij} g_i \frac{\partial g_j}{\partial x_2} p\, dx_2\right\} dx_1 \\ &= E\left[2\pi X_1^k K_{ij} g_i \frac{\partial g_i}{\partial X_2}\right] \end{aligned} \tag{g}$$

where we have used the relationships $\partial p/\partial x_2 = -\phi'(\lambda)\, x_2 p$ and $K_{ij}\partial(g_i g_j)/\partial x_2 = 2K_{ij}g_i(\partial g_j/\partial x_2)$. Substitution of (g) into (e) results in

$$E\left\{X_1^k\left[H - \pi K_{ij}g_i\frac{\partial g_j}{\partial X_2} - u(X_1)\right]\right\} = 0 \tag{h}$$

Equation (h) provides a set of constraints for the determination of the equivalent conservative force of the replacement system.

The expression $H - \pi K_{ij}g_i(\partial g_j/\partial X_2)$ in equation (h) may be separated into two parts:

$$H - \pi K_{ij}g_i\frac{\partial g_j}{\partial X_2} = \nu(X_1, X_2) + \tilde{u}(X_1) \tag{i}$$

where $\tilde{u}(X_1)$ is the part dependent only on X_1, which is clearly the total effective conservative force of the original system, and $\nu(X_1, X_2)$ is the remaining part dependent on X_2. It is logical to choose $u(X_1) = \tilde{u}(X_1)$; that is, the total conservative force of the replacement system is the same as that of the original system. Then equation (h) is reduced to

$$E[X_1^k \nu(X_1, X_2)] = 0 \tag{j}$$

If, in addition, $\nu(X_1, X_2)$ is an odd function of X_2, which is the case with most practical problems, then (j) is identically satisfied for every k. The highly unusual case in which $\nu(X_1, X_2)$ contains even-function terms of X_2 also can be treated (Cai, Lin, and Elishakoff, 1992; To and Li, 1991) but will not be pursued here for lack of practical use.

Next we let $M = X_2^2$ and obtain from (d)

$$E\left\{X_2\left[H - \pi X_2 K_{ij}g_i g_j \phi'(\Lambda) + \pi K_{ij}g_i\frac{\partial g_j}{\partial X_2}\right]\right\} = 0 \tag{k}$$

The approximate probability density $p(x_1, x_2)$, which is of the form of (c), can be substituted into equation (k) to yield

$$\int_{-\infty}^{\infty}\int_{-\infty}^{\infty} \exp[-\phi(\lambda)]\, x_2\left[H - \pi x_2 K_{ij}g_i g_j \phi'(\lambda) + \pi K_{ij}g_i\frac{\partial g_j}{\partial x_2}\right] dx_1\, dx_2 = 0 \tag{l}$$

Let the effective potential energy be denoted by

$$U(x_1) = \int u(x_1)\, dx_1 \tag{m}$$

In general, there exists a unique point $x_1 = x_{10}$ where $U(x_{10})$ is a minimum. Without loss of generality, we let $U(x_{10}) = 0$. We assume that

$$U'(x_1) = u(x_1)\begin{cases} > 0 & x_1 > x_{10} \\ < 0 & x_1 < x_{10} \end{cases} \tag{n}$$

and that for every $\lambda > 0$, the equation $U(x_1) = \lambda$ has two roots, $\mu_{\lambda 1}$ and $\mu_{\lambda 2}$. Such a potential energy function is depicted in Fig. 7.2.1. It is of interest to note that we do not presume $x_{10} = 0$ to allow for possible contribution to $u(x_1)$ from certain multiplicative random excitations.

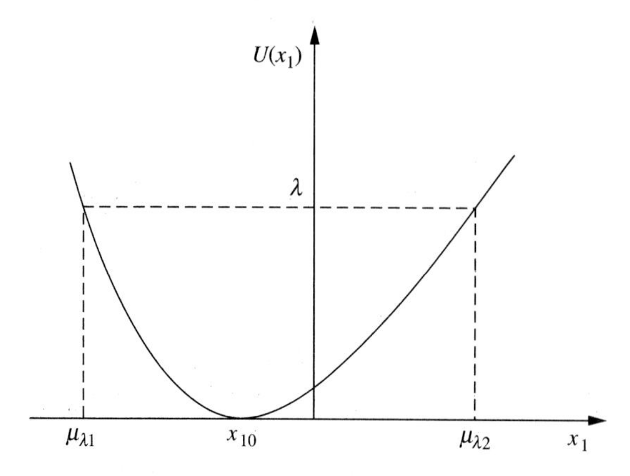

FIGURE 7.2.1
A hypothetical potential energy function $U(x_1)$.

By using the transformation

$$x_1 = x_1 \qquad x_2 = \begin{cases} \sqrt{2\lambda - 2U(x_1)} & x_2 \geq 0 \\ -\sqrt{2\lambda - 2U(x_1)} & x_2 < 0 \end{cases} \tag{o}$$

the integration on x_1 and x_2 in equation (l) can be transformed to that of integration on x_1 and λ as follows:

$$\int_0^\infty \exp[-\phi(\lambda)]\,d\lambda \int_{\mu_{\lambda 1}}^{\mu_{\lambda 2}} \left\{ \left[H - \pi x_2 K_{ij} g_i g_j \phi'(\lambda) + \pi K_{ij} g_i \frac{\partial g_j}{\partial x_2} \right]_{x_2 = \sqrt{2\lambda - 2U(x_1)}} \right.$$
$$\left. - \left[H - \pi x_2 K_{ij} g_i g_j \phi'(\lambda) + \pi K_{ij} g_i \frac{\partial g_j}{\partial x_2} \right]_{x_2 = -\sqrt{2\lambda - 2U(x_1)}} \right\} dx_1 = 0 \tag{p}$$

Since $\phi(\lambda)$ is still unknown, we replace (p) by a stronger sufficient condition, requiring that the integration on x_1 vanishes for every λ:

$$\int_{\mu_{\lambda 1}}^{\mu_{\lambda 2}} \left\{ \left[H - \pi x_2 K_{ij} g_i g_j \phi'(\lambda) + \pi K_{ij} g_i \frac{\partial g_j}{\partial x_2} \right]_{x_2 = \sqrt{2\lambda - 2U(x_1)}} \right.$$
$$\left. - \left[H - \pi x_2 K_{ij} g_i g_j \phi'(\lambda) + \pi K_{ij} g_i \frac{\partial g_j}{\partial x_2} \right]_{x_2 = -\sqrt{2\lambda - 2U(x_1)}} \right\} dx_1 = 0 \tag{q}$$

which leads to the following expression for $\phi'(\lambda)$:

$$\phi'(\lambda) = \frac{\displaystyle\int_{\mu_{\lambda 1}}^{\mu_{\lambda 2}} \left\{ \left[H + \pi K_{ij} g_i \frac{\partial g_j}{\partial x_2} \right]_{x_2 = \sqrt{2\lambda - 2U(x_1)}} - \left[H + \pi K_{ij} g_i \frac{\partial g_j}{\partial x_2} \right]_{x_2 = -\sqrt{2\lambda - 2U(x_1)}} \right\} dx_1}{\displaystyle\pi \int_{\mu_{\lambda 1}}^{\mu_{\lambda 2}} \left\{ \left[x_2 K_{ij} g_i g_j \right]_{x_2 = \sqrt{2\lambda - 2U(x_1)}} - \left[x_2 K_{ij} g_i g_j \right]_{x_2 = -\sqrt{2\lambda - 2U(x_1)}} \right\} dx_1} \tag{r}$$

The approximate stationary probability density is then obtained from

$$p(x_1, x_2) = C \exp\left[-\int_0^\lambda \phi'(\nu)\, d\nu\right] \tag{s}$$

It is interesting to note that by keeping the same total conservative force, $u(X_1) = \tilde{u}(X_1)$, in the replacement system, condition (h) is identically satisfied for any k, and condition (q) becomes the only one that is nontrivial. Moreover, equation (r) can then be used to compute directly the probability potential without a prior knowledge of the $h(X_1, X_2)$ function in the replacement system (7.2.15). This is a very important advantage; it implies that the approximation obtained is indeed the best within the class of generalized stationary potential under condition (q), and possible errors associated with the choice of a form for $h(X_1, X_2)$ are therefore avoided.

The physical implication of constraint (k) is that the average dissipated energy per unit time remains the same for the replaced and replacing systems, while the stronger condition (q) requires that the average dissipated energies be identical at every total energy level λ. Therefore, application of equation (k) or (r) has been referred to as *dissipation energy balancing* (Cai and Lin, 1988b).

Equations (r) and (s) can, at least, be evaluated numerically. A closed-form solution is often possible if the effective conservative force $u(x_1)$ is linear. A general form for a linear $u(x_1)$ function may be written as follows:

$$u(x_1) = c_1(x_1 - x_{10}) \qquad c_1 > 0 \tag{t}$$

Correspondingly,

$$U(x_1) = \tfrac{1}{2}c_1(x_1 - x_{10})^2 \tag{u}$$

and

$$\lambda = \tfrac{1}{2}x_2^2 + \tfrac{1}{2}c_1(x_1 - x_{10})^2 \tag{v}$$

These expressions are in agreement with the previous assumption concerning the general forms of $u(x_1)$ and $U(x_1)$. Letting

$$x_1 = \frac{1}{\sqrt{c_1}}\sqrt{2\lambda}\cos\theta + x_{10} \qquad x_2 = \sqrt{2\lambda}\sin\theta \tag{w}$$

we find, from equation (r),

$$\phi'(\lambda) = \frac{\displaystyle\int_0^{2\pi} \sin\theta \left[H + \pi K_{ij} g_i \frac{\partial g_j}{\partial x_2}\right]_{x_1, x_2 \to \lambda, \theta} d\theta}{\displaystyle\pi\sqrt{2\lambda}\int_0^{2\pi} \sin^2\theta K_{ij}\left[g_i g_j\right]_{x_1, x_2 \to \lambda, \theta} d\theta} \tag{x}$$

where the notation $x_1, x_2 \to \lambda, \theta$ indicates that x_1 and x_2 are expressed in terms of λ and θ according to transformation (w). The integrations in equation (x) can be carried out analytically if $H(x_1, x_2)$ and $g_j(x_1, x_2)$ are polynomials of x_1 and x_2.

If multiplicative random excitations are absent—that is, if the g_i are constants—then (r) is reduced to

$$\phi'(\lambda) = \frac{\displaystyle\int_{\mu_{\lambda 1}}^{\mu_{\lambda 2}} \left\{H\left[x_1, \sqrt{2\lambda - 2U(x_1)}\right] - H\left[x_1, -\sqrt{2\lambda - 2U(x_1)}\right]\right\} dx_1}{\displaystyle 2\pi K_{ij} g_i g_j \int_{\mu_{\lambda 1}}^{\mu_{\lambda 2}} \sqrt{2\lambda - 2U(x_1)}\, dx_1} \tag{y}$$

and (x) is reduced to

$$\phi'(\lambda) = \frac{\int_0^{2\pi} H(\lambda, \theta) \sin\theta \, d\theta}{\pi^2 K_{ij} g_i g_j \sqrt{2\lambda}} \tag{z}$$

where $H(\lambda, \theta)$ is obtained from $H(x_1, x_2)$ by changing the arguments from x_1 and x_2 to λ and θ, according to (w).

Equations (x), (y), and (z) are special forms of equation (r). Equations (h) and (r), or one of the special forms of (r), constitute a set of constraints to determine the equivalent conservative force $u(x_1)$, as well as the probability potential $\phi(\lambda)$. We reiterate that if the total effective conservative force in the original system depends only on the displacement, which is the case with most practical problems, then $u(x_1)$ is known and constraints (h) are trivial. In this case, $\phi(\lambda)$ can be calculated from (r) or one of its special forms alone.

The application of the dissipation energy balancing procedure is illustrated in the following two examples. First, we select a nonlinear system for which closed-form approximations are obtainable from the dissipation energy balancing procedure, as well as two other procedures, equivalent linearization and partial linearization. The equation of motion for this example is given by

$$\ddot{X} + \beta\dot{X} + \gamma\dot{X}^3 + \delta X^3 = W(t) \qquad \gamma, \delta > 0 \tag{aa}$$

The conservative force for this system is clearly $u(X) = \delta X^3$. The method of equivalent linearization leads to

$$p(x_1, x_2) = C_1 \exp\left[-\frac{\beta_e}{2\pi K}\left(x_2^2 + k_e x_1^2\right)\right] \tag{bb}$$

where x_1 and x_2 are possible values of $X(t)$ and $\dot{X}(t)$, respectively, and where

$$\beta_e = \frac{\beta}{2} + \sqrt{\frac{\beta^2}{4} + 3\pi\gamma K} \tag{cc}$$

$$k_e = \sqrt{\frac{3\pi\delta K}{\beta_e}} \tag{dd}$$

The method of partial linearization leads to

$$p(x_1, x_2) = C_2 \exp\left[-\frac{\beta_e}{2\pi K}\left(x_2^2 + \frac{1}{2}\delta x_1^4\right)\right] \tag{ee}$$

with the same β_e given in (cc). In the dissipation energy balancing procedure, equation (y) is applicable since multiplicative excitations are absent, and the same conservative force can be used for the replacement system. The result so obtained is given by

$$p(x_1, x_2) = C_3 \exp\left\{-\frac{1}{2\pi K}\left[\beta\left(x_2^2 + \frac{1}{2}\delta x_1^4\right) + \frac{3\gamma}{7}\left(x_2^2 + \frac{1}{2}\delta x_1^4\right)^2\right]\right\} \tag{ff}$$

The derivation of (ff) will be clear after the reader finishes Exercise 7.3(a).

In Fig. 7.2.2 the computed stationary mean-square values of the displacement X for system (aa) are plotted against the nonlinear damping parameter γ. The values of the other system parameters and the excitation spectral density K used in the computation are indicated in the captions. Results obtained from the three approximation schemes are represented by dotted, dashed, and solid lines, respectively. Also shown in the figure, represented by diamonds, are the Monte Carlo simulation results. As expected, the results obtained on the basis of dissipation energy balancing are the

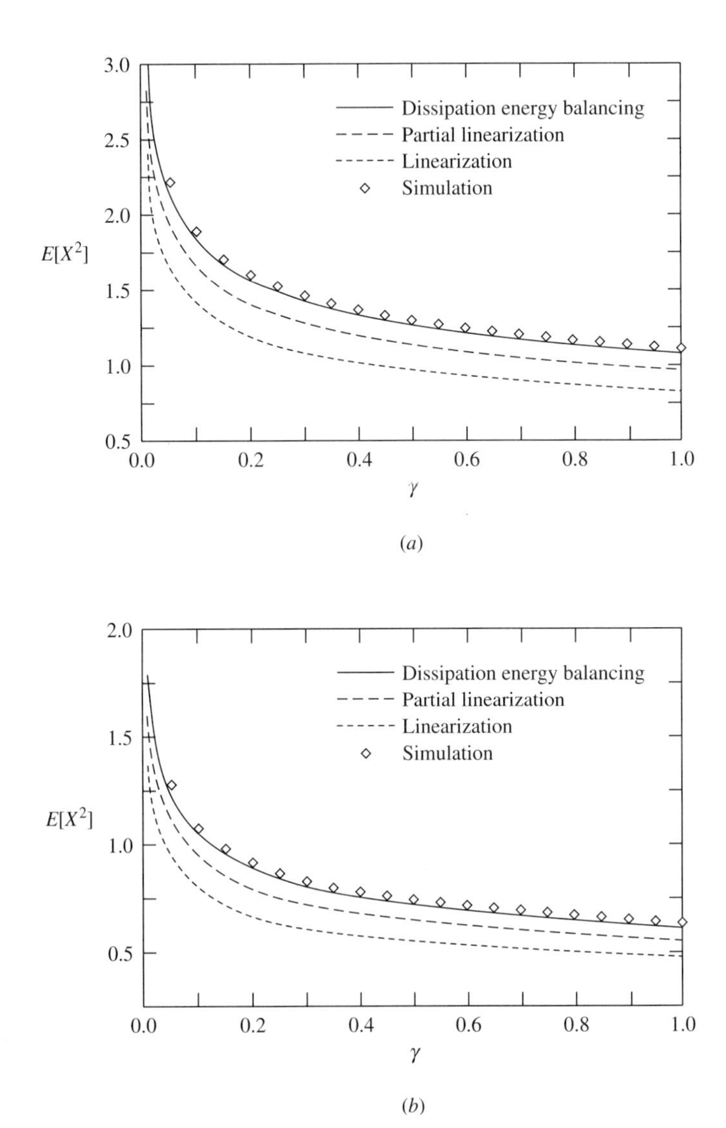

FIGURE 7.2.2
Computed mean-square values of displacement X of a nonlinear system.
(*a*) $\beta = 0.1, \delta = 0.5, K = 1$; (*b*) $\beta = 0.1, \delta = 1.5, K = 1$.

most accurate, and those obtained from equivalent linearization are the least accurate when compared with the simulation results.

Next we consider a van der Pol-type oscillator under combined additive and multiplicative random excitations which are not necessarily independent. Let the equation of motion be

$$\ddot{X} + \alpha\dot{X} + \beta X^2\dot{X} + \gamma\dot{X}^3 + X = XW_1(t) + \dot{X}W_2(t) + W_3(t) \tag{gg}$$

where β and γ are nonnegative constants and $W_1(t)$, $W_2(t)$, and $W_3(t)$ are correlated gaussian white noises with spectral densities K_{ij} $(i, j = 1, 2, 3)$. In comparing equations (gg) and (7.2.14), the following functions can be identified:

$$\begin{gathered} H(X_1, X_2) = \alpha X_2 + \beta X_1^2 X_2 + \gamma X_2^3 + X_1 \\ g_1(X_1, X_2) = X_1 \qquad g_2(X_1, X_2) = X_2 \qquad g_3(X_1, X_2) = 1 \end{gathered} \tag{hh}$$

where $X_1 = X(t)$ and $X_2 = \dot{X}(t)$. From (i),

$$H - \pi K_{ij} g_i \frac{\partial g_j}{\partial x_2} = (\alpha - \pi K_{22}) X_2 + \beta X_1^2 X_2 + \gamma X_2^3 + (1 - \pi K_{12}) X_1 - \pi K_{23} \tag{ii}$$

The total effective conservative force of the replacement system is determined as

$$u(X_1) = (1 - \pi K_{12}) X_1 - \pi K_{23} \tag{jj}$$

It follows that

$$\lambda = \tfrac{1}{2} x_2^2 + \tfrac{1}{2}(1 - \pi K_{12})(x_1 - x_{10})^2 \tag{kk}$$

where

$$x_{10} = \frac{\pi K_{23}}{1 - \pi K_{12}} \tag{ll}$$

In order for the total energy λ to be always positive, it is required that $1 - \pi K_{12} > 0$. Thus $K_{12} < \pi^{-1}$ is a necessary condition for the existence of a stationary solution in probability. Applying equation (x), we obtain

$$\phi'(\lambda) = \frac{2(1 - \pi K_{12})(\alpha + \pi K_{22} + \beta x_{10}^2) + [\beta + 3\gamma(1 - \pi K_{12})]\lambda}{2\pi(1 - \pi K_{12})(K_{11} x_{10}^2 + 2K_{13} x_{10} + K_{33}) + \pi[K_{11} + 3K_{22}(1 - \pi K_{12})]\lambda} \tag{mm}$$

In the case that at least one multiplicative excitation is present so that $K_{11} + 3K_{22}(1 - \pi K_{12}) \neq 0$, the approximate probability density is found to be

$$\begin{aligned} p(x_1, x_2) = {} & C\{4(1 - \pi K_{12})(K_{11} x_{10}^2 + 2K_{13} x_{10} + K_{33}) \\ & + [K_{11} + 3K_{22}(1 - \pi K_{12})][(1 - \pi K_{12})(x_1 - x_{10})^2 + x_2^2]\}^{\sigma} \\ & \exp\left\{ - \frac{3\gamma(1 - \pi K_{12}) + \beta}{2\pi[K_{11} + 3K_{22}(1 - \pi K_{12})]} \left[(1 - \pi K_{12})(x_1 - x_{10})^2 + x_2^2\right]\right\} \end{aligned} \tag{nn}$$

where

$$\sigma = \frac{\begin{array}{c}2(1-\pi K_{12})\{[3\gamma(1-\pi K_{12})+\beta](K_{11}x_{10}^2+K_{33}) \\ -(\alpha+\pi K_{22}+\beta x_{10}^2[K_{11}+3K_{22}(1-\pi K_{12})]\}\end{array}}{\pi[K_{11}+3K_{22}(1-\pi K_{22})]^2} \tag{oo}$$

On the other hand, if only the external excitation $W_3(t)$ is present, then by letting $K_{11} = K_{12} = K_{13} = K_{22} = K_{23} = 0$ in equation (mm), we obtain

$$p(x_1, x_2) = C\exp\left\{-\frac{1}{2\pi K_{33}}\left[\alpha(x_1^2+x_2^2)+\frac{1}{8}(\beta+3\gamma)(x_1^2+x_2^2)^2\right]\right\} \tag{pp}$$

For the following limiting cases, in which the condition for generalized stationary potential is satisfied, the present procedure yields the *exact* results:

1. $K_{12}=K_{13}=K_{23}=0, K_{11}=K_{22}=0, K_{33}\neq 0, \beta=\gamma,$

$$p(x_1, x_2) = C\exp\left\{-\frac{1}{2\pi K_{33}}\left[\alpha(x_1^2+x_2^2)+\frac{\beta}{2}(x_1^2+x_2^2)^2\right]\right\} \tag{qq}$$

2. $K_{12}=K_{13}=K_{23}=0, K_{33}\neq 0, K_{11}=\dfrac{\beta K_{33}}{(\alpha+\pi K_{22})}, K_{22}=\dfrac{\gamma K_{33}}{(\alpha+\pi K_{22})},$

$$p(x_1, x_2) = C\exp\left[-\frac{\alpha+\pi K_{22}}{2\pi K_{33}}(x_1^2+x_2^2)\right] \tag{rr}$$

3. $K_{12}=K_{13}=K_{23}=0,\ K_{11}=K_{22}\neq 0,\ \beta=\gamma,$

$$p(x_1, x_2) = C\left[K_{33}+K_{22}(x_1^2+x_2^2)\right]^{(\beta K_{33}-\pi K_{22}^2-\alpha K_{22})/2\pi K_{22}^2} \times\exp\left[-\frac{\beta}{2\pi K_{22}}(x_1^2+x_2^2)\right] \tag{ss}$$

4. $K_{12}=K_{13}=K_{23}=0,\ K_{11}=K_{33}=0,\ \beta=0,\ \alpha=-\pi K_{22}\neq 0,$

$$p(x_1, x_2) = C\exp\left[-\frac{\gamma}{2\pi K_{22}}(x_1^2+x_2^2)^2\right] \tag{tt}$$

5. $K_{12}=K_{13}=K_{23}=0,\ K_{22}=K_{33}=0,\ K_{11}\neq 0,\ \alpha=\gamma=0,$

$$p(x_1, x_2) = C\exp\left[-\frac{\beta}{2\pi K_{11}}(x_1^2+x_2^2)\right] \tag{uu}$$

It is also of interest to investigate the limiting linear case, $\beta = \gamma = 0$, for which the approximate probability density is given by

$$p(x_1, x_2) = C\{1+\nu[(1-\pi K_{12})(x_1-x_{10})^2+x_2^2]\}^{-\delta} \tag{vv}$$

provided that one or both multiplicative random excitations are present, where

$$\nu = \frac{K_{11}+3K_{22}(1-\pi K_{12})}{4(1-\pi K_{12})(K_{11}x_{10}^2+2K_{13}x_{10}+K_{33})} \tag{ww}$$

$$\delta = \frac{2(1-\pi K_{12})(\alpha+\pi K_{22})}{\pi[K_{11}+3K_{22}(1-\pi K_{12})]} \tag{xx}$$

This approximate probability density exists if $K_{12} < \pi^{-1}$.

The first- and second-order moments computed from equation (vv) are

$$
\begin{aligned}
E[X_1] &= x_{1,0} = \frac{\pi K_{23}}{1 - \pi K_{12}} \\
E[X_2] &= 0 \\
E[X_1^2] &= \frac{\pi(K_{11}x_{10}^2 + 2K_{13}x_{10} + K_{33})}{(\alpha - 2\pi K_{22})(1 - \pi K_{12}) - \pi K_{11}} + x_{10}^2 \\
E[X_1X_2] &= 0 \\
E[X_2^2] &= \frac{\pi(K_{11}x_{10}^2 + 2K_{13}x_{10} + K_{33})(1 - \pi K_{12})}{(\alpha - 2\pi K_{22})(1 - \pi K_{12}) - \pi K_{11}}
\end{aligned}
\tag{yy}
$$

Even though (vv) is approximate, these moments are exact, as can be verified by computing them from the original linear equation of motion. Expressions (yy) imply a first-moment stability condition

$$\pi K_{12} < 1 \tag{zz}$$

and a second-moment stability condition

$$2\pi K_{22} + \frac{\pi K_{11}}{1 - \pi K_{12}} < \alpha \tag{aaa}$$

Note that condition (zz) is a prerequisite for condition (aaa), as expected.

7.3 RANDOMLY EXCITED HYSTERETIC STRUCTURES (Cai and Lin, 1990)

In structural dynamics, the term hysteresis is used to describe a nonconservative system behavior, in which the restoring force depends not only on the instantaneous deformation but also on the past history of the deformation. If motion is cyclic, then a plot of restoring force versus deformation forms a closed loop, the area within the loop being the energy loss per cycle. Reviews of hysteresis characteristics of various structural elements were given by Sozen (1974) and Iwan (1974).

Consider the nondimensional equation of motion for a second-order hysteretic system

$$\ddot{x} + 2\zeta\dot{x} + f(t) = \xi(t) \tag{7.3.1}$$

$$f(t) = \alpha x + (1 - \alpha) z(t) \qquad 0 \le \alpha \le 1 \tag{7.3.2}$$

where $z(t)$ is a hysteretic force and $\xi(t)$ is an excitation. Equation (7.3.2) describes a restoring force, separable into a linear component and a hysteretic component. The hysteretic component z(t) can often be modeled by a first-order differential equation

$$\dot{z}(t) = q(x, \dot{x}, z) \tag{7.3.3}$$

One simple example is the elastoplastic model,

$$\dot{z} = \begin{cases} 0 & z(\operatorname{sgn}\dot{x}) = 1 \\ \dot{x} & |z| < 1 \end{cases} \tag{7.3.4}$$

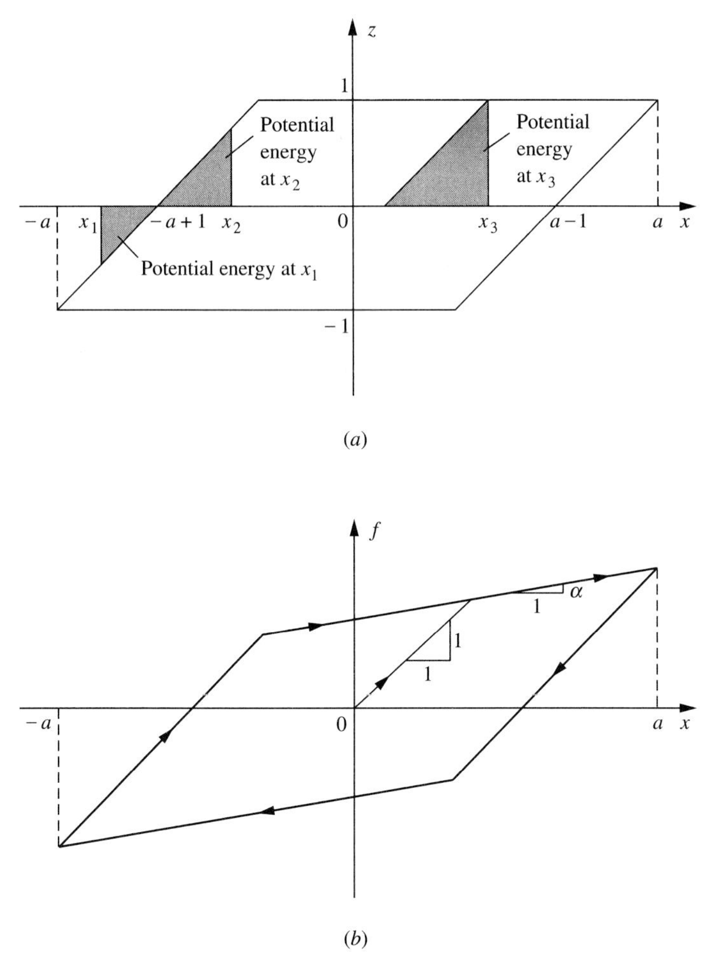

FIGURE 7.3.1
A bilinear hysteresis model. (*a*) Elastoplastic component; (*b*) Restoring force–deformation relationship.

illustrated in Fig. 7.3.1*a*, where the elastic displacement limit has been chosen as the reference length for nondimensionalization. The restoring force $f(t)$, obtained from combining this hysteresis component $z(t)$ and the linear component, constitutes the well-known bilinear model shown in Fig. 7.3.1*b*. Schematically, the bilinear model can be represented by a linear spring and a Jenkin element in parallel, and the latter consists of a linear spring in series with a Coulomb damper, as shown in Fig. 7.3.2.

The abrupt change of slope in the restoring force–deformation relationship in a bilinear model is obviously not realistic. To remedy this deficiency, a smooth hysteretic model was proposed initially by Bouc (1967) and extended by Wen (1976, 1980). This smooth hysteretic model is described by

$$\dot{z} = -\gamma|\dot{x}|z|z|^{n-1} - \beta\dot{x}|z|^n + A\dot{x} \tag{7.3.5}$$

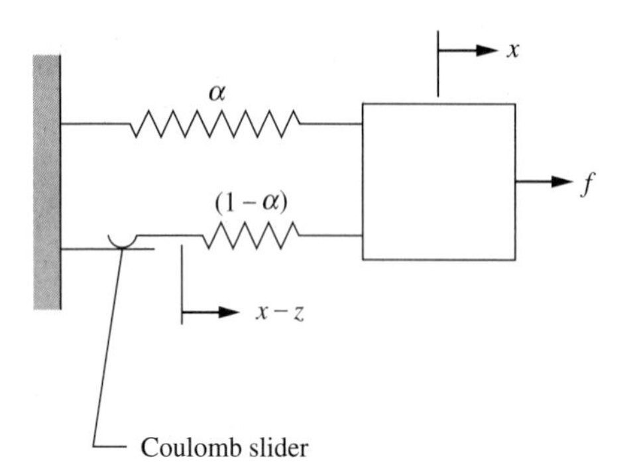

FIGURE 7.3.2
Schematic representation of a bilinear hysteresis model.

Sketches of z versus x and the restoring force f versus x for the Bouc-Wen model are shown in Figs. 7.3.3*a* and 7.3.3*b*, respectively. The smoothness of the force-displacement curve is controlled by value n, the general slope of the curve is controlled by value $\gamma + \beta$, and the slimness of the hysteresis is controlled by γ. By choosing different sets of values for γ, β, n, and A, the behaviors of real hysteretic systems can be approximated closely (Baber and Wen, 1980). This model has become more widely used recently, since it is more amenable to analytical treatments. Further refinements of the model were proposed by Baber and Noori (1984, 1985) and by Casciati (1987).

It can be shown that integration of (7.3.4) or (7.3.5) yields

$$z = \begin{cases} \eta_1(x, \delta) & \dot{x} \geq 0 \\ \eta_2(x, \delta) & \dot{x} \leq 0 \end{cases} \tag{7.3.6}$$

where δ is an integration constant, and $\eta_1 = \eta_2$ when $\dot{x} = 0$. If the excitation $\xi(t)$ in equation (7.3.1) is cyclic, then at the steady state, η_1 would represent one-half the hysteretic loop for increasing x, η_2 would represent the other half for decreasing x, and each value of the integration constant δ would correspond to a given size of the hysteresis loop. Clearly, such a loop size would be determined uniquely by the level of the total energy λ. Thus we can also write, instead of (7.3.6),

$$z = \begin{cases} \bar{\eta}_1(x, \lambda) & \dot{x} \geq 0 \\ \bar{\eta}_2(x, \lambda) & \dot{x} \leq 0 \end{cases} \tag{7.3.7}$$

A larger λ is associated with a larger loop and a longer period.

We now consider a hysteretic system subjected to random excitations, governed by

$$\ddot{X} + 2\zeta\dot{X} + f(t) = g_j(X, \dot{X})W_j(t) \tag{7.3.8}$$

where the $W_j(t)$ are assumed to be gaussian white noises with cross-spectral densities K_{ij}, and the restoring force $f(t)$ is given by equations (7.3.2) and (7.3.3), but with x, $\dot{x}$, and z replaced by X, $\dot{X}$, and Z, respectively, to signify that they are now random. Determination of the probability distribution or statistical properties of the response is difficult for this problem, and no mathematically exact solution is available at the present time. With various degrees of success and limitations, the methods of equivalent linearization (Caughey, 1960; Iwan and Lutes, 1968), equivalent nonlinear system (Lutes, 1970; Lutes and Takemiya, 1974), quasi-conservative averaging

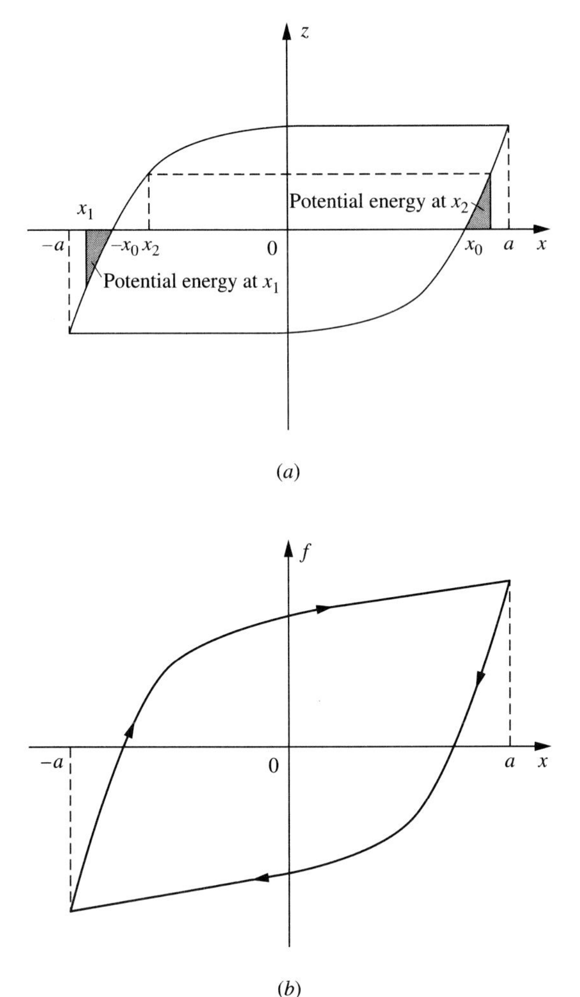

FIGURE 7.3.3
A smooth hysteresis model. (*a*) Elastoplastic component; (*b*) restoring force–deformation relationship.

(Roberts, 1978b; Zhu and Lei, 1987), and moment closure (Kobori, Minai, and Suzuki, 1976) have been applied to the bilinear hysteretic systems, and the methods of Galerkin (Wen, 1976), gaussian closure (Iyenger and Dash, 1978), equivalent linearization (Wen, 1980), and quasi-conservative averaging (Roberts, 1987) have been applied to the smooth Bouc-Wen hysteretic systems.

The preceding references are limited to additive random excitations. One exception is the work by Shih and Lin (1982), in which a multiplicative random excitation was also included in the analysis of a hysteretic column. In this case, the multiplicative excitation arises from the vertical ground motions during an earthquake. The gaussian closure technique was employed to compute the statistics of the structural response.

In this section, the method of dissipation energy balancing is applied to hysteretic systems under gaussian white noise excitations. The excitation can be additive, multiplicative, or both. Certain key issues related particularly to the hysteretic behaviors are explained and resolved.

PROBABILITY DENSITY FUNCTIONS. To apply the dissipation energy balancing approach to a hysteretic system governed by equation (7.3.8), we must calculate the probability potential function ϕ from equation (q) of Section 7.2.3. In the present case, the H function is identified as

$$H = 2\zeta\dot{x} + f(t) = 2\zeta\dot{x} + \alpha x + (1-\alpha)z \tag{7.3.9}$$

where the hysteretic component z is described by (7.3.4) for a bilinear system or by (7.3.5) for a smooth hysteretic system. Under random excitations, δ and λ in (7.3.6) and (7.3.7), respectively, are no longer deterministic. Still, we take the point of view that their one-to-one correspondence remains valid.

Assume that an appropriate expression for the potential energy $U(x)$ can be determined for a hysteretic system. Each total energy level

$$\lambda = \tfrac{1}{2}\dot{x}^2 + U(x) \tag{7.3.10}$$

corresponds to a certain hysteresis loop. The energy consumed by the hysteretic force in one cycle is obviously the area enclosed by the loop:

$$A_r(\lambda) = \oint f\,dx = \int_{-a(\lambda)}^{a(\lambda)} (f)_{\dot{x}>0}\,dx + \int_{a(\lambda)}^{-a(\lambda)} (f)_{\dot{x}<0}\,dx \tag{7.3.11}$$

where $a(\lambda)$ is the amplitude of the loop, that is, the positive root of the equation $U(x) = \lambda$. Substitution of (7.3.9) into (q) of Section 7.2.3 leads to

$$\phi'(\lambda) = \frac{A_r(\lambda) + \int_{-a(\lambda)}^{a(\lambda)} \left[\left(2\zeta\dot{x} + \pi K_{ij} g_i \frac{\partial g_j}{\partial \dot{x}}\right)_{\dot{x}=\sqrt{2\lambda-2U(x)}} - \left(2\zeta\dot{x} + \pi K_{ij} g_i \frac{\partial g_j}{\partial \dot{x}}\right)_{\dot{x}=-\sqrt{2\lambda-2U(x)}}\right] dx}{\pi \int_{-a(\lambda)}^{a(\lambda)} \left[(\dot{x}K_{ij}g_i g_j)_{\dot{x}=\sqrt{2\lambda-2U(x)}} - (\dot{x}K_{ij}g_i g_j)_{\dot{x}=-\sqrt{2\lambda-2U(x)}}\right] dx} \tag{7.3.12}$$

which, in turn, leads to an approximate stationary probability density for X and $\dot{X}$:

$$p(x, \dot{x}) = \rho(\lambda) = C \exp\left[-\int \phi'(\lambda)\,d\lambda\right] \tag{7.3.13}$$

We hasten to add that this expression is not an approximate probability density for Λ. The mean-square value of the displacement may be computed from

$$E[X^2] = \int_{-\infty}^{\infty}\int_{-\infty}^{\infty} x^2 p(x,\dot{x})\,dx\,d\dot{x} = \int_0^{\infty} \rho(\lambda)\,d\lambda \int_{-a(\lambda)}^{a(\lambda)} \frac{2x^2}{\sqrt{2\lambda - 2U(x)}}\,dx \tag{7.3.14}$$

The probability densities for the energy level Λ and the amplitude A, respectively, can be derived from (7.3.13). For the transformation

$$\begin{cases} x = x \\ \lambda = \tfrac{1}{2}\dot{x}^2 + U(x) \end{cases} \quad \text{or} \quad \begin{cases} x = x \\ \dot{x} = \pm\sqrt{2\lambda - 2U(x)} \end{cases} \tag{7.3.15}$$

we obtain the jacobian

$$J = \begin{vmatrix} \dfrac{\partial x}{\partial x} & \dfrac{\partial x}{\partial \lambda} \\ \dfrac{\partial \dot{x}}{\partial x} & \dfrac{\partial \dot{x}}{\partial \lambda} \end{vmatrix} = \pm \frac{1}{\sqrt{2\lambda - 2U(x)}} \tag{7.3.16}$$

The joint probability density $p(x, \lambda)$ of X and Λ is then

$$p_+(x, \lambda) = [p(x, \dot{x})]_{\dot{x}=\sqrt{2\lambda-2U(x)}} |J| = \frac{\rho(\lambda)}{\sqrt{2\lambda - 2U(x)}} \qquad \dot{x} > 0 \tag{7.3.17a}$$

$$p_-(x, \lambda) = [p(x, \dot{x})]_{\dot{x}=-\sqrt{2\lambda-2U(x)}} |J| = \frac{\rho(\lambda)}{\sqrt{2\lambda - 2U(x)}} \qquad \dot{x} < 0 \tag{7.3.17b}$$

The marginal probability density of the total energy Λ follows as

$$p(\lambda) = \int_{-a(\lambda)}^{a(\lambda)} p_+(x, \lambda)\, dx + \int_{-a(\lambda)}^{a(\lambda)} p_-(x, \lambda)\, dx = 2\rho(\lambda) \int_{-a(\lambda)}^{a(\lambda)} \frac{dx}{\sqrt{2\lambda - 2U(x)}} \tag{7.3.18}$$

Since $\lambda = U(a)$, the probability density of the amplitude A is

$$p(a) = p(\lambda) \left| \frac{\partial \lambda}{\partial a} \right| = 2 \left| U'(a) \right| \rho[\lambda(a)] \int_{-a}^{a} \frac{dx}{\sqrt{2\lambda(a) - 2U(x)}} \tag{7.3.19}$$

As indicated, λ must be expressed in terms of the amplitude a in equation (7.3.19).

Once the area A_r within a hysteretic loop and the potential energy $U(x)$ are determined, the probability densities (7.3.13), (7.3.18), and (7.3.19) and the mean-square displacement (7.3.14) can be evaluated numerically. The area A_r can often be obtained in closed form. However, the physical meaning of the potential energy $U(x)$ of a hysteretic system requires clarification. If the deformation of a system is purely elastic, the potential energy is stored as the strain energy, which is fully recoverable when it returns to the original equilibrium configuration. However, if one part of the deformation is hysteretic, then the equilibrium position is no longer unique, but it drifts continuously. A more general definition for the potential energy, applicable to either a purely elastic or a hysteretic system, is that portion of energy associated with the deformation which is recoverable when the system returns to a local equilibrium position. This interpretation is used to obtain analytical expressions for potential energies for various hysteretic systems below.

BILINEAR HYSTERETIC SYSTEMS. For a bilinear hysteretic system, the area of the hysteretic loop, representing the dissipated energy per cycle at amplitude a, can be obtained by inspection:

$$A_r = \begin{cases} 0 & a \le 1 \\ 4(1-\alpha)(a-1) & a \ge 1 \end{cases} \tag{7.3.20}$$

The potential energy of the system is the recoverable part of the total energy if the system is allowed to return to a local equilibrium position upon removal of the external force. Its values for different ranges of x can be determined by referring to the hatched area in Fig. 7.3.1a. Specifically,

$$U(x) = \begin{cases} \frac{1}{2}x^2 & a \le 1 \\ \frac{1}{2}\alpha x^2 + \frac{1}{2}(1-\alpha)(x+a-1)^2 & a \ge 1, \quad -a \le x \le -a+2 \\ \frac{1}{2}\alpha x^2 + \frac{1}{2}(1-\alpha) & a \ge 1, \quad -a+2 \le x \le a \end{cases} \tag{7.3.21}$$

Plots of $U(x)$ versus x for $\alpha = 1/21$ are shown in Fig. 7.3.4a for increasing x ($\dot{x} > 0$). Corresponding plots for decreasing x ($\dot{x} < 0$) are mirror images of those shown in Fig. 7.3.4a. The total energy λ is related to the amplitude as

$$\lambda = \begin{cases} \frac{1}{2}a^2 & a \le 1 \\ \frac{1}{2}(1-\alpha) + \frac{1}{2}\alpha a^2 & a > 1 \end{cases} \tag{7.3.22}$$

Using equations (7.3.21) and (7.3.22), we obtain the following integrals:

$$\int_{-a}^{a} \frac{dx}{\sqrt{2\lambda - 2U(x)}} = \begin{cases} \pi & a \le 1 \\ \frac{\pi}{2}\left(1 + \frac{1}{\sqrt{\alpha}}\right) + \sin^{-1}\frac{1+\alpha-a\alpha}{1-\alpha+a\alpha} + \frac{1}{\sqrt{\alpha}}\sin^{-1}\frac{a-2}{a} & a \ge 1 \end{cases} \tag{7.3.23}$$

$$\int_{-a}^{a} \sqrt{2\lambda - 2U(x)}\, dx = \begin{cases} \frac{1}{2}\pi a^2 & a \le 1 \\ (a-1)(1-\alpha)\sqrt{\alpha(a-1)} + \frac{(1-\alpha+a\alpha)^2}{2}\left(\frac{\pi}{2} + \sin^{-1}\frac{1+\alpha-a\alpha}{1-\alpha+a\alpha}\right) & \\ \quad + \frac{\sqrt{\alpha}}{2}a^2\left(\frac{\pi}{2} + \sin^{-1}\frac{a-2}{a}\right) & a \ge 1 \end{cases} \tag{7.3.24}$$

$$\int_{-a}^{a} \frac{x^2 dx}{\sqrt{2\lambda - 2U(x)}} = \begin{cases} \frac{1}{2}\pi a^2 & a \le 1 \\ \left(4a - 5 - 3\alpha a + 3\alpha + \frac{2-a}{\alpha}\right)\sqrt{\alpha(a-1)} & \\ + \frac{1}{2}[3(a-1)^2(1-\alpha)^2 + a(2-2\alpha-a+2a\alpha)]\left(\frac{\pi}{2} + \sin^{-1}\frac{1+\alpha-a\alpha}{1-\alpha+a\alpha}\right) & \\ + \frac{a^2}{2\sqrt{\alpha}}\left(\frac{\pi}{2} + \sin^{-1}\frac{a-2}{a}\right) & a \ge 1 \end{cases} \tag{7.3.25}$$

Substituting (7.3.20)–(7.3.22) into (7.3.13), (7.3.14), (7.3.18), and (7.3.19) and integrating numerically with respect to λ, we can obtain the stationary probability densities $p(x, \dot{x})$, $p(\lambda)$, and $p(a)$, as well as the mean-square response $E[X^2]$.

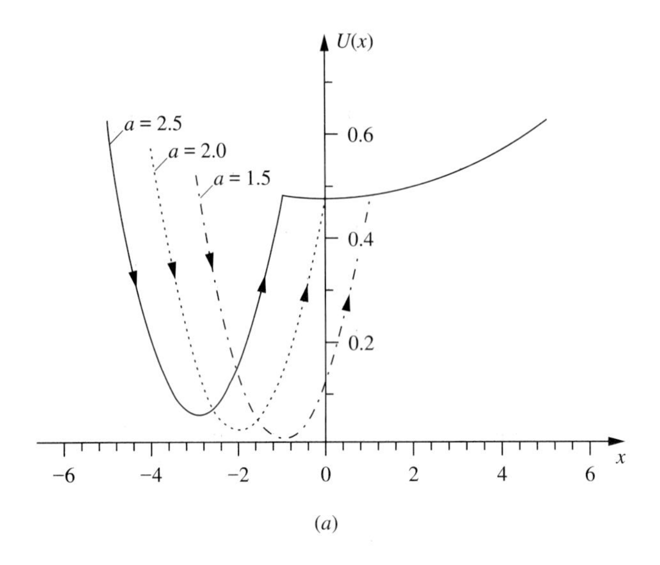

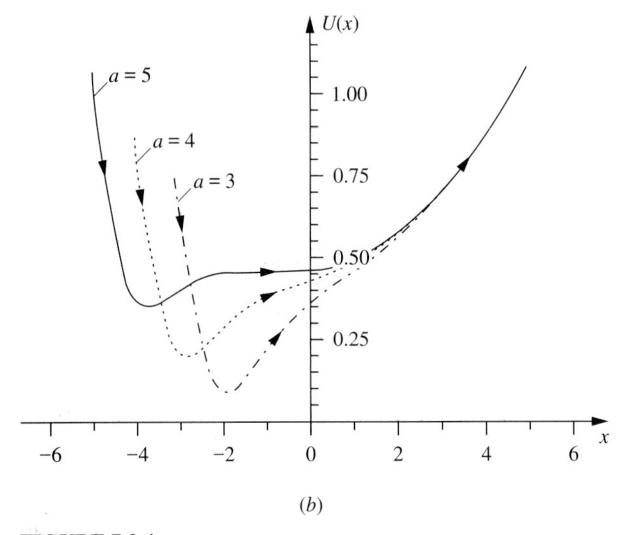

FIGURE 7.3.4
Potential energy of hysteretic systems. (*a*) Bilinear hysteresis model; (*b*) the Bouc-Wen hysteresis model.

SMOOTH HYSTERETIC SYSTEMS. For the Bouc-Wen smooth hysteretic model, equation (7.3.5) can be recast as

$$\dot{z} = \begin{cases} [A + (\gamma - \beta)|z|^n]\,\dot{x} & \dot{x} \geq 0, z \leq 0 \\ [A - (\gamma + \beta)\,z^n]\,\dot{x} & \dot{x} \geq 0, z \geq 0 \\ [A + (\gamma - \beta)\,z^n]\,\dot{x} & \dot{x} \leq 0, z \geq 0 \\ [A - (\gamma + \beta)|z|^n]\,\dot{x} & \dot{x} \leq 0, z \leq 0 \end{cases} \tag{7.3.26}$$

which can be further transformed to

$$\frac{dz}{dx} = \begin{cases} A + (\gamma - \beta)\,|z|^n & \dot{x} \geq 0, z \leq 0 \\ A - (\gamma + \beta)\, z^n & \dot{x} \geq 0, z \geq 0 \\ A + (\gamma - \beta)\, z^n & \dot{x} \leq 0, z \geq 0 \\ A - (\gamma + \beta)\,|z|^n & \dot{x} \leq 0, z \leq 0 \end{cases} \tag{7.3.27}$$

The area of the hysteretic loop corresponding to a given amplitude a is computed from the integral

$$A_r = 2(1 - \alpha)\int_{-a}^{a} z(x)\, dx \tag{7.3.28}$$

The total potential energy, representing the recoverable part of the deformation energy of the system when it returns to a local equilibrium, again consists of two parts: one stored in the linear element, and another in the hysteretic element. The latter may be computed by referring to the hatched areas shown in Fig. 7.3.3*a* for different values of x. For a specific system, the functional relationship between z and x can be obtained by integrating (7.3.27). Then the hysteretic loop area A_r, as a function of amplitude a, and the total potential energy $U(x)$ can also be obtained.

In what follows, the case of $n = 1, A = 1$ is considered in more detail. For this special case, the hysteretic component of the restoring force is found to be

$$z(x) = \begin{cases} \dfrac{1}{\gamma - \beta}[1 - e^{-(\gamma - \beta)(x + x_0)}] & -a \leq x \leq -x_0, \gamma \neq \pm\beta \\ \dfrac{1}{\gamma + \beta}[1 - e^{-(\gamma + \beta)(x + x_0)}] & -x_0 \leq x \leq a, \gamma \neq \pm\beta \end{cases} \tag{7.3.29a}$$

$$z(x) = \begin{cases} x + x_0 & -a \leq x \leq -x_0, \gamma = \beta \\ \dfrac{1}{2\gamma}[1 - e^{-2\gamma(x + x_0)}] & -x_0 \leq x \leq a, \gamma = \beta \end{cases} \tag{7.3.29b}$$

$$z(x) = \begin{cases} \dfrac{1}{2\gamma}[1 - e^{-2\gamma(x + x_0)}] & -a \leq x \leq -x_0, \gamma = -\beta \\ x + x_0 & -x_0 \leq x \leq a, \gamma = -\beta \end{cases} \tag{7.3.29c}$$

The value of x_0, for each of the foregoing expressions, can be determined uniquely for a given amplitude a by using the antisymmetric property $z(a) = -z(-a)$. The area of the hysteretic loop, corresponding to a given amplitude, is given by

$$A_r = \frac{4}{\gamma^2 - \beta^2}(1 - \alpha)\left\{\gamma a - \beta x_0 + \frac{\gamma}{\gamma + \beta}[e^{-(\gamma + \beta)(a + x_0)} - 1]\right\} \qquad \gamma \neq \pm\beta \tag{7.3.30a}$$

$$A_r = (1 - \alpha)\left[-(a - x_0)^2 + \frac{2x_0}{\gamma}\right] \qquad \gamma = \beta \tag{7.3.30b}$$

$$A_r = (1 - \alpha)\left[(a + x_0)^2 - \frac{2x_0}{\gamma}\right] \qquad \gamma = -\beta \tag{7.3.30c}$$

and the potential energy by

$$U(x) = \begin{cases} \frac{1}{2}\alpha x^2 + \frac{1-\alpha}{\gamma-\beta}\Big\{x + x_0 \\ \qquad + \frac{1}{\gamma-\beta}[e^{-(\gamma-\beta)(x+x_0)} - 1]\Big\} & -a \le x \le -x_0, \gamma \ne \pm\beta \\ \frac{1}{2}\alpha x^2 + \frac{1-\alpha}{\gamma^2-\beta^2}\Big\{1 - e^{-(\gamma+\beta)(x+x_0)} - \frac{\gamma+\beta}{\gamma-\beta} \\ \qquad \times \ln\left[1 + \frac{\gamma+\beta}{\gamma-\beta}(1 - e^{-(\gamma+\beta)(x+x_0)})\right]\Big\} & -x_0 \le x \le a, \gamma \ne \pm\beta \end{cases} \tag{7.3.31a}$$

$$U(x) = \begin{cases} \frac{1}{2}\alpha x^2 + \frac{1}{2}(1-\alpha)(x+x_0)^2 & -a \le x \le -x_0, \gamma = \beta \\ \frac{1}{2}\alpha x^2 + \frac{1}{8\gamma^2}(1-\alpha)\left[1 - e^{-2\gamma(x+x_0)}\right]^2 & -x_0 \le x \le a, \gamma = \beta \end{cases} \tag{7.3.31b}$$

$$U(x) = \begin{cases} \frac{1}{2}\alpha x^2 + \frac{1-\alpha}{2\gamma}(x+x_0) \\ \qquad + \frac{1-\alpha}{4\gamma^2}\left[e^{-2\gamma(x+x_0)} - 1\right] & -a \le x \le -x_0, \gamma = -\beta \\ \frac{1}{2}\alpha x^2 + \frac{1-\alpha}{2\gamma}\{x + x_0 \\ \qquad - \frac{1}{2\gamma}\ln[1 + 2\gamma(x+x_0)]\} & -x_0 \le x \le a, \gamma = -\beta \end{cases} \tag{7.3.31c}$$

The potential energy of such a smooth hysteretic system is illustrated in Fig. 7.3.4*b* for increasing x, and for $\alpha = 1/21, \gamma = \beta = 0.5$, and different amplitudes. Plots for decreasing x are mirror images of those shown in Fig. 7.3.4*b*. With the aforementioned results, it is rather simple to compute the stationary probability densities for the state variable X and $\dot{X}$, the total energy Λ, and the amplitude A, as well as the mean-square response using equations (7.3.13), (7.3.14), (7.3.18), and (7.3.19).

It is important to note that this procedure is based on the assumption that the system response can indeed approach the state of statistical stationarity. This is not the case for a bilinear system with $\alpha = 0$ or for a smooth hysteretic system with $\alpha = 0$ and $\gamma + \beta > 0$.

This procedure now is applied to the following two examples.

7.3.1 A Hysteretic System under Additive Excitation

Consider a hysteretic system under an additive gaussian white noise excitation, governed by

$$\ddot{X} + 2\zeta\dot{X} + f(t) = W(t) \tag{a}$$

where the restoring force $f(t)$ is given in (7.3.2). In this case, equation (7.3.12) reduces to

$$\phi'(\lambda) = \frac{2\zeta}{\pi K} + \frac{A_r(\lambda)}{2\pi K \int_{-a(\lambda)}^{a(\lambda)} \sqrt{2\lambda - 2U(x)}\,dx} \tag{b}$$

where K is the spectral density of $W(t)$.

Numerical results have been obtained for both the bilinear and the Bouc-Wen hysteresis models, using a small $\alpha = 1/21$ and several viscous damping coefficients ($\zeta = 0, 0.005, 0.01, 0.02, 0.03, 0.05$) in the computation. The small α value indicates that the system investigated is nearly plastic. The results are described below.

Bilinear model. Figure 7.3.5 depicts the root-mean-square response σ_X computed for different ζ values and normalized with respect to $D = \sqrt{2K}$. Also shown in the figure are results obtained from equivalent linearization and from analogue simulations (Iwan and Lutes, 1968). For all response levels and for all three values of ζ, the results obtained from the dissipation energy balancing procedure are in much better agreement with the analogue simulations than the linearization results, particularly in the range of intermediate excitation levels. Yet it is in this range that the hysteretic component in the restoring force plays a dominant role.

The computed stationary probability density for the total response energy $\Lambda(t)$ is shown in Fig. 7.3.6. It is seen that when the excitation strength is either weak ($D = 0.05$) or strong ($D = 3$), the total energy λ follows approximately an exponential probability distribution, similar to that of a linear system. In such cases, the

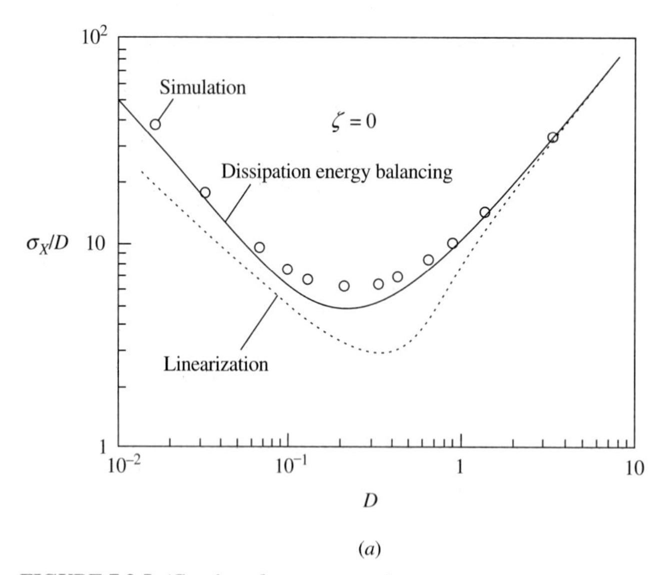

(*a*)

FIGURE 7.3.5 ***(Continued on next page)***
Root-mean-square displacement of a bilinear hysteretic system with $\alpha = 1/21$. (*a*) $\zeta = 0$ [after Cai and Lin, 1990].

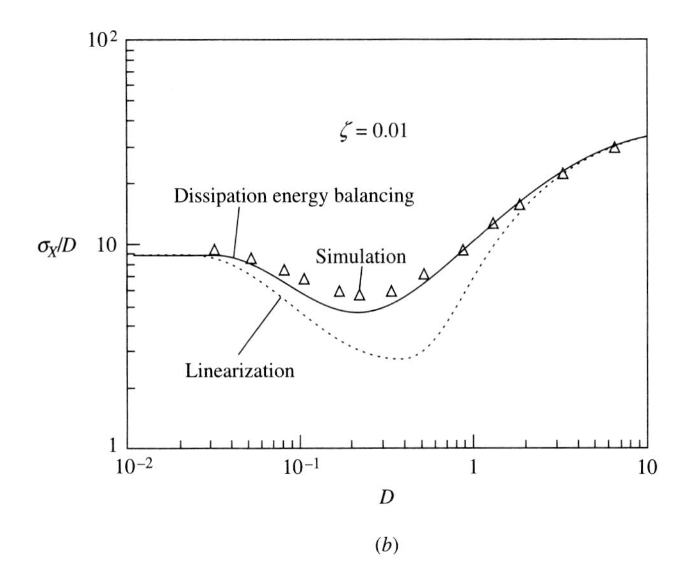

(*b*)

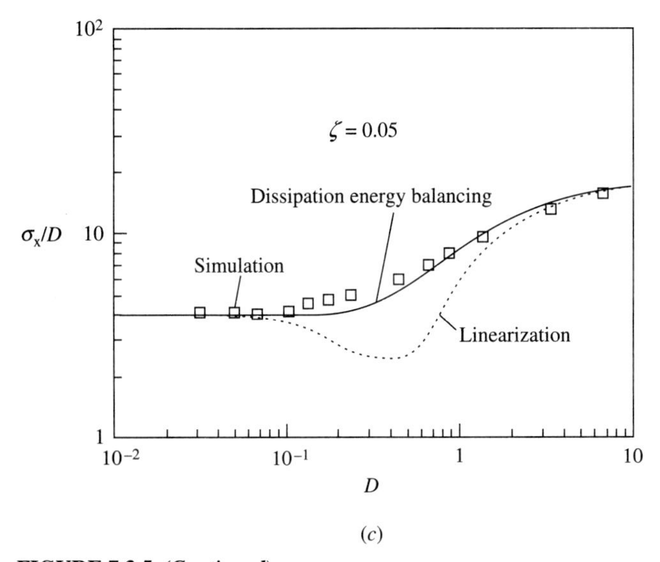

(*c*)

FIGURE 7.3.5 ***(Continued)***
Root-mean-square displacement of a bilinear hysteretic system with $\alpha = 1/21$. (*b*) $\zeta = 0.01$; (*c*) $\zeta = 0.05$ [after Cai and Lin, 1990].

hysteresis effect is unimportant, and the linear viscous effect has a much greater influence on the probability structure of the response. This is also borne out in the computed mean-square properties, shown in Fig. 7.3.5, and explains why the linearization technique can lead to somewhat more accurate results in these ranges. When the excitation strength is intermediate ($D = 0.5$), however, the energy dissipated by the hysteresis component is important. Therefore, the probability distribution of the total energy deviates greatly from being exponential, as shown in Fig. 7.3.6*b*.

Smooth hysteretic model. The computed root-mean-square displacements are shown in Figs. 7.3.7*a* and *b* for $\zeta = 0$ and $\zeta = 0.05$, respectively, for the smooth Bouc-Wen hysteretic model with $n = 1$, $A = 1$, and $\gamma = \beta = 0.5$. Also shown for

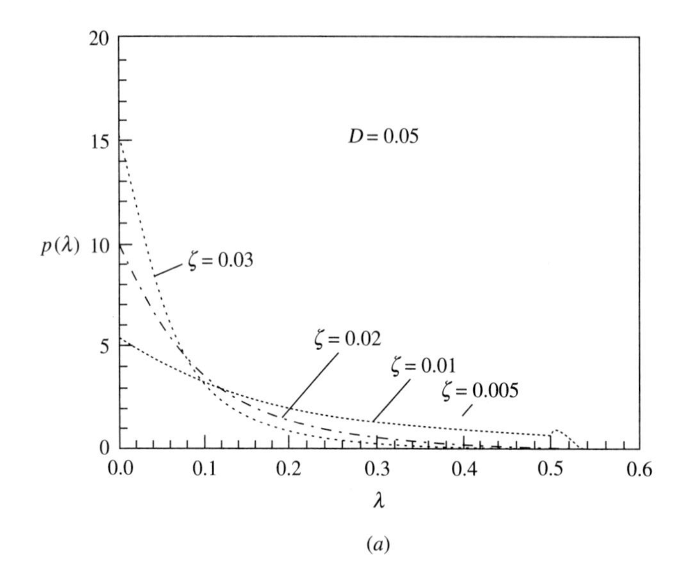

(*a*)

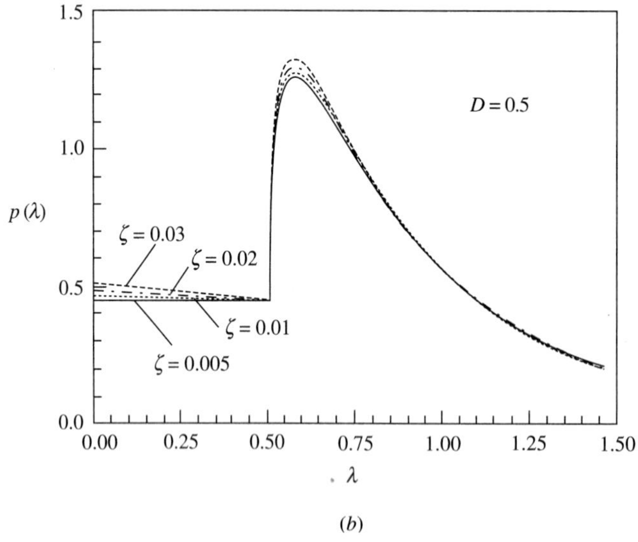

(*b*)

FIGURE 7.3.6 ***(Continued on next page)***
Stationary probability density of the total energy of a bilinear hysteretic system with $\alpha = 1/21$. (*a*) Weak excitation, $D = 0.05$; (*b*) intermediate excitation, $D = 0.5$ [after Cai and Lin, 1990].

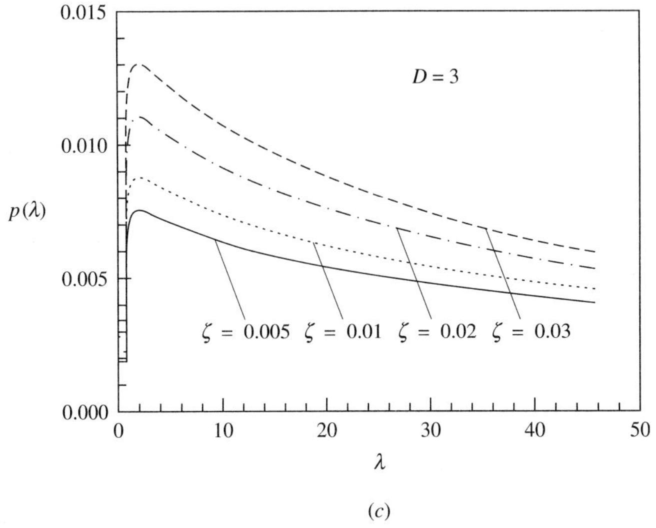

(c)

FIGURE 7.3.6 *(Continued)*
Stationary probability density of the total energy of a bilinear hysteretic system with $\alpha = 1/21$. (*c*) Strong excitation, $D = 3$ [after Cai and Lin, 1990].

comparison are earlier results obtained by Iyenger and Dash (1978) using gaussian closure, and by Wen (1980) using a modified version of equivalent linearization and digital simulation. In Wen's version, only equation (7.3.5) for the hysteretic component z is linearized, instead of the governing equation (7.3.1) for the entire system. Although the root-mean-square response so obtained compares well with the simulation results, the implied approximation of gaussian distribution can be very much in error.

The computed approximate probability density for the response amplitude $A(t)$ is shown in Fig. 7.3.8. When the excitation level is high or low ($D > 2$ or $D < 0.25$), the amplitude is nearly Rayleigh-distributed, indicating that the linear component of the restoring force has the dominant effect. When the excitation level is intermediate ($0.6 < D < 1$), both the linear and hysteretic components contribute importantly, giving rise to two peaks in the amplitude probability density, as shown in Fig. 7.3.8*b*. These figures have been obtained without taking into account the viscous damping effects. Additional calculations have shown that viscous damping does not change the general shape of the probability density, but it causes a shift of the peak to the left, and it flattens the tail of the distribution. These effects are greater when the excitation level is higher.

7.3.2 A Hysteretic Column under Random Ground Excitations

As indicated earlier, the dissipation energy balancing procedure is applicable, even when multiplicative random excitations are present. As an example, consider the

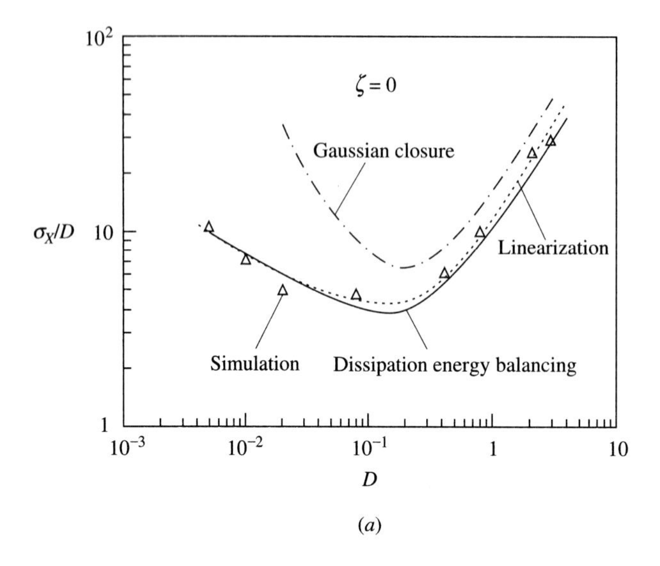

(a)

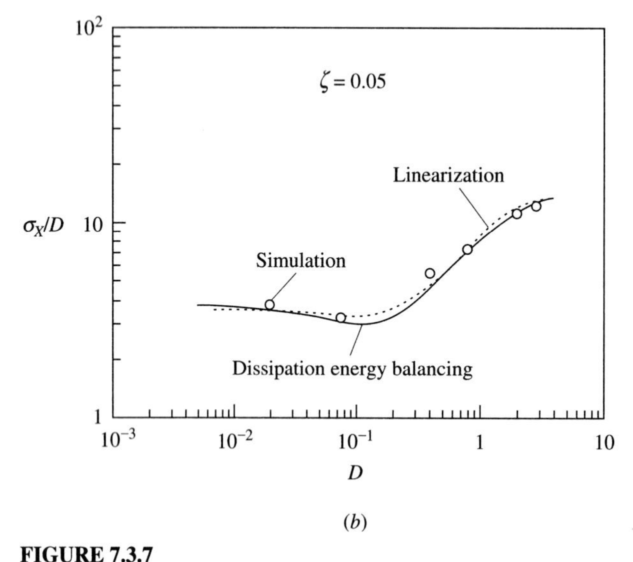

(b)

FIGURE 7.3.7
Root-mean-square displacement of a smooth hysteretic system with $\alpha = 1/21$ and $\gamma = \beta = 0.5$. (a) $\zeta = 0$; (b) $\zeta = 0.05$ [after Cai and Lin, 1990].

massless column shown in Fig. 7.3.9, which is supporting a concentrated mass at the top and is rigidly clamped below in the ground. If the column deformation is linearly elastic and dominated by a single mode, then the governing equation for the horizontal displacement of the concentrated mass may be expressed as

$$\ddot{\Delta} + 2\zeta\omega_0\dot{\Delta} + \omega_0^2\left(1 - \frac{mg}{P_{\text{cr}}} - \frac{m\ddot{V}}{P_{\text{cr}}}\right)\Delta = -\ddot{U} \tag{a}$$

in which P_{cr} is the static buckling load of the column and $\ddot{U}$ and $\ddot{V}$ are, respectively, the horizontal and vertical ground accelerations. The physical meanings of ω_0 and ζ are clear; they are the natural frequency and the modal damping if the effects of gravitation and vertical ground acceleration are not taken into account. Equation (a) may be nondimensionalized in terms of new variables $\tau = \omega_0 t$ and $x = \Delta/\Delta_e$ as

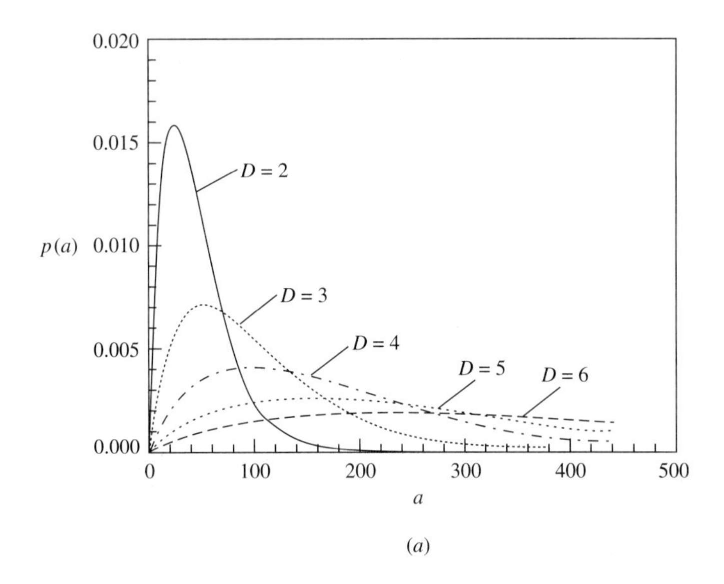

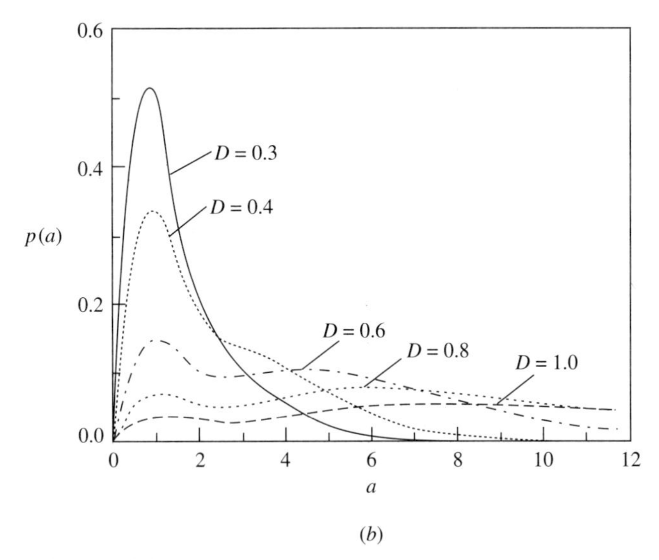

FIGURE 7.3.8 ***(Continued on next page)***
Stationary probability density of the response amplitude of a smooth hysteretic system with $\alpha = 1/21$, $\gamma = \beta = 0.5$, and $\zeta = 0$. (*a*) Strong excitation levels, $D = 2$–6; (*b*) intermediate excitation levels, $D = 0.3$–1.0.

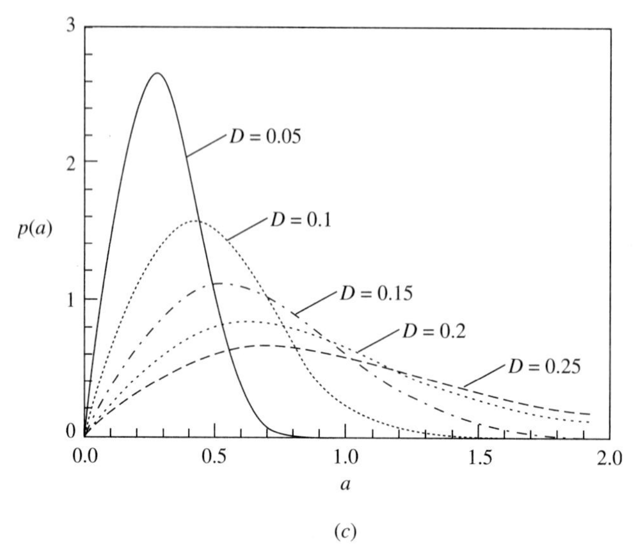

FIGURE 7.3.8 ***(Continued)***
Stationary probability density of the response amplitude of a smooth hysteretic system with $\alpha = 1/21$, $\gamma = \beta = 0.5$, and $\zeta = 0$. (*c*) Weak excitation levels, $D = 0.05$–0.25.

follows:

$$x'' + 2\zeta x' + [1 - \kappa_1 - \kappa_2 \eta(\tau)]\, x = \xi(\tau) \tag{b}$$

in which Δ_e is the elastic limit of the deflection (chosen to be the normalization length), beyond which equation (b) is no longer valid, each prime signifies one differentiation with respect to τ, $\xi = U''/\Delta_e$, $\eta = V''/\Delta_e$, and κ_1 and κ_2 are constants given by $\kappa_1 = mg/P_{\text{cr}}$ and $\kappa_2 = m\omega_0^2/P_{\text{cr}}$.

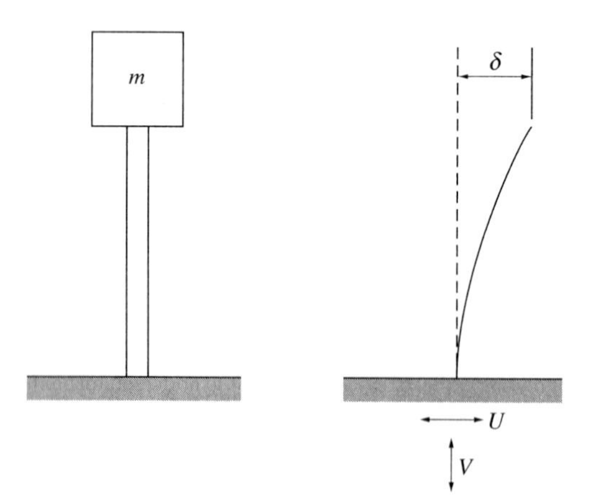

FIGURE 7.3.9
A column excited by vertical and horizontal ground accelerations.

For a hysteretic column under random ground excitations, equation (b) is replaced by

$$X'' + 2\zeta X' + [\alpha - \kappa_1 - \kappa_2\eta(\tau)]X + (1-\alpha)Z(X, X') = \xi(\tau) \tag{c}$$

where Z represents the hysteretic component of the restoring force. In what follows, the excitations $\eta(\tau)$ and $\xi(\tau)$ are idealized as gaussian white noises with the following correlation functions:

$$E[\eta(\tau)\eta(\tau+u)] = 2\pi K_{\eta\eta}\delta(u) \tag{d.1}$$

$$E[\xi(\tau)\xi(\tau+u)] = 2\pi K_{\xi\xi}\delta(u) \tag{d.2}$$

$$E[\eta(\tau)\xi(\tau+u)] = 2\pi K_{\eta\xi}\delta(u) \tag{d.3}$$

The white noise idealization for $\eta(\tau)$ and $\xi(\tau)$ is permissible provided that the correlation times of the excitations are short compared with the relaxation time of the dynamic system. Again, both the bilinear and the Bouc-Wen hysteresis models are considered.

Application of equation (7.3.12) leads to

$$\phi'(\lambda) = \frac{\frac{1}{2}A_r(\lambda) + 2\zeta\int_{-a(\lambda)}^{a(\lambda)}\sqrt{2\lambda - 2U(x)}\,dx}{\pi\int_{-a(\lambda)}^{a(\lambda)}(K_{\xi\xi} + K_{\eta\eta}\kappa_2^2x^2)\sqrt{2\lambda - 2U(x)}\,dx} \tag{e}$$

The area of the hysteretic loop, A_r, still can be computed from equation (7.3.20) for the bilinear model and from equations (7.3.30a–c) for the smooth hysteretic model. However, expressions for the potential energy, (7.3.21) for the bilinear case and (7.3.31a–c) for the smooth case, must be modified. For the bilinear model

$$U(x) = \begin{cases} \frac{1}{2}(1-\kappa_1)x^2 & a \le 1 \\ \frac{1}{2}(\alpha-\kappa_1)x^2 + \frac{1}{2}(1-\alpha)(x+a-1)^2 & a \ge 1, -a \le x \le -a+2 \\ \frac{1}{2}(\alpha-\kappa_1)x^2 + \frac{1}{2}(1-\alpha) & a \ge 1, -a+2 \le x \le a \end{cases} \tag{f}$$

and for the smooth hysteretic model, for example, with $\gamma = \beta$

$$U(x) = \begin{cases} \dfrac{1}{2}(\alpha-\kappa_1)x^2 + \dfrac{1}{2}(1-\alpha)(x+x_0)^2 & -a \le x \le -x_0, \gamma = \beta \\ \dfrac{1}{2}(\alpha-\kappa_1)x^2 + \dfrac{1}{8\gamma^2}(1-\alpha)\left[1 - e^{-2\gamma(x+x_0)}\right]^2 & -x_0 \le x \le a, \gamma = \beta \end{cases} \tag{g}$$

Numerical results have been obtained for hysteretic columns with $\alpha = 0.5$ (a moderate hysteresis) and $\alpha = 0.1$ (a strong hysteresis), and with $\kappa_1 = 0.04$, $\kappa_2 = 0.1$, $\zeta = 0.025$. For the Bouc-Wen smooth hysteretic model, it is assumed that $n = 1$, $A = 1$, and $\gamma = \beta = 0.5$. Two nondimensional quantities, $D_\eta = \sqrt{2K_{\eta\eta}}$ and $D_\xi = \sqrt{2K_{\xi\xi}}$, are introduced to represent the strength of the vertical and horizontal ground excitations. The root-mean-square response would be proportional to D_ξ if the system were linear and if the vertical excitation were absent.

Computed results for the stationary probability density of the response amplitude are shown in Figs. 7.3.10 and 7.3.11 for the Bouc-Wen type hysteretic columns. The strength of the horizontal and vertical excitations, used in the computation, were $D_\xi = 0.1, 0.5, 1, 5$ and $D_\eta = 0, 0.03, 0.034, 0.04, 0.08, 0.1, 0.12$. The low $D_\xi = 0.1$ represents a very weak horizontal excitation, and the high $D_\xi = 5$ a very strong hor-

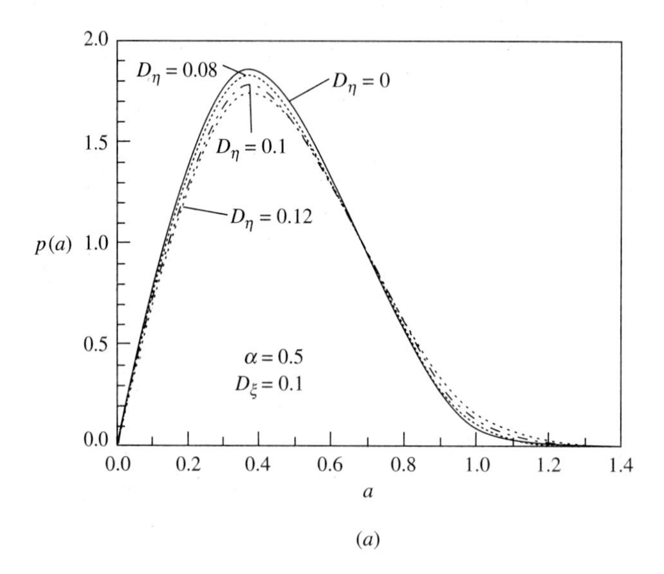

(*a*)

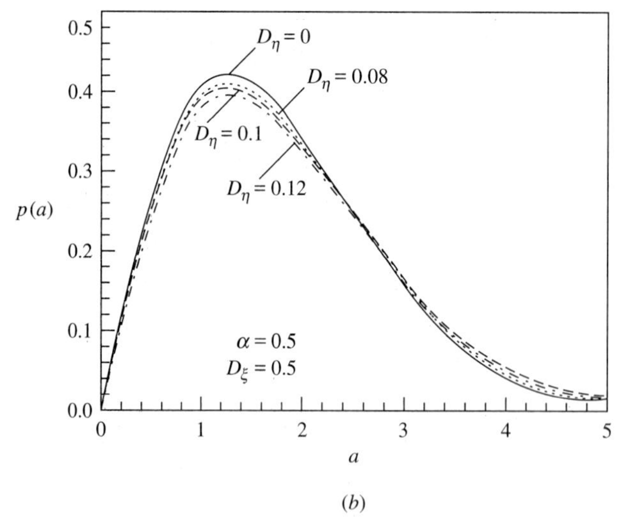

(*b*)

FIGURE 7.3.10 ***(Continued on next page)***
Stationary probability density of the response amplitude of a smooth hysteretic column with moderate hysteresis $\alpha = 0.5$. (*a*) Horizontal excitation spectral level $D_\xi = 0.1$; (*b*) horizontal excitation spectral level $D_\xi = 0.5$.

izontal excitation. The case of $D_\eta = 0$ was included so that the error incurred by neglecting the vertical (multiplicative) excitation could be determined.

Several general trends are revealed in Figs. 7.3.10 and 7.3.11. The general shape of the probability density curve of the response amplitude depends primarily on the relative contribution of the hysteresis component, and to a lesser degree on

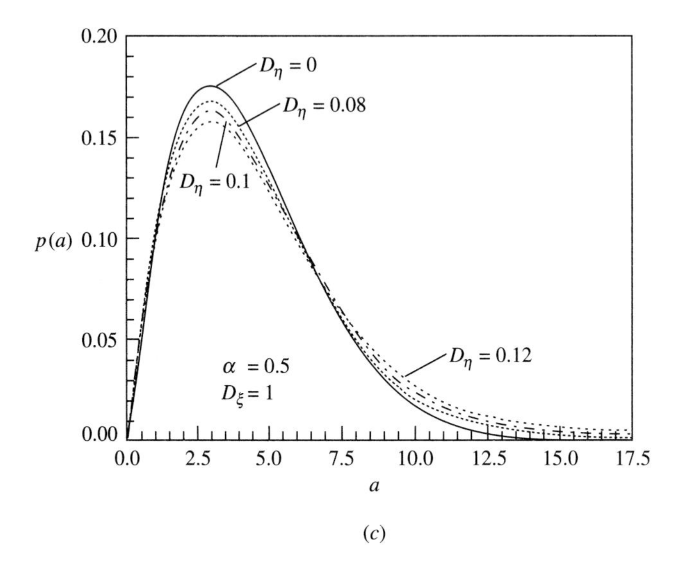

(c)

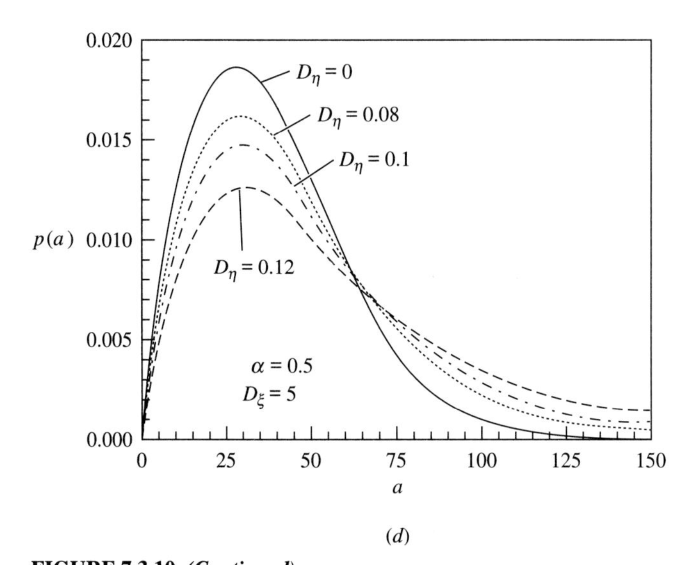

(d)

FIGURE 7.3.10 *(Continued)*
Stationary probability density of the response amplitude of a smooth hysteretic column with moderate hysteresis $\alpha = 0.5$. (*c*) horizontal excitation spectral level $D_\xi = 1$; (*d*) horizontal excitation spectral level $D_\xi = 5$.

the horizontal excitation intensity. In the case of moderate hysteresis ($\alpha = 0.5$), the amplitude is approximately Rayleigh-distributed (Fig. 7.3.10*a–d*), and the strengths of the vertical and horizontal excitations have little influence on the shape of the distribution. In the case of strong hysteresis ($\alpha = 0.1$), the response amplitude is close to being Rayleigh-distributed only if the horizontal excitation is either very

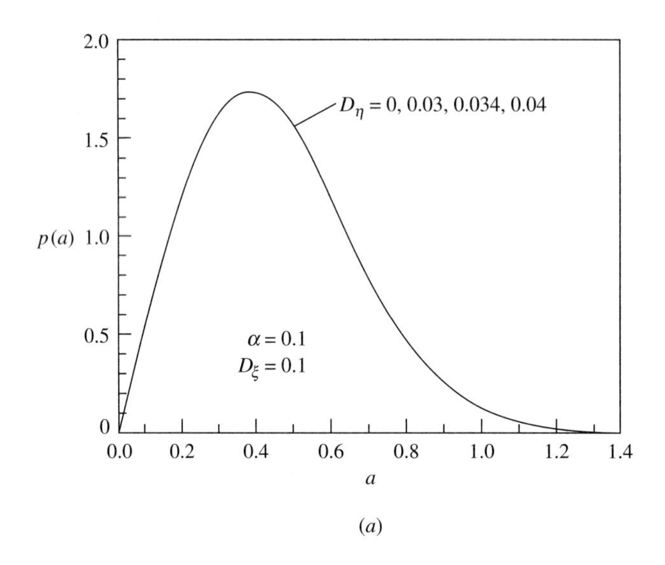

(*a*)

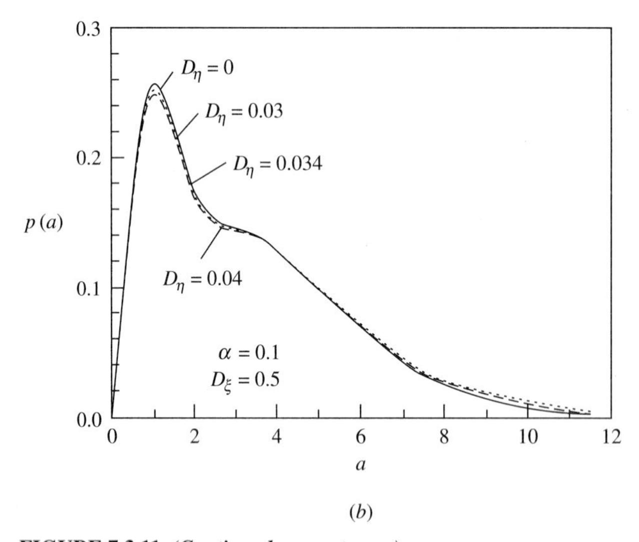

(*b*)

FIGURE 7.3.11 ***(Continued on next page)***
Stationary probability density of the response amplitude of a smooth hysteretic column with strong hysteresis $\alpha = 0.1$. (*a*) Horizontal excitation spectral level $D_\xi = 0.1$; (*b*) horizontal excitation spectral level $D_\xi = 0.5$.

weak (Fig. 7.3.11*a*, $D_\xi = 0.1$) or very strong (Fig. 7.3.11*d*, $D_\xi = 5$). If the strength of the horizontal excitation is intermediate (Fig. 7.3.11*b*, $D_\xi = 0.5$ and Fig. 7.3.11*c*, $D_\xi = 1$), the amplitude distribution is far from Rayleigh, and in some cases it can have two peaks. The presence of a vertical random excitation modifies the amplitude probability density by shifting its peak to the right, decreasing its values in

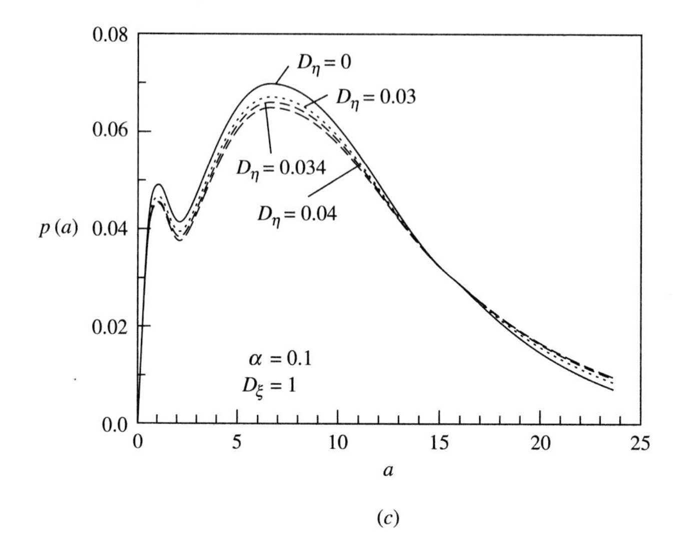

(*c*)

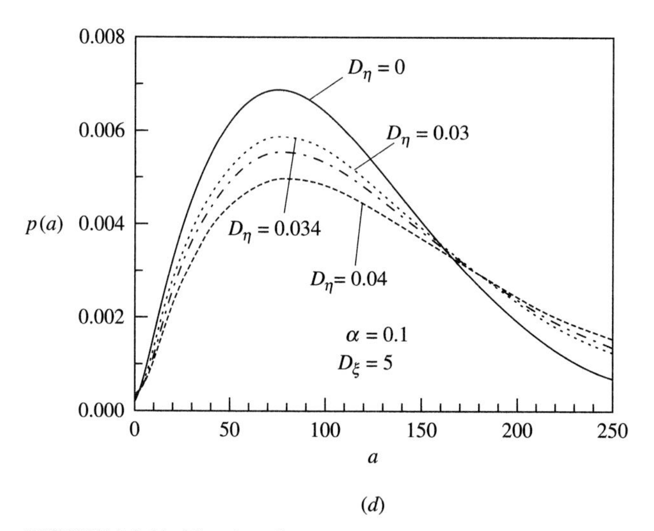

(*d*)

FIGURE 7.3.11 *(Continued)*
Stationary probability density of the response amplitude of a smooth hysteretic column with strong hysteresis $\alpha = 0.1$. (*c*) horizontal excitation spectral level $D_\xi = 1$; (*d*) horizontal excitation spectral level $D_\xi = 5$.

the lower amplitude range, and increasing its values in the higher range. However, these effects are significant only when the horizontal excitation is sufficiently strong (Figs. 7.3.10*d* and 7.3.11*d*). The same characteristics are found also for the bilinear columns.

The root-mean-square displacements σ_X , normalized with respect to D_ξ and computed for the bilinear and the smooth hysteretic models, are shown in Figs. 7.3.12 and 7.3.13, respectively, in solid lines for different values of D_η. The results show

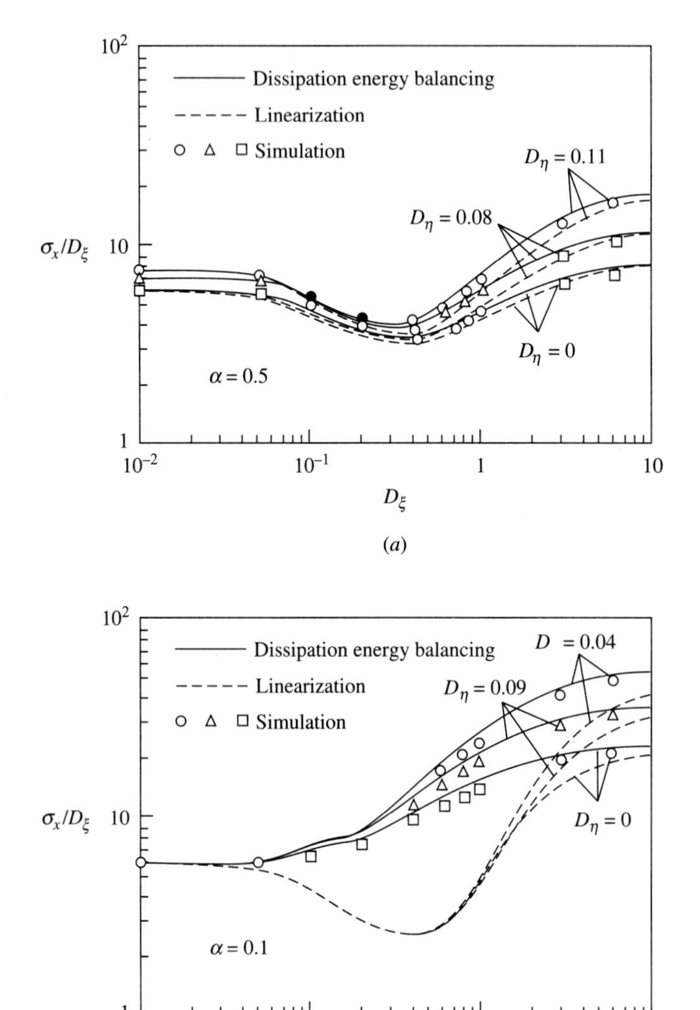

FIGURE 7.3.12
Root-mean-square displacement of a bilinear hysteretic column. (*a*) Moderate hysteresis, $\alpha = 0.5$; (*b*) strong hysteresis, $\alpha = 0.1$.

that the presence of the vertical ground motion increases the displacement response of the column, and that the effect of the vertical ground motion is significant only when the horizontal ground acceleration is sufficiently strong.

To assess the accuracy of the results obtained from the energy dissipation balancing approach, Monte Carlo simulations were carried out for the above hysteretic columns. The root-mean-square displacements obtained from the simulations are

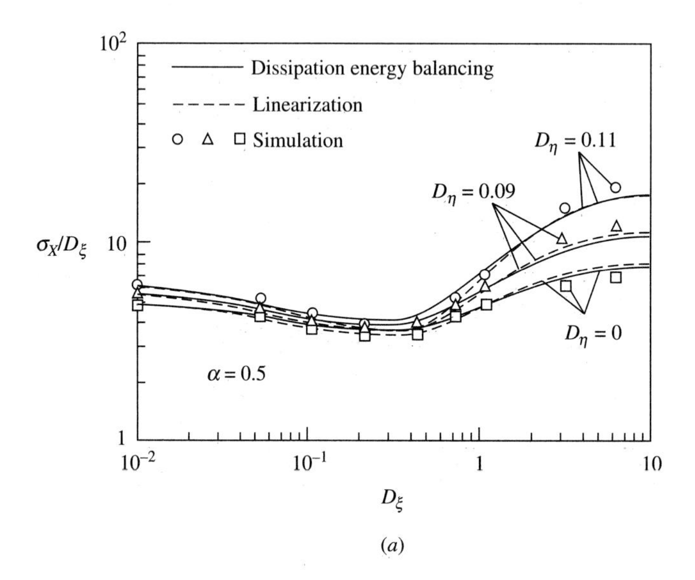

(*a*)

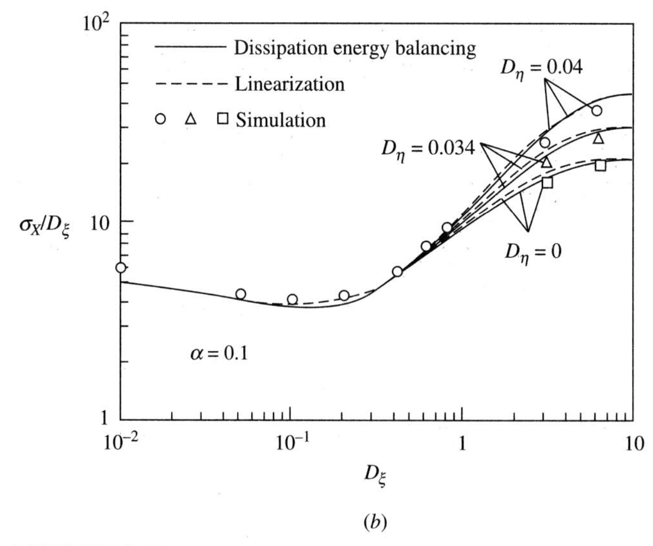

(*b*)

FIGURE 7.3.13
Root-mean-square displacement of a smooth hysteretic column. (*a*) Moderate hysteresis, $\alpha = 0.5$; (*b*) strong hysteresis, $\alpha = 0.1$.

shown also in Figs. 7.3.12 and 7.3.13. These results are represented by circles, triangles, or squares, and they appear to agree very well with the analytical results, obtained from the dissipation energy balancing procedure in all cases investigated, including (1) the bilinear model and the smooth model; (2) a moderate hysteresis ($\alpha = 0.5$) and a strong hysteresis ($\alpha = 0.1$); (3) a wide range of horizontal ground acceleration spectra; and (4) strong, weak, or entirely absent vertical ground excitation.

The dashed lines in Figs. 7.3.12*a* and *b* show the root-mean-square displacements of bilinear columns obtained from equivalent linearization. In the case of moderate hysteresis ($\alpha = 0.5$), the linearization results appear to be in good agreement with those obtained from simulation and from dissipation energy balancing; however, in the case of a nearly elastoplastic system ($\alpha = 0.1$), they deviate far from the simulation results in the range of intermediate horizontal ground excitations, in which the hysteretic component in the restoring force plays an important role. The large discrepancy is attributable to assumptions implied in the linearization method that the response is of a narrow frequency band and that the response amplitude is approximately Rayleigh-distributed. The dashed lines in Figs. 7.3.13*a* and *b* are results for smooth hysteretic columns obtained by using Wen's version of the equivalent linearization technique (1980). They are in good agreement with the results from dissipation energy balancing and from simulations.

7.4 IMPULSIVE NOISE EXCITATION
(Cai and Lin, 1992)

Exact and approximate solution techniques for randomly excited nonlinear systems were developed in Chapter 5 and in Sections 7.1 through 7.3, under the assumption that random excitations were gaussian white noises. A white noise is known to be a special case of shot noise or impulsive noise, which is a sequence of impulses of random amplitude, arriving randomly in time (Lin, 1967, p. 94). The gaussian idealization requires additionally that the average arrival rate of the impulses tends to infinity while the mean-square pulse amplitude tends to zero, such that their product remains finite. In this section the more general excitation model of impulsive noise is considered, and a suitable criterion is established under which the additional gaussian assumption is deemed justifiable.

The impulsive noise model has been used in a variety of engineering problems, such as an airplane tail being buffeted in the "downwash" of a stalled wing (Liepman, 1952), a structure under earthquake excitation (Cornell, 1964; Lin, 1963; Verne-Jones, 1970), a vehicle traveling on a rough road or track (Roberts, 1965, 1966), and a highway bridge under traffic loads (Tung, 1967, 1969). Solutions, in terms of either statistical moments or probability distributions, have been obtained for linear systems under additive impulsive noise excitations (e.g., Iwankiewicz and Sobczyk, 1983; Janssen and Lambert, 1966; Lin, 1965; Roberts, 1965, 1966, 1972; Srinivasan, Subramanian, and Kumaraswamy, 1967). However, if the system is nonlinear or the excitation is multiplicative, the problem becomes much more difficult. Roberts (1972) outlined a perturbation scheme for solving the Fokker-Planck equation for the probability density of the response in a general setting, but the scheme was applied only to a linear system. The equivalent linearization and cumulant-neglect procedure were applied, respectively, by Tylikowski and Marowski (1986) and by Iwankiewicz and Nielsen (1989) to the case of a Duffing oscillator.

In this section the known exact solutions obtained in Chapter 5 for gaussian white noise excitations are used to develop a perturbation scheme for the case in which the excitation is still impulsive but nongaussian. Two specific examples, one with an additive excitation and the other with a multiplicative excitation, are investigated in detail. The results are compared with those obtained from Monte Carlo simulations.

The impulsive noise process belongs to the general class of random pulse trains (Section 3.2) of the form

$$\xi(t) = \sum_{k=1}^{N(T)} Y_k w(t - \tau_k) \qquad t \leq T \tag{7.4.1}$$

where $N(T)$ is a counting process, representing the number of pulses that arrive in the time interval $(-\infty, T]$, Y_k is the random amplitude of the kth pulse arriving at time τ_k, and $w(t - \tau_k)$ is a deterministic function describing the pulse shape. For the present discussion, we assume that $N(T)$ is a Poisson process with a constant average pulse arrival rate λ, and the pulse amplitudes Y_k, for different k, are independent identically distributed random variables with zero mean, and independent of the pulse arrival time τ_k.

The statistical properties of the random excitation $\xi(t)$ described by equation (7.4.1) can be determined simply. The mth cumulant function of $\xi(t)$, given by (3.2.13), is

$$\begin{aligned}\kappa_m & [\xi(t_1), \xi(t_2), \ldots, \xi(t_m)] \\ &= \lambda E[Y^m] \int_{-\infty}^{\min(t_1,t_2,\ldots,t_m)} w(t_1 - \tau)\, w(t_2 - \tau) \cdots w(t_m - \tau)\, d\tau \end{aligned} \tag{7.4.2}$$

where $E[Y^m]$ is the mth moment of any Y_k. It can be shown that this cumulant function remains unchanged upon changing the time origin; therefore, $\xi(t)$ is a stationary random process (Lin, 1967, Eq. 4-88). If the second cumulant of $\xi(t)$ is bounded, then the product $\lambda E[Y^2]$ must be finite, which implies that $E[Y^2]$ is of order λ^{-1} and that $E[Y^n]$ is at most of order $\lambda^{-n/2}$. In the limit as λ tends to infinity, the cumulants of orders higher than 2 tend to zero, and $\xi(t)$ tends to a gaussian random process.

The statistical properties of a random process are characterized by either a complete set of cumulant functions or a complete set of moment functions. The two sets are, of course, related. The first cumulant is the same as the mean, and the second and third cumulants coincide, respectively, with the second and third central moments:

$$\kappa_1[\xi(t)] = E[\xi(t)] = \mu(t) \tag{7.4.3a}$$

$$\kappa_2[\xi(t_1), \xi(t_2)] = E\{[\xi(t_1) - \mu(t_1)][\xi(t_2) - \mu(t_2)]\} \tag{7.4.3b}$$

$$\kappa_3[\xi(t_1), \xi(t_2), \xi(t_3)] = E\{[\xi(t_1) - \mu(t_1)][\xi(t_2) - \mu(t_2)][\xi(t_3) - \mu(t_3)]\} \tag{7.4.3c}$$

A higher order cumulant can also be expressed in terms of moments of the same order and lower orders, as described in Section 7.1.

The case of impulsive noise is one in which the pulse shape function in equation (7.4.1) is a unit impulse; that is, equation (7.4.1) is replaced by

$$\xi(t) = \sum_{k=1}^{N(T)} Y_k \delta(t - \tau_k) \tag{7.4.4}$$

Thus equation (7.4.2) is reduced to

$$\kappa_m[\xi(t_1), \xi(t_2), \ldots, \xi(t_m)] = \lambda E\left[Y^m\right] \delta(t_2 - t_1) \delta(t_3 - t_1) \cdots \delta(t_m - t_1) \tag{7.4.5}$$

Equation (7.4.4) is a mathematical idealization, which is valid if the duration of each random pulse in equation (7.4.1) is much shorter than the so-called relaxation time t_r of the excited dynamical system. For practical purposes, t_r may be defined as the time required for the amplitude of an unexcited free motion of the system to decay by a factor of e^{-1} or increase by a factor of e. Under the assumption that the Poisson process $N(T)$ in equation (7.4.4) has a constant average arrival rate, the impulsive noise model is stationary; thus it is a white noise. If, in addition, the average pulse arrival rate becomes infinite, then its cumulants of orders higher than 2 reduce to zero, and it becomes a gaussian white noise, as indicated earlier.

THE FOKKER-PLANCK EQUATION FOR A SYSTEM UNDER NONGAUSSIAN IMPULSIVE NOISE EXCITATION. Consider the single-degree-of-freedom system

$$\ddot{X} + h(X, \dot{X}) = g(X, \dot{X}) \xi(t) \tag{7.4.6}$$

where the functional forms for h and g are deterministic, and $\xi(t)$ is an impulsive noise described by equation (7.4.4). Since the pulse arrival times τ_k are independent, and Y_k are independent for different k, the response vector $\boldsymbol{X} = \{X_1, X_2\}$, where $X_1 = X$, $X_2 = \dot{X}$, has independent increments in any two nonoverlapping time intervals. Therefore, it satisfies the sufficient condition for a Markov vector. The transition probability density $q = q(\boldsymbol{x}, t \mid \boldsymbol{x}_0, t_0)$ of the response vector $\boldsymbol{X}$ is governed by the Fokker-Planck equation of the more general form of (4.2.15),

$$\begin{aligned}\frac{\partial q}{\partial t} + \frac{\partial}{\partial x_i}(a_i q) - \frac{1}{2!}\frac{\partial^2}{\partial x_i \partial x_j}(b_{ij} q) + \frac{1}{3!}\frac{\partial^3}{\partial x_i \partial x_j \partial x_k}(c_{ijk} q)& \\ - \frac{1}{4!}\frac{\partial^4}{\partial x_i \partial x_j \partial x_k \partial x_l}(d_{ijkl} q) + \cdots = 0&\end{aligned} \tag{7.4.7}$$

where a_i, b_{ij}, c_{ijk}, d_{ijkl}, and so on, are the derivate moments defined as

$$\begin{aligned}a_i &= \lim_{\Delta t \to 0} \frac{1}{\Delta t} E[\Delta X_i \mid \boldsymbol{X}(t) = \boldsymbol{x}] \\ b_{ij} &= \lim_{\Delta t \to 0} \frac{1}{\Delta t} E\left[\Delta X_i \Delta X_j \mid \boldsymbol{X}(t) = \boldsymbol{x}\right] \\ c_{ijk} &= \lim_{\Delta t \to 0} \frac{1}{\Delta t} E\left[\Delta X_i \Delta X_j \Delta X_k \mid \boldsymbol{X}(t) = \boldsymbol{x}\right] \\ d_{ijkl} &= \lim_{\Delta t \to 0} \frac{1}{\Delta t} E\left[\Delta X_i \Delta X_j \Delta X_k \Delta X_l \mid \boldsymbol{X}(t) = \boldsymbol{x}\right] \\ &\;\;\vdots\end{aligned} \tag{7.4.8}$$

and where the subscripts $i, j, k, l, \ldots$, take the values of 1 and 2. In equations (7.4.8), $\Delta X_i = X_i(t + \Delta t) - X_i(t)$, and so on. In the special case in which the impulsive noise is also gaussian, the derivate moments of orders higher than 2 are zero, and equation (7.4.7) reduces to equation (4.2.17).

The derivate moments defined in equation (7.4.8) may be deduced from the equation of motion (7.4.6). For a sufficiently short time interval Δt,

$$\begin{aligned}
[\Delta X_1 \mid \boldsymbol{X}(t) = \boldsymbol{x}] &= x_2 \Delta t + O(\Delta t)^2 \\
[\Delta X_2 \mid \boldsymbol{X}(t) = \boldsymbol{x}] &= -h(x_1, x_2)\Delta t + \int_t^{t+\Delta t} g[X_1(\tau), X_2(\tau)]\,\xi(\tau)\,d\tau + O(\Delta t)^2
\end{aligned} \tag{7.4.9}$$

where the symbol $O(\Delta t)^2$ denotes a remainder of order $(\Delta t)^2$. Substituting (7.4.9) into (7.4.8), and using the properties of the cumulants of an impulsive noise process shown in equation (7.4.5) as well as the relationships between the cumulants and the moments, equations (7.4.3a–c), the derivate moments for system (7.4.6) are found to be

$$\begin{aligned}
a_1 &= x_2 \\
a_2 &= -h(x_1, x_2) + \frac{1}{2}\lambda E[Y^2] g(x_1, x_2) \frac{\partial g(x_1, x_2)}{\partial x_2} \\
b_{22} &= \lambda E[Y^2] g^2(x_1, x_2) \\
c_{222} &= \lambda E[Y^3] g^3(x_1, x_2) \\
d_{2222} &= \lambda E[Y^4] g^4(x_1, x_2) \\
\text{other } b_{ij}, c_{ijk}, &\text{ and } d_{ijkl} = 0 \\
&\cdots\cdots\cdots\cdots
\end{aligned} \tag{7.4.10}$$

These derivate moments do not depend explicitly on time t. If the system is stable, then its response has a stationary probability density $p(x_1, x_2)$ which is governed by the reduced Fokker-Planck equation

$$\begin{aligned}
-x_2 \frac{\partial p}{\partial x_1} &+ \frac{\partial}{\partial x_2}\left\{\left(h - \frac{1}{2}\lambda E[Y^2] g \frac{\partial g}{\partial x_2}\right) p\right\} + \frac{1}{2!}\lambda E[Y^2] \frac{\partial^2}{\partial x_2^2}(g^2 p) \\
&- \frac{1}{3!}\lambda E[Y^3] \frac{\partial^3}{\partial x_2^3}(g^3 p) + \frac{1}{4!}\lambda E[Y^4] \frac{\partial^4}{\partial x_2^4}(g^4 p) - \cdots = 0
\end{aligned} \tag{7.4.11}$$

The foregoing partial differential equation, with an infinite number of terms, cannot be solved even for a simple linear system. However, as presented in Chapter 5, a solution is obtainable if equation (7.4.11) is truncated up to the second-order term, and if the truncated equation belongs to the class of generalized stationary potential. The truncated equation corresponds precisely to the limiting case in which $\xi(t)$ becomes a gaussian white noise, and the third and higher derivate moments in (7.4.10) vanish. This limiting solution now can be used as a basis from which an approximate solution for the case of nongaussian impulsive excitation is obtained by perturbation.

A PERTURBATION SCHEME. We select a perturbation parameter $\varepsilon = \lambda^{-1/2}$, in view of the fact that the excitation process tends to a gaussian process as $\lambda \to \infty$, or $\varepsilon \to 0$. In order that $\lambda E[Y^2]$ remains finite, it is necessary that $E[Y^2]$ be of order ε^2. Thus $E[Y^3]$ is at most of order ε^3, $E[Y^4]$ of order ε^4, and so on. The reduced Fokker-Planck equation (7.4.11) can then be rewritten as

$$-x_2\frac{\partial p}{\partial x_1} + \frac{\partial}{\partial x_2}\left[\left(h - \frac{I_0}{2}g\frac{\partial g}{\partial x_2}\right)p\right] + \frac{I_0}{2!}\frac{\partial^2}{\partial x_2^2}(g^2 p) - \varepsilon\frac{I_1}{3!}\frac{\partial^3}{\partial x_2^3}(g^3 p) + \varepsilon^2\frac{I_2}{4!}\frac{\partial^4}{\partial x_2^4}(g^4 p) - \cdots = 0 \tag{7.4.12}$$

where I_n are defined as

$$\varepsilon^n I_n = \lambda E[Y^{n+2}] \qquad n = 0, 1, 2, \ldots \tag{7.4.13}$$

We seek a solution for equation (7.4.12) in the form of

$$p(x_1, x_2) = p_0(x_1, x_2) + \varepsilon p_1(x_1, x_2) + \varepsilon^2 p_2(x_1, x_2) + \cdots \tag{7.4.14}$$

Upon substituting (7.4.14) into (7.4.12) and grouping terms of the same power of ε, the following set of second-order partial differential equations is obtained:

$$-x_2\frac{\partial p_0}{\partial x_1} + \frac{\partial}{\partial x_2}\left[\left(h - \frac{I_0}{2}g\frac{\partial g}{\partial x_2}\right)p_0\right] + \frac{I_0}{2!}\frac{\partial^2}{\partial x_2^2}(g^2 p_0) = 0 \tag{7.4.15}$$

$$-x_2\frac{\partial p_1}{\partial x_1} + \frac{\partial}{\partial x_2}\left[\left(h - \frac{I_0}{2}g\frac{\partial g}{\partial x_2}\right)p_1\right] + \frac{I_0}{2!}\frac{\partial^2}{\partial x_2^2}(g^2 p_1) = \frac{I_1}{3!}\frac{\partial^3(g^3 p_0)}{\partial x_2^3} \tag{7.4.16}$$

$$-x_2\frac{\partial p_2}{\partial x_1} + \frac{\partial}{\partial x_2}\left[\left(h - \frac{I_0}{2}g\frac{\partial g}{\partial x_2}\right)p_2\right] + \frac{I_0}{2!}\frac{\partial^2}{\partial x_2^2}(g^2 p_2) = \frac{I_1}{3!}\frac{\partial^3(g^3 p_1)}{\partial x_2^3} - \frac{I_2}{4!}\frac{\partial^4(g^4 p_0)}{\partial x_2^4} \tag{7.4.17}$$

Equation (7.4.15), obtained by grouping terms without ε, is the Fokker-Planck equation for the case of gaussian white noise excitation, which is solvable exactly if it belongs to the class of generalized stationary potential. In this case, the solution for equation (7.4.15) may be written in terms of the probability potential function:

$$p_0(x_1, x_2) = C_0 \exp(-\phi) \tag{7.4.18}$$

where C_0 is a normalization constant.

For the second term on the right-hand side of (7.4.14), it is convenient to write

$$p_1(x_1, x_2) = p_0(x_1, x_2)Q_1(x_1, x_2) \tag{7.4.19}$$

Substituting equations (7.4.18) and (7.4.19) into equation (7.4.16) and imposing equation (7.4.15), we obtain

$$-x_2\frac{\partial Q_1}{\partial x_1} + \left(h + \frac{3I_0}{2}g\frac{\partial g}{\partial x_2} - I_0 g^2\frac{\partial \phi}{\partial x_2}\right)\frac{\partial Q_1}{\partial x_2} + \frac{I_0}{2!}g^2\frac{\partial^2 Q_1}{\partial x_2^2} = \frac{I_1}{3!p_0}\frac{\partial^3(g^3 p_0)}{\partial x_2^3} \tag{7.4.20}$$

A general solution for equation (7.4.20) appears difficult for arbitrary h and f functions. However, in most practical cases ϕ is a polynomial of x_1 and x_2. If, in addition,

f is a constant or a linear function of x_1 and x_2, then it can be shown that the equation

$$-x_2\frac{\partial Q_1}{\partial x_1} + (\alpha x_1 + \beta x_2)\frac{\partial Q_1}{\partial x_2} + \frac{I_0}{2!}g^2\frac{\partial^2 Q_1}{\partial x_2^2} = \frac{I_1}{3!p_0}\frac{\partial^3(g^3 p_0)}{\partial x_2^3} \tag{7.4.21}$$

has a solution of the polynomial form. Equation (7.4.21) differs from equation (7.4.20) only in the coefficients of the $\partial Q_1/\partial x_2$ term. Therefore, the solution of equation (7.4.21) may be considered as an approximate solution for equation (7.4.20), and an optional approximation is obtained by minimizing

$$\Delta = \alpha x_1 + \beta x_2 - \left(h + \frac{3I_0}{2}g\frac{\partial g}{\partial x_2} - I_0 g^2\frac{\partial \phi}{\partial x_2}\right) \tag{7.4.22}$$

A reasonable criterion for this purpose is the least-mean-square error

$$E[\Delta^2] = \text{minimum} \tag{7.4.23}$$

which results in

$$\alpha = \frac{E\left[X_1\left(h + \frac{3I_0}{2}g\frac{\partial g}{\partial X_2} - I_0 g^2\frac{\partial \phi}{\partial x_2}\right)\right]}{E[X_1^2]} \tag{7.4.24}$$

$$\beta = \frac{E\left[X_2\left(h + \frac{3I_0}{2}g\frac{\partial g}{\partial X_2} - I_0 g^2\frac{\partial \phi}{\partial X_2}\right)\right]}{E[X_2^2]} \tag{7.4.25}$$

The first-order approximation is then given by

$$p(x_1, x_2) = p_0(x_1, x_2)[1 + \varepsilon Q_1(x_1, x_2)] \tag{7.4.26}$$

For the second-order perturbation analysis, let

$$p(x_1, x_2) = p_0(x_1, x_2)Q_2(x_1, x_2) \tag{7.4.27}$$

and obtain an equation for function Q_2 as follows:

$$-x_2\frac{\partial Q_2}{\partial x_1} + \left(h + \frac{3I_0}{2}g\frac{\partial g}{\partial x_2} - I_0 g^2\frac{\partial \phi}{\partial x_2}\right)\frac{\partial Q_2}{\partial x_2} + \frac{I_0}{2!}g^2\frac{\partial^2 Q_2}{\partial x_2^2}$$
$$= \frac{I_1}{3!p_0}\frac{\partial^3(g^3 Q_1 p_0)}{\partial x_2^3} - \frac{I_2}{4!p_0}\frac{\partial^4(g^4 p_0)}{\partial x_2^4} \tag{7.4.28}$$

Equation (7.4.28) differs from equation (7.4.20) only on the right-hand side; therefore, it can be solved approximately in the same manner. The approximate stationary probability density, corresponding to such a second-order perturbation analysis, is then

$$p(x_1, x_2) = p_0(x_1, x_2)[1 + \varepsilon Q_1(x_1, x_2) + \varepsilon^2 Q_2(x_1, x_2)] \tag{7.4.29}$$

In passing, we note that if the random amplitude Y is symmetrically distributed, then $I_1 = 0$ and equation (7.4.21) gives rise to a trivial solution $Q_1 = 0$. In this case, the second-order term $\varepsilon^2 Q_2$ becomes the leading correction term for nongaussianity.

The perturbation procedure just outlined may be applied iteratively to improve the accuracy of the results. The first term p_0 is obtained by solving equation (7.4.15). The result can then be used to calculate the initial estimates for the optimal α and β values, using equations (7.4.24) and (7.4.25), and in turn the initial estimates of Q_1 and Q_2. The estimates for α and β can be improved when p given in equation (7.4.29), instead of p_0, is used for their calculation. The improved α and β give rise to improved Q_1 and Q_2, and so on. The iterative process continues until it converges. Of course, the resulting probability density must be checked for nonnegativity.

For those cases for which an exact solution for equation (7.4.15) cannot be found analytically, recourse may be sought in an approximate procedure, such as the method of dissipation energy balancing described in Section 7.2.3. In essence, an equivalent nonlinear system is found within the class of generalized stationary potential which can be solved exactly if the excitation is gaussian white noise. Subsequently, the foregoing perturbation procedure can be applied to the equivalent system.

7.4.1 A Duffing Oscillator under Additive Excitation of Impulsive Noise

As the first example, consider the following Duffing oscillator subjected to an additive impulsive noise excitation:

$$\ddot{X} + 2\zeta\omega_0\dot{X} + \omega_0^2(X + \delta X^3) = \xi(t) \tag{a}$$

The zeroth-order term p_0 in the perturbation series corresponds to the limiting case of gaussian white noise excitation, for which the exact stationary probability density is known to be

$$p_0(x_1, x_2) = C\exp\left[-\frac{2\zeta\omega_0}{I_0}\left(x_2^2 + \omega_0^2 x_1^2 + \frac{1}{2}\delta\omega_0^2 x_1^4\right)\right] \tag{b}$$

Function Q_1 of the first-order perturbation term is governed by the equation

$$-x_2\frac{\partial Q_1}{\partial x_1} + [-2\zeta\omega_0 x_2 + \omega_0^2(x_1 + \delta x_1^3)]\frac{\partial Q_1}{\partial x_2} + \frac{I_0}{2}\frac{\partial^2 Q_1}{\partial x_2^2} = \frac{8\zeta^2\omega_0^2 I_1}{I_0^2}\left(x_2 - \frac{4\zeta\omega_0}{3I_0}x_2^3\right) \tag{c}$$

whose analytical solution is not known at the present time. To find an approximate solution, equation (c) is replaced by

$$-x_2\frac{\partial Q_1}{\partial x_1} + (-2\zeta\omega_0 x_2 + \alpha x_1)\frac{\partial Q_1}{\partial x_2} + \frac{I_0}{2}\frac{\partial^2 Q_1}{\partial x_2^2} = \frac{8\zeta^2\omega_0^2 I_1}{I_0^2}\left(x_2 - \frac{4\zeta\omega_0}{3I_0}x_2^3\right) \tag{d}$$

We note that, in the expression multiplying $\partial Q_1/\partial x_2$, only the coefficient α for x_1 needs to be optimized, because the x_2 term in the original equation (c) is already linear. It follows from equation (7.4.24) that

$$\alpha = \omega_0^2(1 + \delta\eta) \qquad \eta = \frac{E[X_1^4]}{E[X_1^2]} \tag{e}$$

The solution for equation (d) can be assumed in the polynomial form

$$Q_1 = c_{10}x_1 + c_{01}x_2 + c_{30}x_1^3 + c_{21}x_1^2x_2 + c_{12}x_1x_2^2 + c_{03}x_2^3 \tag{f}$$

where c_{ij} are constants, which can be determined by substituting (f) into (d) and equating the coefficients on the two sides, resulting in

$$\begin{aligned} Q_1(x_1, x_2) = \frac{8\zeta^2 I_1}{I_0^2(1+\delta\eta+8\zeta^2)}\Bigg[& -\alpha x_1 - \frac{4\zeta\omega_0}{3}x_2 \\ & + \frac{4\zeta\omega_0}{9I_0}(2\alpha^2x_1^3 + 3\alpha x_1x_2^2 + 4\zeta\omega_0x_2^3)\Bigg] \end{aligned} \tag{g}$$

This is an approximate solution for equation (c).

Similarly, by substituting equations (b) and (g) into equation (7.4.28), we obtain the following approximate solution for Q_2:

$$\begin{aligned} Q_2(x_1, x_2) = & \frac{64\zeta^4\omega_0^2 I_1^2}{9I_0^4(1+\delta\eta)}\Big[17\alpha x_1^2 + 28\zeta\omega_0x_1x_2 + 5x_2^2 - \\ & \frac{2\zeta\omega_0}{7I_0}(56x_1^4 + 108\alpha\zeta\omega_0x_1^3x_2 + 98\alpha x_1^2x_2^2 + 184\zeta\omega_0x_1x_2^3 + 7x_2^4)\Big] \\ & - \frac{2\zeta^2\omega_0^2 I_2}{I_0^3}\Big[\alpha x_1^2 + \zeta\omega_0x_1x_2 + x_2^2 - \\ & \frac{\zeta\omega_0}{6I_0}(3\alpha^2x_1^4 + 6\alpha x_1^2x_2^2 + 8\zeta\omega_0x_1x_2^3 + 3x_2^4)\Big] \end{aligned} \tag{h}$$

where the assumption has been made that $\zeta^2 << 1$. An approximate probability density for the system response follows upon substituting equations (b), (g), and (h) into equation (7.4.29). The optimal value for α can be determined by iteration.

The result is greatly simplified when the random pulse amplitude Y is symmetrically distributed. In this case, $I_1 = 0$, $Q_1 = 0$, and the stationary probability density for the system response is given approximately by

$$\begin{aligned} p(x_1, x_2) = C_1\Bigg\{1 - & \frac{2\lambda E[Y^4]\zeta^2\omega_0^2}{I_0^3}\Bigg[\alpha x_1^2 + \zeta\omega_0x_1x_2 + x_2^2 \\ & - \frac{\zeta\omega_0}{6I_0}(3\alpha^2x_1^4 + 6\alpha x_1^2x_2^2 + 8\zeta\omega_0x_1x_2^3 + 3x_2^4)\Bigg]\Bigg\} \\ & \exp\left[-\frac{2\zeta\omega_0}{I_0}\left(x_2^2 + \omega_0^2x_1^2 + \frac{1}{2}\delta\omega_0^2x_1^4\right)\right] \end{aligned} \tag{i}$$

where C_1 is a new normalization constant that is, of course, different from C_0. It can be shown by integrating equation (i) on x_1 or x_2 that the distribution for $X_1(t)$ alone, or $X_2(t)$ alone, is also symmetrical in this case, which is a consequence of the symmetrical distribution of the random pulse amplitude Y. For a nonsymmetrically distributed pulse amplitude Y, the response distribution is also nonsymmetrical, and neither the displacement nor the velocity has a zero mean.

For illustrative purposes, numerical results have been obtained for the preceding Duffing oscillator, with $\zeta = 0.05$ and $\omega_0 = 50$ rad/s. The relaxation time of the system is expected to be near to that of the corresponding linear system ($\delta = 0$), for which $t_r = 1/\zeta\omega_0 = 0.4$ s. The stationary probability densities of the displacement $X(t)$, computed for two different levels of system nonlinearity, $\delta = 0.1$ and 1.0, are shown in Fig. 7.4.1. The random pulse amplitude is assumed to be gaussian-

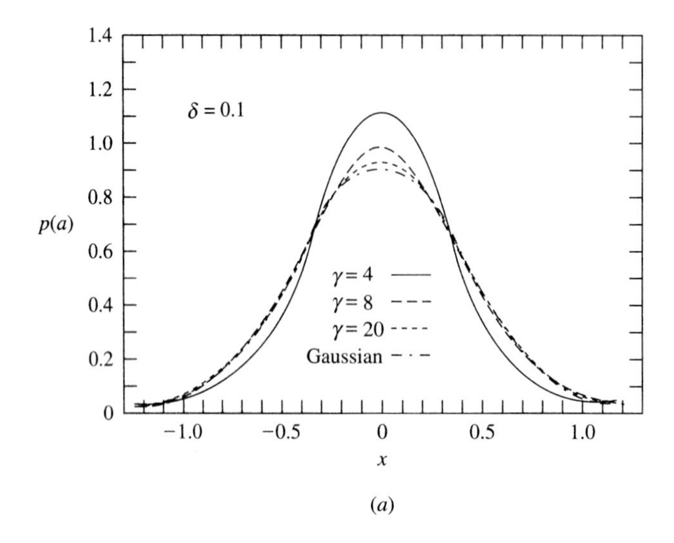

(a)

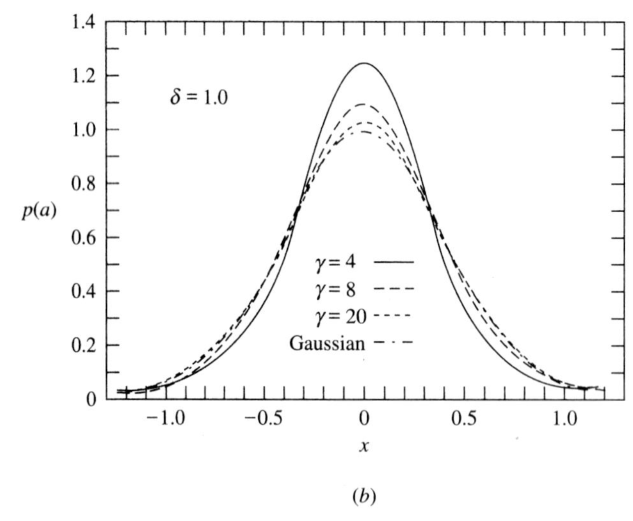

(b)

FIGURE 7.4.1
Stationary probability density of displacement X of a Duffing oscillator. Excitation pulse amplitude symmetrically distributed. (a) $\delta = 0.1$; (b) $\delta = 1.0$ [after Cai and Lin, 1992; Copyright ©1992 Elsevier Science Ltd., reprinted with permission].

distributed with zero mean, and the intensity of the impulsive noise is chosen to be $I_0 = 5000$ for both cases.

It is clear that the average pulse arrival rate in the excitation process, whether fast or slow, should be judged with respect to the relaxation time t_r of the excited system. Therefore, a suitable measure for the closeness to a gaussian excitation is not λ alone, but the product $\gamma = \lambda t_r$. The probability densities computed for three different cases corresponding to $\gamma = 4$, 8, and 20, as well as the case of ideal gaussian white noise excitation, are plotted in Fig. 7.4.1. All the probability densities are symmetric about the zero mean, and the nongaussian ones approach the gaussian one as γ increases. These results suggest that for the particular Duffing oscillator investigated, an additive white noise excitation may be treated as a gaussian white noise if $\gamma > 20$.

Figure 7.4.2 depicts the approximate mean-square values of $X(t)$, computed from the approximate probability densities for the linear case $\delta = 0$ and for the nonlinear cases $\delta = 0.1$ and 1.0. As expected, the mean-square response of the linear system is dependent only on the spectral density of the excitation, either gaussian or nongaussian. The roundabout way in which an approximate probability density is first calculated by perturbation and then used to calculate the mean-square value is of course unnecessary for a linear system. However, it was done for two purposes: to check the accuracy of the perturbation method and to find the range of γ values for which the perturbation analysis might be unreliable. Indeed, the curve associated with $\delta = 0$ in Fig. 7.4.2 is horizontal and straight in the entire range $\gamma > 5$; its departure from the straight line becomes noticeable only in the region $\gamma \leq 3$.

Monte Carlo simulations were performed to further substantiate the present perturbation procedure. When simulating the impulsive noise excitation model, equation (7.4.4), the random arrival times τ_k were generated by calling the IMSL

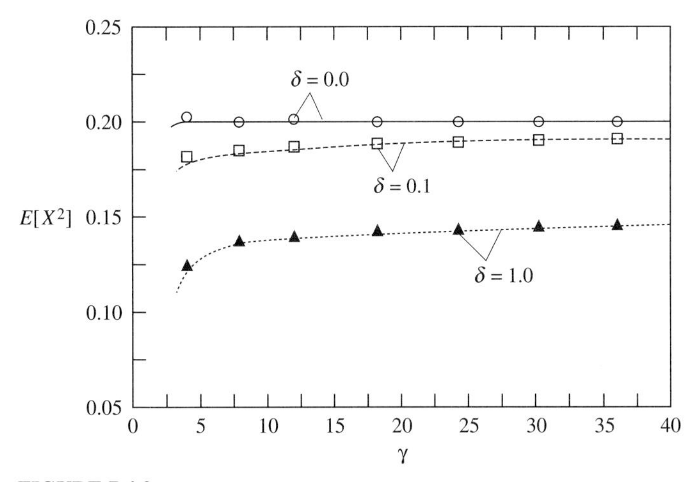

FIGURE 7.4.2
Mean-square response of displacement X for a Duffing oscillator. Excitation pulse amplitude symmetrically distributed [after Cai and Lin, 1992; Copyright ©1992 Elsevier Science Ltd., reprinted with permission].

subroutine RNNPP for each average arrival rate λ, and the samples of the random amplitude Y were generated according to its probability distribution by calling the IMSL subroutine PNNOA. The simulated sample functions of the system response were obtained by solving equation (a) numerically using the fourth-order Runge-Kutta algorithm with a time step of 0.01 s, much shorter than the system relaxation time t_r of 0.4 s. In performing the numerical integration, an instantaneous increment of a randomly generated magnitude was added to the velocity at each impulse arrival time τ_k. The mean and mean-square values were obtained by time-averaging instead of ensemble-averaging, assuming that the response process is ergodic. The simulated mean-square results are shown also in Fig. 7.4.2 as diamonds, squares, and triangles. They agree extremely well with the analytical results obtained from the present perturbation approach in the region $\gamma > 5$.

Next, the effects of a nonsymmetrically distributed pulse amplitude Y are investigated. For this purpose, we choose a shifted Rayleigh distribution

$$p_Y(y) = \begin{cases} \dfrac{y + \sqrt{\pi/2}\,D}{D^2} \exp\left[-\dfrac{\left(y + \sqrt{\pi/2}\,D\right)^2}{2D^2} \right] & y \geq -\sqrt{\pi/2}\,D \\ 0 & y < -\sqrt{\pi/2}\,D \end{cases} \tag{j}$$

where D is a constant, obtainable from a prescribed noise intensity I_0 and the average pulse arrival rate λ. The first four moments of Y are

$$\begin{aligned} E[Y] &= 0 & E[Y^2] &= \left(2 - \frac{\pi}{2}\right) D^2 \\ E[Y^3] &= (\pi - 3)\sqrt{\frac{\pi}{2}}\, D^3 & E[Y^4] &= \left(8 - \frac{3\pi^2}{4}\right) D^4 \end{aligned} \tag{k}$$

Figure 7.4.3 shows the approximate stationary probability densities of the displacement $X(t)$ computed for $\gamma = 4$, 8, and 20, using a shifted Rayleigh distribution for the pulse amplitude and the proposed second-order perturbation procedure. The nonsymmetrical distribution of the pulse amplitude causes the response distribution to also be nonsymmetrical, with its peak shifted slightly to the left. Other features are similar to those of Figs. 7.4.1*a* and *b*. The mean and mean-square values of $X(t)$, computed from these approximate probability densities, are depicted in Figs. 7.4.4 and 7.4.5 for the cases of $\delta = 0$, 0.1, and 1.0. It may be noted in equation (g) that the coefficient of the x_1 term is negative; therefore, the nonsymmetrical distribution of the pulse amplitude Y results in a negative stationary mean value for the response. The effect becomes greater as the level of nonlinearity δ increases or as γ decreases, that is, as the departure from gaussianity in the excitation process increases. That a zero-mean excitation may give rise to a response of nonzero mean for a Duffing oscillator has also been reported in a paper by Iwankiewicz and Nielsen (1989), using a cumulant-neglect closure approach. Again, the mean-square response depends only on the spectral density, not the probability distribution of the white noise excitation if the system is linear. However, if the system is nonlinear, then it decreases as the white noise excitation deviates further from being gaussian.

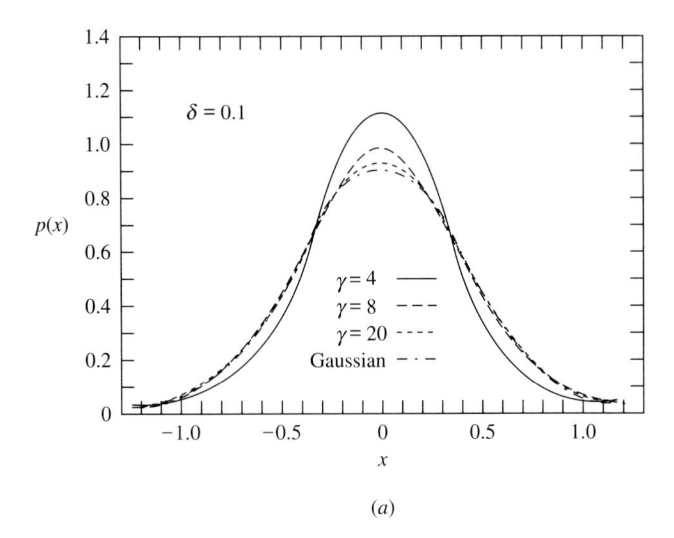

(a)

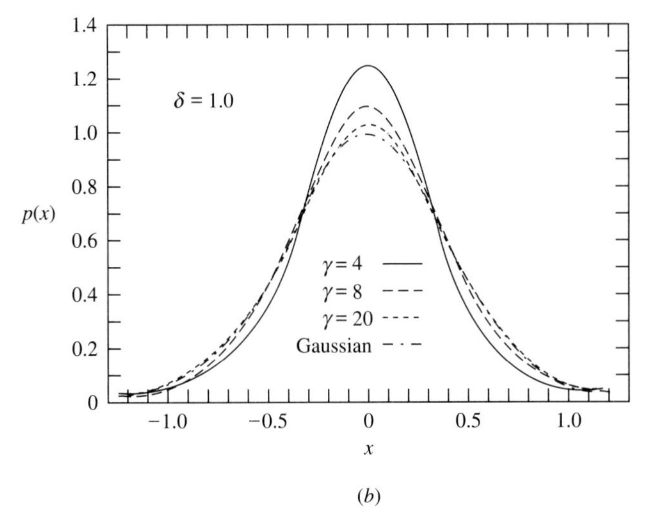

(b)

FIGURE 7.4.3
Stationary probability density of displacement X of a Duffing oscillator. Excitation pulse amplitude nonsymmetrically distributed. (*a*) $\delta = 0.1$ (*b*) $\delta = 1.0$ [after Cai and Lin, 1992; Copyright ©1992 Elsevier Science Ltd., reprinted with permission].

Monte Carlo simulations were also performed for the case of nonsymmetrically distributed pulse amplitude. The samples of the pulse amplitude were obtained from the expression

$$y_k = \sqrt{-2D^2 \ln(1 - z_k)} - \sqrt{\frac{\pi}{2}} D \qquad (l)$$

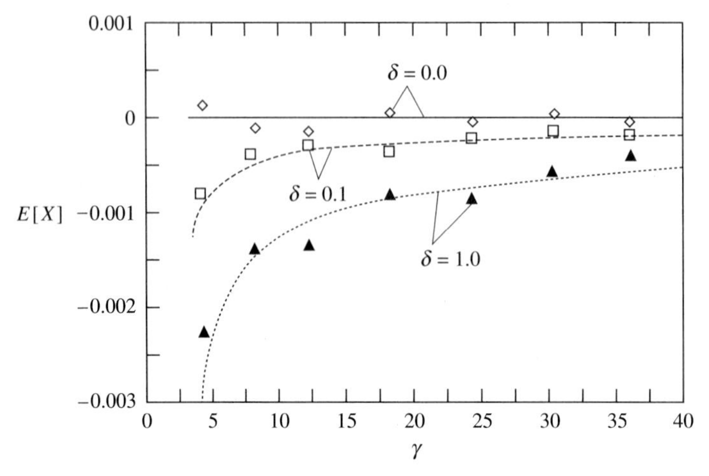

FIGURE 7.4.4
Mean response of displacement X of a Duffing oscillator. Excitation pulse amplitude nonsymmetrically distributed [after Cai and Lin, 1992; Copyright ©1992 Elsevier Science Ltd., reprinted with permission].

where z_k are random numbers uniformly distributed in the interval [0, 1]. It can be verified that the probability density of the simulated pulse amplitude is indeed described by (j). The simulation results are also shown in Figs. 7.4.4 and 7.4.5. It should be noted that the vertical scale used in Fig. 7.4.4. is rather large. Therefore, the seemingly large scattering of the simulated mean values is actually quite small.

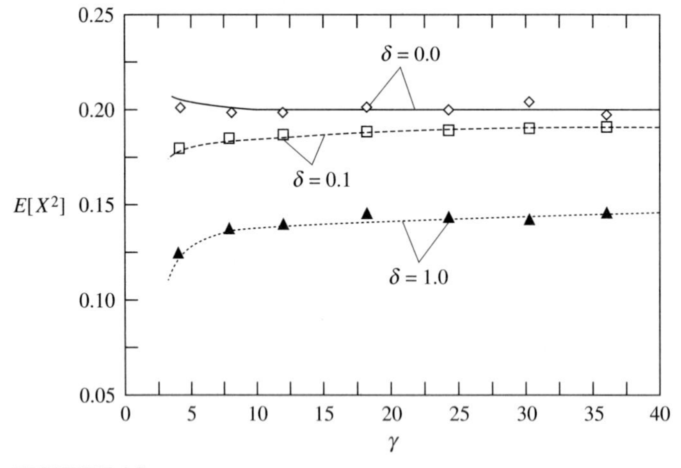

FIGURE 7.4.5
Mean-square response of displacement X of a Duffing oscillator. Excitation pulse amplitude nonsymmetrically distributed [after Cai and Lin, 1992; Copyright ©1992 Elsevier Science Ltd., reprinted with permission].

7.4.2 A Nonlinear System under Multiplicative Excitation of Impulsive Noise

Another example selected for consideration is described by

$$\ddot{X} + 2\zeta\omega_0 X^2 \dot{X} + \omega_0^2(X + \delta X^3) = X\xi(t) \tag{a}$$

which is a nonlinear system with a multiplicative impulsive noise excitation. Note that the trivial solution $X = \dot{X} = 0$ is unstable since the linearized version of equation (a) is undamped. This conclusion can also be reached using the procedure described in Section 6.7. Moreover, if $\xi(t)$ is a gaussian white noise, then the system belongs to the class of generalized stationary potential, and it shares the same stationary probability density, given by equation (b) in Section 7.4.1, for the Duffing oscillator under an additive gaussian white noise excitation.

We now apply the first-order perturbation scheme to obtain an approximate probability density for the response of system (a) when the impulsive excitation is nongaussian. In this case, equation (7.4.20) for $Q_1(x_1, x_2)$ reads

$$\begin{aligned} -x_2\frac{\partial Q_1}{\partial x_1} + \left[-2\zeta\omega_0 x_1^2 x_2 + \omega_0^2(x_1 + \delta x_1^3)\right]\frac{\partial Q_1}{\partial x_2} + \frac{I_0}{2}x_1^2\frac{\partial^2 Q_1}{\partial x_2^2} \\ = \frac{8\zeta^2\omega_0^2 I_1}{I_0^2}x_1^3\left(x_2 - \frac{4\zeta\omega_0}{3I_0}x_2^3\right) \end{aligned} \tag{b}$$

An approximate solution for Q_1 can be obtained by replacing (b) with an equivalent cquation of thc form of (7.4.21) as follows:

$$-x_2\frac{\partial Q_1}{\partial x_1} + (\alpha x_1 + \beta x_2)\frac{\partial Q_1}{\partial x_2} + \frac{I_0}{2}x_1^2\frac{\partial^2 Q_1}{\partial x_2^2} = \frac{8\zeta^2\omega_0^2 I_1}{I_0^2}x_1^3\left(x_2 - \frac{4\zeta\omega_0}{3I_0}x_2^3\right) \tag{c}$$

where the optimal values for α and β are obtained as

$$\begin{aligned} \alpha &= \omega_0^2(1 + \delta\eta) \qquad \eta = \frac{E[X_1^4]}{E[X_1^2]} \\ \beta &= -2\zeta\omega_0 E[X_1^2] \end{aligned} \tag{d}$$

Equation (c) is solved to yield

$$\begin{aligned} Q_1(x_1, x_2) = &-\frac{2\zeta^2\omega_0^2 I_1}{I_0^2}x_1^4 + \frac{32\zeta^3\omega_0^3 I_1}{3I_0^3 D}(c_{60}x_1^6 + c_{51}x_1^5 x_2 + c_{42}x_1^4 x_2^2 + c_{33}x_1^3 x_2^3 \\ &+ c_{24}x_1^2 x_2^4 + c_{15}x_1 x_2^5 + c_{06}x_2^6) \end{aligned} \tag{e}$$

where

$$\begin{aligned} c_{60} = &\left(-5\beta^3 I_0 + \frac{10}{3}\alpha^2\beta^2 - \frac{65}{6\alpha}\beta^2 I_0^2 + \frac{3}{2}\alpha\beta I_0 + \frac{11}{6\beta}\alpha^2 I_0 \right. \\ &\left. + \frac{7}{3}\alpha^3 - \frac{91}{12}I_0^2 - \frac{25}{12\alpha\beta}I_0^3\right) \\ c_{51} = &-\left(7\alpha I_0 + 10\beta^2 I_0 + \frac{15}{\alpha}\beta I_0^2 + \frac{11}{2\beta}I_0^2\right) \end{aligned}$$

$$
\begin{aligned}
c_{42} &= \left(10\alpha\beta^2 + 15\beta I_0 + 7\alpha^2 + \frac{11}{2\beta}\alpha I_0\right) \\
c_{33} &= -\frac{1}{3}\left(14\alpha\beta + 40I_0 + 20\beta^3 + \frac{25}{2\alpha\beta}I_0^2 + \frac{50}{\alpha}\beta^2 I_0\right)
\end{aligned}
\tag{f}
$$

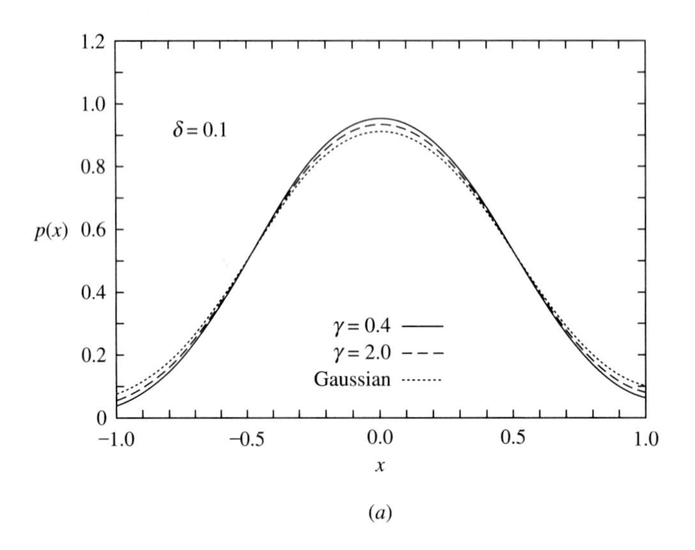

(a)

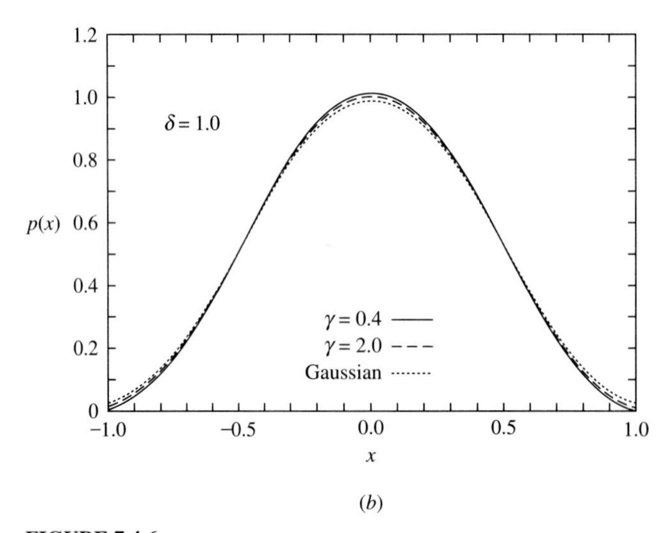

(b)

FIGURE 7.4.6
Stationary probability density of displacement X of a nonlinear system. Excitation pulse amplitude nonsymmetrically distributed. (*a*) $\delta = 0.1$; (*b*) $\delta = 1.0$ [after Cai and Lin, 1992; Copyright ©1992 Elsevier Science Ltd., reprinted with permission].

$$
\begin{aligned}
c_{24} &= -\left(\alpha + 5\beta^2 + \frac{5}{2\beta} I_0 + \frac{25}{2\alpha}\beta I_0\right) \\
c_{15} &= -\left(2\beta + \frac{5}{\alpha} I_0\right) \\
c_{06} &= -\frac{1}{3}\left(1 + \frac{5}{2\alpha\beta} I_0\right)
\end{aligned} \tag{f}
$$

$$
D = \left(170\beta I_0 + 32\alpha^2 + \frac{32}{\beta}\alpha I_0 + 74\alpha\beta^2 + 20\beta^4 + \frac{50}{\alpha}\beta^3 I_0 + \frac{125}{2\alpha} I_0^2\right)
$$

Approximate stationary probability densities of the response $X(t)$, computed for $\delta = 0.1$ and 1.0, respectively, are shown in Fig. 7.4.6. The random pulse amplitude was assumed to have a shifted Rayleigh distribution. It is of interest to note that the computed response probability densities are symmetric even though the distribution of the excitation amplitude is not. A symmetrical distribution implies, of course, a zero mean value. Indeed, this has been confirmed by the Monte Carlo simulation results. Moreover, the nongaussianity of the excitation has little influence on the response distribution even for a highly nongaussian excitation process $\gamma = 0.4$. As shown in Fig. 7.4.7, the mean-square response of system (a) under multiplicative excitation of impulsive noise decreases with a decreasing γ, similar to the case of an additively excited Duffing oscillator, considered in Section 7.4.1. However, the effect of the system nonlinearity and the effect of excitation nongaussianity tend to cancel each other in the present case of multiplicative excitation, while they enhance each other for the Duffing oscillator under additive excitation.

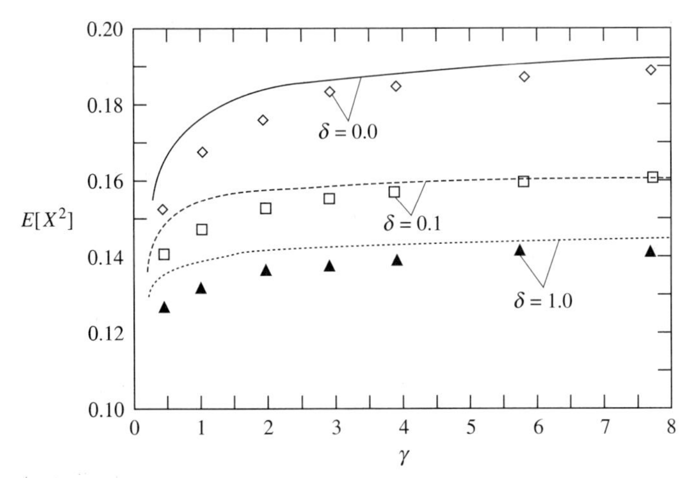

FIGURE 7.4.7
Mean-square response of displacement X of a nonlinear system. Excitation pulse amplitude nonsymmetrically distributed [after Cai and Lin, 1992; Copyright ©1992 Elsevier Science Ltd., reprinted with permission].

The case of symmetrically distributed excitation pulse amplitudes has also been considered for the multiplicatively excited system. In particular, the pulse amplitudes were assumed to be independent and identically distributed gaussian random variables, and both the perturbation analysis and Monte Carlo simulation were carried out for system (a). The effect of the excitation nongaussianity was found to be negligible in the entire range ($\gamma > 0.1$) considered.

7.5 COMBINED HARMONIC AND RANDOM EXCITATIONS

Excitations on a dynamical system need not be entirely random or entirely deterministic. Therefore, the case of combined deterministic and random excitations is of practical interest. In this section we are concerned with an equation of motion of the type

$$\ddot{X} + \varepsilon q(X, \dot{X}) + \omega_0^2 X = \varepsilon f_i(X, \dot{X}) \sin(\omega_i t) + \varepsilon^{1/2} g_j(X, \dot{X}) \xi_j(t) \tag{7.5.1}$$

where ε is a small positive parameter and the $\xi_j(t)$ are weakly stationary random excitations with broadband spectral and cross-spectral densities. As implied in equation (7.5.1), the conservative restoring force is linear, while the damping force and the terms associated with the excitations can be nonlinear. Furthermore, it is assumed that the damping and the harmonic excitation terms are of order ε and the random excitation terms are of order $\varepsilon^{1/2}$, so that their contributions to the system response are commensurable. Under these assumptions, the averaging method is again applicable, as shown by Stratonovich (1967), Ariaratnam and Tam (1976, 1977), Dimentberg (1988), Zhu and Huang (1984), and Ahn (1985).

Without the random excitations, each harmonic excitation term $\sin(\omega_i t)$ in equation (7.5.1) would have a significant contribution to the system response only if ω_i is near the resonance frequency. The location of the resonance frequency depends on the accompanying coefficient $f_i(X, \dot{X})$. For example, it is ω_0 if $f_i(X, \dot{X}) =$ constant, and it is $2\omega_0$ if $f_i(X, \dot{X}) = aX + b\dot{X}$, where a and b are constants. Under both harmonic and random excitations, the resonance and nonresonance cases must be treated separately, as described below.

For the nonresonance case, the harmonic excitations can be ignored. Thus we may use the same transformation as that for the wideband random excitations alone:

$$\begin{aligned} X &= A\cos(\omega_0 t + \phi) \\ \dot{X} &= -A\omega_0 \sin(\omega_0 t + \phi) \end{aligned} \tag{7.5.2}$$

where A and ϕ are the amplitude and phase of the system response, which are slowly varying under the assumption of small damping and weak excitations. After stochastic averaging, the amplitude process A becomes decoupled from the phase process Θ and is approximately a one-dimensional Markov process itself. The corresponding reduced Fokker-Planck equation for the stationary probability density of the amplitude can be obtained readily.

In the resonance case, contributions from the harmonic excitations are significant, and the equations for the amplitude and phase cannot be decoupled by stochastic averaging. Stratonovich (1967) and Ahn (1985) investigated the case of nonlinear damping; however, solutions were provided only for the very special case in which the frequency of the harmonic excitation was tuned *exactly* to the resonance frequency of the system. To allow for a small detuning, it is more convenient to use the alternative transformation

$$\begin{aligned} X &= Ze^{i\nu t} + Z^* e^{-i\nu t} \\ \dot{X} &= i\nu(Ze^{i\nu t} - Z^* e^{-i\nu t}) \end{aligned} \tag{7.5.3}$$

where Z is a complex variable and Z^* is its complex conjugate or, equivalently,

$$\begin{aligned} X &= X_c \cos(\nu t) + X_s \sin(\nu t) \\ \dot{X} &= \nu[-X_c \sin(\nu t) + X_s \cos(\nu t)] \end{aligned} \tag{7.5.4}$$

The new variables Z and Z^* or X_c and X_s are also slowly varying, and the stochastic averaging procedure is applicable. This transformation was used by Ariaratnam and Tam (1976, 1977), Dimentberg (1988), and Zhu and Huang (1984) for obtaining the stability conditions for the statistical moments of linear systems.

In this section, the resonance case is examined for nonlinear systems without the assumption of exact tuning. In particular, we considered the system

$$\ddot{X} + 2\zeta\omega_0\dot{X} + \omega_0^2 X + h(X, \dot{X}) = f_j(X, \dot{X})\sin(j\nu t) + \xi(t) \tag{7.5.5}$$

where $\xi(t)$ is a broadband weakly stationary random process. It is understood that the terms $2\zeta\omega_0\dot{X}$, $h(X, \dot{X})$, and $f_j(X, \dot{X})$ are of a small order ε, and $\xi(t)$ is of order $\varepsilon^{1/2}$, without being so indicated in the equation.

REDUCED FOKKER-PLANCK EQUATION FOR THE AVERAGED SYSTEM. With the transformation (7.5.4), equation (7.5.5) is replaced by an equivalent set of two first-order differential equations for the new variables X_c and X_s:

$$\begin{aligned} \dot{X}_c = {} & 2\zeta\omega_0 \sin(\nu t)[-X_c \sin(\nu t) + X_s \cos(\nu t)] \\ & - \frac{\nu^2 - \omega_0^2}{\nu} \sin(\nu t)[X_c \cos(\nu t) + X_s \sin(\nu t)] + \frac{1}{\nu} \sin(\nu t) h(X_c, X_s) \\ & - \frac{1}{\nu} \sin(\nu t) \sin(j\nu t) f_j(X_c, X_s) - \frac{1}{\nu} \sin(\nu t) \xi(t) \end{aligned} \tag{7.5.6}$$

$$\begin{aligned} \dot{X}_s = {} & -2\zeta\omega_0 \cos(\nu t)[-X_c \sin(\nu t) + X_s \cos(\nu t)] \\ & + \frac{\nu^2 - \omega_0^2}{\nu} \cos(\nu t)[X_c \cos(\nu t) + X_s \sin(\nu t)] - \frac{1}{\nu} \cos(\nu t) h(X_c, X_s) \\ & + \frac{1}{\nu} \cos(\nu t) \sin(j\nu t) f_j(X_c, X_s) + \frac{1}{\nu} \cos(\nu t) \xi(t) \end{aligned} \tag{7.5.7}$$

where $h(X_c, X_s)$ and $f_j(X_c, X_s)$ are obtained from $h(X, \dot{X})$ and $f_j(X, \dot{X})$, respectively, by changing their arguments from X and $\dot{X}$ to X_c and X_s.

When ν is near ω_0, so that $(\nu^2 - \omega_0^2)/\nu$ is small, the right-hand sides of equations (7.5.6) and (7.5.7) are also small, and X_c and X_s are both slowly varying variables. In this case, the stochastic averaging procedure can be applied to eliminate the fast varying terms in equations (7.5.6) and (7.5.7) and to obtain a set of simplified

Itô differential equations for X_c and X_s in the forms

$$dX_c = [-\zeta\omega_0 X_c - \gamma X_s + h_c(X_c, X_s)]\,dt + \sqrt{2\pi K}\,dB_1(t) \tag{7.5.8}$$

$$dX_s = [\gamma X_c - \zeta\omega_0 X_s + h_s(X_c, X_s)]\,dt + \sqrt{2\pi K}\,dB_2(t) \tag{7.5.9}$$

where $B_1(t)$ and $B_2(t)$ are independent unit Wiener processes,

$$\gamma = \frac{\nu^2 - \omega_0^2}{2\nu} \qquad K = \frac{\Phi_{\xi\xi}(\nu)}{2\nu^2} \tag{7.5.10}$$

and $\Phi_{\xi\xi}(\nu)$ is the spectral density of $\xi(t)$ at frequency ν. For $\nu \approx \omega_0$,

$$\gamma = \frac{\nu^2 - \omega_0^2}{2\nu} = \frac{(\nu + \omega_0)(\nu - \omega_0)}{2\nu} \approx \nu - \omega_0 \tag{7.5.11}$$

Thus γ is approximately the difference between ν and ω_0, and it will be referred as the *detuning parameter*. Functions h_c and h_s in equations (7.5.8) and (7.5.9) must be determined in each case for specific $h(X, \dot{X})$ and $f_j(X, \dot{X})$ functions. It can be seen from (7.5.6) and (7.5.7) that whether a harmonic excitation $\sin(j\nu t)$ has a significant contribution toward the averaged system depends on the nature of its accompanying coefficient $f_j(X, \dot{X})$. For instance, excitation $\sin(\nu t)$ is significant if $f_1(X, \dot{X})$ contains a constant term, excitation $\sin(2\nu t)$ is significant if $f_2(X, \dot{X})$ contains a linear term in X and/or $\dot{X}$. The system is said to be in resonance if at least one harmonic excitation contributes significantly to the averaged system.

Equations (7.5.8) and (7.5.9) indicate that, given time, the unknown X_c and X_s will tend to stationary random processes if the averaged system is stable. The reduced Fokker-Planck equation for the stationary probability density $p = p(x_c, x_s)$ is given by

$$\frac{\partial}{\partial x_c}\{[-\zeta\omega_0 x_c - \gamma x_s + h_c(x_c, x_s)]\,p\} + \frac{\partial}{\partial x_s}\{[\gamma x_c - \zeta\omega_0 x_s + h_s(x_c, x_s)]\,p\} - \pi K \frac{\partial^2 p}{\partial x_c^2} - \pi K \frac{\partial^2 p}{\partial x_s^2} = 0 \tag{7.5.12}$$

Equation (7.5.12) is solved subject to the boundary conditions

$$[-\zeta\omega_0 x_c - \gamma x_s + h_c(x_c, x_s)]p - \pi K \frac{\partial p}{\partial x_c} = 0 \qquad |x_c| + |x_s| \to \infty \tag{7.5.13}$$

$$[\gamma x_c - \zeta\omega_0 x_s + h_s(x_c, x_s)]p - \pi K \frac{\partial p}{\partial x_s} = 0 \qquad |x_c| + |x_s| \to \infty \tag{7.5.14}$$

In passing, we note that even when the averaged random processes X_c and X_s reach the stationary state, the original processes X and $\dot{X}$ may remain nonstationary.

EXACT SOLUTION FOR THE AVERAGED SYSTEM. The reduced Fokker-Planck equation (7.5.12) can be cast into the form

$$\frac{\partial}{\partial x_c}\left\{[-\zeta\omega_0 x_c - \gamma x_s + h_c(x_c, x_s)]\,p - \pi K \frac{\partial p}{\partial x_c} - D\frac{\partial p}{\partial x_s}\right\} + \frac{\partial}{\partial x_s}\left\{[\gamma x_c - \zeta\omega_0 x_s + h_s(x_c, x_s)]\,p + D\frac{\partial p}{\partial x_c} - \pi K \frac{\partial p}{\partial x_s}\right\} = 0 \tag{7.5.15}$$

where D is an arbitrary constant. We seek a solution within the class of stationary potential (Stratonovich, 1963) for which equation (7.5.15) is replaced by the following sufficient conditions:

$$[-\zeta\omega_0 x_c - \gamma x_s + h_c(x_c, x_s)]\, p - \pi K \frac{\partial p}{\partial x_c} - D\frac{\partial p}{\partial x_s} = 0 \tag{7.5.16}$$

$$[\gamma x_c - \zeta\omega_0 x_s + h_s(x_c, x_s)]\, p + D\frac{\partial p}{\partial x_c} - \pi K\frac{\partial p}{\partial x_s} = 0 \tag{7.5.17}$$

We note that, as shown in Chapter 5, an oscillatory system does not belong to the class of stationary potential when expressed in the original variables; however, its averaged version in the new variables X_c and X_s may belong to such a class. Now, letting

$$p(x_c, x_s) = C\exp[-\phi(x_c, x_s)] \tag{7.5.18}$$

where ϕ is the probability potential, equations (7.5.16) and (7.5.17) are transformed to

$$\pi K\frac{\partial\phi}{\partial x_c} + D\frac{\partial\phi}{\partial x_s} = \zeta\omega_0 x_c + \gamma x_s - h_c(x_c, x_s) \tag{7.5.19}$$

$$-D\frac{\partial\phi}{\partial x_c} + \pi K\frac{\partial\phi}{\partial x_s} = -\gamma x_c + \zeta\omega_0 x_s - h_s(x_c, x_s) \tag{7.5.20}$$

Equations (7.5.19) and (7.5.20) are two linear algebraic equations for $\partial\phi/\partial x_c$ and $\partial\phi/\partial x_s$ which are solved to yield

$$\frac{\partial\phi}{\partial x_c} = \frac{1}{\pi^2K^2 + D^2}[(\pi K\zeta\omega_0 + \delta D)\, x_c + (\pi K\gamma - \zeta\omega_0 D)\, x_s - \pi K h_c(x_c, x_s) + D h_s(x_c, x_s)] \tag{7.5.21}$$

$$\frac{\partial\phi}{\partial x_s} = \frac{1}{\pi^2K^2 + D^2}[-(\pi K\gamma - \zeta\omega_0 D)\, x_c + (\pi K\zeta\omega_0 - \gamma D)\, x_s - D h_c(x_c, x_s) - \pi K h_s(x_c, x_s)] \tag{7.5.22}$$

A compatible probability potential function ϕ exists provided that

$$\frac{\partial}{\partial x_s}\left(\frac{\partial\phi}{\partial x_c}\right) = \frac{\partial}{\partial x_c}\left(\frac{\partial\phi}{\partial x_s}\right) \tag{7.5.23}$$

Upon substituting equations (7.5.21) and (7.5.22) into equation (7.5.23), we obtain

$$2\pi K\gamma - 2\zeta\omega_0 D - \pi K\left(\frac{\partial h_c}{\partial x_s} - \frac{\partial h_s}{\partial x_c}\right) + D\left(\frac{\partial h_c}{\partial x_c} + \frac{\partial h_s}{\partial x_s}\right) = 0 \tag{7.5.24}$$

If a constant D can be found to satisfy equation (7.5.24), then the averaged system described by equations (7.5.8) and (7.5.9) belongs to the class of stationary potential, and a consistent probability potential function ϕ can be obtained from equations (7.5.21) and (7.5.22).

Three special cases will now be considered.

Case 1. Functions h_c and h_s are linear in x_c and x_s:

$$\begin{aligned} h_c(x_c, x_s) &= c_{00} + c_{10}x_c + c_{01}x_s \\ h_s(x_c, x_s) &= s_{00} + s_{10}x_c + s_{01}x_s \end{aligned} \tag{7.5.25}$$

This case arises when the harmonic excitations are either additive or multiplied to linear terms in X and/or $\dot{X}$. Substituting (7.5.25) into (7.5.24), we obtain

$$D = \frac{\pi K(2\gamma - c_{01} + s_{10})}{2\zeta\omega_0 - c_{10} - s_{01}} \tag{7.5.26}$$

The probability potential function can be obtained by integrating equations (7.5.21) and (7.5.22) to yield

$$\begin{aligned} \phi_l = \frac{1}{\pi^2 K^2 + D^2}\Big\{ & (-\pi K c_{00} + D s_{00})x_c - (D c_{00} + \pi K s_{00})x_s \\ & + \frac{1}{2}[\pi K(-c_{10} + \zeta\omega_0) + D(s_{10} + \gamma)]x_c^2 \\ & + [\pi K(-c_{01} + \gamma) + D(s_{01} - \zeta\omega_0)]x_c x_s \\ & + \frac{1}{2}[\pi K(-s_{01} + \zeta\omega_0) + D(-c_{10} + \gamma)]x_s^2 \Big\} \end{aligned} \tag{7.5.27}$$

Case 2. We now generalize case 1 by seeking an admissible probability potential function ϕ of the form

$$\phi = \phi_l + e_1 x_c^4 + e_2 x_c^2 x_s^2 + e_3 x_c^4 \qquad e_1 \geq 0, e_3 \geq 0, e_2 \geq -2\sqrt{e_1 e_3} \tag{7.5.28}$$

where ϕ_l is given by (7.5.27). From equations (7.5.19) and (7.5.20), it follows that

$$h_c = c_{00} + c_{10}x_c + c_{01}x_s - 4\pi K e_1 x_c^3 - 2De_2 x_c^2 x_s - 2\pi K e_2 x_c x_s^2 - 4De_3 x_s^3 \tag{7.5.29}$$

$$h_s = s_{00} + s_{10}x_c + s_{01}x_s + 4De_1 x_c^3 - 2\pi K e_2 x_c^2 x_s + 2De_2 x_c x_s^2 - 4\pi K e_3 x_s^3 \tag{7.5.30}$$

This probability potential function is admissible if the conditions for e_1, e_2, and e_3 specified in (7.5.28) are satisfied, because it guarantees that the probability density is integrable.

Case 3. The h_c and h_s functions may be generalized further to

$$\begin{aligned} h_c(x_c, x_s) &= c_{00} + c_{10}x_c + c_{01}x_s + c_{30}x_c^3 + c_{21}x_c^2 x_s + c_{12}x_c x_s^2 + c_{03}x_s^3 \\ h_s(x_c, x_s) &= s_{00} + s_{10}x_c + s_{01}x_s + s_{30}x_c^3 + s_{21}x_c^2 x_s + s_{12}x_c x_s^2 + s_{03}x_s^3 \end{aligned} \tag{7.5.31}$$

In this case, equation (7.5.24) becomes

$$\begin{aligned} & 2\pi K\gamma - 2\zeta\omega_0 D - \pi K(c_{01} - s_{10}) + D(c_{10} + s_{01}) \\ & \quad - (\pi K c_{21} - 3\pi K s_{30} - 3Dc_{30} - Ds_{21})x_c^2 \\ & \quad - (2\pi K c_{12} - 2\pi K s_{21} - 2Dc_{21} - 2Ds_{12})x_c x_s \\ & \quad - (3\pi K c_{03} - \pi K s_{12} - Dc_{12} - 3Ds_{03})x_s^2 = 0 \end{aligned} \tag{7.5.32}$$

Equation (7.5.32) is satisfied if constant D is given by equation (7.5.26) and if the following three conditions are met:

$$\begin{aligned} \pi K c_{21} - 3\pi K s_{30} - 3D c_{30} - D s_{21} &= 0 \\ \pi K c_{12} - \pi K s_{21} - D c_{21} - D s_{12} &= 0 \\ 3\pi K c_{03} - \pi K s_{12} - D c_{12} - 3D s_{03} &= 0 \end{aligned} \tag{7.5.33}$$

Of course, a probability density so obtained must be integrable, which is an additional condition.

APPROXIMATE SOLUTION FOR THE AVERAGED SYSTEM. Consider an averaged system governed by the following Itô differential equations:

$$dX_c = [-\zeta\omega_0 X_c - \gamma X_s + H_c(X_c, X_s)]\,dt + \sqrt{2\pi K}\,dB_1(t) \tag{7.5.34}$$

$$dX_s = [\gamma X_c - \zeta\omega_0 X_s + H_s(X_c, X_s)]\,dt + \sqrt{2\pi K}\,dB_2(t) \tag{7.5.35}$$

The corresponding reduced Fokker-Planck equation is given by

$$\frac{\partial}{\partial x_c}\{[-\zeta\omega_0 x_c - \gamma x_s + H_c(x_c, x_s)]\,\tilde{p}\} + \frac{\partial}{\partial x_s}\{[\gamma x_c - \zeta\omega_0 x_s + H_s(x_c, x_s)]\,\tilde{p}\} - \pi K \frac{\partial^2 \tilde{p}}{\partial x_c^2} - \pi K \frac{\partial^2 \tilde{p}}{\partial x_s^2} = 0 \tag{7.5.36}$$

where $\tilde{p} = \tilde{p}(x_c, x_s)$ is the stationary probability density of X_c and X_s. Suppose that condition (7.5.24) is not satisfied upon substituting H_c and H_s for h_c and h_s, respectively. Then an exact analytical solution for equation (7.5.36) is not obtainable from the foregoing procedure of stationary potential, and recourse may be sought again in an approximate solution, using the method of weighted residuals.

Let $p(x_c, x_s)$ be an exact solution for equation (7.5.12), and let it be optimized for the purpose of obtaining an approximate solution for equation (7.5.36). The residual error of approximation is given by

$$\delta = \frac{\partial}{\partial x_c}[(h_c - H_c)\,p] + \frac{\partial}{\partial x_s}[(h_s - H_s)\,p] \tag{7.5.37}$$

Application of the method of weighted residuals (7.2.7) leads to the constraint

$$\Delta_M = E\left[(H_c - h_c)\frac{\partial M}{\partial X_c} + (H_s - h_s)\frac{\partial M}{\partial X_s}\right] = 0 \tag{7.5.38}$$

where M is a selected weighting function. Functions h_c and h_s can be determined from a set of constraints in the form of (7.5.38) corresponding to a set of suitably selected weighting functions.

For some nonlinear structural systems, the H_c and H_s functions may be approximated as being composed of linear and third-order terms in X_c and X_s:

$$H_c(X_c, X_s) = c_{00} + c_{10}X_c + c_{01}X_s + \bar{H}_c(X_c, X_s) \tag{7.5.39}$$

$$H_s(X_c, X_s) = s_{00} + s_{10}X_c + s_{01}Xs + \bar{H}_s(X_c, X_s) \tag{7.5.40}$$

where $\bar{H}_c$ and $\bar{H}_s$ are third-order homogeneous polynomials. Let the approximate probability potential be chosen from the class of (7.5.28) with three coefficients e_1,

e_2, and e_3 to be determined. The functions h_c and h_s of the replacement system are given by (7.5.29) and (7.5.30). Letting $M = X_c^2, X_cX_s$, and X_s^2 and substituting (7.5.29) and (7.5.30) into (7.5.38), we obtain the following three constraints:

$$E\{X_c[\bar{H}_c + 4\pi K e_1 X_c^3 + 2De_2X_c^2X_s + 2\pi K e_2 X_c X_s^2 + 4De_3X_s^3]\} = 0 \tag{7.5.41}$$

$$E\{X_s[\bar{H}_c + 4\pi K e_1 X_c^3 + 2De_2X_c^2X_s + 2\pi K e_2 X_c X_s^2 + 4De_3X_s^3] + X_c[\bar{H}_s - 4De_1X_c^3 + 2\pi K e_2 X_c^2 X_s - 2De_2X_cX_s^2 + 4\pi K e_3 X_s^3]\} = 0 \tag{7.5.42}$$

$$E\{X_s[\bar{H}_s - 4De_1X_c^3 + 2\pi K e_2 X_c^2 X_s - 2De_2X_cX_s^2 + 4\pi K e_3 X_s^3]\} = 0 \tag{7.5.43}$$

where D is obtained from (7.5.26). The preceding ensemble averages are computed from the approximate probability density

$$p(x_c, x_s) = C\exp[-\phi_l - e_1x_c^4 - e_2x_c^2x_s^2 - e_3x_s^4] \tag{7.5.44}$$

where ϕ_l is given by (7.5.27). Equations (7.5.41) through (7.5.43) are a set of algebraic equations for the unknown coefficients e_1, e_2, and e_3; in general, they can be solved numerically.

Application of this procedure is illustrated in the following two examples.

7.5.1 A Linear System under an Additive Random Excitation and a Multiplicative Harmonic Excitation

Consider the equation of motion

$$\ddot{X} + 2\zeta\omega_0\dot{X} + \omega_0^2X[1 + \lambda\sin(2\nu t)] = W(t) \tag{a}$$

where $W(t)$ is a gaussian white noise with a spectral density K_W. Using the transformation (7.5.4) and applying the stochastic averaging procedure, we obtain the following set of averaged Itô stochastic differential equations for the slowly varying variables X_c and X_s:

$$dX_c = (-\zeta\omega_0X_c - \gamma X_s + \zeta\omega_0 r X_c)\,dt + \sqrt{2\pi K}\,dB_1(t) \tag{b}$$

$$dX_s = (\gamma X_c - \zeta\omega_0X_s - \zeta\omega_0 r X_s)\,dt + \sqrt{2\pi K}\,dB_2(t) \tag{c}$$

where
$$\gamma = \frac{\nu^2 - \omega_0^2}{2\nu} \qquad r = \frac{\omega_0\lambda}{4\zeta\nu} \qquad K = \frac{K_w}{2\nu^2} \tag{d}$$

For the near-tuning case of $\nu \approx \omega_0$, we have $\gamma \approx \nu - \omega_0$ and $r \approx \lambda/(4\zeta)$, which is proportional to the ratio between the harmonic excitation amplitude and the damping coefficient. Comparing equations (b) and (c) with equations (7.5.8) and (7.5.9), functions h_c and h_s are seen to be linear. Specifically,

$$h_c = \zeta\omega_0 r X_c \qquad h_s = -\zeta\omega_0 r X_s \tag{e}$$

Thus the problem belongs to case 1 in Section 7.5, and we have from equations (7.5.26) and (7.5.27)

$$D = \pi K\eta \tag{f}$$

where $\eta = \gamma/(\zeta\omega_0)$, which is essentially the ratio of the detuning parameter to the damping coefficient, and

$$\phi_l = \frac{\zeta\omega_0}{2\pi K(1+\eta^2)}[(1+\eta^2-r)x_c^2 - 2r\eta x_c x_s + (1+\eta^2+r)x_s^2] \quad \text{(g)}$$

The quadratic form of the probability potential ϕ_l indicates that the Markov vector (X_c, X_s) is gaussian-distributed, with a probability density

$$p(x_c, x_s) = \frac{\zeta\omega_0\sqrt{1-\mu^2}}{2\pi^2 K}\exp\left\{-\frac{\zeta\omega_0}{2\pi K(1+\eta^2)}[(1+\eta^2-r)x_c^2 - 2r\eta x_c x_s + (1+\eta^2+r)x_s^2]\right\} \quad \text{(h)}$$

where $\mu = r/\sqrt{1+\eta^2}$. Equation (h) is a valid probability density only if

$$\mu = \frac{r}{\sqrt{1+\eta^2}} < 1 \quad \text{(i)}$$

The boundary of the stability region is determined by letting $\mu = 1$, or in terms of original system parameters,

$$\lambda^2 = 4\left(\frac{\nu}{\omega_0}\right)^4 - 8(1-2\zeta^2)\left(\frac{\nu}{\omega_0}\right)^2 + 4 \quad \text{(j)}$$

The critical value λ_{cr} for λ is obtained by minimizing λ with respect to ν/ω_0 to yield

$$\lambda_{\text{cr}} = 4\zeta\sqrt{1-\zeta^2} \quad \text{(k)}$$

This confirms the known fact that the additive excitation has no effect on the stability of a linear system. Figure 7.5.1 shows the stability region in the parameter

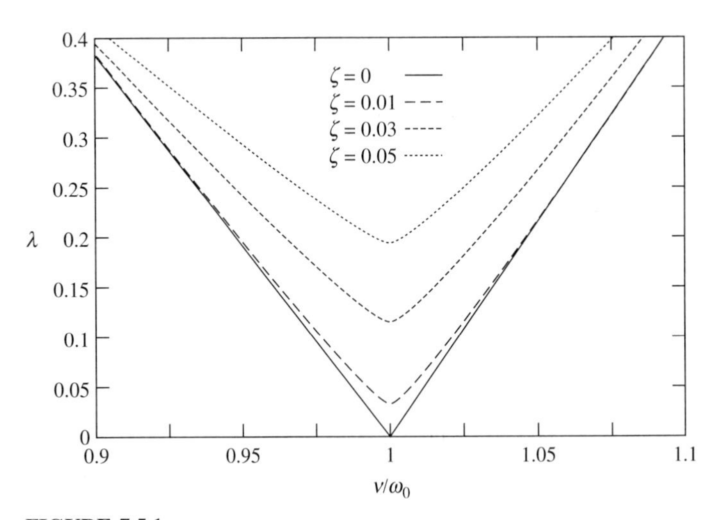

FIGURE 7.5.1
Stability boundary for system $\ddot{X} + 2\zeta\omega_0\dot{X} + \omega_0^2 X(1+\lambda\sin 2\nu t) = W(t)$ [after Cai and Lin, 1994b].

space (λ, ν/ω_0) for different damping coefficients ζ. An unstable system can be stabilized by reducing the intensity of the harmonic excitation, by increasing the detuning parameter γ, or by increasing the damping parameter ζ within the limit $\zeta < 1/\sqrt{2}$.

It is of interest to compare the preceding result with that of the deterministic Mathieu-Hill equation

$$\ddot{u} + 2\beta\dot{u} + (\alpha + \varepsilon \sin 2t)\, u = 0 \tag{l}$$

for which the stability boundary is approximately given by (e.g., Nayfeh and Mook, 1979)

$$\alpha = 1 + \tfrac{1}{2}\left[\varepsilon^2 - 16\beta^2\right]^{1/2} \tag{m}$$

It can be verified that equations (j) and (m) are the same.

The probability density of X and $\dot{X}$ can be converted from (h) using (7.5.4). It can be shown that although X_c and X_s are stationary, X and $\dot{X}$ are nonstationary, except when $\lambda = 0$, that is, when the harmonic excitation is absent.

Additional physical insights can be gained by transforming X_c and X_s to another pair of variables:

$$X_c = A\cos\Theta \qquad X_s = A\sin\Theta \tag{n}$$

which is equivalent to

$$X = A\cos(\nu t - \Theta) \qquad \dot{X} = -\nu A\sin(\nu t - \Theta) \tag{o}$$

The new processes $A(t)$ and $\Theta(t)$ are, respectively, the amplitude and the phase for the narrow-band process $X(t)$, and they are also slowly varying variables. The joint probability density of A and Θ can be obtained readily from equation (h) by change of variables to yield

$$p(a, \theta) = \frac{\zeta\omega_0\sqrt{1-\mu^2}}{2\pi^2 K}\, a \exp\left\{-\frac{\zeta\omega_0 a^2}{2\pi K}\left[1 - \mu\cos(2\theta - 2\psi)\right]\right\} \tag{p}$$

where ψ is determined from

$$\cos(2\psi) = \frac{1}{\sqrt{1+\eta^2}} \qquad \sin(2\psi) = \frac{\eta}{\sqrt{1+\eta^2}} \tag{q}$$

The marginal probability densities for A and Θ are given by

$$p(a) = \frac{\zeta\omega_0\sqrt{1-\mu^2}}{\pi K}\, a\, \mathrm{I}_0\left(\frac{\mu\zeta\omega_0 a^2}{2\pi K}\right)\exp\left(-\frac{\zeta\omega_0 a^2}{2\pi K}\right) \tag{r}$$

and

$$p(\theta) = \frac{\sqrt{1-\mu^2}}{2\pi\left[1 - \mu\cos(2\theta - 2\psi)\right]} \tag{s}$$

obtained from integrating the right-hand side of equation (p) with respect to θ and a, respectively, where I_0 is the modified Bessel function of the zeroth order. As expected, in the limiting case $\lambda = 0$, the amplitude A becomes Rayleigh-distributed, and the phase angle Θ uniformly distributed, which is the case without the harmonic excitation.

Numerical results have been obtained for the linear system (a) with $\xi = 0.05$, $\omega_0 = 10$, and $K_W = 10$. Figure 7.5.2 depicts the stationary probability density of the amplitude process $A(t)$ computed from equation (r) for $\nu = 11$, and the cases $\lambda = 0.05$, 0.1, and 0.2. Also shown in the figure are the results obtained from Monte Carlo simulations of the original system (a). They are in good agreement. Since (r) is the exact solution for the averaged system described by equations (b) and (c),

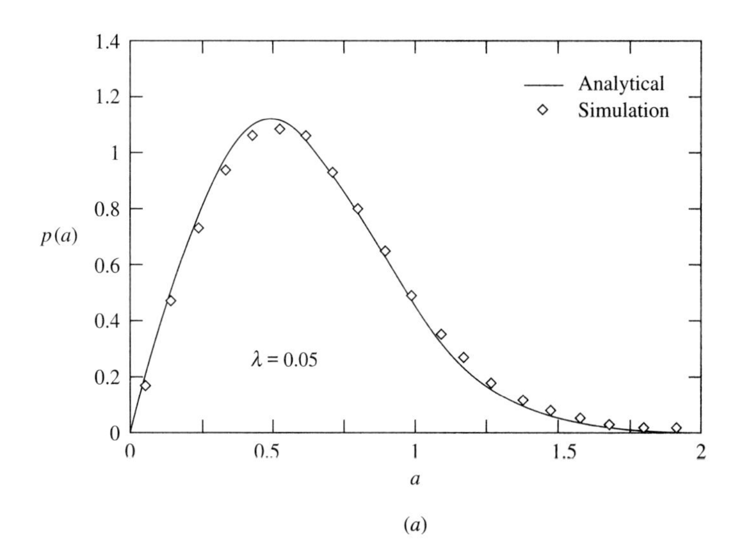

(*a*)

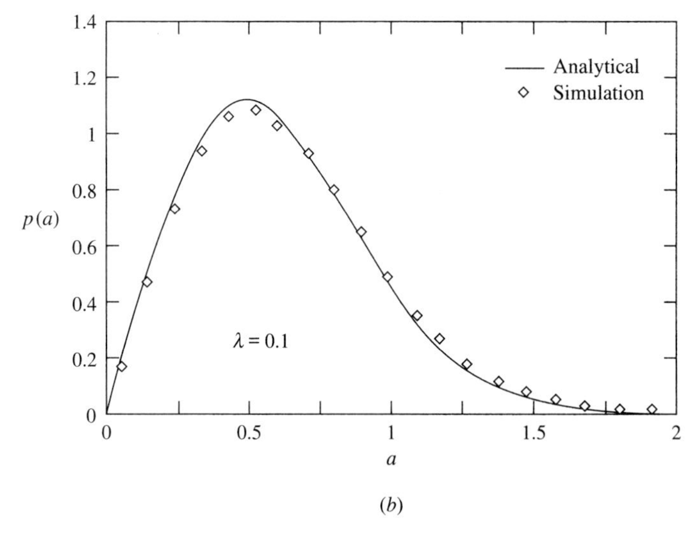

(*b*)

FIGURE 7.5.2 ***(Continued on next page)***
Stationary probability density of the response amplitude A of a linear system (*a*) $\lambda = 0.05$; (*b*) $\lambda = 0.1$ [after Cai and Lin, 1994b].

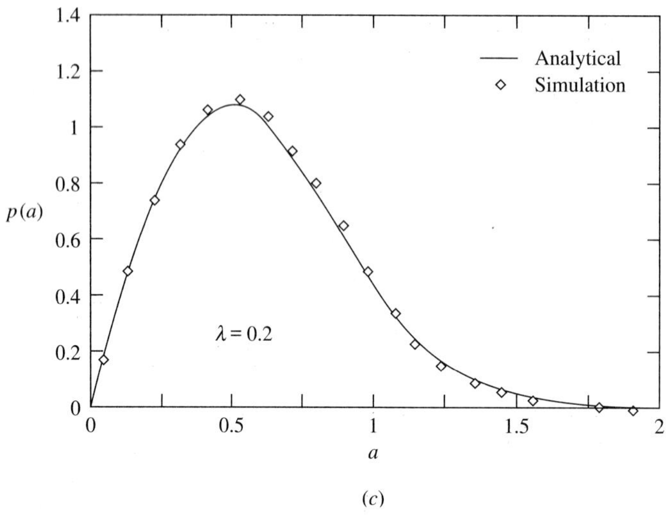

(c)

FIGURE 7.5.2 ***(Continued)***
Stationary probability density of the response amplitude of A of a linear system (c) $\lambda = 0.2$ [after Cai and Lin, 1994b].

the discrepancies between the analytical and simulation results arise mainly from the averaging procedure. The approximate mean values of the amplitude process $A(t)$ computed from (r), as well as those obtained from Monte Carlo simulations, are shown in Fig. 7.5.3 for four different intensity levels of the harmonic excitation, $\lambda = 0.1, 0.15, 0.18$, and 0.19. When it is weak ($\lambda < 0.1$), the harmonic excitation

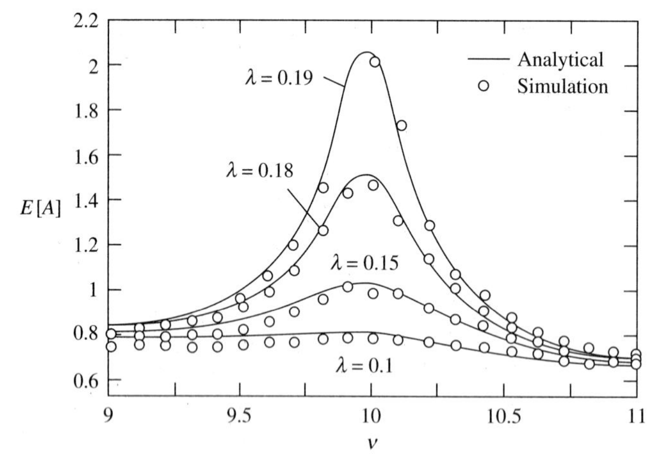

FIGURE 7.5.3
Mean value of the response amplitude A of a linear system [after Cai and Lin, 1994b].

has little influence on the stationary amplitude response. However, it has a dominant effect near the resonant frequency ($\nu \approx \omega_0$), when $\lambda > 0.15$. The system becomes unstable in probability when $\lambda > 0.2$.

7.5.2 A Nonlinearly Damped System under an Additive Random Excitation and a Multiplicative Harmonic Excitation

We now modify the governing equation of the linear system considered in Section 7.5.1 by adding a nonlinear damping term to the left-hand side to read

$$\ddot{X} + 2\zeta\omega_0\dot{X} + \beta X^2\dot{X} + \omega_0^2 X\,[1 + \lambda\sin(2\nu t)] = W(t) \tag{a}$$

Using again the transformation (7.5.4), the averaged Itô equations for the slowly varying $X_c(t)$ and $X_s(t)$ processes are obtained as follows:

$$dX_c = \left[-\zeta\omega_0 - \gamma X_s + \zeta\omega_0 r X_c - \frac{\beta}{8}X_c(X_c^2 + X_s^2)\right]dt + \sqrt{2\pi K}\,dB_1(t) \tag{b}$$

$$dX_s = \left[\gamma X_c - \zeta\omega_0 X_s - \zeta\omega_0 r X_s - \frac{\beta}{8}X_s(X_c^2 + X_s^2)\right]dt + \sqrt{2\pi K}\,dB_2(t) \tag{c}$$

Application of equation (7.5.24) to the present case results in

$$2\pi K\gamma - 2\zeta\omega_0 D - \frac{\beta}{2}D(x_c^2 + x_s^2) = 0 \tag{d}$$

For arbitrary x_c and x_s, equation (d) can be satisfied only if $\gamma = 0$ and $D = 0$, that is, if the frequency of the harmonic excitation is exactly tuned to twice the system natural frequency. In this case, the probability density is given by

$$p(x_c, x_s) = C\exp(-\phi) \tag{e}$$

where
$$\phi = \frac{1}{2\pi K}\left[\zeta\omega_0\left(1 - \frac{\lambda}{4\zeta}\right)x_c^2 + \zeta\omega_0\left(1 + \frac{\lambda}{4\zeta}\right)x_s^2 + \frac{\beta}{16}(x_c^2 + x_s^2)^2\right] \tag{f}$$

The joint stationary probability density for the amplitude process $A(t)$ and the phase process $\Theta(t)$ is obtained by coordinate transformation, to yield

$$p(a, \theta) = Ca\exp\left\{-\frac{\zeta\omega_0 a^2}{2\pi K}[1 - \mu\cos(2\theta)] - \frac{\beta}{32\pi K}a^4\right\} \tag{g}$$

The stationary probability density for the amplitude process $A(t)$ follows upon integrating (g) over θ:

$$p(a) = Ca\,\mathrm{I}_0\left(\frac{\mu\zeta\omega_0 a^2}{2\pi K}\right)\exp\left(-\frac{\zeta\omega_0 a^2}{2\pi K} - \frac{\beta a^4}{32\pi K}\right) \tag{h}$$

where I_0 is the modified Bessel function of the zeroth order. The stationary probability density for the phase process $\Theta(t)$ cannot be obtained in closed form, but it can be obtained by integrating (g) numerically with respect to a. These probability densities exist as long as $\beta > 0$.

When the harmonic excitation is absent, equation (g) is reduced to

$$p(a, \theta) = Ca\exp\left[-\frac{\zeta\omega_0}{2\pi K}\left(a^2 + \frac{\beta}{16\zeta\omega_0}a^4\right)\right] \tag{i}$$

which implies that Θ is uniformly distributed. It can be shown that equation (i) can also be obtained by applying the approximation procedure of dissipation energy balancing.

In practice, the frequency of the harmonic excitation may not be exactly twice the system natural frequency. Then $\gamma \neq 0$ and equation (d) cannot be satisfied for arbitrary x_c and x_s, indicating that the system does not belong to the class of stationary potential. In this case, the approximation described in Section 7.5 is applicable. The functions $\bar{H}_c$ and $\bar{H}_s$ in equations (7.5.41)–(7.5.43) are identified as

$$\bar{H}_c = -\frac{\beta}{8}X_c(X_c^2 + X_s^2) \qquad \bar{H}_s = -\frac{\beta}{8}X_s(X_c^2 + X_s^2) \tag{j}$$

Imposition of the constraints (7.5.41) through (7.5.43) leads to

$$2\pi K m_{40} e_1 + (D m_{31} + \pi K m_{22}) e_2 + 2 D m_{13} e_3 = \frac{\beta}{16}(m_{40} + m_{22}) \tag{k}$$

$$2(-D m_{40} + \pi K m_{31}) e_1 + \pi K (m_{31} + m_{13}) e_2 + 2(\pi K m_{13} + D m_{04}) e_3 = \frac{\beta}{8}(m_{31} + m_{13}) \tag{l}$$

$$-2 D m_{31} e_1 + (\pi K m_{22} - D m_{13}) e_2 + 2\pi K m_{04} e_3 = \frac{\beta}{16}(m_{22} + m_{04}) \tag{m}$$

where $m_{ij} = E[X_c^i X_s^j]$. Equations (k) through (m) are a set of linear algebraic equations for the unknowns e_1, e_2, and e_3; they can be solved numerically in conjunction with equation (7.5.44).

Figure 7.5.4 shows the approximate stationary probability densities of the amplitude process $A(t)$ obtained for four different β values, 0, 0.05, 0.10, and 0.15,

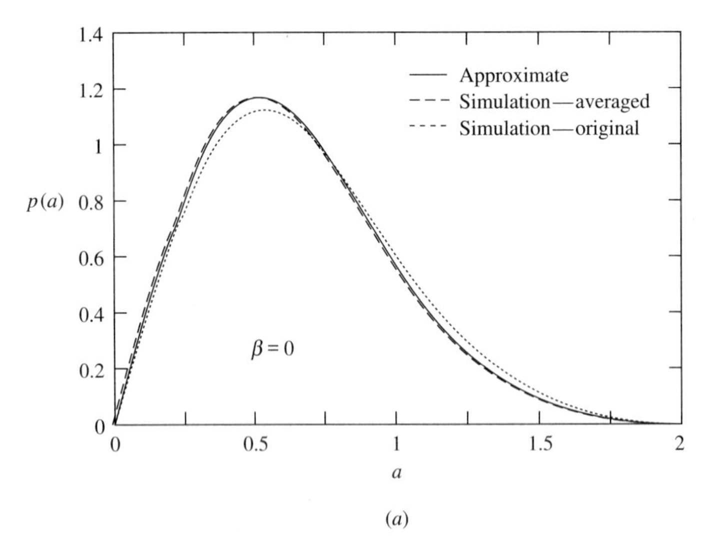

(*a*)

FIGURE 7.5.4 ***(Continued on next page)***
Stationary probability density of the response amplitude A of a nonlinear system (*a*) $\beta = 0.0$ [after Cai and Lin, 1994b].

but keeping the same values $\zeta = 0.05$, $\omega_0 = 10$, $\nu = 11$, $K_W = 10$, and $\lambda = 0.1$. Monte Carlo simulations have also been carried out for both the original system (a) and the averaged system represented by equations (b) and (c). In the case $\beta = 0$ (the linear case) the analytical solution is exact for the averaged system; therefore, the agreement between the analytical and the simulation results of the averaged system is expected and can be used to validate the simulation procedure. For the several

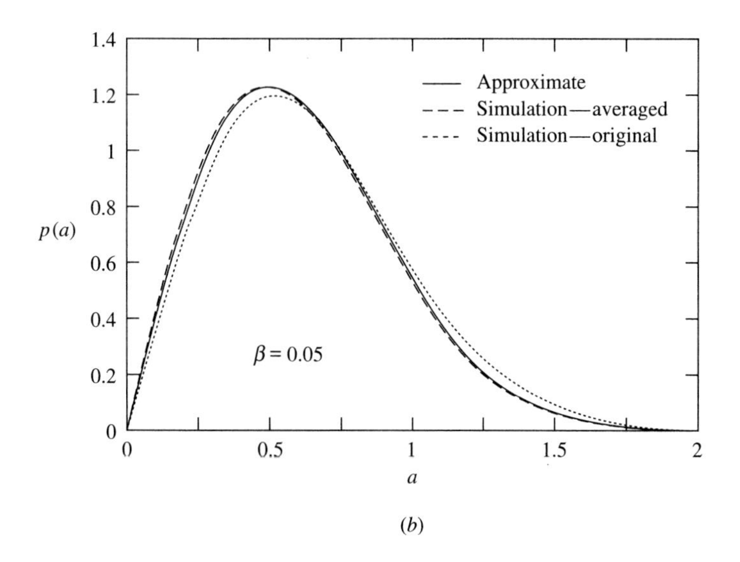

(b)

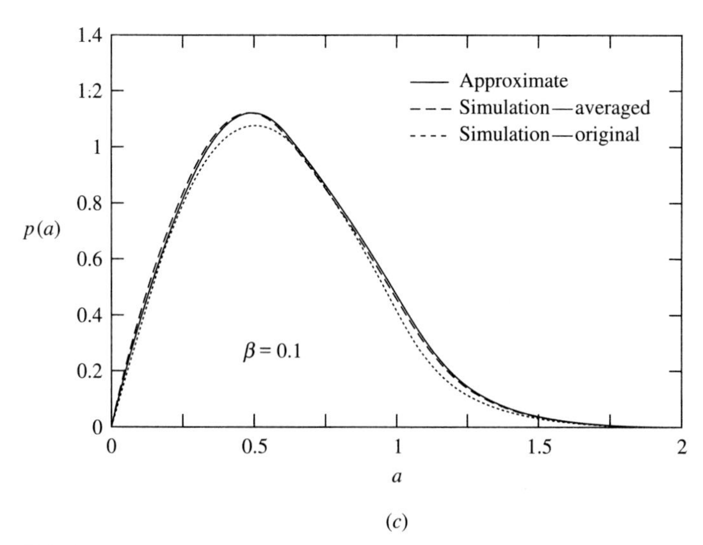

(c)

FIGURE 7.5.4 ***(Continued on next page)***
Stationary probability density of the response amplitude A of a nonlinear system. (*b*) $\beta = 0.05$; (*c*) $\beta = 0.10$ [after Cai and Lin, 1994b].

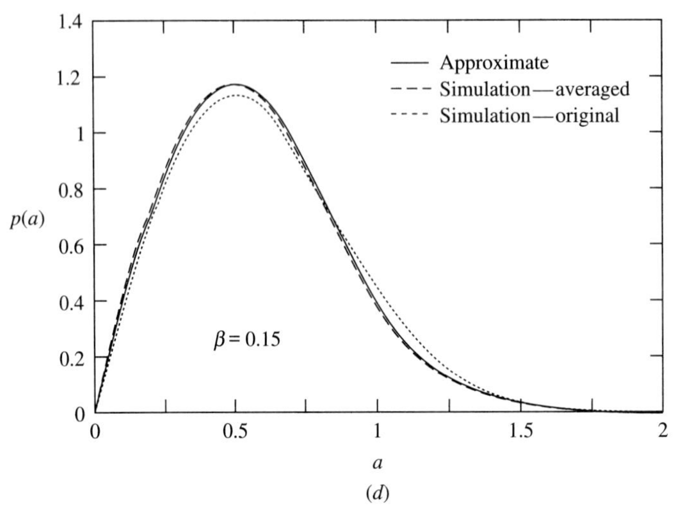

FIGURE 7.5.4 (*Continued*)
Stationary probability density of the response amplitude A of a nonlinear system. (*d*) $\beta = 0.15$ [after Cai and Lin, 1994b].

nonlinearity levels investigated, the analytical results are almost indistinguishable from the simulation results based on the averaged system, but they are distinguishable from the simulation results based on the original system; therefore, the errors in the analytical results are due mainly to the averaging procedure.

7.6 CONCLUDING REMARKS

Several approximate solution techniques have been presented in this chapter for multidimensional nonlinear systems, for which exact probabilistic solutions are presently not obtainable. The method of cumulant-neglect closure is devised to calculate the approximate statistical moments of the system response to gaussian white noise excitations. The procedure is readily applicable to multi-degree-of-freedom systems and is generally reliable when the random excitations are all additive. However, it may lead to erroneous results if multiplicative excitations are present and if the stability boundary is near in some sense.

The method of weighted residuals is a general scheme to obtain an approximate probability density for the response of a system under gaussian white noise excitations. It seeks the best replacement system within a class of exactly solvable systems, under the condition that for each judiciously selected weighting function, the weighted average of the difference between the Fokker-Planck equation for the original system and that for the replacement system is zero. The dissipation energy balancing procedure corresponds to a particular weighting function, such that the requirement for a vanishing weighted residual is equivalent to requiring the average dissipated energy to remain unchanged. The method of equivalent linearization is shown to be another special case of weighted residuals, in which the replacement system is selected from a smaller subclass of solvable systems, namely, the class of linear systems under additive random excitations. The method of partial linearization is similar. An approximate solution is expected to be more accurate if the replacement system is selected from the largest solvable class known.

The nongaussian white noises are the second type of random excitations considered. A white noise is a stationary impulsive noise, and it may be construed as a sequence of impulses of independent, identically distributed random amplitudes, which arrive independently at a constant average arrival rate. In theory, a gaussian white noise requires additionally that the average arrival rate tends to infinity while the mean-square value of the amplitude tends to zero. In practice, the degree in which a white noise tends to be gaussian may be measured by the product γ of the average pulse arrival rate and the relaxation time of the system under excitation.

Under an impulsive noise excitation, the system response vector is still markovian, but its probability density is governed by a Fokker-Planck equation of an infinite number of terms. A perturbation procedure has been developed to obtain an approximate probability density of the response. The numerical results obtained for two *specific* examples suggest that the effects of nongaussianity in the excitation are negligible if the impulsive excitation is multiplicative. On the other hand, if the impulsive excitation is additive, then the effects of nongaussianity are significant only if the product γ of the average impulse arrival rate in the excitation and the relaxation time of the system is less than 10. However, these conclusions are considered tentative before they can be fully substantiated by extensive numerical investigations.

The combination of harmonic and wideband excitations is considered in the last part of the chapter. It is shown that a harmonic excitation is significant only if its frequency is near a resonance frequency of the system, in which case the method of stochastic averaging is applicable. The averaged system, represented by a pair of new variables, can be solved exactly if it belongs to the class of stationary potential. When this is not the case, the method of weighted residuals can again be applied to obtain an approximate solution.

7.7 EXERCISES

7.1. Given a nonlinear system

$$\ddot{X} + \alpha\dot{X} + \gamma\dot{X}^3 + X = W(t)$$

where $W(t)$ is a white noise with a spectral density K, obtain (*a*) an approximate solution for the stationary mean-square value of $X(t)$ using gaussian closure, and (*b*) a set of equations on the basis of the fourth-order cumulant-neglect closure.

7.2. Consider the nonlinearly damped oscillator

$$\ddot{X} + \alpha\dot{X} + \beta X^2\dot{X} + \gamma\dot{X}^3 + \omega_0^2 X = W(t)$$

where $W(t)$ is a white noise with a spectral density K.

(*a*) Obtain $E[X^2]$ and $E[\dot{X}^2]$ in the stationary state, using the equivalent linearization procedure.

(*b*) Obtain an approximate stationary probability density $p(x, \dot{x})$, using the dissipation-energy balancing procedure, and calculate $E[X^2]$ and $E[\dot{X}^2]$ from $p(x, \dot{x})$.

(*c*) Obtain the stationary probability density of the amplitude process $A(t) = [X(t)^2 + \dot{X}(t)^2/\omega_0^2]^{1/2}$ from $p(x, \dot{x})$ in (b), and compare the result with that of Exercise 4.7.

7.3. Obtain and compare the approximate results for the stationary joint probability density of X and $\dot{X}$ for each of the following systems using the equivalent linearization, partial linearization, and dissipation energy balancing procedures:

(*a*) $\ddot{X} + \alpha\dot{X} + \beta X^2\dot{X} + X^3 = W(t)$

(*b*) $\ddot{X} + \alpha\dot{X} + \beta\dot{X}^3 + \gamma X + \delta X^3 = W(t)$

where $W(t)$ is a white noise.

7.4. Consider the following nonlinear system under a multiplicative white noise excitation:

$$\ddot{X} + \alpha\dot{X} + \gamma\dot{X}^3 + X = XW(t)$$

Obtain an approximate stationary probability density $p(x, \dot{x})$ using the dissipation energy balancing procedure, and determine the condition under which $p(x, \dot{x})$ is normalizable.

7.5. Consider the system

$$\ddot{X} + f(X, \dot{X}) + X = X^l W(t)$$

where W(t) is a white noise, and

$$f(X, \dot{X}) = \operatorname{sgn}\dot{X} \sum_{i+j=0}^{N} d_{ij}|X^i\dot{X}^j|$$

(*a*) Obtain an approximate stationary probability density $p(x, \dot{x})$ using the dissipation energy balancing procedure.

(*b*) Determine the condition under which $p(x, \dot{x})$ is normalizable.

(*c*) Show that in the special case

$$f(X, \dot{X}) = \sum_{m=0}^{M} \gamma_m \dot{X}X^{2l}(X^2 + \dot{X}^2)^m$$

the dissipation energy balancing procedure leads to the exact result.

7.6. Obtain an approximate stationary probability density for the system

$$\ddot{X} + f(X, \dot{X}) + u(X) = XW_1(t) + W_2(t)$$

where $W_1(t)$ and $W_2(t)$ are independent white noises with spectral densities K_{11} and K_{22}, respectively, and

$$f(X, \dot{X}) = \operatorname{sgn}\dot{X} \sum_{i+j=0}^{N} d_{ij}|X^i\dot{X}^j|$$

Show that if $f = \alpha\dot{X}(1+K_{11}X^2/K_{22})$, then the system belongs to the class of generalized stationary potential, and the probability density so obtained is the exact solution

$$p(x, \dot{x}) = C\exp\left\{-\frac{\alpha}{\pi K_{22}}\left[\frac{1}{2}\dot{x}^2 + \int u(x)\,dx\right]\right\}$$

7.7 Consider the following system subjected to combined harmonic and random excitations:

$$\ddot{X} + 2\zeta\omega_0\dot{X} + \beta X^2\dot{X} + \omega_0^2 X = \lambda\sin\nu t + W(t)$$

where $W(t)$ is a white noise.

(*a*) Replace the original equation by two first-order equations for X_s and X_c, using the transformation

$$X = X_c \cos \nu t + X_s \sin \nu t$$
$$\dot{X} = \nu(-X_c \sin \nu t + X_s \cos \nu t)$$

(*b*) Obtain the Itô stochastic differential equation for X_s and X_c assuming that $\nu \approx \omega_0$ and employing the smoothed stochastic averaging procedure.

(*c*) Find the condition under which the averaged system belongs to the class of stationary potential, and obtain $p(x_s, x_c)$ under this condition.

(*d*) Obtain an approximate $p(x_s, x_c)$, using the method of weighted residuals, when the preceding condition is not satisfied.

CHAPTER 8

FIRST-EXCURSION FAILURES

A dynamically loaded structure may fail because of one of the following events:

1. The structural motion becomes unstable.
2. The structural response at a critical location exceeds, for the first time, a prescribed safety boundary.
3. The accumulated material damage at a critical location, arising from stress fluctuation, exceeds a tolerance limit.

The first type of failure was considered in Chapter 6 in terms of convergence with probability 1, convergence in probability, and convergence in statistical moments. The second and third types are known as the first-excursion failure and fatigue failure, respectively, and several approximation techniques for obtaining their statistical and probabilistic properties were discussed, e.g., in Lin (1967).

In recent years, the theory of fracture mechanics has been advanced to provide a physical basis for material fatigue, in which fatigue failure is treated as propagation of a dominant crack to a critical size (see, e.g., Frudenthal, 1974; Schütz, 1979). From this point of view, fatigue failure becomes a special case of first-excursion failure. Comprehensive references may be found in a review article by Kozin and Bogdanoff (1989) and a book by Sobczyk and Spencer (1992).

The first-excursion problem is among the most difficult in the theory of probabilistic structural dynamics. At present, a mathematically exact solution is possible only if the random phenomenon in question can be treated as a diffusive Markov process. Still, known solutions are limited to the one-dimensional case (see, e.g., Bharucha-Reid, 1960; Cox and Miller, 1965). Since the state space of a structural system model is generally two-dimensional or higher, a reduction in dimension is necessary in order that the results of the Markov theory can be applied. Fortunately, this is possible using the methods of stochastic averaging and quasi-conservative averaging, as shown in Chapters 4 and 6. These approaches have been applied, for example, by Ariaratnam and Pi (1973), Lennox and Fraser (1974), Roberts (1978a), and Ariaratnam and Tam (1979) for oscillators with linear restoring force, and by Roberts (1976), Spanos (1982), and Zhu and Lei (1989) for the case of nonlinear restoring force.

This chapter is devoted to additional analyses of the first-excursion-type failures, making use of the more thorough treatments of Markov processes given in Chapter 4. The problem of fatigue crack propagation is included as a special case.

8.1 THE GENERALIZED PONTRYAGIN EQUATION

Let a one-dimensional diffusive Markov process, say $Z(t)$, be governed by an Itô-type stochastic differential equation

$$dZ(t) = m(Z)\,dt + \sigma(Z)\,dB(t) \tag{8.1.1}$$

where $B(t)$ is a unit Wiener process and m and σ are the drift and diffusion coefficients, respectively. We consider here only those cases in which m and σ do not depend explicitly on time t, as implied in (8.1.1). We are concerned with the random time T when $Z(t)$ first reaches a critical value z_c, on the condition that $Z(t_0) = z_0$, where $z_0 < z_c$. Let $q(z, t \mid z_0, t_0)$ be the transition probability density of the Markov process, and define a reliability function $R(t, z_c; t_0, z_0)$, that is, the probability of $z_l \leq Z(t) < z_c$ at time t, as follows:

$$R(t, z_c; t_0, z_0) = \int_{z_l}^{z_c} q(z, t \mid z_0, t_0)\,dz = \text{Prob}[T > t - t_0 \mid Z(t_0) = z_0] \tag{8.1.2}$$

$$z_l \leq z_0 < z_c, \quad t \geq t_0$$

where z_l is the left boundary of the process $Z(t)$. We assume that z_l is noncritical. The critical state z_c corresponds to an absorbing boundary, since a sample function is removed from the population of sample functions once it reaches the boundary. With an absorbing boundary, the $Z(t)$ process can never become stationary. Furthermore, the total probability within the region $z_l \leq z \leq z_c$ is no longer conserved; otherwise, the integration shown in (8.1.2) would have resulted in unity.

Clearly, the reliability function R satisfies the same Kolmogorov backward equation as that of $q(z, t \mid z_0, t_0)$:

$$\frac{\partial}{\partial t_0}R + a(z_0)\frac{\partial}{\partial z_0}R + \frac{1}{2}b(z_0)\frac{\partial^2}{\partial z_0^2}R = 0 \tag{8.1.3}$$

where $a(z_0) = m(z_0)$ and $b(z_0) = \sigma^2(z_0)$ are the first and second derivate moments, respectively. In equation (8.1.3), R is treated as a function of t_0 and z_0, while t and z_c are treated as parameters. Alternatively, R may be treated as a function of $\tau = t - t_0$ and z_0, that is, $R(\tau, z_c, z_0)$, and equation (8.1.3) may be rewritten as

$$-\frac{\partial}{\partial \tau}R + a(z_0)\frac{\partial}{\partial z_0}R + \frac{1}{2}b(z_0)\frac{\partial^2}{\partial z_0^2}R = 0 \tag{8.1.4}$$

Equation (8.1.4) is solved subject to the initial and boundary conditions

$$[R(\tau, z_c, z_0)]_{\tau=0} = 1 \qquad z_l \leq z_0 < z_c \tag{8.1.5}$$

$$[R(\tau, z_c, z_0)]_{z_0=z_c} = 0 \tag{8.1.6}$$

$$[R(\tau, z_c, z_0)]_{z_0=z_l} = \text{finite} \tag{8.1.7}$$

The physical meanings of conditions (8.1.5) and (8.1.6) are clear. Condition (8.1.7) is derived from the fact that the transition probability density q in equation (8.1.2) is always integrable.

Equations (8.1.4) through (8.1.7) define an eigenvalue problem; however, closed-form solutions are known only for very few cases. Some such cases are discussed in Section 8.4. Meanwhile, we focus on the more tractable problem of obtaining the statistical moments for the first-excursion time T.

It is clear that the probability distribution function of the first-excursion time T is given by

$$F_T(\tau, z_c, z_0) = \text{Prob}[T < \tau \mid Z(t_0) = z_0] = 1 - R(\tau, z_c, z_0) \tag{8.1.8}$$

and the probability density of T is given by

$$p_T(\tau, z_c, z_0) = \frac{\partial}{\partial \tau} F_T(\tau, z_c, z_0) = -\frac{\partial}{\partial \tau} R(\tau, z_c, z_0) \tag{8.1.9}$$

Thus the nth moment of T can be obtained as follows:

$$\begin{aligned} \mu_n(z_c, z_0) = E[T^n] &= -\int_0^\infty \tau^n \frac{\partial}{\partial \tau} R(\tau, z_c, z_0)\, d\tau \\ &= n \int_0^\infty \tau^{n-1} R(\tau, z_c, z_0)\, d\tau \end{aligned} \tag{8.1.10}$$

where it is assumed that $\tau^n R(\tau, z_c, z_0)$ tends to zero as $\tau \to \infty$. Now, multiplying (8.1.4) by τ^n and integrating on τ, we obtain

$$(n+1)\mu_n + a(z_0)\frac{d}{dz_0}\mu_{n+1} + \frac{1}{2} b(z_0) \frac{d^2}{dz_0^2}\mu_{n+1} = 0 \tag{8.1.11}$$

subject to the boundary conditions

$$[\mu_{n+1}(z_c, z_0)]_{z_0=z_c} = 0 \tag{8.1.12}$$

$$[\mu_{n+1}(z_c, z_0)]_{z_0=z_l} = \text{finite} \tag{8.1.13}$$

Condition (8.1.12) is obvious. Condition (8.1.13) implies that, given time, a first-excursion failure of the system in question will occur, which is an important assumption. It should be noted that equation (8.1.11) is no longer valid when this assumption does not hold. The well-known Pontryagin equation (Andronov, Pontryagin, and Witt, 1933) corresponds to the special case $n = 0$:

$$1 + a(z_0)\frac{d}{dz_0}\mu_1 + \frac{1}{2} b(z_0) \frac{d^2}{dz_0^2}\mu_1 = 0 \tag{8.1.14}$$

Equation (8.1.11) will be called the *generalized Pontryagin equation* (e.g., Ariaratnam and Tam, 1979). In principle, it may be solved recursively, beginning with $n = 0$ and $\mu_0 = 1$.

Although condition (8.1.13) is "qualitative" rather than "quantitative," it is useful to obtain closed-form solutions when possible. However, if equation (8.1.11) must be solved numerically, or numerical solutions are in fact more convenient and preferred, then a quantitative boundary condition is required.

Several cases must now be considered separately. If z_l is nonsingular, that is, if $a(z_l)$ is finite and $b(z_l)$ is nonzero, then it is a regular boundary. In principle, any condition may be imposed on a regular boundary as long as it is consistent with the context of a physical problem. We assumed earlier that z_l is noncritical. Then for the purpose of a first-excursion failure analysis, a conservative assumption is that the left boundary be reflective, in which case we have from (4.2.46)

$$\left[\frac{\partial q(z, t \mid z_0, t_0)}{\partial z_0}\right]_{z_0 = z_l} = 0 \tag{8.1.15}$$

Therefore,
$$\left(\frac{\partial R}{\partial z_0}\right)_{z_0 = z_l} = \left[\int_{z_l}^{z_c} \frac{\partial q}{\partial z_0} dz\right]_{z_0 = z_l} = \int_{z_l}^{z_c} \left(\frac{\partial q}{\partial z_0}\right)_{z_0 = z_l} dz = 0 \tag{8.1.16}$$

provided that the integration and the limiting operation $z_0 \to z_l$ are interchangeable. From equations (8.1.10) and (8.1.16), we have

$$\left[\frac{d\mu_{n+1}}{dz_0}\right]_{z_0 = z_l} = n \int_0^\infty \tau^{n-1} \left[\frac{\partial R(\tau, z_c, z_0)}{\partial z_0}\right]_{z_0 = z_l} d\tau = 0 \tag{8.1.17}$$

If $b(z_l) = 0$, then z_l is a singular boundary of the first kind. It is a shunt if $a(z_l) \neq 0$, or a trap if $a(z_l) = 0$. As shown in Table 4.5.2, a shunt can be regular, an entrance, or an exit; a trap can be regular, an entrance, an exit, or natural. If z_l is a regular, an entrance, or a repulsively natural boundary, then a sample path that begins at or near the boundary can travel inward; thus it is capable of reaching the critical state on the right-hand side. On the other hand, if z_l is an exit, a strictly natural, or an attractively natural boundary, then not every sample path that begins near the boundary is to reach the critical state. Since the generalized Pontryagin equation is derived based on the assumption that passage to the critical state will occur sooner or later for every sample path, a left-hand-side exit, strictly natural, or attractively natural boundary is not admissible. In the case where z_l is a regular shunt or an entrance shunt—that is, $b(z_l) = 0$ and $a(z_l) > 0$—we obtain, directly from equation (8.1.11),

$$\mu'_{n+1}(z_l) = -\frac{(n+1)\mu_n(z_l)}{a(z_l)} \tag{8.1.18}$$

Condition (8.1.18) was suggested by Zhu and Lei (1989). If the left boundary z_l is a regular trap, an entrance trap, or a repulsively natural trap—that is, $b(z_l) = 0$ and $a(z_l) = 0$—we may impose the condition

$$0\left|a(z_0)\mu'_{n+1}(z_0)\right| \sim 0\left|\mu_n(z_0)\right| \qquad z_0 \to z_l \tag{8.1.19}$$

In particular, for $n = 0$,

$$a(z_0)\mu_1'(z_0) = \text{finite} \qquad z_0 \to z_l \tag{8.1.20}$$

Finally, the case in which the first derivate moment $a(z_0)$ becomes unbounded as z_0 tends to z_l is worthy of special attention. In this case, z_l is a singular boundary of the second kind, and we obtain from equation (8.1.11)

$$\frac{d\mu_{n+1}}{dz_0} \sim 0[a(z_0)]^{-1} \to 0 \qquad z_0 \to z_l \tag{8.1.21}$$

indicating that it is a reflective boundary.

Applications of the generalized Pontryagin equation are illustrated below.

8.2 MOMENTS OF FIRST-EXCURSION TIME OF RESPONSE AMPLITUDE

Consider an oscillator with a linear restoring force and light damping, governed by an equation of motion

$$\ddot{Y} + \omega_0^2 Y = \varepsilon f(Y, \dot{Y}) + \varepsilon^{1/2} g_k(Y, \dot{Y})\xi_k(t) \tag{8.2.1}$$

where functions f and g_k may be nonlinear, $\xi_j(t)$ are wideband stationary random excitations, and ε is a small parameter. This is the same equation as (4.7.13). As shown in Section 4.7, equation (8.2.1) may be replaced by two first-order equations for the amplitude process $A(t)$ and the phase process $\phi(t)$, using the transformation

$$\begin{aligned} Y(t) &= A(t)\cos\theta \qquad \theta = \omega_0 t + \phi(t) \\ \dot{Y}(t) &= -\omega_0 A(t)\sin\theta \end{aligned} \tag{8.2.2}$$

to yield

$$\dot{A} = -\frac{\sin\theta}{\omega_0}[\varepsilon f(A\cos\theta, -A\omega_0\sin\theta) + \varepsilon^{1/2} g_k(A\cos\theta, -A\omega_0\sin\theta)\xi_k(t)] \tag{8.2.3}$$

$$\dot{\phi} = -\frac{\cos\theta}{\omega_0 A}[\varepsilon f(A\cos\theta, -A\omega_0\sin\theta) + \varepsilon^{1/2} g_k(A\cos\theta, -A\omega_0\sin\theta)\xi_k(t)] \tag{8.2.4}$$

The right-hand sides of (8.2.3) and (8.2.4) are small; thus the smoothed stochastic averaging procedure may be applied to obtain the Itô-type stochastic differential equation as follows:

$$dA = m_1\,dt + \sigma_{11}\,dB_1(t) + \sigma_{12}\,dB_2(t) \tag{8.2.5}$$

$$d\phi = m_2\,dt + \sigma_{21}\,dB_1(t) + \sigma_{22}\,dB_2(t) \tag{8.2.6}$$

where m_j and σ_{jk} are obtained from equations (4.7.9) and (4.7.10), by identifying $A(t)$ with $X_1(t)$ and $\phi(t)$ with $X_2(t)$. The smoothed stochastic averaging procedure has the advantage of uncoupling the averaged amplitude process $A(t)$ from the phase process $\phi(t)$. The averaged $A(t)$ is approximately a one-dimensional Markov process, and its first-excursion statistics can be calculated by use of the generalized Pontryagin equation. Four examples are given next.

8.2.1 A Linear Oscillator under Both Additive and Multiplicative Random Excitations

The following linear oscillator under both additive and multiplicative random excitations has been studied by Ariaratnam and Tam (1979):

$$\ddot{Y} + \omega_0[2\zeta + \xi_2(t)]\dot{Y} + \omega_0^2[1 + \xi_1(t)]Y = \xi_3(t) \tag{a}$$

where $\xi_1(t)$, $\xi_2(t)$, and $\xi_3(t)$ are wideband stationary processes with zero mean values. It is assumed that ζ is of order ε, and the $\xi_j(t)$ are of order $\varepsilon^{1/2}$. By applying the stochastic averaging procedure, it can be shown that the amplitude process $A(t) = [Y^2(t) + \dot{Y}^2(t)/\omega_0^2]^{1/2}$ is approximately a Markov diffusive process, governed by the Itô equation

$$dA = m(A)\,dt + \sigma(A)\,dB(t) \tag{b}$$

The drift coefficient $m(A)$ and the diffusion coefficient $\sigma(A)$, obtained from equation (4.7.9) and (4.7.10), are given by

$$m(A) = -\alpha A + \frac{\delta}{2A} \tag{c}$$

$$\sigma(A) = (\gamma A^2 + \delta)^{1/2} \tag{d}$$

in which

$$\alpha = \zeta\omega_0 - \frac{\pi\omega_0^2}{8}[2\Phi_{22}(0) + 3\Phi_{22}(2\omega_0) + 3\Phi_{11}(2\omega_0) - 6\Psi_{12}(2\omega_0)] \tag{e}$$

$$\delta = \frac{\pi}{\omega_0^2}\Phi_{33}(\omega_0) \tag{f}$$

$$\gamma = \frac{\pi\omega_0^2}{4}[2\Phi_{22}(0) + \Phi_{22}(2\omega_0) + \Phi_{11}(2\omega_0) + 2\Psi_{12}(2\omega_0)] \tag{g}$$

and

$$\Phi_{ij}(\omega) = \frac{1}{2\pi}\int_{-\infty}^{\infty} E[\xi_i(t)\xi_j(t+\tau)]\cos(\omega\tau)\,d\tau \qquad i, j = 1, 2, 3 \tag{h}$$

$$\Psi_{ij}(\omega) = \frac{1}{2\pi}\int_{-\infty}^{\infty} E[\xi_i(t)\xi_j(t+\tau)]\sin(\omega\tau)\,d\tau \qquad i, j = 1, 2, 3 \tag{i}$$

The generalized Pontryagin equation reads

$$\frac{1}{2}(\delta + \gamma a_0^2)\frac{d^2}{da_0^2}\mu_{n+1} + \left(\frac{\delta}{2a_0} - \alpha a_0\right)\frac{d}{da_0}\mu_{n+1} = -(n+1)\mu_n \tag{j}$$

The left boundary $a_0 = 0$ is singular of the second kind, where the first derivate moment is unbounded and the second derivate moment remains finite. The two boundary conditions are

$$\mu'_{n+1}(a_0) \sim 0(a_0) \qquad a_0 \to 0 \tag{k}$$

$$\mu_{n+1}(a_c) = 0 \tag{l}$$

The first moment is obtained by letting $n = 0$, yielding

$$\mu_1(a_0) = \frac{1}{\eta\gamma}\int_{a_0}^{a_c}\frac{1}{u}[(1+\frac{\gamma}{\delta}u^2)^\eta - 1]\,du \qquad \eta = \frac{\alpha}{\gamma}+\frac{1}{2} \neq 0 \tag{m}$$

$$\mu_1(a_0) = \frac{1}{\gamma}\int_{a_0}^{a_c}\frac{1}{u}\ln(1+\frac{\gamma}{\delta}u^2)\,du \qquad \eta = 0 \tag{n}$$

These results were obtained by Ariaratnam and Tam (1979). For higher order moments, we obtain

$$\mu_{n+1}(a_0) = 2(n+1)\int_{a_0}^{a_c}\frac{(\gamma u^2+\delta)^\eta}{u}\left[\int_0^u \frac{v\mu_n(v)}{(\gamma v^2+\delta)^{\eta+1}}dv\right]du \tag{o}$$

8.2.2 A Linear System under Additive Random Excitation

Return to equation (a) in Section 8.2.1, and let the multiplicative excitations $\xi_1(t)$ and $\xi_2(t)$ be zero. Then $\gamma = 0$, and the results obtained in Section 8.2.1 are no longer valid. In this case, the generalized Pontryagin equation reduces to

$$\frac{\delta}{2}\frac{d^2}{da_0^2}\mu_{n+1} + \left(\frac{\delta}{2a_0} - \zeta\omega_0 a_0\right)\frac{d}{da_0}\mu_{n+1} = -(n+1)\mu_n \tag{a}$$

However, the boundary conditions (k) and (l) in Section 8.2.1 remain valid; their imposition yields

$$\begin{aligned}\mu_1(a_0) &= \frac{1}{\zeta\omega_0}\left[-\ln\frac{a_c}{a_0} + \int_{a_0}^{a_c}\frac{1}{u}\exp\left(\frac{\zeta\omega_0}{\delta}u^2\right)du\right] \\ &= \frac{1}{2\zeta\omega_0}\left[\bar{E}i\left(\frac{\zeta\omega_0}{\delta}a_c^2\right) - \bar{E}i\left(\frac{\zeta\omega_0}{\delta}a_0^2\right) - \ln\left(\frac{a_c}{a_0}\right)^2\right]\end{aligned} \tag{b}$$

where $\bar{E}i$ is an exponential integral function, defined as

$$\bar{E}i(x) = -\int_{-x}^{\infty}\frac{1}{t}e^{-t}dt \qquad x > 0 \tag{c}$$

The higher order moments are obtained as

$$\mu_{n+1}(a_0) = \frac{2(n+1)}{\delta}\int_{a_0}^{a_c}\frac{1}{u}\exp\left(\frac{\zeta\omega_0}{\delta}u^2\right)\left[\int_0^u v\mu_n(v)\exp\left(-\frac{\zeta\omega_0}{\delta}v^2\right)dv\right]du \tag{d}$$

Equation (b) was derived by Ariaratnam and Pi (1973) previously.

8.2.3 A Linear System under Multiplicative Random Excitation

Return again to equation (a) in Section 8.2.1, but assume, instead, that the additive excitation $\xi_3(t)$ is absent. Then $\delta = 0$, and the generalized Pontryagin equation reads

$$\frac{1}{2}\gamma a_0^2 \frac{d^2}{da_0^2}\mu_{n+1} - \alpha a_0 \frac{d}{da_0}\mu_{n+1} = -(n+1)\mu_n \tag{a}$$

Both the first and second derivate moments vanish at $a_0 = 0$; therefore, the left boundary is now a trap. As such, it must be regular, an entrance, or repulsively natural to be compatible with the Pontryagin equation. The diffusion exponent, drift exponent, and character value at $a_0 = 0$ are found to be $\alpha_l = 2$, $\beta_l = 1$, and $c_l = -2\alpha/\gamma$. Referring to Table 4.5.2, the boundary is repulsively natural if $c_l > 1$, that is, if

$$\eta = \frac{\alpha}{\gamma} + \frac{1}{2} < 0 \tag{b}$$

or, according to (e) and (g) in Section 8.2.1,

$$\zeta < \frac{\pi\omega_0}{4}[\Phi_{22}(2\omega_0) + \Phi_{11}(2\omega_0) - 4\Psi_{12}(2\omega_0)] \tag{c}$$

in which case condition (8.1.19) is applicable. The left boundary is attractively natural if $c_l < 1$ or strictly natural if $c_l = 1$. In both cases, it is not an admissible boundary for the first-excursion problem.

Equation (a) is an Euler-type ordinary differential equation, which can be solved easily to yield

$$\mu'_{n+1}(a_0) = a_0^{2\eta-1}\left[\frac{-2(n+1)}{\gamma}\int_0^{a_0}\mu_n(v)v^{-(2\eta+1)}dv + C_{n+1}\right] \tag{d}$$

For $n = 0$, we obtain

$$\mu'_1(a_0) = C_1 a_0^{2\eta-1} + \frac{1}{\eta\gamma a_0} \tag{e}$$

It follows, upon imposing boundary condition (8.1.20), that $C_1 = 0$; thus

$$\mu_1(a_0) = -\frac{1}{\eta\gamma}\ln\frac{a_c}{a_0} \tag{f}$$

which satisfies the condition

$$\mu_1(a_c) = 0 \tag{g}$$

For $n = 1$, we find that

$$\mu'_2(a_0) = C_2 a_0^{2\eta-1} - \frac{2}{\eta^2\gamma^2}\left(\frac{1}{2\eta} + \ln\frac{a_c}{a_0}\right) \tag{h}$$

and that $C_2 = 0$, upon imposing boundary condition (8.1.19). The second moment is obtained by performing another integration and imposing $\mu_2(a_c) = 0$ to yield

$$\mu_2(a_0) = \frac{2}{\eta^2\gamma^2}\ln\frac{a_c}{a_0}\left(-\frac{1}{\eta} + \ln\frac{a_c}{a_0}\right) \tag{i}$$

For $n > 1$, we find again $C_{n+1} = 0$ and

$$\mu_{n+1}(a_0) = \frac{2(n+1)}{\gamma}\int_{a_0}^{a_c} u^{2\eta-1}\left[\int_0^u \mu_n(v)v^{-(2\eta+1)}dv\right]du \tag{j}$$

It is of interest to note that the same result (j) can be obtained from equation (o) in Section 8.2.1 by letting $\delta \to 0$.

Of course, the foregoing results are valid only if the trap boundary at $a_0 = 0$ is repulsively natural, which requires $c_l > 1$, namely, $\eta < 0$ or inequality (c). Under this condition, the system is unstable at $a = 0$ with probability 1; therefore, it is possible for a sample path beginning at near $a_0 = 0$ to travel to the critical state a_c. However, as seen from (f), (i), and (j), the statistical moments of the first-excursion time are very large if a_0 is near the left boundary where both the diffusion and rightward drift are very small.

To show that neither an attractively natural trap nor a strictly natural trap is an admissible left-hand-side boundary for a first-excursion problem, let us return to the generalized Pontryagin equation (a). If $c_l < 1$, then the left boundary is attractively natural, in which case the general solution of (a) for $n = 0$ has the form

$$\mu_1(a_0) = \frac{1}{\eta\gamma} \ln a_0 + C_1 a_0^{2\eta} + D_1 \tag{k}$$

On the other hand, if $c_l = 1$ corresponding to a strictly natural trap, the general solution reads

$$\mu_1(a_0) = -\frac{1}{\gamma}(\ln a_0)^2 + C_1 \ln a_0 + D_1 \tag{l}$$

For any finite C_1 and D_1 values, both (k) and (l) give rise to negative results for $\mu_1(a_0)$ as $a_0 \to 0$. In either case, the Pontryagin equation is no longer physically meaningful.

It is of interest to note that if $c_l < 1$, then the trivial solution $A(t) = 0$ is asymptotically stable with probability 1. In this case, a sample function may never reach the failure boundary, contrary to the very premise on which the generalized Pontryagin equation is derived. In the case of $c_l = 1$, $A(t) = 0$ is neither stable with probability 1 nor unstable with probability 1; therefore, the assumption of eventual first-excursion failure is also violated.

8.2.4 A van der Pol Oscillator

As the last example, consider the following van der Pol-type oscillator under an additive wideband stationary excitation $\xi(t)$:

$$\ddot{Y} + 2\omega_0(\beta + \eta Y^2)\dot{Y} + \omega_0^2 Y = \xi(t) \tag{a}$$

where β and η are assumed to be of order ε, and $\xi(t)$ of order $\varepsilon^{1/2}$. The drift and diffusion coefficients for the amplitude process $A(t)$, obtained from stochastic averaging, are given by

$$m = -\beta\omega_0 A - \frac{1}{4}\eta\omega_0 A^3 + \frac{\delta}{2A} \tag{b}$$

$$\sigma^2 = \delta \tag{c}$$

where $\delta = \pi\Phi_{\xi\xi}(\omega_0)/\omega_0^2$, and $\Phi_{\xi\xi}(\omega)$ is the spectral density of $\xi(t)$. The generalized Pontryagin equation for $A(t)$ is then

$$\frac{1}{2}\delta\frac{d^2}{da_0^2}\mu_{n+1} + \left(-\beta\omega_0 a_0 - \frac{1}{4}\eta\omega_0 a_0^3 + \frac{\delta}{2a_0}\right)\frac{d}{da_0}\mu_{n+1} = -(n+1)\mu_n \qquad \text{(d)}$$

Equation (d) is solved subject to the boundary conditions

$$\mu'_{n+1}(a_0) \sim 0(a_0) \qquad a_0 \to 0 \qquad \text{(e)}$$

$$\mu_{n+1}(a_c) = 0 \qquad \text{(f)}$$

yielding

$$\mu_{n+1}(a_0) = \frac{2(n+1)}{\delta}\int_{a_0}^{a_c}\frac{1}{u}\exp\left[\frac{\omega_0}{\delta}\left(\beta u^2 + \frac{\eta}{8}u^4\right)\right]$$

$$\left\{\int_0^u v\mu_n(v)\exp\left[-\frac{\omega_0}{\delta}\left(\beta v^2 + \frac{\eta}{8}v^4\right)\right]dv\right\}du \qquad \text{(g)}$$

The statistical moments of the first-excursion time can be calculated recursively by carrying out the integrations in (g) or by solving numerically the boundary-value problem defined in (d), (e), and (f).

It is of interest to note that if the first-excursion problem of an oscillator is formulated in terms of the amplitude process, and if an additive wideband excitation is present, then the boundary condition at $a_0 = 0$ for the generalized Pontryagin equation is always reflective, that is, condition (e) holds, irrespective of the damping mechanism and the existence of multiplicative excitations.

8.3 MOMENTS OF FIRST-EXCURSION TIME OF THE RESPONSE ENERGY ENVELOPE

The smoothed version of stochastic averaging is no longer suitable if the restoring force of an oscillator is strongly nonlinear. However, if the random excitations can be modeled as white noises, then the quasi-conservative averaging procedure (Section 4.7) is applicable. Consider the governing equation

$$\ddot{Y} + h(Y) = \varepsilon f(Y, \dot{Y}) + \varepsilon^{1/2} g_j(Y, \dot{Y}) W_j(t) \qquad (8.3.1)$$

where h is assumed to be an odd function of Y, representing a nonlinear restoring force, and $W_j(t)$ are gaussian white noises, with

$$E[W_j(t)W_l(t+\tau)] = 2\pi K_{jl}\delta(\tau) \qquad (8.3.2)$$

The total energy of the oscillator is given by

$$\Lambda(t) = \tfrac{1}{2}\dot{Y}^2 + U(Y) \qquad (8.3.3)$$

where

$$U(Y) = \int_0^Y h(u)\,du \qquad (8.3.4)$$

Differentiate equation (8.3.3) with respect to time and combine the result with equation (8.3.1) to obtain

$$\dot{\Lambda} = \dot{Y}[\varepsilon f(Y, \dot{Y}) + \varepsilon^{1/2} g_j(Y, \dot{Y}) W_j(t)] \qquad (8.3.5)$$

Equation (8.3.5) may be approximated by the Itô equation

$$d\Lambda = m(\Lambda)\,dt + \sigma(\Lambda)\,dB(t) \tag{8.3.6}$$

where the drift and diffusion coefficients may be obtained by use of the quasi-conservative stochastic averaging:

$$m(\Lambda) = \varepsilon\left\langle \dot{Y} f(Y, \dot{Y}) + \pi K_{jl}\dot{Y} g_l(Y, \dot{Y})\frac{\partial}{\partial \Lambda}[\dot{Y} g_j(Y, \dot{Y})]\right\rangle_t \tag{8.3.7}$$

$$\sigma^2(\Lambda) = \varepsilon 2\pi K_{jl}\left\langle \dot{Y}^2 g_j(Y, \dot{Y}) g_l(Y, \dot{Y})\right\rangle_t \tag{8.3.8}$$

in which $\langle[\cdot]\rangle_t$ denotes the time-averaging of $[\cdot]$, and $\dot{Y}$ is treated as a function of Λ and Y:

$$\dot{Y} = \pm\sqrt{2[\Lambda - U(Y)]} \tag{8.3.9}$$

In the quasi-conservative averaging procedure, each time-averaging is substituted by averaging on Y over a pseudo-period,

$$\langle[\cdot]\rangle = \frac{1}{T}\left\{\int_{-A}^{A} \frac{[\cdot]_{\dot{Y}=\sqrt{2[\Lambda-U(Y)]}} + [\cdot]_{\dot{Y}=-\sqrt{2[\Lambda-U(Y)]}}}{\sqrt{2[\Lambda - U(Y)]}}\,dY\right\} \tag{8.3.10}$$

where

$$T = 4T_{1/4} = 4\int_0^A \frac{1}{\sqrt{2[\Lambda - U(Y)]}}\,dY \tag{8.3.11}$$

is a quasi-period, and A is the would-be amplitude of free oscillation if the oscillator, possessing a total energy Λ, were to undergo an undamped oscillation or, symbolically,

$$A = U^{-1}(\Lambda) \tag{8.3.12}$$

The same generalized Pontryagin equation (8.1.11) and the boundary conditions (8.1.12) and (8.1.13) apply, with z_0 replaced by the initial state variable λ_0 of the total energy. When appropriate, the qualitative boundary condition (8.1.13) may be replaced by one of the quantitative conditions (8.1.17), (8.1.18), (8.1.19), and (8.1.21).

The foregoing procedures are illustrated in the following examples.

8.3.1 A Nonlinear Oscillator under Additive White Noise Excitation

Roberts (1976) has considered the oscillator

$$\ddot{Y} + 2\zeta\dot{Y} + k\left|Y\right|^{\rho}\operatorname{sgn}(Y) = W(t) \qquad k, \rho > 0 \tag{a}$$

which involves only one additive white noise excitation. In equation (a), it is assumed that the damping coefficient ζ and the spectral density K of the excitation $W(t)$ are of order ε, so that the quasi-conservative averaging procedure is applicable. Using equations (8.3.7) and (8.3.8), we obtain, after some algebra,

$$m(\Lambda) = -2\delta\zeta\Lambda + \pi K \tag{b}$$

$$\sigma^2(\Lambda) = 2\pi K \delta \Lambda \tag{c}$$

where $\delta = 2(\rho + 1)/(\rho + 3)$. The generalized Pontryagin equation is now

$$\pi K \delta \lambda_0 \frac{d^2}{d\lambda_0^2} \mu_{n+1} + (\pi K - 2\delta\zeta\lambda_0) \frac{d}{d\lambda_0} \mu_{n+1} = -(n+1)\mu_n \tag{d}$$

The left boundary at $\lambda_0 = 0$ is singular of the first kind. It can be shown, by using Table 4.5.2, that the left boundary is either an entrance shunt or a regular shunt; therefore, the boundary conditions for equation (d) are

$$\mu'_{n+1}(0) = -\frac{(n+1)\mu_n(0)}{\pi K} \tag{e}$$

$$\mu_{n+1}(\lambda_c) = 0 \tag{f}$$

It follows that

$$\mu'_{n+1}(\lambda_0) = -\frac{n+1}{\pi K \delta} e^{(2\zeta/\pi K)\lambda_0} \lambda_0^{-1/\delta} \int_0^{\lambda_0} \mu_n(v) v^{1/\delta - 1} e^{-(2\zeta/\pi K)v} \, dv \tag{g}$$

and $$\mu_{n+1}(\lambda_0) = -\int_{\lambda_0}^{\lambda_c} \mu'_{n+1}(u) \, du \tag{h}$$

It is of interest to compare the present result with that of the linear case, by letting $\delta = \rho = 1$. Upon carrying out the integration, we obtain

$$\mu_1 = \frac{1}{2\zeta} \left[\bar{E}i\left(\frac{2\zeta}{\pi K}\lambda_c\right) - \bar{E}i\left(\frac{2\zeta}{\pi K}\lambda_0\right) - \ln\left(\frac{\lambda_c}{\lambda_0}\right) \right] \tag{i}$$

Equation (i) agrees with equation (b) in Section 8.2.2, although the behaviors of the drift and diffusion coefficients are different at the left boundary for the two Pontryagin equations, and the respective boundary conditions are also different. This is remarkable, since in the linear case energy Λ is exactly the amplitude squared, that is, $\Lambda = A^2$, and the asymptotic stability conditions for $E[A(t)]$ and $E[A^2(t)]$ are known to be different.

It is also remarkable that if the first-excursion problem is formulated in terms of energy, then the boundary condition (e) is shared by all cases as long as an additive noise is present, whether the system is linear or nonlinear, and none of the system parameters ζ, k, and ρ appears in the condition. Moreover, this is true irrespective of whether a multiplicative noise is also present. We note that Kozin (1982) obtained a very simple result for an undamped linear system under additive gaussian white noise excitation, which, in the present notation, reads

$$\mu_1 = \frac{\lambda_c - \lambda_0}{\pi K} \tag{j}$$

Kozin's result was obtained by applying the so-called Dynkin's formula (Dynkin, 1965) and, in fact, it is also valid when the undamped system is nonlinear. Equation (j) shows that the slope of μ_1 with respect to λ_0 remains unchanged at $-(\pi K)^{-1}$ if the system is undamped. Therefore, in terms of the first excursion of the total energy to a critical energy level, all systems under additive random excitation begin with the same slope for μ_1 at the boundary $\lambda_0 = 0$, and damping can change the slope only after departing from the boundary.

8.3.2 A Duffing Oscillator under Multiplicative White Noise Excitation

Next consider the following system:

$$\ddot{Y} + 2\zeta\dot{Y} + Y + \eta Y^3 = YW(t) \tag{a}$$

The drift and diffusion coefficients for the energy envelope process $\Lambda(t)$, obtained from equations (8.3.7) and (8.3.8), respectively, are given by

$$(m(\Lambda)) = \frac{A^2\left[\pi K\int_0^{\pi/2}\cos^2\phi(1-\bar{k}^2\sin^2\phi)^{-1/2}d\phi - 2\zeta(1+\eta A^2)\int_0^{\pi/2}\sin^2\phi(1-\bar{k}^2\sin^2\phi)^{1/2}d\phi\right]}{\displaystyle\int_0^{\pi/2}(1-\bar{k}^2\sin^2\phi)^{-1/2}d\phi} \tag{b}$$

$$\sigma^2(\Lambda) = 2\pi K A^4(1+\eta A^2)\frac{\displaystyle\int_0^{\pi/2}\sin^2\phi\cos^2\phi(1-\bar{k}^2\sin^2\phi)^{1/2}d\phi}{\displaystyle\int_0^{\pi/2}(1-\bar{k}^2\sin^2\phi)^{-1/2}d\phi} \tag{c}$$

where
$$\bar{k} = A\sqrt{\frac{\eta}{2(1+\eta A^2)}} \tag{d}$$

and where A is treated as a function of Λ and is the positive root of the equation

$$\eta A^4 + 2A^2 - 4\Lambda = 0 \tag{e}$$

To determine the nature of the left boundary, the limiting behavior of the drift and diffusion coefficients near $\Lambda = 0$ must be investigated. It can be shown from equations (b), (c), and (d) that as $\Lambda \to 0$, $A \to \sqrt{2\Lambda}$, $\bar{k} \to 0$, and

$$m(\Lambda) \to (\pi K - 2\zeta)\Lambda \tag{f}$$

$$\sigma^2(\Lambda) \to \pi K \Lambda^2 \tag{g}$$

Thus the diffusion exponent, drift exponent, and character value of the left boundary are $\alpha_l = 2$, $\beta_l = 1$, and $c_l = 2 - 4\zeta/\pi K$, respectively. The boundary is repulsively natural if

$$c_l = 2 - \frac{4\zeta}{\pi K} > 1 \tag{h}$$

in which case the boundary condition (8.1.19) is applicable. Since the drift and diffusion coefficients, given by equations (b) and (c), are complicated, an analytical solution for the generalized Pontryagin equation seems difficult. Therefore, a direct numerical solution of equation (d) is preferable, for which the behavior of the solution near the left boundary must first be examined.

When λ_0 is very small, the drift and diffusion coefficients may be approximated by (f) and (g). Thus the solution of the generalized Pontryagin equation may be written as

$$\mu'_{n+1}(\lambda_0) \approx -\frac{2(n+1)}{\pi K}\lambda_0^{(4\zeta/\pi K)-2}\left[\int_0^{\lambda_0}\mu_n(u)u^{-4\zeta/\pi K}du + C_{n+1}\right] \qquad \lambda_0 << 1 \qquad \text{(i)}$$

For $n = 0$, equation (i) reduces to

$$\mu'_1(\lambda_0) \approx \frac{2}{(4\zeta - \pi K)}\left[\frac{1}{\lambda_0} + C_1\lambda_0^{(4\zeta/\pi K)-2}\right] \qquad \lambda_0 << 1 \qquad \text{(j)}$$

Upon imposing boundary condition (8.1.20) and condition (h), we obtain $C_1 = 0$ and

$$\mu'_1(\lambda_0) \approx \frac{2}{(4\zeta - \pi K)\lambda_0} \qquad \lambda_0 << 1 \qquad \text{(k)}$$

Condition (k) may be used in conjunction with the condition $\mu_1(\lambda_c) = 0$ in carrying out the numerical solution of the Pontryagin equation for μ_1. For the second moment μ_2, we obtain, from integrating (i) by parts,

$$\mu'_2(\lambda_0) \approx \frac{4}{(4\zeta - \pi K)\lambda_0}\left[\mu_1(\lambda_0) + \frac{2\pi K}{(4\zeta - \pi K)^2}\right] - \frac{4}{\pi K}C_2\lambda_0^{(4\zeta/\pi K)-2} \qquad \lambda_0 << 1 \qquad (l)$$

Upon imposing (8.1.19) and (h), we conclude that $C_2 = 0$, and (l) reduces to

$$\mu'_2(\lambda_0) \approx \frac{4}{(4\zeta - \pi K)\lambda_0}\left[\mu_1(\lambda_0) + \frac{2\pi K}{(4\zeta - \pi K)^2}\right] \qquad \lambda_0 << 1 \qquad \text{(m)}$$

Condition (m) may be used in conjunction with $\mu_2(\lambda_c) = 0$ to obtain μ_2. Calculation of higher order moments can proceed analogously.

Numerical results have been obtained, in three different ways, for the mean first-excursion time μ_1 of the energy envelope for the Duffing oscillators, with $\eta = 0.1$ and several combinations of ζ and K values. First, the Pontryagin equation was solved numerically with boundary conditions $\mu_1(\lambda_c) = 0$ and that given approximately in (k). When applying the approximate boundary condition (k), two trial initial values, 0.01 and 0.005, were selected for λ_0 to begin the computation, and the results were found to be practically the same. These results are shown in Figs. 8.3.1 and 8.3.2 as solid and dashed lines. Next, Monte Carlo simulations were performed on the basis of the original governing equation for a Duffing oscillator, and the results so obtained are represented by the crosses in Figs. 8.3.1 and 8.3.2. Finally, the Itô equation for the energy envelope of the averaged system was converted to its corresponding Stratonovich-type equation, which was then used as the basis to perform another set of Monte Carlo simulations. The results of the latter set of simulations

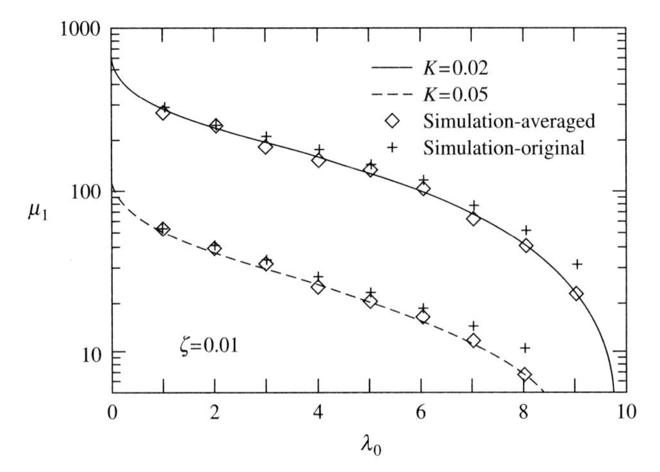

FIGURE 8.3.1
The mean first-excursion time of a Duffing oscillator under parametric white noise excitation. Damping ratio $\zeta = 0.01$, white noise spectral density $K = 0.02, 0.05$ [after Cai and Lin, 1994a].

are shown as diamonds in Figs. 8.3.1 and 8.3.2. As expected, the simulation results for the averaged system are in excellent agreement with the results obtained by numerically solving the Pontryagin equation. The difference between the simulation results for the original and the averaged systems indicates the order of magnitude of the error arising from the averaging procedure. This error is significant only when the initial energy level λ_0 is relatively large.

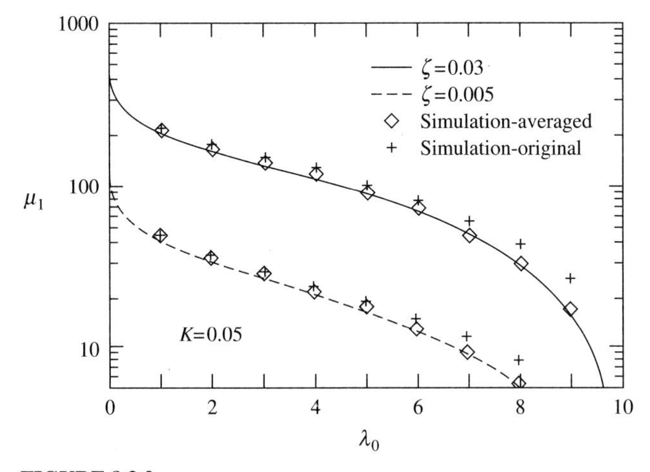

FIGURE 8.3.2
The mean first-excursion time of a Duffing oscillator under parametric white noise excitation. Damping ratio $\zeta = 0.03, 0.005$, white noise spectral density $K = 0.05$ [after Cai and Lin, 1994a].

8.3.3 Toppling of a Rigid Block under Random Base Excitation (Dimentberg, Lin, and Zhang, 1993)

Shown in Fig. 8.3.3 is a freestanding rigid block undergoing a rocking motion, caused by the shaking of its supporting base in both the horizontal and vertical directions. The rigid block represents a simplified model for a piece of computer equipment, a tall building, or a storage tank, and the shaking of the supporting base may represent the more intense portion of a strong earthquake. It is assumed that friction is sufficiently strong so that no sliding takes place between the block and the supporting base. If this is not the case, then a separate analysis of the sliding motion (see Section 4.7.2) or a combined rocking-sliding motion will be necessary.

The rocking motion is described by the rotational angle θ, of which the equation of motion may be written as

$$I_0\ddot{\theta} + (\operatorname{sgn}\theta)MRg(1 + \ddot{y}_G)\sin(\theta_c - |\theta|) = -MRg\ddot{x}_G\cos(\theta_c - |\theta|) \qquad \text{(a)}$$

where I_0 = mass moment of inertia about O (or O'), M = mass, $R = (H^2 + B^2)^{1/2}$ = distance between the centroid and O (or O'), g = gravitational acceleration, $\ddot{x}_G$ = horizontal base acceleration in unit g, $\ddot{y}_G$ = vertical base acceleration in unit g, $\operatorname{sgn}\theta$ = sign of θ, and $\theta_c = \tan^{-1}(B/H)$ = the critical $|\theta|$ value at which toppling occurs. The rotational angle θ is considered positive when rotation is centered about O and negative when it is about O'. Transition from a positive to a negative θ, or vice versa, is accompanied by an impact which, in turn, is accompanied by a loss of energy. The amount of energy loss can be determined theoretically if impacts are inelastic and totally without bouncing (Housner, 1963). However, an ideal inelastic impact for a rocking block is generally not possible, and the accompanying energy

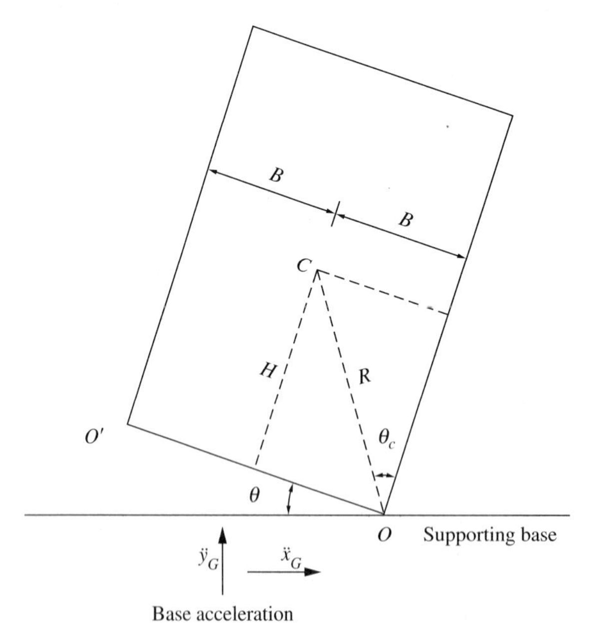

FIGURE 8.3.3
Rocking of a freestanding rigid block.

loss must be determined by experiments. As a result of such an energy loss, the angular velocity $\dot{\theta}$ is discontinuous at $\theta = 0$. The ratio ν between the two angular velocities, immediately after and before an impact, called the restitution coefficient, is given by

$$\nu = \frac{\dot{\theta}_+}{\dot{\theta}_-} \qquad 0 < \nu \le 1 \tag{b}$$

For example, Aslam, Godden, and Scalise (1980) found that ν is about 0.925 for a concrete block rocking on a steel base.

Dividing (a) by I_0 and taking into account the energy loss due to impacts at $\theta = 0$, we obtain

$$\ddot{\theta} + (1 - \nu)\dot{\theta}\left|\dot{\theta}\right|\delta(\theta) + (\operatorname{sgn}\theta)\Omega^2(1 + \ddot{y}_G)\sin(\theta_c - |\theta|) = -\Omega^2\ddot{x}_G\cos(\theta_c - |\theta|) \tag{c}$$

where $\Omega^2 = MRg/I_0$. The impact damping term in (c) is obtained by noting that (Dimentberg and Menyailov, 1979)

$$(\dot{\theta}_- - \dot{\theta}_+)\delta(t - t_*) = (1 - \nu)\dot{\theta}\left|\dot{\theta}\right|\delta(\theta) \tag{d}$$

where t_* is the time of impact.

Equation (c) represents a second-order strongly nonlinear system. Application of the quasi-conservative averaging procedure to this system requires the knowledge of its undamped free motion, for which (c) is reduced to

$$\ddot{\theta} + f_b(\theta) = 0 \tag{e}$$

where

$$f_b(\theta) = (\operatorname{sgn}\theta)\Omega^2\sin(\theta_c - |\theta|) \tag{f}$$

Equation (e) can be integrated once to yield

$$\frac{\dot{\theta}^2}{2} + U(\theta) = \Lambda \tag{g}$$

where $U(\theta)$ is the potential energy, obtained from

$$U(\theta) = \int_0^{\theta} f_b(\theta')\,d\theta' = \Omega^2[\cos(\theta_c - |\theta|) - \cos\theta_c] \tag{h}$$

and the integration constant Λ is the conservative total energy level for the undamped free rocking. The undamped free motion is periodic. Since the motion trajectory on the phase plane of θ and $\dot{\theta}$ is doubly symmetric for a given Λ, the length of a quarter-period may be computed within the region of $\theta > 0$ and $\dot{\theta} > 0$ as follows:

$$T_{1/4}(\Lambda) = \int_0^{\theta_m} \frac{d\theta}{\sqrt{2[\Lambda - U(\theta)]}} = \frac{1}{\sqrt{2}\Omega}\int_{\theta_c-\theta_m}^{\theta_c} \frac{d\phi}{\sqrt{\cos(\theta_c - \theta_m) - \cos\phi}} \tag{i}$$

where θ_m is the maximum magnitude of θ for a given Λ, that is, $\Lambda = U(\theta_m)$. A complete period is, of course, four times $T_{1/4}(\Lambda)$.

The restoring force $f_b(\theta)$, the potential energy $U(\theta)$, and the trajectory of undamped free motion on the phase plane are illustrated in Figs. 8.3.4 through 8.3.6.

We now return to equation (c) and take the impact damping and base excitations into consideration. Again we focus on the total energy

$$\Lambda(t) = \frac{\dot{\theta}^2}{2} + U(\theta) \tag{j}$$

where $U(\theta)$ is given by equation (h). However, the total energy is now time-dependent. Differentiating (j) with respect to t and combining the result with (c), we obtain

$$\begin{aligned}\dot{\Lambda} = &- \dot{\theta}[(1-\nu)\dot{\theta}\,|\dot{\theta}|\,\delta(\theta) \\ &+ (\operatorname{sgn}\theta)\Omega^2 \sin(\theta_c - |\theta|)W_v(t) + \Omega^2 \cos(\theta_c - |\theta|)W_h(t)]\end{aligned} \tag{k}$$

where the horizontal and vertical base excitations, $\ddot{x}_G$ and $\ddot{y}_G$, have been idealized as gaussian white noises. These idealizations are acceptable, under the assumption that the correlation times of the physical excitations are short compared with the dominant quasi-period of the rocking motion. The spectral densities of the base accelerations, in the unit of the gravitational acceleration g per radian per second, are generally much smaller than unity. If, in addition, the restitution coefficient ν in equation (k) is close to unity, which is usually the case, then $\Lambda(t)$ varies slowly with t. In this case, the quasi-conservative averaging procedure is applicable. The averaged $\Lambda(t)$ is approximately a Markov process, governed by the following Itô-type stochastic differential equation:

$$d\Lambda(t) = m(\Lambda)\,dt + \sigma(\Lambda)\,dB(t) \tag{l}$$

where $B(t)$ is a unit brownian motion process, and $m(\Lambda)$ and $\sigma(\Lambda)$ are the drift and diffusion coefficients, respectively. They are determined according to

$$\begin{aligned}m(\Lambda) = \Big\langle &-(1-\nu)\dot{\theta}^2\,|\dot{\theta}|\,\delta(\theta) + \pi\Omega^4 \frac{d}{d\Lambda}\Big\{ K_v\dot{\theta}^2 \sin^2(\theta_c - |\theta|) \\ &+ K_h\dot{\theta}^2 \cos^2(\theta_c - |\theta|) + \sqrt{K_v K_h}\,\gamma_{vh}\dot{\theta}^2 \operatorname{sgn}\theta \sin[2(\theta_c - |\theta|)]\Big\}\Big\rangle_t\end{aligned} \tag{m}$$

$$\begin{aligned}\sigma^2(\Lambda) = 2\pi\Omega^4 \Big\langle &K_v\dot{\theta}^2 \sin^2(\theta_c - |\theta|) + K_h\dot{\theta}^2 \cos^2(\theta_c - |\theta|) \\ &+ \sqrt{K_v K_h}\,\gamma_{vh}\dot{\theta}^2 \operatorname{sgn}\,\theta\, \sin[2(\theta_c - |\theta|)]\Big\rangle_t\end{aligned} \tag{n}$$

in which K_v = spectral density of $W_v(t)$, K_h = spectral density of $W_h(t)$, and $\sqrt{K_v K_h}\,\gamma_{vh}$ = cross-spectral density of $W_v(t)$ and $W_h(t)$. In the quasi-conservative averaging procedure, each time-averaging is substituted by averaging on θ over a pseudo period. In view of the symmetrical properties in the θ domain of the expressions on the right-hand sides of (m) and (n), all the terms except one may be averaged over a quarter-period:

$$\langle[\cdot]\rangle_t = \frac{1}{T_{1/4}(\Lambda)} \int_0^{\theta_m} \frac{[\cdot]d\theta}{\sqrt{2[\Lambda - U(\theta)]}} \tag{o}$$

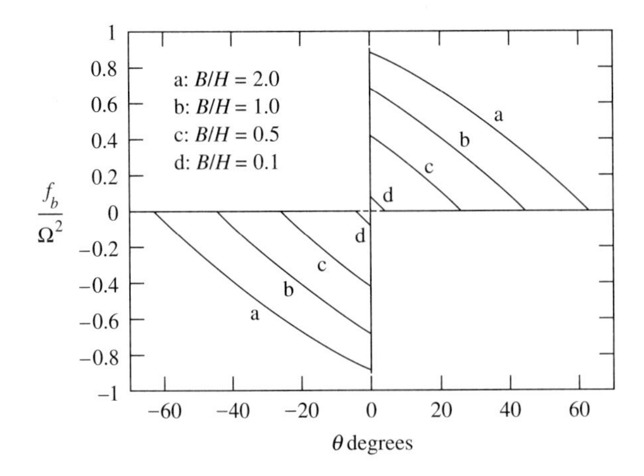

FIGURE 8.3.4
Restoring moment.

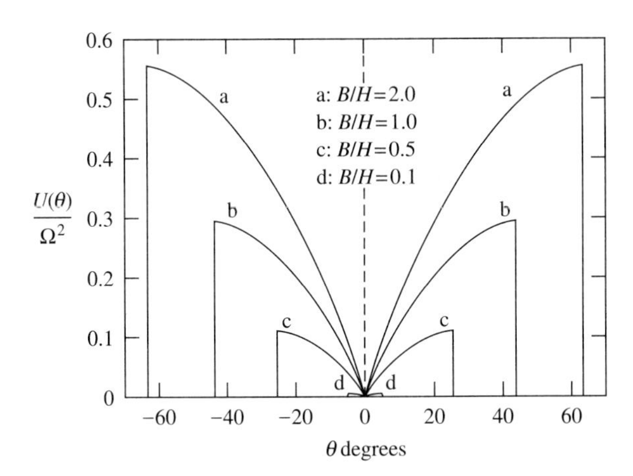

FIGURE 8.3.5
Potential energy.

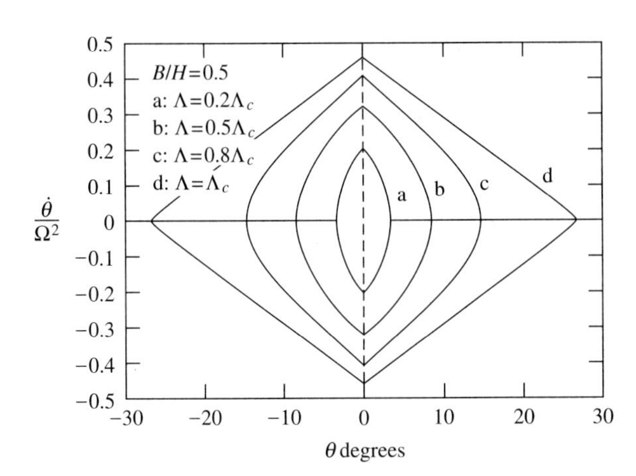

FIGURE 8.3.6
Phase plane trajectory of undamped free rocking motion.

The impulsive damping term in equation (m) should be averaged over a half-period, since there are two impacts in each period. Upon carrying out the procedure, we obtain

$$m(\Lambda) = \frac{1}{T_{1/4}(\Lambda)}\left\{-Q(\Lambda) + \pi\Omega^4[K_v S_s'(\Lambda) + K_h S_c'(\Lambda)]\right\} \tag{p}$$

$$\sigma^2(\Lambda) = \frac{2\pi}{T_{1/4}(\Lambda)}\Omega^4[K_v S_s(\Lambda) + K_h S_c(\Lambda)] \tag{q}$$

where

$$Q(\Lambda) = \tfrac{1}{2}(1-\nu)\int_0^{\theta_m} 2[\Lambda - U(\theta)]\delta(\theta)\,d\theta = (1-\nu)\Lambda \tag{r}$$

$$S_s(\Lambda) = \int_0^{\theta_m}\sqrt{2[\Lambda - U(\theta)]}\sin^2(\theta_c - \theta)\,d\theta \tag{s}$$

$$S_s'(\Lambda) = \frac{d}{d\Lambda}S_s(\Lambda) = \int_0^{\theta_m}\frac{\sin^2(\theta_c - \theta)}{\sqrt{2[\Lambda - U(\theta)]}}\,d\theta \tag{t}$$

$$S_c(\Lambda) = \int_0^{\theta_m}\sqrt{2[\Lambda - U(\theta)]}\cos^2(\theta_c - \theta)\,d\theta \tag{u}$$

$$S_c'(\Lambda) = \frac{d}{d\Lambda}S_c(\Lambda) = \int_0^{\theta_m}\frac{\cos^2(\theta_c - \theta)}{\sqrt{2[\Lambda - U(\theta)]}}\,d\theta \tag{v}$$

As can be inferred from equation (h), the upper integration limit θ_m in (r) through (v) is a function of Λ:

$$\theta_m = \theta_c - \cos^{-1}\left(\cos\theta_c + \frac{\Lambda}{\Omega^2}\right) \tag{w}$$

The terms associated with S_s' and S_c' on the right-hand side of (p) are the Wong-Zakai correction terms. It is interesting to note that the cross-spectral density of the horizontal and vertical excitations does not appear in the expressions for the drift and diffusion coefficients, (p) and (q); therefore, it does not affect the averaged total energy $\Lambda(t)$.

The generalized Pontryagin equation for $\Lambda(t)$ is given by

$$\frac{1}{2}b(\lambda_0)\frac{d^2}{d\lambda_0^2}\mu_{n+1} + a(\lambda_0)\frac{d}{d\lambda_0}\mu_{n+1} = -(n+1)\mu_n \tag{x}$$

where $a(\lambda_0)$ and $b(\lambda_0)$ are obtained, respectively, from the right-hand sides of (p) and (q), with Λ replaced by λ_0. Equation (x) is solved subject to two boundary conditions, one of which is

$$\mu_{n+1}(\lambda_c) = 0 \tag{y}$$

where

$$\lambda_c = \Omega^2(1 - \cos\theta_c) \tag{z}$$

Another boundary condition, associated with $\lambda_0 = 0$, depends on the behaviors of $a(\lambda_0)$ and $b(\lambda_0)$ as $\lambda_0 \to 0$. We now digress to examine these behaviors.

In the neighborhood of $\Lambda = 0$, the upper limit θ_m for the integrations on the right-hand sides of equations (s) through (v) are small. Within the range $-\theta_m \leq \theta \leq \theta_m$, the magnitude of θ is also small. Thus the potential energy may be approximated as

$$U(\theta) \approx (\Omega^2 \sin\theta_c)\theta \tag{aa}$$

It can be shown that as $\lambda_0 \to 0$, $b(\lambda_0) \to 0$ and $a(\lambda_0) \to \Delta$, where

$$\Delta = \pi\Omega^4(K_v \sin^2\theta_c + K_h \cos^2\theta_c) \tag{bb}$$

Thus the left boundary $\lambda_0 = 0$ is either a regular shunt or an entrance shunt. According to equation (8.1.18), the left-hand-side boundary condition is given by

$$\mu'_{n+1}(0) = -\frac{(n+1)}{\Delta}\mu_n(0) \tag{cc}$$

The statistical moments of the toppling time T can be calculated recursively, beginning from the first moment. For the present block toppling problem under consideration, an analytical solution appears difficult; however, with the knowledge of precise boundary conditions, direct numerical solution of equation (x) is straightforward.

In Fig. 8.3.7, the average toppling time μ_1 is plotted against λ_0, both in nondimensional units. Note that μ_1 decreases slowly in the region of small λ_0, although it drops rapidly to zero in the vicinity of λ_c. Since λ_0 is expected to be small in most practical cases, the choice of $\lambda_0 = 0$ is considered justified, and has been so assumed when obtaining other numerical results.

The variation of μ_1 with the base excitation levels is shown in Fig. 8.3.8. The solid and dotted lines correspond to the cases of horizontal excitation alone and combined horizontal–vertical excitations, respectively. In the latter case, the spectral density of the vertical excitation is assumed to be one-half the spectral density of the horizontal excitation.

The standard deviation σ of the toppling time is plotted against the excitation spectral densities in Fig. 8.3.9. Interestingly, σ and μ_1 follow the same trend, decreasing in about the same order of magnitude with increasing excitation levels.

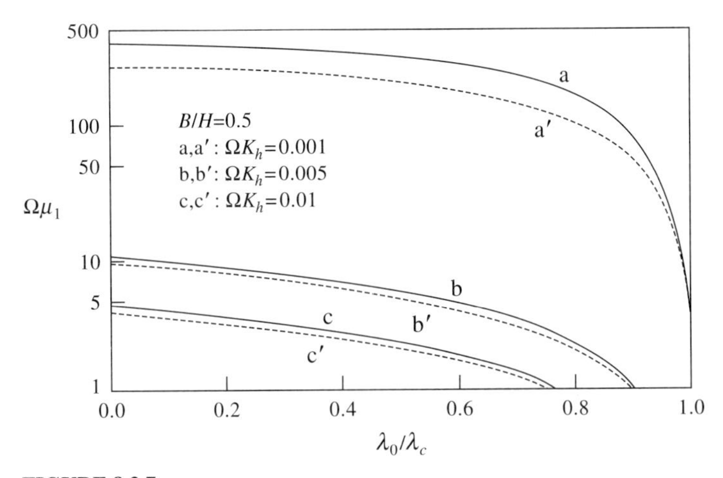

FIGURE 8.3.7
Average toppling time vs. initial energy level. Solid line: horizontal excitation only; dotted line: combined horizontal and vertical excitations [after Dimentberg et al., 1993; Copyright ©1993 ASCE, reprinted with permission].

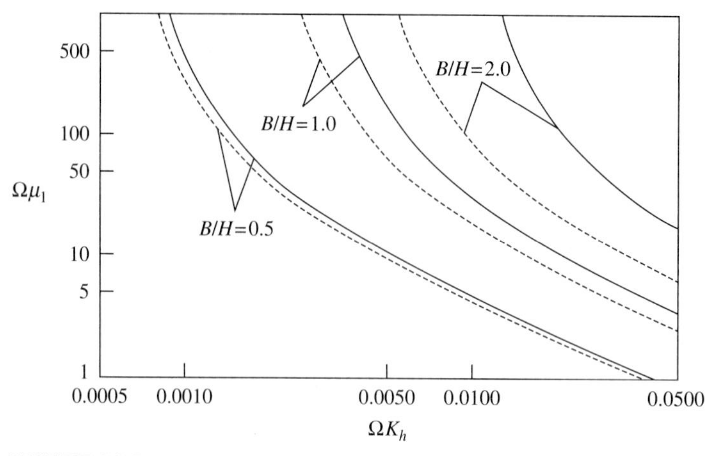

FIGURE 8.3.8
Average toppling time vs. base excitation level, $\lambda_0 = 0$. Solid line: horizontal excitation only; dotted line: combined horizontal and vertical excitations [after Dimentberg et al., 1993; Copyright © 1993 ASCE, reprinted with permission].

It has been suggested that blocks of the same geometrical proportion are not equally stable against toppling, and larger blocks are more stable than smaller ones, known as the "scale effect" (Housner, 1963). Indeed, a similar scale effect has been found in the present analysis. As shown in Fig. 8.3.10, the average toppling time increases with R, which is the distance between the block centroid and the center of rotation.

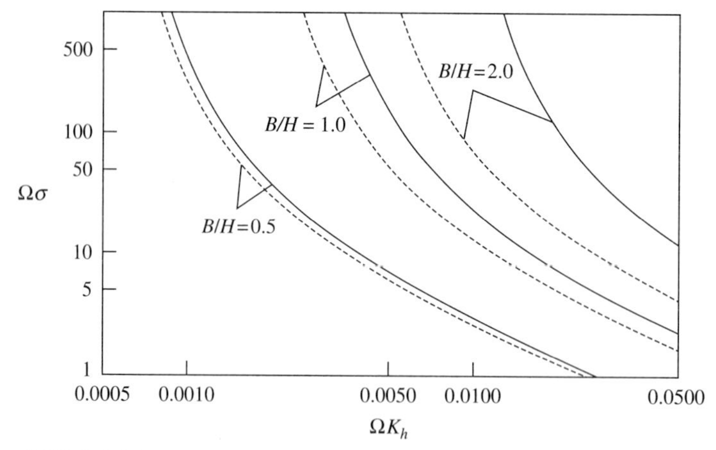

FIGURE 8.3.9
Standard deviation of toppling time vs. base excitation level, $\lambda_0 = 0$. Solid line: horizontal excitation only; dotted line: combined horizontal and vertical excitations [after Dimentberg et al., 1993; Copyright © 1993 ASCE, reprinted with permission].

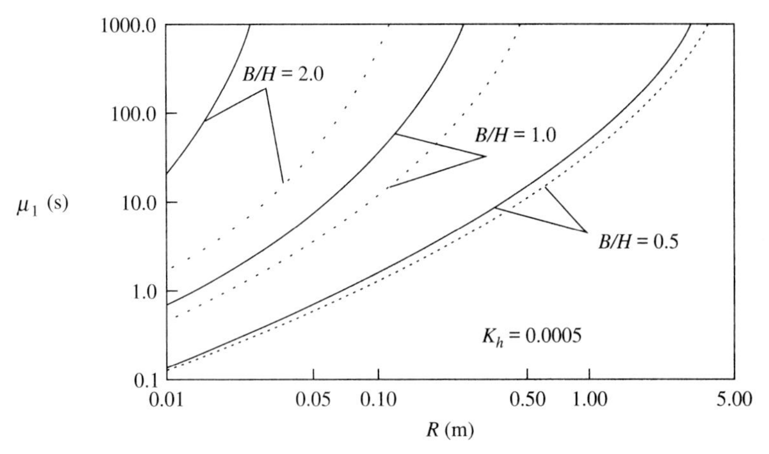

FIGURE 8.3.10
Average toppling time vs. size scale. $\lambda_0 = 0$. Solid line: horizontal excitation only; dotted line: combined horizontal and vertical excitations [after Dimentberg et al., 1993; Copyright © 1993 ASCE, reprinted with permission].

When it is desirable and feasible, additional resistance against toppling may be provided by anchoring the block to the supporting system with elastic ties. This is done in some instances for computer equipment. The potential energy in the system will then include the contribution of the elastic ties, and the critical toppling angle can be greater than $\tan^{-1}(B/H)$. Except for these differences, the analysis of such an anchored block is quite similar to that of a freestanding block given previously. Interested readers are referred to the paper by Dimentberg, Lin, and Zhang (1993).

8.4 THE RELIABILITY FUNCTION

As indicated earlier, the governing equation (8.1.4) for the reliability function R and its associated initial and boundary conditions (8.1.5) through (8.1.7) constitute an eigenvalue problem. Its solution is illustrated in the following examples.

8.4.1 A Linear Oscillator under Additive Random Excitation

Consider the following linear oscillator under an additive excitation

$$\ddot{Y} + 2\zeta\omega_0\dot{Y} + \omega_0^2 Y = \xi_3(t) \tag{a}$$

obtained by dropping the multiplicative excitations in equation (a) of Section 8.2.1. The statistical moments of the first-excursion time were obtained for this system in Section 8.2.2. Now we focus on the reliability function R, which is governed by

$$-\frac{\partial}{\partial\tau}R + \left(\frac{\delta}{2a_0} - \zeta\omega_0 a_0\right)\frac{\partial}{\partial a_0}R + \frac{1}{2}\delta\frac{\partial^2}{\partial a_0^2}R = 0 \tag{b}$$

where, as given in Section 8.2.1,

$$\delta = \frac{1}{2\omega_0^2}\int_{-\infty}^{\infty} E[\xi_3(t)\xi_3(t+\tau)]\cos\omega\tau\, d\tau = \frac{\pi}{\omega_0^2}\Phi_{33}(\omega_0) \tag{c}$$

Equation (b) is solved subject to the initial and boundary conditions

$$[R(\tau, a_c, a_0)]_{\tau=0} = 1 \qquad a_0 < a_c \tag{d}$$

$$[R(\tau, a_c, a_0)]_{a_0=a_c} = 0 \tag{e}$$

$$\frac{\partial R(\tau, a_c, a_0)}{\partial a_0} \sim 0(a_0) \qquad a_0 \to 0 \tag{f}$$

Equation (f) is obtained from equation (b) by noting that as a_0 tends to 0, the first derivate moment increases as a_0^{-1}. We seek a solution of equation (b) in the form of

$$R(\tau, a_c, a_0) = e^{-\lambda\tau}P(\lambda, a_0) \tag{g}$$

and obtain an ordinary differential equation for $P(\lambda, a_0)$ as follows:

$$\frac{1}{2}\delta\frac{d^2P}{da_0^2} + \left(\frac{\delta}{2a_0} - \zeta\omega_0 a_0\right)\frac{dP}{da_0} + \lambda P = 0 \tag{h}$$

where λ is treated as a parameter. Equation (h) is solved subject to the boundary conditions

$$(P)_{a_0=a_c} = 0 \tag{i}$$

$$\frac{dP}{da_0} \sim 0(a_0) \qquad a_0 \to 0 \tag{j}$$

Equation (h) can be simplified by a change of variable:

$$z = \frac{\zeta\omega_0 a_0^2}{\delta} = \frac{a_0^2}{2\sigma_y^2} \qquad \sigma_y^2 = \frac{\delta}{2\zeta\omega_0} = \frac{\pi\Phi_{33}(\omega_0)}{2\zeta\omega_0^3} \tag{k}$$

where σ_y^2 is the variance of the stationary displacement response. Then, equation (h) is transformed to

$$z\frac{d^2P}{dz^2} + (1-z)\frac{dP}{dz} + \frac{\lambda}{2\alpha}P = 0 \tag{l}$$

and the boundary conditions (i) and (j) are transformed to

$$(P)_{z=z_c} = 0 \tag{m}$$

$$\left[\frac{dP}{dz}\right]_{z=0} = \left[\frac{dP}{da_0}\frac{da_0}{dz}\right]_{a_0=0} = \text{finite} \tag{n}$$

where $z_c = \zeta\omega_0 a_c^2/\delta$. Equation ($l$) is a *degenerate hypergeometric equation* (Whittaker and Watson, 1952), which has only one solution satisfying the condition of finite derivative at $z = 0$, given by

$$P = M\left(-\frac{\lambda}{2\zeta\omega_0}, 1; z\right) \tag{o}$$

where M is known as the *confluent* or *degenerate hypergeometric function*. Using the boundary condition (m), we obtain an equation for the eigenvalues λ_n:

$$M\left(-\frac{\lambda_n}{2\zeta\omega_0}, 1; z_c\right) = 0 \tag{p}$$

Thus the reliability function R can be expressed in a series as follows:

$$R(\tau, a_c, a_0) = \sum_{n=1}^{\infty} C_n M\left(-\frac{\lambda_n}{2\zeta\omega_0}, 1; \frac{\zeta\omega_0}{\delta} a_0^2\right) e^{-\lambda_n \tau} \tag{q}$$

where C_n are constants. Substituting (q) into the initial condition (d) and using the orthogonality property of the eigenfunctions,

$$\int_0^{z_c} e^{-z} M\left(-\frac{\lambda_m}{2\zeta\omega_0}, 1, z\right) M\left(-\frac{\lambda_n}{2\zeta\omega_0}, 1, z\right) dz = \begin{cases} \neq 0 & m = n \\ = 0 & m \neq n \end{cases} \tag{r}$$

we obtain

$$C_n = \frac{\displaystyle\int_0^{z_c} e^{-z} M\left(-\frac{\lambda_n}{2\zeta\omega_0}, 1, z\right) dz}{\displaystyle\int_0^{z_c} e^{-z} M^2\left(-\frac{\lambda_n}{2\zeta\omega_0}, 1, z\right) dz} \tag{s}$$

The foregoing results were obtained by Lennox and Fraser (1974).

Numerical results have been obtained for three critical values $z_c = 0.5$, 2, and 4.5 (corresponding to $a_c = \sigma_y$, $2\sigma_y$, and $3\sigma_y$, respectively). The first nine eigenvalues are listed in Table 8.4.1. Figure 8.4.1 shows the probability distribution function of the first-excursion time T, computed for an initial value $a_0 = 0$ and three critical levels $a_c = \sigma_y$, $2\sigma_y$, and $3\sigma_y$. Figure 8.4.2 illustrates the probability density function of T for one critical level $a_c = 3\sigma_y$ but two different initial values $a_0 = 0$ and $1.5\sigma_y$.

8.4.2 A Linear Oscillator under Both Additive and Multiplicative Random Excitations

Return to the linear oscillator under both additive and multiplicative random excitations, considered in Section 8.2.1, and governed by

$$\ddot{Y} + \omega_0[2\zeta + \xi_2(t)]\dot{Y} + \omega_0^2[1 + \xi_1(t)]Y = \xi_3(t) \tag{a}$$

where $\xi_1(t)$, $\xi_2(t)$, and $\xi_3(t)$ are wideband stationary processes with zero mean values. We assume again that the first-excursion failure occurs when the amplitude process $A(t) = [Y^2(t) + \dot{Y}^2(t)/\omega_0^2]^{1/2}$ exceeds, for the first time, a critical value a_c. We now investigate the reliability function $R(\tau, a_c, a_0)$, which is governed by

$$-\frac{\partial}{\partial \tau} R + \left(\frac{\delta}{2a_0} - \alpha a_0\right)\frac{\partial}{\partial a_0} R + \frac{1}{2}(\delta + \gamma a_0^2)\frac{\partial^2}{\partial a_0^2} R = 0 \tag{b}$$

where, as shown in Section 8.2.1,

$$\alpha = \zeta\omega_0 - \frac{\pi\omega_0^2}{8}[2\Phi_{22}(0) + 3\Phi_{22}(2\omega_0) + 3\Phi_{11}(2\omega_0) - 6\Psi_{12}(2\omega_0)] \tag{c}$$

$$\delta = \frac{\pi}{\omega_0^2}\Phi_{33}(\omega_0) \tag{d}$$

TABLE 8.4.1
Eigenvalues λ_n for equation (p)[†]

	λ_n		
n	$a_c = \sigma_y$	$a_c = 2\sigma_y$	$a_c = 3\sigma_y$
1	2.418811	0.329478	0.040987
2	14.774574	3.465318	1.548872
3	36.984065	9.023553	4.030491
4	69.061214	17.044586	7.597266
5	111.007448	27.531894	12.258870
6	162.823090	40.486194	18.016587
7	224.508247	55.907712	24.870726
8	296.062962	73.796536	32.821393
9	377.487360	94.152710	41.868630

[†] After Lennox and Fraser (1974).

$$\gamma = \frac{\pi\omega_0^2}{4}[2\Phi_{22}(0) + \Phi_{22}(2\omega_0) + \Phi_{11}(2\omega_0) + 2\Psi_{12}(2\omega_0)] \tag{e}$$

and

$$\Phi_{ij}(\omega) = \frac{1}{2\pi}\int_{-\infty}^{\infty} E[\xi_i(t)\xi_j(t+\tau)]\cos(\omega\tau)\,d\tau \qquad i, j = 1, 2, 3 \tag{f}$$

$$\Psi_{ij}(\omega) = \frac{1}{2\pi}\int_{-\infty}^{\infty} E[\xi_i(t)\xi_j(t+\tau)]\sin(\omega\tau)\,d\tau \qquad i, j = 1, 2, 3 \tag{g}$$

Equation (b) is solved subject to the initial and boundary conditions,

$$[R(\tau, a_c, a_0)]_{\tau=0} = 1 \qquad a_0 < a_c \tag{h}$$

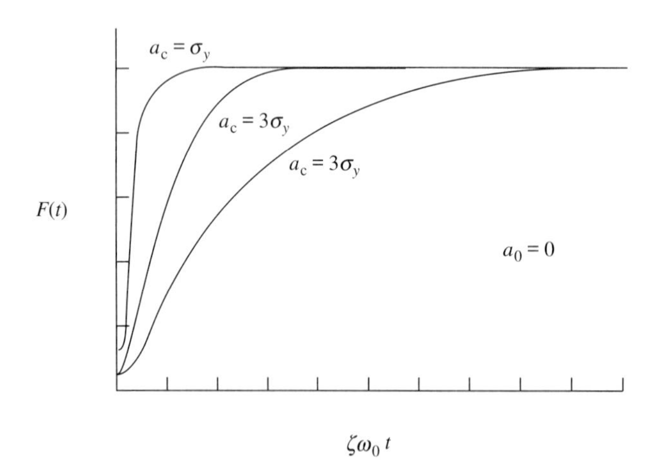

FIGURE 8.4.1
The probability distribution of the first-excursion time of a linear oscillator. Initial condition: $a_0 = 0$; critical amplitude: $a_c = \sigma_y, 2\sigma_y, 3\sigma_y$ [after Lennox and Fraser, 1974].

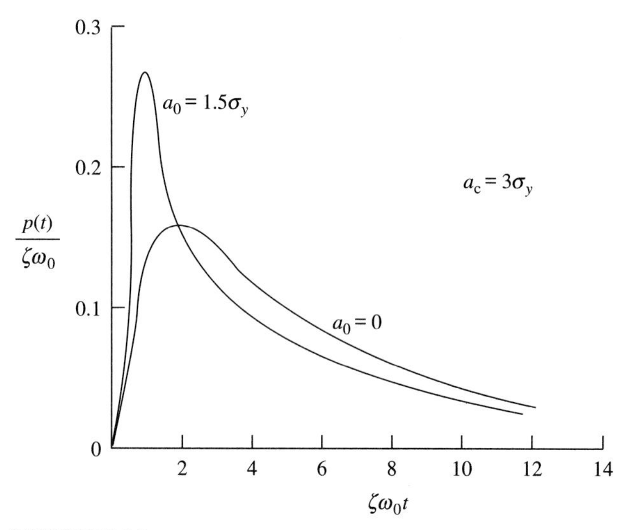

FIGURE 8.4.2
The probability density of the first-excursion time of a linear oscillator. Initial condition: $a_0 = 0, 1.5\sigma_y$, critical amplitude: $a_c = 3\sigma_y$ [after Lennox and Fraser, 1974].

$$[R(\tau, a_c, a_0)]_{a_0=a_c} = 0 \tag{i}$$

$$\frac{\partial R(\tau, a_c, a_0)}{\partial a_0} \sim (a_0) \qquad a_0 \to 0 \tag{j}$$

Letting
$$R(\tau, a_c, a_0) = e^{-\lambda\tau} P(\lambda, a_0) \tag{k}$$

we obtain an ordinary differential equation for $P(\lambda, a_0)$:

$$\frac{1}{2}(\delta + \gamma a_0^2)\frac{d^2P}{da_0^2} + \left(\frac{\delta}{2a_0} - \alpha a_0\right)\frac{dP}{da_0} + \lambda P = 0 \tag{l}$$

Equation (l) is solved subject to the boundary conditions

$$(P)_{a_0=a_c} = 0 \tag{m}$$

$$\frac{dP}{da_0} \sim 0(a_o) \qquad a_0 \to 0 \tag{n}$$

In this case, the change of variable $z = \alpha a_0^2/\delta$ used in Section 8.4.1 is no longer suitable. Instead, letting

$$z = -\frac{\gamma}{\delta}a_0^2 \tag{o}$$

equations (l), (m), and (n) become, respectively,

$$z(1-z)\frac{d^2P}{dz^2} + \left[1 - \left(\frac{1}{2} - \frac{\alpha}{\gamma}\right)z\right]\frac{dP}{dz} - \frac{\lambda}{2\gamma}P = 0 \tag{p}$$

$$(P)_{z=z_c} = 0 \tag{q}$$

$$\left[\frac{dP}{dz}\right]_{z=0} = \text{finite} \tag{r}$$

where $z_c = -\gamma a_c^2/\delta$. Equation (p) is a hypergeometric equation, which possesses only one solution as follows satisfying boundary condition (r) (Whittaker and Watson, 1952):

$$P(\lambda, z) = \begin{cases} F(c, d, 1, z) & -1 \le z \le 0 \\ \dfrac{\Gamma(c-d)}{\Gamma(d)\Gamma(c-1)}(-z^{-1})^c F(c, c, c+1-d, z^{-1}) & \\ +\dfrac{\Gamma(d-c)}{\Gamma(c)\Gamma(d-1)}(-z^{-1})^d F(d, d, d+1-c, z^{-1}) & z < -1 \end{cases} \tag{s}$$

where F is a hypergeometric function and the parameters c and d are given by

$$c = -\frac{1}{2}\eta + \frac{1}{2}\sqrt{\eta^2 - \frac{2\lambda}{\gamma}} \qquad d = -\frac{1}{2}\eta - \frac{1}{2}\sqrt{\eta^2 - \frac{2\lambda}{\gamma}} \tag{t}$$

in which

$$\eta = \frac{1}{2} + \frac{\alpha}{\gamma} \tag{u}$$

Imposing the boundary condition (q) on (s), we obtain an equation for the eigenvalues λ_n:

$$P(\lambda_n, z_c) = 0 \tag{v}$$

Thus a complete solution for $R(\tau, a_c, a_0)$ may be written as

$$R(\tau, a_c, a_0) = \sum_{n=1}^{\infty} C_n P\left(\lambda_n, -\frac{\gamma}{\delta}a_0^2\right)e^{-\lambda_n \tau} \tag{w}$$

where constants C_n are determined from

$$C_n = \frac{\displaystyle\int_0^{z_c} (1-z)^{\eta-1} P(\lambda_n, z)\, dz}{\displaystyle\int_0^{z_c} (1-z)^{\eta-1} P^2(\lambda_n, z)\, dz} \tag{x}$$

The foregoing results were obtained previously by Ariaratnam and Tam (1979) in different notations.

8.5 FATIGUE CRACK GROWTH DUE TO RANDOM LOADING

From a fracture mechanics point of view, the fatigue damage of a structural or machine component, subjected to dynamic loads, can be measured by the size of a dominant crack, and failure occurs when this crack reaches a critical magnitude. The growth of a fatigue crack is a random phenomenon, since the stress fluctuation that causes the crack growth and the material resistance to the growth are generally random in time.

Several stochastic models have been proposed for the analysis of fatigue crack growth under cyclic loading of deterministic amplitude, or programmed loading of other deterministic forms, including those of Madsen (1982), Lin and Yang (1983, 1985), Ditlevsen and Olesen (1986), and Ortis and Kiremidjian (1988). Comprehensive references may be found in a review article by Kozin and Bogdanoff (1989) and a monograph by Sobczyk and Spencer (1992). These stochastic models were mostly randomized versions of known deterministic models, for example, multiplying the well-known Paris-Erdogan equation (1963) by a nonnegative random variable or a nonnegative random function of time. The parameters in such randomized models were inferred from published experimental data, such as those by Virkler, Hillberry, and Goel (1978) in the case of constant-amplitude cyclic loading and by Norohna, Henslee, Gordon, Wolanski, and Yee (1978) in the case of programmed loading.

A typical deterministic model for fatigue crack growth under cyclic loading has the general form of (e.g., Hoeppner and Krupp, 1974; Miller and Gallagher, 1981),

$$\frac{da}{dn} = f(k, \Delta k, a, R) \tag{8.5.1}$$

where a = crack size (half the length for a through crack), n = number of stress cycles, f = a nonnegative function, k = stress intensity factor, Δk = stress intensity factor range, and R = stress ratio. It is generally recognized that the effect of the stress intensity range Δk is dominant.

Experimental evidence indicates that the da/dn versus Δk relationship follows a general trend shown in Fig. 8.5.1. There exist a threshold Δk_0 for the stress intensity factor range below which the crack growth is negligible and a maximum stress intensity factor k_{ft}, known as the fracture toughness, beyond which the crack growth becomes unstable. Away from these two boundaries, the relationship between da/dn and Δk may be approximated by a straight line on a log-log plot, suggesting an approximate power law. The well-known Paris-Erdogan model (1963) is given by

$$\frac{da}{dn} = \alpha(\Delta k)^\beta \tag{8.5.2}$$

$$\Delta k = h(a)\Delta x \tag{8.5.3}$$

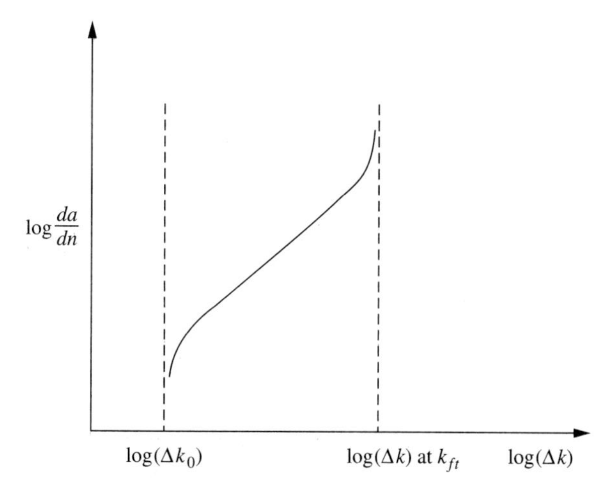

FIGURE 8.5.1
Schematic relationship between da/dn and Δk.

in which α and β are nonnegative material constants, Δx is the stress range, and $h(a)$ is a nonnegative function whose functional form depends on the geometrical shapes of the crack and of the structural or mechanical component. Clearly, equation (8.5.2) simulates only the power law portion of the crack growth phenomenon.

Neglecting the secondary factors in equation (8.5.1), and noting the dependence of Δk on the stress range Δx and a, one may write

$$\frac{da}{dt} = \eta g(a, \Delta x) \tag{8.5.4}$$

where the discrete independent variable n has been replaced by the continuous time variable t, η is the number of cycles per unit time, and $g(\cdot, \cdot)$ is again a nonnegative function. To account for the random material resistance, equation (8.5.4) may be modified to read

$$\frac{dA}{dt} = \eta g(A, \Delta x)\psi(t) \tag{8.5.5}$$

where $\psi(t)$ is a random process accounting for such effects as temperature and corrosion, and the symbol for the crack size is capitalized to signify that it is now also a random process. The stress range remains deterministic. In most cases, $\psi(t)$ may be modeled as a slowly varying stationary random process with a long correlation time. In the limiting case in which the correlation time tends to infinity, $\psi(t)$ reduces to a random variable, which is a model proposed by Yang, Salivar, and Annis (1983).

Complication arises when random stresses are taken into consideration. The usual definition for stress range is no longer meaningful. Sobczyk (1986) suggested the replacement of the stress intensity factor range in cyclic loading with the root-mean-square value in random loading. Among other investigators, Tsurui and Ishikawa (1986) obtained a generalized Fokker-Planck equation for the crack length distribution, and Bolotin (1989) derived an asymptotic expression for the crack length, making use of the central limit theorem. Zhu, Lin, and Lei (1992) proposed another version of the Markov approximation, based on the argument that the random variation of the stress range had a much shorter correlation time than the relaxation time of crack growth. This latter argument is adopted in the following analysis.

Let $X(t)$ be the random stress causing the growth of a dominant fatigue crack, and let $\Delta X(t)$ be the *stress range*, defined as the difference between a local maximum and the following local minimum:

$$\Delta X(t) = \left|X(t_n) - X(t_{n+1})\right| \qquad t_n \leq t < t_{n+1} \tag{8.5.6}$$

where t_n and t_{n+1} are the times at which the two consecutive extrema occur. We assume that the fatigue crack growth under random loading can still be described by equation (8.5.5), with Δx replaced by a random process $\Delta X(t)$. The modified version now reads

$$\frac{dA}{dt} = \eta g[A(t), \Delta X(t)]\, \psi(t) \tag{8.5.7}$$

where η is the number of the local maxima or local minima per unit time.

To proceed, we further assume that fatigue crack size $A(t)$ is a slowly varying random process compared with the stress range process $\Delta X(t)$. This is a reasonable assumption, since the correlation time τ_c of $\Delta X(t)$ is expected to be much shorter than the life of a component, defined as the time spent for the crack to grow to the critical size. According to the Stratonovich-Khasminskii limit theorem described in Section 4.7, the crack size $A(t)$ is approximately a diffusive Markov process governed by the Itô equation

$$dA = m(A)\,dt + \sigma(A)\,dB(t) \tag{8.5.8}$$

where the drift coefficient $m(A)$ and the diffusion coefficient $\sigma(A)$ are obtained from the stochastic averaging procedure as follows:

$$m(A) = \eta E[g]E[\psi] + \eta \int_{-\infty}^{0} \text{cov}\left[\frac{\partial g(t)}{\partial A}, g(t+\tau)\right] \text{cov}[\psi(t), \psi(t+\tau)]\,d\tau \tag{8.5.9}$$

$$\sigma^2(A) = \eta^2 \int_{-\infty}^{\infty} \text{cov}[g(t), g(t+\tau)]\,\text{cov}[\psi(t), \psi(t+\tau)]\,d\tau \tag{8.5.10}$$

It is implied that the $\Delta X(t)$ and $\psi(t)$ processes are independent. Note that the integrals in equations (8.5.9) and (8.5.10) involve the covariance functions, instead of the correlation functions as in equations (4.7.3) and (4.7.4). The difference stems from the fact that the ensemble averages of both function g and $\psi(t)$ are generally nonzero. Moreover, an unsmoothed stochastic averaging procedure is carried out in (8.5.9) and (8.5.10), without additional time-averaging; therefore, the terms on the right-hand sides of (8.5.9) and (8.5.10) need not be small.

We are concerned with the random time T for the crack size $A(t)$ to reach a critical value a_c. The nth moment μ_n of the random time T is governed by the generalized Pontryagin equation

$$\frac{1}{2}\sigma^2(a_0)\frac{d^2}{da_0^2}\mu_{n+1} + m(a_0)\frac{d}{da_0}\mu_{n+1} = -(n+1)\mu_n \tag{8.5.11}$$

Since the Wiener process $B(t)$ in (8.5.8) has an unbounded variation, the idealized Markov model does not guarantee that $A(t_1) \leq A(t_2)$ for every $t_1 < t_2$. However, if the drift coefficient m is sufficiently large and the diffusion coefficient σ is sufficiently small, then the rightward movement dominates. Furthermore, by imposing a reflective boundary at $A = a_i$ where a_i is the initial crack length, the possibility of $A < a_i$ is eliminated. Thus the boundary conditions for equation (8.5.11) are

$$\mu_{n+1}(a_c) = 0 \tag{8.5.12}$$

$$\mu'_{n+1}(a_i) = 0 \tag{8.5.13}$$

The reliability function $R(\tau, a_c, a_0)$ defined in (8.1.2) is governed by

$$-\frac{\partial}{\partial \tau}R + m(a_0)\frac{\partial}{\partial a_0}R + \frac{1}{2}\sigma^2(a_0)\frac{\partial^2}{\partial a_0^2}R = 0 \tag{8.5.14}$$

subject to the initial and boundary conditions

$$[R(\tau, a_c, a_0)]_{\tau=0} = 1 \qquad a_i \leq a_0 < a_c \tag{8.5.15}$$

$$[R(\tau, a_c, a_0)]_{a_0=a_c} = 0 \tag{8.5.16}$$

$$\left[\frac{\partial}{\partial a_0}R(\tau, a_c, a_0)\right]_{a_0=a_i} = 0 \tag{8.5.17}$$

With the knowledge of the functional forms of $g[A(t), \Delta X(t)]$, the initial crack size a_i, the critical crack size a_c, and the statistical properties of the stress range process $\Delta X(t)$ and the random process $\psi(t)$, the moments of the fatigue life, as well as the reliability function, can be obtained at least numerically.

PARIS-ERDOGAN FATIGUE MODEL. The randomized version of the Paris-Erdogan fatigue model (8.5.2) has the form of

$$\frac{dA}{dt} = \eta Q(A)(\Delta X)^{\beta}\psi(t) \tag{8.5.18}$$

where

$$Q(A) = \alpha[h(A)]^{\beta} \tag{8.5.19}$$

In this case, the problem can be simplified by introducing another random process

$$Z[A(t)] = \int_{a_i}^{A} \frac{du}{Q(u)} \tag{8.5.20}$$

which is monotonically increasing with $A(t)$ since $Q(A)$ is a nonnegative function. It is assumed implicitly that the right-hand side of equation (8.5.20) exists at least in probability. With this transformation, equation (8.5.18) becomes

$$\frac{dZ}{dt} = \eta(\Delta X)^{\beta}\psi(t) \tag{8.5.21}$$

Note that the right hand-side of equation (8.5.21) is independent of Z, which is an advantage. Now $Z(t)$ is also approximately a diffusive Markov process. The drift and diffusion coefficients of process $Z(t)$ can be obtained by applying equations (8.5.9) and (8.5.10), with A replaced by Z, and g by $(\Delta X)^{\beta}$. Since the right-hand side of equation (8.5.21) is independent of Z, the second term in equation (8.5.9) vanishes, resulting in

$$m(Z) = \eta E[(\Delta X)^{\beta}]E[\psi(t)] \tag{8.5.22}$$

$$\sigma^2(Z) = \eta^2 \int_{-\infty}^{\infty} \operatorname{cov}[(\Delta X_t)^{\beta}, (\Delta X_{t+\tau})^{\beta}]\operatorname{cov}[\psi(t), \psi(t+\tau)]\, d\tau \tag{8.5.23}$$

The stress process $X(t)$ is assumed to be stationary, as is the process $(\Delta X)^{\beta}$. Thus $m(Z)$ and $\sigma^2(Z)$ are both constants, denoted by m_Z and σ_Z^2, respectively.

The reliability function $R(\tau, z_c, z_0)$ is now governed by

$$-\frac{\partial}{\partial \tau}R + m_Z\frac{\partial}{\partial z_0}R + \frac{1}{2}\sigma_Z^2\frac{\partial^2}{\partial z_0^2}R = 0 \tag{8.5.24}$$

subject to the initial and boundary conditions

$$[R(\tau, z_c, z_0)]_{\tau=0} = 1 \qquad 0 \le z_0 < z_c \tag{8.5.25}$$

$$[R(\tau, z_c, z_0)]_{z_0=z_c} = 0 \tag{8.5.26}$$

$$\left[\frac{\partial}{\partial z_0}R(\tau, z_c, z_0)\right]_{z_0=0} = 0 \tag{8.5.27}$$

where

$$z_c = \int_{a_i}^{a_c} \frac{du}{Q(u)} \tag{8.5.28}$$

Note that z_c depends on both the initial crack size a_i and the critical crack size a_c, and that the initial crack size a_i corresponds to $z_0 = 0$.

Assuming that the solution for equation (8.5.24) has the form of

$$R(\tau, z_c, z_0) = e^{-\lambda\tau} P(z_0) \tag{8.5.29}$$

we obtain from (8.5.24), (8.5.26), and (8.5.27)

$$\frac{1}{2}\sigma_Z^2 \frac{d^2 P}{dz_0^2} + m_Z \frac{dP}{dz_0} + \lambda P = 0 \tag{8.5.30}$$

$$P(z_c) = 0 \tag{8.5.31}$$

$$P'(0) = 0 \tag{8.5.32}$$

Equations (8.5.30) through (8.5.32) define an eigenvalue problem. It can be shown that the eigenfunctions for the problem are

$$P_n = e^{-m_z z_0/\sigma_Z^2}\left(\sin \omega_n z_0 + \frac{\sigma_Z^2}{m_Z}\omega_n \cos \omega_n z_0\right) \qquad n = 1, 2, \ldots \tag{8.5.33}$$

where the ω_n are determined by the transcendental equation

$$\tan \omega_n z_c = -\frac{\sigma_Z^2}{m_Z}\omega_n \tag{8.5.34}$$

The eigenvalues λ_n are related to ω_n through

$$\lambda_n = \frac{1}{2\sigma_Z^2}\left(m_Z^2 + \sigma_Z^4 \omega_n^2\right) \tag{8.5.35}$$

Now equation (8.5.30) can be changed into a self-adjoint form by multiplying a common factor $\exp(2m_Z z_0/\sigma_Z^2)$. It then becomes clear that the eigenfunctions P_n are orthogonal with respect to the weighting function $\exp(2m_Z z_0/\sigma_Z^2)$:

$$\int_0^{z_c} e^{2m_Z z_0/\sigma_Z^2} P_n(z_0) P_m(z_0)\, dz_0 = \begin{cases} \neq 0 & m = n \\ = 0 & m \neq n \end{cases} \tag{8.5.36}$$

The general solution for equation (8.5.24) can then be constructed by superposition as follows:

$$R(\tau, z_c, z_0) = e^{-m_Z z_0/\sigma_Z^2} \sum_{n=1}^{\infty} d_n \left(\sin \omega_n z_0 + \frac{\sigma_Z^2}{m_Z}\omega_n \cos \omega_n z_0\right) e^{-\lambda_n \tau} \tag{8.5.37}$$

The coefficients d_n are determined by imposing the initial condition (8.5.25) and the orthogonality condition (8.5.36), to yield

$$d_n = \frac{2 m_Z \sigma_Z^2 e^{m_Z z_c/\sigma_Z^2} \sin \omega_n z_c}{z_c(m_Z^2 + \sigma_Z^4 \omega_n^2) + m_Z \sigma_Z^2} \tag{8.5.38}$$

The reliability function is obtained from equation (8.5.37) by letting $z_0 = 0$:

$$R(\tau, z_c, 0) = \frac{\sigma_Z^2}{m_Z} \sum_{n=1}^{\infty} d_n \omega_n e^{-\lambda_n \tau} \tag{8.5.39}$$

The probability density of the fatigue life follows as

$$p_T(\tau, a_c, a_i) = -\frac{\partial}{\partial \tau} R(\tau, z_c, 0) = \frac{\sigma_Z^2}{m_Z} \sum_{n=1}^{\infty} d_n \lambda_n \omega_n e^{-\lambda_n \tau} \tag{8.5.40}$$

If only the statistical moments of the fatigue life are of interest, then they can be obtained directly from the simpler generalized Pontryagin equation, given by

$$\frac{1}{2}\sigma_Z^2 \frac{d^2}{dz_0^2}\mu_{n+1} + m_Z \frac{d}{dz_0}\mu_{n+1} = -(n+1)\mu_n \tag{8.5.41}$$

subject to the boundary conditions

$$\mu_{n+1}(z_c) = 0 \tag{8.5.42}$$

$$\mu'_{n+1}(0) = 0 \tag{8.5.43}$$

Equation (8.5.41) can easily be integrated to yield

$$\mu_{n+1}(z_0) = \frac{2(n+1)}{\sigma_Z^2} \int_{z_0}^{z_c} \exp\left(-\frac{2m_Z}{\sigma_Z^2}u\right) du \int_0^u \mu_n(v) \exp\left(\frac{2m_Z}{\sigma_Z^2}v\right) dv \tag{8.5.44}$$

where boundary conditions (8.5.42) and (8.5.43) have been imposed. The moments of the fatigue life of the process $A(t)$ are obtained by letting $z_0 = 0$:

$$\mu_{n+1}(0) = \frac{2(n+1)}{\sigma_Z^2} \int_0^{z_c} \exp\left(-\frac{2m_Z}{\sigma_Z^2}u\right) du \int_0^u \mu_n(v) \exp\left(\frac{2m_Z}{\sigma_Z^2}v\right) dv \tag{8.5.45}$$

In particular, we obtain for $n = 0$ the mean fatigue life

$$(\mu_1)_{a_0=a_i} = \frac{z_c}{m_Z} + \frac{\sigma_Z^2}{2m_Z^2}\left[\exp\left(-\frac{2m_Z}{\sigma_Z^2}z_c\right) - 1\right] \tag{8.5.46}$$

and for $n = 1$ the mean square of the fatigue life

$$(\mu_2)_{a_0=a_i} = \frac{1}{m_Z^2}\left(z_c^2 - \frac{\sigma_Z^4}{m_Z^2}\right) + \frac{\sigma_Z^2}{m_Z^3}\left(3z_c + \frac{\sigma_Z^2}{2m_Z}\right)\exp\left(-\frac{2m_Z}{\sigma_Z^2}z_c\right) + \frac{\sigma_Z^4}{2m_Z^4}\exp\left(-\frac{4m_Z}{\sigma_Z^2}z_c\right) \tag{8.5.47}$$

STATIONARY GAUSSIAN NARROW-BAND STRESS PROCESSES. One important step in the preceding analysis is to obtain the covariance functions required in equations (8.5.9) and (8.5.10) for a general fatigue model or those in equation (8.5.23) for the randomized Paris-Erdogan model. In the special case of a narrow-band stress process with zero mean, nearly all the maxima are positive, all the minima are negative, and a zero crossing occurs between two consecutive extrema. Such a stress range process may be approximated by two times the envelope process, and the parameter η in equation (8.5.5) or (8.5.18) may be approximated by the average number of the up (or down) zero crossings.

In general, the statistical analysis of an envelope process is also difficult. However, for a stationary gaussian process, the first- and second-order probability density functions of its envelope process can be obtained analytically. Assume that the stress process $X(t)$ is a stationary gaussian process with zero mean. It is known from Section 2.2 that $X(t)$ may be represented by

$$X(t) = \int_{-\infty}^{\infty} e^{i\omega t} dZ(\omega) \tag{8.5.48}$$

where $Z(\omega)$ is an orthogonal increment random process satisfying

$$E[dZ(\omega)\, dZ^*(\omega')] = \begin{cases} \Phi_{XX}(\omega)\, d\omega & \omega = \omega' \\ 0 & \omega \neq \omega' \end{cases} \tag{8.5.49}$$

in which the asterisk denotes the complex conjugate, and $\Phi_{XX}(\omega)$ is the spectral density of $X(t)$. Now define another real stationary process $\hat{X}(t)$:

$$\hat{X}(t) = \int_{-\infty}^{\infty} h(\omega) e^{i\omega t} dZ(\omega) \tag{8.5.50}$$

where

$$h(\omega) = \begin{cases} i & \omega > 0 \\ 0 & \omega = 0 \\ -i & \omega < 0 \end{cases} \tag{8.5.51}$$

It can be shown that

$$R_{XX}(\tau) = R_{\hat{X}\hat{X}}(\tau) = \int_{-\infty}^{\infty} e^{i\omega\tau} \Phi_{XX}(\omega)\, d\omega = 2\int_0^{\infty} \Phi_{XX} \cos \omega\tau\, d\omega \tag{8.5.52}$$

$$R_{X\hat{X}}(\tau) = -R_{\hat{X}X}(\tau) = \int_{-\infty}^{\infty} h^*(\omega) e^{i\omega\tau} \Phi_{XX}(\omega)\, d\omega = 2\int_0^{\infty} \Phi_{XX} \sin \omega\tau\, d\omega \tag{8.5.53}$$

The envelope process of $X(t)$ is defined by

$$A(t) = [X^2(t) + \hat{X}^2(t)]^{1/2} \tag{8.5.54}$$

which is equivalent to

$$X(t) = A(t) \cos \Theta(t) \qquad \hat{X}(t) = A(t) \sin \Theta(t) \tag{8.5.55}$$

The first- and second-order probability density functions of the envelope process $A(t)$ are given by

$$p(a) = \frac{a}{\sigma_X^2} \exp\left(-\frac{a^2}{2\sigma_X^2}\right) \tag{8.5.56}$$

$$p(a_1, a_2, \tau) = \frac{a_1 a_2}{\sigma_X^2(1-\rho^2)} \mathrm{I}_0\left[\frac{a_1 a_2 \rho}{\sigma_X^2(1-\rho^2)}\right] \exp\left[-\frac{a_1^2 + a_2^2}{2\sigma_X^2(1-\rho^2)}\right] \tag{8.5.57}$$

where σ_X^2 is the variance of $X(t)$, I_0 is the modified Bessel function of the zeroth order, and ρ is the correlation coefficient of $A(t)$ and $A(t+\tau)$. Equations (8.5.56) and (8.5.57) were first obtained by Rice (1944, 1945) under the additional assumption that $X(t)$ is a narrow-band process. However, as shown by Cramer and Leadbetter (1967), they are valid without the narrow-band assumption. Given the spectral density $\Phi_{XX}(\omega)$ of the stress process $X(t)$,

$$\sigma_X^2 = \int_{-\infty}^{\infty} \Phi_{XX}(\omega)\, d\omega \tag{8.5.58}$$

$$\rho(\tau) = \frac{[R_{XX}^2(\tau) + R_{X\hat{X}}^2(\tau)]^{1/2}}{\sigma_X^2} \tag{8.5.59}$$

The first- and second-order probability densities of the stress range process are then approximately

$$p(\Delta x) = \frac{\Delta x}{4\sigma_X^2} \exp\left(-\frac{\Delta x^2}{8\sigma_X^2}\right) \tag{8.5.60}$$

$$p(\Delta x_1, \Delta x_2, \tau) = \frac{\Delta x_1 \Delta x_2}{16\sigma_X^2(1-\rho^2)} \mathrm{I}_0\left[\frac{\Delta x_1 \Delta x_2 \rho}{4\sigma_X^2(1-\rho^2)}\right] \exp\left[-\frac{\Delta x_1^2 + \Delta x_2^2}{8\sigma_X^2(1-\rho^2)}\right] \tag{8.5.61}$$

Making use of the series expansion (Gradshteyn and Ryzhik, 1980),

$$\frac{(uvw)^{-\alpha/2}}{1-w} \exp\left(-w\frac{u+v}{1-w}\right) \mathrm{I}_\alpha\left(\frac{2\sqrt{uvw}}{1-w}\right) = \sum_{n=0}^{\infty} n! \frac{L_n^\alpha(u) L_n^\alpha(v)}{\Gamma(n+\alpha+1)} w^n \tag{8.5.62}$$

where L_n^α are the Laguerre polynomials of order α, the second-order probability density (8.5.61) may be expressed in a series as follows:

$$p(\Delta x_1, \Delta x_2, \tau) = \tfrac{1}{2} s_1 s_2 \exp[-(s_1^2 + s_2^2)] \sum_{n=0}^{\infty} L_n^0(s_1^2) L_n^0(s_2^2) \rho^{2n}(\tau) \tag{8.5.63}$$

where

$$s_1 = \frac{\Delta x_1}{2\sqrt{2}\sigma_X} \qquad s_2 = \frac{\Delta x_2}{2\sqrt{2}\sigma_X} \tag{8.5.64}$$

The average number of up zero crossings of a stationary gaussian process $X(t)$ is a classical result, given by (Rice, 1944, 1945)

$$\eta_0 = \frac{1}{2\pi}\left[\frac{\int_0^\infty \omega^2 \Phi_{XX}(\omega)\, d\omega}{\int_0^\infty \Phi_{XX}(\omega)\, d\omega}\right]^{1/2} \tag{8.5.65}$$

Thus given the spectral density $\Phi_{XX}(\omega)$ of $X(t)$ and the mean and covariance functions of $\psi(t)$, the drift and diffusion coefficients of the crack size $A(t)$ can be obtained from equations (8.5.9) and (8.5.10).

If the Paris-Erdogan model is employed, we obtain from equations (8.5.60) and (8.5.61)

$$\begin{aligned} E[\Delta X^\beta] &= \int_0^\infty \Delta x^\beta p(\Delta x)\, d(\Delta x) \\ &= (2\sqrt{2}\sigma_X)^\beta \Gamma\left(1 + \frac{\beta}{2}\right) \end{aligned} \tag{8.5.66}$$

$$\begin{aligned} \mathrm{cov}[\Delta X^\beta(t), \Delta X^\beta(t+\tau)] &= \int_0^\infty \int_0^\infty \Delta x_1^\beta \Delta x_2^\beta p(\Delta x_1, \Delta x_2, \tau)\, d(\Delta x_1)\, d(\Delta x_2) \\ &= (8\sigma_X^2)^\beta \sum_{n=1}^{\infty} c_n^2 \rho^{2n}(\tau) \end{aligned} \tag{8.5.67}$$

where
$$c_n = \sum_{j=0}^{n} \frac{(-1)^j n!}{(n-j)!(j!)^2} \Gamma\left(1 + \frac{j}{2} + \frac{\beta}{2}\right) \tag{8.5.68}$$

In obtaining (8.5.66) and (8.5.67), use has been made of

$$L_n^0(u) = \sum_{j=0}^{n} (-1)^j \frac{n!}{(j!)^2(n-j)!} u^j \tag{8.5.69}$$

Thus the drift and diffusion coefficients of process $Z(t)$ can be calculated from equations (8.5.22) and (8.5.23), and the probability density and the statistical moments of the fatigue life can be calculated from equations (8.5.40) and (8.5.45), respectively.

8.5.1 Fatigue Crack Growth in a Plate

For illustration, consider a thin square plate of size $l \times l$, shown in Fig. 8.5.2, with an initial central crack of length $2a_0$ and supporting an infinitely rigid heavy mass M at its end. The plate is idealized to be massless, made of homogeneous and isotropic material, and with linear viscous damping. The mass M is subjected to a gaussian wideband excitation $\xi(t)$ perpendicular to the crack. We are interested in the statistical properties of the first-excursion time when the fatigue crack length reaches a critical value a_c. This problem has been studied by Grigoriu (1990), taking into account a slow degradation of the plate stiffness due to crack growth. It is reexamined here using the stochastic averaging procedure described above. However, we assume that the critical crack size $2a_c$ is much smaller than l; therefore, the stiffness degradation is negligible prior to failure. The random variation of the material resistance, represented by function $\psi(t)$ in equation (8.5.7), is also ignored.

First, we investigate the properties of the stress process. Let $Y(t)$ be the displacement of the plate, governed by differential equation

$$M\ddot{Y} + c\dot{Y} + kY = \xi(t) \tag{a}$$

where damping c and stiffness k are assumed to be constants. If stress is a linear function of the displacement, then it is also governed by equation (a), provided that

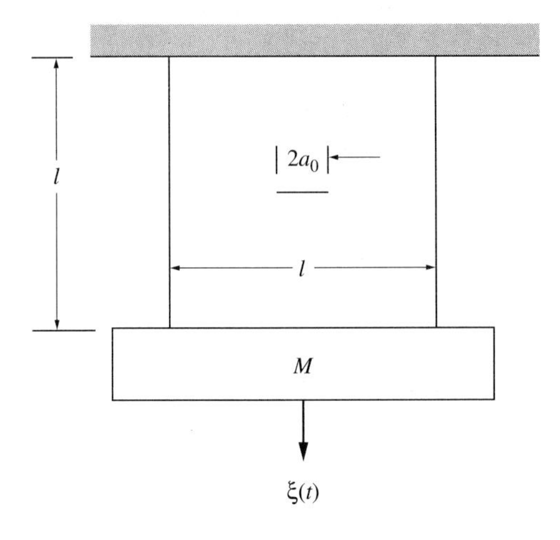

FIGURE 8.5.2
A square plate with initial central crack of length $2a_0$ subject to random load $\xi(t)$.

$\xi_1(t)$ is suitably rescaled:

$$\ddot{X} + 2\zeta\omega_0\dot{X} + \omega_0^2 X = \xi_1(t) \tag{b}$$

where X is now the stress process, $\zeta = c/(2\omega_0 M)$, $\omega_0^2 = k/M$, and $\xi_1(t)$ is another gaussian wideband process. The spectral density of the stress process is given by

$$\Phi_{XX}(\omega) = \frac{\Phi_{\xi_1\xi_1}(\omega)}{(\omega_0^2 - \omega^2)^2 + 4\zeta\omega_0^2\omega^2} \tag{c}$$

where $\Phi_{\xi_1\xi_1}(\omega)$ is the spectral density of $\xi_1(t)$. For a small ζ, which is often the case, $X(t)$ tends to a narrow-band stationary gaussian process, and the stress range process may be approximated by two times the envelope process.

The function $h(A)$ in equation (8.5.19) for the present case is assumed to be

$$h(A) = \gamma\sqrt{A} \tag{d}$$

where γ is a constant, resulting in

$$Q(A) = \alpha\gamma^\beta A^{\beta/2} \tag{e}$$

The drift and diffusion coefficients for process $Z(t)$ are then obtained as

$$m_Z = \eta_0 E[\Delta X^\beta] = \eta_0(2\sqrt{2}\sigma_X)^\beta \Gamma\left(1 + \frac{\beta}{2}\right) \tag{f}$$

$$\sigma_Z^2 = \eta_0^2(8\sigma_X^2)^\beta \sum_{n=1}^{\infty} c_n^2 \int_{-\infty}^{\infty} \rho^{2n}(\tau)\,d\tau \tag{g}$$

Given the material parameters α, β, and γ and the spectral density $\Phi_{XX}(\omega)$ of the stress process $X(t)$, we calculate σ_X^2, $\rho(t)$, and η_0 from equations (8.5.58), (8.5.59), and (8.5.65). We then obtain m_Z and σ_Z^2 from (f) and (g) and finally find $p_T(\tau)$ and μ_n from (8.5.40) and (8.5.45).

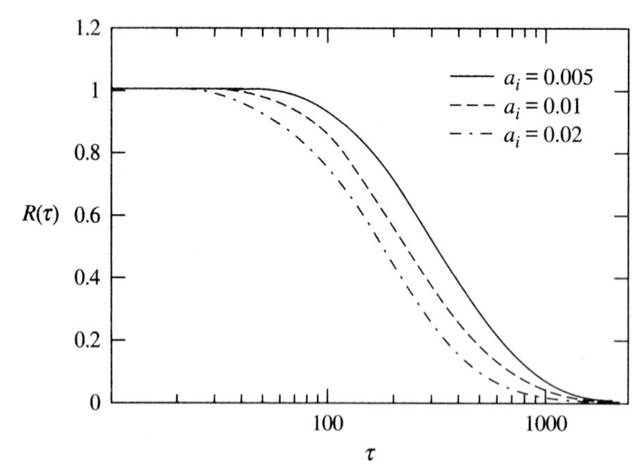

FIGURE 8.5.3
Reliability function, excitation level $K = 0.1$.

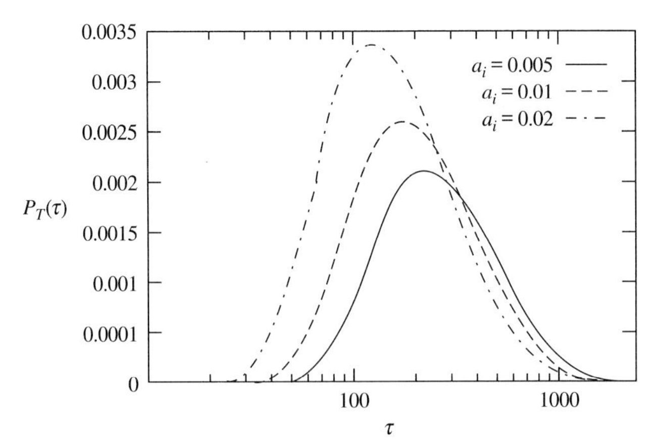

FIGURE 8.5.4
Probability density of the fatigue life, excitation level $K = 0.1$.

In numerical calculations, the parameters in (d) and (e) for the fatigue model are chosen to be $\alpha = 0.01$, $\beta = 2$, and $\gamma = 2$, and those in (c) are chosen to be $\omega_0 = 1, \zeta = 0.1$. The excitation $\xi_1(t)$ is assumed to be a white noise with a spectral density $\Phi_{\xi_1\xi_1}(\omega) = K$. The critical crack length is $a_c = 1$. Figures 8.5.3 and 8.5.4 depict the reliability and the probability density of the fatigue life with respect to time τ, for $K = 0.1$ and three different initial crack lengths, $a_i = 0.005$, 0.01, and 0.02. Figure 8.5.5 shows values of the mean μ_1 and the standard deviation σ_T of the fatigue life versus the initial crack size a_i for three different values of spectral density, $K = 0.01$, 0.02, and 0.05. It is expected that the excitation level plays a dominant role in the fatigue failure since the magnitude of the stress range process depends mainly on the excitation level. As shown in Figure 8.5.5*a*, a change in the excitation spectral level indeed leads to an enormous difference in the mean fatigue life. Figure 8.5.5*a* shows that the mean fatigue life decreases sharply in the range of small initial crack length, indicating that most of the fatigue life is spent in the early stage of crack propagation after it is initiated. The probability density depicted in Fig. 8.5.4 shows that the fatigue life has a wide range distribution, thus a large standard deviation.

8.6 CONCLUDING REMARKS

The approximation schemes discussed in Lin (1967) for the probability of first-excursion failure were based on statistical properties of structural response in the stationary state. While these approximations provided some indication of structural vulnerability in this regard, they were conceptually inconsistent with the very notion of failure. If failure does occur in the first-excursion mode, then the critical state represents an absorbing boundary, in the sense that a sample function is removed from the population of sample functions once it touches the boundary. Therefore, the total probability within the safe domain cannot be preserved, and a stationary state can never be attained. The presumption of a stationary response is equivalent to

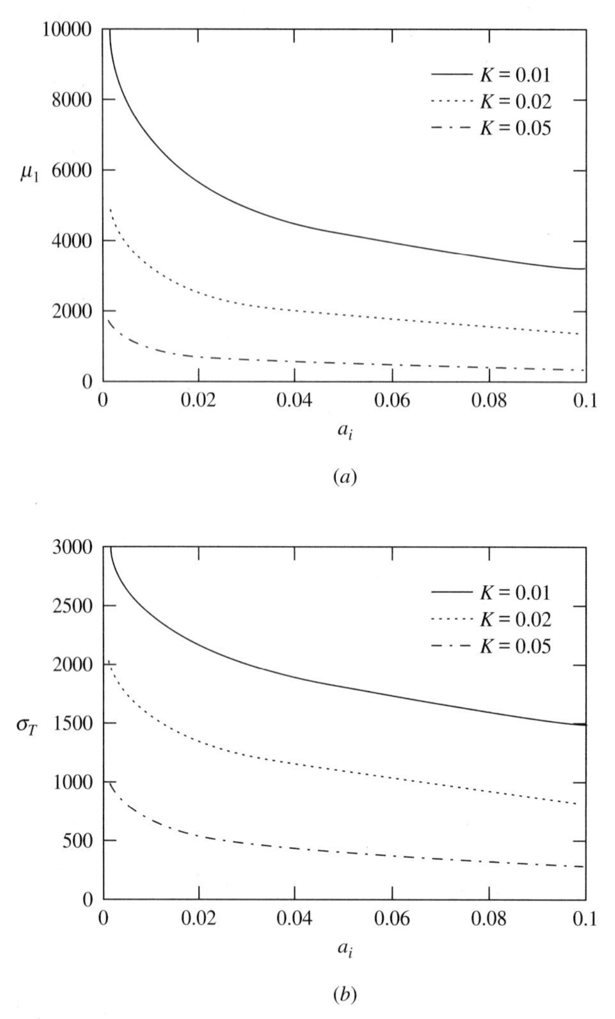

FIGURE 8.5.5
(*a*) Mean and (*b*) standard deviation of the fatigue life.

allowing a removed sample function to return to the safe domain, which is generally unconservative.

An analytically rigorous and consistent treatment of the first-excursion failure must rely again on the theory of Markov processes. In particular, if the response variable in question can be approximated as a one-dimensional Markov process, then the reliability function R is governed by the Kolmogorov backward equation, and the statistical moments of the first-excursion time are governed by the generalized Pontryagin equation. Of course, the Markov approximation must be justified in each case by comparing the correlation time of the excitations and the relaxation time of the system. The Kolmogorov backward equation and the associated initial and boundary conditions constitute an eigenvalue problem. In the context of first-excursion failure, at least one boundary must be absorbing, and the boundary at the trivial (zero) state, which is generally singular, must not be an exit, attractively natural, or strictly natural. Such an eigenvalue problem can sometimes be solved analytically, using the standard procedure of separation of variables. If it must be solved numerically, then a quantitative boundary condition must first be found at or near the trivial state. The generalized Pontryagin equation for the statistical moments is an ordinary differential equation of the recursive type. It can be solved either analytically or numerically, beginning from the first moment. A numerical solution requires again a quantitative boundary condition at or near the trivial state.

From a fracture mechanics point of view, fatigue failure of a structural or mechanical component results from the growth of a dominant crack to a critical size; thus it may also be formulated as a first-excursion problem. If the stress range in such a component is random, with a reasonably short correlation time, then the crack size is approximately a Markov process, since its change is generally much slower in time. However, determination of the drift and diffusion coefficients for the Markov model requires the knowledge of the mean and the covariance functions of the random stress range, which is not presently available for a general stress process. In the special case of a stationary gaussian narrow-band stress, the stress range is approximately twice the stress envelope process, for which the first- and second-order probability densities are known. Application of the procedure is illustrated in the example of a square plate with an initial central crack.

8.7 EXERCISES

8.1. A model for the genetics of a population is described by the following Itô stochastic differential equation:

$$dX = (1 + X)\,dt + X\,dB(t)$$

Find the statistical moments of the first-excursion time for process $X(t)$ to reach a critical state x_c.

8.2. Given a nonlinearly damped oscillator

$$\ddot{X} + 2\zeta\omega_0\dot{X} + \beta\dot{X}^3 + \omega_0^2 X[1 + W_1(t)] = W_2(t)$$

where $W_1(t)$ and $W_2(t)$ are gaussian white noises with correlation functions

$$E[W_j(t)W_k(t+\tau)] = 2\pi K_{jk}\delta(\tau) \qquad j, k = 1, 2$$

(*a*) Using the transformation

$$X = A\cos\theta \qquad \theta = \omega_0 t + \phi$$
$$\dot{X} = -A\omega_0 \sin\theta$$

and the smoothed stochastic averaging procedure, obtain an Itô equation for the amplitude process $A(t)$.

(*b*) Find the statistical moments of the first-excursion time for the amplitude process to reach a critical value a_c, in the special case of $W_1(t) = 0$ and $W_2(t) \neq 0$.

(*c*) If $W_2(t) = 0$ and $W_1(t) \neq 0$, then under what condition is the first-excursion problem meaningful? Obtain the statistical moments of the first-excursion time when such a condition is satisfied.

8.3. Consider a nonlinear oscillator under a multiplicative excitation as follows:

$$\ddot{X} + \eta|\dot{X}|^\delta \operatorname{sgn}\dot{X} + \omega_0^2 X = \dot{X}W(t) \qquad \eta > 0,\ \delta \geq 0$$

(*a*) Obtain an Itô stochastic differential equation for the amplitude process

$$A(t) = \left[X^2(t) + \frac{\dot{X}^2(t)}{\omega_0^2}\right]^{1/2}$$

using the smoothed stochastic averaging procedure.

(*b*) Determine the condition under which $a = 0$ is an admissible boundary for a first-excursion problem.

(*c*) Obtain the statistical moments of the first-excursion time for the amplitude process to reach a critical level a_c when the preceding condition is satisfied.

8.4. Consider the linear oscillator

$$\ddot{X} + \omega_0[2\zeta + W_2(t)]\dot{X} + \omega_0^2[1 + W_1(t)]X = W_3(t)$$

where $W_1(t)$, $W_2(t)$, and $W_3(t)$ are gaussian white noises and

$$E[W_j(t)W_k(t+\tau)] = 2\pi K_{jk}\delta(\tau) \qquad j, k = 1, 2$$

(*a*) Obtain an Itô stochastic differential equation for the energy process

$$\Lambda(t) = \tfrac{1}{2}\dot{X}^2(t) + \tfrac{1}{2}\omega_0^2 X^2(t)$$

using the quasi-conservative averaging procedure.

(*b*) Find the statistical moments of the first-excursion time T for the energy process $\Lambda(t)$ to reach a critical value λ_c.

(*c*) Obtain the reliability function defined by

$$R(\tau, \lambda_c, \lambda_0) = \text{Prob}[T > \tau \mid \Lambda(0) = \lambda_0]$$

(*d*) Repeat (b) and (c) for the special case in which $W_1(t) = 0$ and $W_2(t) = 0$, namely, only the additive excitation $W_3(t)$ is present.

(*e*) For the case of $W_3(t) = 0$, $W_1(t) \neq 0$, and $W_2(t) \neq 0$, determine the condition under which the first-excursion problem is meaningful, and find the statistical moments of the first-excursion time for the energy process.

(*f*) Compare the foregoing results with those obtained in Sections 8.2.1 through 8.2.3.

8.5. An oscillator which has a nonlinear spring and is subjected to a multiplicative gaussian white noise excitation is governed by

$$\ddot{X} + 2\eta\dot{X} + k|X|^{\delta}\,\text{sgn}\,X = \dot{X}W(t) \qquad k > 0, \;\; \delta \geq 0$$

(*a*) Derive an Itô stochastic differential equation for the energy process

$$\Lambda(t) = \frac{1}{2}\dot{X}^2(t) + \frac{k}{\delta+1}|X(t)|^{\delta+1}$$

using the quasi-conservative averaging procedure.

(*b*) Find the statistical moments of the first-excursion time for the energy process to reach a critical level λ_c.

CHAPTER 9

DISORDERED STRUCTURES

In Chapters 2 through 8, the excitations on a dynamical system are treated as being random, while the constitutive properties of the system are assumed to be deterministic. In reality, a physical system can never be exactly the same as the ideal mathematical model in an analysis, because measurements cannot be absolutely accurate, manufacturing processes cannot be absolutely perfect, materials cannot be absolutely homogeneous, and so on. The departure of a physical system from its ideal model is called disorder.

The need for statistical analyses of disordered systems was first advocated by physicists (e.g., Born, 1955). In the engineering literature, Brillouin (1946) and Chenea and Bogdanoff (1958) were among the earliest to document quantitative studies on the effects of variability of system parameters. A review of related works to the mid-1970s was given by Soong and Cozzarelli (1976). More recent reviews were provided by Vanmarcke et al. (1986), and by Shinozuka and Yamazaki (1988).

The problems of disordered systems may be classified according to the presumed knowledge of the random system parameters. If the joint probability distribution of the random parameters is given, then the problem often can be solved as one of probability space transformation or on the basis of total probability. If only the statistical moments of the random system parameters are given, then a perturbation approach is generally required. The majority of published works belong to the second category, including those under the general headings of stochastic eigenvalues (e.g., Collins and Thomson, 1969; Shinozuka and Astill, 1972) and stochastic finite elements (e.g., Cambou, 1975; Chang and Yang, 1991; Deotatis, 1991; Deotatis and Shinozuka, 1991; Ghanem and Spanos, 1991; Liu et al., 1986; Nakagiri and Hisada, 1981; Shinozuka and Deotatis, 1988a; Vanmarcke and Grigoriu, 1983; Zhu and Wu, 1991). In a certain sense, the analyses of wave propagation in stochastic media also belong to this category (Sobczyk, 1984). One general drawback of the perturbation approach is the difficulty with which the accuracy of the results can be estimated, especially when random dynamic loads are involved. Therefore, most published works were restricted to the statistical properties of eigenvalues and eigenvectors, or responses to static loads. In many such works, Monte Carlo simulation results were also included for comparison.

In this chapter, the case of known probability distribution of the system parameters is discussed in a general setting, with the purpose of obtaining the probability distribution or statistical properties of the response of relatively simple systems. Next, the case of known statistical properties of the system parameters, for which the perturbation procedure is suitable, is treated. Our main emphasis is on a special type of structure—disordered periodic structures—which we analyze using two approaches, the perturbation approach and the wave propagation approach.

9.1 RANDOM SYSTEM PARAMETERS WITH KNOWN PROBABILITY DISTRIBUTION

Consider an n-dimensional dynamical system governed by the equation

$$\dot{X}_j = f_j[\boldsymbol{X}(t), \boldsymbol{Y}, \boldsymbol{F}(t)] \qquad j = 1, 2, \ldots, n \tag{9.1.1}$$

subject to the initial conditions

$$\boldsymbol{X}(0) = \boldsymbol{X}_0 \tag{9.1.2}$$

where $\boldsymbol{Y}$ is an m-dimensional vector of random parameters and $\boldsymbol{F}(t)$ is a vector of excitations, either random or deterministic. In a broader sense, the initial conditions $\boldsymbol{X}_0$ may be treated also as random system parameters. We assume that the joint probability density $p_{\boldsymbol{Y}\boldsymbol{X}_0}(\boldsymbol{y}, \boldsymbol{x}_0)$ of $\boldsymbol{Y}$ and $\boldsymbol{X}_0$ is known.

Solution can now proceed in two steps. First we obtain the probability density $\tilde{p}_{\boldsymbol{X}}(\boldsymbol{x}, t \mid \boldsymbol{y}, \boldsymbol{x}_0)$ of the system response, conditional on the knowledge of system parameters and the initial conditions. Then the unconditional probability density follows as

$$p_{\boldsymbol{X}}(\boldsymbol{x}, t) = \underbrace{\int \cdots \int}_{(n+m)\text{-fold}} \tilde{p}_{\boldsymbol{X}}(\boldsymbol{x}, t \mid \boldsymbol{y}, \boldsymbol{x}_0) p_{\boldsymbol{Y}\boldsymbol{X}_0}(\boldsymbol{y}, \boldsymbol{x}_0)\, d\boldsymbol{y}\, d\boldsymbol{x}_0 \tag{9.1.3}$$

Of course, the method is applicable only if $\tilde{p}_{\boldsymbol{X}}(\boldsymbol{x}, t \mid \boldsymbol{y}, \boldsymbol{x}_0)$ is obtainable. If only certain conditional statistics, say, $E\{g[\boldsymbol{X}(t)] \mid \boldsymbol{y}, \boldsymbol{x}_0\}$, are obtainable, then equation (9.1.3) is replaced by

$$E\{g[\boldsymbol{X}(t)]\} = \underbrace{\int \cdots \int}_{(n+m)\text{-fold}} E\{g[\boldsymbol{X}(t)] \mid \boldsymbol{y}, \boldsymbol{x}_0\} p_{\boldsymbol{Y}\boldsymbol{X}_0}(\boldsymbol{y}, \boldsymbol{x}_0)\, d\boldsymbol{y}\, d\boldsymbol{x}_0 \tag{9.1.4}$$

The application of equation (9.1.4) is straightforward. For example, consider a simple linear oscillator under the excitation of an additive gaussian white noise, governed by

$$\ddot{X} + 2\zeta\omega_0\dot{X} + \omega_0^2 X = W(t) \tag{9.1.5}$$

The variance of $X(t)$ at the stationary state, conditional on the knowledge of system parameters ζ and ω_0, is given by

$$\tilde{\sigma}_X^2(\zeta, \omega_0) = E[X^2(t) \mid \zeta, \omega_0] = \frac{\pi K}{2\zeta\omega_0^3} \tag{9.1.6}$$

where K is the spectral density of $W(t)$. Given the joint probability density $p(\zeta, \omega_0)$ of the random parameter ζ and ω_0, the unconditional mean-square response follows simply as

$$\sigma_X^2 = E[X^2(t)] = \int\int \frac{\pi K}{2\zeta\omega_0^3} p(\zeta, \omega_0)\, d\zeta\, d\omega_0 \tag{9.1.7}$$

In particular, let ζ and ω_0 be independent and uniformly distributed in (ζ_1, ζ_2) and (ω_1, ω_2), respectively. Then the right-hand side of (9.1.7) is integrated to yield

$$\sigma_X^2 = \frac{\pi K(\omega_1 + \omega_2)}{4(\zeta_2 - \zeta_1)\omega_1^2\omega_2^2} \ln\left(\frac{\zeta_2}{\zeta_1}\right) \tag{9.1.8}$$

The application of equation (9.1.3) is also conceptually simple, but closed-form solutions are generally difficult to obtain. For example, the conditional stationary probability density of $X(t)$ governed by equation (9.1.5) is the well-known

$$\tilde{p}_X(x \mid \zeta, \omega_0) = \frac{1}{\sqrt{2\pi}\tilde{\sigma}_X(\zeta, \omega_0)} \exp\left[-\frac{x^2}{2\tilde{\sigma}_X^2(\zeta, \omega_0)}\right] \qquad -\infty < x < \infty \tag{9.1.9}$$

where $\tilde{\sigma}_X^2(\zeta, \omega_0)$ is the conditional variance given in equation (9.1.6). Computation of the unconditional probability density according to equation (9.1.3) must be carried out numerically even when the probability distributions of ζ and ω_0 are quite simple. It is also clear that the result obtained by substituting the unconditional variance σ_X^2 given in equation (9.1.8) for the conditional variance $\tilde{\sigma}_X^2(\zeta, \omega_0)$ in equation (9.1.9) can be very much in error.

When the excitation $\boldsymbol{F}(t)$ in equation (9.1.1) is deterministic, the conditional probability density $\tilde{p}_X(\boldsymbol{x}, t \mid \boldsymbol{y})$ of the system response $\boldsymbol{X}(t)$ is governed by the Liouville equation

$$\frac{\partial \tilde{p}}{\partial t} + \frac{\partial}{\partial x_j}(f_j \tilde{p}) = 0 \tag{9.1.10}$$

or

$$\frac{\partial \tilde{p}}{\partial t} + \tilde{p}\frac{\partial}{\partial x_j} f_j + f_j \frac{\partial}{\partial x_j}\tilde{p} = 0 \tag{9.1.11}$$

where f_j are the drift coefficients obtained from equation (9.1.1) by replacing $\boldsymbol{X}(t)$ and $\boldsymbol{Y}$ by their state variables:

$$f_j = f_j[\boldsymbol{x}, \boldsymbol{y}, \boldsymbol{F}(t)] \tag{9.1.12}$$

Equation (9.1.10) is solved subject to the random initial conditions

$$[\tilde{p}_X(\boldsymbol{x}, t \mid \boldsymbol{y})]_{t=0} = \tilde{p}_{X_0}(\boldsymbol{x}_0 \mid \boldsymbol{y}) \tag{9.1.13}$$

The solution for (9.1.10) may be constructed from the solution of the following $n+1$ subsidiary equations:

$$\frac{dt}{1} = -\frac{d\tilde{p}}{\tilde{p}\dfrac{\partial}{\partial x_j} f_j} = \frac{dx_1}{f_1} = \frac{dx_2}{f_2} = \cdots = \frac{dx_n}{f_n} \tag{9.1.14}$$

The n subsidiary equations

$$\frac{dt}{1} = \frac{dx_j}{f_j} \qquad j = 1, 2, \ldots, n \tag{9.1.15}$$

are exactly the deterministic counterparts of equations (9.1.1),

$$\dot{x}_j = f_j[\boldsymbol{x}, \boldsymbol{y}, \boldsymbol{F}(t)] \tag{9.1.16}$$

subject to initial conditions $\boldsymbol{x}(0) = \boldsymbol{x}_0 = (x_{10}, x_{20}, \ldots, x_{n0})$. We assume that equation (9.1.16) can be solved to yield

$$x_j = \eta_j[\boldsymbol{y}, \boldsymbol{x}_0, \boldsymbol{F}(t)] \tag{9.1.17}$$

The remaining subsidiary equation

$$\frac{dt}{1} = -\frac{d\tilde{p}}{\tilde{p}\dfrac{\partial}{\partial x_j} f_j} \tag{9.1.18}$$

has a general solution

$$\tilde{p} = C \exp\left[-\int_0^t \frac{\partial}{\partial x_j} f_j(\tau)\, d\tau\right] \tag{9.1.19}$$

The general solution to (9.1.10) now can be written as

$$\tilde{p} = \psi[\eta_1, \eta_2, \ldots, \eta_n] \exp\left[-\int_0^t \frac{\partial}{\partial x_j} f_j(\tau)\, d\tau\right] \tag{9.1.20}$$

where ψ represents an arbitrary functional form. Imposing the initial condition (9.1.13),

$$\tilde{p}_X(\boldsymbol{x}, t \mid \boldsymbol{y}) = \left\{ \tilde{p}_{X_0}(\boldsymbol{x}_0 \mid \boldsymbol{y}) \exp\left[-\int_0^t \frac{\partial}{\partial x_j} f_j(\tau)\, d\tau\right]\right\}_{\boldsymbol{x}_0 = \eta^{-1}(\boldsymbol{x})} \tag{9.1.21}$$

In equation (9.1.21), $\boldsymbol{x}_0$ should be expressed in terms of $\boldsymbol{x}$ by inverting (9.1.17), as indicated. Finally, the probability density of $\boldsymbol{X}(t)$ is obtained as

$$\begin{aligned}
\tilde{p}_X(\boldsymbol{x}, t) &= \underbrace{\int \cdots \int}_{m\text{-fold}} \left\{ \tilde{p}_{X_0}(\boldsymbol{x}_0 \mid \boldsymbol{y}) \exp\left[-\int_0^t \frac{\partial}{\partial x_j} f_j(\tau)\, d\tau\right]\right\}_{\boldsymbol{x}_0 = \eta^{-1}(\boldsymbol{x})} p_Y(\boldsymbol{y})\, dy_1\, dy_2 \cdots dy_m \\
&= \underbrace{\int \cdots \int}_{m\text{-fold}} \left\{ p_{X_0 Y}(\boldsymbol{x}_0 \mid \boldsymbol{y}) \exp\left[-\int_0^t \frac{\partial}{\partial x_j} f_j(\tau)\, d\tau\right]\right\}_{\boldsymbol{x}_0 = \eta^{-1}(\boldsymbol{x})} dy_1\, dy_2 \cdots dy_m
\end{aligned} \tag{9.1.22}$$

where $p_{X_0 Y}(\boldsymbol{x}_0, \boldsymbol{y})$ is the given joint probability density of vector $\boldsymbol{X}_0$ of random initial values and vector $\boldsymbol{Y}$ of random system parameters.

It is of interest to note that equation (9.1.22) may be viewed as mapping of one probability space to another, namely, from an $(n+m)$-dimensional space of $\boldsymbol{x}_0$ and $\boldsymbol{y}$ to an n-dimensional space of $\boldsymbol{x}$. Making use of Eq. 2-50 in Lin (1967), we may write

$$p_X(\boldsymbol{x}, t) = \underbrace{\int \cdots \int}_{m\text{-fold}} \left\{ p_{X_0 Y}(\boldsymbol{x}_0, \boldsymbol{y}) \left| J_n \right| \right\}_{\boldsymbol{x}_0 = \eta^{-1}(\boldsymbol{x})} dy_1 dy_2 \cdots dy_m \tag{9.1.23}$$

where J_n is the jacobian of transformation, constructed as follows:

$$J_n = \left| \frac{\partial x_{j0}}{\partial x_k} \right| \tag{9.1.24}$$

Comparing equations (9.1.22) and (9.1.23), we find that

$$\left| J_n \right| = \exp\left[-\int_0^t \frac{\partial}{\partial x_j} f_j(\tau)\, d\tau \right] \tag{9.1.25}$$

This relationship was first obtained by Soong (1973).

As an example, consider an undamped linear oscillator under a deterministic sinusoidal excitation

$$\ddot{X} + \Omega_0^2 X = A \sin \omega t \qquad \omega \neq \Omega_0 \tag{9.1.26}$$

Letting $X = X_1$ and $\dot{X} = X_2$, we can replace equation (9.1.26) by

$$\frac{d}{dt} \begin{Bmatrix} X_1 \\ X_2 \end{Bmatrix} = \begin{bmatrix} 0 & 1 \\ -\Omega_0^2 & 0 \end{bmatrix} \begin{Bmatrix} X_1 \\ X_2 \end{Bmatrix} + \begin{Bmatrix} 0 \\ A \sin \omega t \end{Bmatrix} \tag{9.1.27}$$

Solution to equation (9.1.27), subject to the initial conditions $X_1(0) = X_0$ and $X_2(0) = V_0$, is found to be

$$\begin{Bmatrix} X_1 \\ X_2 \end{Bmatrix} = \begin{bmatrix} \cos \Omega_0 t & \dfrac{\sin \Omega_0 t}{\Omega_0} \\ -\Omega_0 \sin \Omega_0 t & \cos \Omega_0 t \end{bmatrix} \begin{Bmatrix} X_0 \\ V_0 \end{Bmatrix} + \frac{A}{\Omega_0^2 - \omega^2} \begin{Bmatrix} \sin \omega t - \dfrac{\omega}{\Omega_0} \sin \Omega_0 t \\ \omega(\cos \omega t - \cos \Omega_0 t) \end{Bmatrix} \tag{9.1.28}$$

Equation (9.1.28) can be easily converted to yield

$$\begin{Bmatrix} X_0 \\ V_0 \end{Bmatrix} = \begin{bmatrix} \cos \Omega_0 t & \dfrac{-\sin \Omega_0 t}{\Omega_0} \\ \Omega_0 \sin \Omega_0 t & \cos \Omega_0 t \end{bmatrix} \begin{Bmatrix} X_1 \\ X_2 \end{Bmatrix} - \frac{A}{\Omega_0^2 - \omega^2} \begin{Bmatrix} \cos \Omega_0 t \sin \omega t - \dfrac{\omega}{\Omega_0} \sin \Omega_0 t \cos \omega t \\ \Omega_0 \sin \Omega_0 t \sin \omega t + \omega \cos \Omega_0 t \cos \omega t - \omega \end{Bmatrix} \tag{9.1.29}$$

The jacobian of transformation from (X_0, V_0) to (X_1, X_2), or vice versa, is clearly equal to 1. Given a joint probability density of the random initial conditions X_0 and V_0 and the random system parameter Ω_0, the probability density for X_1 and X_2 may be obtained from equation (9.1.23).

Specifically, let us assume that the initial conditions are known to be $X_0 = V_0 = 0$, and that the random system parameter Ω_0 has a discrete distribution,

$$p_{X_0V_0\Omega_0}(x_0, v_0, \omega_0) = \delta(x_0)\,\delta(v_0)\sum_{j=i}^{n} p_j\,\delta(\omega_0 - \omega_j) \tag{9.1.30}$$

where

$$\sum_{j=1}^{n} p_j = 1 \tag{9.1.31}$$

Then from (9.1.23)

$$\begin{aligned} &p_{X_1X_2}(x_1, x_2, t) \\ &= \sum_{j=1}^{n} p_j\delta\left[x_1 \cos\omega_j t - x_2\frac{\sin\omega_j t}{\omega_j} - \frac{A}{\omega_j^2 - \omega^2}\right. \\ &\times\left.\left(\cos\omega_j t \sin\omega_j t - \frac{\omega}{\omega_j}\sin\omega_j t\cos\omega t\right)\right] \\ &\times\delta\left[x_1\omega_j \sin\omega_j t + x_2\cos\omega_j t - \frac{A}{\omega_j^2 - \omega^2}(\omega_j\sin\omega_j t\sin\omega t + \omega\cos\omega_j t\cos\omega t - \omega)\right] \end{aligned} \tag{9.1.32}$$

Equation (9.1.31) may be used to compute the statistics of $X_1(t)$ and $X_2(t)$. By letting $\omega = 0$, we obtain the results for the free vibration of a linear oscillator, considered earlier by Kozin (1961).

9.2 FREE VIBRATION OF A DISORDERED PERIODIC BEAM (Lin and Yang, 1974)

As an example of the perturbation procedure, consider the free vibration of an N-span beam on $N + 1$ "simple" supports, as shown in Fig. 9.2.1. The beam is designed to be spatially periodic from one span to the next. Due to variabilities of material and manufacturing processes, disorder in periodicity is expected.

Supposing that the beam shown in Fig. 9.2.1 is undergoing an undamped free vibration of a natural frequency ω, the state of motion at an arbitrary location on the

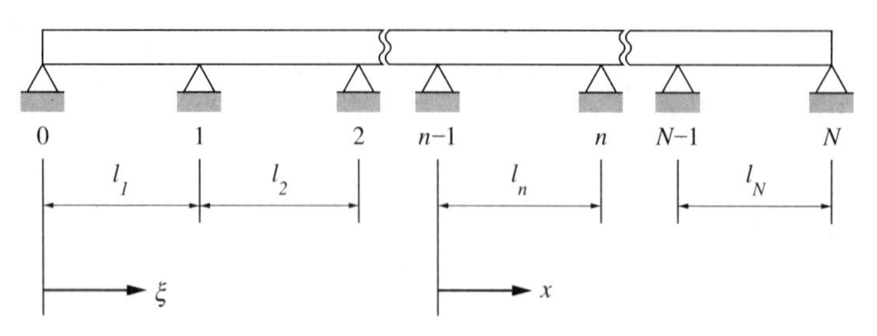

FIGURE 9.2.1
An N-span beam on simple supports.

beam is described by $\boldsymbol{Z}\exp(i\omega t) = \{w, \theta, M, V\}'\exp(i\omega t)$ where w, θ, M, and V are the complex amplitudes of the deflection, slope, bending moment, and transverse shear, respectively. Let us focus our attention on two neighboring support stations, say $n-1$ and n, and write

$$\boldsymbol{Z}_n = \boldsymbol{F}_n \boldsymbol{Z}_{n-1} \tag{9.2.1}$$

where $\boldsymbol{F}_n$ is a transfer matrix, which may be computed from the material and geometrical constants of the structural element, connecting stations $n-1$ and n. It can be shown that $\boldsymbol{F}_n$ is given by (e.g., Pestel and Leckie, 1963)

$$\boldsymbol{F}_n = \begin{bmatrix} (C_0)_n & (S_1)_n & \dfrac{(C_2)_n}{(EI)_n} & \dfrac{(S_3)_n}{(EI)_n} \\ \sigma_n^4 (S_3)_n & (C_0)_n & \dfrac{(S_1)_n}{(EI)_n} & \dfrac{(C_2)_n}{(EI)_n} \\ \sigma_n^4 (EI)_n (C_2)_n & \sigma_n^4 (EI)_n (S_3)_n & (C_0)_n & (S_1)_n \\ \sigma_n^4 (EI)_n (S_1)_n & \sigma_n^4 (EI)_n (C_2)_n & \sigma_n^4 (S_3)_n & (C_0)_n \end{bmatrix} \tag{9.2.2}$$

where $(EI)_n$ = bending rigidity of the beam between supports $n-1$ and n, $\sigma_n = [\omega^2 m_n/(EI)_n]^{1/4}$, m_n = mass of beam per unit length between supports $n-1$ and n, l_n = distance between supports $n-1$ and n, and where

$$\begin{aligned} (C_0)_n &= \frac{1}{2}(\cosh\sigma_n l_n + \cos\sigma_n l_n) & (C_2)_n &= \frac{1}{2\sigma_n^2}(\cosh\sigma_n l_n - \cos\sigma_n l_n) \\ (S_1)_n &= \frac{1}{2\sigma_n}(\sinh\sigma_n l_n + \sin\sigma_n l_n) & (S_3)_n &= \frac{1}{2\sigma_n^3}(\sinh\sigma_n l_n - \sin\sigma_n l_n) \end{aligned} \tag{9.2.3}$$

For the present special case, the deflections are zero at the supports. We then have, from the first row of equation (9.2.1),

$$V_{n-1} = -\frac{(f_{12})_n \theta_{n-1} + (f_{13})_n M_{n-1}}{(f_{14})_n} \tag{9.2.4}$$

where $(f_{ij})_n$ are the elements of $\boldsymbol{F}_n$ on the ith row and the jth column. We see that by restricting $w = 0$ at the supports, the other three components of the state vector $\boldsymbol{Z}$ are no longer independent. Therefore, equation (9.2.1) can be simplified to involve just the slopes and the bending moments, θ and M,

$$\begin{Bmatrix} \theta \\ M \end{Bmatrix}_n = \boldsymbol{T}_n \begin{Bmatrix} \theta \\ M \end{Bmatrix}_{n-1} \tag{9.2.5}$$

that is, only two-dimensional state vectors are required to describe the motion of the beam. A simple calculation will show that the elements of $\boldsymbol{T}_n$ are given by

$$\begin{aligned} (t_{11})_n = (t_{22})_n &= \frac{\sinh\sigma_n l_n \cos\sigma_n l_n - \cosh\sigma_n l_n \sin\sigma_n l_n}{\sinh\sigma_n l_n - \sin\sigma_n l_n} \\ (t_{12})_n &= \frac{\cosh\sigma_n l_n \cos\sigma_n l_n - 1}{(EI)_n \sigma_n (\sinh\sigma_n l_n - \sin\sigma_n l_n)} \\ (t_{21})_n &= \frac{2(EI)_n \sigma_n \sinh\sigma_n l_n \sin\sigma_n l_n}{\sinh\sigma_n l_n - \sin\sigma_n l_n} \end{aligned} \tag{9.2.6}$$

The matrices $\boldsymbol{F}_n$ and $\boldsymbol{T}_n$ were introduced in Section 3.3.2 when modeling a surface-mounted pipeline. To compare, we equate ω with kc and note the additional subscript n, which identifies a particular span n.

AN IDEAL PERIODIC BEAM. To provide the required reference for the subsequent perturbation analysis, a brief discussion of an ideal periodic beam without disorder is presented. For such a beam, all the spans are identically constructed, that is, $l_n = l$, $(EI)_n = EI$, $m_n = m$, and so on, for all n; therefore, the subscript n associated with σ, C_0, C_2, S_1, S_3, $\boldsymbol{F}$, f_{ij}, $\boldsymbol{T}$, and t_{ij} in equations (9.2.1) through (9.2.6) can be removed. The relation between the state vectors at the two ends of the beam can be obtained by repeated application of equation (9.2.5), resulting in

$$\begin{Bmatrix} \theta \\ M \end{Bmatrix}_N = \boldsymbol{T}^N \begin{Bmatrix} \theta \\ M \end{Bmatrix}_0 \tag{9.2.7}$$

The computation of $\boldsymbol{T}^N$ can be simplified by recalling that the eigenvalues of the transfer matrix $\boldsymbol{T}$ are a reciprocal pair. Denoting these eigenvalues by $\exp(\pm i\beta)$, it can be shown that (Lin and McDaniel, 1969)

$$\boldsymbol{T}^N = \frac{\cos N\beta}{\cos\beta}\begin{bmatrix} t_{11} & 0 \\ 0 & t_{22} \end{bmatrix} + \frac{\sin N\beta}{\sin\beta}\begin{bmatrix} 0 & t_{12} \\ t_{21} & 0 \end{bmatrix} \tag{9.2.8}$$

where β can be computed from

$$\cos\beta = t_{11} = t_{22} = \frac{\sinh\sigma l\cos\sigma l - \cosh\sigma l\sin\sigma l}{\sinh\sigma l - \sin\sigma l} \tag{9.2.9}$$

We now impose the boundary conditions $M_N = M_0 = 0$ in equation (9.2.7), and from the second row of this matrix equation we obtain

$$\frac{t_{21}\sin N\beta}{\sin\beta} = 0 \tag{9.2.10}$$

Since $|\boldsymbol{T}| = t_{11}t_{22} - t_{12}t_{21} = \cos^2\beta - t_{12}t_{21} = 1$,

$$t_{21} = -\frac{\sin^2\beta}{t_{12}} \tag{9.2.11}$$

Equations (9.2.10) and (9.2.11) may be combined to obtain

$$\sin\beta\sin N\beta = 0 \tag{9.2.12}$$

provided $t_{12} \neq 0$. That t_{12} is indeed nonzero will be verified later. Equation (9.2.12) is the frequency equation derived independently by Krein (1933) and Miles (1956).

The roots of equation (9.2.12) may be expressed as

$$\beta_{k0} = \pi\left(1 + \frac{k-1}{N}\right) \qquad k = 1, 2, \ldots \tag{9.2.13}$$

The corresponding σ_{k0} can be obtained through equation (9.2.9), and finally the corresponding ω_{k0} can be obtained through

$$\omega = \sigma^2\left(\frac{EI}{m}\right)^{1/2} \tag{9.2.14}$$

It can be shown (Miles, 1956) that the natural frequencies ω_{k0} are clustered in groups, with exactly N frequencies in each group. The first (lowest) frequency in the sth group, where $s = 1, 2, \ldots,$ corresponds to $k = N(s-1)+1$. For such a k-value

$$\beta_{k0} = s\pi \tag{9.2.15a}$$

$$\sigma_{k0} = \frac{s\pi}{l} \tag{9.2.15b}$$

Therefore, these first natural frequencies in the individual groups are the same as those of a single-span simply supported beam.

The condition $t_{12} \neq 0$ will now be verified. For those roots corresponding to $\sin N\beta = 0$ but $\sin\beta \neq 0$, we have

$$\sin^2\beta = -t_{12}t_{21} \neq 0 \tag{9.2.16}$$

which guarantees that both t_{12} and t_{21} are nonzero. For those roots corresponding to $\sin\beta = 0$, namely, $\beta = s\pi$, we find that $t_{21} = 0$, but

$$t_{12} = \frac{(-1)^s \cosh s\pi - 1}{EI\sigma \sinh s\pi} \neq 0 \tag{9.2.17}$$

Thus the roots given in (9.2.13) are all valid.

RANDOM BENDING STIFFNESS AND SPAN LENGTHS. Refer again to Fig. 9.2.1 and let the bending stiffness of the beam at the global coordinate ξ be $EI + \varepsilon\alpha(\xi)$ where EI is the ideal (design) value of the beam stiffness, $\alpha(\xi)$ is a random process with zero mean, and $|\varepsilon| \ll 1$. Since the bending stiffness varies along the beam, the transfer relationship, equation (9.2.1), no longer holds. To derive an expression replacing equation (9.2.1), we divide the nth span into M equal subintervals of length Δx which are small enough so that the bending stiffness in each subinterval may be treated as a constant. Then write

$$\boldsymbol{Z}_n = \prod_{j=1}^{M} \boldsymbol{F}_{nj}(\Delta x)\boldsymbol{Z}_{n-1} \tag{9.2.18}$$

where each $\boldsymbol{F}_{nj}(\Delta x)$ is constructed in the same way as in equation (9.2.2), but using the bending stiffness of that subinterval and replacing l_n by Δx. Expand $\boldsymbol{F}_{nj}(\Delta x)$ in a Taylor series about $\varepsilon = 0$,

$$\boldsymbol{F}_{nj}(\Delta x) = \boldsymbol{F}_n(\Delta x) + \varepsilon\alpha(\xi_n)\frac{\partial}{\partial(EI)}\boldsymbol{F}_n(\Delta x) + \cdots \tag{9.2.19}$$

where ξ_n represents the global coordinate in the nth span and $\boldsymbol{F}_n(\Delta x) = [\boldsymbol{F}_{nj}(\Delta x)]_{\varepsilon=0}$, that is, $\boldsymbol{F}_n(\Delta x)$ is computed in the same way as equation (9.2.2), but using a constant EI as the bending stiffness and replacing l_n by the augment of $\boldsymbol{F}_n(\cdot)$. This notation is implicit for the remainder of the section. Substitution of equation (9.2.19) into equation (9.2.18) yields, to an accuracy of order ε,

$$\boldsymbol{Z}_n = \left[\boldsymbol{F}_n^M(\Delta x) + \varepsilon\sum_{j=1}^{M}\alpha(\xi_n)\boldsymbol{F}_n^{M-j}(\Delta x)\frac{\partial \boldsymbol{F}_n(\Delta x)}{\partial(EI)}\boldsymbol{F}_n^{j-1}(\Delta x)\right]\boldsymbol{Z}_{n-1} \tag{9.2.20}$$

From the physical meaning of matrix $\boldsymbol{F}_n(\cdot)$, it is clear that $\boldsymbol{F}_n^m(\Delta x) = \boldsymbol{F}_n(m\Delta x)$, m = integer. Then in the limit, as $\Delta x \to 0$ accompanied by $M \to \infty$, the summation in equation (9.2.20) tends to an integral, and $[\partial/\partial(EI)]\boldsymbol{F}_n(\Delta x) \to \boldsymbol{A}\,dx$, that is,

$$\boldsymbol{Z}_n = \left[\boldsymbol{F}_n(l_n) + \varepsilon \int_0^{l_n} \alpha(\xi_n)\boldsymbol{F}_n(l_n - x)\boldsymbol{A}\boldsymbol{F}_n(x)\,dx\right]\boldsymbol{Z}_{n-1} \tag{9.2.21}$$

where we have made the substitutions $M\Delta x = l_n$, $(M - j)\Delta x = l_n - x$ and $(j-1)\Delta x = x$. From equation (9.2.2), it can be found that all elements in matrix $\boldsymbol{A}$ are zero, except the (2, 3) element $a_{23} = -1/(EI)^2$.

Denote the integral in equation (9.2.21) by $\boldsymbol{Q}_n(l_n)$:

$$\boldsymbol{Q}_n(l_n) = \int_0^{l_n} \alpha(\xi)\boldsymbol{F}_n(l_n - x)\boldsymbol{A}\boldsymbol{F}_n(x)\,dx \tag{9.2.22}$$

The components of $\boldsymbol{Q}_n(l_n)$ are given by

$$(q_{ij})_n = -\frac{1}{(EI)^2}\int_0^{l_n} \alpha(\xi) f_{i2,n}(l_n - x) f_{3j,n}(x)\,dx \tag{9.2.23}$$

We now impose $w_{n-1} = w_n = 0$ in equation (9.2.21) to obtain

$$V_{n-1} = -\frac{[(f_{12})_n + \varepsilon(q_{12})_n]\theta_{n-1} + [(f_{13})_n + \varepsilon(q_{13})_n]M_{n-1}}{(f_{14})_n + \varepsilon(q_{14})_n} \tag{9.2.24}$$

Again, equation (9.2.24) shows that only two-dimensional state vectors are required to describe the motion of the beam. Thus equation (9.2.21) may be replaced by

$$\begin{Bmatrix} \theta \\ M \end{Bmatrix}_n = [\boldsymbol{T}_n(l_n) + \varepsilon \boldsymbol{Y}_n(l_n)] \begin{Bmatrix} \theta \\ M \end{Bmatrix}_{n-1} \tag{9.2.25}$$

where the elements of $\boldsymbol{T}_n(l_n)$ are computed as in equation (9.2.6), except that $(EI)_n$ is replaced by EI, and the elements of $\boldsymbol{Y}_n(l_n)$ are given by

$$(y_{ij})_n = (q_{i+1,j+1})_n - \frac{(f_{1,j+1})_n(q_{i+1,4})_n + (f_{i+1,4})_n(q_{1,j+1})_n}{(f_{14})_n} + \frac{(f_{1,j+1})_n(f_{i+1,4})_n(q_{14})_n}{(f_{14})_n^2} \tag{9.2.26}$$

At this point we are ready to consider the additional complexity due to the uncertainty in span lengths l_n. Let

$$l_n = l + \varepsilon X_n \qquad n = 1, 2, \ldots, N \tag{9.2.27}$$

where l is the design span length and X_n are random variables of zero mean. The matrices $\boldsymbol{T}_n(l_n)$ and $\boldsymbol{Y}_n(l_n)$ in equation (9.2.25) can be expanded into

$$\boldsymbol{T}_n(l_n) = \boldsymbol{T} + \varepsilon X_n \frac{d\boldsymbol{T}}{dl} + \cdots \tag{9.2.28a}$$

$$\boldsymbol{Y}_n(l_n) = \boldsymbol{Y}_n(l) + \varepsilon X_n \frac{d\boldsymbol{Y}_n(l)}{dl} + \cdots \tag{9.2.28b}$$

where matrices $\boldsymbol{T}$ and $\boldsymbol{Y}_n(l)$ are constructed according to (9.2.6) and (9.2.26), respectively, but l_n is replaced by l. However, the subscript n for matrix $\boldsymbol{Y}_n(l)$ cannot be dropped, since the random variation in the bending stiffness in the nth span must be included in its computation. Then, to the first order in ε, equation (9.2.25) is approximated as follows:

$$\begin{Bmatrix} \theta \\ M \end{Bmatrix}_n = \left[\boldsymbol{T} + \varepsilon X_n \frac{d\boldsymbol{T}}{dl} + \varepsilon \boldsymbol{Y}_n(l) \right] \begin{Bmatrix} \theta \\ M \end{Bmatrix}_{n-1} \tag{9.2.29}$$

The transfer relationship between the state vectors at the two ends of the beam is now

$$\begin{aligned} \begin{Bmatrix} \theta \\ M \end{Bmatrix}_N &= \prod_{n=1}^{N} \left[\boldsymbol{T} + \varepsilon X_n \frac{d\boldsymbol{T}}{dl} + \varepsilon \boldsymbol{Y}_n(l) \right] \begin{Bmatrix} \theta \\ M \end{Bmatrix}_0 \\ &= \left[\boldsymbol{T}^N + \varepsilon \sum_{j=1}^{N} X_j \boldsymbol{T}^{N-j} \frac{d\boldsymbol{T}}{dl} \boldsymbol{T}^{j-1} + \varepsilon \sum_{j=1}^{N} \boldsymbol{T}^{N-j} \boldsymbol{Y}_j(l) \boldsymbol{T}^{j-1} \right] \begin{Bmatrix} \theta \\ M \end{Bmatrix}_0 \end{aligned} \tag{9.2.30}$$

Imposing the boundary conditions $M_N = M_0 = 0$, we obtain from the second row of equation (9.2.30),

$$\gamma_{21}(\omega, N) + \varepsilon \sum_{j=1}^{N} X_j \eta_{21}(\omega, N, j) + \varepsilon \sum_{j=1}^{N} \zeta_{21}(\omega, N, j) = 0 \tag{9.2.31}$$

where γ_{21}, η_{21}, and ζ_{21} are the (2,1) elements of matrices $\boldsymbol{T}^N$, $\boldsymbol{T}^{N-j}(d\boldsymbol{T}/dl)\boldsymbol{T}^{j-1}$, and $\boldsymbol{T}^{N-j}\boldsymbol{Y}_j(l)\boldsymbol{T}^{j-1}$, respectively.

Let ω_k be the kth root of equation (9.2.31) and let

$$\omega_k = \omega_{k0} + \varepsilon \omega_{k1} \tag{9.2.32}$$

Then

$$\gamma_{21}(\omega_k, N) = \gamma_{21}(\omega_{k0}, N) + \varepsilon \omega_{k1} \frac{\partial \gamma_{21}(\omega_{k0}, N)}{\partial \omega_{k0}} \tag{9.2.33a}$$

$$\eta_{21}(\omega_k, N, j) = \eta_{21}(\omega_{k0}, N, j) + \varepsilon \omega_{k1} \frac{\partial \eta_{21}(\omega_{k0}, N, j)}{\partial \omega_{k0}} \tag{9.2.33b}$$

$$\zeta_{21}(\omega_k, N, j) = \zeta_{21}(\omega_{k0}, N, j) + \varepsilon \omega_{k1} \frac{\partial \zeta_{21}(\omega_{k0}, N, j)}{\partial \omega_{k0}} \tag{9.2.33c}$$

Substituting equations (9.2.32) and (9.2.33a–c) into equation (9.2.31), we obtain the equations for ω_{k0} and ω_{k1}:

$$\gamma_{21}(\omega_{k0}, N) = \frac{t_{21} \sin N\beta_{k0}}{\sin \beta_{k0}} = 0 \tag{9.2.34}$$

$$\omega_{k1} = \frac{-\sum_{j=1}^{N} [X_j \eta_{21}(\omega_{k0}, N, j) + \zeta_{21}(\omega_{k0}, N, j)]}{\partial \gamma_{21}(\omega_{k0}, N)/\partial \omega_{k0}} \tag{9.2.35}$$

As expected, equation (9.2.34) is identical to equation (9.2.10) for a perfectly periodic beam, and the relationship between ω_{k0} and β_{k0} (indirectly through σ_{k0}) here is the same as that of a perfectly periodic beam. Furthermore, within the scope of a first-order perturbation theory (accurate to the first order in ε), the frequency perturbation $\varepsilon\omega_{k1}$ depends linearly on the effect of span length variation (the $X_j\eta_{21}$ terms) and the effect of stiffness variation (the ζ_{21} terms). In most practical situations it is reasonable to assume that the two types of variation are statistically independent. Then each type may be considered separately and the resultant variance of the frequency perturbation can be obtained as the sum of two variances computed on the basis of one type of variation being present at a time.

RANDOM BENDING STIFFNESS ALONE. Letting $X_j = 0$ in equation (9.2.35), we obtain

$$\omega_1 = -\frac{\sum_{j=1}^{N} \zeta_{21}(\omega_0, N, j)}{\partial\gamma_{21}(\omega_0, N)/\partial\omega_0} \tag{9.2.36}$$

where we have dropped the subscript k for simplicity. This omission is understood unless stated otherwise. We recall that ω_0 corresponds to a natural frequency for the perfectly periodic beam, and it is determined indirectly by use of equation (9.2.34) for β_0, equation (9.2.9) for σ_0, and equation (9.2.14) for ω_0.

It is convenient to consider separately the case of $\beta_0 = s\pi$ (hence $\sigma_0 = s\pi/l$) and the case of $\beta_0 \neq s\pi$. The former corresponds to the lowest natural frequencies in the individual frequency groups, which are the same as the natural frequencies of a single-span beam. It can be shown that the denominator in equation (9.2.36) is given by

$$\frac{\partial\gamma_{21}(\omega_0, N)}{\partial\omega_0} = \begin{cases} \dfrac{(-1)^{N_s+1} NEI\sigma_0^2 l}{\omega_0} & \beta_0 = s\pi \\[2ex] \dfrac{NEI\sigma_0^2 l \cos N\beta_0[2\sinh\sigma_0 l \sin\sigma_0 l + \cos\beta_0(\cosh\sigma_0 l - \cos\sigma_0 l)]}{2\omega_0(1 - \cosh\sigma_0 l \cos\sigma_0 l)} & \beta_0 \neq s\pi \end{cases} \tag{9.2.37}$$

and that

$$\zeta_{21}(\omega_0, N, j) = \sum_{k=1}^{2}\sum_{m=1}^{2} \tilde{\tau}_{km}(\omega_0, N, j) y_{km}(\omega_0, j) \tag{9.2.38}$$

where $y_{km}(\omega_0)$ are the elements of matrix $\boldsymbol{Y}_j(l)$; that is, they are computed as in equation (9.2.26), and

$$\begin{aligned} \tilde{\tau}_{11} &= -\frac{1}{t_{12}} \sin\beta_0 \sin(N-j)\beta_0 \cos(j-1)\beta_0 \\ \tilde{\tau}_{12} &= -\frac{t_{21}}{t_{12}} \sin(N-j)\beta_0 \sin(j-1)\beta_0 \\ \tilde{\tau}_{21} &= \cos(N-j)\beta_0 \cos(j-1)\beta_0 \\ \tilde{\tau}_{22} &= -\frac{1}{t_{12}} \sin\beta_0 \cos(N-j)\beta_0 \sin(j-1)\beta_0 \end{aligned} \tag{9.3.39}$$

Within the scope of a first-order perturbation theory, the variance of each random natural frequency is simply

$$\sigma_\omega^2 = \varepsilon^2 E[\omega_1^2] \tag{9.2.40}$$

Upon substituting equation (9.2.36) into equation (9.2.40),

$$\sigma_\omega^2 = \varepsilon^2 \left(\frac{\partial \gamma_{21}}{\partial \omega_0}\right)^{-2} \sum_{i=1}^{N}\sum_{j=1}^{N}\sum_{k=1}^{2}\sum_{l=1}^{2}\sum_{m=1}^{2}\sum_{n=1}^{2} \tilde{\tau}_{kl}(i)\tilde{\tau}_{mn}(j)E[y_{kl}(i)y_{mn}(j)] \tag{9.2.41}$$

where the dependence of various matrix elements on N and ω_0 is tacitly implied but dependences of $\tilde{\tau}$ and the y elements on i or j are explicitly indicated to show how the summations are carried out.

The calculation of the ensemble averages $E[y_{kl}y_{mn}]$ on the right-hand side of equation (9.2.41) requires the statistical information about the random process $\alpha(\xi)$. Although not essential for the present analysis, we assume that $\alpha(\xi)$ is statistically homogeneous along the entire length of the beam. Then

$$E[\alpha(\xi_i)\alpha(\xi_j)] = \sigma_\alpha^2 \rho_\alpha(\xi_i - \xi_j) = \sigma_\alpha^2 \rho_\alpha(il - jl + x_i - x_j) \tag{9.2.42}$$

where the x's are local coordinates, σ_α is the standard deviation of $\alpha(\xi)$, ρ_α is the correlation coefficient of $\alpha(\xi)$, and where we have made use of the fact that $\alpha(\xi)$ has a zero mean. Of course, for the bending stiffness itself, the correlation coefficient is also ρ_α, but the standard deviation is $\varepsilon\sigma_\alpha$. Substituting (9.2.23), (9.3.26), and (9.2.42) into equation (9.2.41), we obtain

$$\begin{aligned} R_1^2 &= \left(\frac{\sigma_\omega/\omega_0}{\varepsilon\sigma_\alpha/EI}\right)^2 \\ &= \frac{\Phi^2}{N^2 l^2} \sum_{i=1}^{N}\sum_{j=1}^{N}\sum_{k=1}^{2}\sum_{l=1}^{2}\sum_{m=1}^{2}\sum_{n=1}^{2} \tau_{kl}(i)\tau_{mn}(j) \\ &\quad \times \int_0^l \int_0^l \rho_\alpha(il - jl + x_i - x_j)\delta_{kl}(x_i)\delta_{mn}(x_j)\, dx_i\, dx_j \end{aligned} \tag{9.2.43}$$

where

$$\Phi = \begin{cases} 1 & \beta_0 = s\pi \\ \dfrac{2(\cosh\sigma_0 l \cos\sigma_0 l - 1)}{\cos N\beta_0[2\sinh\sigma_0 l \sin\sigma_0 l + \cos\beta_0(\cosh\sigma_0 l - \cos\sigma_0 l)]} & \beta_0 \neq s\pi \end{cases}$$

$$\tau_{11} = \frac{\tilde{\tau}_{11}}{\sigma_0 EI} \qquad \tau_{12} = \frac{\tilde{\tau}_{12}}{(\sigma_0 EI)^2} \qquad \tau_{21} = \tilde{\tau}_{21} \qquad \tau_{22} = \frac{\tilde{\tau}_{22}}{\sigma_0 EI}$$

$$\begin{aligned} \delta_{11}(x) &= c_0(l-x)s_3(x) - \frac{1}{s_3(l)}[c_0(l-x)s_1(x)s_1(l) + s_1(l-x)s_3(x)c_2(l)] \\ &\quad + \frac{1}{s_3^2(l)} s_1(l-x)s_1(x)s_1(l)c_2(l) \\ \delta_{12}(x) &= c_0(l-x)c_0(x) - \frac{1}{s_3(l)}[c_0(l-x)s_1(x)c_2(l) + s_1(l-x)c_0(x)c_2(l)] \\ &\quad + \frac{1}{s_3^2(l)} s_1(l-x)s_1(x)c_2^2(l) \end{aligned}$$

$$\delta_{21}(x) = s_3(l-x)s_3(x) - \frac{1}{s_3(l)}[s_3(l-x)s_1(x)s_1(l) + s_1(l-x)s_3(x)s_1(l)] + \frac{1}{s_3^2(l)}s_1(l-x)s_1(x)s_1^2(l)$$

$$\delta_{22}(x) = s_3(l-x)c_0(x) - \frac{1}{s_3(l)}[s_3(l-x)s_1(x)c_2(l) + s_1(l-x)c_0(x)s_1(l)] + \frac{1}{s_3^2(l)}s_1(l-x)s_1(x)s_1(l)c_2(l)$$

(9.2.44)

and where σ_0 is related to ω_0 according to (9.2.14), and functions c_0, c_2, s_1, and s_3 are given by

$$c_0(y) = \tfrac{1}{2}(\cosh\sigma_0 y + \cos\sigma_0 y) \qquad c_2(y) = \tfrac{1}{2}(\cosh\sigma_0 y - \cos\sigma_0 y)$$
$$s_1(y) = \tfrac{1}{2}(\sinh\sigma_0 y + \sin\sigma_0 y) \qquad s_3(y) = \tfrac{1}{2}(\sinh\sigma_0 y - \sin\sigma_0 y) \tag{9.2.45}$$

The left-hand side of equation (9.2.43) has been expressed as the squared ratio of the variation coefficient of a natural frequency to the variation coefficient of the bending stiffness. The variation coefficient of a random variable is the ratio of its standard deviation to its mean value.

Equation (9.2.43) is greatly simplified in the case of $\beta_0 = s\pi$, that is, for the first natural frequency in each frequency group. In this case, $\tau_{11}(j) = \tau_{12}(j) = \tau_{22}(j) = 0$, $\tau_{21}(j) = (-1)^{(N-1)S}$, and $\delta_{21}(x) = (-1)^s \sin^2(s\pi x/l)$; therefore, equation (9.2.43) reduces to

$$R_1^2 = \frac{1}{N^2 l^2}\sum_{i=1}^{N}\sum_{j=1}^{N}\int_0^l\int_0^l \rho_\alpha(il - jl + x_i - x_j)\sin^2\left(\frac{s\pi x_i}{l}\right)\sin^2\left(\frac{s\pi x_j}{l}\right)dx_i\,dx_j$$
$$\beta_0 = s\pi \tag{9.2.46}$$

To proceed further, knowledge of the correlation coefficient function ρ_α is required. In a practical situation, this function can be estimated from actual measurements. For illustrative purposes we consider an idealized case where $\alpha(\xi)$ is exponentially correlated within each span but uncorrelated in different spans:

$$\rho_\alpha = \begin{cases} \exp(-a|x_i - x_j|) & a > 0,\ i = j \\ 0 & i \neq j \end{cases} \tag{9.2.47}$$

Then from equation (9.2.46),

$$R_1^2 = \frac{al[3(al)^2 + 8(s\pi)^2][(al)^2 + 4(s\pi)^2] - 64(s\pi)^4(1 - e^{-al})}{4N(al)^2[(al)^2 + 4(s\pi)^2]^2} \tag{9.2.48}$$

It is interesting to consider two limiting situations. The first is when $a = 0$, representing the case in which the bending stiffness is uniform in each span, but its value varies from the ideal value EI in a random manner, uncorrelated from span to span. Taking the limit on the right-hand side of equation (9.2.48) as $al \to 0$ we obtain $R^2 = 1/(4N)$. However, this simple result could have been obtained directly from equation (9.2.46) by letting $\rho_\alpha = 1$ and replacing the double sum by a simple sum. The second extreme is the opposite case of $al \to \infty$, representing completely uncorrelated fluctuation in bending stiffness along the length. The second limiting case gives rise to $R_1 = 0$, showing that the effects of uncorrelated random variation in the bending stiffness cancel each other. It may be noted that the value a^{-1} is a measure of the correlation length or scale of randomness. If this scale is much less than the span length l, that is, if $a^{-1} \ll l$ or $al \gg 1$, then R_1 is expected to be negligible.

RANDOM SPAN LENGTHS ALONE. For this case, equation (9.2.35) reduces to

$$\omega_{k1} = -\frac{\sum_{j=1}^{N} X_j \eta_{21}(\omega_{k0}, N, j)}{\partial \gamma_{21}(\omega_{k0}, N)/\partial \omega_{k0}} \tag{9.2.49}$$

where, unlike equation (9.2.36), the subscript k has been retained for reasons that will become obvious later. The denominator of this expression is obtainable from equation (9.2.37). After some algebraic manipulation it can be shown that

$$\omega_{k1} = \sum_{j=1}^{N} X_j W(\omega_{k0}, N, j) \tag{9.2.50}$$

where $$W(\omega_{k0}, N, j) = -\frac{2\omega_{k0}}{Nl}[1 + \Psi_k \cos(N + 1 - 2j)\beta_{k0}] \tag{9.2.51}$$

$$\Psi_k = \begin{cases} 0 & \beta_{k0} = s\pi \\ \left[\cos N\beta_{k0}\left(\cos\beta_{k0} + \dfrac{2\sinh\sigma_{k0}l \sin\sigma_{k0}l}{\cosh\sigma_{k0}l - \cos\sigma_{k0}l}\right)\right]^{-1} & \beta_{k0} \neq s\pi \end{cases} \tag{9.2.52}$$

The covariance of the natural frequencies ω_k and ω_m can be written readily as follows:

$$\begin{aligned} \operatorname{cov}(\omega_k, \omega_m) = \varepsilon^2 \left(\frac{e}{Nl}\right)^2 \omega_{k0}\omega_{m0} \sum_{i=1}^{N}\sum_{j=1}^{N} &[1 + \Psi_k \cos(N + 1 - 2i)\beta_{k0}] \\ &\times [1 + \Psi_m \cos(N + 1 - 2j)\beta_{m0}] E[X_i X_j] \end{aligned} \tag{9.2.53}$$

The variance of ω_k follows from equation (9.2.53) by letting $m = k$.

Two special cases are considered next.

Case 1: X_i are independent and identically distributed random variables. For this case, $E[X_i X_j] = 0$ when $i \neq j$; hence we obtain

$$R_2^2 = \left(\frac{\sigma_{\omega_k}/\omega_{k0}}{\varepsilon\sigma_X/l}\right)^2 = \frac{4}{N}\left(1 + \frac{\Psi_k^2}{2}\right) \tag{9.2.54}$$

where, as indicated, R_2 is the ratio of the variation coefficient of a natural frequency to that of the span length and σ_X is the standard deviation of each X_i. It is seen that R_2^2 is inversely proportional to the number of span N, indicating a cancellation effect of uncorrelated variations in individual span length.

Case 2: All X_i are identical random variables. For this case, we let $E[X_iX_j] = \sigma_X^2$ and again obtain an extremely simple result $R_2^2 = 4$. For this second special case, the covariance of any two natural frequencies computed from equation (9.2.53) is equal to the product of their standard deviations; therefore, they are perfectly correlated, as expected. Since this simple result is independent of N, it must be applicable to the case of a single span. Indeed, the same result can be obtained by letting $N = 1$ and $\Psi_k = 0$ in equation (9.2.54). In fact, $R_2^2 = 4$ in the case of a single-span beam is an obvious result, since each natural frequency is then inversely proportional to the square of the span length.

FREQUENCY POSITION INTERCHANGE. Within the scope of a first-order perturbation theory, the deviation of a natural frequency from its ideal value, which is the same as the ensemble average here, is a linear function of random imperfections, as seen in equation (9.2.50) or more generally in equation (9.2.35). Take equation (9.2.50), for example, and assume that the joint probability distribution of the X_j is given. Then, in principle, the joint probability distribution of any two natural frequencies, say ω_k and ω_m, can be determined, as well as the probability for these two natural frequencies to reverse the order of their positions. The latter event can be stated as that of $\omega_k \geq \omega_m$ when $\omega_{k0} < \omega_{m0}$.

We restrict our present discussion to a sufficient condition under which the event of frequency position reversal cannot occur. Denoting this event by Q, we obtain from equation (9.2.50)

$$Q = \left\{ \varepsilon \sum_{j=1}^{N} X_j \Delta W_{k,m,j} \geq \Delta_{k,m} \right\} \tag{9.2.55}$$

where

$$\begin{aligned} \Delta W_{k,m,j} &= W(\omega_{k0}, N, j) - W(\omega_{m0}, N, j) \\ \Delta_{k,m} &= \omega_{m0} - \omega_{k0} \qquad\qquad m > k \end{aligned} \tag{9.2.56}$$

Let X_j, $j = 1, 2, \ldots, N$ be independent and identically distributed random variables, and let the range of each X_j be restricted within the interval $(-a_L\sigma_X, a_U\sigma_X)$ where a_L and a_U are positive. Then a sufficient condition for Prob$[Q] = 0$ is

$$\varepsilon\sigma_X a_{\max} \sum_{j=1}^{N} \left|\Delta W_{k,m,j}\right| \leq \Delta_{k,m} \qquad m > k \tag{9.2.57}$$

where $a_{\max} = \max(a_L, a_U)$. For example, in the case of uniformly distributed X_j, we have $a_{\max} = a_L = a_U = \sqrt{3}$. Inequality (9.2.57) may be expressed more meaningfully in terms of the coefficient of variation of the span length. Using equation (9.2.51) in (9.2.57) and regrouping, we obtain the following sufficient condition for nonreversal of frequency positions:

$$\begin{aligned} \frac{\varepsilon\sigma_X}{l} \leq N\Delta_{k,m} \Bigg[2a_{\max} \sum_{j=1}^{N} \Big| \Delta_{k,m} &- \omega_{k0}\Psi_k \cos(N + 1 - 2j)\beta_{k0} \\ &+ \omega_{m0}\Psi_m \cos(N + 1 - 2j)\beta_{m0} \Big| \Bigg]^{-1} \qquad m > k \end{aligned} \tag{9.2.58}$$

This condition is rather crude since it requires only the knowledge of a_{max}. A sharper condition must come from additional knowledge of the X_j distribution.

However, the position order can never be interchanged among the first natural frequencies in different frequency groups, regardless of the X_j distribution. For these first natural frequencies, $\beta_{k0} = s\pi$ and $\Psi_k = 0$. Therefore, from equations (9.2.50) through (9.2.52)

$$\omega_k = \omega_{k0} - \frac{2\varepsilon\omega_{k0}}{Nl} \sum_{j=1}^{N} X_j \tag{9.2.59}$$

or, equivalently,

$$\frac{\omega_k}{\omega_{k0}} = 1 - \frac{2\varepsilon}{Nl} \sum_{j=1}^{N} X_j \tag{9.2.60}$$

The right-hand side of this equation is independent of k, enabling us to write

$$\frac{\omega_k}{\omega_{k0}} = \frac{\omega_m}{\omega_{m0}} \tag{9.2.61}$$

Equation (9.2.61) shows that, to the accuracy of a first-order perturbation analysis, the ratio between the first natural frequencies in any two frequency groups remains the same as that of an ideal periodic beam; thus their position order cannot be reversed.

NUMERICAL EXAMPLES. For illustration, numerical results have been obtained for a beam with the following physical properties:

$E = 10.5 \times 10^6$ psi; thickness $= 0.05$ in
l = design span length $= 6.5$ in; width $= 1$ in
m = unit mass $= 2.616 \times 10$ lb $\cdot$ sec^2/in^2

Table 9.2.1 summarizes the results for a four-span beam, considering random span lengths alone. It has been pointed out that the effect of random bending stiffness is unimportant for a multispan beam if its correlation length is short compared with the span length. The deviations of individual span lengths from the design value l are assumed to be uncorrelated but identically distributed random variables.

Columns $\sigma_0 l$ and ω_0 show the ideal (= mean) values for the frequency parameter σl and frequency ω, respectively, disregarding the random imperfections. Column R_2 gives the ratio of the variation coefficient of a natural frequency to that of the random span length. The correlation coefficients between neighboring natural frequencies are listed under column ρ. Each value in the last column is the sufficiency bound computed from inequality (9.2.58), which guarantees the nonreversal in the position order for a pair of neighboring natural frequencies. It is of interest to note that the R_2 value for the first natural frequency in each group is always the same. This is intimately related to the fact that the $\sigma_0 l$ values differ by a multiple of π between two first natural frequencies. The same observation can be made for the third (this number depends on the total number of spans) natural frequencies in

TABLE 9.2.1
Numerical results for a four-span beam with random span lengths

Frequency Group	$\sigma_0 l$	ω_0 rad/sec	R_2	ρ	$\varepsilon\sigma_X/l$
1	π	151.03	1.0000	0.85655	0.06843
	3.3932	176.19	1.1675	0.85655	0.16247
	3.9266	235.93	1.0000	0.86513	0.11256
	4.4633	304.84	1.1559	0.86513	0.28868
2	2π	604.12	1.0000	0.86643	0.03933
	6.5454	655.60	1.1542	0.86643	0.08328
	7.0685	764.59	1.0000	0.86606	0.06654
	7.5916	881.92	1.1547	0.86603	0.27068
3	3π	1359.2	1.0000	0.86601	0.02666
	9.6865	1435.8	1.1547	0.86601	0.05551
	10.210	1595.2	1.0000	0.86602	0.04759
	10.733	1763.0	1.1547		

individual groups. The other R_2 values in the first and second groups do not show exact repetitions, but after the second group their values begin to form a fixed pattern. Indeed, beyond the second group the $\sigma_0 l$ values become sufficiently large and their values in a higher group may be approximated by adding a multiple of π to those in the third group, as first noted by Miles (1956).

9.3 WAVE MOTION IN DISORDERED PERIODIC STRUCTURES

As indicated earlier on several occasions, the natural frequencies of an ideal periodically supported beam are grouped in distinctive bands, and the number of natural frequencies in each band is equal to the number of spans. Let us suppose that the number of spans is increased to infinity. Then every frequency in each band is a natural frequency. In other words, the values of natural frequencies are no longer discrete; they become continuous in distinctive bands. This same conclusion applies to other periodic structures of infinite length, provided that each periodic unit is a continuous structure itself.

When a structure is infinitely long, it is often convenient to treat the structural motion from a wave propagation point of view. The result so obtained, however, is not purely academic, since damping generally exists in a physical structure, and a wave motion generated at one location on the structure becomes negligible when allowed to propagate to a far enough distance. Therefore, an infinitely long structure is often a good model for a sufficiently long structure. If an infinitely long structure is ideally periodic, then the bands of natural frequencies become the so-called *wave-passage bands* (Brillouin, 1946), in the sense that disturbance at a frequency within a wave-passage band may be propagated to a great distance (unlimited distance if damping is ignored). The remainder of the frequency domain is referred to collectively as the *wave-stoppage bands,* and disturbance at a frequency within a wave-stoppage band will attenuate in a short distance from the source even without damping.

When referring to a periodic structure, the term *wave propagation* is interpreted in a broader sense. In general, each of the periodic units (or periodic cells) may consist of bars, plates, and other continuous elements. The wave motions within each unit are extremely complicated, since wave scattering occurs wherever two elements meet at a joint. A simpler point of view is to regard a periodic unit as possessing n degrees of freedom at its interface with a neighboring unit. By focusing our attention on these degrees of freedom at the unit-to-unit interface, we bypass the need to know the detailed motions within each unit. Such a periodic unit may then be regarded as being capable of transmitting n types of waves through the interface to its neighboring unit. For example, a multispan beam is an ideal periodic structure if the cross section is uniform, the material is homogeneous, and the supports are identical and uniformly spaced. Each support may be considered as an interface between two periodic units. If all the supports are rigid in the transverse direction, then only the rotational degree of freedom is permitted at each interface. In this case, the periodic units are said to be monocoupled, and the periodic structure may be regarded as being capable of transmitting only one type of wave motion.

Disorder in a periodic structure has a profound effect on wave propagation. If a structure were exactly periodic and undamped, then a structural wave, at a frequency within a wave-passage band, would pass through the interface of every pair of adjoining units without attenuation. However, in a disordered system, two adjoining units are generally different; therefore, some reflection generally takes place at the interfaces, and a total transmission is no longer possible. Therefore, disorder in a periodic structure causes additional attenuation for wave motion. This is known as the localization effect, first shown by Anderson (1958) for atomic lattices.

Analyses of deterministically disordered periodic structures conducted, for example, by Valero and Bendiksen (1986), Cornwell and Bendiksen (1987), and Pierre and his co-workers (1987, 1989) have shown that the normal modes which would be periodic along the length of a perfectly periodic structure are localized in a small region when periodicity is disrupted. This phenomenon can be explained in terms of wave localization.

The average exponential rate per periodic unit, at which a structural wave decays with respect to the wave propagation distance attributable to disorder, is known as the localization factor. Hodges and Woodhouse (1983) and Pierre (1990) obtained the localization factors for some specific disordered periodic systems, using perturbation procedures. The first analysis of a generic disordered periodic system, given by Kissel (1988), made use of the concept of wave transmission and reflection and the limit theorem of Furstenberg (1963). Additional results were obtained by Ariaratnam and Xie (1991). Extension to the case of multicoupling was considered recently by Kissel (1991) and by Ariaratnam and Xie (1994).

The ability to treat a generic disordered periodic system is clearly desirable, since the analysis is applicable to a general class of problems, not restricted to a specific type of governing equation. In this section, three different systematic procedures are presented for calculating the localization factor for a generic disordered periodic structure. The first one, called the method of multiple reflection (Cai and

Lin, 1991a), is to take into account reflections from nearby periodic cells. It permits successive improvement of the accuracy by including more and more cells in the analysis. Account can also be taken of structural damping. The second procedure, called the method of invariant phase probability (Cai and Lin, 1991b), is applicable to undamped disordered periodic structures. In such a structure the probability distribution of the phase difference between two waves of the same type, but traveling in opposite directions, is identical at every cell-to-cell interface. With the knowledge of this probability distribution, the localization factor can be obtained simply. In the third procedure, the concept of localization factor is related to the Lyapunov exponent. The key to the successful development of these three methods, under a rather general setting, is the assumption of spatial ergodicity, thus permitting the substitution of ensemble averages for sequential averages of certain statistical properties of the structure.

WAVE TRANSMISSION AND REFLECTION MATRICES. Consider two typical cell units in a disordered periodic structure, denoted as cells n and $n+1$ and shown in Fig. 9.3.1a. The relation between the state vector at the left end of cell unit n and that at the left end of cell unit $n+1$ may be written as

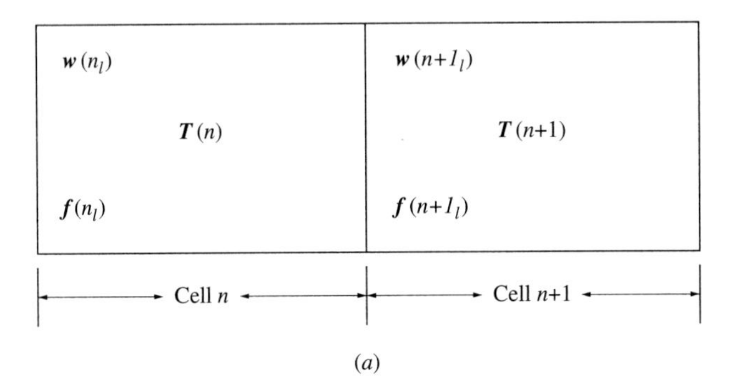

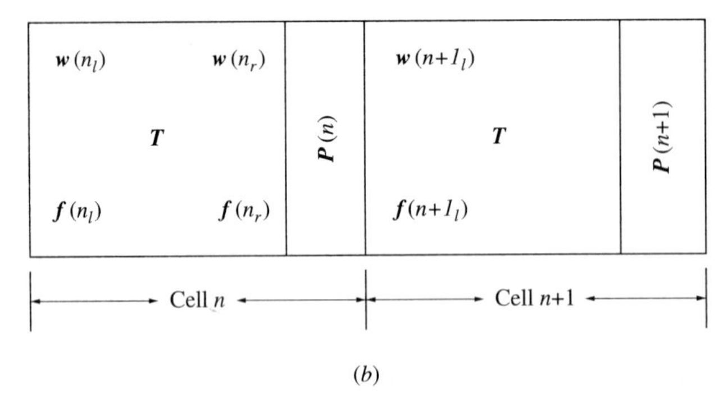

FIGURE 9.3.1
State vector representation. (a) Original system; (b) equivalent system.

$$\begin{Bmatrix} \boldsymbol{w}(n+1_l) \\ \boldsymbol{f}(n+1_l) \end{Bmatrix} = \boldsymbol{T}(n) \begin{Bmatrix} \boldsymbol{w}(n_l) \\ \boldsymbol{f}(n_l) \end{Bmatrix} \tag{9.3.1}$$

where each $\boldsymbol{w}$ is a p-dimensional vector of generalized displacements, each $\boldsymbol{f}$ is a p-dimensional vector of generalized forces, and $\boldsymbol{T}(n)$ is a $2p \times 2p$ transfer matrix associated with the nth cell. It is implied in equation (9.3.1) that the motion is a timewise sinusoidal, and the displacements and forces are interpreted as their complex-valued amplitudes. Because of disorder, $\boldsymbol{T}(n)$ is generally different from the ideal transfer matrix $\boldsymbol{T}$ associated with the ideal design condition of exact periodicity. Nevertheless, it is always possible to write

$$\boldsymbol{T}(n) = \boldsymbol{P}(n)\boldsymbol{T} \tag{9.3.2}$$

which means that the effect of disorder in the nth cell can be lumped at its right end and represented by a "point" transfer matrix $\boldsymbol{P}(n)$. In other words, the original system shown in Fig. 9.3.1*a* is replaced by an equivalent system shown in Fig. 9.3.1*b*. In typical matrix analyses (e.g., Pestel and Leckie, 1963), a point transfer matrix is used to relate state vectors on the two sides of a structural discontinuity which may arise from a concentrated mass, spring, damper, and so on. However, $\boldsymbol{P}(n)$ in equation (9.3.2) is merely a mathematical device, and it may even correspond to, say, a negative mass. With this device, equation (9.3.1) is replaced by

$$\begin{Bmatrix} \boldsymbol{w}(n_r) \\ \boldsymbol{f}(n_r) \end{Bmatrix} = \boldsymbol{T} \begin{Bmatrix} \boldsymbol{w}(n_l) \\ \boldsymbol{f}(n_l) \end{Bmatrix} \tag{9.3.3a}$$

$$\begin{Bmatrix} \boldsymbol{w}(n+1_l) \\ \boldsymbol{f}(n+1_l) \end{Bmatrix} = \boldsymbol{P}(n) \begin{Bmatrix} \boldsymbol{w}(n_r) \\ \boldsymbol{f}(n_r) \end{Bmatrix} \tag{9.3.3b}$$

We recall that the eigenvalues of every transfer matrix are reciprocal pairs. Thus we can denote the eigenvalues of $\boldsymbol{T}$ as $\nu_1, \nu_2, \ldots, \nu_p$ and $\nu_1^{-1}, \nu_2^{-1}, \ldots, \nu_p^{-1}$, where $|\nu_1| \le |\nu_2| \le \cdots \le |\nu_p| \le 1$. The eigenvectors corresponding to these eigenvalues constitute a transformation matrix $\boldsymbol{D}$, with which the state vectors in equation (9.3.3a) are transformed to wave vectors as follows:

$$\begin{Bmatrix} \boldsymbol{w}(n_l) \\ \boldsymbol{f}(n_l) \end{Bmatrix} = \boldsymbol{D} \begin{Bmatrix} \boldsymbol{\mu}^r(n_l) \\ \boldsymbol{\mu}^l(n_l) \end{Bmatrix} \tag{9.3.4a}$$

$$\begin{Bmatrix} \boldsymbol{w}(n_r) \\ \boldsymbol{f}(n_r) \end{Bmatrix} = \boldsymbol{D} \begin{Bmatrix} \boldsymbol{\mu}^r(n_r) \\ \boldsymbol{\mu}^l(n_r) \end{Bmatrix} \tag{9.3.4b}$$

where the superscript r or l associated with a wave vector $\boldsymbol{\mu}$ indicates the direction of wave propagation, either rightgoing or leftgoing.

Substituting equations (9.3.4a) and (9.3.4b) into equation (9.3.3a), we obtain

$$\begin{Bmatrix} \boldsymbol{\mu}^r(n_r) \\ \boldsymbol{\mu}^l(n_r) \end{Bmatrix} = \boldsymbol{D}^{-1}\boldsymbol{T}\boldsymbol{D} \begin{Bmatrix} \boldsymbol{\mu}^r(n_l) \\ \boldsymbol{\mu}^l(n_l) \end{Bmatrix} = \boldsymbol{Q} \begin{Bmatrix} \boldsymbol{\mu}^r(n_l) \\ \boldsymbol{\mu}^l(n_l) \end{Bmatrix} \tag{9.3.5}$$

where $\boldsymbol{Q}$ is also a transfer matrix, but it transfers wave vectors, instead of state vectors. This wave transfer matrix is clearly diagonal. In fact,

$$\boldsymbol{Q} = \boldsymbol{D}^{-1}\boldsymbol{T}\boldsymbol{D} = \begin{bmatrix} \boldsymbol{\nu} & \boldsymbol{0} \\ \boldsymbol{0} & \boldsymbol{\nu}^{-1} \end{bmatrix} \tag{9.3.6}$$

where
$$\boldsymbol{\nu} = \begin{bmatrix} \nu_1 & & & \\ & \cdot & & 0 \\ & & \cdot & \\ & 0 & & \cdot \\ & & & \nu_p \end{bmatrix} \tag{9.3.7}$$

Equation (9.3.5) can be rewritten in another form:

$$\begin{Bmatrix} \boldsymbol{\mu}^l(n_l) \\ \boldsymbol{\mu}^r(n_r) \end{Bmatrix} = \boldsymbol{S} \begin{Bmatrix} \boldsymbol{\mu}^r(n_l) \\ \boldsymbol{\mu}^l(n_r) \end{Bmatrix} \tag{9.3.8}$$

with
$$\boldsymbol{S} = \begin{bmatrix} 0 & \boldsymbol{\nu} \\ \boldsymbol{\nu} & 0 \end{bmatrix} \tag{9.3.9}$$

Equation (9.3.8) is written to suggest that $\boldsymbol{\mu}^r(n_l)$ and $\boldsymbol{\mu}^l(n_r)$ are input waves and $\boldsymbol{\mu}^l(n_l)$ and $\boldsymbol{\mu}^r(n_r)$ are output waves, as "viewed" by the structural unit, which is characterized by $\boldsymbol{S}$.

Similar transformation of the state vectors in equation (9.3.3b) to wave vectors results in

$$\begin{Bmatrix} \boldsymbol{\mu}^r(n + 1_l) \\ \boldsymbol{\mu}^l(n + 1_l) \end{Bmatrix} = \boldsymbol{Q}(n) \begin{Bmatrix} \boldsymbol{\mu}^r(n_r) \\ \boldsymbol{\mu}^l(n_r) \end{Bmatrix} \tag{9.3.10}$$

where
$$\boldsymbol{Q}(n) = \boldsymbol{D}^{-1}\boldsymbol{P}(n)\boldsymbol{D} = \boldsymbol{D}^{-1}\boldsymbol{T}(n)\boldsymbol{D}\begin{bmatrix} \boldsymbol{\nu}^{-1} & \boldsymbol{0} \\ \boldsymbol{0} & \boldsymbol{\nu} \end{bmatrix} \tag{9.3.11}$$

Equation (9.3.10) can also be cast in the form of equation (9.3.8):

$$\begin{Bmatrix} \boldsymbol{\mu}^l(n_r) \\ \boldsymbol{\mu}^r(n + 1_l) \end{Bmatrix} = \boldsymbol{S}(n) \begin{Bmatrix} \boldsymbol{\mu}^r(n_r) \\ \boldsymbol{\mu}^l(n + 1_l) \end{Bmatrix} \tag{9.3.12}$$

The elements in $\boldsymbol{S}(n)$ can be obtained from those of $\boldsymbol{Q}(n)$ as follows:

$$\boldsymbol{s}_{11} = -\boldsymbol{q}_{22}^{-1}\boldsymbol{q}_{21} \tag{9.3.13a}$$

$$\boldsymbol{s}_{12} = \boldsymbol{q}_{22}^{-1} \tag{9.3.13b}$$

$$\boldsymbol{s}_{21} = \boldsymbol{q}_{11} - \boldsymbol{q}_{12}\boldsymbol{q}_{22}^{-1}\boldsymbol{q}_{21} \tag{9.3.13c}$$

$$\boldsymbol{s}_{22} = \boldsymbol{q}_{12}\boldsymbol{q}_{22}^{-1} \tag{9.3.13d}$$

In equations (9.3.13a–d), $\boldsymbol{s}_{ij}$ and $\boldsymbol{q}_{ij}$ are $p \times p$ submatrices of $\boldsymbol{S}(n)$ and $\boldsymbol{Q}(n)$, respectively, and the argument (n) is omitted in denoting the submatrices for simplicity.

It is physically meaningful to rename the submatrices of $\boldsymbol{S}(n)$ as $\boldsymbol{r}^r(n) = \boldsymbol{s}_{11}$, $\boldsymbol{t}^l(n) = \boldsymbol{s}_{12}$, $\boldsymbol{t}^r(n) = \boldsymbol{s}_{21}$, and $\boldsymbol{r}^l(n) = \boldsymbol{s}_{22}$ and to rewrite equation (9.3.12) as

$$\begin{Bmatrix} \boldsymbol{\mu}^l(n_r) \\ \boldsymbol{\mu}^r(n + 1_l) \end{Bmatrix} = \begin{bmatrix} \boldsymbol{r}^r(n) & \boldsymbol{t}^l(n) \\ \boldsymbol{t}^r(n) & \boldsymbol{r}^l(n) \end{bmatrix} \begin{Bmatrix} \boldsymbol{\mu}^r(n_r) \\ \boldsymbol{\mu}^l(n + 1_l) \end{Bmatrix} \tag{9.3.14}$$

The submatrices $\boldsymbol{t}$ and $\boldsymbol{r}$ are called the transmission and reflection matrices, respectively. The superscript for each of these submatrices signifies the direction of an incoming wave group to be transmitted or reflected. For example, $\boldsymbol{t}^r(n)$ is a transmission matrix for an incoming wave group traveling in the right direction.

Matrix $\boldsymbol{S}(n)$, which contains the transmission and reflection submatrices, is known as a scattering matrix. This particular scattering matrix characterizes the

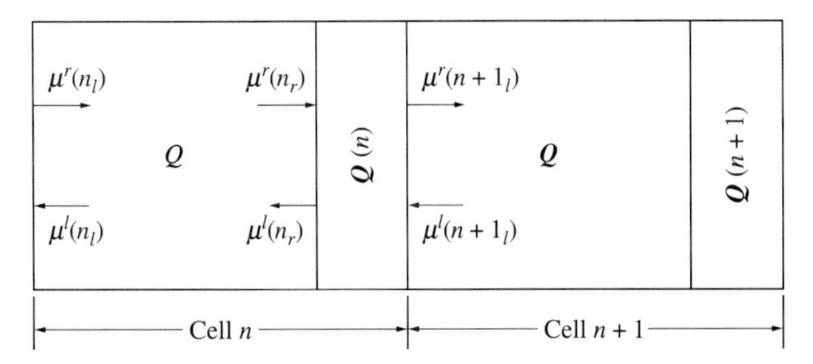

FIGURE 9.3.2
Wave vector representation. Equivalent system.

effects of disorder in cell n which are lumped at the right end of cell n, as shown schematically in Fig. 9.3.2. In wave mechanics, the concept of scattering matrix is quite general; it characterizes the behavior of an identifiable structural element. In fact, matrix $\boldsymbol{S}$ in equation (9.3.8) is also a scattering matrix, in which the transmission and reflection matrices are $\boldsymbol{t}^r = \boldsymbol{t}^l = \boldsymbol{\nu}$ and $\boldsymbol{r}^r = \boldsymbol{r}^l = \boldsymbol{0}$. These submatrices of $\boldsymbol{S}$ are also related to those of a matrix $\boldsymbol{Q}$ according to equations (9.3.13a–d). It is of interest to note that all scattering matrices are symmetric, a consequence of the reciprocity theorem for linear systems.

We now consider the case of monocoupling in more detail, which means that only one type of wave motion (rightgoing, leftgoing, or both) is permitted at every cell-to-cell interface. In this case, the wave transfer matrices $\boldsymbol{Q}$ and the scattering matrices $\boldsymbol{S}$ become 2×2, and the transmission and reflection matrices reduce to the transmission and reflection coefficients. Moreover, $t^r(n) = t^l(n)$, as can be proven from equations (9.3.13b) and (9.3.13c) and from the fact $|\boldsymbol{Q}(n)| = 1$.

The exponential decay rate of wave amplitude due to passage through cell n is given by

$$\gamma(n) = -\ln\left|\frac{\mu^r(n+1_l)}{\mu^r(n_l)}\right| \tag{9.3.15}$$

The localization factor is defined as

$$\gamma = \lim_{N\to\infty} \frac{1}{N} \sum_{n=1}^{N} \gamma(n) \tag{9.3.16}$$

METHOD OF MULTIPLE REFLECTIONS. As indicated earlier, wave reflection takes place at every interface between two nonidentical cell units and contributes to the wave decay along its path. Therefore, the localization factor can be computed on the basis of such reflections. Several specific cases are considered in the order of increasing complexity.

Case 1: Small reflections. If all the reflection coefficients due to disorder are small, such that the effect of multiple reflection is negligible, then from equations (9.3.8) and (9.3.14),

$$\mu^r(n+1_l) \approx vt^r(n)\mu^r(n_l) \tag{9.3.17}$$

Moreover, since matrix $\boldsymbol{Q}(n)$ is 2×2 with a determinant equal to 1, it can be shown that $t^r(n) = t^l(n) = t(n) = q_{22}^{-1}(n)$, and that

$$\gamma = -\ln|\nu| + \lim_{N\to\infty} \frac{1}{N} \sum_{n=1}^{N} \ln|q_{22}(n)| \tag{9.3.18}$$

where ν is the one eigenvalue of $\boldsymbol{T}$ satisfying $|\nu| \leq 1$. Since $\boldsymbol{T}$ is a 2×2 transfer matrix, the other eigenvalue must be ν^{-1} and $|\nu^{-1}| \geq 1$. The sequential average in equation (9.3.18) may be replaced by an ensemble average as follows:

$$\gamma = -\ln|\nu| + \int \ln|q_{22}(\boldsymbol{x})|\, p(\boldsymbol{x})\, d\boldsymbol{x} \tag{9.3.19}$$

where $p(\boldsymbol{x})$ is the probability density of a random vector $\boldsymbol{X}(n)$ of disordered parameters in a typical cell n. It is assumed that $p(\boldsymbol{x})$ is independent of n, and that $\boldsymbol{X}(n)$ for different n constitute an ergodic sequence. Given the forms of $q_{22}[\boldsymbol{x}(n)]$ and the probability density $p(\boldsymbol{x})$ of $\boldsymbol{X}(n)$, the integral can be evaluated at least numerically. In simple cases, it can often be carried out in closed form, as shown later in an example.

If the system disorder is also small, then equation (9.3.19) may be further approximated by

$$\gamma = -\ln|\nu| + \frac{\sigma_i^2}{2}\left[\frac{\partial^2}{\partial x_i^2} \ln|q_{22}(\boldsymbol{x})|\right]_{\boldsymbol{x}=0} \tag{9.3.20}$$

where σ_i^2 are the variance of $X_i(n)$. In deducing (9.3.20), use has been made of $|q_{22}(\boldsymbol{x})|_{\boldsymbol{x}=0} = 1$. If, in addition, the structure is undamped and γ is evaluated at a wave-passage frequency, then $|\nu| = 1$, and equation (9.3.20) reduces to

$$\gamma = \frac{\sigma_i^2}{2}\left[\frac{\partial^2}{\partial x_i^2} \ln|q_{22}(\boldsymbol{x})|\right]_{\boldsymbol{x}=0} \tag{9.3.21}$$

Equation (9.3.21) has been obtained previously by Kissel (1988).

It should be emphasized that equation (9.3.19) is valid only for small reflection coefficients, which imply that localization is weak. Equation (9.3.20) requires both small reflection coefficients and small disorder, and equation (9.3.21) requires additionally a zero damping and a wave-passage frequency.

Case 2: Moderate reflections. Equation (9.3.19) may not be sufficiently adequate if the reflection coefficients due to disorder are not sufficiently small. For improved accuracy, the effects of multiple reflections must be taken into account. As shown in Fig. 9.3.3, the rightgoing wave $\mu^r(n_r)$ gives rise to a rightgoing transmitted wave $t(n)\mu^r(n_r)$ and a leftgoing reflected wave $r^r(n)\mu^r(n_r)$. The reflected wave travels through cell n and is itself partially reflected back at the interface between cells n and $n-1$. This back-reflected wave travels also through cell n, whereupon it splits again into a reflected part and a transmitted part. The latter is clearly $t(n)\nu^2 r^l(n-1)r^r(n)\mu^r(n_r)$, the former travels back through cell n, and so on. Summing up all the rightgoing waves which pass through the interface between cells n and $n+1$, but taking into account only the reflections at the interfaces between cells n and $n+1$,

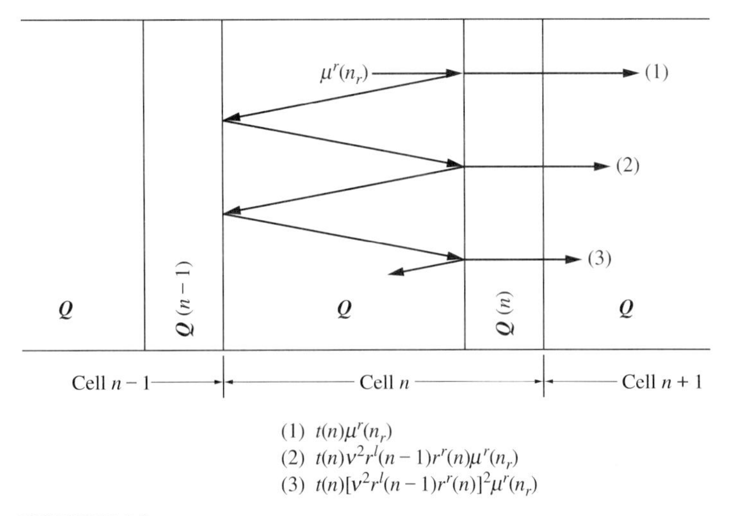

FIGURE 9.3.3
Transmitted waves through the interface between cells n and $n + 1$.

and between cells $n - 1$ and n, we obtain

$$\begin{aligned}\mu^r(n + 1_l) &= t(n)\{1 + \nu^2 r^l(n - 1)r^r(n) + [\nu^2 r^l(n - 1)r^r(n)]^2 + \cdots\}\mu^r(n_r) \\ &= \frac{t(n)}{1 - \nu^2 r^l(n - 1)r^r(n)}\mu^r(n_r)\end{aligned} \tag{9.3.22}$$

Thus

$$\gamma(n) = \frac{t(n)\nu}{1 - \nu^2 r^l(n - 1)r^r(n)} \tag{9.3.23}$$

It follows from equations (9.3.9) and (9.3.13a–d) and the ergodicity assumption that

$$\begin{aligned}\gamma = &- \ln|\nu| + \int \ln\left|q_{22}(\boldsymbol{x})\right| p(\boldsymbol{x})\,d\boldsymbol{x} \\ &+ \int\!\!\int \ln\left|1 + \frac{\nu^2 q_{12}(\boldsymbol{x}_2)q_{21}(\boldsymbol{x}_1)}{q_{22}(\boldsymbol{x}_2)q_{22}(\boldsymbol{x}_1)}\right| p(\boldsymbol{x}_1)p(\boldsymbol{x}_2)\,d\boldsymbol{x}_1\,d\boldsymbol{x}_2\end{aligned} \tag{9.3.24}$$

The third term is seen to provide additional accuracy over equation (9.3.19). Equation (9.3.24) is valid for the case of moderate localization.

Case 3: Strong reflections. Equation (9.3.24) may still be inadequate if the reflection coefficients are relatively large (although each cannot have a magnitude greater than 1). More accurate results, however, can always be obtained by including more reflection terms in the formulation. It can be shown that if the effect of reflection at the interface between cells $n - 1$ and $n - 2$ is also taken into consideration, then

$$\mu^r(n + 1_l) = t(n)\left[\frac{1}{1 - a} + \frac{c}{(1 - a)(1 - b)(1 - c)}\right]\mu^r(n_r) \tag{9.3.25}$$

where
$$a = \nu^2 r^l(n-1) r^r(n) \tag{9.3.26a}$$
$$b = \nu^2 r^l(n-2) r^r(n-1) \tag{9.3.26b}$$
$$c = t^2(n-1)\nu^4 r^l(n-2) r^r(n) \tag{9.3.26c}$$

Again, using equations (9.3.13a–d) and invoking the ergodicity assumption,

$$\gamma = -\ln|\nu| + \int \ln\left|q_{22}(\boldsymbol{x})\right| p(\boldsymbol{x})d\boldsymbol{x} - \int\int\int \ln\left|\frac{1}{1+f_1(\boldsymbol{x}_1,\boldsymbol{x}_2)} - \frac{f_2(\boldsymbol{x}_1,\boldsymbol{x}_2,\boldsymbol{x}_3)}{[1+f_1(\boldsymbol{x}_1,\boldsymbol{x}_2)][1+f_1(\boldsymbol{x}_2,\boldsymbol{x}_3)][1+f_2(\boldsymbol{x}_1,\boldsymbol{x}_2,\boldsymbol{x}_3)]}\right| p(\boldsymbol{x}_1)p(\boldsymbol{x}_2)p(\boldsymbol{x}_3)d\boldsymbol{x}_1 d\boldsymbol{x}_2 d\boldsymbol{x}_3 \tag{9.3.27}$$

where
$$f_1(\boldsymbol{x}_1,\boldsymbol{x}_2) = \frac{\nu^2 q_{12}(\boldsymbol{x}_2) q_{21}(\boldsymbol{x}_1)}{q_{22}(\boldsymbol{x}_1) q_{22}(\boldsymbol{x}_2)} \tag{9.3.28a}$$
$$f_2(\boldsymbol{x}_1,\boldsymbol{x}_2,\boldsymbol{x}_3) = \frac{\nu^4 q_{12}(\boldsymbol{x}_3) q_{21}(\boldsymbol{x}_1)}{q_{22}(\boldsymbol{x}_1) q_{22}^2(\boldsymbol{x}_2) q_{22}(\boldsymbol{x}_3)} \tag{9.3.28b}$$

Formulas of still higher accuracies can be derived in an analogous manner.

METHOD OF INVARIANT PHASE DISTRIBUTION. Since disorder has an effect similar to damping on wave propagation in a periodic structure, it is reasonable to neglect damping if the objective of an analysis is to estimate the effect due to disorder alone (Hodges and Woodhouse, 1983; Kissel, 1988; Pierre, 1990). Consider, therefore, an undamped disordered chain of N cells, and focus attention on the portion from cell n through cell N. External excitations, if any, are assumed to be on the left of cell n; thus they are not located on the segment of the chain under investigation. Since energy must be conserved within this segment, we have

$$|\mu^r(n_l)|^2 + |\mu^l(N_l)|^2 = |\mu^l(n_l)|^2 + |\mu^r(N_l)|^2 \tag{9.3.29}$$

The left- and right-hand sides of equation (9.3.29) represent, respectively, the average energy inflow and energy outflow of the segment within one period of time $2\pi/\omega$. As $N \to \infty$, the waves $\mu^r(N_l)$ and $\mu^l(N_l)$ vanish due to localization: $|\mu^r(N_l)| \to 0$ and $|\mu^l(N_l)| \to 0$. It follows that $|\mu^r(n_l)|^2 = |\mu^l(n_l)|^2$ and

$$\left|\eta(n)\right| = \frac{|\mu^r(n_l)|}{|\mu^l(n_l)|} = 1 \tag{9.3.30}$$

where $\eta(n)$ is the ratio of the amplitudes of the rightgoing and leftgoing waves. Equation (9.3.30) implies that $\eta(n)$ can be expressed as

$$\eta(n) = e^{i\Theta(n)} \tag{9.3.31}$$

where $\Theta(n)$ is a purely real random variable; it represents the random phase difference between the two waves $\mu^r(n_l)$ and $\mu^l(n_l)$.

It can be shown from equations (9.3.5) and (9.3.10) that $\Theta(n)$ and $\Theta(n+1)$ are related as follows:

$$e^{i\Theta(n)} = \frac{q_{12}(n) - e^{i\Theta(n+1)}q_{22}(n)\nu^2}{e^{i\Theta(n+1)}q_{21}(n)\nu^2 - q_{11}(n)} \qquad n = N-1, \ldots, 2, 1 \qquad (9.3.32)$$

where ν may be taken to be either one of the two eigenvalues of transfer matrix $\boldsymbol{T}$. However, when $N \to \infty$, the probability density $p[\Theta(n)]$ of $\Theta(n)$ must be independent of n, for any finite n. In other words, the probability distribution for the phase difference between the rightgoing and leftgoing waves becomes invariant in an infinite undamped disordered chain. The existence of such an invariant probability density (also referred to as invariant measure) is guaranteed by the Furstenberg theorem (1963). Due to the functional dependence of the random variables $\Theta(n)$ and $\Theta(n+1)$ shown in equation (9.3.32), the invariant probability density $p(\theta)$ must satisfy the relation

$$\begin{aligned} p[\theta(n)]_{\theta(n)=\theta} &= \left\{ \int p[\theta(n+1)]p[\boldsymbol{x}(n)] \left| \frac{\partial\theta(n+1)}{\partial\theta(n)} \right| d\boldsymbol{x}(n) \right\}_{\theta(n)=\theta} \\ &= \{p[\theta(n+1)]\}_{\theta(n+1)=\theta} \end{aligned} \qquad (9.3.33)$$

In obtaining equation (9.3.33), use has been made of the independence of the two random variables $\Theta(n)$ and $\boldsymbol{X}(n)$:

$$p[\theta(n+1), \boldsymbol{x}(n)] = p[\theta(n+1)p[\boldsymbol{x}(n)] \qquad (9.3.34)$$

due to the fact that $\Theta(n+1)$ is related only to $\Theta(n+2)$ and random vector $\boldsymbol{X}(n+1)$, as can be seen from the recursive relationship, equation (9.3.32).

An explicit analytical expression for the invariant probability density $p(\theta)$ is generally difficult to obtain; however, for specific cases, it may be calculated approximately from equation (9.3.33) by using a perturbation approach, as shown later in an example. In any case, it can always be obtained quite simply by Monte Carlo simulation. With the knowledge of the probability density $p(\theta)$, the localization factor can be obtained as follows.

We begin by recasting equation (9.3.10) as

$$\begin{Bmatrix} \mu^r(n_r) \\ \mu^l(n_r) \end{Bmatrix} = \boldsymbol{V}(n) \begin{Bmatrix} \mu^r(n+1_l) \\ \mu^l(n+1_l) \end{Bmatrix} \qquad (9.3.35)$$

where $\boldsymbol{V}(n) = \boldsymbol{Q}(n)^{-1}$. The first row of equation (9.3.35) reads

$$\mu^r(n_r) = v_{11}(n)\mu^r(n+1_l) + v_{12}(n)\mu^l(n+1_l) \qquad (9.3.36)$$

where $v_{ij}(n)$ are elements of $\boldsymbol{V}(n)$. Comparing equations (9.3.5) and (9.3.36), we obtain

$$\mu^r(n_l) = v_{11}(n)\nu^{-1}\mu^r(n+1_l) + v_{12}(n)\nu^{-1}\mu^l(n+1_l) \qquad (9.3.37)$$

Dividing both sides of equation (9.3.37) by $\mu^r(n+1_l)$ and expressing the wave ratio $\eta(n+1)$ as $e^{i\Theta(n+1)}$, we obtain

$$\frac{\mu^r(n_l)}{\mu^r(n+1_l)} = v_{11}(n)\nu^{-1} + v_{12}(n)\nu^{-1}e^{i\Theta(n+1)} \qquad (9.3.38)$$

It is known that if the frequency of the wave motion is within a wave-passage band, then $|\nu| = 1$ and $\nu^{-1} = \nu^*$ for an undamped periodic structure. Thus the exponential decay rate of wave amplitude due to passage through cell n is given by

$$\begin{aligned}\gamma(n) &= \ln\left|\frac{\mu^r(n_l)}{\mu^r(n+1_l)}\right| \\ &= \frac{1}{2}\ln\left\{|v_{11}(n)|^2 + |v_{12}(n)|^2 + 2\mathrm{Re}[v_{11}(n)v_{12}^*(n)e^{i\Theta(n+1)}]\right\} \end{aligned} \tag{9.3.39}$$

where Re[·] denotes the real part of a complex quantity and an asterisk denotes the complex conjugate. Being a function of the disordered parameters, each $v_{ij}(n)$ in equation (9.3.39) may be denoted by

$$v_{ij}(n) = v_{ij}[\boldsymbol{X}(n)] \tag{9.3.40}$$

As shown in equation (9.3.34), the random variables $\boldsymbol{X}(n)$ and $\Theta(n+1)$ are independent. Hence $p[\theta(n+1), \boldsymbol{x}(n)] = p[\theta]p[\boldsymbol{x}(n)]$. The localization factor can then be obtained as

$$\begin{aligned}\gamma &= \lim_{N\to\infty}\frac{1}{N}\sum_{n=1}^{N}\gamma(n) = E[\gamma(n)] \\ &= \frac{1}{2}\int\int \ln\left\{|v_{11}(\boldsymbol{x})|^2 + |v_{12}(\boldsymbol{x})|^2 + 2\mathrm{Re}[v_{11}(\boldsymbol{x})v_{12}^*(\boldsymbol{x})e^{i\theta}]\right\} p(\boldsymbol{x})p(\theta)\,d\boldsymbol{x}\,d\theta\end{aligned} \tag{9.3.41}$$

in which the sequential averaging has been replaced by ensemble averaging on the basis of the ergodicity assumption for the random sequence $\boldsymbol{X}(n)$, $n = 1, 2, \ldots$.

Equation (9.3.41) is an exact expression, applicable to any undamped and randomly disordered monocoupled chain, whether localization is weak, moderate, or strong. Given the probability distribution of the disorder, the accuracy of the computed localization factor depends only on the accuracy of the computed invariant probability density $p(\theta)$.

METHOD OF LYAPUNOV EXPONENTS. Combining equations (9.3.5) and (9.3.10), we have the following wave transfer relationship:

$$\boldsymbol{\mu}_{n+1} = \boldsymbol{Q}_n\boldsymbol{\mu}_n \tag{9.3.42}$$

where

$$\boldsymbol{\mu}_{n+1} = \begin{Bmatrix}\boldsymbol{\mu}^r(n+1_l)\\ \boldsymbol{\mu}^l(n+1_l)\end{Bmatrix} \qquad \boldsymbol{\mu}_n = \begin{Bmatrix}\boldsymbol{\mu}^r(n_l)\\ \boldsymbol{\mu}^l(n_l)\end{Bmatrix} \tag{9.3.43}$$

and

$$\boldsymbol{Q}_n = \boldsymbol{Q}(n)\boldsymbol{Q} \tag{9.3.44}$$

which is also a wave transfer matrix. For a disordered periodic structure consisting of N cell units, repeated application of equation (9.3.42) yields

$$\boldsymbol{\mu}_{N+1} = \boldsymbol{Q}(N,1)\boldsymbol{\mu}_1 \tag{9.3.45}$$

where $\boldsymbol{Q}(N, 1)$ is the wave transfer matrix of the entire structure given by

$$\boldsymbol{Q}(N,1) = \boldsymbol{Q}_N\boldsymbol{Q}_{N-1}\cdots\boldsymbol{Q}_1 \tag{9.3.46}$$

Recall that in Chapter 6 on stability of stochastic systems, we introduced the concept of the Lyapunov exponent, which characterizes the average exponential growth rate per unit time of the system motion, given a set of initial conditions. Motion stability requires that the maximum Lyapunov exponent be negative. Clearly, the same concept applies to the wave propagation phenomenon in a disordered periodic structure governed by equation (9.3.42), with the continuous time parameter t replaced by a discrete spatial parameter n. It should be noted, however, that wave motions occur in two directions, one following the increasing sequence of n, and another following the decreasing sequence of n. In the latter case, a wave motion appears growing with increasing n, and the associated Lyapunov exponent is positive. Physically, of course, a wave motion can never grow without energy input.

Since the numbering of spatial orders is reversible, the Lyapunov exponents associated with a wave transfer matrix $\boldsymbol{Q}(N, 1)$ must be pairwise equal in magnitude but opposite in sign. In other words, for every positive Lyapunov exponent $\lambda_j^{(N)}$, associated with $\boldsymbol{Q}(N, 1)$, there exists another Lyapunov exponent $-\lambda_j^{(N)}$. A transfer matrix of order $2p \times 2p$ possesses a total of p pairs of Lyapunov exponents. Each negative Lyapunov exponent characterizes the average exponential decay rate per unit cell of one type of wave. Let the negative Lyapunov exponents of $\boldsymbol{Q}(N, 1)$ be ordered as $-\lambda_1^{(N)} \geq -\lambda_2^{(N)} \geq \cdots \geq -\lambda_p^{(N)}$. Then, the largest negative Lyapunov exponent $-\lambda_1^{(N)}$ characterizes the least decaying wave, which is the one of significance when the chain is long. Thus we obtain

$$\gamma = \lim_{N\to\infty} \lambda_1^{(N)} \tag{9.3.47}$$

Note that $-\lambda_1^{(N)}$ is numerically the smallest positive Lyapunov exponent. For a monocoupled disordered chain, only one pair of Lyapunov exponents exists, in which case the smallest positive Lyapunov exponent is, in fact, the largest one. Therefore,

$$\gamma = \lim_{N\to\infty} \lambda_T^{(N)} \qquad \text{for a monocoupling chain} \tag{9.3.48}$$

Assume that $\boldsymbol{Q}_1, \boldsymbol{Q}_2, \ldots, \boldsymbol{Q}_N$ are independent and identically distributed random matrices. According to the multiplicative ergodic theorem (Oseledec, 1968), the behavior of the wave vector $\boldsymbol{\mu}_{N+1}$ is dominated by the largest Lyapunov exponent λ_T as N tends to infinity:

$$\lambda_T = \lim_{N\to\infty} \frac{1}{N} \ln \left\| \boldsymbol{\mu}_{N+1} \right\| \tag{9.3.49}$$

where $\left\| \boldsymbol{\mu}_{N+1} \right\|$ is the euclidean norm of vector $\boldsymbol{\mu}_{N+1}$ defined by

$$\left\| \boldsymbol{\mu}_{N+1} \right\|^2 = (\boldsymbol{\mu}_{N+1}^*)' \boldsymbol{\mu}_{N+1} = (\boldsymbol{\mu}_1^*)' [\boldsymbol{Q}^*(N, 1)]' \boldsymbol{Q}(N, 1) \boldsymbol{\mu}_1 \tag{9.3.50}$$

in which an asterisk denotes the complex conjugate and a prime denotes a matrix transposition. The procedure proposed by Khasminskii (1967) can, again, be applied to obtain λ_T. Let

$$\Upsilon = \ln \left\| \boldsymbol{\mu}_n \right\| \qquad \boldsymbol{U}_n = \frac{\boldsymbol{\mu}_n}{\left\| \boldsymbol{\mu}_n \right\|} \qquad n = 1, 2, \ldots \tag{9.3.51}$$

The random variables $\boldsymbol{U}_1, \boldsymbol{U}_2, \ldots, \boldsymbol{U}_n, \ldots,$ constitute a Markov chain on a $2p$-dimensional unit sphere $\|\boldsymbol{U}\| = 1$. Then,

$$\Upsilon_{n+1} = \ln \|\boldsymbol{\mu}_{n+1}\| = \ln \|\boldsymbol{Q}_n \boldsymbol{\mu}_n\| = \ln \Big\| \boldsymbol{Q}_n \boldsymbol{U}_n \|\boldsymbol{\mu}_n\| \Big\| = \Upsilon_n + \ln \|\boldsymbol{Q}_n \boldsymbol{U}_n\| \quad (9.3.52)$$

The recursive relationship (9.3.52) leads to

$$\Upsilon_{n+1} = \Upsilon_1 + \sum_{n=1}^{N} \ln \|\boldsymbol{Q}_n \boldsymbol{U}_n\| \quad (9.3.53)$$

Under a set of rather general conditions (Khasminskii, 1980), we have from (9.3.53)

$$\begin{aligned}\lambda_T &= \lim_{N\to\infty} \frac{\Upsilon_{N+1}}{N} = \lim_{N\to\infty} \frac{1}{N} \sum_{n=1}^{N} \ln \|\boldsymbol{Q}_n \boldsymbol{U}_n\| \\ &= E[\ln \|\boldsymbol{Q}(X)\boldsymbol{U}\|] = \iint \ln \|\boldsymbol{Q}(\boldsymbol{x})\boldsymbol{u}\| p(\boldsymbol{x}) p(\boldsymbol{u})\, d\boldsymbol{x}\, d\boldsymbol{u} \end{aligned} \quad (9.3.54)$$

where $p(\boldsymbol{x})$ is the probability density of the disordered system parameters $\boldsymbol{X}$, and $p(\boldsymbol{u})$ is the probability density of $\boldsymbol{U}$, both of which are independent of n.

Now equation (9.3.54) can be used to calculate the localization factor for a monocoupled and undamped disordered periodic structure. In this case, the wave transfer matrix can be written as

$$\boldsymbol{Q}(\boldsymbol{X}) = \begin{bmatrix} q_{11}(\boldsymbol{X}) & q_{12}(\boldsymbol{X}) \\ q_{21}(\boldsymbol{X}) & q_{22}(\boldsymbol{X}) \end{bmatrix} \quad (9.3.55)$$

For a 2×2 wave transfer matrix (Kissel, 1988),

$$q_{11}(\boldsymbol{X}) = q_{22}^{*}(\boldsymbol{X}) \qquad q_{12}(\boldsymbol{X}) = q_{21}^{*}(\boldsymbol{X}) \quad (9.3.56)$$

$$|q_{22}(\boldsymbol{X})|^2 - |q_{21}(\boldsymbol{X})|^2 = 1 \quad (9.3.57)$$

Furthermore, a vector on a unit circle may be expressed as

$$\boldsymbol{U} = \frac{1}{\sqrt{2}} \begin{Bmatrix} e^{i\phi} \\ e^{-i\phi} \end{Bmatrix} \quad (9.3.58)$$

Then, a simple calculation leads to

$$\|\boldsymbol{Q}\boldsymbol{U}\| = |q_{22}(\boldsymbol{X}) + q_{21}(\boldsymbol{X}) e^{-2i\phi}| \quad (9.3.59)$$

Substituting (9.3.59) into (9.3.54) results in

$$\gamma = \lambda_T = \int_0^{2\pi} p(\phi)\, d\phi \int \ln |q_{22}(\boldsymbol{x}) + q_{21}(\boldsymbol{x}) e^{-2i\phi}| p(\boldsymbol{x})\, d\boldsymbol{x} \quad (9.3.60)$$

Equation (9.3.60) is an exact result for a monocoupled chain. In general, $p(\phi)$ can be evaluated from Monte Carlo simulation, for which an algorithm has been devised by Xie (1990). In the case of small disorder, ϕ is approximately uniformly distributed (Kissel, 1988), in which case equation (9.3.60) is reduced to

$$\gamma = \int \ln[|q_{22}(\boldsymbol{x})| + |q_{21}(\boldsymbol{x})|]p(\boldsymbol{x})\,d\boldsymbol{x} \tag{9.3.61}$$

If, in addition, the reflection is weak, then equation (9.3.13a) implies $|q_{21}(\boldsymbol{x})| << |q_{22}(\boldsymbol{x})|$, and (9.3.61) further reduces to

$$\gamma = \int \ln|q_{22}(\boldsymbol{x})|\, p(\boldsymbol{x})\,d\boldsymbol{x} \tag{9.3.62}$$

Equation (9.3.62) is in agreement with equation (9.3.19) without the first damping term on the right-hand side.

We reiterate that the foregoing procedure is applicable only for the case of monocoupling, for which there exists only one positive Lyapunov exponent. In the case of multiwave coupling, the localization factor is identified with the smallest positive Lyapunov exponent (Kissel, 1991; Ariaratnam and Xie, 1994).

9.3.1 A Multispan Beam with Random Torsional Springs at Supports

For illustration, the three different procedures just described are applied to obtain the localization factor for an Euler-Bernoulli beam on evenly spaced hinge supports, and with an additional torsional spring at each support, as shown in Fig. 9.3.4. In order to focus our attention on certain key theoretical issues, without being obscured by numerical complexities, it is assumed that the beam is undamped and that only the torsional spring stiffnesses are random and are described by

$$k_n = k_0[1 + X(n)] \qquad n = 1, 2, \ldots \tag{a}$$

where k_0 is the average k_n, and $X(n)$ are independent and identically distributed random variables with zero mean. The other physical parameters are assumed to be deterministic, including the distances between neighboring supports l, the bending rigidity of the beam EI, and the mass of the beam per unit length m. All three methods are, of course, applicable when several or all parameters of the system are random.

Several choices can be made for a typical cell unit in the analysis. The one selected for the following discussion is a typical beam element between two neighboring supports plus the entire right spring. The entire left spring is treated as belonging to the preceding cell. The state vector at each cell-to-cell interface is composed of a rotational angle and a bending moment. Accordingly, the transfer matrix for a typical

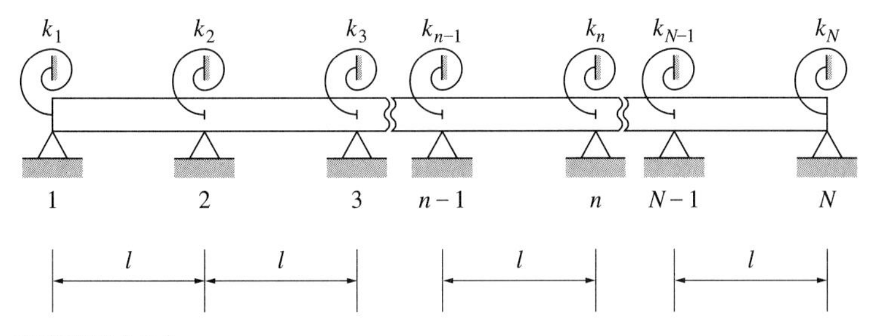

FIGURE 9.3.4
A beam on hinge supports with additional torsional spring at each support.

cell, say the nth cell, is given by

$$\boldsymbol{T}(n) = \begin{bmatrix} \beta & \dfrac{\alpha}{\sigma EI} \\ -\dfrac{\sigma EI(1-\beta^2)}{\alpha} + \dfrac{\beta}{\delta}[1+X(n)] & \beta + \dfrac{\alpha}{\delta\sigma l}[1+X(n)] \end{bmatrix} \tag{b}$$

where $\delta = EI/lk_0$, $\sigma = (\omega^2 m/EI)^{1/4}$, ω = frequency,

$$\alpha = \frac{\cosh\sigma l \cos\sigma l - 1}{\sinh\sigma l - \sin\sigma l} \tag{c}$$

and

$$\beta = \frac{\sinh\sigma l \cos\sigma l - \cosh\sigma l \sin\sigma l}{\sinh\sigma l - \sin\sigma l} \tag{d}$$

The nondimensional δ, known as the internal (cell-to-cell) coupling parameter, is the ratio between EI/l, the resistance against rotation provided by the beam, and k_0, the resistance against rotation provided by the torsional spring at a support. If k_0 is very large such that δ is very small, then there is little cell-to-cell coupling, and strong localization is expected to occur.

The transfer matrix $\boldsymbol{T}$ for the ideal cell unit without disorder and the point transfer matrix $\boldsymbol{P}(n)$, lumping the effect of disorder, are obtained as follows:

$$T = \begin{bmatrix} \beta & \dfrac{\alpha}{\sigma EI} \\ -\dfrac{\sigma EI(1-\beta^2)}{\alpha} + \dfrac{\beta}{\delta} & \beta + \dfrac{\alpha}{\delta\sigma l} \end{bmatrix} \tag{e}$$

$$P(n) = \begin{bmatrix} 1 & 0 \\ \dfrac{X(n)}{\delta} & 1 \end{bmatrix} \tag{f}$$

The eigenvalues of transfer matrix $\boldsymbol{T}$ are a reciprocal pair, which may be denoted conveniently by $\nu = e^{i\psi}$ and $\nu^{-1} = e^{-i\psi}$. It can be shown that ψ can be found from

$$\cos\psi = \beta + \frac{\alpha}{2\delta\sigma l} \tag{g}$$

and the wave-passage frequency bands are determined by the inequality

$$-1 \le \beta + \frac{\alpha}{2\delta\sigma l} \le 1 \tag{h}$$

The wave transfer matrix $\boldsymbol{Q}(n)$ is obtained as

$$Q(n) = \begin{bmatrix} 1 - iAX(n) & -iAX(n) \\ iAX(n) & 1 + iAX(n) \end{bmatrix} \tag{i}$$

where $A = \alpha/(2\delta\sigma l \sin\psi)$.

The following physical properties have been used in the numerical calculations: $m = 1.8043$ kg/m, $EI = 0.3140\ \text{N}\cdot\text{m}^2$, and $l = 0.1651$ m. In addition, the $X(n)$ are assumed to be uniformly distributed between $-\sqrt{3}\sigma_X$ and $\sqrt{3}\sigma_X$, where σ_X is the common standard deviation of $X(n)$.

The invariant probability density of the wave-phase difference for an infinite undamped disordered chain must satisfy equation (9.3.33). For the present multispan beam, equation (9.3.33) has the specific form

$$[p(\theta')]_{\theta'=\theta} = \left\{ \frac{1}{2\sqrt{3}\sigma_X} \int_{-\sqrt{3}\sigma_X}^{\sqrt{3}\sigma_X} \left| \frac{\partial\theta}{\partial\theta'} \right| p[\theta(\theta', x)]\, dx \right\}_{\theta'=\theta} = p(\theta) \qquad \text{(j)}$$

where $\left| \dfrac{\partial\theta}{\partial\theta'} \right| = G(\theta', x) = \{1 - 2Ax \sin(\theta' + 2\psi) + 2A^2x^2[1 + \cos(\theta' + 2\psi)]\}^{-1}$

(k)

and where $\theta(\theta', x)$ represents an implicit relation described by the following pair of equations:

$$\sin\theta = \{\sin(\theta' + 2\psi) - 2Ax[1 + \cos(\theta' + 2\psi)]\}G(\theta', x) \qquad (l.1)$$

$$\cos\theta = \{\cos(\theta' + 2\psi) + 2Ax \sin(\theta' + 2\psi) - 2A^2x^2[1 + \cos(\theta' + 2\psi)]\}G(\theta', x) \qquad (l.2)$$

Since $|x|$ is bounded by $\sqrt{3}\sigma_X$, a perturbation procedure, applicable when $|A\sigma_X|$ is small, can be devised as follows. Let

$$p(\theta) = \frac{1}{2\pi} + \sum_{j=1}^{\infty} (C_j \sin j\theta + D_j \cos j\theta) \qquad \text{(m)}$$

where

$$C_j = C_{j0} + C_{j1}A\sigma_X + C_{j2}(A\sigma_X)^2 + \cdots \qquad \text{(n.1)}$$

$$D_j = D_{j0} + D_{j1}A\sigma_X + D_{j2}(A\sigma_X)^2 + \cdots \qquad \text{(n.2)}$$

In equation (m), the probability density $p(\theta)$ is expanded into a Fourier series because it is a periodic function with a period 2π. The constant term in this expression is known to be $1/2\pi$; otherwise, the normalization condition for $p(\theta)$ will not be satisfied. The coefficients C_{jk} and D_{jk} are obtained by substituting equations (m), (n.1), and (n.2) into equation (j) and equating terms of the same power in $A\sigma_X$. The final expression for $p(\theta)$ can be expressed as follows:

$$p(\theta) = \frac{1}{2\pi} \left\{ 1 + A^2\sigma_X^2 \frac{\sin(\theta + \psi)}{\sin\psi} \left[1 + \frac{\cos(\theta + \psi)}{\cos\psi}\right] \right\} + O(A\sigma_X)^4 \qquad \text{(o)}$$

$$\cos 2n\psi \neq 1, \qquad -\pi \leq \theta \leq \pi$$

The preceding perturbation scheme is not applicable when $\cos 2n\psi = 1$.

Figure 9.3.5 depicts the computed approximate invariant probability densities $p(\theta)$ for the cases $\sigma_X = 0.01$, 0.1, and 0.2. The results are in good agreement with those obtained from Monte Carlo simulations, also shown in the figure as diamonds, triangles, and squares. The simulations were carried out according to equation (9.3.32), starting from an arbitrarily chosen right boundary. The distribution of θ became essentially invariant after 1000 cells, and the simulation results shown in Fig. 9.3.5 correspond to the relative frequencies calculated for the subsequent 10^6 cells.

For the present case, equation (9.3.41) is reduced to

$$\gamma = \frac{1}{2\sqrt{3}\sigma_X} \int_{-\pi}^{\pi} p(\theta)\, d\theta \int_{-\sqrt{3}\sigma_X}^{\sqrt{3}\sigma_X} \ln[1 + 2Ax \sin\theta + 2A^2x^2(1 + \cos\theta)]\, dx \qquad \text{(p)}$$

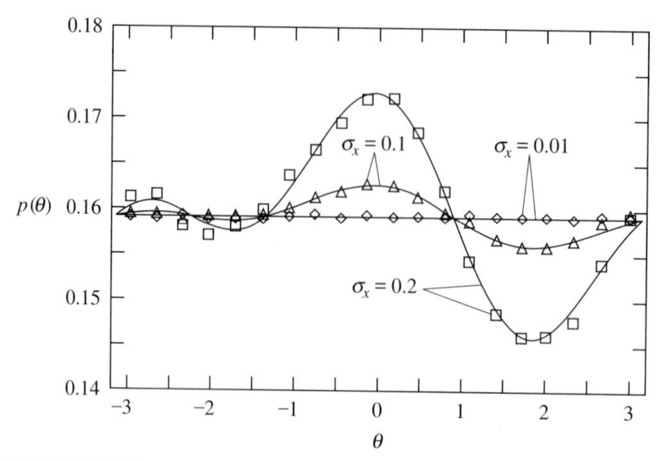

FIGURE 9.3.5
Invariant probability density of phase difference θ between the rightgoing and leftgoing waves, $\omega = 250$ rad/s, $\delta = 0.1$.

The localization factor γ can be computed numerically from equations (p) and (o). Although equation (p) is exact, the computed γ is approximate, because equation (o) is approximate. We recall that equation (o) is valid only for small $|A\sigma_X|$, which implies a weak localization.

Substituting equation (i) into equation (9.3.19), we obtain an approximate localization factor as follows for the case of a wave-passage frequency and weak localization on the basis of multiple reflections:

$$\gamma = \frac{1}{2}\ln(1 + 3A^2\sigma_X^2) - 1 + \frac{1}{\sqrt{3}A\sigma_X}\tan^{-1}(\sqrt{3}A\sigma_X) \tag{q}$$

where σ_X^2 is the variance of $X(n)$, which is assumed to be the same for every n. If, in addition, the disorder is small, we obtain from equation (9.3.21)

$$\gamma = \frac{\alpha^2\sigma_X}{8\delta^2\sigma^2 l^2 \sin^2\psi} \tag{r}$$

Equation (r) is valid only if the reflection coefficient at the cell-to-cell interface is small; otherwise, the use of equation (9.3.24) or (9.3.27) is more appropriate.

The computed localization factors in the first four wave-passage frequency bands are shown in Fig. 9.3.6*a* and *b* for $\delta = 1$, $\sigma_X = 0.1$ and for $\delta = 0.1$, $\sigma_X = 0.1$, respectively. Except for frequencies very close to the lower boundary of a wave-passage frequency band (not shown in Fig. 9.3.6), all seven equations (9.3.41), (9.3.19), (9.3.21), (9.3.24), (9.3.27), (9.3.61), and (9.3.62) yield essentially the same results, and in excellent agreement with those obtained from Monte Carlo simulations, also shown in the figure. That the values of the localization factors are below 0.02 is an indication of weak localization in these cases. At the lower boundary of a wave-passage frequency band, the deformations of two neighboring spans tend to be out of phase (exactly out of phase for an ideal periodic structure)

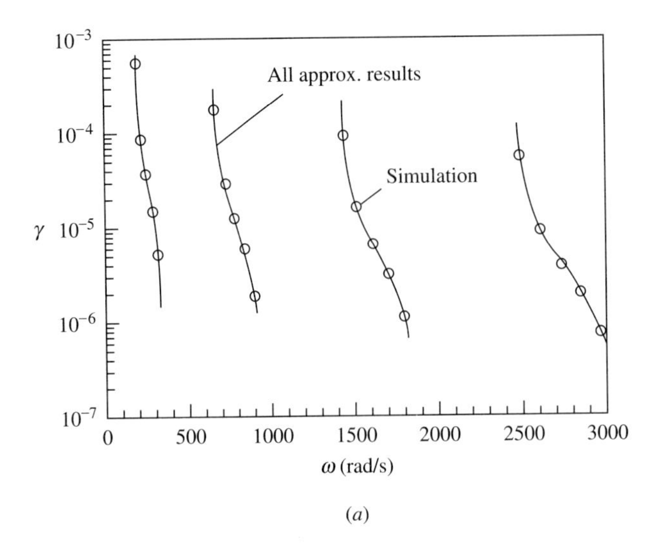

(*a*)

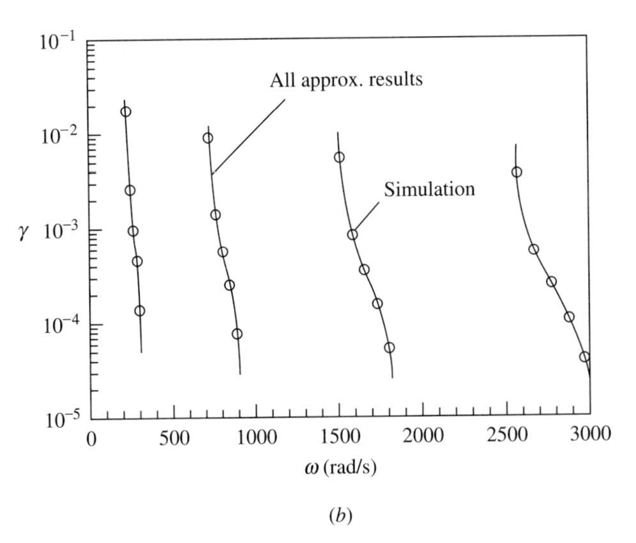

(*b*)

FIGURE 9.3.6
Localization factors at frequencies in wave-passage frequency bands.
(*a*) $\delta = 1, \sigma_X = 0.1$; (*b*) $\delta = 0.1, \sigma_X = 0.1$.

and the torsional springs at the supports influence the beam motion to a great extent. Therefore even a small disorder of the torsional springs may give rise to strong localization. On the other hand, the deformations of two neighboring spans are nearly in phase at the upper boundary of a wave-passage frequency band, and the torsional springs have little effect on the beam motion. Hence their disorder is relatively unimportant.

The Monte Carlo simulations, referred to previously, were carried out by multiplying a large number of random wave transfer matrices at a given frequency to obtain one realization of the wave transfer matrix for an overall N-span system,

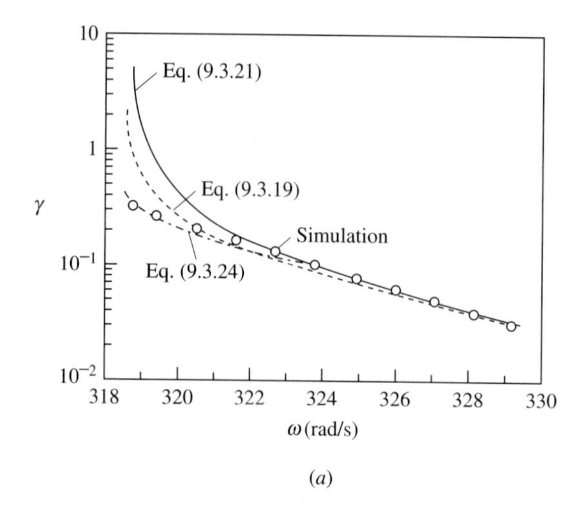

(*a*)

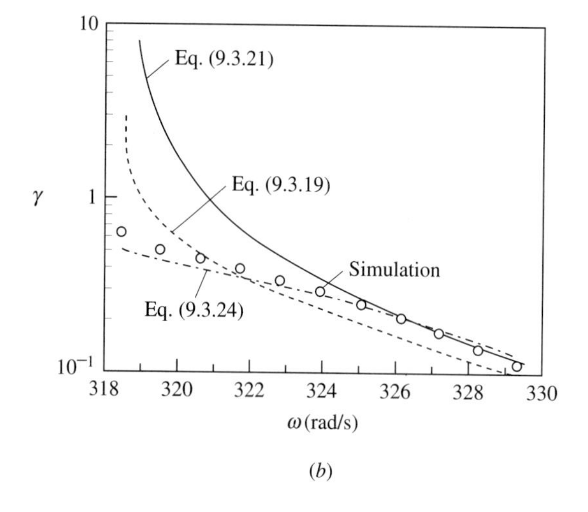

(*b*)

FIGURE 9.3.7
Localization factors at frequencies in lower half of the first wave-passage frequency band [after Cai and Lin, 1991a; Copyright ©1990 AIAA, reprinted with permission]. (*a*) $\delta = 0.01$, $\sigma_X = 0.25$; (*b*) $\delta = 0.01$, $\sigma_X = 0.5$.

and thus one realization of the overall transmission coefficient $t^r(N, 1)$. The values of $\ln |t^r(N, 1)| /N$ for different realizations were averaged over a sufficiently large number of realizations to obtain an estimate for the localization factor. A total number of 300 cells were used in each realization, and the average was taken over 2000 realizations.

Figure 9.3.7*a* and *b* depicts the computed localization factors, using the method of multiple reflections, for two disordered systems, one with $\delta = 0.01$, $\sigma_X = 0.25$ and the other with $\delta = 0.01$, $\sigma_X = 0.5$, and for frequencies falling within the lower part of the first wave-passage frequency band. When compared with the simulation results, it is seen that the use of equations (9.3.19) and (9.3.21) greatly overestimates the localization factor near the lower boundary of the wave-passage frequency band, where the reflection coefficients are larger, and the effect of multiple reflection is no longer negligible. Much more accurate results are obtained by taking into account the reflections from one additional cell on the immediate left, that is, by using equation (9.3.24). These figures show that for the disordered beam systems with the given δ and σ_X investigated, localization is strong at a frequency near the lower boundary of a wave-passage frequency band, and it becomes weaker as the frequency increases. Thus the disorder to internal coupling ratio (σ_X/δ) is not the only factor affecting the intensity of localization, as once suggested; the frequency location relative to a wave-passage band is at least another factor. Other factors might be important in more complicated systems.

Results obtained from equation (9.3.19) are generally more accurate than those obtained from equation (9.3.21). However, since both equations are approximate, exceptions can occur, as shown in the higher frequency region in Fig. 9.3.7. Although equation (9.3.19) effectively retains more terms in the approximation, it does not follow that those additional terms will increase the accuracy consistently at all frequencies.

For the particular example under consideration, disorder arises from the random variation of torsional spring over the supports. In this case, localization may be strong at frequencies near the lower boundaries of wave-passage frequency bands, even when the disorder is relatively small. Three cases—(a) $\delta = 0.1$, $\omega = 232.5$ rad/s; (b) $\delta = 0.01$, $\omega = 318.7$ rad/s; and (c) $\delta = 0.001$, $\omega = 339.7$ rad/s—have been examined in more detail, using the method of multiple reflections. The results obtained are shown in Fig. 9.3.8*a*–*c*, respectively, with varying standard deviation of disorder σ_X. In all these cases equation (9.3.24) is clearly inadequate. Equation (9.3.27) leads to satisfactory results for cases (b) and (c) . For case (a), in which

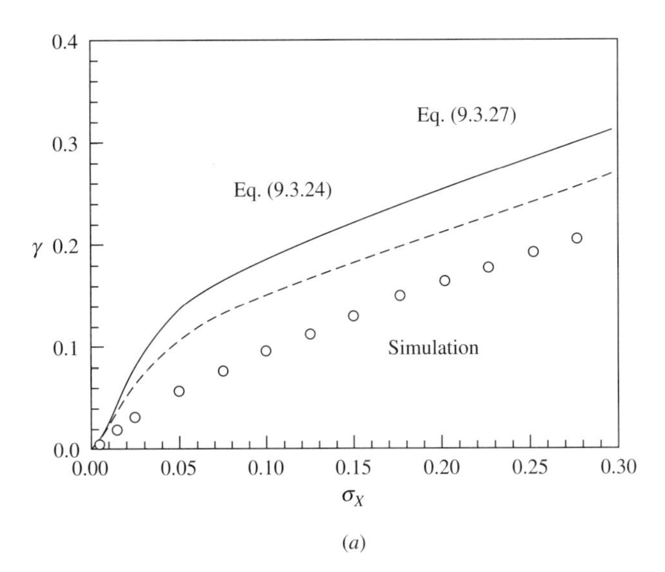

FIGURE 9.3.8
Localization factors at frequencies near the lower boundary of the first wave-passage frequency band [after Cai and Lin, 1991a; Copyright ©1990 AIAA, reprinted with permission]. (*a*) $\delta = 0.1$, $\omega = 232.5$ rad/s.

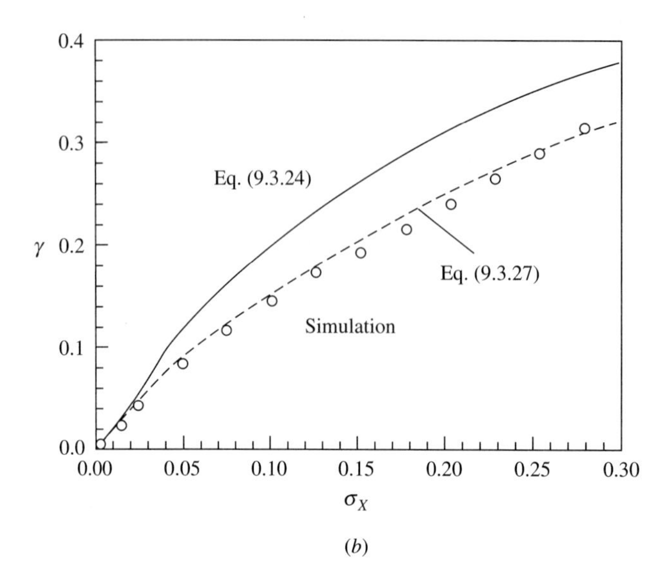

(*b*)

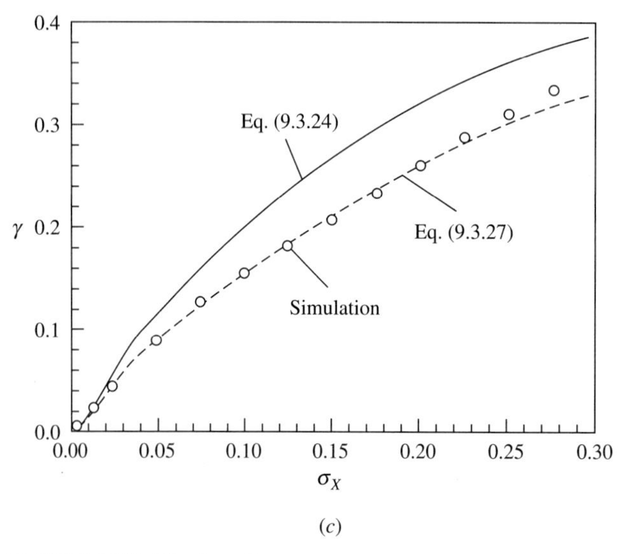

(*c*)

FIGURE 9.3.8 (*Continued*)
(*b*) $\delta = 0.01, \omega = 318.7$ rad/s; (*c*) $\delta = 0.001, \omega = 339.7$ rad/s.

the disorder is also greater, equation (9.3.27) is still inadequate, although it is much better than equation (9.3.24). The use of higher order approximations appears warranted.

9.4 FREQUENCY RESPONSE OF DISORDERED PERIODIC STRUCTURES

The variability of system response to external excitations due to disorder was considered in general terms in Section 9.1. In the present section, our attention is focused on disordered periodic structures, for which the general approach is not suitable. Early publications on the response of disordered periodic structures were related to specific systems, for example, a chain of linear oscillators by Soong and Bogdanoff (1964), a multispan beam by Yang and Lin (1975), and a mistuned bladed disk by Huang (1982), using perturbation analyses. Here we treat instead a generic monocoupled disordered chain, employing a procedure similar to the method of invariant phase distribution discussed in Section 9.3.

EXTERNAL EXCITATION AT THE INTERFACE BETWEEN TWO CELLS. Consider a disordered chain composed of N cell units, subjected to an external sinusoidal force with a complex amplitude f_e acting at the interface between cells n and $n+1$, as shown in Fig. 9.4.1. The external force f_e and the displacement at the excitation point, denoted by w_e, are related to the state variables at stations n_r and $n+1_l$ as follows:

$$w_e = w(n+1_l) = w(n_r) \tag{9.4.1}$$

$$f_e = f(n+1_l) - f(n_r) \tag{9.4.2}$$

Equation (9.4.2) implies that f_e is expressed in the same unit as the generalized forces on the two sides of the interface. The frequency response function $H(\omega)$ is given by

$$H(\omega) = \frac{w_e}{f_e} = \frac{w(n_r)}{f(n+1_l) - f(n_r)} = \left[\frac{f(n+1_l)}{w(n+1_l)} - \frac{f(n_r)}{w(n_r)}\right]^{-1} \tag{9.4.3}$$

By making use of equations (9.3.4a) and (9.3.4b), equation (9.4.3) may be changed to read

$$H(\omega) = \left[\frac{d_{21}\eta(n+1_l) + d_{22}}{d_{11}\eta(n+1_l) + d_{12}} - \frac{d_{21}\eta(n_r) + d_{22}}{d_{11}\eta(n_r) + d_{12}}\right]^{-1} \tag{9.4.4}$$

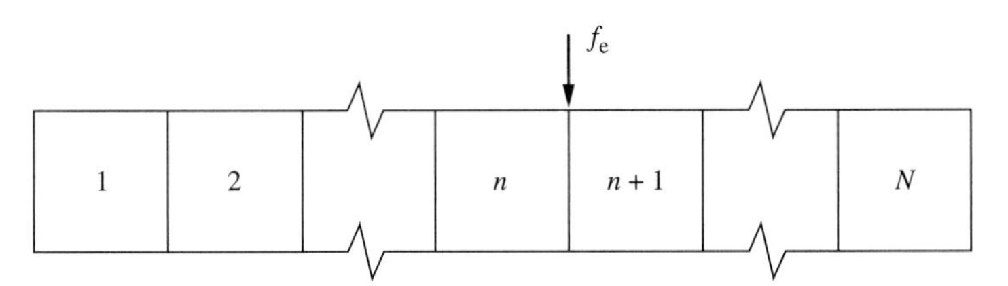

FIGURE 9.4.1
External excitation at the interface between cells n and $n+1$.

where d_{ij} are elements of matrix $\boldsymbol{D}$, which is composed of the eigenvectors of the ideal transfer matrix $\boldsymbol{T}$, and where η is the ratio of the rightgoing and leftgoing waves:

$$\eta(n+1_l) = \frac{\mu^r(n+1_l)}{\mu^l(n+1_l)} \qquad \eta(n_r) = \frac{\mu^r(n_r)}{\mu^l(n_r)} \tag{9.4.5}$$

Since each μ is complex-valued, each wave ratio η is also complex-valued. The modulus of η is the magnitude ratio of the rightgoing and leftgoing waves, while the phase angle of η is the phase difference between the two waves. Note that $\eta(n+1_l) \neq \eta(n_r)$ because an external force f_e is present, and their moduli are generally not equal to 1, on account of damping in the system.

Since no excitation is present on the right of cell n, we have for $m > n$,

$$\begin{Bmatrix} \mu^r(m_r) \\ \mu^l(m_r) \end{Bmatrix} = \boldsymbol{D}^{-1}\boldsymbol{T}(m)\boldsymbol{D}\begin{Bmatrix} \mu^r(m_l) \\ \mu^l(m_l) \end{Bmatrix} = \boldsymbol{Q}(m)\begin{Bmatrix} \mu^r(m_l) \\ \mu^l(m_l) \end{Bmatrix} \tag{9.4.6}$$

$$\begin{Bmatrix} \mu^r(m+1_l) \\ \mu^l(m+1_l) \end{Bmatrix} = \begin{Bmatrix} \mu^r(m_r) \\ \mu^l(m_r) \end{Bmatrix} \tag{9.4.7}$$

Note that matrix $\boldsymbol{Q}(m)$ in equation (9.4.6) is defined differently from that of equation (9.3.11). It can be shown that the wave ratios, $\eta(m_l)$ and $\eta(m_r)$, at two consecutive stations are related as follows:

$$\eta(m_l) = \frac{q_{12}(m) - \eta(m_r)q_{22}(m)}{\eta(m_r)q_{21}(m) - q_{11}(m)} \qquad m = n+1, \ldots, N \tag{9.4.8}$$

$$\eta(m_r) = \eta(m+1_l) \qquad m = n+1, \ldots, N \tag{9.4.9}$$

where $q_{ij}(m)$ are elements of matrix $\boldsymbol{Q}(m)$. Similarly, the wave ratios at two consecutive stations on the left-hand side of cell $n+1$ have the relationship

$$\eta(m_r) = \frac{\eta(m_l)q_{11}(m) + q_{12}(m)}{\eta(m_l)q_{21}(m) + q_{22}(m)} \qquad m = n, n-1, \ldots, 1 \tag{9.4.10}$$

$$\eta(m_l) = \eta(m-1_r) \qquad m = n, n-1, \ldots, 2 \tag{9.4.11}$$

Note that the wave ratios, $\eta(N_r)$ and $\eta(1_l)$, at the right and left boundaries are dependent only on the respective boundary conditions. For example, if a boundary is clamped (that is, $w = 0$), then $\eta = -d_{12}/d_{11}$; and if a boundary is free ($f = 0$), then $\eta = -d_{22}/d_{21}$.

Equations (9.4.4) and (9.4.8) through (9.4.11) provide a simple scheme with which the frequency response function $H(\omega)$ can be calculated for a disordered chain. Beginning with $\eta(N_r)$ determined by the right boundary condition, we compute progressively $\eta(N_l), \eta(N-1_r), \ldots, \eta(n+1_l)$ using equations (9.4.8) and (9.4.9). Similarly, we obtain $\eta(n_r)$ beginning from the left boundary condition, using equations (9.4.10) and (9.4.11). Finally, the frequency response function is calculated from equation (9.4.4). The scheme is numerically efficient, due to the recursive nature of equations (9.4.8) through (9.4.11). It can be applied to a long chain with a large number of cells and to a large number of sample chains for the purpose of Monte Carlo simulation.

We digress to note that if the structure is perfectly periodic, and if no damping mechanism exists at the two boundaries, then

$$\eta(n+1_l) = \eta(N_r)\nu^{-2(N-n)} \qquad \eta(n_r) = \eta(1_l)\nu^{2n} \tag{9.4.12}$$

where ν is of the smaller modulus of the two eigenvalues of matrix $\boldsymbol{T}$, and $|\nu| < 1$ because of damping. Substitution of equation (9.4.12) into equation (9.4.4) leads to the frequency response function for an ideal periodic structure.

Return now to the disordered chain. Let each of the random parameters in a disordered cell unit be represented by a mean (or the nominal design value) plus a random variable with zero mean. For cell m, these random variables constitute a random vector denoted by $\boldsymbol{X}(m) = \{X_1(m), X_2(m), \ldots, X_k(m)\}$. It is reasonable to assume that $X_j(1), X_j(2), \ldots, X_j(N)$, for each j, are independent and identically distributed random variables. Being functions of the disordered parameters, $q_{ij}(m)$ may be denoted by

$$q_{ij}(m) = q_{ij}[\boldsymbol{X}(m)] \tag{9.4.13}$$

Thus equations (9.4.4) and (9.4.8) through (9.4.11) describe the functional relationships among random variables H and $q_{ij}[\boldsymbol{X}(m)]$, $m = 1, 2, \ldots, N$. If the probability distribution for $\boldsymbol{X}(m)$ is known, then in principle the probability distribution and the statistical moments of H are obtainable.

Specifically, we obtain from equations (9.4.6) and (9.4.7)

$$\begin{Bmatrix} \mu^r(N_r) \\ \mu^l(N_r) \end{Bmatrix} = \boldsymbol{Q}(N, n+1) \begin{Bmatrix} \mu^r(n+1_l) \\ \mu^l(n+1_l) \end{Bmatrix} \tag{9.4.14}$$

and

$$\begin{Bmatrix} \mu^r(n_r) \\ \mu^l r(n_r) \end{Bmatrix} = \boldsymbol{Q}(n, 1) \begin{Bmatrix} \mu^r(1_l) \\ \mu^l(1_l) \end{Bmatrix} \tag{9.4.15}$$

where

$$\boldsymbol{Q}(i, j) = \boldsymbol{Q}(i)\boldsymbol{Q}(i-1)\cdots\boldsymbol{Q}(j) \qquad i > j \tag{9.4.16}$$

It follows that

$$\eta(n+1_l) = \frac{q_{12}(N, n+1) - \eta(N_r)q_{22}(N, n+1)}{\eta(N_r)q_{21}(N, n+1) - q_{11}(N, n+1)} \tag{9.4.17}$$

$$\eta(n_r) = \frac{\eta(1_l)q_{11}(n, 1) + q_{12}(n, 1)}{\eta(1_l)q_{21}(n, 1) + q_{22}(n, 1)} \tag{9.4.18}$$

Equations (9.4.17) and (9.4.18) are analogous to equations (9.4.8) and (9.4.10), respectively. Since $\eta(N_r)$ is a constant determinable from the boundary condition at N_r, $\eta(n+1_l)$ is a function of random vectors $\boldsymbol{X}(n+1), \boldsymbol{X}(n+2), \ldots, \boldsymbol{X}(N)$. Similarly, $\eta(n_r)$ is a function of random vectors $\boldsymbol{X}(1), \boldsymbol{X}(2), \ldots, \boldsymbol{X}(n)$. Therefore, the frequency response H, given by equation (9.4.4), is a function of $\boldsymbol{X}(1), \boldsymbol{X}(2), \ldots, \boldsymbol{X}(N)$. The lth moments of $Y = | H |$ are then obtained from ensemble-averaging:

$$E[|H|^l] = \int y^l p[\boldsymbol{x}(1)]\, p[\boldsymbol{x}(2)] \cdots p[\boldsymbol{x}(N)]\, d\boldsymbol{x}(1)\, d\boldsymbol{x}(2) \cdots d\boldsymbol{x}(N) \tag{9.4.19}$$

The domain of the $(N \times k)$-fold integration in equation (9.4.19) is a superrectangle. The product $p[\boldsymbol{x}(1)] \cdots p[\boldsymbol{x}(N)]$ implies that $\boldsymbol{X}(1), \ldots, \boldsymbol{X}(N)$ are mutually independent, and all $p[\boldsymbol{x}(m)] (m = 1, 2, \ldots, N)$ are of the same form under the assumption of identical distribution. In general, the integration on the right-hand side of (9.4.19) must be carried out numerically, and it becomes impractical when the number of cells in the chain is large. Analytical determination of the probability distribution of H from those of $\boldsymbol{X}(m)$ is even less tractable, although conceptually equally simple.

However, the recurrence relationship, equations (9.4.8) through (9.4.11), provides an efficient framework for Monte Carlo simulation. With a large enough sample size, accurate probability distribution and statistical moments can be obtained by averaging the simulated results. In general, a larger sample size is required for the computation of probability density than what is needed for the mean and mean-square values.

EXTERNAL EXCITATION AT AN INTERIOR POINT OF A CELL. The problem becomes slightly more complicated if the external force f_e is applied at an interior point P of cell n, as shown in Fig. 9.4.2. In this case, the state vectors at the two ends of cell n may be written as

$$\begin{Bmatrix} w(n_r) \\ f(n_r) \end{Bmatrix} = \boldsymbol{T}(n) \begin{Bmatrix} w(n_l) \\ f(n_l) \end{Bmatrix} + \begin{Bmatrix} g_1 \\ g_2 \end{Bmatrix} f_e \tag{9.4.20}$$

where g_1 and g_2 are obtainable from the knowledge of the nth cell and the position of point P. Furthermore, the external force f_e need not be expressed in the same unit as the generalized forces $f(n_r)$ and $f(n_l)$, since any unit conversion, if required, may be incorporated in g_1 and g_2. Transforming the state vectors in equation (9.4.20) into the wave vectors, we obtain

$$\begin{Bmatrix} \mu^r(n_r) \\ \mu^l(n_r) \end{Bmatrix} = \boldsymbol{Q}(n) \begin{Bmatrix} \mu^r(n_l) \\ \mu^l(n_l) \end{Bmatrix} + \begin{Bmatrix} \gamma_1 \\ \gamma_2 \end{Bmatrix} f_e \tag{9.4.21}$$

where

$$\begin{Bmatrix} \gamma_1 \\ \gamma_2 \end{Bmatrix} = \boldsymbol{D}^{-1} \begin{Bmatrix} g_1 \\ g_2 \end{Bmatrix} \tag{9.4.22}$$

Equation (9.4.21) can be rearranged to read

$$\mu^l(n_r)\eta(n_r) = [q_{11}(n)\eta(n_l) + q_{12}(n)]\mu^l(n_l) + \gamma_1 f_e \tag{9.4.23a}$$

$$\mu^l(n_r) = [q_{21}(n)\eta(n_l) + q_{22}(n)]\mu^l(n_l) + \gamma_2 f_e \tag{9.4.23b}$$

Solving for $\mu^l(n_l)$,

$$\mu^l(n_l) = \frac{[\gamma_1 - \gamma_2\eta(n_r)]f_e}{[q_{21}(n)\eta(n_l) + q_{22}(n)]\eta(n_r) - [q_{11}(n)\eta(n_l) + q_{12}(n)]} \tag{9.4.24}$$

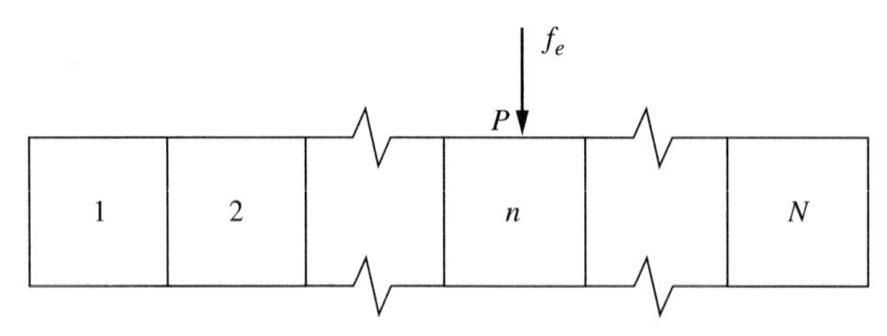

FIGURE 9.4.2
External excitation at an interior point of cell n.

Now, let w_e be the displacement at the excitation point P of interest. It need not be in the same unit as $w(n_r)$ and $w(n_l)$, and it is expressible in terms of $w(n_l)$ and $f(n_l)$ as follows:

$$w_e = h_1 w(n_l) + h_2 f(n_l) \tag{9.4.25}$$

where h_1 and h_2, similar to g_1 and g_2, can also be obtained from the knowledge of cell n and the position of point P. Transforming $w(n_l)$ and $f(n_l)$ in terms of the wave vector $\mu^l(n_l)$ and the wave ratio $\eta(n_l) = \mu^r(n_l)/\mu^l(n_l)$,

$$w_e = \{h_1[d_{11}\eta(n_l) + d_{12}] + h_2[d_{21}\eta(n_l) + d_{22}]\}\mu^l(n_l) \tag{9.4.26}$$

Finally, the frequency response function is obtained by combining equations (9.4.24) and (9.4.26):

$$H(\omega) = \frac{w_e}{f_e} = \frac{[\gamma_1 - \gamma_2\eta(n_r)]\{h_1[d_{11}\eta(n_1) + d_{12}] + h_2[d_{21}\eta(n_l) + d_{22}]\}}{[q_{21}(n)\eta(n_l) + q_{22}(n)]\eta(n_r) - [q_{11}(n)\eta(n_l) + q_{12}(n)]} \tag{9.4.27}$$

where $\eta(n_r)$ can be calculated by using equations (9.4.8) and (9.4.9) beginning from the right boundary, and $\eta(n_l)$ can be calculated by using equations (9.4.10) and (9.4.11) beginning from the left boundary. Following a procedure similar to that leading to (9.4.17) and (9.4.18), we obtain

$$\eta(n_r) = \frac{q_{12}(N, n+1) - \eta(N_r)q_{22}(N, n+1)}{\eta(N_r)q_{21}(N, n+1) - q_{11}(N, n+1)} \tag{9.4.28}$$

$$\eta(n_l) = \frac{\eta(1_l)q_{11}(n-1, 1) + q_{12}(n-1, 1)}{\eta(1_l)q_{21}(n-1, 1) + q_{22}(n-1, 1)} \tag{9.4.29}$$

Thus the probabilistic and statistical properties of the frequency response can be obtained from Monte Carlo simulation, using equations (9.4.27) and (9.4.8) through (9.4.11). If the number of cells is small, the statistical moments can also be obtained analytically using equations (9.4.19) and (9.4.27) through (9.4.29).

In the ideal case without disorder and without energy dissipation at the two boundaries, equations (9.4.28) and (9.4.29) reduce to

$$\eta(n_r) = \eta(N_r)\nu^{-2(N-n)} \qquad \eta(n_l) = \eta(1_l)\nu^{2(n-1)} \tag{9.4.30}$$

9.4.1 A Multispan Beam with Random Span Lengths

For illustration, the foregoing procedure is applied to an Euler-Bernoulli beam on multiple hinge supports, as shown in Fig. 9.4.3. For simplicity, it is assumed that the number of spans N is $2n - 1$, and an external transverse force is applied at the middle of the beam, that is, at the center of the nth span. The response of interest is the transverse displacement at the same point. A typical cell unit is chosen to be a beam element between two neighboring supports. The generalized displacement and the corresponding generalized force at the cell-to-cell interface are the rotational

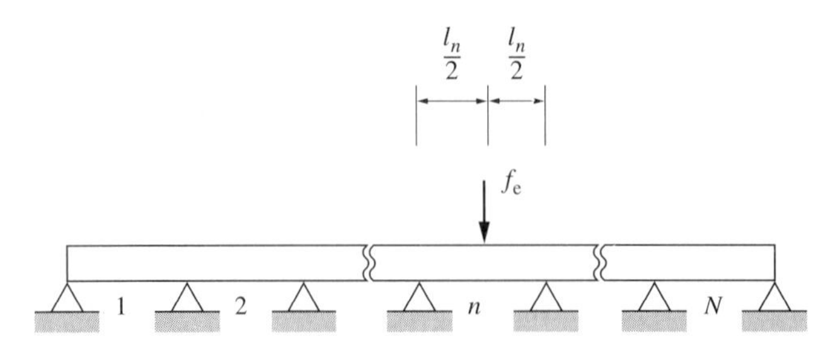

FIGURE 9.4.3
A multispan beam with random span lengths, $N = 2n - 1$.

angle and the bending moment, respectively. Since the external force is applied at an interior point of the nth cell, the frequency response function is given by equation (9.4.27).

To focus our attention on certain key issues and avoid numerical complexities, only the spacings between neighboring supports are assumed to be random and described by

$$l_m = l[1 + X(m)] \qquad m = 1, 2, \ldots, N \tag{a}$$

where l is the average l_m, and the $X(m)$ are independent and identically distributed random variables with zero mean. For numerical calculations, the $X(m)$ are further assumed to be uniformly distributed between $-\sqrt{3}\sigma_X$ and $\sqrt{3}\sigma_X$, where σ_X is the standard deviation. The other physical parameters are taken to be deterministic, including the bending rigidity of the beam EI and the mass of the beam per unit length M. The structural damping is introduced by adding an imaginary part to Young's modulus E_0: $E = E_0(1 + i\zeta \operatorname{sgn} \omega)$, where ζ is the loss factor.

The transfer matrix for the mth cell is given by

$$\boldsymbol{T}(m) = \begin{bmatrix} \beta_m & \dfrac{\alpha_m}{\sigma EI} \\ -\dfrac{\sigma EI(1 - \beta_m^2)}{\alpha_m} & \beta_m \end{bmatrix} \tag{b}$$

where $\sigma = (\omega^2 M/EI)^{1/4}$, ω = frequency,

$$\alpha_m = \frac{\cosh \sigma l_m \cos \sigma l_m - 1}{\sinh \sigma l_m - \sin \sigma l_m} \tag{c}$$

and

$$\beta_m = \frac{\sinh \sigma l_m \cos \sigma l_m - \cosh \sigma l_m \sin \sigma l_m}{\sinh \sigma l_m - \sin \sigma l_m} \tag{d}$$

The transfer matrix $\boldsymbol{T}$ of the ideal periodic structure is given by

$$\boldsymbol{T} = \begin{bmatrix} \beta & \dfrac{\alpha}{\sigma EI} \\ -\dfrac{\sigma EI(1 - \beta^2)}{\alpha} & \beta \end{bmatrix} \tag{e}$$

where α and β are constants obtained from equations (c) and (d), respectively, by letting $l_m = l$. The eigenvalues of transfer matrix $\boldsymbol{T}$ are

$$\nu = \beta \pm \sqrt{\beta^2 - 1} \tag{f}$$

The transformation matrix $\boldsymbol{D}$ and the wave transfer matrix $\boldsymbol{Q}(m)$ can then be easily constructed. The functions g_1, g_2 in equation (9.4.20) and h_1, h_2 in equation (9.4.25) are found to be

$$g_1 = \frac{1}{2EI\sigma^2}\left(\cosh\frac{\sigma l_n}{2} - \cos\frac{\sigma l_n}{2}\right) \tag{g}$$

$$g_2 = \frac{1}{2\sigma}\left(\sinh\frac{\sigma l_n}{2} + \sin\frac{\sigma l_n}{2}\right) \tag{h}$$

$$h_1 = \frac{1}{2\sigma}\left[\sinh\frac{\sigma l_n}{2} + \sin\frac{\sigma l_n}{2} - \frac{\left(\cosh\frac{\sigma l_n}{2} - \cos\frac{\sigma l_n}{2}\right)(\sinh\sigma l_n + \sin\sigma l_n)}{\sinh\sigma l_n - \sin\sigma l_n}\right] \tag{i}$$

$$h_2 = \frac{1}{2EI\sigma^2}\left[\cosh\frac{\sigma l_n}{2} - \cos\frac{\sigma l_n}{2} - \frac{\left(\sinh\frac{\sigma l_n}{2} - \sin\frac{\sigma l_n}{2}\right)(\cosh\sigma l_n - \cos\sigma l_n)}{\sinh\sigma l_n - \sin\sigma l_n}\right] \tag{j}$$

The following physical properties have been used in the numerical calculations: $M = 1.8043$ kg/m, $E_0 I = 0.3140$ N $\cdot$ m^2, and $l = 0.1651$ m.

As indicated earlier, the Monte Carlo simulation procedure is efficient for generating the sample functions of the frequency response. The probability density functions of the frequency response magnitude, so obtained and normalized with respect to the reference value y_0 of the corresponding ideal periodic system, are shown in Figs. 9.4.4 and 9.4.5 for a 21-span disordered beam. The selected excitation frequency $\omega = 200$ rad/s is within the first wave-passage frequency band of the ideal periodic system. Figure 9.4.4 is plotted for a fixed loss factor $\zeta = 0.01$ and three different values for the standard deviation of the random span length, $\sigma_X = 0.01$, 0.02, and 0.05. Figure 9.4.5, however, is plotted for a fixed standard deviation of the random span length $\sigma_X = 0.01$ and three different values for the loss factor,

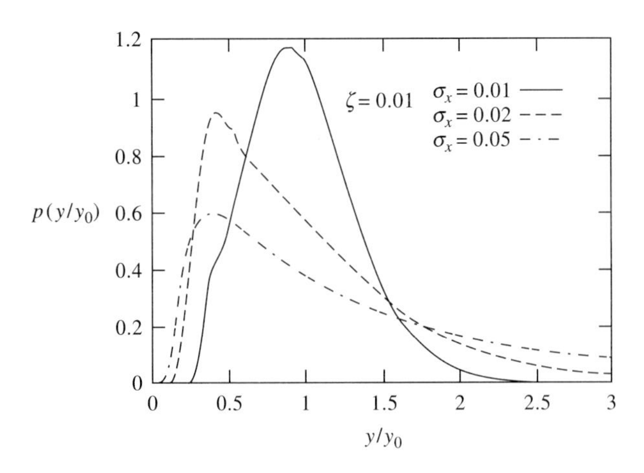

FIGURE 9.4.4
Probability density of nondimensional frequency response magnitude, disordered 21-span beam, $\omega = 200$ rad/s, standard deviation of disorder $\sigma = 0.01, 0.02, 0.05$.

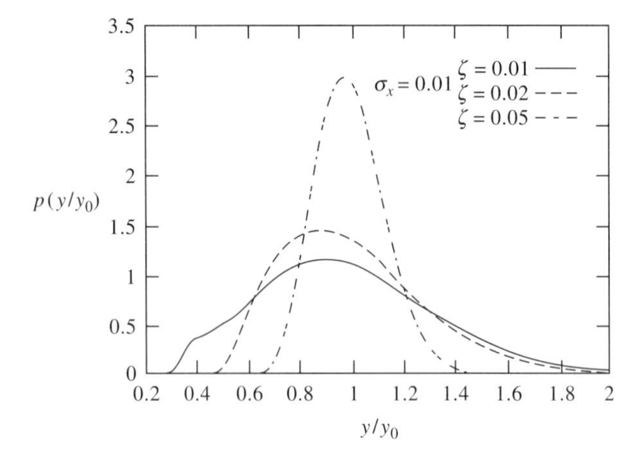

FIGURE 9.4.5
Probability density of nondimensional frequency response magnitude, disordered 21-span beam, $\omega = 200$ rad/s, loss factor $\zeta = 0.01, 0.02, 0.05$.

$\zeta = 0.01, 0.02$, and 0.05. A total of 10^6 samples were generated for each case. It is clear from the shapes of the probability density functions that the frequency response magnitude of a sample disordered structure can be either larger or smaller than that of the corresponding ideal periodic counterpart. Moreover, while both damping and disorder play a similar role of causing decay in wave propagation, their effects on the frequency response magnitude are different. When damping is increased, the range over which the response magnitude is distributed is reduced, resulting in a smaller standard deviation. Therefore, damping is beneficial in two ways: it reduces the response magnitude and it makes a disordered structure behave more like an ideal structure. On the other hand, with a higher level of disorder, represented by a larger standard deviation of the random parameter, the peak location of the probability density is shifted lower, but it is accompanied by a higher probability in the higher range of y/y_0. Thus disorder can be harmful from a structure reliability point of view—it increases the probability for the response magnitude to exceed a tolerant limit.

The mean and the standard deviation of the magnitude of the frequency response have also been calculated for the 21-span beam, using samples generated from Monte Carlo simulations. The results are shown in Fig. 9.4.6 for three different levels of damping. A total of 10^5 samples were used to obtain a mean value or a standard deviation. At an excitation frequency $\omega = 200$ rad/s, the mean of the frequency response magnitude of a disordered beam is mostly higher than the frequency response magnitude of the corresponding ideal periodic structure. Only in the case of weak damping ($\zeta = 0.01$) and small disorder ($\sigma_X < 0.015$) is the mean of the frequency response magnitude slightly lower than that of the ideal periodic counterpart. As expected, the standard deviation of the response magnitude increases with the standard deviation of the disorder parameter. A higher damping reduces the standard deviation of the response magnitude and brings the mean value closer to the ideal reference value y_0.

Figure 9.4.7 depicts the exact mean value and standard deviation of the frequency response magnitude for a five-span disordered beam, calculated by

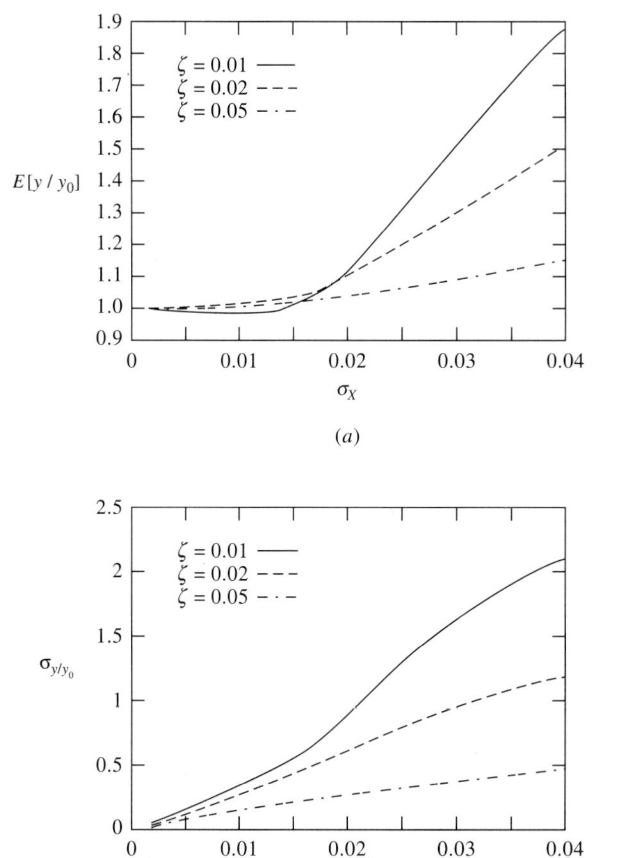

FIGURE 9.4.6
(*a*) Mean and (*b*) standard deviation of nondimensional frequency response magnitude, disordered 21-span beam, $\omega = 200$ rad/s, loss factor $\zeta = 0.01$, 0.02, 0.05.

ensemble-averaging, that is, by numerically integrating equation (9.4.19). Shown as diamonds, squares, and triangles in the figure are the results from Monte Carlo simulations. It is seen again that disorder generally increases the mean as well as the standard deviation of the response magnitude, when compared with those of the ideal periodic structure, and that damping reduces the effect of disorder on the response magnitude. It is interesting to note that the computation time required for numerically integrating equation (9.4.19) increases dramatically with the number of cells in the system, while the time spent for Monte Carlo simulation increases only linearly. The CPU times spent on a DEC station 5000/200 to obtain a mean value of the response magnitude by exact ensemble-averaging and by Monte Carlo simulation with 10^5 samples are compared in Table 9.4.1. It is seen that when the number of spans is small (≤ 7), ensemble-averaging is more economical than Monte Carlo simulation. In this particular example, ensemble-averaging is no longer practical when the number of spans is greater than 9.

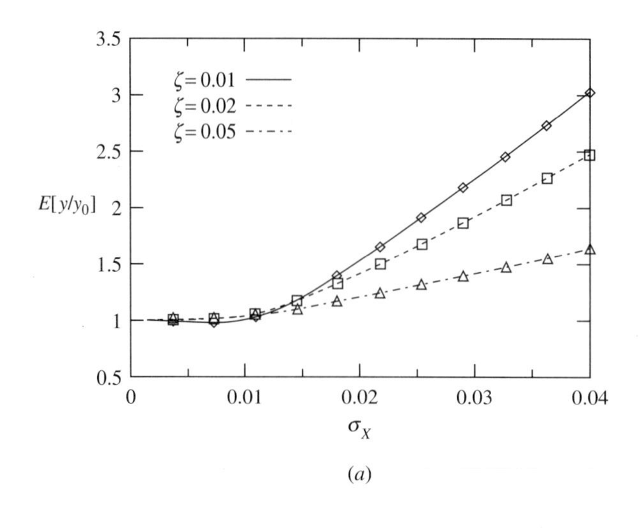

(*a*)

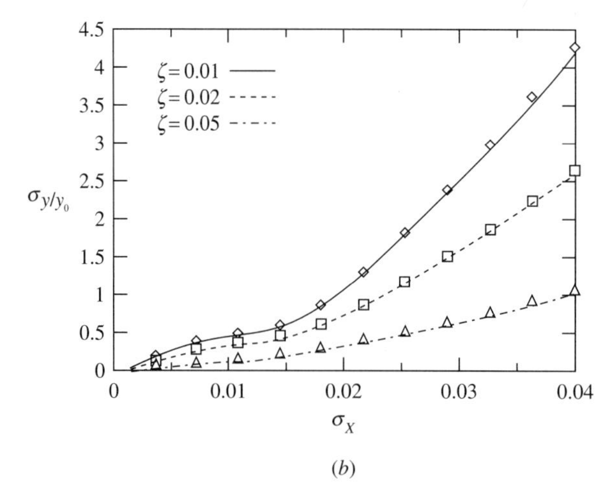

(*b*)

FIGURE 9.4.7
(*a*) Mean and (*b*) standard deviation of nondimensional frequency response magnitude, disordered five-span beam, ω = 200 rad/s, loss factor ζ = 0.01, 0.2, 0.05.

TABLE 9.4.1
CPU time for computing a mean value by two methods (in seconds)

Cell number	5	7	9	11
Time for integrating (9.4.19)	<1	26	$>10^4$	
Time for simulation	113	153	196	234

9.5 CONCLUDING REMARKS

The problem of disordered parameters in a dynamical system is conceptually simple. If the probability distribution of the disordered parameter is known, then in principle the problem can be solved in two steps. First, the probability distribution or the statistical moments of the system response to external random excitation can be obtained on the condition of the knowledge of the disordered parameters. Second, the results are averaged over the distribution of the parameters. If the external excitations are deterministic, then the problem can be solved alternatively by way of the Liouville equation, which is a Fokker-Planck equation without the diffusion term. However, these general procedures are practical only for simple systems, such as a single-degree-of-freedom oscillator.

For complicated systems, the perturbation scheme is generally applicable. In most cases, the objective of a perturbation analysis is to obtain the statistical properties of a system, given the statistical properties of the random parameters. The greatest drawback of the approach, however, is that the accuracy of results so obtained is difficult to estimate.

Special methods of analysis can sometimes be devised for special types of systems. This is the case with disordered periodic structures. By using the concepts of wave propagation and wave scattering when motion is propagated across the interface of two nonidentical cell units, the additional attenuation due to disorder, known as the localization effect, can be evaluated. These concepts have also been applied in this chapter in the development of a recursive procedure to compute the frequency response function for a disordered periodic structure. The procedure is useful for obtaining analytically the statistical moments of the frequency response function. More important, it can also be used to obtain the probability density by Monte Carlo simulation, especially when the disordered chain is long.

9.6 EXERCISES

9.1. Given a linear oscillator under a harmonic excitation,

$$\ddot{X} + 2\zeta\omega_0\dot{X} + \omega_0^2 X = A \sin \nu t$$

in which ζ, ν, and A are constants, $0 < \zeta < 1$, and ω_0 is a random variable uniformly distributed in $[\nu - a, \nu + a]$ where $0 < a \leq \nu$. Find the mean $E[X]$ and mean-square $E[X^2]$ of the random process $X(t)$. Investigate how the results vary as the ratio a/ν decreases.

9.2. Given a linear system

$$\ddot{X} + 2\zeta\omega_0\dot{X} + \omega_0^2 X = W(t)$$

in which $0 < \zeta < 1$ is a constant, $W(t)$ is a gaussian white noise with a spectral density K, and ω_0 is a random variable uniformly distributed in the interval $[\nu - a, \nu + a]$ where $0 < a \leq \nu$. Obtain the stationary solutions for the mean $E[X]$, mean-square $E[X^2]$, and probability density $p_X(x)$ of the random process $X(t)$. Investigate how the results vary as the ratio a/ν decreases.

9.3. Consider a linear system

$$\ddot{X} + 2\zeta\omega_0\dot{X} + \omega_0^2 X = A \sin \nu t$$

ζ, ω_0, ν, and A are positive constants and $\zeta < 1$. The initial conditions $X(0)$ and $\dot{X}(0)$ are assumed to be correlated random variables. Determine the mean and mean-square functions of the random process $X(t)$ in terms of the statistical properties of $X(0)$ and $\dot{X}(0)$.

9.4. An undamped oscillator subjected to a harmonic excitation is governed by

$$\ddot{X} + \omega_0^2 X = A \sin \nu t \qquad \omega_0 \neq \nu$$

where ω_0, ν, and A are positive constants. The oscillator is initially at rest, that is, $\dot{X}(0) = 0$; however, the initial position $X(0)$ is assumed to be a gaussian-distributed random variable. Find the joint probability density $p(x, \dot{x})$ of the random processes $X(t)$ and $\dot{X}(t)$.

9.5. Using the method of multiple reflections, find the localization factor for a disordered chain shown in Fig. P9.5, where $k_1, k_2, k_3, \ldots$, are independent positive random variables, uniformly distributed in $[k_0 - b, k_0 + b]$, where $0 < b < k_0$. Assume that the reflection is weak.

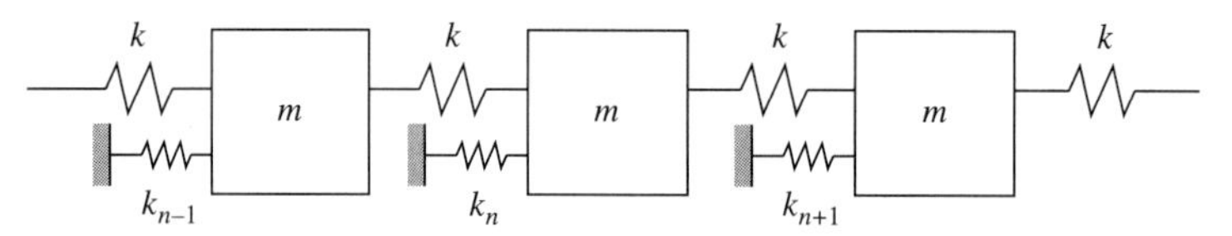

FIGURE P9.5
A disordered periodic structure with one random parameter.

9.6. A disordered mass-spring chain is shown in Fig. P9.6. The masses $m_1, m_2, m_3, \ldots$, are a set of independent random variables, uniformly distributed in $[m - a, m + a]$ where $0 < a < m$. The springs $k_1, k_2, k_3, \ldots$, are another set of independent random variables, uniformly distributed in $[k - b, k + b]$ where $0 < b < k$, and are independent of the masses. Obtain the equations of motion, the transfer matrix, and the wave transfer matrix for the nth cell unit. Find the localization factor for the three cases (a) $b = 0$, (b) $a = 0$, and (c) $a \neq 0, b \neq 0$, under the assumption that the reflection is weak.

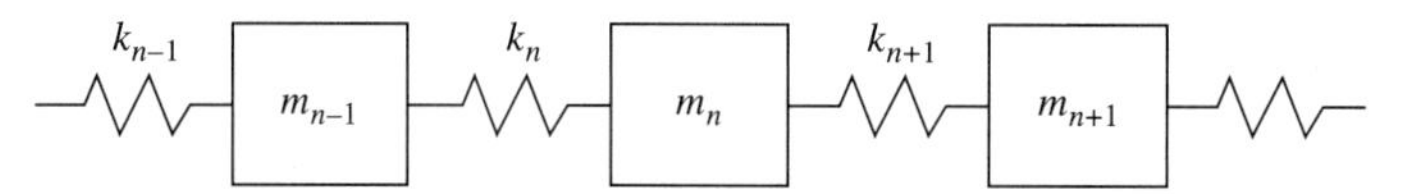

FIGURE P9.6
A disordered mass-spring chain with two random parameters.

9.7. Consider a disordered mass-spring chain shown in Fig. P9.7, where $m_2 = m_3 = m_4 = m$, and m_1 is a random variable, uniformly distributed in $[m - a, m + a]$ and where $0 < a < m$. An external force f_e is applied at mass m_2, and the displacement response w_e of mass m_2 is of interest. Obtain the mean and mean-square value of the magnitude of the frequency response.

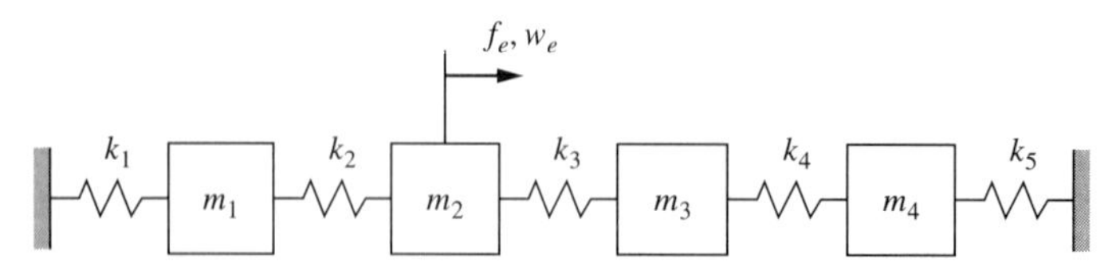

FIGURE P9.7
A disordered mass-spring system under external excitation.

REFERENCES

Abramowitz, M., and Stegun, I. A. 1972. *Handbook of Mathematical Functions*. Dover, New York.

Ahn, N. D. 1985. "On the Study of Random Oscillations in Non-Autonomous Mechanical Systems Using the Fokker-Planck-Kolmogorov Equations." *PMM U.S.S.R.,* vol. 49, no. 3, pp. 392–397.

Aki, K., Bouchon, M., Chouet, B., and Das, S. 1977. "Quantitative Prediction of Strong Motion for a Potential Earthquake Fault." *Annali di Geofisica,* vol. 30, pp. 341–368.

Aki, K., and Richards, P. G. 1980. *Quantitative Seismology: Theory and Methods.* W. H. Freeman, New York.

Amin, M., and Ang, A. H. S. 1968. "Nonstationary Stochastic Model of Earthquake Motions." *ASCE Journal of the Engineering Mechanics Division,* vol. 94, no. 2, pp. 559–583.

Anderson, P. W. 1958. "Absence of Diffusion in Certain Random Lattices." *Physical Review,* vol. 109, no. 5, pp. 1492–1505.

Andronov, A., Pontryagin, L., and Witt, A. 1933. "On the Statistical Investigation of Dynamical Systems." *Zh. Eksp. Teor. Fiz.,* vol. 3, pp. 165–180. (in Russian)

Ariaratnam, S. T. 1967. "Dynamic Stability of a Column under Random Loading." In *Dynamic Stability of Structures*, ed. G. Herrmann. Pergamon Press, New York, pp. 255–265.

Ariaratnam, S. T. 1980. "Bifurcation in Non-Linear Stochastic Systems." In *New Approaches to Non-Linear Problems in Dynamics*, ed. P. J. Holmes. SIAM Publications, Philadelphia, pp. 470–473.

Ariaratnam, S. T., and Ly, B. L. 1989. "Almost-Sure Stability of Some Linear Stochastic Systems." *ASME Journal of Applied Mechanics,* vol. 56, no. 1, pp. 175–178.

Ariaratnam, S. T., and Pi, H. N. 1973. "On the First-Passage Time for Envelope Crossing for a Linear Oscillator." *International Journal of Control,* vol. 18, no. 1, pp. 89–96.

Ariaratnam, S. T., and Srikantaiah, T. K. 1978. "Parametric Instabilities in Elastic Structures under Stochastic Loading." *Journal of Structural Mechanics,* vol. 6, no. 4, pp. 349–365.

Ariaratnam, S. T., and Tam, D. S. F. 1976. "Parametric Random Excitation of a Damped Mathieu Oscillator." *Z. Angew. Math. Mech.,* vol. 56, no. 11, pp. 449–452.

Ariaratnam, S. T., and Tam, D. S. F. 1977. "Moment Stability of Coupled Linear Systems under Combined Harmonic and Stochastic Excitation." In *Stochastic Problems in Dynamics,* ed. B. L. Clarkson. Pitman, London, pp. 90–103.

Ariaratnam, S. T., and Tam, D. S. F. 1979. "Random Vibration and Stability of a Linear Parametrically Excited Oscillator." *Z. Angew. Math. Mech.,* vol. 59, no. 2, pp. 79–84.

Ariaratnam, S. T., and Xie, W. C. 1988. "Effect of Derivative Process on the Almost-Sure Asymptotic Stability of Second-Order Linear Stochastic Systems." *Dynamics and Stability of Systems,* vol. 3, no. 1/2, pp. 69–78.

Ariaratnam, S. T., and Xie, W. C. 1989. "Effect of Correlation on Almost-Sure Asymptotic Stability of Second-Order Linear Stochastic Systems." *ASME Journal of Applied Mechanics,* vol. 56, no. 3, pp. 685–690.

Ariaratnam, S. T., and Xie, W. C. 1990. "Lyapunov Exponent and Rotation Number of a Two-Dimensional Nilpotent Stochastic System." *Dynamics and Stability of Systems,* vol. 5, no. 1, pp. 1–9.

Ariaratnam, S. T., and Xie, W. C. 1991. "On the Localization Phenomenon in Randomly Disordered Engineering Structures." In *Nonlinear Stochastic Mechanics,* eds. N. Bellomo and F. Casciati. Springer-Verlag, Berlin, pp. 13–24.

Ariaratnam, S. T., and Xie, W. C. 1992. "Lyapunov Exponents and Stochastic Stability of Coupled Linear Systems under Real Noise Excitation." *ASME Journal of Applied Mechanics,* vol. 59, no. 3, pp. 664–673.

Ariaratnam, S. T., and Xie, W. C. 1994. "Localization of Stress Wave Propagation in Disordered Multi-Wave Structures." In *Structural Safety and Reliability,* vol. 1, eds. G. I. Schuëller, M. Shinozuka, and J. T. P. Yao. Balkema, Rotterdam, pp. 77–83.

Arnold, L. 1984. "A Formula Connecting Sample and Moment Stability of Linear Stochastic Systems." *SIAM Journal of Applied Mathematics,* vol. 44, no. 4, pp. 793–802.

Arnold, L., and Wihstutz, V. 1984. "Lyapunov Exponent: A Survey," In *Lyapunov Exponents,* Lecture Notes in Mathematics, 1186, eds. L. Arnold and V. Wihstutz. Springer-Verlag, Berlin, pp. 1–26.

Aslam, M., Godden, W. G., and Scalise, D. T. 1980. "Earthquake Rocking Response of Rigid Bodies." *ASCE Journal of the Structural Division,* vol. 106, no. 2, pp. 377–392.

Assaf, S. A., and Zirkie, L. D. 1976. "Approximate Analysis of Non-Linear Stochastic Systems." *International Journal of Control,* vol. 23, no. 4, pp. 477–492.

Atalik, T. S., and Utku, S. 1976. "Stochastic Linearization of Multi-Degree-of-Freedom Systems." *Earthquake Engineering and Structural Dynamics,* vol. 4, pp. 411–420.

Baber, T. T., and Noori, M. N. 1984. "Random Vibration of Pinching Systems." *ASCE Journal of Engineering Mechanics,* vol. 110, no. 7, pp. 1036–1049.

Baber, T. T., and Noori, M. N. 1985. "Random Vibration of Degrading Pinching Systems." *ASCE Journal of Engineering Mechanics,* vol. 111, no. 8, pp. 1010–1026.

Baber, T. T., and Wen, Y. K. 1980. " Stochastic Equivalent Linearization for Hysteretic, Degrading, Multistory Structures." In *Civil Engineering Studies SRS No. 471,* Department of Civil Engineering, University of Illinois, Urbana.

Beck, J. L., and Hall, J. F. 1986. "Engineering Features of the Recent Mexican Earthquake." *Engineering and Science,* January, pp. 2–9.

Beliveau, J.-G., Vaicaitis, R., and Shinozuka, M. 1977. "Motion of Suspension Bridge Subject to Wind Loads." *ASCE Journal of the Structural Division,* vol. 103, no. ST6, pp. 1189–1205.

Beran, M. J. 1968. *Statistical Continuum Theories.* Interscience, New York.

Bertram, J. E., and Sarachik, P. E. 1959. "On the Stability of Systems with Random Parameters." *Transaction, IRE-PGCT-5 Special Supplement,* pp. 260–270.

Bharucha-Reid, A. T. 1960. *Elements of Markov Processes and Their Applications.* McGraw-Hill, New York.

Bisplinghoff, R. L., and Ashley, H. 1962. *Principles of Aeroelasticity.* Wiley, New York.

Bogdanoff, J. L., Goldberg, J. E., and Bernard, M. C. 1961. "Response of a Simple Structure to Random Earthquake-Type Disturbance." *Bulletin of the Seismological Society of America,* vol. 51, pp. 293–310.

Bogoliubov, N. N., and Mitropolsky, Y. A. 1961. *Asymptotic Methods in the Theory of Nonlinear Oscillations.* Gordon and Breach, New York.

Bolotin, V. V. 1964. *The Dynamic Stability of Elastic Systems.* Holden-Day, San Francisco.

Bolotin, V. V. 1965. *Statistical Methods in Structural Mechanics.* Holden-Day, San Francisco.

Bolotin, V. V. 1989. *Prediction of Service Life of Machines and Structures.* ASME Press, New York.

Booton, R. C. 1954. "Nonlinear Control Systems with Random Inputs." *IRE Transactions on Circuit Theory,* CT-1, vol. 1, pp. 9–19.

Born, M. 1955. "Continuity, Determination and Reality." *Matematisk-Fysiske Meddelelser,* vol. 30, pp. 3–9.

Bouc, R. 1967. "Forced Vibration of Mechanical System with Hysteresis." In *Proceedings of the 4th Conference on Nonlinear Oscillation,* Prague, Czechoslovakia. (abstract)

Bouchon, M. 1979. "Discrete Wave Number Representation of Elastic Wave Field in Three-Space Dimensions." *Journal of Geophysical Research,* vol. 84, no. B7, pp. 3609–3614.

Bouchon, M. 1981. "A Simple Method to Calculate Green's Function for Elastic Layered Media." *Bulletin of the Seismological Society of America,* vol. 71, pp. 959–971.

Box, M. J. 1965. "A New Method of Constrained Optimization and a Comparison with Other Methods." *Computer Journal,* vol. 8, no. 1, pp. 42–52.

Brillouin, L. 1946. *Wave Propagation in Periodic Structures.* McGraw-Hill, New York.

Brückner, A., and Lin, Y. K. 1987. "Generalization of the Equivalent Linearization Method for Non-Linear Random Vibration Problems." *International Journal of Non-Linear Mechanics,* vol. 22, no. 3, pp. 227–235.

Bucher, C. G., and Lin, Y. K., 1988. "Effect of Spanwise Correlation of Turbulence Field on the Motion Stability of Long-Span Bridges." *Journal of Fluids and Structures,* vol. 2, pp. 437–451.

Burridge, R., and Knopoff, L. 1964. "Body Force Equivalents for Seismic Dislocations." *Bulletin of the Seismological Society of America,* vol. 54, pp. 1875–1888.

Cai, G. Q., and Lin, Y. K. 1988a. "On Exact Stationary Solutions of Equivalent Non-Linear Stochastic Systems." *International Journal of Non-Linear Mechanics,* vol. 23, no. 4, pp. 315–325.

Cai, G. Q., and Lin Y. K. 1988b. "A New Approximate Solution Technique for Randomly Excited Non-Linear Oscillators." *International Journal of Non-Linear Mechanics,* vol. 23, no. 5/6, pp. 409–420.

Cai, G. Q., and Lin, Y. K. 1990. "On Randomly Excited Hysteretic Structures." *ASME Journal of Applied Mechanics,* vol. 57, no. 2, pp. 442–448.

Cai, G. Q., and Lin, Y. K. 1991a. "Localization of Wave Propagation in Disordered Periodic Structures." *AIAA Journal,* vol. 29, no. 3, pp. 450–456.

Cai, G. Q., and Lin, Y. K., 1991b. "Wave Attenuation in Disordered Periodic Structures." In *Stochastic Structural Dynamics, vol. 1: New Theoretical Developments,* eds. Y. K. Lin and I. Elishakoff. Springer-Verlag, Berlin, pp. 33–61.

Cai, G. Q., and Lin, Y. K. 1992. "Response Distribution of Non-Linear Systems Excited by Non-Gaussian Impulsive Noise." *International Journal of Non-Linear Mechanics,* vol. 27, no. 6, pp. 955–967.

Cai, G. Q., and Lin, Y. K. 1994a. "On Statistics of First-Passage Failure." *ASME Journal of Applied Mathematics,* vol. 61, no. 1, pp. 93–99.

Cai, G. Q., and Lin, Y. K. 1994b. "Nonlinearly Damped Systems under Simultaneous Harmonic and Random Excitations." *Nonlinear Dynamics,* vol. 5.

Cai, G. Q., Lin, Y. K., and Elishakoff, I. 1992. "A New Approximate Solution Technique for Randomly Excited Non-Linear Oscillators, Part II." *International Journal of Non-Linear Mechanics,* vol. 27, no. 6, pp. 969–979.

Cambou, B. 1975. "Applications of First-Order Uncertainty Analysis in the Finite Elements Method in Linear Elasticity." In *Proceedings of the 2nd International Conference on Applications of Statistics and Probability in Soil and Structural Engineering,* Aachen, Germany, pp. 67–87.

Casciati, F. 1987. "Nonlinear Stochastic Dynamics of Large Structural Systems by Equivalent Linearization." In *Proceedings of the 5th International Conference on Applications of Statistics and Probability in Soil and Structural Engineering,* Vancouver, Canada, pp. 1165–1172.

Casciati, F., and Faravelli, L. 1991. *Fragility Analysis of Complex Structural Systems.* Wiley, New York.

Caughey, T. K. 1959a. "Response of a Nonlinear String to Random Loading." *ASME Journal of Applied Mechanics,* vol. 26, no. 3, pp. 341–344.

Caughey, T. K. 1959b. "Response of van der Pol Oscillator to Random Excitation." *ASME Journal of Applied Mechanics,* vol. 26, no. 3, pp. 345–348.

Caughey, T. K. 1960. "Random Excitation of a System with Bilinear Hysteresis." *ASME Journal of Applied Mechanics,* vol. 27, no. 4, pp. 649–652.

Caughey, T. K. 1964. "On the Response of a Class of Nonlinear Oscillator to Stochastic Excitation." In *Proc. Colloq. Intern. du Centre National de la Recherche Scientifique,* No. 148, Marseilles, September, pp. 393–402.

Caughey, T. K. 1971. "Nonlinear Theory of Random Vibrations." In *Advances in Applied Mechanics,* vol. 11, ed. C. S. Yih. Academic Press, New York, pp. 209–253.

Caughey, T. K. 1986. "On the Response of Non-Linear Oscillators to Stochastic Excitation." *Probabilistic Engineering Mechanics,* vol. 1, no. 1, pp. 2–4.

Caughey, T. K., and Dienes, J. K. 1961. "Analysis of a Nonlinear First-Order System with a White Noise Input." *Journal of Applied Physics,* vol. 23, no. 11, pp. 2476–2479.

Caughey, T. K., and Gray, A. H., Jr. 1965. "On the Almost Sure Stability of Linear Dynamic Systems with Stochastic Coefficients." *ASME Journal of Applied Mechanics,* vol. 32, no. 2, pp. 365–372.

Caughey, T. K., and Ma, F. 1982a. "The Exact Steady-State Solution of a Class of Nonlinear Stochastic Systems." *International Journal of Non-Linear Mechanics,* vol. 17, no. 3, pp. 137–142.

Caughey, T. K., and Ma, F. 1982b. "The Steady-State Response of a Class of Dynamical Systems to Stochastic Excitations." *ASME Journal of Applied Mechanics,* vol. 49, no. 3, pp. 629–632.

Chang, C. C., and Yang, H. T. Y. 1991. "Random Vibration of Flexible, Uncertain Beam Element." *ASCE Journal of Engineering Mechanics,* vol. 117, no. 10, pp. 2329–2350.

Chenea, P. F., and Bogdanoff, J. L. 1958. "Impedance of Some Disordered Systems." In *Colloquium on Mechanical Impedance Methods,* ASME, pp. 125–128.

Chetayev, N. G. 1961. *The Stability of Motion.* Pergamon Press, New York.

Clough, R. W., and Penzien, J. 1975. *Dynamics of Structures.* McGraw-Hill, New York.

Collins, J. D., and Thomson, W. T. 1969. "Eigenvalue Problem for Structural Systems with Statistical Properties." *AIAA Journal,* vol. 7, no. 4, pp. 642–648.

Copson, E. T. 1965. *Asymptotic Expansions.* Cambridge University Press, New York.

Corcos, G. M. 1963. "Resolution of Pressure in Turbulence," *Journal of the Acoustical Society of America,* vol. 35, no. 2, pp. 192–199.

Cornell, C. A. 1964. "Stochastic Process Models in Structural Engineering." *Technical Report 34,* Department of Civil Engineering, Stanford University, Stanford, Calif.

Cornwell, P. J., and Bendiksen, O. O. 1987. "Localization of Vibrations in Large Space Reflectors." AIAA paper 87-0949. In *Proceedings of the 28th AIAA/ASME/ASCE/AHS Structures, Structural Dynamics and Materials Conference,* Monterey, Calif., pp. 925–935.

Cox, D. R., and Miller, H. D. 1965. *The Theory of Stochastic Processes.* Chapman and Hall, New York.

Cramer, H., and Leadbetter, M. R. 1967. *Stationary and Related Stochastic Processes.* Wiley, New York.

Crandall, S. H. 1978. "Heuristic and Equivalent Linearization Techniques for Random Vibration of Non-Linear Oscillators." In *Proceedings of the 8th International Conference on Non-Linear Oscillation,* vol. 1. Academia, Prague, pp. 211–226.

Crandall, S. H. 1980. "Non-Gaussian Closure for Random Vibration of Non-Linear Oscillators." *International Journal of Non-Linear Mechanics,* vol. 15, no. 4/5, pp. 303–313.

Das, S., and Aki, K. 1977. "Fault Plane with Barriers: A Versatile Earthquake Model." *Journal of Geophysical Research,* vol. 82, no. 36, pp. 5658–5670.

Davenport, A. G. 1962. "Buffeting of a Suspension Bridge by Stormy Winds." *ASCE Journal of the Structural Division,* vol. 1, no. 3, pp. 233–268.

Davis, P. J., and Rabinowitz, P. 1984. *Methods of Numerical Integration.* Academic Press, New York.

Deodatis, G. 1991. "Weighted Integral Method. I: Stochastic Stiffness Matrix." *ASCE Journal of Engineering Mechanics,* vol. 117, no. 8, pp. 1851–1864.

Deodatis, G., and Shinozuka, M. 1991. "Weighted Integral Method. II: Response Variability and Reliability." *ASCE Journal of Engineering Mechanics,* vol. 117, no. 8, pp. 1865–1877.

Dimentberg, M. F. 1982. "An Exact Solution to a Certain Nonlinear Random Vibration Problem." *International Journal of Non-Linear Mechanics,* vol. 17, no. 4, pp. 231–236.

Dimentberg, M. F. 1988. *Statistical Dynamics of Nonlinear and Time-Varying Systems.* Wiley, New York.

Dimentberg, M. F., Lin, Y. K., and Zhang, R. C. 1993. "Toppling of Computer-Type Equipment under Base Excitation." *ASCE Journal of Engineering Mechanics,* vol. 119, no. 1, pp. 145–160.

Dimentberg, M. F., and Menyailov, A. I. 1979. "Response of a Single-Mass Vibroimpact System to White Noise Random Excitation." *Z. Angew. Math. Mech.,* vol. 59, pp. 709–716.

Ditlevsen, O., and Olesen, R. 1986. "Statistical Analysis of the Virkler Data on Fatigue Crack Growth." *Engineering Fracture Mechanics,* vol. 25, no. 2, pp. 177–195.

Dryden, H. L. 1961. "A Review of the Statistical Theory of Turbulence." In *Turbulence,* ed. S. K. Friedlander and L. Topper. Interscience, New York, pp. 115–150.

Dynkin, E. B. 1965. *Markov Processes,* vols. 1 and 2. Springer-Verlag, New York.

Elishakoff, I., and Cai, G. Q. 1992. "Approximate Solution for Nonlinear Random Vibration Problems by Partial Stochastic Linearization." In *Nonlinear Vibrations,* ASME Winter Annual Meeting AMD, vol. 144, eds. R. A. Ibrahim, N. Sri Namachchivaya, and A. K. Bajaj. ASME, New York, pp. 117–121.

Feller, W. 1952. "The Parabolic Differential Equation and the Associated Semigroups of Transformations." *Annals of Mathematics,* vol. 55, no. 3, pp. 468–519.

Feller, W. 1954. "Diffusion Process in One Dimension." *Transactions of the American Mathematical Society,* vol. 77, July–December, pp. 1–31.

Feller, W. 1957. *An Introduction to Probability Theory and Its Applications,* vol. 1. John Wiley and Sons, New York.

Finlayson, B. A. 1972. *The Method of Weighted Residuals and Variational Principles.* Academic Press, New York.

Freudenthal, A. M. 1974. "New Aspects of Fatigue and Fracture Mechanics." *Engineering Fracture Mechanics,* vol. 6, no. 4, pp. 775–794.

Fujimori, Y., and Lin, Y. K. 1973. "Analysis of Airplane Response to Nonstationary Turbulence Including Wing Bending Flexibility." *AIAA Journal,* vol. 11, no. 3, pp. 334–339.

Fujimori, Y., Lin, Y. K., and Ariaratnam, S. T. 1979. "Rotor Blade Stability in Turbulence Flow, Part II." *AIAA Journal,* vol. 17, no. 7, pp. 673–678.

Fung, Y. C. 1953. "Statistical Aspects of Dynamic Loads." *Journal of Aeronautical Sciences,* vol. 20, pp. 317–330.

Furstenberg, H. 1963. "Noncommuting Random Products." *Transactions of the American Mathematical Society,* vol. 108, no. 3, pp. 377–428.

Gardiner, C. W. 1983. *Handbook of Stochastic Methods for Physics, Chemistry and the Natural Sciences.* Springer-Verlag, Berlin.

Gear, C. W. 1971. *Numerical Initial Value Problems in Ordinary Differential Equations.* Prentice-Hall, Englewood Cliffs, N.J.

Ghanem, R. G., and Spanos, P. D. 1991. *Stochastic Finite Elements: A Spectral Approach.* Springer-Verlag, Berlin.

Gradshteyn, I. S., and Ryzhik, I. M. 1980. *Tables of Integrals, Series and Products.* Academic Press, New York.

Graham, R., and Haken, H. 1971. "Generalized Thermo-Dynamic Potential for Markoff Systems in Detailed Balance and Far from Thermal Equilibrium." *Zeitschrift für Physik,* vol. 243, pp. 289–302.

Grigoriu, M. 1990. "Reliability of Degrading Dynamic Systems." *Structural Safety,* vol. 8, no. 1/4, pp. 345–351.

Haskell, N. A. 1964. "Radiation Pattern of Surface Waves from Point Sources in a Multilayered Medium." *Bulletin of the Seismological Society of America,* vol. 54, pp. 377–393.

Hodges, C. H., and Woodhouse, J. 1983. "Vibration Isolation from Irregularity in a Nearly Periodic Structure: Theory and Measurements." *Journal of the Acoustic Society of America,* vol. 74, no. 3, pp. 894–905.

Hoeppner, D. W., and Krupp, W. E. 1974. "Prediction of Component Life by Application of Fatigue Crack Growth Knowledge." *Engineering Fracture Mechanics,* vol. 6, no. 1, pp. 47–70.

Housner, G. W. 1947. "Characteristics of Strong Motion Earthquakes." *Bulletin of the Seismological Society of America,* vol. 37, pp. 19–31.

Housner, G. W. 1963. "The Behavior of Inverted Pendulum Structures during Earthquakes." *Bulletin of the Seismological Society of America,* vol. 53, no. 2, pp. 407–417.

Howell, L. J., and Lin, Y. K. 1971. "Response of Flight Vehicles to Nonstationary, Atmospheric Turbulence." *AIAA Journal,* vol. 9, no. 11, pp. 2201–2207.

Hsu, L. C. 1948. "A Theorem on the Asymptotic Behavior of a Multiple Integral." *Duke Mathematical Journal,* vol. 15, pp. 623–632.

Huang, W. 1982. "Vibration of Some Structures with Periodic Random Parameters." *AIAA Journal,* vol. 18, no. 3, pp. 318–323.

Hudson, J. A. 1969. "A Quantitative Evaluation of Seismic Signals at Teleseismic Distances. I: Radiation from Point Sources." *Geophysical Journal of the Royal Astronomical Society,* vol. 18, no. 3, pp. 233–249.

Huston, D. R. 1986. "The Effect of Upstream Gusting on the Aeroelastic Behavior of Long Suspended-Span Bridges." Ph.D. dissertation, Princeton University, Princeton, N.J.

Ibrahim, R. A., Soundararajan, A., and Heo, H. 1985. "Stochastic Response of Non-Linear Dynamic Systems Based on a Non-Gaussian Closure." *ASME Journal of Applied Mechanics,* vol. 52, no. 4, pp. 965–970.

Infante, E. F. 1968. "On the Stability of Some Linear Nonautonomous Systems." *ASME Journal of Applied Mechanics,* vol. 35, no. 1, pp. 7–12.

Itô, K., 1951a. "On Stochastic Differential Equations." *Memoirs of the American Mathematical Society,* no. 4, pp. 289–302.

Itô, K., 1951b. "On a Formula Concerning Stochastic Differentials." *Nagoya Mathematical Journal,* vol. 3, pp. 55–65.

Itô, K., and McKean, H. P., Jr. 1965. *Diffusion Processes and Their Sample Paths.* Academic Press, New York.

Iwan, W. D. 1974. "Application of Nonlinear Analysis Techniques." In *Applied Mechanics in Earthquake Engineering.* ASME Annual Meeting, AMD, vol. 8, ed. W. D. Iwan, ASME, New York, pp. 135–162.

Iwan, W. D., and Lutes, L. D. 1968. "Response of the Bilinear Hysteretic System to Stationary Random Excitation." *Journal of the Acoustic Society of America,* vol. 43, no. 3. pp. 545–552.

Iwankiewicz, R., and Nielsen, S. R. K. 1989. "Dynamic Response of Non-Linear Systems to Poisson-Distributed Random Impulses." Institute of Material Science and Applied Mechanics, Report No. 11, Technical University of Wroclaw, Poland.

Iwankiewicz, R., and Sobczyk, K. 1983. "Dynamic Response of Linear Structures to Correlated Random Impulses." *Journal of Sound and Vibration,* vol. 86, no. 3, pp. 303–317.

Iyengar, R. N., and Dash, P. K. 1978. "Study of the Random Vibration of Nonlinear Systems by the Gaussian Closure Technique." *ASME Journal of Applied Mechanics,* vol. 45, no. 2, pp. 393–399.

Jancauskas, E. C., and Melbourne, W. H. 1985. "The Aerodynamic Admittance of Two-Dimensional Rectangular Section Cylinders in Turbulent Flow." In *Proceedings of the 5th U.S. National Conference on Wind Engineering,* Lubbock, Texas, November 6-8, 1985, pp. 4A 65–72.

Janssen, R. A., and Lambert, R. F. 1966. "Numerical Calculation of Some Response Statistics for a Linear Oscillator under Impulsive-Noise Excitation." *Journal of the Acoustical Society of America,* vol. 41, no. 4, pp. 827–835.

Kameda, H. 1975. "Evolutionary Spectra of Seismogram by Multifilter." *ASCE Journal of the Engineering Mechanics Division,* vol. 101, no. 6, pp. 787–801.

Kanai, K. 1957. "Seismic-Empirical Formula for the Seismic Characteristics of the Ground." *Bulletin of the Earthquake Research Institute* (Tokyo University), vol. 35, pp. 309–325.

Karlin, S., and Taylor, H. M. 1975. *A First Course in Stochastic Processes.* Academic Press, New York.

Karlin, S., and Taylor, H. M. 1981. *A Second Course in Stochastic Processes.* Academic Press, New York.

Kennett, B. L. N. 1985. *Seismic Wave Propagation in Stratified Media.* Cambridge University Press, New York.

Khasminskii, R. Z. 1955, "Probability Distribution for the Functionals of Trajectory of Stochastic Diffusion Process." *Sov. Math. Dokl.*, vol. 104, no. 1, pp. 22–25. (in Russian)

Khasminskii, R. Z. 1964. "On the Behavior of a Conservative System with Small Friction and Small Random Noise." *Prikladnaya Matematika i Mechanica* (Applied Mathematics and Mechanics), vol. 28, no. 5, pp. 1126–1130. (in Russian)

Khasminskii, R. Z. 1966. "A Limit Theorem for the Solution of Differential Equations with Random Right Hand Sides." *Theory of Probability and Application,* vol. 11, no. 3, pp. 390–405.

Khasminskii, R. Z. 1967. "Sufficient and Necessary Conditions of Almost Sure Asymptotic Stability of a Linear Stochastic System." *Theory of Probability and Application,* vol. 12, no. 1, pp. 144–147.

Khasminskii, R. Z. 1980. *Stochastic Stability of Differential Equations.* Kluwer Academic Pubs., Norwell, MA.

Kissel, G. J. 1988. "Localization in Disordered Periodic Structures." Ph.D. dissertation, Massachusetts Institute of Technology, Cambridge, Mass.

Kissel, G. J. 1991. "Localization Factor for Multichannel Disordered Systems." *Physical Review A,* vol. 44, no. 2, pp. 1008–1014.

Kobori, T., Minai, R., and Suzuki, Y. 1976. "Stochastic Seismic Response of Hysteretic Structures." *Bulletin of the Disaster Prevention Research Institute* (Kyoto University), vol. 26, part 1, no. 236, pp. 57–70.

Kolmogorov, A. 1931. "Über die analytischen Methoden in der Wahrscheinlichkeitsrechnung." *Mathematische Annalen,* vol. 104, pp. 415–458.

Kozin, F. 1961. "On the Probability Densities of the Output of Some Random Systems." *ASME Journal of Applied Mechanics,* vol. 28, no. 2, pp. 161–165.

Kozin, F. 1963. "On Almost Sure Stability of Linear Systems with Random Coefficients." *Journal of Mathematics and Physics,* vol. 42, no. 1, pp. 59–67.

Kozin, F. 1969. "A Survey of Stability of Stochastic Systems." *Automatica,* vol. 5, no. 1, pp. 95–112.

Kozin, F. 1977. "An Approach to Characterizing, Modeling and Analyzing Earthquake Excitation Records." In *Random Excitation of Structures by Earthquakes and Atmospheric Turbulence,* CISM Courses and Lectures, 225, ed. H. Parkus. Springer-Verlag, Wien-New York, pp. 77–109.

Kozin, F. 1982. "First-Passage Times: Some Results." In *Proceedings of The International Workshop on Structural Mechanics,* Innsbruck, Austria, November 6–8, 1982, pp. 28–32.

Kozin, F., and Bogdanoff, J. L. 1989. "Recent Thought on Probabilistic Fatigue Crack Growth." *Applied Mechanics Review,* vol. 42, no. 11, pp. s121–s127.

Kozin, F., and Prodromou, S. 1971. "Necessary and Sufficient Conditions for Almost Sure Sample Stability of Linear Itô Equations." *SIAM Journal of Applied Mathematics,* vol. 21, no. 3, pp. 413–424.

Kozin, F., and Sugimoto, S. 1977. "Relation between Sample and Moment Stability for Linear Stochastic Differential Equations." In *Proceedings, Conference on Stochastic Differential Equations,* ed. D. Mason. Academic Press, New York, pp. 145–162.

Kozin, F., and Sunahara, Y. 1987. "An Application of the Averaging Method to Noise Stabilization of Nonlinear Systems." In *Proceedings of the 20th Midwestern Mechanics Conference,* West Lafayette, Indiana, vol. 14(a), pp. 291–298.

Kozin, F., and Wu, C. M. 1973. "On the Stability of Linear Stochastic Differential Equations." *ASME Journal of Applied Mechanics,* vol. 40, no. 1, pp. 87–92.

Kozin, F., and Zhang, Z. Y. 1990. "On Almost Sure Sample Stability of Nonlinear Itô Differential Equations." *Stochastic Structural Dynamics, vol 1: New Theoretical Developments,* eds. Y. K. Lin and I. Elishakoff. Springer-Verlag, Berlin, pp. 147–154.

Kramers, H. A. 1940. "Brownian Motion in a Field of Force and Diffusion Model of Chemical Reactions." *Physica,* vol. 7, pp. 284–304.

Krein, M. G. 1933. "Vibration Theory of Multispan Beams." *Vestnic Inzhenerov i Teknikov,* vol. 4, pp. 142–145. (in Russian)

Landa, P. S., and Stratonovich, R. L. 1962. "Theory of Stochastic Transitions of Various Systems between Different States." *Vestnik MGU* (Proc. of Moscow University), series III(1), pp. 33–45. (in Russian)

Lee, E. H. 1962. "Viscoelasticity." In *Handbook of Engineering Mechanics,* ed. W. Flügge. McGraw-Hill, New York, Chapter 53.

Lee, E. H., and Kanter, I. 1953. "Wave Propagation in Finite Rods of Viscoelastic Materials." *Journal of Applied Physics,* vol. 24, no. 9, pp. 1115–1122.

Lennox, W. C., and Fraser, D. A. 1974. "On the First-Passage Distribution for the Envelope of a Nonstationary Narrow-Band Stochastic Process." *ASME Journal of Applied Mechanics,* vol. 41, no. 3, pp. 793–797.

Lévy, P. 1948. *Processus Stochastiques et Mouvement Brownien.* Gauthier-Villars, Paris.

Levy, R., and Kozin, F. 1968. "Processes for Earthquake Simulation." *ASCE Journal of the the Engineering Mechanics Division,* vol. 94, no. 6, pp. 1597–1601.

Li, Q. C. 1993. "Theoretical and Experimental Investigations of Motion Stability of Long-Span Bridges in Turbulent Flow." Ph.D. dissertation, Florida Atlantic University, Boca Raton.

Liepman, H. W. 1952. "On the Application of Statistical Concepts to the Buffeting Problem." *Journal of Aeronautical Sciences,* vol. 19, pp. 793–800.

Lin, Y. K. 1963. "Application of Nonstationary Shot Noise in the Study of System Response to a Class of Nonstationary Excitations." *ASME Journal of Applied Mechanics,* vol. 30, no. 4, pp. 555–558.

Lin, Y. K. 1965. "Nonstationary Excitation in Linear Systems Treated as Sequences of Random Pulses." *Journal of the Acoustical Society of America,* vol. 38, no. 3, pp. 453–460.

Lin, Y. K. 1977. "Structural Response under Turbulence Flow Excitations." In *Random Excitation of Structures by Earthquakes and Atmospheric Turbulence,* ed. H. Parkus. Springer-Verlag, Wien-New York, pp. 238–307.

Lin, Y. K. 1986. "On Random Pulse Train and Its Evolutionary Spectral Representation." *Probabilistic Engineering Mechanics,* vol. 1, no. 4, pp. 219–223.

Lin, Y. K. 1967. *Probabilistic Theory of Structural Dynamics.* McGraw-Hill, New York. Reprint R. E. Krieger, Melbourne, Fla., 1976.

Lin, Y. K., and Ariaratnam, S. T. 1980. "Stability of Bridge Motion in Turbulent Winds." *Journal of Structural Mechanics,* vol. 8, no. 1, pp. 1–15.

Lin, Y. K., and Cai, G. Q. 1988a. "Exact Stationary-Response Solution for Second-Order Nonlinear Systems under Parametric and External White-Noise Excitations: Part II." *ASME Journal of Applied Mechanics,* vol. 55, no. 3, pp. 702–705.

Lin, Y. K., and Cai, G. Q. 1988b. "Equivalent Stochastic Systems." *ASME Journal of Applied Mechanics,* vol. 55, no. 4, pp. 918–922.

Lin, Y. K., Dimentberg, M. F., Zhang, R. C., Cai, G. Q., and Holung, J. A. 1994. "Sliding Motion of Anchored Rigid Block under Random Base Excitations." *Probabilistic Engineering Mechanics,* vol. 9, no. 1, pp. 33–38.

Lin, Y. K., Li, Q. C., and Su, T. C. 1993. "Application of a New Wind Turbulence Model in Predicting Motion Stability of Wind-Excited Long-Span Bridges." *Journal of Wind Engineering and Industrial Aerodynamics,* vol. 49, pp. 507–516.

Lin, Y. K., and Maekawa, S. 1977. "Decomposition of Turbulence Forcing Field and Structural Response." *AIAA Journal,* vol. 15, no. 5, pp. 608–610.

Lin, Y. K., Maekawa, S., Nijim, H., and Maestrello, L. 1976. "Response of Periodic Beam to Supersonic Boundary-Layer Pressure Fluctuations." In *Stochastic Problems in Dynamics,* ed. B. L. Clarkson. Pitman, London, pp. 468–486.

Lin, Y. K., and McDaniel, T. J. 1969. "Dynamics of Beam-Type Periodic Structures." *Journal of Engineering for Industry,* vol. 91, no. 4, pp. 1131–1141.

Lin, Y. K., and Wu, W. F. 1984a. "A Closed Form Earthquake Response Analysis of Multistory Building on Compliance Soil." *Journal of Structural Mechanics,* vol. 12, no. 1, pp. 87–110.

Lin, Y. K., and Wu, W. F. 1984b. "Along-Wind Motion of Building on Compliant Soil." *ASCE Journal of Engineering Mechanics,* vol. 110, no. 1, pp. 1–19.

Lin, Y. K., and Yang, J. N. 1974. "Free Vibration of Disordered Periodic Beam." *ASME Journal of Applied Mechanics,* vol. 41, no. 2, pp. 383–391.

Lin, Y. K., and Yang, J. N. 1983. "On Statistical Moments of Fatigue Crack Propagation." *Engineering Fracture Mechanics,* vol. 18, no. 2, pp. 243–256.

Lin, Y. K., and Yang, J. N. 1985. "A Stochastic Theory of Fatigue Crack Propagation." *AIAA Journal,* vol. 23, no. 1, pp. 117–124.

Lin, Y. K., and Yong, Y. 1987. "Evolutionary Kanai-Tajimi Earthquake Models." *ASCE Journal of Engineering Mechanics,* vol. 113, no. 8, pp. 1119–1137.

Lin, Y. K., and Yong, Y. 1989. "Discussions of Evolutionary Kanai-Tajimi Earthquake Models." *ASCE Journal of Engineering Mechanics,* vol. 115, no. 4, pp. 884–887.

Lin, Y. K., Zhang, R. C., and Yong, Y. 1990. "Multiply Supported Pipeline under Seismic Excitation." *ASCE Journal of Engineering Mechanics,* vol. 116, no. 5, pp. 1094–1108.

Liu, S. C. 1970. "Dynamics of Correlated Random Pulse Trains." *ASCE Journal of the Engineering Mechanics Division,* vol. 96, no. 3, pp. 455–475.

Liu, W. K., Belytschko, T., and Mani, A. 1986. "Probabilistic Finite Elements for Nonlinear Structural Dynamics." *Journal of Computer Methods in Applied Mechanics and Engineering,* vol. 56, pp. 61–81.

Lumley, J. L., and Panofsky, H. A. 1964. *The Structure of Atmospheric Turbulence.* Wiley, New York.

Lutes, L. D. 1970. "Approximate Technique for Treating Random Vibration of Hysteretic Systems." *Journal of the Acoustical Society of America,* vol. 48, no. 1, pp. 299–306.

Lutes, L. D., and Takemiya, M. 1974. "Random Vibration of Yielding Oscillator." *ASCE Journal of the Engineering Mechanics Division,* vol. 100, no. 2, pp. 345–357.

Lyapunov, A. M. 1892. "Problème générale de la stabilité du mouvement." *Comm. Soc. Math. Kharkov,* vol. 2, pp. 265–272. Reprinted in *Annals of Mathematical Studies,* vol. 17, Princeton University Press, Princeton, 1947.

Madsen, H. O. 1982. *Deterministic and Probabilistic Models for Damage Accumulation due to Time-Varying Loading.* Dialog 5-82, Danish Engineering Academy, Lyngby, Denmark.

Maruyama, T. 1963. "On the Force Equivalents of Dynamic Elastic Dislocation with Reference to the Earthquake Mechanism." *Bulletin of the Earthquake Research Institute* (Tokyo University), vol. 41, pp. 467–486.

Mead, D. J. 1971. "Vibration Response and Wave Propagation in Periodic Structures." *Journal of Engineering for Industry,* vol. 93, no. 3, pp. 783–792.

Metler, E. 1962. "Dynamic Buckling." In *Handbook of Engineering Mechanics,* ed. W. Flügge. McGraw-Hill, New York, Chapter 62.

Miles, J. W. 1956. "Vibration of Beams on Many Supports." *ASCE Journal of the Engineering Mechanics Division,* vol. 82, no. 1, pp. 1–9.

Miller, M. S., and Gallagher, J. P. 1981. "An Analysis of Several Fatigue Growth Rate (FCGR) Descriptions." In *Fatigue Crack Growth Measurements and Data Analysis,* ASTM-STP 738, pp. 205–251.

Mitchell, R. R., and Kozin, F. 1974. "Sample Stability of Second-Order Linear Differential Equations with Wide-Band Noise Coefficients." *SIAM Journal of Applied Mathematics,* vol. 27, no. 4, pp. 571–604.

Moyal, J. E. 1949. "Stochastic Processes and Statistical Physics." *Journal of the Royal Statistical Society, Series B,* vol. 11, pp. 150–210.

Naess, A., and Johnsen, J. M. 1993. "Response Statistics of Nonlinear, Compliant Offshore Structures by the Path Integral Solution Method." *Probabilistic Engineering Mechanics,* vol. 8, no. 2, pp. 91–106.

Nakagiri, S., and Hisada, T. 1981. "Finite Element Stress Analysis Extended to Stochastic Treatment in Problems of Structural Safety and Reliability." *Transactions of the 6th International Conference on Structural Mechanics in Reactor Technology,* Paris, France, August 17–21, 1981.

Nayfeh, A. H., and Mook, D. T. 1979. *Nonlinear Oscillations.* Wiley, New York.

Nishioka, K. 1976. "On the Stability of Two-Dimensional Linear Stochastic Systems." *Kodai Mathematics Seminar Reports,* vol. 27, pp. 211–230.

Norohna, P. J., Henslee, S. P., Gordon, D. E., Wolanski, Z. R., and Yee, B. G. W. 1978. *Fastener Hole Quality,* vol. I. Air Force Flight Dynamics Laboratory, Wright-Patterson Air Force Base, Dayton, Ohio, AFFDL-TR-78-206.

Ortis, K., and Kiremidjian, A. S. 1988. "Stochastic Modeling of Fatigue Crack Growth." *Engineering Fracture Mechanics,* vol. 29, no. 3, pp. 317–334.

Oseledec, V. I. 1968. "A Multiplicative Ergodic Theorem: Lyapunov Characteristic Number for Dynamical Systems." *Transactions of the Moscow Mathematical Society,* vol. 19, pp. 197–231.

Papageorgiou, A., and Aki, K. 1983. "A Specific Barrier Model for the Quantitative Description of Inhomogeneous Faulting and the Prediction of Strong Ground Motion." *Bulletin of the Seismological Society of America,* vol. 73, pp. 693–722, 953–978.

Papanicolaou, G. C., and Kohler, W. 1974. "Asymptotic Theory of Mixing Stochastic Ordinary Differential Equations." *Communications on Pure and Applied Mathematics,* vol. 27, pp. 641–668.

Papanicolaou, G. C., and Kohler, W. 1975. "Asymptotic Analysis of Deterministic and Stochastic Ordinary Differential Equations." *Communications in Mathematical Physics,* vol. 45, pp. 217–232.

Pardoux, E., and Wihstutz, V. 1988. "Lyapunov Exponent and Rotation Number of Two-Dimensional Linear Stochastic Systems with Small Diffusion." *SIAM Journal of Applied Mathematics,* vol. 48, no. 2, pp. 442–457.

Paris, P. C., and Erdogan, F. 1963. "A Critical Analysis of Crack Propagation Laws." *Journal of Basic Engineering,* vol. 85, pp. 528–534.

Parzen, E. 1962. *Stochastic Processes.* Holden-Day, San Francisco.

Pestel, E. C., and Leckie, F. A. 1963. *Matrix Methods in Elastomechanics.* McGraw-Hill, New York.

Pierre, C. 1990. "Weak and Strong Vibration Localization in Disordered Structures: A Statistical Investigation." *Journal of Sound and Vibration,* vol. 139, no. 1, pp. 549–564.

Pierre, C., and Cha, P. D. 1989. "Strong Mode Localization in Nearly Periodic Disordered Structures." *AIAA Journal,* vol. 27, no. 2, pp. 227–241.

Pierre, C., and Dowell, E. H. 1987. "Localization of Vibrations by Structural Irregularity." *Journal of Sound and Vibration,* vol. 114, no. 3, pp. 549–564.

Pierre, C., Tang, D. M., and Dowell, E. H. 1987. "Localized Vibrations of Disordered Multispan Beams: Theory and Experiment." *AIAA Journal,* vol. 25, no. 9, pp. 1249–1257.

Priestley, M. B. 1965. "Evolutionary Spectra and Non-Stationary Processes." *Journal of the Royal Statistical Society, Series B,* vol. 27, pp. 204–237.

Rayleigh, J. W. S. 1887. "On Waves Propagated along the Plane Surface of an Elastic Solid." *Proceedings, London Mathematical Society,* vol. 17, pp. 4–11.

Rice, S. O. 1944, 1945. "Mathematical Analysis of Random Noise." *Bell System Technical Journal,* vol. 23, pp. 282–332; vol. 24, pp. 46–156. Reprinted in *Selected Papers on Noise and Stochastic Processes,* ed. N. Wax, Dover, New York, 1954.

Richardson, J. A., and Kuester, J. L. 1973. "The Complex Method for Constrained Optimization." *Communications of the ACM,* vol. 16, no. 8, pp. 487–489.

Roberts, J. B., 1965. "The Response of Linear Vibratory Systems to Random Impulses." *Journal of Sound and Vibration,* vol. 2, no. 3, pp. 375–390.

Roberts, J. B. 1966. "On the Response of a Simple Oscillator to Random Impulses." *Journal of Sound and Vibration,* vol. 4, no. 1, pp. 51–61.

Roberts, J. B. 1972. "System Response to Random Impulses." *Journal of Sound and Vibration,* vol. 24, no. 1, pp. 23–34.

Roberts, J. B. 1976. "First Passage Probability for Nonlinear Oscillators." *ASCE Journal of the Engineering Mechanics Division,* vol. 102, no. 5, pp. 851–866.

Roberts, J. B. 1978a. "First-Passage Time for Oscillators with Nonlinear Damping." *ASME Journal of Applied Mechanics,* vol. 45, no. 1, pp. 175–180.

Roberts, J. B., 1978b. "The Response of an Oscillator with Bilinear Hysteresis to Stationary Random Excitation." *ASME Journal of Applied Mechanics,* vol. 45, no. 4, pp. 923–928.

Roberts, J. B. 1987. "Application of Averaging Methods to Randomly Excited Hysteretic Systems." In *Proceedings of IUTAM Symposium on Nonlinear Stochastic Engineering Systems,* eds. F. Ziegler and G. I. Schuëller. Springer-Verlag, Berlin, pp. 361–374.

Roberts, J. B., and Spanos, P. D. 1990. *Random Vibration and Statistical Linearization.* Wiley, New York.

Scanlan, R. H. 1978. "The Action of Flexible Bridges under Wind. I: Flutter Theory; II. Buffeting Theory." *Journal of Sound and Vibration,* vol. 60, no. 2, pp. 187–211.

Scanlan, R. H., Beliveau, J.-G., and Budlong, K. S. 1974. "Indicial Aerodynamic Functions for Bridge Decks." *ASCE Journal of the Engineering Mechanics Division,* vol. 100, no. 4, pp. 657–672.

Scanlan, R. H., and Tomko, J. J. 1971. "Airfoil and Bridge Deck Flutter Derivatives." *ASCE Journal of the Engineering Mechanics Division,* vol. 97, no. 6, pp. 1717–1737.

Scherer, R. J., Riera, J. D., and Schuëller, G. I. 1982. "Estimation of the Time Dependent Frequency Content of Earthquake Accelerations." *Nuclear Engineering and Design,* vol. 71, no. 3, pp. 301–310.

Scheurkogel, A., and Elishakoff, I. 1988. "Non-Linear Random Vibration of a Two-Degree-of-Freedom System." In *Non-Linear Stochastic Engineering Systems,* eds. F. Ziegler and G. I. Schuëller. Springer-Verlag, Berlin, pp. 285–299.

Schuëller, G. I., and Scherer, R. J. 1985. "A Stochastic Earthquake Loading Model." In *Proceedings of PRC-U.S.-Japan Trilateral Symposium/Workshop on Engineering for Multiple Natural Hazard Mitigation.* Beijing, China, January 7–12, pp. G-2-1–G-2-16.

Schütz, W. 1979. "The Prediction of Fatigue Life in the Crack Initiation and Propagation Stages—A State of the Art Survey." *Engineering Fracture Mechanics,* vol. 11, no. 2, pp. 405–421.

Sears, W. R. 1941. "Some Aspects of Nonstationary Airfoil Theory and Its Practical Applications." *Journal of Aeronautical Sciences,* vol. 8, pp. 104–108.

Shih, T. Y., and Lin, Y. K. 1982. "Vertical Seismic Load Effect on Hysteretic Columns." *ASCE Journal of the Engineering Mechanics Division,* vol. 88, no. 2, pp. 242–254.

Shinozuka, M., and Astill, C. J. 1972. "Random Eigenvalue Problems in Structural Analysis." *AIAA Journal,* vol. 10, no. 4, pp. 456–462.

Shinozuka, M., and Brant, P. 1969. "Application of the Evolutionary Power Spectrum in Structural Dynamics." In *Proceedings, ASCE-EMD Specialty Conference on Probabilistic Concepts and Methods in Engineering,* Purdue University, West Lafayette, Ind., November 12–14, pp. 42–46.

Shinozuka, M., and Deodatis, G. 1988a. "Response Variability of Stochastic Finite Element Systems." *ASCE Journal of Engineering Mechanics,* vol. 114, no. 39, pp. 499–519.

Shinozuka, M., and Deodatis, G. 1988b. "Stochastic Wave Models of Seismic Ground Motion." *International Workshop on Spatial Variation of Earthquake Ground Motion,* Princeton University, Dunwalke, N. J., November 7–9. Reprinted in *Structural Safety,* vol. 10, no. 1/3, 1991, pp. 235–246.

Shinozuka, M., and Sato, Y. 1967. "Simulation of Nonstationary Random Processes." *ASCE Journal of the Engineering Mechanics Division,* vol. *93,* no. 1, pp. 11–40.

Shinozuka, M., and Yamazaki, F. 1988. "Stochastic Finite Element Analysis: An Introduction." In *Stochastic Structural Dynamics: Progress in Theory and Applications,* eds. S. T. Ariaratnam, G. I. Schuëller, and I. Elishakoff. Elsevier Applied Science, London, pp. 241–291.

Simiu, E., and Scanlan, R. H. 1986. *Wind Effects on Structures: An Introduction to Wind Engineering.* Wiley, New York.

Sneddon, I. N. 1951. *Fourier Transforms.* McGraw-Hill, New York.

Sobczyk, K. 1984. *Stochastic Wave Propagation.* PWN-Polish Scientific Publishers, Warsaw.

Sobczyk, K. 1986. "Modeling of Random Fatigue Crack Growth." *Engineering Fracture Mechanics,* vol. 24, no. 4, pp. 609–623.

Sobczyk, K., and Spencer, B. F. 1992. *Random Fatigue.* Academic Press, San Diego.

Soize, C. 1988. "Steady-State Solution of Fokker-Planck Equation in High Dimension." *Probabilistic Engineering Mechanics,* vol. 3, no. 4, pp. 196–206.

Soong, T. T. 1973. *Random Differential Equations in Science and Engineering.* Academic Press, New York.

Soong, T. T., and Bogdanoff, J. L. 1964. "On the Impulsive Admittance and Frequency Response of a Disordered Linear Chain of *N* Degrees of Freedom." *International Journal of Mechanical Sciences,* vol. 6, no. 3, pp. 225–237.

Soong, T. T., and Cozzarelli, F. A. 1976. "Vibration of Disordered Structural Systems." *Shock and Vibration Digest,* vol. 8, no. 5, pp. 21–35.

Sozen, M. A. 1974. "Hysteresis in Structural Elements." In *Applied Mechanics in Earthquake Engineering,* ASME Annual Meeting, AMD, vol. 8, ed. W. D. Iwan, ASME, New York, pp. 63–68.

Spanos, P.-T. D. 1981. "Stochastic Linearization in Structural Dynamics." *Applied Mechanics Review,* vol. 34, no. 1, pp. 1–8.

Spanos, P.-T. D. 1982. "Approximate Analysis of Random Vibration through Stochastic Averaging." In *Proceedings, IUTAM Symposium on Random Vibration and Reliability, Frankfurt/Oder,* ed. K. Hennig. Akademie-Verlag, Berlin, pp. 327–337.

Spencer, B. F., and Bergman, L. A. 1993. "On the Numerical Solution of the Fokker-Planck Equation for Nonlinear Stochastic Systems." *Nonlinear Dynamics,* vol. 4, pp. 357–372.

Sri Namachchivaya, N. 1989. "Instability Theorem Based on the Nature of the Boundary Behavior for One-Dimensional Diffusion." *Solid Mechanics Archives,* vol. 14, no. 3/4, pp. 131–142.

Sri Namachchivaya, N., and Lin, Y. K. 1988. "Application of Stochastic Averaging for Nonlinear Systems with High Damping." *Probabilistic Engineering Mechanics,* vol. 3, no. 3, pp. 159–167.

Srinivasan, S. K., Subramanian, R., and Kumaraswamy, S. 1967. "Response of Linear Vibratory Systems to Non-Stationary Stochastic Impulses." *Journal of Sound and Vibration,* vol. 6, no. 2, pp. 169–179.

Stratonovich, R. L. 1963. *Topics in the Theory of Random Noise,* vol. 1. Gordon and Breach, New York.

Stratonovich, R. L. 1967. *Topics in the Theory of Random Noise,* vol. 2. Gordon and Breach, New York.

Su, T. C., and Lian, Q. X. 1992. Personal communication.

Sun, J.-Q., and Hsu, C. S. 1987. "Cumulant-Neglect Closure Method for Nonlinear Systems under Random Excitations." *ASME Journal of Applied Mechanics,* vol. 54, no. 3, pp. 649–655.

Sun, J.-Q., and Hsu, C. S. 1990. "The Generalized Cell Mapping Method in Nonlinear Random Vibration Based upon Short-Time Gaussian Approximation." *ASME Journal of Applied Mechanics,* vol. 57, no. 4, pp. 1018–1025.

Tajimi, H. 1960. "A Statistical Method of Determining the Maximum Response of a Building Structure during an Earthquake." In *Proceedings of the 2nd World Conference on Earthquake Engineering,* Tokyo-Kyoto, Japan, July, 1960, pp. 781–798.

Tanaka, H. 1957. "On Limiting Distributions for One-Dimensional Diffusion Processes." *Bulletin of Mathematical Statistics,* vol. 7, pp. 84–91.

To, W. S., and Li, D. M. 1991. "Equivalent Nonlinearization of Nonlinear Systems to Random Excitations." In *Stochastic Structural Dynamics, vol. 1: New Theoretical Developments,* eds. Y. K. Lin and I. Elishakoff. Springer-Verlag, Berlin, pp. 245–266.

Tsurui, A., and Ishikawa, H. 1986. "Application of the Fokker-Planck Equation to a Stochastic Fatigue Crack Growth Model." *Structural Safety,* vol. 4, no. 1, pp. 15–29.

Tung, C. C. 1967. "Random Response of Highway Bridges to Vehicle Loads." *ASCE Journal of the Engineering Mechanics Division,* vol. 93, no. 5, pp. 79–94.

Tung, C. C. 1969. "Response of Highway Bridges to Renewal Traffic Loads." *ASCE Journal of the Engineering Mechanics Division,* vol. 95, no. 1, pp. 41–57.

Tylikowski, A., and Marowski, W. 1986. "Vibration of a Non-Linear Single-Degree-of-Freedom System due to Poissonian Impulse Excitation." *International Journal of Non-Linear Mechanics,* vol. 21, no. 3, pp. 229–238.

Valero, N. A., and Bendikson, O. O. 1986. "Vibration Characteristics of Mistuned Shrouded Blade Assemblies." *Journal of Engineering for Gas Turbines and Power,* vol. 108, no. 2, pp. 293–299.

Van Kampen, N. G. 1957. "Derivation of the Phenomenological Equations from the Master Equation. II: Even and Odd Variables." *Physica,* vol. 23, no. 9, pp. 816–829.

Van Kampen, N. G. 1981. "Itô versus Stratonovich." *Journal of Statistical Physics,* vol. 24, no. 1, pp. 175–187.

Vanmarcke, E., and Grigoriu, M. 1983. "Stochastic Finite Element Analysis of Simple Beams." *ASCE Journal of Engineering Mechanics,* vol. 109, no. 5, pp. 1203–1214.

Vanmarcke, E., Shinozuka, M., Nakagiri, S., Schuëller, G. I., and Grigoriu, M. 1986. "Random Field and Stochastic Finite Elements." *Structural Safety,* vol. 3, no. 3/4, pp. 143–166.

Verne-Jones, D. 1970. "Stochastic Models for Earthquake Occurrence." *Journal of the Royal Statistical Society, Series B,* vol. 32, no. 1, pp. 1–62.

Virkler, D. A., Hillberry, B. M., and Goel, P. K. 1978. *The Statistical Nature of Fatigue Crack Growth.* Air Force Flight Dynamics Laboratory, Wright-Patterson Air Force Base, Dayton, Ohio, AFFDL-TR-78-43.

Von Flotow, A. H. 1986. "Disturbance Propagation in Structural Networks." *Journal of Sound and Vibration,* vol. 106, no. 3, pp. 433–450.

Von Karman, T. 1948. "Progress in the Statistical Theory of Turbulence." In *Proceedings of the National Academy of Sciences, National Academy Press,* Washington, D.C., pp. 530–539.

Wang, M. C., and Uhlenbeck, G. E. 1945. "On the Theory of the Brownian Motion II." *Reviews of Modern Physics,* vol. 17, pp. 323–342. Reprinted in *Selected Papers on Noise and Stochastic Processes,* ed. N. Wax, Dover, New York, 1954.

Wedig, W. V. 1988. "Lyapunov Exponents of Stochastic Systems and Related Bifurcation Problems." *Stochastic Structural Dynamics: Progress in Theory and Applications,* eds. S. T. Ariaratnam, G. I. Schuëller, and I. Elishakoff. Elsevier Applied Science, London, pp. 315–327.
Wedig, W. V. 1989. "Analysis and Simulation of Nonlinear Stochastic Systems." In *Nonlinear Dynamics in Engineering Systems,* ed. W. Schiehlen. Springer-Verlag, Berlin, pp. 337–344.
Wedig, W. V. 1990. "Dynamic Stability of Beams under Axial Forces: Lyapunov Exponents for General Fluctuating Loads." In *Structural Dynamics,* eds. W. B. Krätzig et al. Balkema, Rotterdam, pp. 141–148.
Wedig, W. V., Lin, Y. K., and Cai, G. Q. 1990. "Necessary and Sufficient Conditions for Existence of Stationary Solutions of Some Nonlinear Stochastic Systems." *Journal of Nonlinear Dynamics,* vol. 1, no. 1, pp. 75–90.
Wen, Y. K. 1976. "Method for Random Vibration of Hysteretic Systems." *ASCE Journal of the Engineering Mechanics Division,* vol. 103, no. 2, pp. 249–263.
Wen, Y. K. 1980. "Equivalent Linearization for Hysteretic Systems under Random Excitation." *ASME Journal of Applied Mechanics,* vol. 47, no. 1, pp. 150–154.
Whittaker, E. T., and Watson, G. N. 1952. *A Course of Modern Analysis.* University Press, Cambridge.
Wong, E., and Zakai, M. 1965. "On the Relation between Ordinary and Stochastic Equations." *International Journal of Engineering Sciences,* vol. 3, no. 2, pp. 213–229.
Wu, W. F., and Lin, Y. K. 1984. "Cumulant-Neglect Closure for Non-Linear Oscillators under Random Parametric and External Excitations." *International Journal of Non-Linear Mechanics,* vol. 19, no. 4, pp. 349–362.
Xie, W. C. 1990. "Lyapunov Exponents and Their Applications in Structural Dynamics." Ph.D. thesis, University of Waterloo, Waterloo, Ontario, Canada.
Yang, J. N., and Lin, Y. K. 1975. "Frequency Response Functions of a Disordered Periodic Beam." *Journal of Sound and Vibration,* vol. 38, no. 3, pp. 317–340.
Yang, J. N., Salivar, G. C., and Annis, C. G. 1983. "Statistical Modeling of Fatigue Crack Growth in a Nickel-Base Superalloy." *Engineering Fracture Mechanics,* vol. 18, no. 2, pp. 257–270.
Yong, Y., and Lin, Y. K. 1987. "Exact Stationary-Response Solution for Second-Order Nonlinear Systems under Parametric and External White-Noise Excitations." *ASME Journal of Applied Mechanics,* vol. 54, no. 2, pp. 414–418.
Yong, Y., and Lin, Y. K. 1989. "Propagation of Decaying Wave in Periodic and Piecewise Periodic Structures." *Journal of Sound and Vibration,* vol. 129, no. 2, pp. 99–118.
Zhang, R. C., Yong, Y, and Lin, Y. K. 1991a. "Earthquake Ground Motion Modeling. I: Deterministic Point Source." *ASCE Journal of Engineering Mechanics,* vol. 117, no. 9, pp. 2114–2132.
Zhang, R. C., Yong, Y., and Lin, Y. K. 1991b. "Earthquake Ground Motion Modeling. II: Stochastic Line Source." *ASCE Journal of Engineering Mechanics,* vol. 117, no. 9, pp. 2133–2148.
Zhang, Z. Y. 1991. "New Developments in Almost Sure Sample Stability of Nonlinear Stochastic Dynamical Systems," Ph.D. dissertation, Polytechnic University, New York.
Zhang, Z. Y., and Kozin, F. 1990. "On Almost Sure Sample Stability of Nonlinear Stochastic Dynamic Systems." In *Proceedings of the IEEE International Conference on System Engineering,* Pittsburgh, Pa., August 9–11.
Zhu, W. Q. 1990. "The Exact Stationary Response Solution of Several Classes of Nonlinear Systems to White Noise Parametric and/or External Excitations." *Applied Mathematics and Mechanics,* vol. 11, no. 2, pp. 165–175.
Zhu, W. Q., Cai, G. Q., and Lin, Y. K. 1990. "On Exact Stationary Solutions of Stochastically Perturbed Hamiltonian Systems." *Probabilistic Engineering Mechanics,* vol. 5, no. 2, pp. 84–87.
Zhu, W. Q., Cai, G. Q., and Lin, Y. K. 1991. "Stochastically Excited Hamiltonian Systems." *Proceedings of IUTAM Symposium on Nonlinear Stochastic Mechanics*, eds. N. Bellomo and F. Casciati, pp. 543–552.
Zhu, W. Q., and Huang, T. C. 1984. "Dynamic Instability of a Liquid-Free Surface in a Container with an Elastic Bottom under Combined Harmonic and Stochastic Longitudinal Excitation." In *Random Vibrations,* ASME Winter Annual Meeting, AMD, vol. 65, eds. T. C. Huang and P. D. Spanos, ASME, New York, pp. 195–220.

Zhu, W. Q., and Lei, Y. 1987. "Stochastic Averaging of Energy Envelope of Bilinear Hysteretic Systems." In *Proceedings of IUTAM Symposium on Nonlinear Stochastic Dynamic Engineering Systems,* Springer-Verlag, Berlin, pp. 381–391.

Zhu, W. Q., and Lei, Y. 1989. "First Passage Time for State Transition of Randomly Excited Systems." *Proceedings of the 47th International Statistical Institute Meeting,* Paris, August 29–September 6, pp. 517–531.

Zhu, W. Q., Lin, Y. K., and Lei, Y. 1992. "On Fatigue Crack Growth under Random Loading." *Engineering Fracture Mechanics,* vol. 43, no. 1, pp. 1–12.

Zhu, W. Q., and Wu, W. Q. 1991. "A Stochastic Finite Element Method for Real Eigenvalue Problems." *Stochastic Structural Dynamics, vol. 2: New Practical Applications,* eds. I. Elishakoff and Y. K. Lin. Springer-Verlag, Berlin, pp. 337–351.

Zhu, W. Q., Yu, J. S., and Lin, Y. K. 1994. "On Improved Stochastic Averaging Procedure." *Probabilistic Engineering Mechanics,* vol. 9, no. 3, pp. 203–211.

NAME INDEX

SUBJECT INDEX